OSA Trends in Optics and Photonics Series

Series Editor
Tingye Li *AT&T Labs.*

Volume 21

ADVANCES IN OPTICAL IMAGING AND PHOTON MIGRATION

Trends in Optics and Photonics Series

Tingye Li, Series Editor

Vol. 23 Symposium on Electro-Optics: Present and Future
Hermann Haus, ed.

Vol. 22 joint volume - Biomedical Optical Spectroscopy and Diagnostics / Therapeutic Laser Applications (1998)
Eva Sevick-Muraca, Joseph A. Izatt, and Marwood N. Ediger, eds.

Vol. 21 Advances in Optical Imaging & Photon Migration (1998)
James G. Fujimoto and Michael S. Patterson, eds

Vol. 20 Optical Networks and Their Applications
Richard A. Barry, ed..

Vol. 19 Advanced Solid State Lasers (1998)
Walter R. Bosenberg and Martin M. Fejer, eds.

Vol. 18 Radiative Processes and Dephasing in Semiconductors
D. S. Citrin, ed.

Vol. 17 Diode Pumped Solid State (DPSS) Lasers: Applications and Issues
Mark W. Dowley, ed.

Vol. 16 Optical Amplifiers and Their Applications (1997)
Michalis N. Zervas, Alan E. Willner, and Shinya Sasaki, eds.

Vol. 15 Advances in Vertical Cavity Surface Emitting Lasers
Connie J. Chang-Hasnain, ed.

Vol. 14 Spatial Light Modulators
Geoffrey Burdge and Sadik C. Esener, eds.

Vol. 13 Ultrafast Electronics and Optoelectronics
Martin Nuss and John Bowers, eds.

Vol. 12 System Technologies
Curtis R. Menyuk and Alan E. Willner, eds.

Vol. 11 Noninvasive Assessment of the Visual System
Dean Yager, ed.

Vol. 10 Advanced Solid State Lasers (1997)
Clifford R. Pollock and Walter R. Bosenberg, eds.

Vol. 9 Lasers and Optics for Manufacturing
Andrew C. Tam, ed.

Vol. 8 Environmental Monitoring and Instrumentation
Frank Bucholtz, ed.

Volumes 1-7 Contact OSA Customer Service for more information

OSA Trends in Optics and Photonics Series
Volume 21

ADVANCES IN OPTICAL IMAGING AND PHOTON MIGRATION

Edited by
James G. Fujimoto
and
Michael S. Patterson

From the Topical Meeting
Advances in Optical Imaging
and Photon Migration
March 8-11, 1998
Orlando, Florida

Sponsored by
Optical Society of America

In cooperation with
Laser Institute of America

Optical Society of America
2010 Massachusetts Avenue, NW
Washington, DC 20036-1023
(202) 223-8130

Articles in this publication may be cited in other publications. In order to facilitate access to the original publication source, the following form for the citation is suggested:

> Name of Author(s), Title of Paper, OSA *Trends in Optics and Photonics Vol. 21, Advances in Optical Imaging and Photon Migration*, James G. Fujimoto and Michael S. Patterson, eds. (Optical Society of America, Washington, DC 1998), pp. xx-xx.

Cover art: Cellular level optical coherence tomography (OCT) imaging of a xenopus tadpole in vivo. Cellular level OCT could have future clinical applications for the detection of early neoplastic changes in humans. (See article on pp. 317-320).

ISSN Number 1094-5695

ISBN Number 1-55752-546-3
LC Number 98-86315

Printed in the USA

Advances in Optical Imaging and Photon Migration

Contents

Optical characterization of tissue and other scattering media

Diffuse optical imaging: Theory, simulation, and experiment

Diffuse optical imaging and functional assessment: Clincal and *in vivo* studies

Optical coherence tomography and coherence imaging

Imaging and microscopy including nonlinear optics and light/sound interaction

Preface

We are pleased to present this volume of the *Trends in Optics and Photonics* series covering the exciting field of biomedical optical imaging and photon migration. This volume contains manuscripts from the Optical Society of America Topical Meeting on Advances in Optical Imaging and Photon Migration (AOIPM), which was held in Orlando, Florida, March, 1998. The AOIPM conference had 53 talks and 33 poster presentations. Invited talks from 17 speakers highlighted key advances by leading research groups. The conference was attended by approximately 150 participants, with a large fraction of participants from Europe and Asia.

This AOIPM conference was the third in a biennial series. The first AIOPM conference, held in March 1994, was organized under the direction of Dr. Robert Alfano and was one of the first major peer-reviewed conferences devoted entirely to biomedical optical imaging and photon propagation in scattering tissues. The creation of this conference in 1994 also marked the first OSA topical meeting in the biomedical optics area. In 1996, a second topical meeting, Biomedical Optical Spectroscopy and Diagnostics (BOSD) was organized and was co-located with AOIPM. This year, the Biomedical Optical Spectroscopy and Diagnostics Topical Meeting, under the direction of Drs. Eva Sevick and Joseph Izatt, continues as the second in its series. In addition, a third topical meeting was held for the first time. Therapeutic Laser Applications (TLA) was organized under the direction of Drs. Woody Ediger, Reginald Birngruber, and A. J. Welch. These three co-located conferences cover complementary themes and form the OSA Biomedical Topical Meeting series. The Biomedical Topical Meetings taken together had 189 total presentations and were attended by close to 300 participants.

The field of optical imaging and photon migration brings together basic physics of photon propagation, optical engineering, and clinical sciences. In this sense, it is perhaps the most vertically integrated topical area in biomedical optics. The field of biomedical optics in general has undergone tremendous growth in the past few years. This growth has been stimulated not only by the discovery of important new technologies, techniques, and applications, but also by the entry of many talented researchers and research groups into the field. In addition, we have also witnessed the evolution of many concepts from basic science to clinical practice. These proceedings feature a chapter on clinical and *in vivo* applications. It is especially striking that the number of manuscripts in this chapter has increased several fold from 1994, to 1996, to 1998. This increase is evidence for the rapid evolution of our field. Each advance moves us closer toward integrating basic science, technology, and clinical care.

We would like to thank the many people and organizations who have contributed to the success of this conference and to the publication of this volume. We gratefully acknowledge the members of the Technical Program Committee, whose guidance shaped the direction of the conference. They reviewed the many manuscripts that were submitted and also carefully crafted the technical program. The conference was sponsored and organized under the direction of the Optical Society of America. We would like to especially acknowledge the contributions of Dr.

Thomas Baer, who, along with Barbara Hicks, organized the new OSA Division on Optics in Medicine and Biology. Dr. Baer and Ms. Hicks were instrumental in the creation of the OSA Topical Meetings on Biomedical Optics. The skill, dedication, and hard work of the OSA staff – including Rosemary Dwyer, Amy Hutto, Kari Amundsen, Naomi Chavez, and Kelly Cohen – were essential in bringing both the conference and this proceedings volume into reality.

The conferences were supported in part by funding from the U.S. Air Force Office of Scientific Research, the Whitaker Foundation, Humphrey Instruments, SpectRx Inc., and Hamamatsu Corporation. Finally, we wish to acknowledge the researchers who participated in this conference. Through their insight, innovation, and continued efforts, our field advances toward its ultimate objective: to create new technologies, diagnostics, and therapies that will improve health care and the quality of life for humankind.

James G. Fujimoto
Massachusetts Institute of Technology
Cambridge, Massachusetts

Michael S. Patterson
Hamilton Regional Cancer Centre and McMaster University
Ontario, Canada

1998

Light propagation in scattering media: Theory, simulation, and experiment

Simulation of 3D light propagation in a layered head model including a clear CSF layer

Shuichi TAKAHASHI [1)] Yukio YAMADA [2)]
[1)] MEC Laboratory, Daikin Industries. LTD.
[2)] Biomechanics Division, Mechanical Engineering Laboratory, AIST, MITI
1-2 Namiki, Tsukuba-city, Ibaraki 305-8564 Japan
phone: +81-298-58-7055 fax: +81-298-58-7291
E-mail: xiu@mec.daikin.co.jp[1)] yamada@mel.go.jp[2)]

Abstract

For the purpose of understanding the effect of the clear CSF (cerebrospinal fluid) layer on the light distribution in human heads, we have used 3D FEM (Finite Element Method) to solve the photon diffusion equation combined with the radiosity concept to predict the light propagation in the scattering and non-scattering layers. The numerical model for the calculation is a hemisphere which simulates human heads more realistically than slabs which have been used for numerical simplicity. The numerical model consists of a hemispherical shell and a hemisphere and they are connected by a clear layer. The distributions of the fluence rate and the light path probability have been obtained and shown by using the technique of computer graphics. As the result, the clear CSF layers have been found to have a great effect on the light propagation in human heads if the thickness of the clear CSF layer is thick.

Key Words: (170.3660) Light propagation in tissues, (170.3880) Medical and biological imaging, (170.5280) Photon Migration, (999.9999) Radiosity.

Introduction

We previously reported calculation results using 3D (three dimensional) FEM (the Finite Element Method) to solve the photon diffusion equation for a hemispherical head model consisting of four layers; scalp, skull, gray and white matters[1]. We presented the simulation results of 3D time-resolved pulse light propagation in the hemispherical model. However, in the calculation we neglected the existence of the clear CSF layer because of the difficulty of incorporating the light propagation in the clear layer where there is no light scattering and the photon diffusion approximation does not hold.

Recently, it is discussed that light propagation through the clear CSF (cerebrospinal fluid) layer may have a great effect on redistribution of light propagation in human heads showing the possibility of light channeling through the clear layer. However, previous investigators [2,3] have calculated the cases of simple geometry such as slabs or cylinders because of complexities involved in numerical calculation. They used a Monte Carlo method for slabs and cylinders, or an analytical solution for slabs combined with the radiosity concept. Monte Carlo methods require a very long computation time to obtain results with a satisfactory accuracy. The analytical solution is not appropriate to be applied to complicated geometry or inhomogeneous distribution of optical properties such as human heads.

In this paper we have extended our previous FEM model of a hemisphere simulating a human head to include a clear CSF layer. The model consists of an outer hemispherical shell, an inner hemispherical shell and an inner hemisphere. The photon diffusion equation is applied to the outer hemispherical shell and the inner hemisphere where the FEM meshing is made, and the radiosity concept [2] is applied to the inner hemispherical shell. The FEM results in the outer hemispherical shell and in the inner hemisphere are combined with the radiosity concept which describes the light propagation in clear media. This hybrid method combining the photon diffusion approximation and radiosity concept gives the results of the fluence rate distribution and the distribution of the light path probability. Finally, the effect of the clear CSF layer on

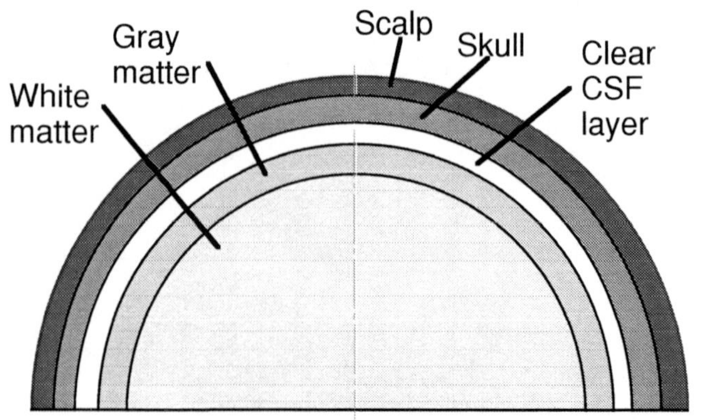

Fig.1 Cross section of a hemispherical head model including a clear CSF layer.

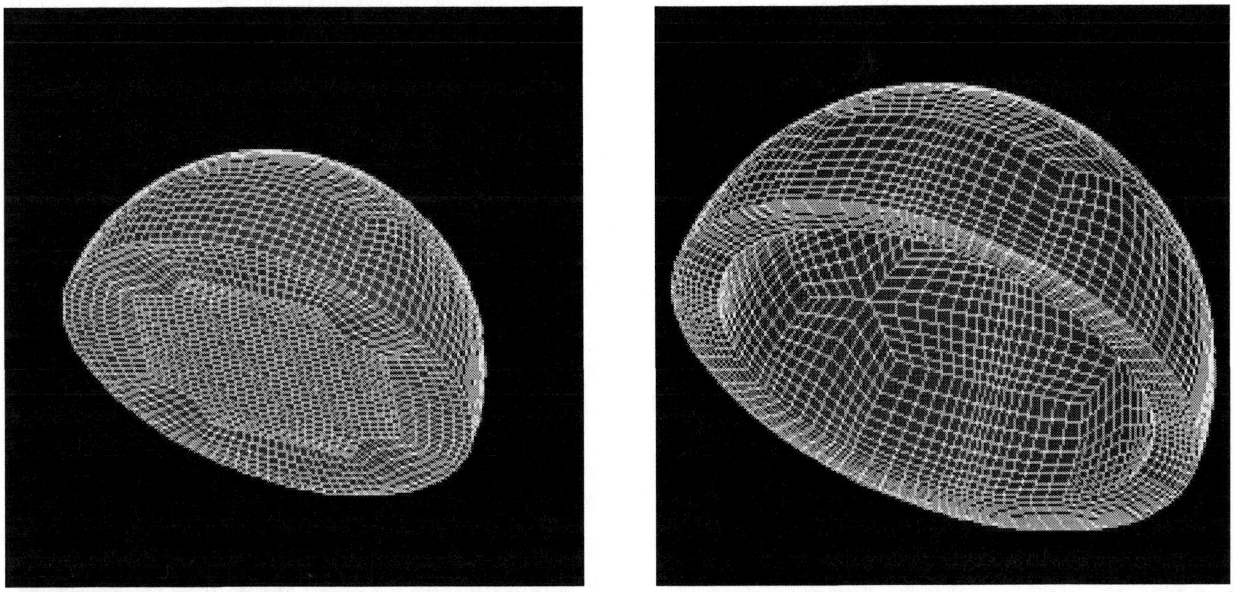

Fig. 2. 3D FEM meshing of the scattering regions of the gray and white matters (left hemisphere), and the scalp and skull layers (right hemispherical shell).

the light propagation in the head model are shown by using a computer graphics technique.

Model Construction

We simplifies a human head as a hemisphere consisting of the outer hemispherical shell, inner hemispherical shell and inner hemisphere as shown in Fig. 1. The outer hemispherical scattering shell represents the combined scalp and skull, the inner scattering hemisphere representing the combined gray and white matters, and the inner hemispherical clear shell representing the CSF layer, respectively.

Each scattering region is divided into about twenty thousand elements for FEM calculation to solve the photon diffusion equation as shown in Fig. 2, where both regions are observed obliquely from the bottom.

Radiation Exchange in the Clear Layer

In the clear layer light propagates without scattering and with very weak absorption. So, the photon diffusion equation cannot be applied. In this study, radiosity concept is applied[2] to estimate the light propagation in the clear layer. Thus the incoming radiative flux Γ at the point q on the surface **D** in Fig. 3 is given by the following equation (1) based on the radiosity concept [2].

$$\Gamma = \iint_{\mathbf{S}} \frac{I(p,\mathbf{R})\cos\theta\cos\psi}{R^2}\exp(-R\mu_a)\,d\mathbf{S} \qquad (1)$$

where **S** expresses the radiating surface, $I(p, \mathbf{R})$ the intensity in the direction **R** at the point p on **S**, **R** the directional vector from p to q, θ the angle between **R** and the normal at p, ψ the angle between **R** and the normal at q, R the distance between p and q, μ_a the absorption coefficient in the clear layer. In our model, the outer surface of the gray matter is irradiated by the inner surface of the skull. And the inner surface of the skull is irradiated by both the outer surface of the gray matter and the inner surface of the skull itself. By considering the radiation exchange in the clear layer, light propagation between the two scattering regions are connected.

Light Propagation in the Scattering Regions

In the scattering regions of the outer hemispherical shell (scalp, skull) and the inner hemisphere (white and gray matters), the photon diffusion equation (2) is applied [1].

$$\frac{1}{c}\frac{\partial\phi(\mathbf{r},t)}{\partial t} = \nabla[D(\mathbf{r})\nabla\phi(\mathbf{r}t)] - \mu_a(\mathbf{r})\phi(\mathbf{r},t) + Q(\mathbf{r},t) \qquad (2)$$

where ϕ is the fluence rate, t time, **r** the position vector, c the

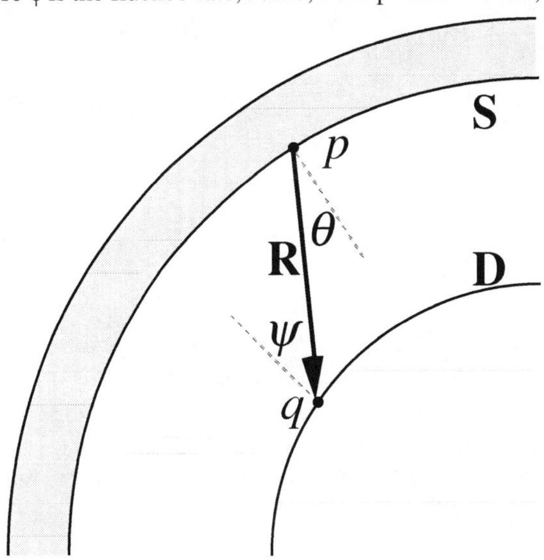

Fig. 3. Radiation exchange in the clear layer.

speed of light, $D = 1/3\mu_s'$ the diffusion coefficient, μ_s' the reduced scattering coefficient, Q the light source.

The boundary condition for equation (2) is given by the following equation (3), which is derived from the continuity of the photon flux at the boundary.

$$-D(\boldsymbol{r})\frac{\partial\phi(\boldsymbol{r},t)}{\partial\boldsymbol{n}} = \frac{1}{2}\phi(\boldsymbol{r},t) - 2\Gamma \text{ : at the surface} \quad (3)$$

where $\boldsymbol{n}$ is the outward unit vector normal to the boundary, and the reflection due to the refractive index mismatch at the boundary is neglected. This boundary condition connects the photon diffusion in the scattering regions and the radiation exchange in the clear layer. Once the fluence rate is calculated the path probability distribution (or photon hitting density) between a source and a detector at specified positions can be calculated by making a product of two fluence rates for the cases of light injection at the source and the detector positions [4, 5].

Light Propagation in the Total Hemisphere

Because of the difficulty in calculating the temporal variation of ultra-short pulse light propagation by the radiosity concept, only continuous wave (CW) light is assumed. CW light source is incident on a point at the periphery of the bottom plane of the outer hemisphere representing scalp. Initial 3D distribution of the fluence rate is appropriately given by an analytical solution for a semi-infinite homogeneous medium. Using the initial distribution of the fluence rate, the intensity at the inner surface of the skull is obtained by assuming that the radiation is isotropic because the error induce d by this assumption has been found small [2]. Then the incoming fluxes Γ to the outer surface of the gray matter is calculated by equation (1). Similarly, the incoming fluxes Γ to the inner surface of the skull is calculated.

New distribution of the fluence rate is then calculated by 3D FEM to solve the photon diffusion equation with the above obtained Γ used in the boundary condition (3). Obtained distribution is used as the new initial distribution for calculating the incoming fluxes by the radiosity concept. This process is iterated until a convergence is reached.

Although the optical properties of scalp, skull, gray matter and white matter are different each other, they are equally given as $\mu_s' = 1.0$ mm^{-1} and $\mu_a = 0.001$ mm^{-1} because the main purpose of this study is to see the effect of the clear CSF layer on the light propagation. The clear CSF layer does not have scattering obviously and its absorption coefficient was also neglected. The size of a neonatal head was referred to the model. The outer and inner diameters of the outer hemisphere were given as 70 mm and 64 mm, with the thickness of 3 mm for both the scalp and skull. The thickness of the CSF layer was changed from 3 mm to 6 mm which was rather thicker than the realistic human head to emphasize the effect of the CSF layer. Therefore, the outer diameter of the inner hemisphere was changed from 58 mm to 52 mm with the fixed thickness of 3 mm for the gray matter layer.

Results and Discussion

The distribution of the fluence rate for homogeneous

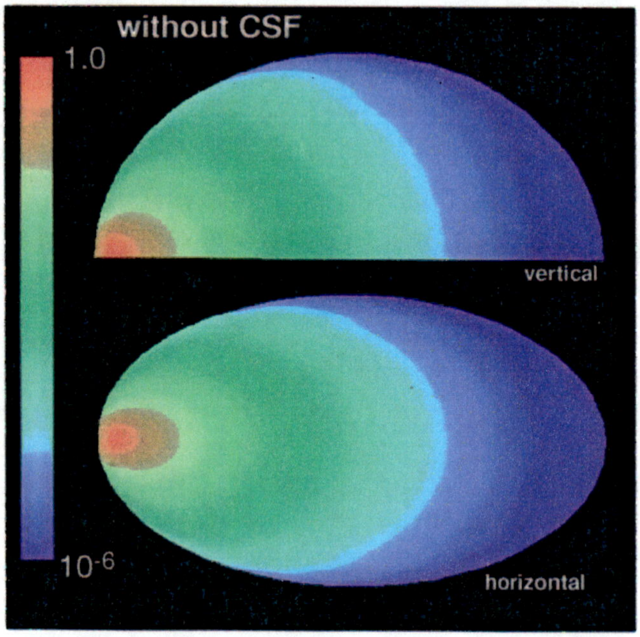

Fig. 4. Distribution of the fluence rate for a homogeneous hemisphere without clear CSF layer. $\mu_s' = 1.0$ mm^{-1} and $\mu_a = 0.001$ mm^{-1}, and the diameter of the hemisphere is 70 mm.

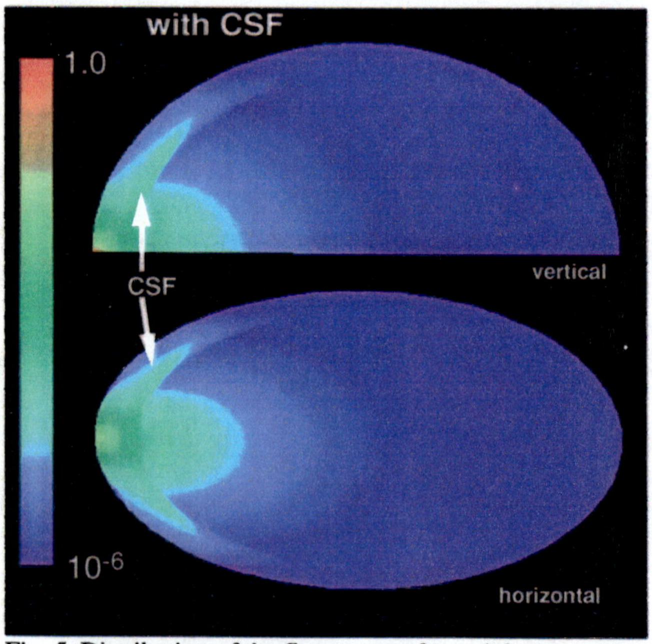

Fig. 5. Distribution of the fluence rate for an inhomogeneous hemisphere with a clear CSF layer. The hemispherical shell of the clear CSF layer has 3 mm-thickness with the outer diameter of 64 mm. The oprical properties of the scattering regions are $\mu_s' = 1.0$ mm^{-1} and $\mu_a = 0.001$ mm^{-1}.

hemisphere without including the CSF layer is shown in Fig. 4. We see a monotonous decrease of the fluence rate from the source point. On the other hand the distribution of the fluence rate for the case including the clear layer in Fig. 5 clearly shows the effect of the clear layer. In addition to the light propagation into the inner hemisphere, the propagation through the clear layer is extending further and looking like branches. The extension of the branch increases with the increase of the thickness of the clear layer to 6 mm as shown in Fig. 6.

Figure 7 shows the path probability distribution for the 3 mm thick clear layer with the angle of (a) 90° between the source and detector, and (b) 180° between the source and detector. Two light paths are seen, one through the inner hemisphere showing a so-called banana shape, and the other through the clear layer. The light channeling through the clear layer is clearly seen. Figure 8 shows the case of the 6 mm thick clear layer with the angle of (a) 90° between the source and detector, and (b) 180° between the source and detector. The effect of the clear layer is more pronounced and the path for the 180° shows a "ϕ" pattern.

In order to clarify the effect of the clear CSF layer, we intentionally specified the thickness of the layer thicker than that of a healthy human infant head. The results of the 3D FEM-radiosity calculation in this paper are very similar to those obtained previously, suggesting a strong effect of the clear layer on the light propagation in human heads. On the other hand, some people who have been working on brain oxygen monitoring by NIRS have a strong feeling that the effect of the clear layer is not so strong as suggested by this

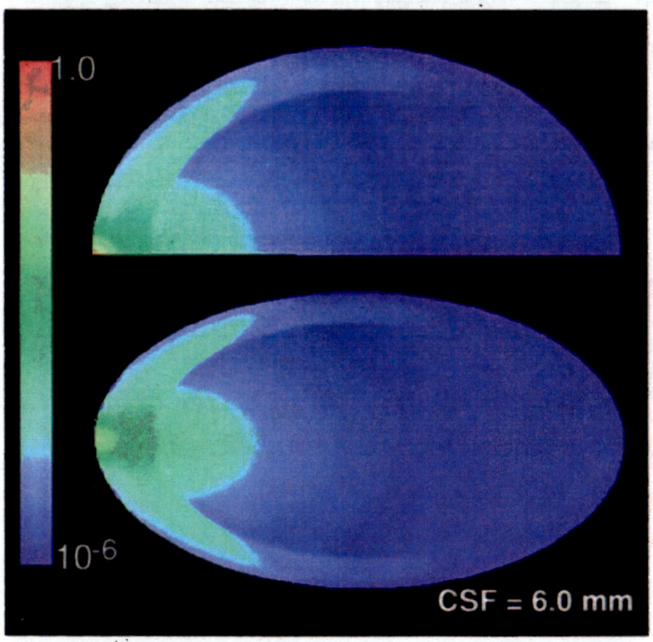

Fig. 6. Distribution of the fluence rate for an inhomogeneous hemisphere with a clear CSF layer. The hemispherical shell of the clear CSF layer has 6 mm-thickness with the outer diameter of 64 mm. The oprical properties of the scattering regions are $\mu_s' = 1.0\ mm^{-1}$ and $\mu_a = 0.001\ mm^{-1}$.

and the previous reports. One reason of the strong effect may be the overespecified thickness of the clear layer. The calculation models of these reports also assume smooth surfaces containing the clear layer, but usually the real surfaces of the inner surface of the skull and the brain surface are not smooth but rugged. In addition, there exist some membranes such as dura and arachnoid mater between the skull and the brain. These ragged surfaces and membranes may block the light traveling for a long distance straight through the clear CSF layer. The model of this study should be modified and improved to include these rugged surfaces and membranes which make the calculations more complicated but more realistic.

Conclusions

The hybrid 3D FEM and radiosity calculation of CW light propagation through a layered scattering hemisphere including a clear layer has suggested a strong effect of the clear layer on the light propagation by the channeling phenomenon. However, the model simplifies the geometry and more elaboration of the model may be necessary to include the rugged surfaces and membranes in the clear layer space.

Acknowledgment

The computation in this paper was done in RIPS (Research Information Processing System) of AIST (Agency of Industrial Science and Technology), MITI (Ministry of International Trade and Industry), Japan.

References

1. S. Takahashi, D. Imai, Y. Tanikawa-Takahashi, and Y. Yamada, "Fundamental 3D FEM analysis of light propagation in head model toward 3D optical tomography," Proc. of SPIE, Vol. 2979, pp. 250-260 (1997).
2. M. Firbank, S. R. Arridge, M. Schweiger, and D. T. Delpy, "An investigation of light transport through scattering bodies with non-scattering regions," Phys. Med. Biol., Vol. 41, pp. 767-783 (1996).
3. E. Okada, M. Firbank, M. Schweiger, S. R. Arridge, M. Cope, and D. T. Delpy, "Theoretical and experimental investigation of near-infrared light propagation in a model of the adult head," Appl. Opt., Vol. 36, pp. 21-31 (1997).
4. J. Haselgrove, J. Leigh, C. Yee, N-G. Wang, M. Maris, and B. Chance, "Monte Carlo and diffusion calculations of photon migration in non-infinite highly scatterign media," Proc. of SPIE, Vol. 1431, pp. 30-41 (1991).
5. J. C. Schotland, J. C. Haselgrove, and J. S. Leigh, "Photon hitting density," Appl. Opt., Vol. 32, pp. 448-453 (1993).

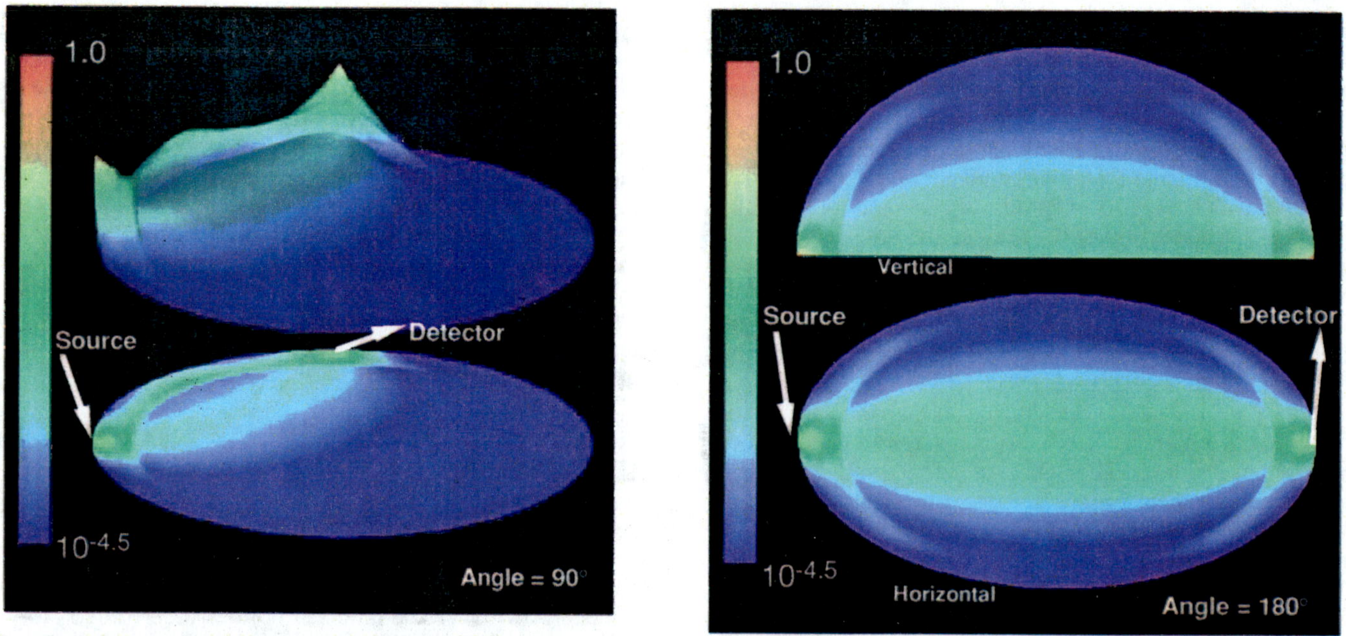

Fig. 7. Path probability distribution for an inhomogeneous hemisphere with a clear CSF layer, corresponding to Fig. 5 with the clear layer thickness of 3 mm. The angle between the source and detector is 90° (left figure) and 180° (right figure).

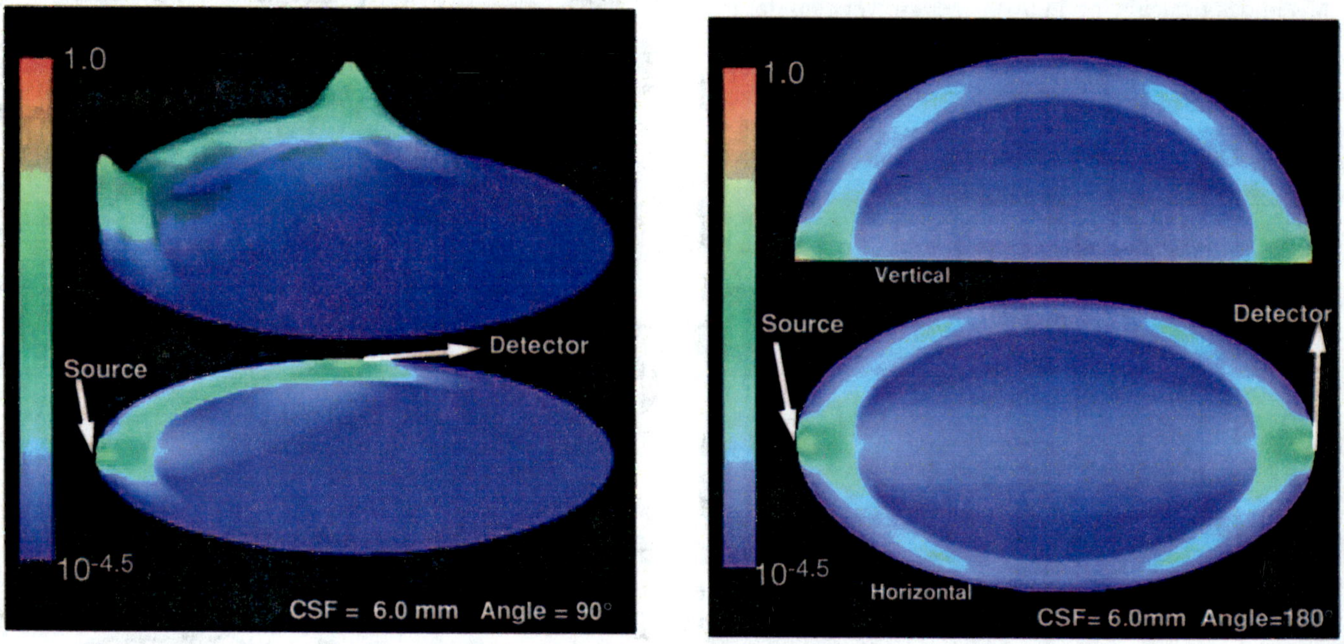

Fig. 8. Path probability distribution for an inhomogeneous hemisphere with a clear CSF layer, corresponding to Fig. 6 with the clear layer thickness of 6 mm. The angle between the source and detector is 90° (left figure) and 180° (right figure).

Investigation of the Direct and Indirect Signal Contributions of Brain Haematoma in Near Infrared Spectroscopy

Eiji Okada, Yukari Tanigawa*, Yukio Yamada*, Michael Firbank** and David T. Delpy***

Department of Electronics and Electrical Engineering, Keio University
3-14-1, Hiyoshi, Kohoku-ku, Yokohama 223 Japan

* Biomedical Division, Mechanical Engineering Laboratory, AIST-MITI
1-2, Namiki, Tsukuba, Ibaragi 305 Japan

** University Department of Radiology, Imaging Laboratory, Royal Victoria Infirmary
Newcastle upon Tyne NE1 4LP, UK

*** Department of Medical Physics and Bioengineering, University College London
1st Floor Shropshire House, 11-20 Capper Street, London WC1E 6JA, UK

Abstract

The direct and indirect contribution of brain haematoma to the near infrared spectroscopy instrument signal has been investigated by both experiment and Monte Carlo prediction. Absorption change by the haematoma (the direct effect) strongly contributes to change in both intensity and mean time of flight of the detected signal when the haematoma is located at or near the surface of the brain. In cases where the haematoma is located more than 10 mm below the brain surface the direct signal effect is negligible. However the haematoma is still indirectly detectable if it causes brain swelling because the detected light intensity is very sensitive to changes in the CSF thickness.

Key Words: (170.3660) light propagation in tissue, (170.7050) turbid media, near-infrared spectroscopy, Monte Carlo prediction.

1. Introduction

The heterogeneity of the tissue of the head strongly affects the propagation of light in the brain and causes ambiguity in the volume of tissue sampled by near infrared spectroscopy (NIRS) instruments. Recent studies on light propagation in layered media reveal that the clear cerebrospinal fluid (CSF) strongly affects light propagation in the adult brain and the sampled volume tends to be confined to the surface tissues of the cortex [1,2]. On the other hand published experimental data on the optical detection of brain haematoma shows that haematoma located in deeper regions of the brain are detectable from the change in the NIRS signal [3]. These results seem to be contradictory.

In this study, the direct and indirect contributions of haematoma to the NIRS signal are investigated to try and resolve this contradiction. The direct contribution of the haematoma to the NIRS signal is assumed to arise from an absorption change in the brain. However, since cerebral haematoma can result in several other effects such as swelling of the brain or leakage of blood into the cerebrospinal fluid (CSF), the effect of these as indirect contributions of the haematoma to the NIRS signal must also be considered. In this paper, the indirect contribution of the haematoma is assumed to be caused by a decrease in the CSF thickness due to brain swelling. Both the direct and indirect effects of the haematoma have been incorporated into a model of the adult head and the resulting changes in the detected light intensity and mean time of flight predicted by a Monte Carlo method.

2. Model and Method

2.1 Head Model

The geometry of the simplified adult head model is shown in Fig. 1. The model mainly consists of a homogeneous scattering medium and a clear layer. The surface component of the scattering medium imitates scalp and skull, and the deeper part imitates the brain. The clear layer mimics the cerebrospinal fluid. The

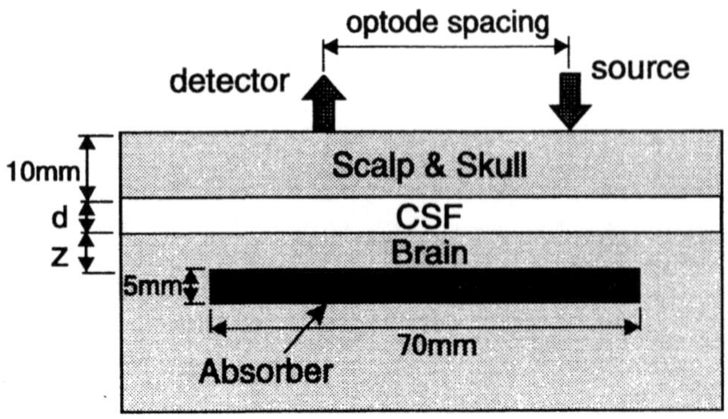

Fig. 1 Schematic diagram of head model.

optical properties of the scattering medium [4] are $\mu_s' = 2.0\ mm^{-1}$ and $\mu_a = 0.025\ mm^{-1}$ and of the clear layer are $\mu_s' = 0.01\ mm^{-1}$ and $\mu_a = 0.003\ mm^{-1}$.

A rectangular solid block of absorber, 70 mm in width, 5 mm in thickness, 100 mm in depth is embedded in the deeper part of the scattering medium to evaluate the direct effect of haematoma. The μ_a of the absorber is $0.5\ mm^{-1}$ and the μ_s' is the same as that of the scattering medium. The top surface of absorber is placed 0, 5 or 10 mm below the surface of the brain and the CSF thickness is kept constant at 2 mm. In order to imitate the indirect effect of the haematoma which could arise from brain swelling, the CSF thickness of the model is reduced from 2.0 mm to 1.5 or 1.0 mm, but no added absorber imitating the haematoma is embedded in the brain part so that the independent effects of added absorber or CSF thickness change on the NIRS signal can be assessed. Photon propagation in the model was calculated by a previously described Monte Carlo method [5].

In order to evaluate the contribution of the direct or indirect effects, the detected light intensity and mean time of flight for the model with and without the effects of the haematoma were predicted as a function of optode spacing. The attenuation change (ΔOD) caused by the effect of the haematoma was calculated from the detected light intensity for the model with haematoma (I_h) and without haematoma (I_0) as follows:

$$\Delta OD = \ln (I_0 / I_h) \qquad (1)$$

The change in mean time of flight caused by the effects of the haematoma was similarly calculated as follows:

$$rmt = \langle t_h \rangle / \langle t_0 \rangle \qquad (2)$$

Where $\langle t_h \rangle$ and $\langle t_0 \rangle$ are mean times of flight for the model with and without the haematoma, respectively.

2.2 Experiment

The phantom for the experiment was a water tank filled with intralipid suspension and ink. The wall of the water tank, mimicking the surface tissues and CSF layer was made of epoxy resin. The surface layer contains TiO_2 scattering particles and an IR absorbing dye and the optical properties were adjusted to $\mu_s' = 2.0\ mm^{-1}$ and $\mu_a = 0.025\ mm^{-1}$ at 780 nm. The thickness of surface layer was 10 mm and of CSF layer was 0, 1.0, 1.5, 2.0 mm for each wall. A rectangular opaque black plastic plate, 70 mm in width, 5 mm in thickness was immersed in the intralipid and was able to be positioned at a variable distance away from the wall. A picosecond pulsed laser at 780 nm illuminated the phantom and the reflected light was detected with an optical fibre bundle connected to the detector. The optical fibre bundle was connected to optical power meter to measure the reflected light intensity or to a streak camera to measure the temporal point spread function for calculating mean time of flight.

3. Results and Discussion

The direct contributions to detected intensity and mean time of flight are shown in Fig. 2 and 3, respectively. Lines represent Monte Carlo prediction and symbols represent experimental results. The Monte Carlo prediction and experimental resluts show the same tendency. In the case where the haematoma is located at the surface of the brain, both the intensity and mean time are significantly affected by the additional absorption of the haematoma. From the experimental results the difference in detected intensity for the model with or without the haematoma is about 1.5 OD at an optode spacing of 40 mm. The mean time of flight is less sensitive to the absorption change caused by the haematoma. The difference in mean time for the models is about 10 %. When the haematoma is located deeper in the brain, neither the detected intensity nor mean time tend to be affected by the absorption change caused by the haematoma.

Figure 4 shows the partial mean optical pathlength in the brain layer predicted by Monte Carlo method. In the homogeneous case (CSF thickness, d = 0.0) the partial optical pathlength in the brain steeply increases

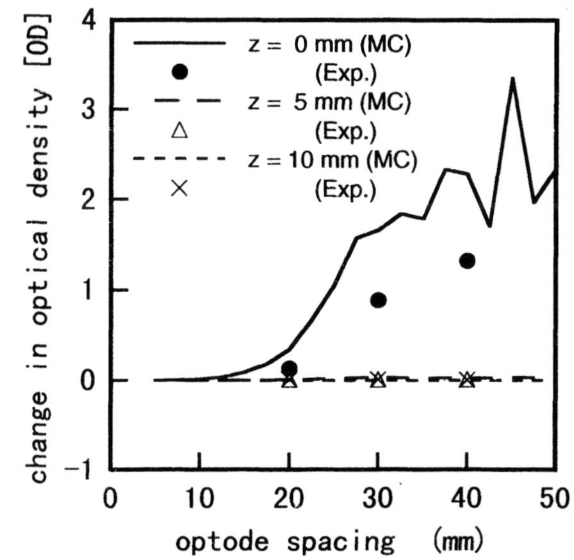

Fig. 2 The attenuation change caused by the direct effect of haematoma.

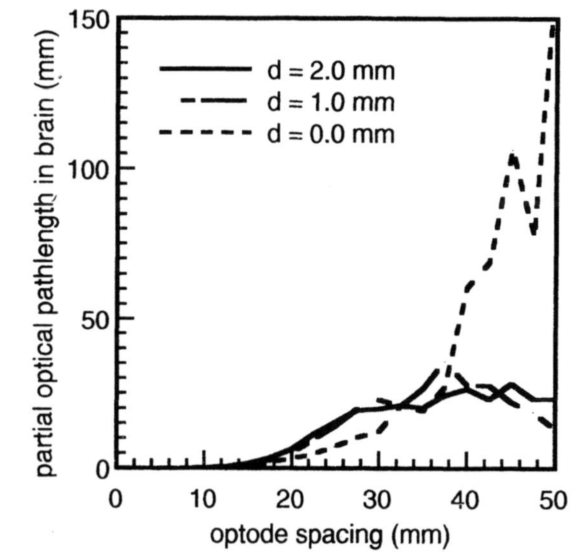

Fig. 4 Partial optical pathlength in the brain predicted by Monte Carlo method.

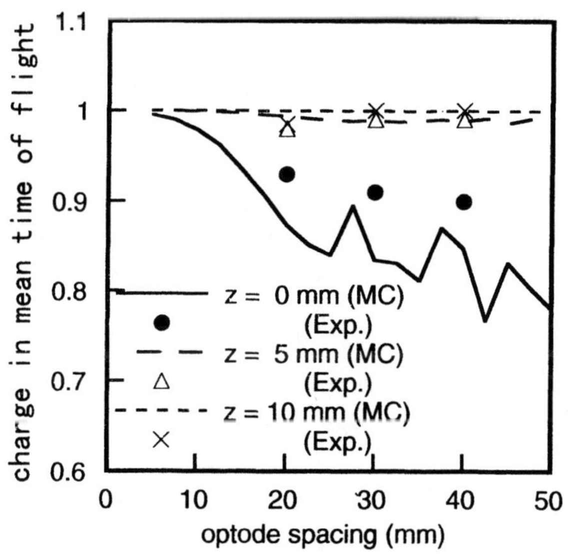

Fig. 3 The change in mean time of flight caused by the direct effect of haematoma.

once the optode spacing exceed 40 mm. On the other hand in the model with CSF layer, the partial optical pathlength in the brain is almost constant once the optode spacing is greater than 30 mm. This result means that large optode spacings do not enable one to probe the absoption change at a deeper level. It should be noted that the late arriving photons, which have travelled a long path in the head do tend to penetrate deeper into the brain even if CSF layer exists and hence continue to carry information about the absorption change even if the haematoma is located deeper. However the overall contribution of these late photons to the detected intensity and mean time is very slight and they are probably often below the detectable level. In order to detect the absorption change caused by a haematoma located deep in the brain, the late photons should be selectively detected using a time gate system.

The indirect contribution of the haematoma caused by a change in CSF thickness are shown in Fig. 5 and 6 for detected intensity and mean time of flight respectively. The CSF thickness significantly affects the detected intensity, and easily measurable changes can be observed even if the CSF thickness decreases by only 0.5 mm. The attenuation change caused by a decrease in CSF thickness of 0.5 mm is about 0.5 OD at an optode spacing of 40 mm. The mean time of flight is however only slightly changed by the CSF thickness variation. The difference in the mean time between the model with and without the haematoma is less than the statistical error of the Monte Carlo calculation. These results suggest that the observed considerable change in the light intensity caused by haematomas located deep in the brain does not probably reflect the absorption change in the brain and more probably reflects changes in CSF thickness due to

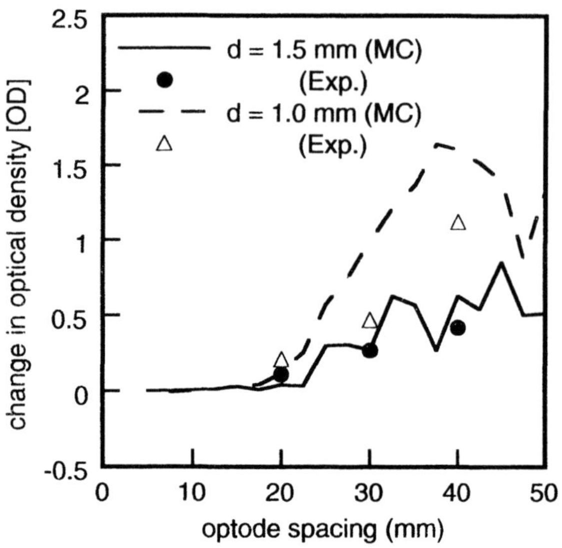

Fig. 5 The attenuation change caused by the CSF thickness variation.

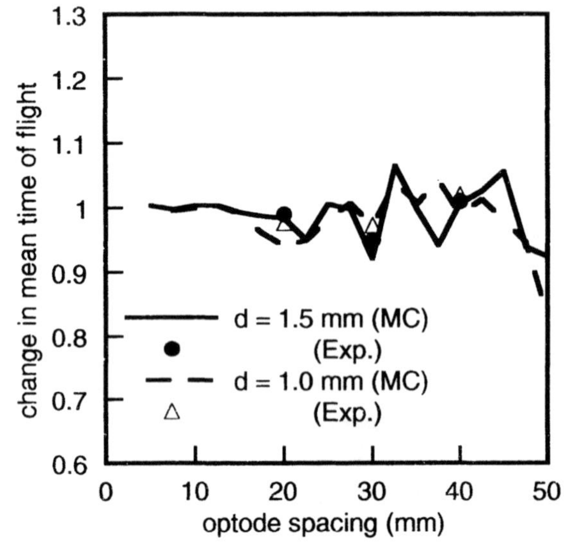

Fig. 6 The change in mean time of flight caused by the CSF thickness variation.

brain swelling. However, it must be noted that because of the effect of the CSF on the light distribution in the adult head, a very small change in the μ_a of the CSF (due to blood leakage) or a swelling of the pial vessels on the brain surface would also have a significant effect on detected intensity. This effect has not been modelled here, but could be anothermajor contributor to the indirect detection of haematomas.

4. Conclusions

In this study the contribution to the NIRS signal of an absorption change in the adult brain and of a change in CSF thickness caused by a brain haematoma has been estimated. In the case where the haematoma is located at or near the surface of the brain, the direct effect of the haematoma i.e. absorption, causes considerable change in both the optical density and mean time of flight. On the other hand, if the haematoma is located more than 10 mm below the brain surface it is difficult to directly detect the NIRS signal change caused by the added absorption. However the haematoma is still indirectly detectable if it causes brain swelling because the detected light intensity is very sensitive to the CSF thickness.

This work was supported by TEPCO Research Foundation, the Ministry of Education, Science and Culture of Japan under a Grant-in-Aid for Developmental Scientific Research No. 012-10091047, and the Engineering and Physical Sciences Research Council (UK).

References

1. E. Okada *et al.*: "Theoretical and experimental investigation of near-infrared light propagation in a model of the adult head", Appl. Opt., 36-1, 21/31 (1997).

2. E. Okada *et al.*: "The effect of overlying tissue on NIR light propagation in neonatal brain", OSA TOPs on Advances in Optical Imaging and Photon Migration, 2, 338/343 (1996).

3. S. P. Gopinath *et al.*: "Near infrared spectroscopic localisation of intracranial hematomas", J. Neurosurg., 79, 43/47 (1993).

4. P. van der Zee *et al.*: "Optical properties of brain tissue", Proc. SPIE, 1888, 454/465 (1993)

5. M. Hiraoka *et at.*: "A Monte Carlo investigation of optical pathlength in inhomogeneous tissue and its application to near-infrared spectroscopy", Phys. Med. Biol., 38, 1859/1876 (1993).

Accuracy of the diffusion approximation in determining the optical properties of a two-layer turbid medium

George Alexandrakis, Thomas J. Farrell, and Michael S. Patterson
Department of Medical Physics, Hamilton Regional Cancer Center and McMaster University, 699 Concession Street, Hamilton, Ontario, Canada L8V 5C2

Abstract
The purpose of this study was to examine the possibility of determining the optical properties of a two-layer turbid medium using a diffusion approximation radiation transport model. Continuous wave (CW) and frequency domain (FD) low noise Monte Carlo (M.C.) data were fitted to the model. Marquardt-Levenberg and a simulated annealing algorithm were used, and compared as optimization techniques. Our particular choice of optical properties for the two-layer model was consistent with skin and underlying fat in the presence of an exogenous chromophore (table 1). It was found that the CW diffusion solution could never estimate all optical properties reliably. The combined CW and FD solutions could not estimate some of the top layer optical properties to better than 10%, although the absorption and transport scattering coefficients of the bottom layer could be estimated to within 0.5 and 7% respectively. No improvement was found from simultaneously fitting M.C. data at three different modulation frequencies. These results point to the need for a more accurate radiation transfer model at small source-detector separations.

Key words
(290.1990) diffusion; two-layer turbid medium; Monte Carlo simulations; frequency domain.

Introduction

Optical reflectance measurements offer the potential to gain diagnostic information about biological tissue by determination of the absorption and scattering coefficients. The absorption coefficient depends on the concentration of endogenous chromophores such as hemoglobin and may reflect the physiological status of tissue (1). Pharmacokinetic information about exogenous chromophores can also be obtained in this manner (2). The scattering coefficient has been exploited as an indicator of tissue structure (3) and the relative indices of refraction of tissue components (4). Estimation of the absolute absorption (μ_a) and scattering (μ_s) coefficients requires a model of light propagation between the light source and detectors on the tissue surface. To date, most in vivo applications of these models have relied on the diffusion approximation of light transport and the assumption that the tissue is macroscopically uniform (5). This made μ_a and μ_s spatially invariant, while the preferential forward scattering of photons was equivalently described by an isotropic transport scattering coefficient μ_s'.

It is clear that in most situations the tissue itself is optically inhomogeneous or physiological changes occur only in one tissue component. The goal of our work was to determine whether it is possible in principle to estimate the optical properties of two-layer tissues (μ_a, μ_s' in both layers and the thickness of the top layer, for 'skin' and underlying fat) within 10% accuracy. Hence, from fitting an appropriate two-layer approximate model to measurements of surface reflectance, the drug concentration in each layer could be determined.

Methods

The diffusion approximation radiation transport model for a two-layer medium as reported by Kienle *et al.* (6) was computed. A pencil beam of light enters the top layer of the medium (thickness l) along the z-direction. It is assumed that all photons scatter at one transport mean free path length $z_o = 1/\mu_s'$ into the top layer. The two-layer medium diffusion equations for the fluence rates due to the intensity modulated source (circular frequency ω) are then given by:

$$D_1\nabla^2\Phi_1(r,\omega) - \mu_{a1}\Phi_1(r,\omega) = -\delta(x,y,z-z_o)e^{j\frac{\omega}{c_m}r}, \quad 0 \le z \le l \qquad (1)$$

$$D_1\nabla^2\Phi_2(r,\omega) - \mu_{a2}\Phi_2(r,\omega) = 0, \quad l \le z \qquad (2)$$

where $r = \sqrt{x^2+y^2+z^2}$. Also, $D_i = \frac{1}{3}(\mu_{ai} + \mu_{si}')^{-1}$ are the diffusion constants, and Φ_i the fluence rates for layers i=1, and 2 respectively. The boundary conditions implemented were: (a) Zero fluence rate at an extrapolated boundary at $z= -z_b= -2D_1$. (b) Finite photon fluence rate for all large r. (c) Continuity of fluence rate and flux at the boundary between the two layers. It was shown by Kienle *et al.* (6) that the fluence rate at the surface (z=0) can be obtained by a spatial Fourier inversion of eq.(1), giving:

$$\Phi_1(\rho, z=0) = \frac{1}{2\pi}\int_0^\infty \phi_1(z=0) s I_o(s\rho) ds \tag{3}$$

where

$$\phi_1(z=0) = \frac{\sinh(\alpha_1(z_b+z_0))}{D_1\alpha_1} \times \frac{D_1\alpha_1\cosh(\alpha_1 l) + D_2\alpha_2\sinh(\alpha_1 l)}{D_1\alpha_1\cosh(\alpha_1(l+z_b)) + D_2\alpha_2\sinh(\alpha_1 l + z_b))} - \frac{\sinh(\alpha_1 z_0)}{D_1\alpha_1} \tag{4}$$

$\rho = \sqrt{x^2+y^2}$, 's' is a spatial frequency dummy variable, $I_0(s\rho)$ the zeroth order Bessel function, and $\alpha_i = (D_i s^2 + \mu_{ai} + j\omega / c_m) / D_i$ (i=1,2). For a refractive index-matched boundary, the surface reflectance can then be calculated by:

$$R(\rho,\omega) = \frac{1}{4}\Phi_1(\rho, z=0,\omega) + \frac{1}{2}\left.\frac{\partial\Phi_1(\rho,z,\omega)}{\partial z}\right|_{z=0} \tag{5}$$

A low noise M.C. reflectance calculation for a two-layer medium with known optical properties was used as a tool to test how much information the CW and FD two-layer medium diffusion approximation solutions could extract under ideal circumstances. Marquardt-Levenberg ('Mrqrt', non-linear least squares) and Simulated Annealing ('S.A.', combinatorial optimization) were used and compared as algorithms in fitting the appropriate diffusion approximation solutions to M.C. generated reflectance measurements. The latter were fitted as relative or absolute quantities, in order to find the optimal combination of measurements that is readily realizable experimentally. A reflectance scaling factor (q_r), and an additive phase constant (q_{ph}) that minimize χ^2 at each iteration were used to match the relative M.C. measurements with the absolute predictions of the diffusion approximation. For absolute reflectance and phase fittings, q_r= 1, and q_{ph}= 0 respectively. Results from 'Mrqrt' fittings using the 'true' parameters of the two-layer medium as an initial guess (table 1) are indicated 'Mrqrt-t', and those using a 'false' initial guess are indicated 'Mrqrt-f '. Similarly for 'S.A.' the conventions are 'S.A.-t' and 'S.A.-f ' respectively. Several different 'false' initial guesses were tested, but one (table 1) was consistently used, in order to compare the performance of the different fitting algorithms.

Results

The CW reflectance as calculated in the diffusion approximation for a two-layer medium, and the corresponding M.C. results are shown in fig. 1. M.C. results for a semi-infinite medium having the optical properties of the top layer and the semi-infinite medium diffusion solutions for the top and bottom layers are also shown. It can be seen that for ρ>6mm the agreement between diffusion and M.C. results is good. Not so for smaller ρ, though the two-layer diffusion solution becomes identical to the semi-infinite medium solution for the top layer, and similarly for the M.C. results. Fig.1 also illustrates the interesting point that the two-layer diffusion approximation was no less accurate at small ρ compared to the semi-infinite medium solution. This leads to the conclusion that the discrepancy between the two-layer diffusion solution and corresponding M.C. results at small ρ does not stem from the more complicated geometry of the two-layer medium relative to the semi-infinite medium one. Rather, it comes from the breakdown of the diffusion approximation at small ρ, and hence is a fundamental limitation of the model.

The AC amplitude and phase delay (fig. 2) for an amplitude modulated incident beam were also calculated. The phase delay as calculated by the diffusion model was close to the M.C. results though it consistently underestimated them. The AC amplitude offered information very similar to that shown in fig.1 for the CW reflectance. The agreement of these FD diffusion calculations with M.C. results became worse with increasing modulation frequencies for all ρ. It was also found that including M.C. reflectance measurements for ρ<6mm in any fitting procedure reduced the quality of fit and the accuracy of the retrieved optical properties.

The fitted parameter percentage errors for 'Mrqrt-t', Mrqrt-f ', and 'S.A.-f ' from fitting M.C. CW reflectances as relative measurements for 6mm≤ρ≤20mm are shown in fig. 3. It was found that 'Mrqrt' was sensitive to initial guesses of the true optical properties and was consistently trapped at local minima, whereas 'S.A.' was found to be insensitive to initial parameter guesses. In all cases the errors in fitted optical parameters were considerably larger than our goal of 10%. Even when fitting the M.C. results as absolute CW reflectance measurements (fig. 4), the difference in fitted optical parameters obtained form 'Mrqrt-t ' and 'S.A.-f ' was very large even though the minima in χ^2 found by the two methods were very close to each other. This demonstrated the ill-conditioned nature of the problem.

To increase the amount of information fed into the fitting procedures, all possible combinations of CW, AC reflectance amplitude and phase delay M.C. data were fitted as relative or absolute measurements. The most informative, yet the hardest measurements to obtain experimentally are absolute ones. Fig. 5 shows the results obtained from fitting CW reflectance and phase as absolute quantities. A considerable improvement is seen compared to fitting CW reflectance alone. The top layer thickness was determined to 5%, μ_{a2} to 0.5%, and μ_{s1}' to 7%, while the errors in μ_{a1} and μ_{s1}' were slightly over 10%. Between the two extremes of having all measurements relative or absolute there are two more possibilities: Absolute phase and relative CW reflectance fittings to the M.C. data produced results almost as bad as having both measurements relative. Absolute reflectance and relative phase on the other hand, produced results almost as good as when both measurements were absolute.

In the hope that the accuracy of fitted parameters could be further improved, AC reflectance amplitude information was included in the fittings. The AC reflectance amplitude curves look very much like the CW ones shown in fig. 1. The difference is that they drop faster than the CW ones with increasing ρ, more so for higher modulation frequencies. No improvement was observed in the accuracy of the estimated optical properties by the addition of such information.

The next logical step was to simultaneously fit all of CW, AC reflectance amplitude and phase delays as absolute quantities at several widely different modulation frequencies. This would assess the potential of the time domain method, where δ-like light pulses are used, to improve parameter estimates. Fig. 6 shows the results from fitting measurements at three modulation frequencies (195MHz, 300MHz, 1GHz). The errors in fitted parameters were found to be almost identical to those from fitting only CW reflectance and AC phase as absolute quantities at a single modulation frequency.

Conclusions

The aim of our work was to find if, and how accurately, the diffusion approximation could determine the true optical properties of a two-layer medium from low noise M.C. data. It was shown that the CW diffusion solution alone could never estimate optical properties and layer thickness reliably, irrespective of whether data were fitted as relative or absolute quantities. The combined CW and FD diffusion solution could estimate μ_{a2}, μ_{s2}' and l to an accuracy better than 10%, but the errors in μ_{a1} and μ_{s1}' were slightly over that. This was only true though, if all of the fitting data were absolute quantities. The inclusion of AC reflectance at a single modulation frequency did not improve the accuracy of the estimated parameters, nor did the simultaneous fitting of absolute CW, AC reflectance and phase at three different modulation frequencies. Hence it is concluded that a limit was reached in the amount of useful information that can be extracted using a diffusion model. It is thought that there exists useful information in the M.C. data that the diffusion approximation cannot exploit because:

(a) At short source-detector distances where the reflectance is characteristic of the top layer, the agreement between diffusion and M.C. is not good.

(b) As modulation frequency increases the diffusion approximation becomes less accurate.

Developing a two-layer turbid medium light transport model for regimes where diffusion fails is our next goal.

Acknowledgments

This research was supported by the National institutes of Health grant P01-CA43892.

References

1. B. Chance, Q. Luo, S. Nioká, D.C. Alsop, and J.A. Detre, "Optical investigation of physiology: A study of biomedical contrast: Intrinsic and extrinsic," Phil. Trans. Roy. Soc. Lond. *B* 352 707-716 (1997).
2. R. A. Weersink, J. E. Hayward, K. R. Diamond, and M. S. Patterson , "Accuracy of non-invasive *in vivo* measurements of photosensitizer uptake based on a diffusion model of reflectance spectroscopy," J. Photochem. Photobiol. 66, 326-335 (1997).
3. J. R. Mourant, I. J. Bigio, J. Boyer, R. L. Conn, T. Johnson, and T. Simada, "Spectroscopic diagnosis of bladder cancer with elastic light scattering," Lasers Surg. Med. 17, 350-357 (1995).
4. J. T. Bruulsema, J. E. Hayward, T. J. Farrell, M. S. Patterson, M. Essenpreis, G. Scmolzeisen-Redeker, D. Bocker, L. Heinemann, M. Berger, T. Koschinsky, J. Sandahl-Christiansen, and H. Orskov, " Correlation between blood glucose concentration in diabetics andnon-invasively measured tissue optical scattering coefficient," Opt. lett. 22, 190-192 (1997).
5. T. J. Farrell, and M. S. Patterson, " A diffusion theory model of spatially resolved, steady state diffuse reflectance for the non-invasive determination of tissue optical properties *in vivo*," Med. Phys. 19, 879-888 (1992).
6. A. Kienle, M. S. Patterson, N. Utke, R. Rays, G. Wagnieres, and H. Van Den Bergh, "Determination of the optical properties of two-layer turbid media," Appl. Opt. 37, 779-791 (1998).

Table 1 The 'true' set of optical properties used to generate M.C. data, and a selected 'false' guess of the parameters used as input to all fitting routines.

Parameters	'True'	'False'
μ_{a1} / mm^{-1}	0.0775	0.0175
μ_{s1}' / mm^{-1}	1.49	1.0
μ_{a2} / mm^{-1}	0.007	0.013
μ_{s2}' / mm^{-1}	1.0	1.1
l / mm	1.5	3.5

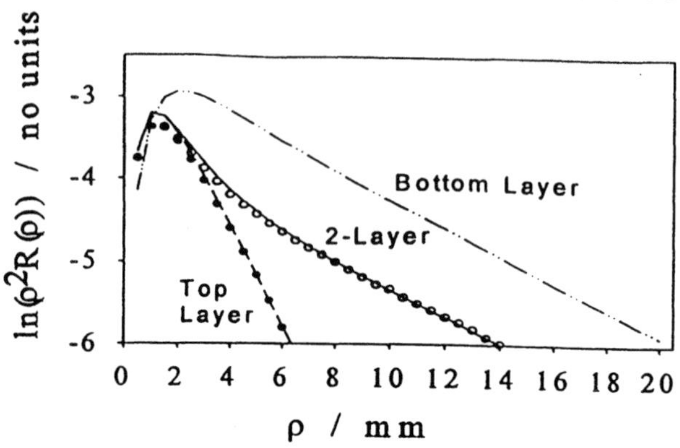

Fig. 1 M.C. data (circles) and the corresponding diffusion approximation solution (lines) for CW reflectance.

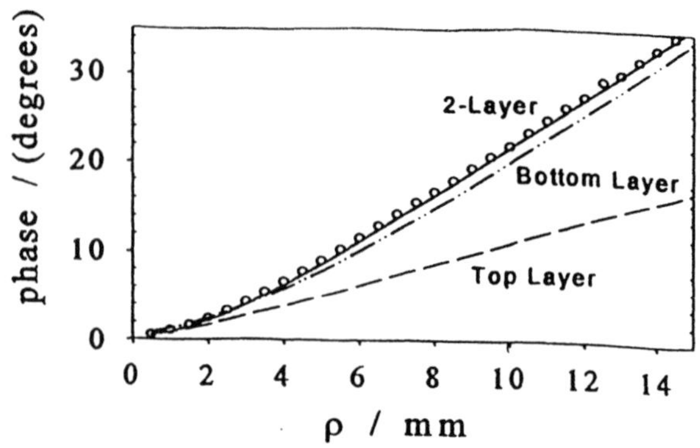

Fig. 2 M.C. data (circles) and the corresponding diffusion approximation solution (lines) for AC phase at 195MHz.

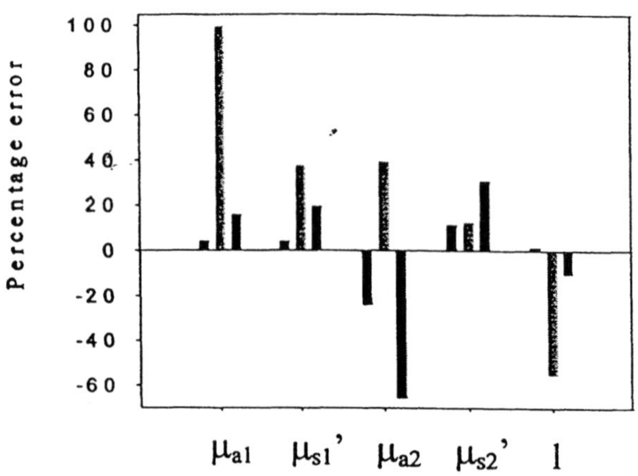

Fig. 3 Fitted % errors for M.C. data fitted as relative CW reflectance measurements (1st column from the left: 'Mrqrt-t', 2nd column: 'Mrqrt-f', 3rd column: 'S.A.-t').

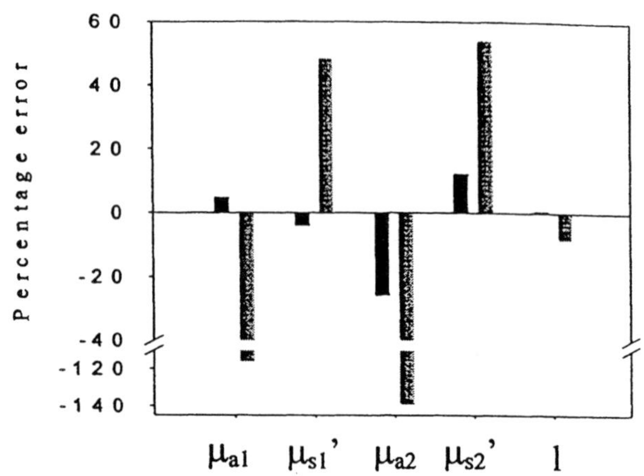

Fig. 4 Fitted % errors for M.C. data fitted as absolute CW reflectance measurements (1st column from the left: 'Mrqrt-t', 2nd column: 'S.A.-f', 'Mrqrt-f ' did not converge).

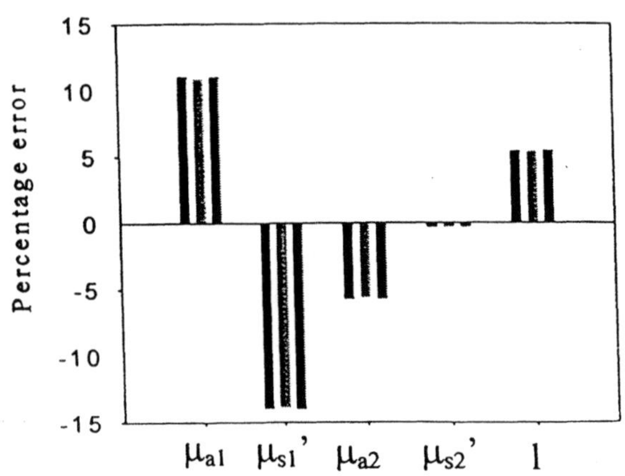

Fig. 5 Fitted % parameter errors for M.C. data fitted as absolute CW reflectances and AC phase measurements (1st column from the left: 'Mrqrt-t', 2nd column: 'S.A.-f', 3rd column: 'S.A.-t')

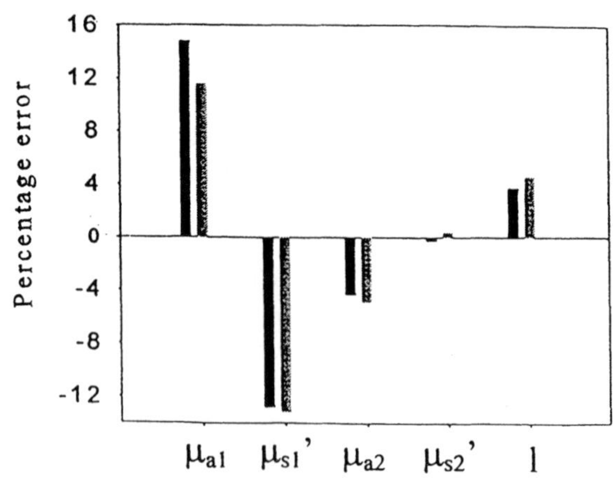

Fig. 6 Fitted % parameter errors for M.C. data simultaneously fitted as absolute CW reflectance, AC reflectance and phase measurements at 195MHz, 300MHz, and 1GHz (1st column from the left: 'Mrqrt-t', 2nd column: 'S.A.-f')

Density of Photon Paths in Turbid Medium for Short Times of Flight

L. T. Perelman, J. N. Winn, K. Chen, J. Wu, R. R. Dasari, and M. S. Feld

G. R. Harrison Spectroscopy Laboratory
Massachusetts Institute of Technology
Cambridge, Massachusetts 02139

The major obstacle to employing laser-based methods for imaging disease inside the body is the turbidity of biological tissue [1]. The theory of radiative transfer is usually used to treat this problem [2], but analytic solutions cannot be obtained for the equation of transfer for most realistic problems. The solutions which do exist are often based on the isotropic diffusion approximation, which assumes that enough scattering events have occurred so that light is uniformly scattered in all directions. This approach fails near boundaries or sources, and (when each scattering event itself is not isotropic, as is the case in biological tissue) does not hold for early times, before multiple scattering has randomized photon trajectories. Yet these early-arriving photons would be useful for tomography, since they preserve information about where they came from.

We designed an experiment to measure the distribution of photon paths in a turbid medium, for short times of flight. The scattering medium was a glass tank (15" x 15" x 5") filled with an aqueous solution of polystyrene beads, of diameter $1.06\mu m$. Figure 1 shows the view looking down into the tank from above. By varying the concentration of the beads, we could adjust the scattering coefficient μ_s of the medium, as measured by a spectrophotometer. The anisotropy factor g (the average cosine of the scattering angle) for these polystyrene beads was computed to be 0.91 using Mie theory, permitting calculation of the reduced scattering coefficient $\mu_s' = \mu_s(1-g)$.

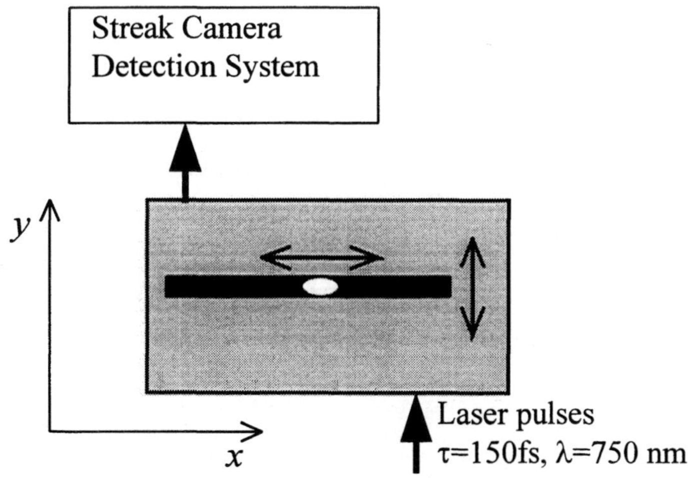

Figure 1. Schematic view of the tank, from above. To be detected, photons must travel from the source to the hole in the screen to the collection fibers. The source and detector are separated by $\Delta x = 10cm$ and $\Delta y = 12.7cm$.

Entering the tank on one side were laser pulses with a duration of ~150 fs and a wavelength of 750 nm, which were generated by a Coherent Mira 900 mode-locked Ti:sapphire laser pumped by a Coherent Innova 400 multiline argon-ion laser. On the other side of the tank they were collected by a bundle of eight optical fibers.

To separate photon paths by time of flight, we focused the light from the fiber bundle onto the slit of a Hamamatsu streak camera. To separate these photon paths spatially, we dangled a black screen with a small (0.5 cm) hole into the solution. By adjusting the location of the screen and the hole in two dimensions (on a square grid with spacing 0.5 cm), we could map out the temporal signal $P(t)$ for each point (x, y) in the tank, giving the density of photon paths $P(t, x, y)$.

To describe these paths theoretically, we borrow the central result of the "path integral approach"(PI) to photon migration [3] derived in [4], which is the probability of finding a photon at location $\mathbf{r}_2$ which began at $\mathbf{r}_1$ after time t. To describe our experiment, we divide a photon's propagation into two steps: propagation from the source at $\mathbf{r}_s$ to the hole at $\mathbf{r}_h$ in time τ, and then propagation from $\mathbf{r}_h$ to the collection fibers $\mathbf{r}_f$ at $\mathbf{r}_0$ in time $t-\tau$. For the proper treatment of a boundary at each step we use the imaginary source similar to one described for diffusion approximation in [5] and then we integrate over all possible times t. Calling $\mathbf{r}_s - \mathbf{r}_h = \mathbf{r}_1$ and $\mathbf{r}_f - \mathbf{r}_h = \mathbf{r}_2$, the result is:

$$P_{PI}(t, \mathbf{r}_1, \mathbf{r}_2) = \left(\frac{3\mu}{4\pi}\right)^3 y_1 y_2 \int_{r_1}^{t-r_2} d\tau \, \frac{1}{\tau^{5/2}} \left(1 - \frac{r_1^2}{\tau^2}\right)^{(3\mu\tau/4)-1} \cdot \frac{1}{(t-\tau)^{5/2}} \left(1 - \frac{r_2^2}{(t-\tau)^2}\right)^{(3\mu(t-\tau)/4)-1} . \quad (2)$$

where μ is the reduced scattering length. We have used units in which $c = 1$. This formula is simple enough to be quickly computed and compared to the experimental data. Meanwhile, the corresponding prediction of diffusion theory can be obtained by replacing the factors $\left(1 - r^2/t^2\right)^{3\mu t/4-1}$ with the exponentials $\exp(-3\mu r^2/4t)$, in which case the integral can be computed by hand:

$$P_{diff}(t, \mathbf{r}_1, \mathbf{r}_2) = \frac{(3\mu)^{3/2}}{8\sqrt{\pi}} \frac{y_1 y_2 (r_1 + r_2)}{4\pi^2 r_1^3 r_2^3 t^{5/2}} \exp\left(-\frac{(r_1 + r_2)^2}{t^2}\right) \left[2(r_1 - r_2)^2 + 2r_1 r_2 \left(\frac{3\mu(r_1 + r_2)^2}{2t} + 1\right)\right] . \quad (3)$$

The PI formula (2) agrees with the data for early times, unlike diffusion theory (3).

To provide a more detailed comparison of theory and experiment, in figure 2 the contours of the entire PI probability distribution (2) are plotted on top of the experimental data. These contours line up nicely with the dots representing the data.

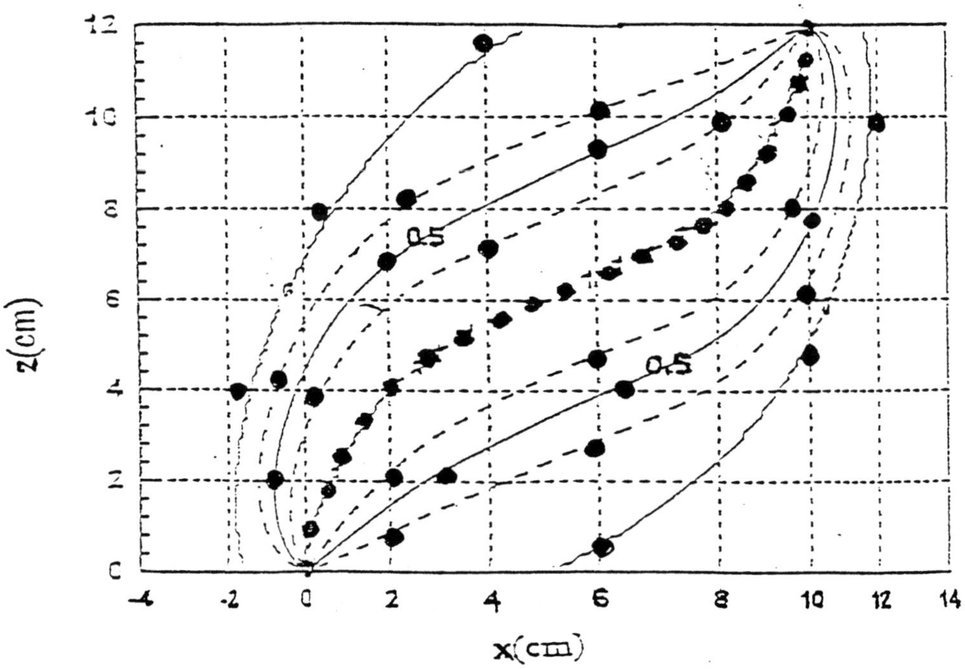

Figure 2. The distribution of photon paths arriving at time $t = (850 \pm 5)ps$, as a function of position, for the case $\mu_s' = 0.51cm^{-1}$. The dots represents the measured intensity (normalized along each horizontal line), the contour lines predicted by the PI theory (2).

In summary, a direct analytic formula can be derived for the density of photon paths in a turbid medium, for very early times of flight, using the path integral approach. The formula agrees with the experimentally-measured photon paths at early times, unlike diffusion theory.

References

1. A. G. Yodh and B. Chance, *Physics today*, **48**(3), 34-40, (1995).
2. A. Ishimaru, *Wave Propagation and Scattering in Random Media*, (Academic Press, Orlando, 1978).
3. L. T. Perelman, J. Wu, I. Itzkan, and M. S. Feld, *Phys. Rev. Lett.* **72**, 1341-1344, (1994).
4. L. T. Perelman, J. Wu, I. Itzkan, Y. Wang, R. R. Dasari and M. S. Feld, *Phys. Rev. E*, **51**, 6134-6141 (1995).
5. M. S. Patterson, B. Chance, and B. C. Wilson, *Applied Optics* **28**, 2331 (1989).

Dependence of Time and Space Resolved Reflectance on the anisotropy factor g and on the shape of the phase function

Sigrid AVRILLIER [a], Bernard GELEBART [a], Jean-Michel TUALLE [a], Eric TINET [a] and Georges WAGNIERES [b]
[a] *Laboratoire de Physique des Lasers, Université Paris Nord, 93430 Villetaneuse, France*
Tel : (33) 1 49 40 33 97, Fax : (33) 1 49 40 32 00, E-mail : avrillier@lpl.univ-paris13.fr
[b] *Institute of Environmental Engineering, Swiss Federal Institute of Technology (EPFL)*
CH-1015 Lausanne, Switzerland

Abstract

We have performed experiments to explore the dependence of time and space resolved reflectance measurements on the anisotropy factor g and on the shape of the phase function for a given g. The experimental results have been compared to Monte-Carlo simulations. We have deduced some rules to obtain the g value of turbid media by analyzing the backscattered light.

Key words

(170.5280) Photon migration ; (170.4580) Optical diagnostics for medecine ; (290.7050) Turbid media, (290.1350) Backscattering.

Introduction

Average optical coefficients measured in the near infrared range on relatively uniform volumes of tissues can yield useful information about their structure and their functions [1]. Clinically, it is also very useful to obtain *in vivo* optical parameters directly on the patient during laser surgery or therapy, since these parameters can vary during the treatment and thus lead to an evolution of the laser efficiency. There is therefore the need for non invasive real time determination of tissues optical properties.
Various methods, in different laboratories, are classically used to obtain the absorption coefficient μ_a and the reduced scattering coefficient $\mu_s'=(1-g)\,\mu_s$ of thick tissues from reflectance measurements [2,3,4] but, to our knowledge, no measurements of tissues anisotropy factor g has ever been performed with this type of technique.

Since g is related to the size distribution of the particles in the tissue, an evaluation of g could have a diagnostic interest [5,6,7]. For the same reasons another goal is to explore the influence of the phase function shape in these conditions [5,7].
The aim of this work is to determine if time and space resolved reflectance measurements R(r,t) can be used to explore the different moments of the scattering phase function.

Dependence of R(r,t) on g in case of a Mie phase function

In a recent work [8] we have shown that time and space resolved reflectance measurements at early times and close to the light source, could be used to measure the anisotropy factor g of latex spheres suspensions in water.

Monte Carlo simulations

We first explored this possibility by the means of Monte-Carlo numerical simulations. The algorithm we used has been described in details elsewhere [9]. The sample was assumed to be an homogeneous semi-infinite medium (depth is infinite and the upper surface is planar) with a refraction index equal to 1.33. The reduced scattering coefficient μ_s' was set to be 1 mm^{-1}, which is an usual value for biological tissues, and the g value was successively chosen to be : 0, 0.5, 0.715, 0.859 and 0.95. The results of the simulations (figure 1) clearly show that R(r,t) depends independently on g and on μ_s at early times when the distance r from the source is of the order of $1/\mu_s'$. Significant intensity variations with g can be seen in the early times range where the reflectance maxima can be observed. But these differences become negligible at longer times or at larger distances r from the source.

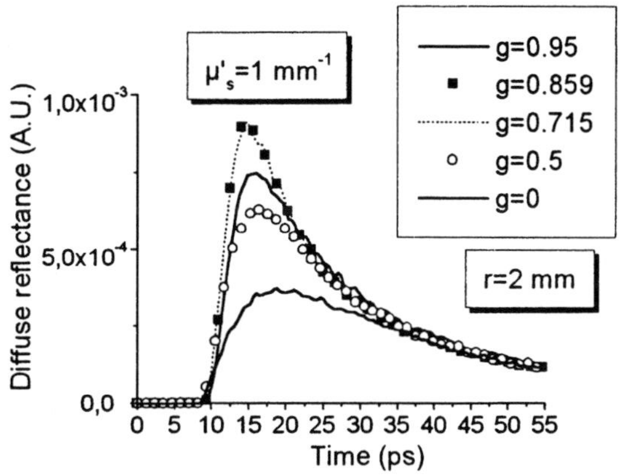

Figure 1. Monte Carlo simulations of a section of the reflectance map R(r,t) at r=2mm for different values of the anisotropy factor g. μ_s'=1mm^{-1}, and the scattering angle is determined from the Mie phase function. The illumination beam is supposed to be normally incident on a point of the sample surface. The backscattered light is detected, with a numerical aperture of 0.141.

Experimental measurements

As a consequence, experimental measurements of R(r,t) in these time and space ranges should allow separate evaluations of g and μ_s especially if μ_a and μ_s' have been previously determined by measurements performed in the diffusion approximation spatial and temporal ranges [10,11].Unfortunately, as can be seen in figure 1, tissue measurements, for which μ_s' is of the order of 1mm^{-1} and μ_s is of the order of 10 mm^{-1} in the near infrared range, would require a time resolution of about 1 picosecond, which is still difficult to reach experimentally.

To overcome this difficulty we performed experiments on aqueous suspensions of latex spheres with a μ_s value of 0.3mm^{-1} , that is about ten to a hundred times smaller than in biological tissues. In these conditions, the time and space ranges of interest are respectively a few hundreds picoseconds and a few milimetres. Various values of the spheres diameter (125nm, 363nm and 760nm) were alternately used in order to vary the anisotropy factor g of the solutions (respectively g=0.074, 0.619 and 0.857).

The apparatus used for these experiments is shown in figure 2 and has been previously described in detail [6-10] .Ultra-fast laser pulses are generated from a home-built self mode-locked Ti-Sa laser centred at λ=800 nm with a bandwidth of 10 nm. The temporal pulse width of the laser is about 100 fs and the repetition rate is 100MHz.

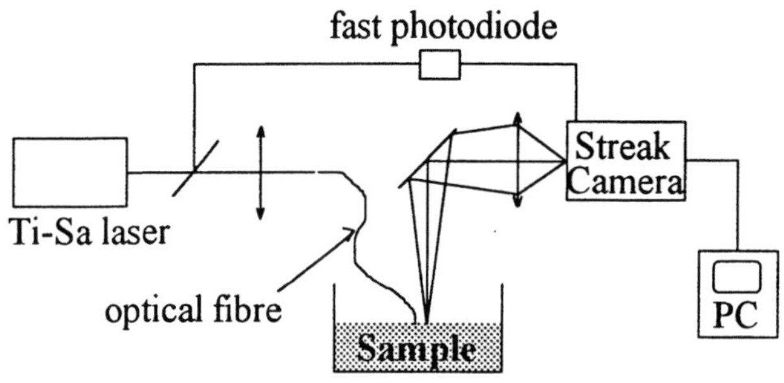

Figure 2. Experimental set-up.

The short light pulses produced by the laser reach the medium through an optical fibre. A synchroscan streak camera (ARP Photonetics) directly measures time and space-resolved reflectance along a segment of the sample's surface and the images are stored in a PC. The streak camera is triggered externally by a fast photodiode activated by the laser.

The R(r,t) maps obtained with the streak camera (figure 3) have been fitted by MC simulations which can easily lead to a measurement of g.

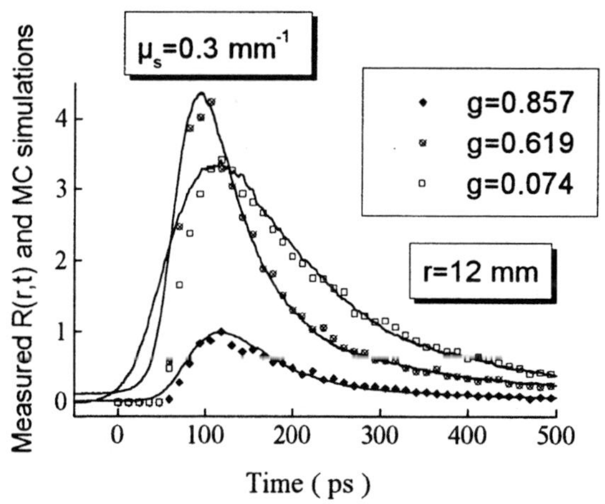

Figure 3. Experimental results (solid lines) and Monte Carlo simulations (symbols) for a section of the reflectance map R(r,t) at r=12mm for different values of the anisotropy factor ,g. A constant value of the scattering coefficient, μ_s=0.3mm^{-1}, was obtained for each g value by adjusting the latex spheres concentration in water.

The relatively poor signal to noise ratio in the Monte-Carlo (MC) results (symbols) is due to the fact that the reception numerical aperture is very small which drastically reduces the number of useful « photons » in rapid simulations. However one can observe that the raising time on the left part of the experimental curve (solid lines) is increased by the experimental time resolution of about 30ps. A better agreement between the experimental and the MC results has been found by performing a convolution of the MC curves by a 30ps large gate.

A possible method for g measurements

From these preliminary results it becomes reasonable to think of a possible method for the experimental determination of g from time and space resolved reflectance measurements. First, one could measure μ_a and μ_s' using the part of the map which can be described by the diffusion approximation. Then knowing these coefficients, the g value can be determine with Monte Carlo numerical simulations for r of the order of $1/\mu_s$'or less. The time resolution of about 1ps required for this type of experiments in case of tissues should be possible to reach in the near future with commercially available techniques.

Dependence of R(r,t) on the shape of the phase function for a given g

To explore the influence of the phase function shape in time and space resolved reflectance measurements we used suspensions of a mixture of latex spheres with different diameters.

Theoretical background

We have shown in a previous paper [5] that in case of a mixture of spheres with different diameters d_i , the scattering coefficient of the suspension is given by:

$$\mu_{s_{eq}} = \sum_i \rho_i \, \sigma_{s_i} = \sum_i \mu_{si}$$

where ρ_i is the number of spheres of diameter d_i per unit volume, σ_{si} is the scattering cross section of the spheres of diameter d_i and μ_{si} is the corresponding scattering coefficient.

The phase function of the suspension is then given by :

$$p_{eq}(\theta) = \sum_i \mu_{s_i} p_i(\theta) / \sum_i \mu_{s_i}$$

where $p_i(\theta)$ is the Mie phase function corresponding to the spheres of diameter d_i . Consequently, the g value of the mixture is also a linear combination of the individual g_i values:

$$g_{eq} = \sum_i \mu_{s_i} g_i / \sum_i \mu_{s_i}$$

It is therefore possible to modulate g and the shape of the phase function by using various mixtures of spheres.

In this study we decided to compare time and space resolved reflectance measurements performed on two different sphere suspensions having the same g value and the same μ_s value. The first one is a suspension of latex spheres with a diameter of 760nm corresponding to a g value of 0.857. The second one is a mixture of 4 types of latex spheres which diameters are 125nm, 214nm, 760nm and 2.032 µm. The relative concentrations of the spheres, which are respectively : 79%, 17%, 4.3% and 0.25%, have been chosen to be proportional to $d_i^{-\alpha}$ where α=2.895 is the fractal dimension corresponding to a g value of 0.857 for such a mixture [5].

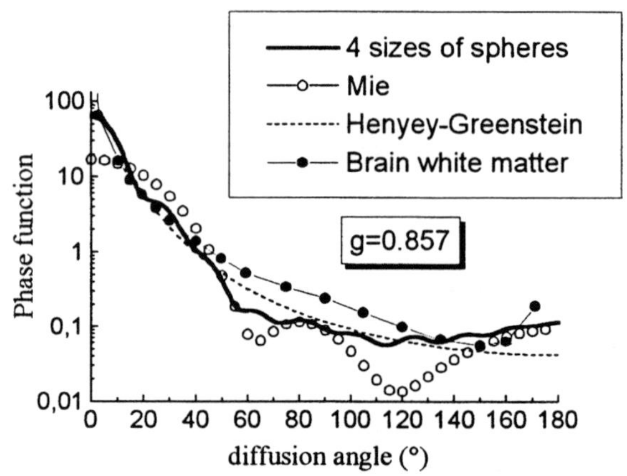

Figure 4. Phase functions for g=0.857.

Figure 4 allows a comparison between the Mie phase function corresponding to the suspension of 760nm spheres, the calculated phase function of the 4 sizes mixture, the Henyey-Greenstein phase function [13] and a measurement performed by Van der Zee on brain white matter at a wavelength of 750nm [14]. By chance the g value corresponding to these measurements is also 0.85.

From these curves, it is obvious that a mixture of spheres of different diameters is a better way to fit a tissue phase function than the use of a single size. No strong oscillations are observed as in the Mie phase function and the value obtained at 180° is better than the one obtained with the Henyey-Greenstein phase function.

Experimental measurements

R(r,t) measurements where performed with the same apparatus on the suspension of 760nm spheres and on the 4 sizes mixture.

The concentrations where adjusted in order to obtain μ_s =0.3mm^{-1} as in the previous experiments. Sections of the R(r,t) measured reflectance map at r= 12mm are shown in figure 5 and compared to Monte Carlo simulations. Here r=12mm corresponds to r=0.5/ μ_s '.

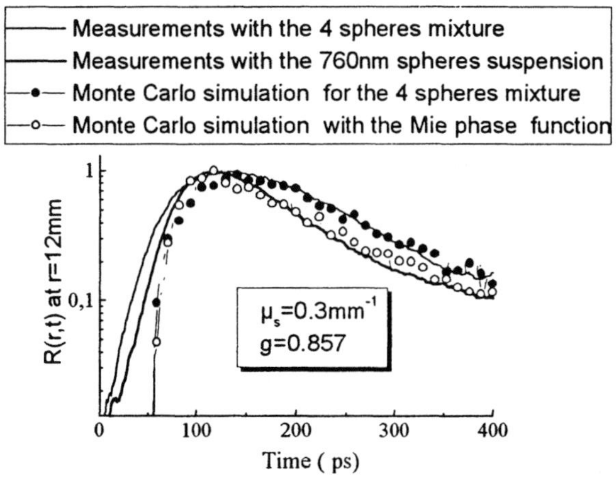

Figure 5. Measured phase functions for g=0.857 compared to MonteCarlo simulations.

These results clearly show that R(r,t) depends on the shape of the phase function at short distances from the source. This dependence could be used to have some information on the shape of tissue phase function and on the size distribution of the scatterers.

Conclusion

We have experimentally demonstrated that time and space resolved reflectance depends on g at short times when the distance from the source is of the order of $1/\mu_s$'.

This dependence can be used to determine g by the means of Monte Carlo simulations when μ_a and μ_s' have been previously measured from the results obtained in the diffusion approximation region.

The experimental time resolution has to be increased for biological tissue experiments. The 1ps time resolution required for tissues measurements should be possible to reach by performing a better synchronization between the laser and the streak camera.

We have also experimentally demonstrated that R(r,t) at a given g depends on the shape of the phase function at short times when the distance from the source is of the order of $0.5/\mu_s$'.

An evaluation of g and of the shape of the phase function could have a diagnostic interest and could give information on the particles size distribution in small volumes of tissues located near the surface. Such a technique could therefore contribute to the detection of early superficial cancerous lesions. These quite flat lesions are often asymptomatic and/or invisible under white light endoscopy and their detection remains a challenge since their presence is clearly related to tumor recurrence and subsequent cancer progression.

Acknowledgments

The authors are grateful to Photonetics for the supply of the streak camera and to the French Army Health Authority (Service de Santé des Armées) for financial support.

References

1. B Chance, J.S. Leigh, H Miyake, D.S. Smith, S Nioka, R Greenfeld, M Finander, K Kaufman, W Levy, M Young, P Cohen, H Yoshioka and R Boretsky « *Comparison of time resolved and unresolved measurements of deoxyhemoglobin in brain* » Proc. Natl. Acad. Sci. USA **85,** 4971-4975 (1988).
2. S J Matcher, M Cope, D T Delpy « *In vivo measurements of the wavelength dependence of tissue-scattering coefficients between 760 and 900nm measured with time-resolved spectroscopy* », Applied Optics **36**, 386-396, (1997).
3. F Bevilacqua, P Marquet, C Depeursinge "*Determination of biological tissues optical coefficients from the spatial or temporal scattered profile of a collimated pulsed light source*", SPIE *Proc.* **2326**, 173-184 (1994).
4. H Liu, B Chance, F.K. Tittel, S.L. Jacques « *Time-resolved photon emission from layered turbid media* », Applied Optics **35**, 719-728 (1996).
5. B Gélébart, E Tinet, J-M Tualle, S Avrillier « *Phase function simulation in tissue phantoms : a fractal approach* », Pure and Applied Optics **5**, 377-388 (1996).
6. L.T. Perelman, V. Backman, M. Wallace, G. Zonios, R. Manoharan, A. Nustrat, S. Shields, M. Seiler, C. Lima, T. Hamano, I. Itzkan, J. Van Dam, J.M. Crawford and M.S. Feld « *Observation of fine structure in reflectance from biological tissue : a new technique for measuring nuclear size distribution* », Phys. Rev. Lett. **80**, 627-630 (1998).

7. J. Mourant, A.H. Hielscher, J.F. Freyer, A.A. Eick, D. Shen and T.M. Jonhson « *Scattering properties of biological cells* » OSA TOPS (1998).
8. S. Avrillier, B. Gélébart, E. Tinet, J.-M. Tualle and J.P. Ollivier « *Does turbid media time and space resolved reflectance depend on the different moments of the scattering phase function ?* », SPIE *Proc.* **3251** (1998) (to be published).
9. E Tinet, S Avrillier, J-M Tualle « *Fast semi-analytical Monte-Carlo simulation for time resolved light propagation in turbid media* », J. Opt. Soc. Am. A **13**, 1903-1915 (1996).
10. J-M Tualle, B Gélébart, E Tinet, S Avrillier, J.P Ollivier « *Real time optical coefficients evaluation from time and space resolved reflectance measurements in biological tissues* », Optics Communications **124**, 216-221 (1996).
11. J-M Tualle, B Gélébart, S Avrillier « *Time and space resolved reflectance for the measurement of optical coefficients of tissues* », Ultrafast Processes in Spectroscopy O Svelto, S de Silvestri and G Denardo Ed., 479-482 (1996).
12. B Gélébart, J-M Tualle, E Tinet, S Avrillier, J-P Ollivier « *Time and space resolved reflectance from multi-layered turbid media* », SPIE **3195** (1997).
13. Henyey L, Greenstein J « *Diffuse radiation in the Galaxy* », Astrophysics Journal **93**, 70-83 (1941).
14. Van Der Zee P, Essenpreis M and Delpy D T « *optical properties of brain tissue* » SPIE **1888**, 454-65 (1993).

Discrete-Ordinate Transport Simulations of Light Propagation in Highly Forward Scattering Heterogeneous Media

Andreas H. Hielscher[1,2] and Raymond E. Alcouffe[2]

State University of New York Health Science Center
Department of Pathology, Box 25,
450 Clarkson Ave., Brooklyn, NY 11203
Tel. 718 270 4562, Fax 718 270 1305, E-mail: ahielscher@netmail.hscbklyn.edu

Los Alamos National Laboratory, Transport Methods,
XTM MS E535, Los Alamos, New Mexico 87545,
Tel. 505 667 7270, Fax 505 665 4637, E-mail: rea@lanl.gov

Abstract

We implemented a delta-Eddington scheme in conjunction with a finite-difference, discrete-ordinate, radiative-transport code to accurately and efficiently simulate light propagation in highly-forward-scattering media. It is demonstrated that light propagation does only weakly depend on the anisotropy factor, g, as long as the reduced scattering coefficient, μ_s', is much larger than the absorption coefficient, μ_a. When $\mu_a \sim \mu_s'$, the fluence rate decays faster in media with high g-values compared to isotropically scattering (g = 0) media. In heterogeneous media that contain void-like spaces, such as the cerebrospinal-fluid-filled ventricles in the brain, the choice of g barely affects the predicted light propagation. However, the diffusion approximation does not yield an accurate description of the light transport in these cases.

Key Words

(170.3660) Light propagation in tissues 170.5280 Photon migration; 170.7050 Turbid media.

1. Introduction

In previous works [1-3] we reported on the development of a finite-difference discrete-ordinate code, which numerically solves the equation of radiative transfer for photon propagation in highly scattering media. We used this code to show that the widely employed diffusion approximation leads to large errors in the predicted fluence rates in media where the absorption coefficient is comparable to the value of the scattering coefficients. Furthermore, the diffusion approximations leads to erroneous results in heterogeneous media that contain void-like regions, such as the cerebrospinal fluid filled ventricles in the brain. These differences occurred even though isotropic scattering (g = 0) was assumed in the transport calculations.

In this work we address in particular the problem of highly forward scattering media and what type of additional effects and difference between diffusion and transport simulation can be observed, when anisotropy values of g > 0.9 are assumed. To this end we implemented the delta-Eddington scheme. This scheme allows to efficiently use the finite-difference, discrete-ordinate transport code to accurately calculate fluence rates in optically thick media, with highly forward peaked scattering phase functions.

2. Mathematical Background

For the calculation of the photon propagation in biomedical tissues we employ a finite-difference radiative-transport code called DANTSYS (Diffusion Accelerated Neutral Particle Transport Code System), which was originally developed to calculate neutron transport in nuclear materials. The time independent transport equation is given by:

$$\Omega \bullet \nabla\Psi(\mathbf{r},\Omega) + (\mu_a(\mathbf{r}) + \mu_s(\mathbf{r}))\Psi(\mathbf{r},\Omega) = S(\mathbf{r},\Omega) + \mu_s \int_{4\pi} \Psi(\mathbf{r},\Omega')p(\Omega \bullet \Omega')d\Omega' \quad (1)$$

Here **r** is the position vector and Ω a unit vector pointing in the direction of photon propagation. $\Psi(\mathrm{r},\Omega)$ is the energy radiance in units of $\mathrm{Wcm^{-2}sr^{-1}}$.The source term $S(\mathrm{r},\Omega)$ represents power injected into a solid angle centered on Ω in a unit volume at **r**. The absorption and scattering coefficient μ_a and μ_s, are the inverse of the mean free path (mfp) for absorption and scattering, respectively. The phase function $p(\Omega\bullet\Omega')$ describes the probability that during a scattering event a photon with direction Ω' is scattered in the direction Ω. The goal of any algorithm that solves Eq. (1) is to determine $\Psi(\mathbf{r},\Omega)$, or the derived parameter

$$\Phi(\mathrm{r}) = \int_{4\pi} \Psi(\mathrm{r},\Omega)d\Omega, \quad (2)$$

which is the energy fluence rate with units of $\mathrm{W/cm^2}$.

OSA TOPS Vol. 21 Advances in Optical Imaging and Photon Migration, James G. Fujimoto and Michael S. Patterson (eds.)

Equation (1) is spatial discretized by a diamond differencing scheme, also called Crank-Nicolson method [4,5]. For the discretization of the direction or angle the method of discrete ordinates is employed [6,7]. This method divides the radiation field into a number of discrete directions Ω_m. In this way the transport equation is transformed into a set of coupled integro-differential equations. Each of these equation is further transformed into a linear differential equation by expanding the phase function, $p(\omega = \Omega \bullet \Omega')$, in a series of Legendre polynomials $P_n(\omega)$,

$$p(\omega) = (1/4\pi) \sum_{n=0}^{\infty} (2n+1) b_n P_n(\omega) \tag{3}$$

and replacing the scattering integral by a quadrature formula. In this work a Gauss-Legendre quadrature was used. The set of linear differential equations is commonly solved iteratively [8,9].

In general, finite-difference, discrete-ordinate techniques provide an efficient way of numerically solving the transport equation for large, optically thick, heterogeneous media, such as the brain or the breast. Monte Carlo simulations are virtually impossible in these practical cases. The computation times easily exceed several weeks or months, because large numbers of photons have to be launched and traced to obtain reasonable signal to noise ratios. The discrete-ordinate techniques provide results in minutes.

This advantage in speed for large media is compromised if systems with highly forward peaked phase functions, $p(\omega)$, are considered. The slower convergence in these cases is caused by the fact that $p(\omega)$ is expanded in Legendre polynomials. In case of isotropic scattering only one term is need and linearly varying phase function can be described by two Legendre terms. For highly forward peaked phase function, many orders of Legendre expansions have to be evaluated. For example to accurately describe a Henyey-Greenstein phase function described by an anisotropy factor of g = 0.99, over 100 Legendre orders are necessary. The calculations of these many orders considerably slows down the convergence of finite-difference, discrete-ordinate transport codes.

In this work we overcome this problem by expressing the phase function for highly forward scattering media as a sum of a new phase function $p'(\omega)$ and a delta function, $\delta(1-\omega)$:

$$p(\omega) = (1-f)p'(\omega)) + f\delta(1-\omega)/(2\pi), \tag{4}$$

with $0 < f < 1$. This procedure is often called the delta-Eddington approach [10,11]. The new phase function $p'(\omega)$ can also be expanded in a series of Legendre polynomials and the new coefficients, b'_n, are given by

$$b'_n = (b_n - f)/(1 - f). \tag{5}$$

This follows directly from combining Eq. (4) and (3) and using the fact that $\delta(1-\omega)/2\pi$ is represented by Eq.(3) with $b_n = 1$ for all n.

Inserting Eq. (4) into Eq.(1) the transport equation can now be written as

$$\Omega \bullet \nabla\Psi(\mathbf{r},\Omega) + (\mu_a(\mathbf{r}) + \mu_s(\mathbf{r})(1-f))\Psi(\mathbf{r},\Omega) = S(\mathbf{r},\Omega) + \mu_s(1-f) \int_{4\pi} \Psi(\mathbf{r},\Omega')p'(\Omega \bullet \Omega')d\Omega' \tag{6}$$

Since the delta function has been taken out of the scattering phase function, a much smaller number of orders in the Legendre expansion are necessary to accurately describe a highly forward peaked phase function. In the case of g = 0.99, the number of expansion coefficient is reduced from over 100 to less than 10, depending on the allowable residual error. This leads to an increase in the speed of convergence for the DANTSYS discrete-ordinate code by approximately a factor of 1000.

3. Results

To study the effects of high g-values on the photon transport, we compare transport calculations of media with the same reduced scattering coefficients, $\mu'_s = (1-g)\mu_s$, but different g-values. For isotropic scattering (g = 0) we solved Eq. 1. To simulate a highly forward scattering medium with g =0.99, we chose $f = g^3$, $b'_0 = 1$, $b'_1 = (g - g^3)/(1 - g^3)$, $b'_2 = (g^2 - g^3)/(1 - g^3)$. Solving Eq. 6 with these paramters, is identical to solving Eq. (1) with the expansion coefficient $b_0 = 1$, $b_1 = g$, $b_2 = g^2$, $b_n = g^3$ for all n>2.

3.1. Homogeneous Media

As a first example we consider a simple homogenous cube (10x10x10cm) with $\mu_s' = 5.0\ cm^{-1} >> \mu_a = 0.05\ cm^{-1}$. The point source is spatially centered on one face of the cube. We calculate the fluence rates for g = 0 and g = 0.99. To better display the difference between these two case, we divide the result obtained with g = 0.99 by the result obtained with g = 0. Figure 1a shows this ratio in the plane that contains the source (arrow on left side). Fig. 1b shows this ratio on the dashed line indicated in Fig. 1a.

It can be seen that the ratios are close to one for all points considered. Differences between the calculations obtained with g = 0.99 and g = 0 are smaller than 4% throughout the medium. In general we found that as long as $\mu_s' >> \mu_a$, differences in the g-value only lead to negligible difference in the fluence rates in media larger than 20 transport mean free path (mfp' = $1/\mu_s'$). This is expected because for these cases the diffusion theory is valid and the photon migration process is dominated by μ_s'.

As a second example, we consider the same medium only with a 50 times larger absorption coefficient, $\mu_a = 2.5\ cm^{-1}$. Therefore, in this case μ_a is not much smaller than μ_s'. .Again, we divide the result for the fluence rates obtained with g = 0.99 by the result obtained with g = 0. Figure 2a shows this ratio in the plane that contains the source (arrow on left side). Fig. 2b shows this ratio on the dashed line indicated in Fig. 1a.

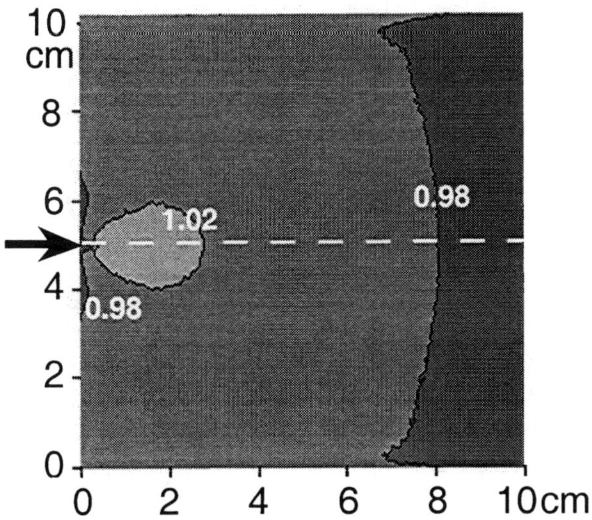

Figure 1a. Ratio of the fluence rates calculated with g = 0.99 and g = 0 ($\mu_s' = 5$ cm^{-1}, $\mu_a = 0.05$ cm^{-1}) shown for a 2-dimensional slice through a cube. The slice contains the pencil beam source as illustrated on the left face of the cube.

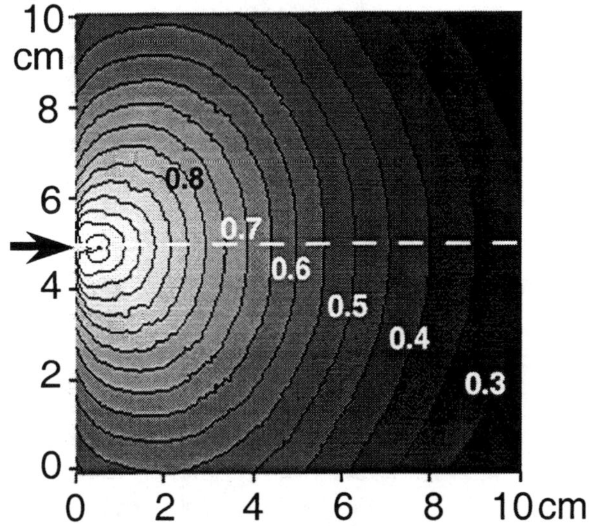

Figure 2a. Ratio of the fluences as calculated with g = 0.99 and g = 0 ($\mu_s' = 5$ cm^{-1}, $\mu_a = 0.05$ cm^{-1}) shown for a 2-dimensional slice through a cube, The slice contains the pencil beam source as illustrated on the left face of the cube.

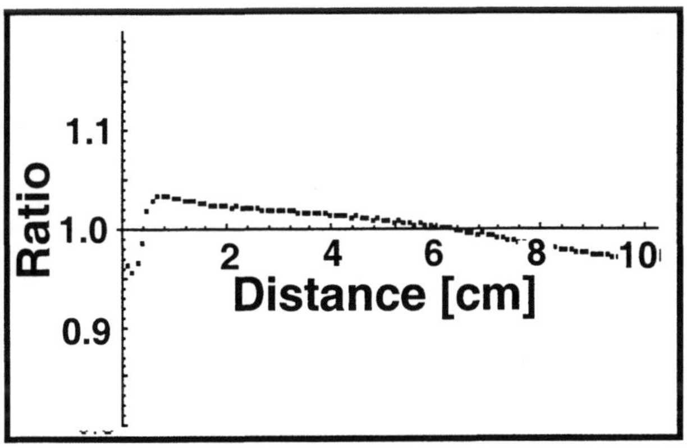

Fig. 1b. Values of the ratio of the fluence rates as calculated with g = 0.99 and g = 0 shown for a the dashed line in Fig. 1a.

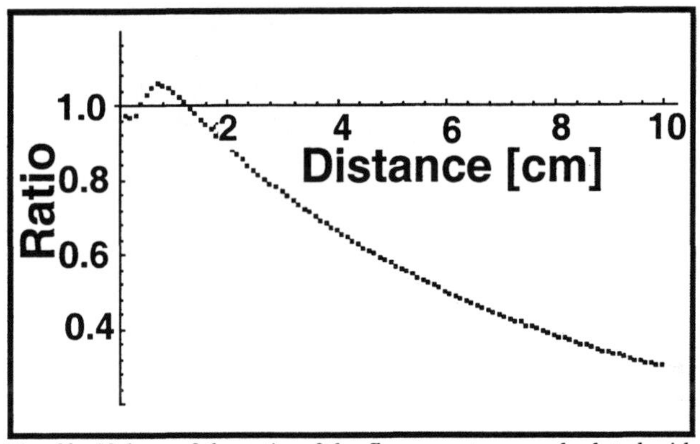

Fig. 2b: Values of the ratio of the fluence rates as calculated with g = 0.99 and g = 0 shown for a the dashed line in Fig. 1a.

This time, the ratios are not close to 1, indicating large differences in the calculated fluence rates for the cases of g = 0.99 and g = 0. In general we found that the larger g-value leads to an increase in the fluence rates around 2 mfp' from the source and a decrease in the fluence rates further away from the source, when compared to the case of isotropic scattering, g = 0.

3.2. Simple Heterogeneous Media

Next we studied a heterogeneous system, for which we observed in the past large differences between diffusion and transport calculations. In Figure 3 the geometry of this problem is displayed. We consider a 10x10x10-cm cube with μ_s'=5 cm^{-1} and $\mu_a = 0.05$ cm^{-1}. This cube contains a void-like structure ($\mu_s' = 0.1$ cm^{-1} and $\mu_a = 0.001$ cm^{-1}), which consists of a 4-mm-thick box, which surrounds an 72x72x72-mm inner cube. This model mimics the cerebro-spinal-fluid filled region that surrounds the brain.

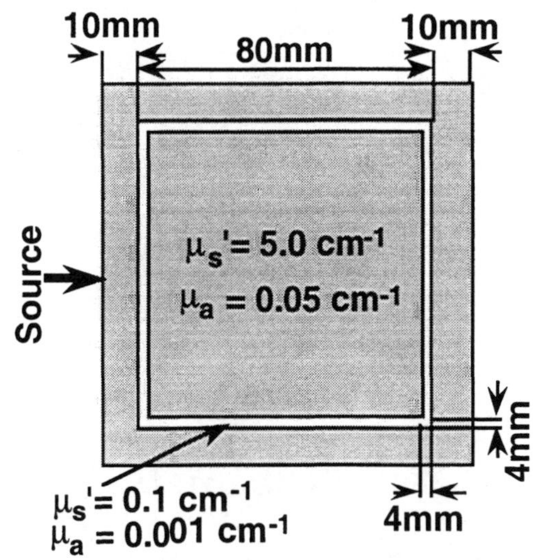

Figure 3. Geometry of simple heterogeneous medium.

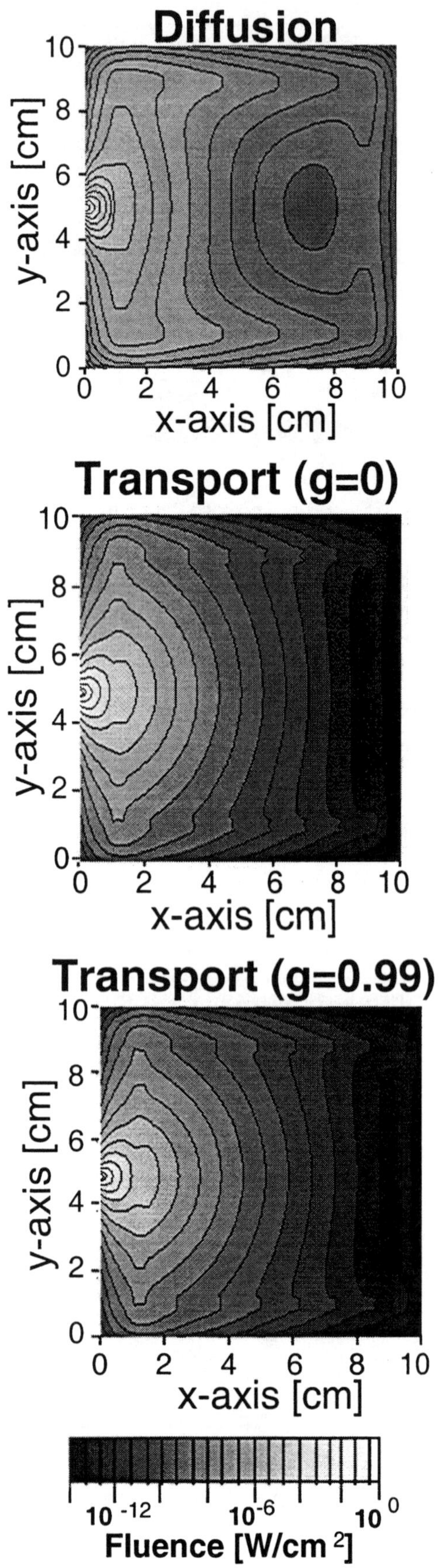

Figure 4. (a-top) Fluence rates, as calculated with the diffusion approximation, on a plane through the cube shown in Fig. 3 The plane displayed contains the source at position x = 0 cm and y = 5 cm. (b-middle) Fluence rates as calculated with a finite-difference transport code (Eq. 1, g = 0). (c-bottom) Fluence rates as calculated with finite difference transport code(Eq. 6, g = 0.99)

Figures 4a, 4b, and 4c show the finite-difference diffusion and transport simulations (g = 0 and g = 0.99) respectively for a plane through the cube, which contains the source on the left. Big differences between the diffusion and transport results can be observed, which are caused by a light channeling effect within the almost clear layer as predicted by diffusion theory. The differences between the cases of g = 0 and g = 0.99 are much smaller. Throughout the medium $\mu_s' \gg \mu_a$ and therefore different values of g do not have a strong influence on the outcome of the simulation.

3.3. Brain

As a final example a 2-dimensional slice through the brain is considered. We took an MRI scan of the brain and segmented it into different areas with various optical properties. MRI imaging techniques allow to distinguish between fatty white matter, the more watery gray matter, the CSF filled ventricles, the skull, the skin and blood. These different tissues and fluids appear in MRI scan with different densities. From this one obtains an optical property map (μ_a , μ_s') by assigning different optical properties to different density values. Fig. 5 shows such a segmented scan for a slice that was taken just above the eyebrows through a human head. For example, clearly visible are the butterfly like ventricles in the center, which are filled with cerebrospinal fluid (CSF). The optical

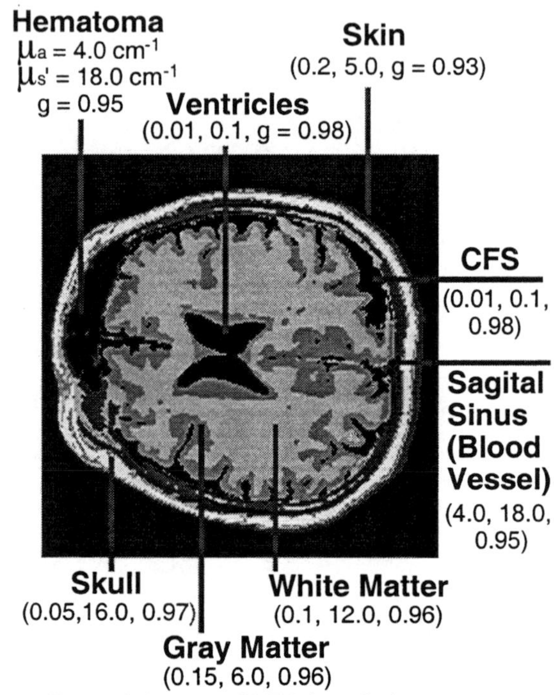

Figure 5. Segmented MRI data of a human brain.

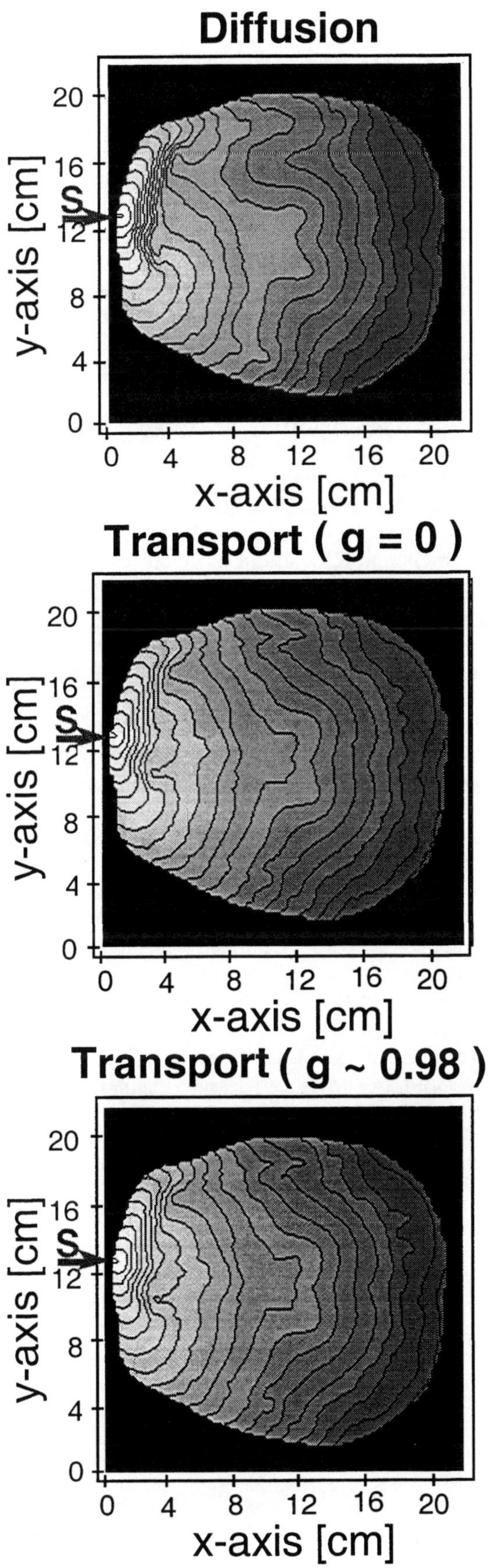

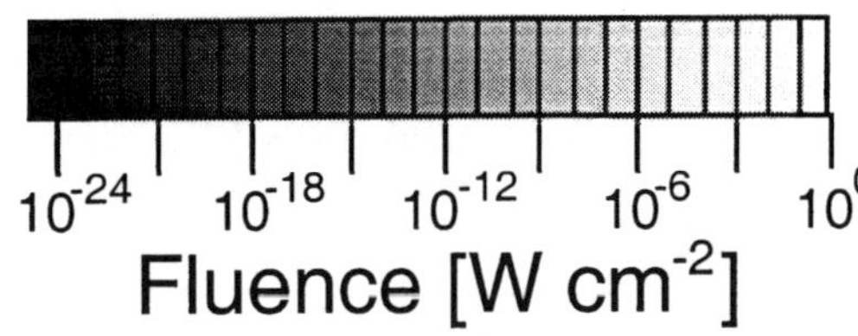

Figure 6. (a-top) Fluence rates as calculated by finite-difference diffusion code for 2-dimensional slice through human brain shown in Fig. 5. (b-middle) Fluence rates as calculated with finite- difference transport code (Eq. 1, g = 0), (c-bottom) Fluence rates as calculated with finite-difference transport code (Eq. 6, g = 0.99).

properties chosen for the different tissues and fluids are representative values for light in the near-infrared (600-900nm) [1].

The results of the 2-dimensional, finite-difference transport and diffusion simulations are shown in Figs. 6a, 6b, and 6c. The light source is place on the forehead (x = 0.4, y=12.8) just above the hematoma. The different gray values represent different fluence-rate levels. The iso-fluence-rate lines indicate a one-order-of-magnitude drop in the fluence rate between two lines. Large differences can be observed between the diffusion and transport simulations, while the transport calculations with g = 0 and g ~ 0.98 are almost identical.

This is even more evident when the values of the fluence rate are shown on a line that contains the source and crosses through the hematoma and one of the ventricles (Fig. 7). Diffusion and transport calculations of the fluence rates differ as much as 3 order of magnitude right behind the hematoma, while differences between the calculations with isotropic and anisotropic scattering never excite more than 40%.

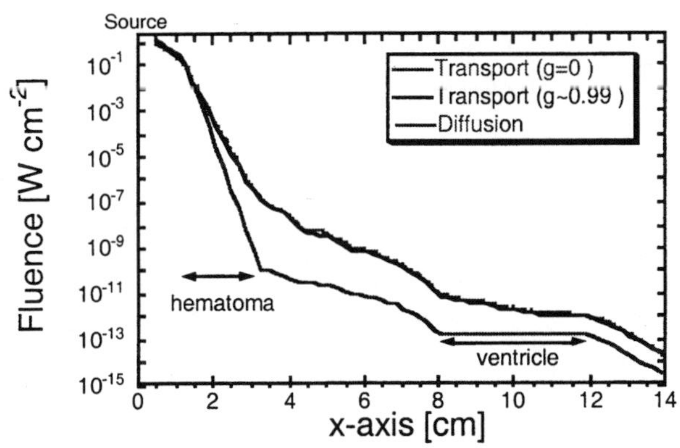

Figure 7. Finite-difference diffusion and transport results of fluence calculations for line through the human brain shown in Fig. 5. The line includes the source at x = 0.4 cm y=12.8 cm (see Fig. 6), and is parallel to the x-axis.

4. Conclusions and Summary

In this work we employed a delta-Eddington scheme in conjunction with a finite-difference, discrete-ordinate code to simulate light propagation in highly-forward-scattering, optically thick (>20 transport mean free path) media. We showed that as long as $\mu_s' >> \mu_a$ the differences between Boltzmann transport calculations for isotropic media ($g = 0$) and for highly forward scattering media ($g \sim 0.99$) are negliable. When μ_a is comparable in magnitude to μ_s' we found that the medium with a higher g-value displays a faster decay in the fluence rate as the distance from the source is increased. However, these differences between media of different g-values are small compared to difference between the results obtained with transport theory and diffusion theory. Especially we showed that in the brain the fluence rates calculated under the diffusion approximation are up to 3 orders of magnitude smaller than the fluence rates calculated with the transport equation. In contrast, transport simulations with $g = 0$ throughout the head differ less than 40% from transport simulations with $g \sim 0.98$ throughout the head. In simulations on breast tissue we observed even smaller effects.

We can conclude that in media for which the diffusion approximation holds ($\mu_s' >> \mu_a$), the details of the scattering phase functions can be neglected. For the cases of $\mu_s \sim \mu_a$ and media containing void-like regions, such as cerebro-spinal-fluid filled areas in the brain or cysts in the breast, transport-theory based simulation with isotropic scattering yield substantial improvement over diffusion-theory based solution. Further improvement of the results are possible if anisotropic scattering is considered. However, these additional improvements are small and may not warrant in all cases the additional complexity and time needs of these simulations.

Acknowledgments
This work was supported in part by an NIH grant, CA66184-A02, and a fellowship from the director's office at Los Alamos National Laboratory. We would like to thank Prof. Randall Barbour, from the State University of New York, Health Science Center at Brooklyn, for providing the MRI data of the brain.

References

1. A. H. Hielscher, R. E. Alcouffe, R. L. Barbour, "Comparison of finite-difference transport and diffusion calculations for photon migration in homogeneous and heterogeneous tissues," Phys. Med. Biol., in press.
2. A. H. Hielscher, R. E. Alcouffe, R. L. Barbour, "Transport and diffusion calculations on MRI generated data," in Optical Tomography and Spectroscopy of Tissue: Theory, Instrumentation, Model, and Human Studies, B. Chance and R.R. Alfano, eds., SPIE-The International Society for Optical Engineering, Proc. 2979, pp. 500-508, 1997.
3. A.H. Hielscher, R.E. Alcouffe, "Non-diffusive photon migration in homogeneous and heterogeneous tissues," in Photon Propagation in Tissue II, B. Chance, D. Benaron, and G.J. Müller, eds.,SPIE-The Internat. Society of Optical Engineering , Proc. 2925, pp. 22-30, 1996.
4. R.E. Alcouffe, "An adaptive weighted diamond differencing method for three-dimensional xyz geometry," Trans. Am. Nuc. Soc. 68 Part A, pp. 206-212 (1993).
5. W.H. Press, B.P. Flannery, S.A. Teukolsky, and W.T. Vetterling, Numerical Recipes in C (Cambridge University Press, New York, NY, 1992)
6. B.G. Carlson and K.D. Lathrop, "Transport Theory - The Method of Discrete Ordinates," in Computational Methods in Reactor Physics, H. Greenspan, C.N Kelber and D. Okrent, eds. (Gordon and Breach Science Publishers, New York, NY, 1968), pp 171-270.
7. G.I. Bell and S. Glasstone, "Discrete ordinates and discrete S_N methods," in Nuclear Reactor Theory (Van Nostrand Reinhold, New York, NY, 1970), pp. 214-251.
8. R.E. Alcouffe, "Diffusion Synthetic Acceleration: Method for the Diamond Differenced Discrete Ordinates Equation," Nuc. Sci. Eng. 64, pp. 344-352 (1977).
9. J.E. Morel, "A synthetic acceleration method for discrete ordinates calculations with highly anisotropic scattering," Nuclear Science and Engineering 82, pp. 34-46 (1982).
10. A.J. Welch, M.J.C. van Gemert, eds., Optical-Thermal Response of Laser-Irradiated Tissue (Plenum Press, New York, NY, 1995), pp. 144-148.
11. J. H. Joseph, W.J. Wiscombe, J.H. Weinman, "The delta-Eddington approximation for radiative flux transfer," J. Atmos. Sci. 33, pp. 2452-2459 (1976).

Analytic Transfer Function of the Forward Propagation of Diffuse Photon Density Waves in Turbid Media

Deborah L. Lasocki, Peter J. Collins and Charles L. Matson[†]
Department of Electrical and Computer Engineering, Air Force Institute of Technology,
Wright-Patterson Air Force Base, Ohio 45433-7765
[†]Optical Sensing Division, Air Force Phillips Laboratory, Kirtland AFB, NM 87117-6008
email: dlasocki@afit.af.mil, voice: (937) 255-3636 x4733, fax: (937) 656-4055
email: pcollins@afit.af.mil, voice: (937) 255-6565 x4304, fax: (937) 656-4055
email: matson@plk.af.mil, voice: (505) 846-2049, fax: (505) 846-2045

1. Introduction

Interest in using optical radiation to image and diagnose human breast cancer is motivating researchers to clearly understand light propagation through human tissue. The forward propagation behavior of light has been explored to better understand the wave nature in human tissue, particularly tissue containing tumors.[3] Frequency-domain methods are a promising method to localize and characterize inhomogeneous objects within turbid media.[6] These methods use diffuse photon density waves (DPDW) which are damped, spherical waves propagating outwards from the source. Analytic solutions have been determined to exactly model the forward propagation of the DPDW in a turbid media that contains an embedded spherical inhomogeneity.[2] Recently, this propagation has been approximated by a Fourier optics model using diffraction tomographic methods which assumes weak perturbations from the background medium but arbitrary inhomogeneity structure.[4] In this paper, an exact transfer function is developed to characterize the forward propagation behavior of the DPDW using Fourier optics theory. Although this new model assumes a spherical inhomogeneity, it can be used to improve the diffraction tomography approach by permitting strong perturbations from the background medium.

2. Transfer Function Development

In linear systems, for a given input and output pair, there is a transfer function that can be applied to the input to yield the given output. For this development, the system in Fig. 1(a) can be modeled as a linear system. The input to the system is a point source of intensity modulated light, and the output can be modeled in the form of an analytic solution.[2] Since the total field outside the sphere is a superposition of the incident and scattered waves, then via the linearity property of Fourier transforms (FT), the inverse FT of the total wave, represented by $\mathcal{F}^{-1}\{\Phi\}$, is simply a superposition of the individual inverse FTs of the incident (Φ_{AC}) and scattered (Φ_{SC}) waves. Meaning, $\mathcal{F}^{-1}\{\Phi\} = \mathcal{F}^{-1}\{\Phi_{AC}\} + \mathcal{F}^{-1}\{\Phi_{SC}\}$. Rearranging terms, this expression can be rewritten as

$$\mathcal{F}^{-1}\{\Phi\} = \mathcal{F}^{-1}\{\Phi_{AC}\}\left\{1 + \frac{\mathcal{F}^{-1}\{\Phi_{SC}\}}{\mathcal{F}^{-1}\{\Phi_{AC}\}}\right\}, \tag{1}$$

provided $1/\mathcal{F}^{-1}\{\Phi_{AC}\}$ exists. The transfer function is seen to be in Eqn. (1) as the right bracketed term.

Assuming that the diffusion approximation to the transport equation is valid, the analytic solution used in this development is an exact solution. The field outside the inhomogeneity is a superposition of the

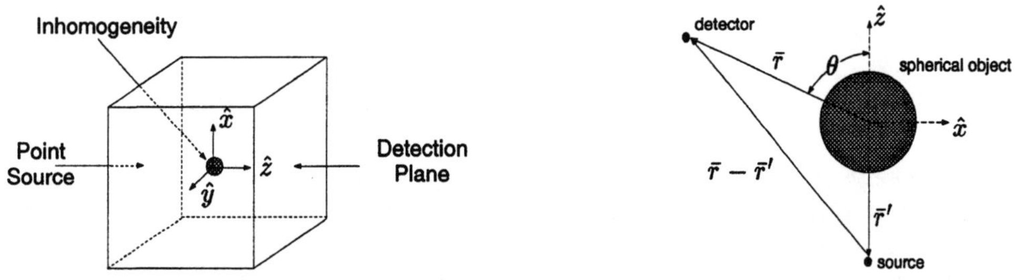

(a) (b)

Figure 1: (a) A point source (-$\hat{z}$) and a detector plane (+$\hat{z}$) is referenced to the object center located at the system origin. (b) The radial distances are indicated for the source, detector and object.[2]

incident DPDW plus the scattered wave from the spherical object,[2,3] i.e.,

$$\Phi(\bar{r},\bar{r}') = \frac{vS_{AC}}{4\pi D_{out}|\bar{r}-\bar{r}'|}e^{jk_{out}|\bar{r}-\bar{r}'|} + \sum_{l=0}^{\infty} A_l h_l^{(1)}(k_{out}r)Y_l^{(0)}(\theta,\phi), \qquad r \geq a, \tag{2}$$

where S_{AC} is the modulation amplitude, v is the speed of light in the media, and f is the modulation frequency. k_{out} is the complex photon density wavenumber of the surrounding media given by $k_{out}^2 = (-v\mu_a + j\omega)/D$, where $D_{out} = v/[3(\mu_a + \mu_s')]$ is the photon diffusion coefficient, μ_s' is the reduced scattering coefficient, and μ_a is the absorption coefficient in the medium. $h_l^{(1)}$ is the spherical Hankel function of the first kind, type l, and $Y_l^{(0)}$ are spherical harmonics of order $m = 0$, degree l. The amplitude of the multipole moments of the scattered wave is given by A_l (see Ref. [2]).

The difficulty of the problem is that while the incident DPDW has spherical symmetry, the boundary (detector) is a plane. It is natural to express the spherical wave in terms of plane waves in the desired Cartesian coordinate system in order to evaluate the wave on the planar boundary.

Using a double integral representation of the scalar Green's Function, the homogeneous wave solution (the first term of Eqn. (2)) can be rewritten as[1]

$$\Phi_{AC}(\bar{r},\bar{r}') = \frac{vS_{AC}}{8\pi^2 D}\int_{-\infty}^{\infty}\int_{-\infty}^{\infty}\frac{1}{\gamma}e^{-\gamma(z-z')+j[\xi(x-x')+\eta(y-y')]}d\xi d\eta, \qquad z-z' > 0. \tag{3}$$

The system geometry dictates that the source and detector will be on opposite sides of the inhomogeneity, and the wave travels in the positive $\hat{z}$ direction. The unprimed variable ($\bar{r}$) indicates the distance to the detector and the primed ($\bar{r}'$) to the source as seen in Fig. 1(b). The path of integration for each of the two complex variables ξ and η, is the corresponding real axis. The attenuating factor, γ, is defined as,

$$\begin{aligned} \gamma &= \sqrt{\xi^2+\eta^2-k^2}, && \text{if } \xi^2+\eta^2 > k^2, \\ \gamma &= -j\sqrt{k^2-\xi^2-\eta^2}, && \text{if } \xi^2+\eta^2 \leq k^2. \end{aligned} \tag{4}$$

This definition requires the real part of the square root to be greater than zero and keeps the function analytic over a strip region outside the spherical inhomogeneity. The constraints on γ are valid in that the right side of Eqn. (3) satisfies the Helmholtz wave equation and gives the correct value of the field at $z = z'$. The plane waves are homogeneous if $\xi^2+\eta^2 \leq k^2$ and evanescent if $\xi^2+\eta^2 > k^2$. The elegant result of Eqn. (3) is a decomposition of a spherical wave into a superposition of elementary plane waves in the direction determined by ξ and η as indicated by the exponent. The form for a two-dimensional inverse FT of the incident wave, U_{AC}, is readily apparent, e.g.

$$U_{AC}(\xi,\eta;z') = \frac{vS_{AC}}{8\pi^2 D}\frac{1}{\gamma}e^{-\gamma(z-z')}, \qquad z-z' > 0. \tag{5}$$

It is interesting to note that even though the incident wave is singular at the source location, the non-zero imaginary component of k and the real frequencies, ξ and η, keep γ from becoming zero.

We now want to rewrite the scattered wave into a plane wave expansion in Cartesian coordinates. To that end, the scattered wave outside the inhomogeneity from the second term of Eqn. (2) can be expressed in an integral format as[5]

$$\Phi_{SC}(\bar{r}) = \sum_{l=0}^{\infty} A_l \frac{1}{4\pi j^l}\int_0^{2\pi} d\beta \int_B e^{j\bar{k}\cdot\bar{r}} P_l(\cos\alpha)\sin\alpha d\alpha, \qquad r \geq a, \tag{6}$$

where the spherical harmonic of order $m = 0$, degree l, has been replaced with the equivalent Legendre polynomial, P_l. The contour B is chosen to be from $0 \longrightarrow \frac{\pi}{2} - j\infty$. The vector $\bar{k}$ is of length k and has the spherical angles α and β. Using the spherical coordinate to rectilinear transform variables, the dot product is replaced with $\bar{k}\cdot\bar{r} = k[z\cos\alpha + x\sin\alpha\cos\beta + y\sin\alpha\sin\beta]$. From this spherical representation in transform space, a conformal transformation into polar coordinates is defined as $\lambda = k\sin\alpha$, $d\lambda = k\cos\alpha d\alpha$; $\gamma = -\sqrt{\lambda^2-k^2} = -jk\cos\alpha$, where the path of integration in the λ plane is restricted along the positive half of the real axis, e.g. $0 \leq \lambda < \infty$. Making the conformal transformation, the scattered wave expression becomes

$$\Phi_{SC}(\bar{r}) = \sum_{l=0}^{\infty} A_l \frac{1}{4\pi j^l} \frac{1}{jk} \int_0^{2\pi} \mathrm{d}\beta \int_0^{\infty} \frac{1}{\gamma} e^{-\gamma z} e^{j(x\lambda\cos\beta + y\lambda\sin\beta)} P_l(\frac{j\gamma}{k}) \lambda \mathrm{d}\lambda. \quad (7)$$

The result in Eqn. (7) is in the same form as the double integral of Eqn. (3). Let the frequency variables $\xi = \lambda\cos\beta$ and $\eta = \lambda\sin\beta$, where $\lambda = \sqrt{\xi^2 + \eta^2}$ and $\lambda \mathrm{d}\lambda \mathrm{d}\beta = \mathrm{d}\xi \mathrm{d}\eta$. Note that ξ and η are real transform variables, so the limits of integration are now over the entire real axis. Using these transform variables, the scattered wave can now be expanded into a superposition of ordinary plane waves in Cartesian coordinates, namely,

$$\Phi_{SC}(\bar{r}) = \sum_{l=0}^{\infty} \frac{A_l}{4\pi k j^{l+1}} \int_{-\infty}^{\infty} \int_{-\infty}^{\infty} \frac{1}{\gamma} e^{-\gamma z} e^{j(\xi x + \eta y)} P_l(\frac{j\gamma}{k}) \mathrm{d}\xi \mathrm{d}\eta, \qquad |z| \geq a, \quad (8)$$

where the plane wave expansion in Eqn. (8) has a different region of validity than the multipole expansion. The inverse FT of the scattered wave is

$$U_{SC}(\xi, \eta; z) = \sum_{l=0}^{\infty} \frac{1}{4\pi j^l} \frac{1}{jk} A_l \frac{1}{\gamma} e^{-\gamma z} P_l(\frac{j\gamma}{k}). \quad (9)$$

Using the results in Eqns. (5) and (9), the analytic form of the transfer function can be determined from Eqn. (1), e.g.

$$H(\xi, \eta; z, z') = 1 + \sum_{l=0}^{\infty} \frac{2\pi D}{v k S_{AC} j^{l+1}} A_l P_l(\frac{j\gamma}{k}) e^{-\gamma z'}, \qquad z - z' > 0, \quad |z| \geq a. \quad (10)$$

3. Conclusions

In this paper, an analytic transfer function was determined for a system of infinite homogeneous turbid media with a spherical inhomogeneity situated at the origin. The transfer function completely characterizes this system so that any input can be applied to get a corresponding perturbed output wave at a detector plane. The constraints on γ in Eqn. (4) indicate that the two-dimensional Fourier components of Eqn. (10) about which the information is carried is in the homogeneous waves, i.e. $\xi^2 + \eta^2 \leq k^2$, and defines the range of accessible Fourier components of the system, acting like a spatial frequency filter. The transfer function shows that some of the three-dimensional Fourier components of the system may be determined directly from knowing the two-dimensional Fourier components in the detector plane. It is then possible to determine the three-dimensional structure of the inhomogeneity with the measurements in the detector plane.

References

[1] Baños, A. *Dipole Radiation in the Presence of a Conducting Half-Space.* Oxford: Pergamon Press, 1966.

[2] Boas, D. A., et al. "Scattering of diffuse photon density waves by spherical inhomogeneities within turbid media: Analytic solution and applications," *Proc. Natl. Acad. Sci. USA*, *91*:4887–4891 (May 1994).

[3] Feng, S., et al. "Photon migration in the presence of a single defect: a perturbation analysis," *Appl. Opt.*, *34*:3826 (July 1995).

[4] Matson, C. L. "A diffraction tomographic model of the forward problem using diffuse photon density waves," *Optics Express*, *1*:6–11 (1997).

[5] Morse, P. M. and H. Feshbach. *Methods of Theoretical Physics, I & II.* New York: McGraw-Hill, 1953.

[6] O'Leary, M. A., et al. "Experimental images of heterogeneous turbid media by frequency-domain diffusing-photon tomography," *Opt. Lett.*, *20*:426–428 (March 1995).

Time-dependent equation for the intensity in the diffusion limit

Meir Gershenson
Coastal Systems Station, Code R22, Panama City, Florida 32407
Gershenson_Meir@ccmail.ncsc.navy.mil

Submitted March, 1998

Abstract

The first two terms in the spherical-harmonics expansion (the P_1 approximation) of the radiative transfer equation yield the diffusion equation. This approximation applies to multiple scattering and results in a solution for the energy density, the gradient of which is proportional to the light intensity. In this work a higher-order spherical-harmonics expansion of the radiative transfer equation is developed. Using the Fourier transform, an approximation based on expanding at small wave vectors k leads to an equation similar to the diffusion equation. This equation applies to the radiant intensity rather than the energy density. The equation is expected to predict the intensity for multiple scattering at earlier times and shorter distances than the DE can. The notion of an equivalent wave field is introduced.

1. Introduction

The radiative transfer equation (RTE) and the diffusion equation (DE) to which it reduces in the case of many scatterings provide the basis for describing many physical phenomena. (In this paper I consider the phenomenon of unpolarized light transport.) Analytical solutions for the RTE exist only for restricted cases and one usually has to resort to numerical methods. One difficulty of the problem is the dimensionality of the system. Besides the position vector and time, each point in space has an associated vector describing the direction of propagation of light at that point. Thus while the diffusion equation has three spatial dimensions and one of time, the RTE has two additional dimensions.

The problem is usually to investigate the angular distribution using some limiting approximation. The most common approximations are the uniform intensity approximation, leading to the diffusion equation, and the forward–backward dominant scattering approximation, leading to the two-stream model. The DE follows from the RTE under the assumption that after a sufficiently large number of collisions the intensity is almost isotropic, and then provides an asymptotic approximation applicable to later times. Large numbers of physical systems obey the DE independent of the details of the RTE. The diffusion approximation fails for intermediate scattering ranges[1] that are appropriate for many practical applications such as shallow medical imaging, longer-range underwater imaging, and electron transport in small semiconductor devices. Second-order corrections to the DE appear occasionally in the literature,[2–4] but the validity of these corrections has been debated for the last 20 years.[5–9] A leading reason for the confusion in this matter is the lack of a general solution of the RTE.

2. Theory

In this paper I derive an equation analogous to the DE that includes the angular distribution of the light intensity. Such an equation is expected to be valid at shorter distances and earlier times than the diffusion equation as it does not assume times or distances sufficient to form a uniform intensity distribution. The outline of the derivation is as follows: First the RTE will be represented using the spherical-harmonics expansion (SHE) for the angular intensity with coefficients that are time and space dependent. The resulting representation—the coefficient matrix—will be inverted and then expanded to second order in the spatial derivatives. Summation of the terms will eliminate the SHE, leading to a new differential equation independent of the intermediate SHE.

We start from the time-dependent radiative transfer equation for the intensity I at time t, position r, in the direction Ω

$$\left[\frac{1}{c}\frac{\partial}{\partial t} + \Omega\cdot\nabla + \sigma_t(r)\right] I(r,t,\Omega) - \\ -\frac{\sigma_s}{4\pi}\int I(r,t,\Omega')\,p(r,\Omega\cdot\Omega')\,d\Omega' = \varepsilon(r,t,\Omega)\,, \tag{1}$$

where σ_s and σ_a are the scattering and absorption coefficients, $\sigma_t = \sigma_s + \sigma_a$, and $p(\Omega\cdot\Omega')$ is the phase function. The first term (the term in the large parenthesis) on the left-hand side of Eq. (1) describes the time evolution of attenuated ballistic photons, and the second term is the scattering integral, and the term on the right-hand side is the source term.

The SHE has the form[10]

$$I_l^m(r,t) = \int I(r,t,\Omega) Y_l^{m\,*}(\Omega)\, d\Omega \tag{2a}$$

$$I(r,t,\Omega) = \sum_{l=0}^{\infty} \sum_{m=-l}^{m=l} \frac{2l+1}{4\pi} I_l^m(r,t) Y_l^m(\Omega) \tag{2b}$$

with

$$p(r,\Omega\cdot\Omega') = \sum_{l=0}^{\infty} \frac{4\pi}{2l+1} p_l(r) P_l(\Omega\cdot\Omega'). \tag{3}$$

$P_l(\Omega\cdot\Omega')$ is the Legendre polynomial of the cosine of the angle between Ω and Ω'. Using the addition theorem for spherical harmonics we get

$$p(r,\Omega\cdot\Omega') = \sum_{l=0}^{\infty} \sum_{m=-l}^{m=l} p_l(r) Y_l^m(\Omega) Y_l^{m\,*}(\Omega'). \tag{4}$$

Using the identity $\Omega = (\sin\theta\cos\phi, \sin\theta\sin\phi, \cos\theta)$, we obtain from Eq. (1) of Ref. 11 the equivalent set of coupled differential equations

$$\left[\frac{1}{c}\frac{\partial}{\partial t} + \sigma_t(r)\, p_l(r)\right] I_l^m(r,t) + \\ +\frac{1}{2l+1}\left[\begin{array}{l} \sqrt{(l-m+1)(l+m+1)}\,\frac{\partial}{\partial z} I_{l+1}^m(r,t) + \sqrt{(l-m)(l+m)}\,\frac{\partial}{\partial z} I_{l-1}^m(r,t) - \\ -\frac{1}{2}\sqrt{(l+m-1)(l+m)}\left[\frac{\partial}{\partial x} - i\cdot\frac{\partial}{\partial y}\right] I_{l-1}^{m-1}(r,t) + \\ +\frac{1}{2}\sqrt{(l-m+1)(l-m+2)}\left[\frac{\partial}{\partial x} - i\cdot\frac{\partial}{\partial y}\right] I_{l+1}^{m-1}(r,t) + \\ +\frac{1}{2}\sqrt{(l-m-1)(l-m)}\left[\frac{\partial}{\partial x} + i\cdot\frac{\partial}{\partial y}\right] I_{l-1}^{m+1}(r,t) - \\ -\frac{1}{2}\sqrt{(l+m+1)(l+m+2)}\left[\frac{\partial}{\partial x} + i\cdot\frac{\partial}{\partial y}\right] I_{l+1}^{m+1}(r,t) \end{array}\right] = \varepsilon_l^m(r,t). \tag{5}$$

Spherical harmonics are eigenvalues of all the terms except the propagation term, which couples adjacent l's and m's. Recognizing that, we expect that any approximation based on spherical harmonics will be a good approximation for scattering-dominated propagation, but will perform poorly in the ballistic regime. This conclusion is consistent with the inefficacy of the SHE in describing transport in the case of a strongly peaked phase function.

Using matrix notation and applying the Fourier transform we write

$$D(s,s_t)\,\hat{I}(s,s_t) = \hat{\varepsilon}(s,s_t,\Omega) \tag{6a}$$

where $s_t = i\omega$ and $\mathbf{s} = (s_x,s_y,s_z) = i(k_x,k_y,k_z)$, bold characters represent a vector or a matrix, and the hat represents the Fourier (or, the Laplace) transform; special notation will be used for functions that are defined only in the transform domain. The indices of the matrix elements are combinations of all possible l's and m's. (An effective way of counting l's and m's will be introduced later.) The operator $D(s,s_t)$ in equation 6a can be writen as

$$D(s,s_t) = O(s_t) + B(s), \tag{6b}$$

where the matrix

$$O(s_t)_{(l,m),\,(l',m')} = \\ = \left[\frac{s_t}{c} + \hat{\sigma}_t(s) - \hat{\sigma}_s(s)\; p_l(s)\right] \delta(l,l')\,\delta(m,m') \tag{7}$$

is diagonal and B is a sparse off-diagonal matrix representing $\Omega\cdot\nabla$, which is the spatial ballistic part of the propagation. A formal solution for $\hat{I}$ is given by

$$\hat{I} = D^{-1}\hat{\boldsymbol{\varepsilon}}\,. \tag{8}$$

The determinant of the matrix represents a differential equation applied to $I(r,t)$. In calculating the determinant we can take advantage of some properties of Eq. (5). First, as a result of symmetry with respect to rotation, the only terms that are functions of s are of the form $s^2 = s_x^2 + s_y^2 + s_z^2$; therefore it is sufficient to calculate the s_z contribution and deduce the rest. Equation (5) simplifies to

$$\left[\frac{1}{c}\frac{\partial}{\partial t} + \sigma_t(r)\, p_l(r)\right] I_l^m(r,t) + \\ +\frac{1}{2l+1}\left[\begin{array}{l} \sqrt{(l-m+1)(l+m+1)}\,\frac{\partial}{\partial z} I_{l+1}^m(r,t) + \\ +\sqrt{(l-m)(l+m)}\,\frac{\partial}{\partial z} I_{l-1}^m(r,t) \end{array}\right] \\ = \varepsilon_l^m(r,t). \tag{9}$$

The matrix associated with the transform of Eq. (9) can be decomposed into tridiagonal block matrices by grouping the elements according to the same m, while l values ranging from $|m|$ to L, where L is an upper bound of the approximation

The submatrices of an m value have the form

$$D_{m,L} = O_{m,L} + B_{m,L} \tag{10a}$$

where

$$O_{m,L} = \begin{bmatrix} O_{|m|} & 0 & 0 & \cdots & \cdots & \cdots \\ 0 & O_{|m|+1} & \cdots & \cdots & \cdots & \cdots \\ 0 & \cdots & \cdots & \cdots & 0 & 0 \\ \cdots & \cdots & \cdots & O_{L-2} & 0 & 0 \\ \cdots & \cdots & 0 & 0 & O_{L-1} & 0 \\ \cdots & \cdots & 0 & 0 & 0 & O_L \end{bmatrix} \tag{10b}$$

and

$$B_{m,L} = \begin{bmatrix} 0 & \frac{1}{\sqrt{2|m|+1}} & 0 & \cdots & \cdots & \cdots \\ \frac{1}{\sqrt{2|m|+1}} & 0 & \cdots & \cdots & \cdots & \cdots \\ 0 & \cdots & \cdots & \cdots & 0 & 0 \\ \cdots & \cdots & \cdots & 0 & \frac{\sqrt{(L-1)^2-m^2}}{2L-3} & 0 \\ \cdots & \cdots & 0 & \frac{\sqrt{(L-1)^2-m^2}}{2L-1} & 0 & \frac{\sqrt{(L-1)^2-m^2}}{2L-1} \\ \cdots & \cdots & 0 & 0 & \frac{\sqrt{(L-1)^2-m^2}}{2L+1} & 0 \end{bmatrix} \cdot s_z \tag{10c}$$

From Eqs. (10) by expanding through the last column and row we find the relation

$$\det(D_{m,\,L}) = \det(D_{m,\,L-1})\, O_L - \frac{L^2-m^2}{4L^2-1} \det(D_{m,\,L-2})\, s_z^2 \,. \tag{11}$$

where $\det(A)$ is the determinant of A.

If we define

$$A_{m,\,L} = \frac{L^2-m^2}{4L^2-1} \frac{1}{O_L O_{L-1}}\,, \tag{12}$$

then

$$\det(D_{m,\,m+n}) = \left[\sum_{n=0}^{N} C_{m,n}(-1)^n s_z^{2n}\right] \prod_{l=|m|}^{|m|+n} O_l + 1. \tag{13}$$

$C_{m,n}$ is the sum of all the product of n $A_{m,l}$-terms with no two l's being of adjacent values and $2N \le L - |m|$.

The total determinant up to order L is given by

$$\det(D_L) = \prod_{m=-L}^{L} \det(D_{m,L}) \tag{14a}$$

$$\det(D_L) = \prod_{m=-L}^{L} \det(D_{m,L}) \tag{14b}$$

.

Equation (8) can be written as

$$\det(D_L)\, I = Cd_L \cdot \hat{\varepsilon}\,. \tag{15}$$

where Cd_L is the Lth codeterminant of D. We will not attempt to calculate Cd_L in this paper. $\det(D_L)$ is a polynomial in s_t and s_z^2 which corresponds to a differential operator in the space–time domain. The left-hand side of Eq. (15) is a generalized diffusion operator of order L (the diffusion equation itself is first order). The high order of s in the polynomial makes the equation hard to deal with for most cases. A simpler approximation is necessary. Consistent with the diffusion or small s (but not s_t). We will expand the determinant approximation, we will assume small spatial derivatives keeping zeroth- and second-order terms while keeping track of the fourth-order residual term. Using the first two terms of the expansion in s_z^2 in Eq. (13), which we call $d_{m,L}$, we get

$$d_{m,L} = \left[1 + s_z^2 \sum_{l=|m|}^{L} \frac{l^2-m^2}{4l^2-1} \frac{1}{O_l O_{l-1}}\right] \prod_{l=|m|}^{L} O_l\,, \tag{16}$$

and the fourth-order residual term is

$$\left[\sum_{\substack{l,\,\lambda=|m| \\ |l-\lambda|\ge 2}}^{L} \frac{l^2-m^2}{4l^2-1} \frac{\lambda^2-m^2}{4\lambda^2-1} \frac{1}{O_l O_{l-1}} \frac{1}{O_\lambda O_{\lambda-1}}\right] \cdot s_z^4 \prod_{l=|m|}^{L} O_l\,.$$

A weak requirement for the approximation to be valid is that $|s_z| \ll |O_l|$ or that

$$|s_z| \ll |s_t/c + \sigma_t(r) - \sigma_s(r) p_l(r)|\,, \tag{17}$$

which applies for distances larger than the attenuation length or early times. The second-order approximation in s_z for $di_L = \det(D_L)$ is obtained through second-order expansion of the products $\prod_{m=-L}^{L} d_{m,L}$:

$$di_L = \left[1 - \frac{1}{3}\sum_{l=1}^{L} \frac{1}{O_l O_{l-1}} s_z^2\right] O_0 \prod_{l=1}^{L} (O_l)^{2l}. \tag{18}$$

The relation between the time and space derivatives is given by

$$d_L = 1 - \frac{1}{3}\sum_{l=1}^{L} \frac{1}{O_l O_{l-1}} s_z^2 \equiv 1 - \Psi(s_t) s_z^2. \tag{19}$$

Equation (19) is the basis for an extended DE which accounts for scattering up to order L.

$$[1 - \Psi(r, s_t)\, \nabla^2]\ \tilde{I}(r, s_t, \Omega) = \tilde{\Xi}(r, s_t, \Omega) \tag{20a}$$

$$\Psi = \frac{1}{3}\sum_{l=1}^{L} \frac{l}{\left[\frac{s_t}{c} + \sigma_t - \sigma_s\, p_l(r)\right]\left[\frac{s_t}{c} + \sigma_t - \sigma_s\, p_{l-1}\right]} \tag{20b}$$

Equation (20) is similar to the DE in that it includes only second-order derivatives in r but is more general than the DE because it accounts for higher-order terms of the expansion of the phase function and applies to the intensity I rather than to the density.

3. Discussion

The first observation we make is that in the transform domain, in all the equations derived from the RTE, including the RTE itself, the term s_t appears in the form $s_t/c + \sigma_t$. Using the properties of the Laplace transform we note that if $\tilde{X}(s)$ is the transform of $X(t)$, then the inverse transform of $\tilde{X}(s_t/c + \sigma_t)$ is $c\, X(ct) \exp(-\sigma_t ct)$. As a result, the decay of the intensity is dominated by the exponential factor $\exp(-\sigma_t ct)$.

Considering the universality of the DE and the derivation approach used here which is based on similar assumptions, the merit of the results should be judged relative to the DE. As a first check, we compare the result to the DE ($L = 1$).

For isotropic scattering

$$\left[1 - \frac{1}{3} \frac{1}{\left(\frac{s_t}{c} + \sigma_t - \sigma_s\, p_1\right)\left(\frac{s_t}{c} + \sigma_t - \sigma_s\, p_0\right)} \nabla^2\right] \cdot \tilde{I}(r, s_t, \Omega) = \tilde{\Xi}(r, s_t, \Omega)$$

This is the Ishimaru[2] version of the DE. Using only the first term of Eq. (29b) we get

$$\left[1 - \frac{1}{3} \frac{c}{\sigma_s \cdot (p_0 - p_1)} \frac{1}{\frac{s_t}{c} + \sigma_t - \sigma_s\, p_0} \nabla^2\right] \cdot \tilde{I}(r, s_t, \Omega) = \tilde{\Xi}(r, s_t, \Omega)$$

which is Furutsu's[5] version of the DE. The right-hand side is slightly different because it accounts for the intensity rather than the density. For orders higher than $L = 1$ the complete solution includes the Laplacian operator with powers greater than one (∇^2); Eq. (20) ignores these terms. The coefficient of ∇^2, $\Psi(r, s_t)$, is described by Eq. (20b). This series does not necessarily converge, in particular because of the l factor, which is part of the coefficient. The series must be terminated at a certain L value, which has to be determined on the basis of the details of the problem. We expect the equation to be a good description for cases where the DE is almost a good description, i.e., late times, long distances, and small-l–dominated phase function and source. The equation is expected to be a good description of "snake photons"[12] which are early-arrived scattered-photons used to improve image details.

ACKNOWLEDGMENTS

This work was supported by Coastal System Station independent research.

I would like to thank P. Schippnick for his help in manuscript preparation and G. Sammelmann for discussion regarding symmetry properties.

Appendix A: Equivalent wave field

Equation (20) has the form

$$[1 - \Psi(s_t)\nabla^2]\, \tilde{I}(r, s_t, \Omega) = \tilde{\Xi}(r, s_t, \Omega), \tag{A1}$$

which, can be shown, using arguments similar to those used above, is quite common as a second order approximation in many physical phenomenon. We will show that the solution $I(r, t, \Omega)$ can be related to the solution of an equivalent wave equation:

$$\left[1 - \frac{1}{p_t^2}\nabla^2\right] \tilde{I}_w(r, p_t, \Omega) = \tilde{\Xi}_w(r, p_t, \Omega) \tag{A2}$$

Such a relation is very beneficial due to the mathematical properties of the wave equation that make it useful in remote sensing and also relatively easy to solve in both direct and inverse problems. For the sake of simplicity, we ignore some of the mathematical subtleties. For compactness, we omit the arguments r and Ω:

$$\Psi\left(g(p_t)\right) = \frac{1}{p_t^2} \tag{A3}$$

$$\tilde{I}\left(g(p_t)\right) = \tilde{I}_w(p_t), \qquad \tilde{\Xi}\left(g(p_t)\right) = \tilde{\Xi}_w(p_t). \tag{A4}$$

Using the forward and inverse Laplace transform

$$\tilde{I}_w(p_t) = \int_0^\infty I(t) \exp\left[-g(p_t)\, t\right] dt \tag{A5}$$

$$I_w(t) = \frac{1}{2\pi i}\int_{c-i\infty}^{c+i\infty} \tilde{I}_w(p_t) \exp(p_t \cdot t)\, dp_t . \tag{A6}$$

and performing the two integrations rearranging the order of integration we obtain

$$I_w(t) = \int_0^\infty I(\tau)\, K(t,\tau)\, d\tau \tag{A7a}$$

$$\Xi_w(t) = \int_0^\infty \Xi(\tau)\, K(t,\tau)\, d\tau\,, \tag{A7b}$$

$$K(t,\tau) = \frac{1}{2\pi i}\int_{c-i\infty}^{c+i\infty} \exp\left[-g(p_t)\,\tau\right] \exp(p_t\, t)\, dp_t . \tag{A8}$$

If the boundary conditions are frequency-dependent, they should be accounted for. Equations (A7) and (A8) usually represent an ill-posed problem, and one should exercise caution when using them. However, with proper care it is possible to obtain an optimal form of the kernel $K(t,\tau)$ for practical use. An inverse relation of the form

$$I(t) = \int_0^\infty I_w(t)\, K(t,\tau)\, d\tau \tag{A9}$$

can also be derived. (For a discussion of the equivalent wave field in the DE limit see Ref. 13.)

1. K. M. Yoo, F. Liu, and R. Alfano, "When does the diffusion equation approximation fail to describe photon transport in random media," Phys. Rev. Lett. **64**, 2647–2650 (1990).
2. A. Ishimaru, "Diffusion of a pulse in densely distributed scatterers," J. Opt. Soc. Am. **68**, 1045–1050 (1978).
3. D. J. Durian and J. Rudnick, "Photon migration at short times and distances and in the case of strong absorption," J. Opt. Soc. Am. A **14**, 235–245 (1997).
4. S. Kumar, K. Mitra, and Y. Yamada, "Hyperbolic damped-wave model for transient light-pulse propagation in scattering media," Appl. Opt. **35**, 3372 (1976). [No. 19]
5. K. Furutsu, "Diffusion equation derived from the space–time transport equation," J. Opt. Soc. Am. **70**, 360–365 (1980).
6. S. Ito, "Comparison of diffusion theories for optical pulse waves propagated in discrete random media," J. Opt. Soc. Am. A **1**, 502 (1984).
7. A. Ishimaru, "Difference between Ishimaru's and Furutsu's theories on pulse propagation in discrete random media," J. Opt. Soc. Am. A **1**, 506–XXX (1984).
8. J. M. Porra, J. Masoliver, and G. H. Weiss, "When the telegrapher's equation furnishes a better approximation to the transport equation than the diffusion approximation," Phys. Rev. E **55**, 7771–7774 (1997).
9. S. Godoy and L. S. García-Colín, "Nonvalidity of the telegrapher's diffusion equation in two and three dimensions for crystalline solids," Phys. Rev. E **55**, 2127–2131 (1997).
10. K. M. Case and P. F. Zweifel, *Linear Transport Theory* (Addison-Wesley, Reading, Massachusetts, 1967), YEAR), p. 7.
11. Ibid., p. 219.
12. L. Wang, P. P. Ho, C. Liu, G. Zhang, and R. R. Alfano, "Ballistic 2-D imaging through scattering walls using an ultrafast optical Kerr gate," Science **253**, 769 (1991).
13. M. Gershenson, "Simple interpretation of time-domain electromagnetic sounding using similarities between wave and diffusion propagation," Geophysics **62**, 763–774 (1997).

The Fourier Diffraction Theorem for Turbid Media

Charles L. Matson
Optical Sensing Division, Air Force Phillips Laboratory, Kirtland AFB, NM 87117-6008

Hanli Liu
Department of Biomedical Engineering, University of Texas at Arlington, Arlington, TX 76019

Abstract

Diffraction tomography theory describes both the forward and inverse problems in media where electromagnetic radiation propagation must be modeled in terms of its wave nature. In general, this theory assumes media in which the radiation is not attenuated. An important result from this theory is the Fourier diffraction theorem, which relates the two-dimensional spatial Fourier transform of the radiation which is detected in a plane to the three-dimensional spatial Fourier transform of the inhomogeneous structure imbedded in the background medium. In this paper we present a generalization of the forward problem in diffraction tomography to attenuating media, such as turbid media, with an emphasis on the turbid media version of the Fourier diffraction theorem.

Key Words

(170.3880) Medical and biological imaging; (100.3020) Image reconstruction/restoration; (170.3660) Light propagation in tissue.

Introduction

The use of optical radiation for imaging human tissue has many desirable advantages over other imaging methodologies.[1] One promising category of approaches, called frequency-domain methods,[2,3] is being explored for use in localizing and characterizing inhomogeneous structures imbedded in turbid media. Recently, this problem has been explored in the context of diffraction tomography, with preliminary results indicating that significant speed improvements are possible[4] along with decreasing the number of views needed.[5] We have previously developed a diffraction tomographic theory for imaging absorptive inhomogeneities in turbid media.[6] In this paper, we extend this theory to both absorbing and scattering inhomogeneities. In addition, we develop the turbid media version of the Fourier diffraction theorem.

The forward problem

The physical geometry of our observation scenario is shown in Figure 1, where the object is imbedded in a homogeneous and infinite turbid medium. The illumination plane is located at $z=z_1$, the detection plane is located at $z=z_o$, where $z_o>z_1$, and the object is located between the two planes. We assume plane wave illumination, where the illumination plane is defined as the location where the plane wave has unit amplitude and zero phase. We approach our analysis by deriving the Fourier transform of the wave scattered from the object as measured in the detection plane. We are interested in this

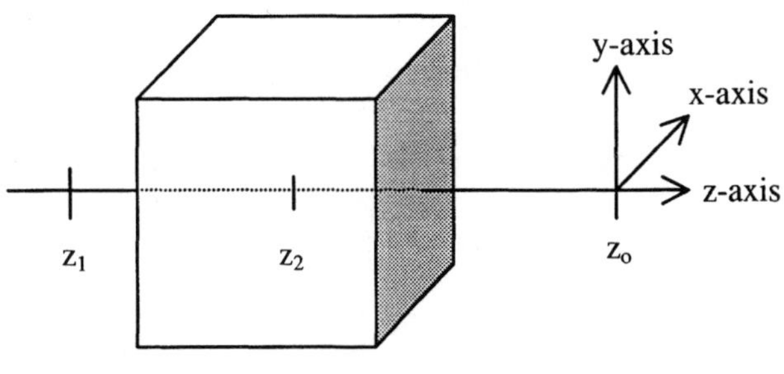

Figure 1. Conceptual diagram showing the geometry for the diffraction tomography development. The inhomogeneity is represented by the cube, which is assumed to be imbedded in an infinite homogeneous turbid medium. The illumination plane is located at z_1, the center of the inhomogeneity is located at x=0, y=0, $z=z_2$, and the detection plane is located at z_o. The illumination and detection apparatus are not shown.

Fourier transform because it is needed to obtain the Fourier diffraction theorem for turbid media. We present the solution first for absorptive inhomogeneities, then for scattering inhomogeneities, and finally put the two solutions together and analyze the properties of the overall solution.

Absorptive inhomogeneities

In this subsection we present an expression for the Fourier transform of the detected scattered wave, $u_{Ba}(x,y,z)$, when our inhomogeneity structure consists purely of weak absorptive perturbations when compared to the background medium. We employ the first Born solution[7] which is given by

$$u_{Ba}(x,y,z) = \iiint o_a(x',y',z')u_o(x',y',z') \times g(x-x',y-y',z-z')dx'dy'dz' \quad (1)$$

where $u_{Ba}(x,y,z)$ is the solution of the wave equation for absorptive inhomogeneities, $o_a(x,y,z)$ is related to the absorptive inhomogeneity that we are trying to reconstruct,[8] $u_o(x,y,z)$ is the solution of the wave equation without the inhomogeneities, $g(x,y,z)$ is the Green's function associated with the homogeneous infinite medium, and k is the complex wavenumber of the homogeneous medium. By using one version of the angular spectrum form of the Green's function,[9] assuming plane wave illumination, Fourier transforming Eq.(1), and simplifying, we can show that

$$\begin{aligned} U_{Ba}(\omega_x,\omega_y;z_o) = {} & \frac{1}{2\gamma_\omega}\exp[-(z_o-z_1)k_i] \\ & \times\exp(-iz_1k_r - iz_o\gamma_{\omega i})\iiint o_a(x',y',z') \\ & \times\exp[-(z_o-z')(\gamma_{\omega r}-k_i)] \\ & \times\exp[-ix'\omega_x - iy'\omega_y - iz'(-\gamma_{\omega i}-k_r)] \\ & \times dx'dy'dz' \end{aligned} \quad (2)$$

where $U_{Ba}(\omega_x,\omega_y;z_o)$ is the two-dimensional Fourier transform of $u_{Ba}(x,y,z)$ in the $z=z_o$ plane, k_r and k_i are the real and imaginary parts of the wavenumber, and

$$\begin{aligned} \gamma_\omega & \equiv \gamma_{\omega r} + i\gamma_{\omega i} \\ & = \mathrm{Re}\left(\sqrt{\omega_x^2+\omega_y^2-k^2}\right) + i\,\mathrm{Im}\left(\sqrt{\omega_x^2+\omega_y^2-k^2}\right) \end{aligned} \quad (3)$$

where Re() and Im() denote the real and imaginary parts of a complex number, respectively.

Scattering inhomogeneities

In this subsection we present an expression for the Fourier transform of the detected scattered wave when our inhomogeneity structure consists purely of weak scattering perturbations when compared to the background medium. It has been previously shown that the integral solution for the scattered wave, denoted by $u_{Bs}(x,y,z)$, is given by

$$\begin{aligned} u_{Bs}(x,y,z) = {} & \iiint o_s(x',y',z')\nabla u_o(x',y',z')' \\ & \cdot\nabla g(x-x',y-y',z-z')dx'dy'dz \end{aligned} \quad (4)$$

where $o_s(x,y,z)$ is related to the scattering inhomogeneity that we are trying to reconstruct.[8] By using an approach similar to the one used for the results in Eq.(2), we can show that the Fourier transform of the scattered wave measured in the detection plane, $U_{Bs}(\omega_x,\omega_y;z_o)$, is given by

$$\begin{aligned} U_{Bs}(\omega_x,\omega_y;z_o) = {} & \frac{-ik}{2}\exp[-(z_o-z_1)k_i] \\ & \times\exp(-iz_1k_r - iz_o\gamma_{\omega i})\iiint o_s(x',y',z') \\ & \times\exp[-(z_o-z')(\gamma_{\omega r}-k_i)] \\ & \times\exp[-ix'\omega_x - iy'\omega_y - iz'(-\gamma_{\omega i}-k_r)] \\ & \times dx'dy'dz' \end{aligned} \quad (5)$$

Because Eq.(4) contains the gradients of the illumination function and the Green's function, the method of solution for Eq.(5) is more dependent upon these functions than for the absorptive inhomogeneity case.

Combined absorption and scattering solution

In this subsection, we combine Eqs.(2) and (5) to obtain the Fourier transform of the detected scattered wave due to both absorptive and scattering inhomogeneities. This gives us the turbid media version of the Fourier diffraction theorem. We will then compare our results to the standard Fourier diffraction theorem.

Our combined solution, denoted by $U_B(\omega_x,\omega_y;z_o)$, is the sum of $U_{Ba}(\omega_x,\omega_y;z_o)$ and $U_{Bs}(\omega_x,\omega_y;z_o)$:

$$\begin{aligned} U_B(\omega_x,\omega_y;z_o) = {} & \frac{1}{2\gamma_\omega}\exp[-(z_o-z_1)k_i] \\ & \times\exp(-iz_1k_r - iz_o\gamma_{\omega i})\iiint [o_a(x',y',z') \\ & - ik\gamma_\omega o_s(x',y',z')]\exp[-(z_o-z') \\ & \times(\gamma_{\omega r}-k_i)]\exp[-ix'\omega_x - iy'\omega_y \\ & - iz'(-\gamma_{\omega i}-k_r)]dx'dy'dz' \end{aligned} \quad (6)$$

Eq.(6) is a statement of the Fourier diffraction theorem for turbid media, where we have moved the $1/\gamma_\omega$ factor outside the integral to facilitate comparison with the standard Fourier diffraction theorem. A two-dimensional version of this theorem can be found in Ref.[7]. Below we rewrite the standard Fourier diffraction theorem in our notation:

$$U_B(\omega_x,\omega_y;z_o) = \frac{1}{2\gamma_\omega}\exp(-iz_o\gamma_{\omega i}) \times O(\omega_x,\omega_y,-\gamma_{\omega i}-k_r) \qquad (7)$$

We notice, when comparing Eq.(7) with Eq.(6), that the turbid media version of the Fourier diffraction theorem differs from the standard Fourier diffraction theorem primarily by additional exponential terms. There is an overall exponential attenuation outside the integral in Eq.(6) which is a function of the separation of the illumination and detection planes. There is also an additional phase term outside the integral, $\exp(-iz_1k_r)$, which is due to our definition of the illuminating plane wave. The triple integral term in Eq.(6) is the Fourier transform of the object multiplied by a frequency-dependent attenuating exponential, where we define the object as the sum of $o_a(x,y,z)$ with $ik\gamma_\omega o_s(x,y,z)$. In the limit of no attenuation, the attenuating exponential function inside the integral becomes unity because both k_i and $\gamma_{\omega r}$ become zero. This can be seen by referring back to Eq.(3) as will now be discussed. If the medium is not attenuating, the imaginary part of k becomes zero and the real part of k is positive.[7] In turn, this makes the term inside the radical in Eq.(3) negative for $\omega_x^2+\omega_y^2<k_r^2$ (the usual assumption in standard diffraction tomography), which forces $\gamma_{\omega r}$ to zero. Thus the triple integral term becomes the Fourier transform of the object, the attenuating exponential term outside the integral becomes unity, and Eq.(6) becomes equal to Eq.(7) with the addition of the phase term resulting from our definition of the plane wave illumination. In addition, an interesting point about the object function is that the part of the object which corresponds to the scattering inhomogeneity is multiplied by $ik\gamma_\omega$, which is a function of ω. It can be shown that this makes characterizing scattering inhomogeneities potentially easier than for absorptive inhomogeneities.

Conclusions

We have developed the Fourier diffraction theorem for turbid media in this paper. The difference between the turbid media version and the standard diffraction tomography result are additional exponential terms, most of which are attenuating. We have also extended our previous results for purely absorptive inhomogeneities to include scattering inhomogeneities. Because of the frequency-dependent term multiplying the portion of the object function which is due to the scattering nature of the inhomogeneity, scattering inhomogeneities are potentially easier to characterize than absorptive inhomogeneities. Applications of this new theory include guidance to develop optimal systems for probing turbid media, and extending existing reconstruction algorithms that were developed under the standard diffraction tomographic theory to turbid media applications.

References

[1] *Medical Optical Tomography: Functional Imaging and Monitoring*, R.F. Potter, ed., SPIE Optical Engineering Press **IS11** (1993)

[2] M.A. O'Leary, D.A. Boas, B. Chance, and A.G. Yodh, "Experimental images of heterogeneous turbid media by frequency-domain diffusing-photon tomography," *Optics Letters* **20**, 426-428 (1995)

[3] H.B. Jiang, K.D. Paulsen, U.L. Osterberg, M.S. Patterson, "Frequency-domain optical-image reconstruction in turbid media – an experimental study of single-target detectability," *Applied Optics* **36**, 52-63 (1997)

[4] X.D. Li, T. Durduran, A.G. Yodh, B. Chance, and D.N. Pattanayak, "Diffraction tomography for biochemical imaging with diffuse-photon density waves," *Optics Letters* **22**, 573-575 (1997)

[5] C.L. Matson, N. Clark, L. McMackin, and J.S. Fender, "Three-dimensional tumor localization in thick tissue with the use of diffuse photon-density waves," *Applied Optics* **36**, 214-220 (1997)

[6] C.L. Matson, "A diffraction tomographic model of the forward problem using diffuse photon density waves," *Optics Express* **1**, 6-11 (1997)

[7] A. Kak and M. Slaney, *Principles of Computerized Tomographic Imaging*, IEEE Press, New York (1988)

[8] M.A. O'Leary, D.A. Boas, B. Chance, and A.G. Yodh, "Simultaneous scattering and absorption images of heterogeneous media using diffuse waves within the Rytov approximation," in *Optical Tomography, Photon Migration, and Spectroscopy of Tissue and Model Media: Theory, Human Studies, and Instrumentation*, Proceedings SPIE **2389**, 320-327 (1995)

[9] A. Baños, Jr., *Dipole Radiation in the Presence of a Conducting Half-Space*, Pergamon Press, Oxford (1966)

Accuracy Tests of Frequency Domain Immersion Technique

Eamon Briggs and Gregory W. Faris
Molecular Physics Laboratory, SRI International, 333 Ravenswood Avenue, Menlo Park, CA 94025

Abstract

We use comparison with conventional techniques and Mie theory to test the accuracy of the frequency domain immersion technique for absorption and scattering property measurements.

Key Words

(290.7050) Turbid media; (170.3660) Light propagation in tissues; (010.4450) Ocean Optics; (170.5270) Photon density waves.

The accurate determination of absorption and scattering coefficients in turbid media is important for applications in medicine, ocean optics, and other areas. Such property determinations may be made using continuous wave, time domain, or frequency domain measurements. We are investigating frequency domain measurements because of the simplicity and high potential accuracy of the technique.

The accuracy of frequency domain measurements can be enhanced by using immersion in a matched scattering medium.[1] There are several reasons immersion can provide improved accuracy for noninfinite scattering samples. First, the effects of boundary conditions are removed when a sample is immersed in a medium with nearly the same absorption and scattering coefficients. Second, the effects of probe coupling and motion and eliminated. Third, the effects of sample geometry are greatly reduced. And finally, the measurements are performed as a differential measurement against a well characterized background. The immersion technique is an alternative to multi-distance[2] and multi-frequency[3] techniques.

Our frequency domain system operating at 100 MHz and 671 nm, has been described previously.[1] Fiber optic bundles carry light in and out of the immersion medium. As an immersion medium, we use a mixture of Ropaque and dye, with concentration adjusted to provide the proper amount of scatter and absorption. The optical properties of the medium are determined by measuring the variation of the phase and amplitude as a function of distance. An example of such a measurement is shown in Figure 1. As is clear from the Figure, the high quality and number of the multi-distance measurements allow the immersion medium to be characterized to very high accuracy.

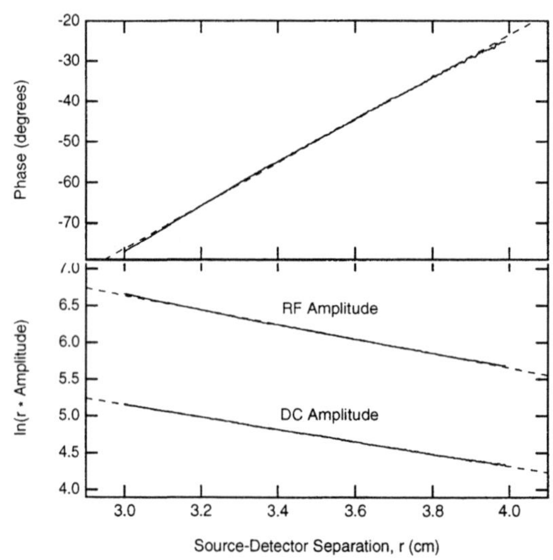

Figure 1. Variation of phase and amplitude in a Ropaque immersion medium.

Once the immersion medium has been characterized, the only measurement required on a sample is the difference in phase and amplitude with and without the sample between the optical fiber bundles. An example of such a measurement in shown in Figure 2. Using these differential measurements together with the multi-distance measurements allows direct determination of the scattering and absorption properties of the sample.[1]

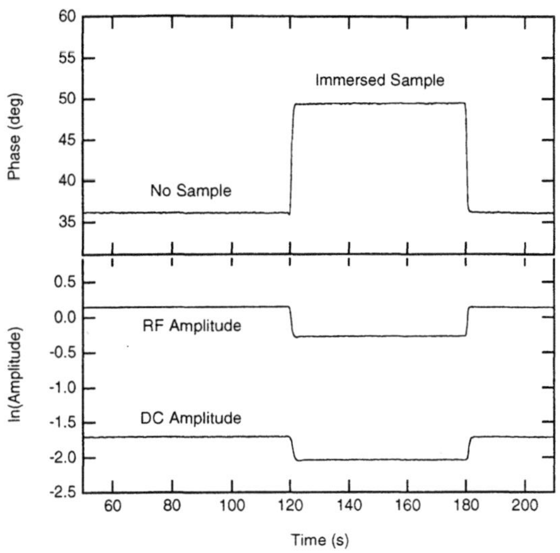

Figure 2. Measurement of differential phase and amplitude during immersion measurement.

We have tested the accuracy of the immersion frequency domain technique in two ways: comparison of multi-distance and immersion measurements on the same sample, and comparison with Mie theory. For the first measurement, the immersed sample consisted of diluted Ropaque, which was immersed in a lower concentration of Ropaque. The resulting absorption and scattering coefficients were compared with those determined by multi-distance measurements on the higher concentration. The results of the two types of measurements agreed to within 4 and 3% for the absorption and scattering coefficients, respectively.

For the second test, the immersed samples consisted of diluted polystyrene microspheres. Based on the known concentrations and diameter of the microspheres, we calculated the scattering coefficient using Mie theory calculations.[4] The measured scattering coefficient agreed with the calculated coefficient to within 5%.

We have performed two tests on the accuracy of the frequency domain immersion technique. These tests demonstrate an accuracy of 5%.

This work was supported by the Office of Naval Research. E. Briggs was supported by the Research Experiences for Undergraduates Program of the National Science Foundation.

References

1. X. Wu., L. Stinger, and G. W. Faris, "Determination of Tissue Properties by Immersion in a Matched Scattering Fluid," Proc. SPIE **2979**, 300-306 (1997).
2. S. Fantini, M. A. Franceschini, J. B. Fishkin, B. Barbieri, and E. Gratton, "Quantitative Determination of the Absorption and Spectra of Chromophores in Strongly Scattering Media: A Light-Emitting-Diode Based Technique," Appl. Opt. **33** 5204 (1994).
3. B. J. Tromberg, L. O. Svaasand, T.-T. Tsay, and R. C. Haskell, "Properties of Photon Density Waves in Multiple-Scattering Media," Appl. Opt. **32**, 607 (1993).
4. C. F. Bohren and D. R. Huffman, Absorption and Scattering of Light by Small Particles, (J. Wiley & Sons, New York, 1983) pp. 477-482

The influence of the FM tuning characteristic on wavelength modulation spectroscopy

L. Sandström, H. Ahlberg, T. Andersson*, S. Lundqvist*, and R. Tell*

Chalmers University of Technology

S-412 96 Göteborg
Sweden
Phone +46 31 772 1598
Fax +46 31 772 1540
E-mail lars@elm.chalmers.se

* AltOptronic AB
Box 8910
S-402 37 Göteborg
Sweden
Phone +46 31 22 21 10
Fax +46 31 22 21 19
E-mail info@altoptronic.se
Home page http://www.altoptronic.se

Introduction

Demands on efficient process and emission control have increased the need for on-line *in situ* concentration monitoring of several gases in industrial processes. Therefore, several efforts to transfer optical gas analysis techniques, working well within the laboratory, to the real world have been made [1,2]. However, reliability problems including sensitivity to environmental conditions, stability and drift have to be successfully eliminated. The component normally determining the stability and the quality of the measurement is the DFB laser that is used in the majority of such systems. Therefore, we have investigated the influence of several laser parameters to identify those crucial to the measurement quality for instruments based on Wavelength Modulation Spectroscopy (WMS). Earlier work has to large extent focused on laser noise and Residual Amplitude Modulation (RAM) [3]. Our work indicates that other laser parameters will limit the performance of practical systems.

Measurements and Results

Three DFB lasers with different design-wavelength were chosen for the investigation. Laser 1 is manufactured by IMC in Stockholm and has threshold current I_{th} = 30 mA, Laser 2 and 3 are manufactured by Sarnoff / Sensors Unlimited having an I_{th} of 20 mA and 25 mA, respectively.

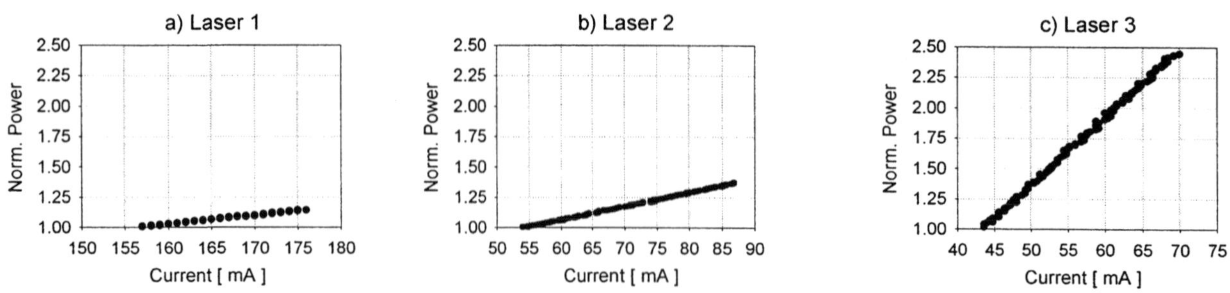

Figure 1. The L/I characteristics of the investigated lasers.

Reprinted from Advances in Optical Imaging & Photon Migration, 1998 Technical Digest, © 1998 Optical Society of America

Figure 1 shows the laser light-current (L/I) characteristic for the investigated lasers and clearly illustrates the large differences in the efficiency. The output power increases for Laser 1, Laser 2 and Laser 3 with 0.7 %/mA, 1 %/mA, and 5 %/mA, respectively.

For the tunability measurements we used an F-P etalon with a FSR of 300 MHz and finesse better than 300. The accuracy of the measurements of the frequency tuning versus current was better than 10 MHz/mA when using this F-P etalon. Each laser was modulated with a current ramp giving a 14 GHz optical tuning from the center-wavelength.

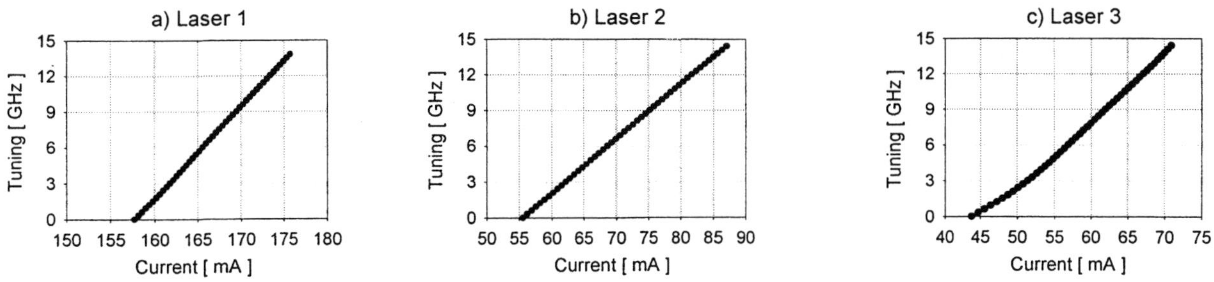

Figure 2. Laser current tuning characteristic.

The laser current and temperature setting were selected so that the center-wavelength for each laser coincided with a strong gas-absorption-line to get a good signal to noise ratio in the final WMS measurements. In Figure 2 we have plotted the frequency tuning of the laser versus injection current. If we adapt the characteristic to a straight line we obtain the FM-index for Laser 1, Laser 2 and Laser 3 of 0.77 GHz/mA, 0.46 GHz/mA and 0.55 GHz/mA respectively. Laser 1 and 3 shows a clear variation in FM-index over the tuning rage.

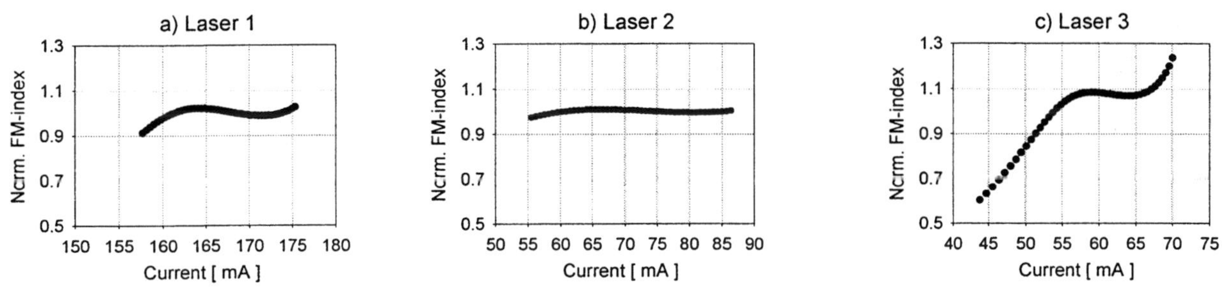

Figure 3. Normalized FM index.

We propose that the normalized FM index is the parameter important to the WMS performance.

$$Normalized\ FM-index = \frac{\Delta\upsilon_{i,i+1}}{\Delta I_{i,i+1}} \Big/ Mean\left(\frac{\Delta\upsilon_{i,i+1}}{\Delta I_{i,i+1}}\right)$$

Figure 3 reveals that Laser 3 exhibits the most nonlinear FM-index, vary, from 0.6 to 1.3 over the tested tuning range, while the Laser 1 and Laser 2 varies less than 10% and 5%, respectively. The absolute difference from the linear characteristic for Laser 3 is as much as 0.9 GHz that is substantial compared to the typical 3 GHz linewidth of a measured absorption line in a gas at atmospheric pressure.

Finally, we evaluated these lasers in a WMS diode laser system [4]. Figure 4 shows the detected second harmonic (2f) using NH_3 with Laser 1 and 3 and CO with Laser 2. The difference in absorption linewidth reflects the actual linewidth of the measured absorption lines. Laser 3 produces a heavily distorted absorption line. Moreover, we find that the received signal and distortion is very sensitive to the detection phase. The correct phase for synchronous detection will vary along the sweep due to the large variation in FM-index. This will make the measurement setup prone to long-term drift and instability. The other two lasers produce significantly more symmetric and undistorted absorption-lines whose distortion is not sensitive to limited variations in the detection phase.

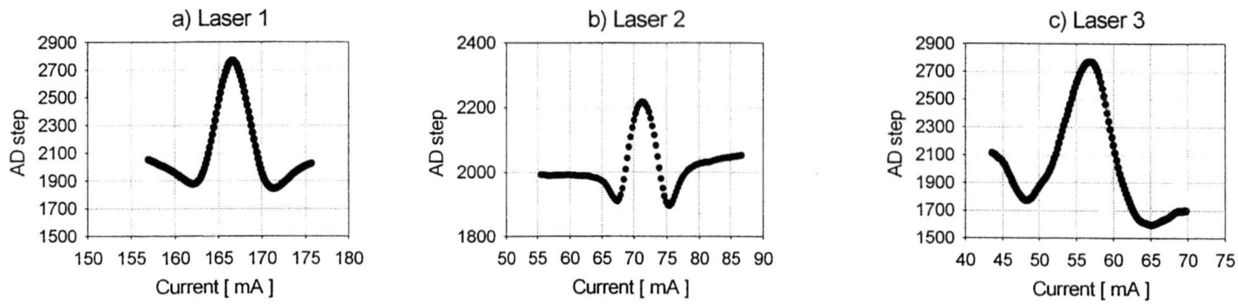

Figure 4. Second harmonic signal for the three lasers.

Discussion and Conclusion

This study has shown the need to identify the important characteristics of DFB lasers optimized for spectroscopic applications. The detected 2f signals have all different types of distortion. The WMS theory explains the observed amplitude difference of the left and the right minimum of the 2f-signal [3]. This effect can be seen for all three lasers. However, Laser 3 has a completely different type of phase-sensitive distortion that must be due to another mechanism. It is possible to correlate the measurements with the large difference in the normalized FM-index and the observed large nonlinearity in the FM sweep. This nonlinearity will give raise to overtones to the fundamental modulation frequency distorting the detected 2f-signal. Furthermore, the phase of the overtones will vary over the scan and consequently this type of distortion will be very sensitive to the detection phase and virtually impossible to suppress over the entire. In addition, the modulation amplitude will also change during the scan resulting in a distortion of the shape of the 2f signal.

In conclusion, we have shown that DFB lasers intended for WMS application should be optimized for linear frequency sweep and not for high efficiency. A nonlinear FM-index, resulting in a very phase sensitive distortion, is often the dominating source limiting system performance while the laser AM-index is of secondary importance. We will also present simulations that will support this conclusion.

1. H. Ahlberg, S. Lundqvist, R. Tell, and T. Andersson, "Industrialized High-Sensitivity Fiber Optic Near-IR Diode-Laser-Based Gas Analysis System," *SPIE Proceedings* Vol. 2112, p. 118-129, 1993.
2. D. Bomse, "Diode lasers; Finding Trace Gases in the Lab and the Plant," Photonic Spectra June 1995.
3. J. A. Silver, "Frequency-modulation spectroscopy for trace species detection: theory and comparison among experimental methods," Applied Optics, Vol. 31, No. 6, pp. 707-717, 1992.
4. LDS3000 instrument for process monitoring and control," *Technical documentation AltOptronic AB* 1996.

Image Processing by Holographic Gratings in Polydiacetylene

Hossin Abdeldayem, Mark S. Paley, William Witherow, and Donald O. Frazier
ES 76, Building 4481, Space Science Laboratory, NASA-Marshall Space Flight Center,
Huntsville, AL 35812
Phone (205)544-3494, Fax (205)544-2102
E-mail: hossin.abdeldayem@msfc.nasa.gov

Polydiacetylenes are a very promising class of polymers for both photonic and electronic applications[1-3] because of their highly conjugated structures. We, recently, have discovered the possibility of depositing a permanent holographic grating of information, made of polydiacetylene on a glass substrate. A novel technique for obtaining high quality thin films of a polydiacetylene derivative of 2-methyl-4-nitroaniline (PDAMNA) using photodeposition from monomer solutions onto UV transparent substrates was discovered by members of our group a few years back[4]. This compound was one of several asymmetric diacetylenes that were first studied extensively for their optical and electronic properties by Garito and co-workers in the late 1970s; however, their investigations did not include behavior in solutions[5,6]. PDAMNA films can be obtained readily from solutions of DAMNA in 1,2 -dichloroethane by irradiation with UV light through a quartz or glass window, which serves as the substrate. This simple straightforward process yields transparent films with thickness on the order of 1μm[7]. To obtain PDAMNA thin films, a solution of DAMNA in 1,2-dichloroethane is placed inside a chamber shown in figure 1. DAMNA monomer exhibits strong absorption at 366nm. A hand-held 15W UV lamp placed directly in front of the substrate (approximate radiation intensity of 800 $\mu W/cm^2$ at 6 in.), a film of approximately 0.6 μm thickness is obtained after 24 hours of exposure.

Initially, we realized that any substrate which is sufficiently transparent to UV light can be used; thus far we have grown PDAMNA films onto glass, quartz, mica, indium tin oxide coated glass, polyethylene teraphthalate , KBr, and NaCl. In the deposition process, we have conducted masking experiments in which a portion of the substrate is blocked from exposure to the UV light during film deposition. The mask is placed on the exterior surface of the substrate (opposite to the side on which the film is grown) and thus is not in contact with the solution ; it serves merely to protect part of the substrate from the light. Interestingly, the result is that film deposition occurs only where the substrate is directly exposed to the light (figure 2).

Photodeposition was also achieved using a UV-Argon ion laser at 364m. The chamber was mounted onto a computer-controlled translation stage, which had been programmed to trace out a desired pattern. The laser was then focused onto the substrate as the translation stage traced out the test pattern shown in figure 3. The line thickness in the pattern is controlled by the intensity of the laser, the translation rate, and the number of times the pattern is traced. The width of the line is determined by the focal width of the laser beam.

Most recently, a holographic grating of information made of PDAMNA was constructed on a glass substrate. This holographic recording on the polydiacetylene film

Reprinted from Advances in Optical Imaging & Photon Migration, 1998 Technical Digest, © *1998 Optical Society of America*

Most recently, a holographic grating of information made of PDAMNA was constructed on a glass substrate. This holographic recording on the polydiacetylene film was an analogue recording, which is necessary for a number of vital application, such as associative memory, pattern recognition, holographic optical elements, holographic interconnection for neural networks, etc.. In the process, two coherent beams of UV at 364nm from an Ar^+ laser were made to interfere at the substrate which is an optical window of the monomer solution in figure 1. One of the beams was carrying certain information. The interference of the two beams at the substrate induces an interference grating which carries the same information that was carried by one of the beams. The information on the grating was later retrieved by a cw a He-Ne laser at 633nm as shown in figure 4 (23X magnification). The intensity of the interference pattern formed by the grating induces changes in the property of the recording medium through polymerization. Consequently the polymer will be deposited on the substrate according to the grating intensity pattern. The complex index of refraction n of the deposited polymer film will be given by :

$$n = n_0 - i\kappa \qquad (2)$$

where n_0 is the real index of refraction and κ is the imaginary index of refraction which depends on the absorption coefficient of the solution to the UV light. The hologram obtained by PDA is a combination of both amplitude and phase gratings.

The hologram was distinguished as an intermediate grating between a thick (3-D) or thin (2-D). This was determined by the Q-parameter criterion given by Klein as

$$Q = 2\pi\lambda d / n\Lambda^2 \qquad (3)$$

where λ is the wavelength of the illumination reference beam (366nm), n is the refractive index of the recording medium(1.7) and Λ is the grating period of the holographic grating, which is ~364nm, where the angle of incidence is ~ 30^0. More investigations are under way to estimate the maximum resolution possible of these holograms, and consequently the maximum storage density.

References:

1. Polydiacetylenes; Bloor, D., Chance, R. R. , Eds.; Martinus Nijhoff Publishers: Dordrecht, The Netherlands, 1985.
2. Nonlinear Optical Properties of Organic Molecules and Crystals ;Chemla, D. S. , Zyss, J. Eds; Academic Press: Orlando, FL , 1987; vol. 2.
3. Introduction to Nonlinear Optical Effects in Molecules and Polymers; Prasad, P. N., Williams, D. J., Eds..; John wiley and Sons, Inc. : New York, 1991.
4. Paley, M. S. ; Frazier, D.O.; Abdeldayem, H. ; McManus, S. P. Chem. Mater. 1994, 6(12), 2213.
5. Horner, C. J.; Garito, A. F. Makromol. Chem. 1981, 182, 19.
6. Garito, A. F. ; Singer, K. D. ; Teng, C. C. In Nonlinear Optical Properties of Organic and Polymeric Materials ; Williams, d. J. , Ed.: ACS Symposium Series, vol. 233; American Chemical Society: Washington, DC. 1983; pp.1-26.
7. Paley, M. S. ; Frazier, D.O.; Abdeldayem, H. ; Armstrong, S.; and McManus, S. P., J. Am. Chem. Soc. Vol. 117, No. 17, pp. 4775-4780, 1995.

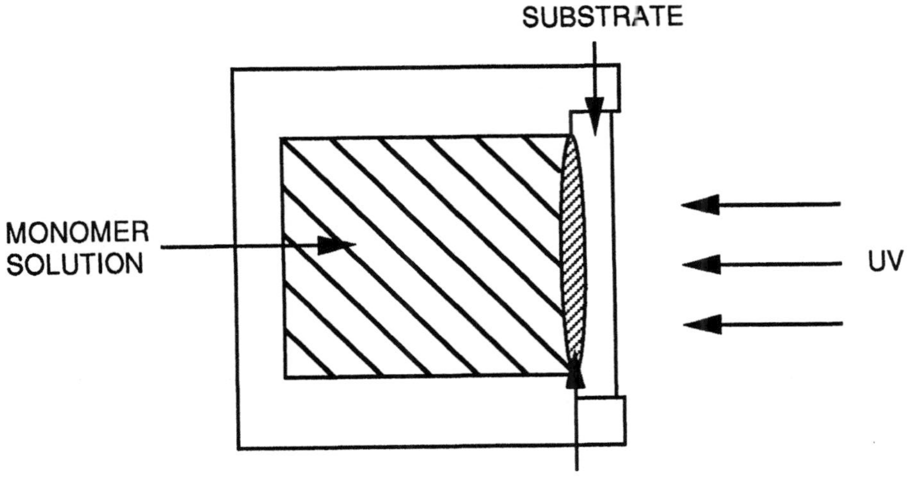

Figure 1. Diagram of the apparatus used for the photodeposition of PDAMNA thin films from solution substrate.

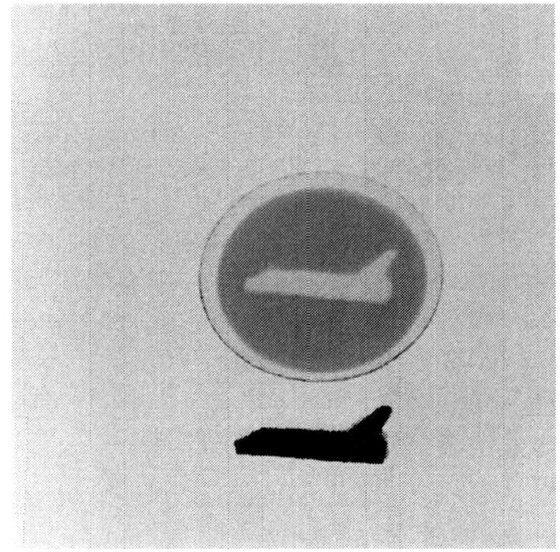

Figure 2. PDAMNA film formed by photodeposition from solution onto a thin glass. A portion of the substrate was blocked from exposure to the UV light by the mask shown below the film.

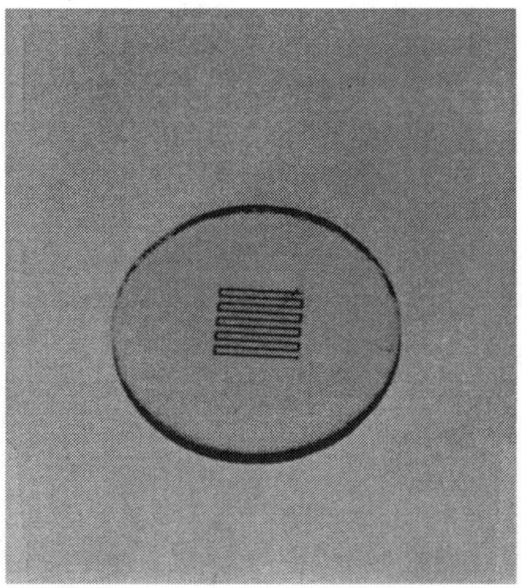

Figure 3. PDAMNA circuit photodeposited onoto quartz using the UV laser.

Figure 4. Retrieved information from the holographic grating of PDAMNA.

Generation of diffuse density waves from a coherently modulated source: definition of the spatial validity of parameters extraction for collimated source.

X. INTES, B. LE JEUNE, F. PELLEN, Y. GUERN, J. CARIOU, J. LOTRIAN

L.S.O.L - University of Brest
6 avenue Le Gorgeu - B.P. 809 - 29285 Brest CEDEX - France.

Phone : (33) 02-98-01-66-30 - Fax : (33) 02-98-01-61-31
E-mail : intes@univ-brest.fr

We present here a frequency domain experiment which deals with the photon propagation in a highly scattering medium.

This experiment showed that, when two continuous optical waves of frequencies υ_1 and υ_2 respectively, with a constant phase relationship, are injected in an infinite scattering medium through a fiber, they generate a diffuse wave which propagates at the frequency modulation $\Delta\upsilon = \upsilon_1 - \upsilon_2$ (140MHz in our case) [1]. We treated this diffuse wave within the P_1 approximation [2] of the transport equation:

$$D\nabla^2\Phi(r,t) - \mu_a\Phi(r,t) - \frac{1}{v}\frac{\partial\Phi(r,t)}{\partial t} = -S_0(r,t) \qquad (1)$$

where $\Phi(r,t)$ is the photon density, $D = \left\{3\left[\mu_a + \mu_s'\right]\right\}^{-1}$ is the diffusion coefficient, μ_a is the absorption coefficient, μ_s' is the reduced scattering coefficient, v is the velocity of the light in the medium, $S_0(r,t)$ is our source distribution and is considered as an intensity-modulated isotropic point source in this theoretical approach. The solution of this equation is of the following form:

$$\Phi(r,t) = \Phi_{DC}(r) + \Phi_{AC}(r)e^{-i\omega t} \qquad (2)$$

where $\Phi_{DC}(r)$ is the continuous part and $\Phi_{AC}(r)$ is the modulation part of the diffuse wave. In addition, $\omega = 2\pi\Delta v$ is the angular frequency of modulation.

The macroscopic optical parameters of absorption (μ_a) and scattering (μ_s') characterizing several turbid media constituted by various concentrations of milk diluted in water (Table 1) were then extracted from our experimental data (continuous intensity $I_{DC}(r)$, modulated intensity $I_{AC}(r)$ and phase $\varphi(r)$ of the diffuse wave), fitted to the analytical solution expression of (1).

The good agreement of the extracted parameters with those reported in the literature [3,4] validates the use of the diffuse photon density wave (DPDW) formalism. The estimated error on the extraction of these parameters was 10%.

The DPDW is created by the coherent interference between the two waves injected in the medium, although, at such a depth in the medium, these waves cannot not be considered as coherent.

concentration (%)	5	10	20	40
μ_a (10^{-4} mm^{-1})	9.4	9.6	12.5	17.2
μ'_s (mm^{-1})	0.20	0.45	0.92	1.62

Table 1: Macroscopic optical parameters of several concentrations of milk diluted in pure water.

To spatially define the validity of parameters extraction, we carried out polar detection in the medium. The diffuse wave was acquired for various r, the distance between the source and the detector, and different θ, the angle between the axis of the source fiber and the detector fiber. Figure 1 presents an example of acquired data corresponding to $ln[r \times I_{DC}(r,\theta)]$ for a 5% concentration in milk.

The numerical aperture of the emission fiber was experimentally determined to be $N = 0.038$ in water. It can be considered as a quasi-collimated source.

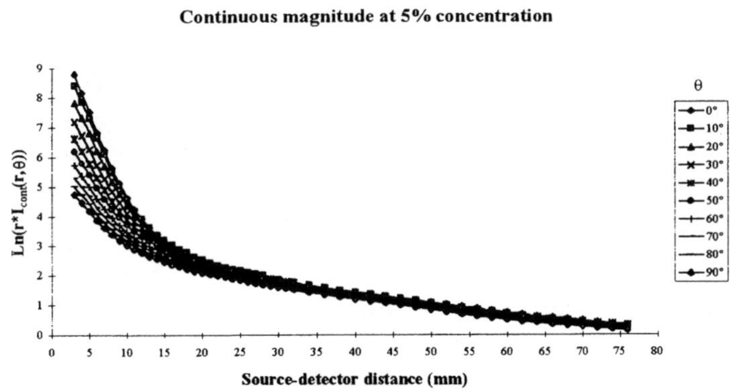

Figure 1: $Ln[r \times I_{DC}(r,\theta)]$ **for a 5% concentration of milk in water.** $Ln[r \times I_{AC}(r,\theta)]$ **and** $\varphi(r,\theta)$ **were extracted from the same data. This three quantities are linear functions of the distance** r**, when the theoretical approach is valid.**

The optical parameters of the turbid media were extracted with respect to the distance between the first considered point of detection and the source. This distance was expressed in units of $l_{tr}^* = \frac{1}{\mu_s'}$ where l_{tr}^* is the reduced mean free path (Figures 2, 3). The parameters were extracted with reliability in the 10% error scale for source-detector distances greater than $10\,l_{tr}^*$ in axial detection ($\theta = 0°$), and $5\,l_{tr}^*$ in a 90°-diffusion detection ($\theta = 90°$) respectively. These results are applicable to a collimated source geometry and show that setting the detector far from the source is important to determine optical parameters accurately.

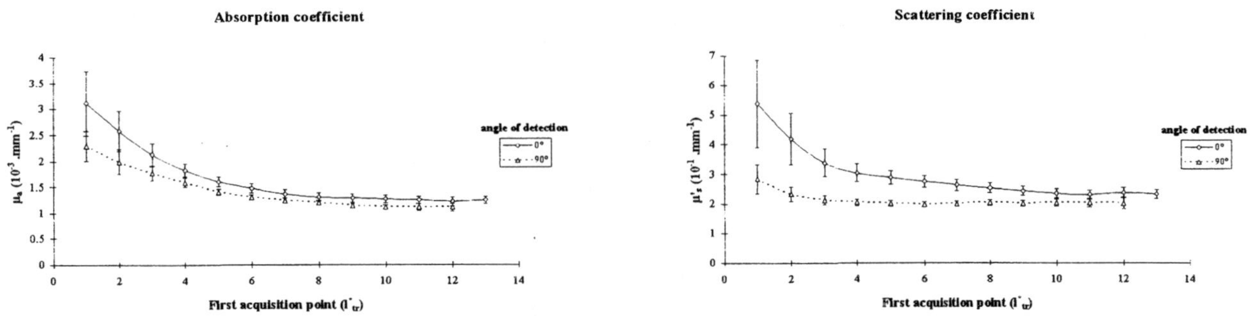

2 (a) **2 (b)**

Figure 2: (a) μ_a**, (b)** μ_s' **for a 5% milk concentration. The parameters were extracted as a function of the distance between the first acquisition point and the detector. Uncertainty bars were determined from uncertainties on the slopes of the linear functions that fit the quantities experimentally measured.**

This experiment showed that the algorithms based on the use of ordinates at the origin of the straight lines associated with the measured quantities do not provide accurate results [**5**, **6**]. In the region where $r < 10l^*_{tr}$, the source term cannot be considered as an isotropic source term. From the results presented in Figure 1 we constructed iso-intensities (Figure 3). They show that a collimated beam can be approximated by an isotropic point source only when detection is made far from the source ($r > 10l^*_{tr}$). Moreover, this point source must be located deeper in the medium, at one l^*_{tr}.

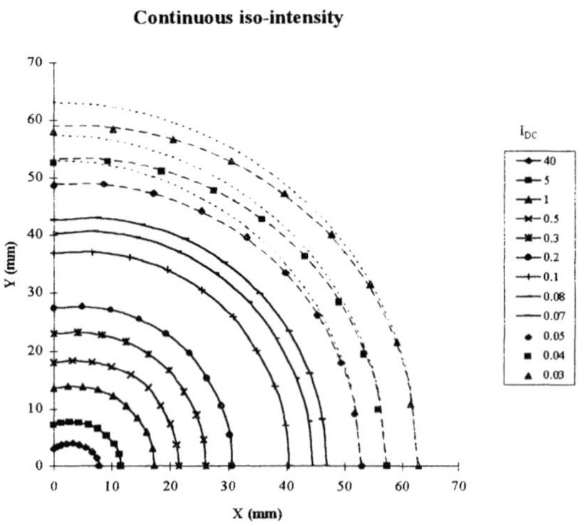

Figure 3: Iso-intensities of $I_{cont}(X,Y)$ where X is the axis of the emission fiber. The dotted curves represent the circles centered on the tip of the emissive fiber with a radius equal to the value of the abscissa of the considered iso-intensity on the X-axis. The dashed curves indicate the circles centered on one l^*_{tr} on the X-axis, with a radius equal to the value of the abscissa of the considered iso-intensity on the X-axis minus one l^*_{tr}. The points represent experimentally-acquired data. This experiment was made with a 5% milk concentration.

These results confirm experimentally the validity of utilizing this common truncation. For distances smaller than $10l^*_{tr}$, an "effective source term" must be used to apply the diffusion equation (1) [7]. Obtaining the analytical expression of this source term will be one of our next tasks. Furthermore, the originality of our modulation technique makes possible the coupling of DPDW generation with coherent techniques such as heterodyne tomography or polarization detection.

[1] X.Intes, B.Le Jeune, F.Pellen, Y.Guern, J.Cariou, and J.Lotrian, « *Determination of optical properties of multiple-scattering media by using a coherent-modulated source* », Journal of Optics, Vol. 28, pp. 1-7 (1997)

[2] J.Fishkin and E.Gratton, « *Propagation of photon-density waves in strongly scattering media containing an absorbing semi-infinite plane bounded by a straight edge* », J. Opt. Soc. Am. A., Vol. 10, n°1, pp. 127-140 (1993)

[3] H.Wabnitz and H.Rinneberg, « *Imaging in turbid media by photon density waves: spatial resolution and scaling relations* », Appl. Opt., Vol. 36, n°1, pp. 67-73 (1997)

[4] K.M.Yoo, B.B.Das and R.R.Alfano, «*Imaging of a translucent object hidden in a highly scattering medium from the early portion of the diffuse component of a transmitted ultrafast laser pulse* », Opt. Lett., Vol. 17, n°13, pp. 958-960 (1992)

[5] M.Franceschini, S.Fantini, S.Walker, J.Maier, W.Mantulin, and E.Gratton, « *Multi-channel optical instrument for near-infrared imaging of tissues* », SPIE, Vol. 2389, pp. 264-273 (1995)

[6] H.Liu, D.Boas, Y.Zhang, A.Yodh, and B.Chance, « *A simplified approach to characterize optical properties an blood oxygenation in tissue using continuous near-infrared light* », SPIE, Vol. 2389, pp. 469-502 (1995)

[7] S.Fantini, M.Franceschini, and E.Gratton, « *Effective source term in the diffusion equation for transport in turbid media* », Appl. Opt., Vol. 36, n°1, pp. 156-163 (1997)

Approximate Scalar Equations for Polarized Radiative Transfer

N. J. McCormick
University of Washington
Department of Mechanical Engineering, Seattle, Washington 98195-2600
Tel: (206) 543-4355; Fax: (206) 685-8047; E-mail: mccor@u.washington.edu
G. C. Pomraning
University of California, Los Angeles
Engineering and Applied Science, Los Angeles, California 90095-1597
Tel: (310) 825-1744; Fax: (310) 206-2302; E-mail: pom@seas.ucla.edu

Abstract—An asymptotic analysis of the radiative transfer equation with polarization gives a renormalized scalar equation for the Stokes parameter I and quadrature results for Q, U, V. The only change in the approximate transfer equation for I is the scattering phase function has expansion coefficients that depend on three elements of the scattering phase matrix rather than one. The approximate equations enable estimation, with purely scalar equations, of the propagation in optically thick media of the four polarized radiation components.

Key Words: (030.5620) Radiative transfer; (260.5430) Polarization; (170.5280) Photon migration

Introduction

A general treatment of polarized radiative transfer requires the solution for four radiation components, often taken to be the four Stokes parameters I (the total specific intensity of radiation), Q, U, and V; see for example, Chandrasekhar.[1] The Stokes vector, a column vector with components I, Q, U, and V, is a convenient representation and can be defined in terms of time averages of the electric field.[1,2] The element I is the total intensity and V is the degree of circular polarization; Q and U account for the linear polarization effects.

Many elementary radiative transfer analyses simply assume that $Q = U = V = 0$ in order to avoid the necessity of accounting for the changes between different polarization states. Here we derive an appropriate scalar radiative transfer equation for I that approximately incorporates the effects of polarization in three dimensional geometries with general anisotropic scattering. The equation developed is valid to order $O(\epsilon)$, where the smallness parameter ϵ is such that the total, absorption, and scattering parameters ($\sigma = \sigma_a + \sigma_s$, σ_a, and σ_s) all are of order $O(1/\epsilon)$. The parameter ϵ also measures the departure of the radiation field from isotropy. What this means in practice is that the analysis is asymptotically valid for media of large optical thicknesses. We also develop equations to compute the polarization components Q and U with an error of $O(\epsilon^2)$; the component $V = 0$ to this order.

Analysis Procedure

The time-independent equation for the Stokes vector $\mathbf{\Psi}(\mathbf{r}, \mu, \phi) = [I, Q, U, V]^T$ at position $\mathbf{r}$ (suppressing the spatial dependence of σ, σ_s, and $\mathbf{P}$) is[1]

$$\mathbf{\Omega} \cdot \nabla \mathbf{\Psi}(\mathbf{r}, \mu, \phi) + \sigma \mathbf{\Psi}(\mathbf{r}, \mu, \phi) = \mathbf{S}(\mathbf{r}, \mu, \phi) + \frac{\sigma_s}{4\pi} \int_0^{2\pi} \mathrm{d}\phi' \int_{-1}^{1} \mathrm{d}\mu' \mathbf{P}(\mu, \mu', \phi - \phi') \mathbf{\Psi}(\mathbf{r}, \mu', \phi'), \quad (1)$$

where $S(\mathbf{r}, \mu, \phi)$ denotes a spatially distributed interior source. The two angular variables μ and ϕ are those that define the unit vector $\mathbf{\Omega}$. For the source we take the thermal emission column vector $\mathbf{S}(\mathbf{r}, \mu, \phi) = [\sigma_a \mathcal{B}(\mathbf{r}), 0, 0, 0]^T$, where $\mathcal{B}$ is the local Planck blackbody radiation function and T denotes the transpose, so that the row vector in this equation is a column vector.

The form of the scattering phase function $\mathbf{P}(\mu, \mu', \phi-\phi')$ derived by Siewert[3,4] is used because it facilitates the use of a Fourier series expansion of $\mathbf{\Psi}(\mathbf{r}, \mu, \phi)$ in ϕ. The phase function is valid in various situations, such as: (i) randomly oriented particles having a plane of symmetry, like spheres and ellipsoids; (ii) particles and their mirror particles in equal numbers with random orientation; and (iii) randomly oriented optically inactive particles exhibiting Rayleigh scattering with or without depolarization effects. (Mie scattering, for example, is a special case.) The phase function can be expressed in terms of the associated Legendre functions $P_\ell^m(\mu)$ that generalize the Legendre polynomials $P_\ell(\mu)$ and the functions $R_\ell^m(\mu)$ and $T_\ell^m(\mu)$ that are defined[3,4] in terms of the generalized spherical functions $P_{m,n}^\ell(\mu)$. The expansion coefficients of the scattering phase function are contained in the matrix[3,4]

$$\mathbf{B}_\ell = \begin{bmatrix} \beta_\ell & \gamma_\ell & 0 & 0 \\ \gamma_\ell & \alpha_\ell & 0 & 0 \\ 0 & 0 & \zeta_\ell & -\varepsilon_\ell \\ 0 & 0 & \varepsilon_\ell & \delta_\ell \end{bmatrix}, \quad \ell = 0 \text{ to } L, \qquad (2)$$

where the scattering coefficients β_ℓ, α_ℓ, γ_ℓ, ζ_ℓ, δ_ℓ, and ε_ℓ can be computed with a Mie scattering program.[5]

Deep in the interior of a medium in near thermodynamic equilibrium the intensity $I(\mathbf{r}, \mu, \phi)$ is essentially isotropic and very nearly equal to the local Planck function, i.e., $I(\mathbf{r}, \mu, \phi) = \mathcal{B}(\mathbf{r}) + O(\epsilon)$; in complete thermodynamic equilibrium, $I = \mathcal{B}$ and $Q = U = V = 0$. This near-isotropy assumption, together with a spatial smoothness assumption on $\mathbf{\Psi}(\mathbf{r}, \mu, \phi)$, constitutes the underlying basis of the asymptotic analysis.

Results

We used an expansion of the Stokes vector in an asymptotic series and equated coefficients of like powers of ϵ to obtain[6]

$$\begin{bmatrix} Q(\mathbf{r}, \mu, \phi) \\ U(\mathbf{r}, \mu, \phi) \\ V(\mathbf{r}, \mu, \phi) \end{bmatrix} = \epsilon \frac{\omega}{2\pi} \sum_{\ell=0}^{L} \sum_{\ell=m+2}^{L} \frac{(2\ell+1)(\Gamma_\ell^m)^2 \gamma_\ell}{(1+\delta_{m,0})(2\ell+1-\omega\alpha_\ell)} \times \int_0^{2\pi} \int_{-1}^{1} P_\ell^m(\mu') \begin{bmatrix} R_\ell^m(\mu) \cos m(\phi-\phi') \\ -T_\ell^m(\mu) \sin m(\phi-\phi') \\ 0 \end{bmatrix} \times I(\mathbf{r}, \mu', \phi') \mathrm{d}\mu' \mathrm{d}\phi' + \mathrm{O}(\epsilon^2) \qquad (3)$$

where $(\Gamma_\ell^m)^2 = (\ell-m)!/(\ell+m)!$.

Equation (3), which allows the computation of Q and U once I has been determined, shows that the effective scattering coefficient for the conversion of polarization component I to components Q and U is $(2\ell+1)\gamma_\ell/(2\ell+1-\omega\alpha_\ell)$. The factor γ_ℓ is not surprising, in view of the "1–2" and "2–1" elements of $\mathbf{B}_\ell$ in Eq. (2), but the "polarization enhancement factor" $[1-\omega\alpha_\ell/(2\ell+1)]^{-1}$ acts to enhance that conversion process when $\omega < 1$ because $|\alpha_\ell| \leq (2\ell+1)$.

The final form of the renormalized equation for the scalar intensity I is[6]

$$\mathbf{\Omega} \cdot \nabla I(\mathbf{r}, \mu, \phi) + \sigma I(\mathbf{r}, \mu, \phi) = \sigma_a \mathcal{B}(\mathbf{r}) + \frac{\sigma_s}{4\pi} \int_0^{2\pi} \int_{-1}^{1} \mathcal{P}(\mu, \mu', \phi, \phi') I(\mathbf{r}, \mu', \phi') \mathrm{d}\mu' \mathrm{d}\phi' + \mathrm{O}(\epsilon). \qquad (4)$$

Here the effective scalar scattering phase function is given by

$$\mathcal{P}(\mu, \mu', \phi, \phi') = \sum_{\ell=0}^{L} \sum_{m=0}^{\ell} \frac{2(\Gamma_\ell^m)^2}{1+\delta_{m,0}} \beta_\ell^{polar} P_\ell^m(\mu) P_\ell^m(\mu') \cos m(\phi-\phi'), \qquad (5)$$

where the "polarization scattering coefficient" β_ℓ^{polar} is defined by

$$\beta_\ell^{polar} = \beta_\ell + \frac{\omega\gamma_\ell^2}{2\ell+1-\omega\alpha_\ell}. \qquad (6)$$

The "polarization enhancement term" $\beta_\ell^{polar} - \beta_\ell$ of the polarization scattering coefficient is the contribution of Q to I and is always a nonnegative number like the polarization enhancement factor for Q and U.

Equation (4) is the renormalized radiative transfer equation to be used to approximately account for polarization effects. Numerical tests[6] have shown that the equation gives very good results for the diffuse attenuation coefficient deep within a medium, much better than the use of an elementary radiative transfer analysis with $Q = U = V = 0$. More details are in Ref. 6.

References

1. S. Chandrasekhar, *Radiative Transfer*, (Dover, New York, 1960).
2. M. Kerker, *The Scattering of Light and Other Electromagnetic Radiation*, (Academic, New York, 1969).
3. C. E. Siewert, *Astrophys. J.* **245**, 1080-1086 (1981).
4. C. E. Siewert, *Astron. Astrophys.* **109**, 195-200 (1982).
5. W. A. de Rooij and C. C. A. H. van der Stap, "Expansion of Mie scattering matrices in generalized spherical functions," *Astron. Astrophys.* **131**, 237-248 (1984).
6. G. C. Pomraning and N. J. McCormick, *J. Opt. Soc. Am. A* (in press).

Time-resolved optical backscattering from highly scattering media

W. Cai, B. Luo, M. Lax, R. R. Alfano

Institute for Ultrafast Spectroscopy and Lasers
New York State Center of Advanced Technology for Ultrafast Photonic Materials and Applications
Department of Physics
The City College and Graduate Center of City University of New York
New York, NY 10031

Abstract

A time-resolved backscattering model, which combines a single large-angle scattering with multiple small-angle scattering,' is used to produce scattered light profile about the medium. While a multiple small-angle scattering leads to light at near-forward direction, a single large-angle scattering transfers the direction of light to the backward hemisphere. This model can be served as a forward model for image reconstruction of highly scattering media, since inhomogeneity of the medium is included in the model. Some multidimensional integrals can be evaluated analytically, which greatly saves computational time.

Key Words

(170.5280) Photon migration; (170.6920) Time-resolved imaging; backscattering.

Introduction

Probing the internal structure of an inhomogeneous scattering medium using backscattering light from a medium has broad applications in medical and non-medical areas. The backscattering signal from the earth's atmosphere using a laser source located in an air plane or a satellite provides information to monitor the earth's atmospheric structure, such as clouds and ozone layers. The time-resolved intensity profile collected by the detector-fibers near the illuminating source-fiber at early time (10 - 300 psec) provides important information of the structure inside tissues below the sub-surface. It has been shown that optical backscattering images from different skin surfaces have high resolution. With an increase of depth under the skin surface the image resolution is reduced. One of the reasons is that multiple scattering in the medium smears the image.

The diffusion approximation of photon migration in highly scattering media is often used in developing optical tomography algorithms. It is not suitable for the back-scattering case where highly anisotropic scattering for early times plays an important role. The Monte Carlo method can be used to simulate photon migration; however, due to its complexity it is rather inconvenient when one wants to treat an inverse problem to obtain a image from light scattering data. Recently, Polishchuk et al[1] and Perelman et al[2] have suggested different models for photon migration. They consider only multiple small-angle scattering, based on the fact that phase function (angular distribution of the scattering cross-section) in many media has a very sharp forward peak. This assumption, generally speaking, is true, but not for the backscattering case. It requires the accumulation of many small-angle scattering events to obtain scattered light at angles near π to the incident light direction.

In this paper we introduce a time-resolved backscattering model, combining a single large-angle scattering with small-angle multiple scatterings, for image reconstruction of scattering medium. While a multiple small-angle scattering leads to light at the near-forward direction, a single large-angle scattering transfers the direction of light to the backward hemisphere. This idea was first employed in 1971[3] and recently used by Katsev et al[4] for estimating the static backscattering light power in the atmosphere. We generalize this concept to the time-dependent case. Inhomogeneity of the medium with a real phase function can be included in our model, so that this model can serve as a forward model for image reconstruction of highly scattering media. Some multidimensional integrals can be evaluated analytically to save computational time. This model reduces to the conventional diffusion model at great depths from the source and detectors.

Formulation of the model

A schematic diagram of our model is shown in Fig. 1.

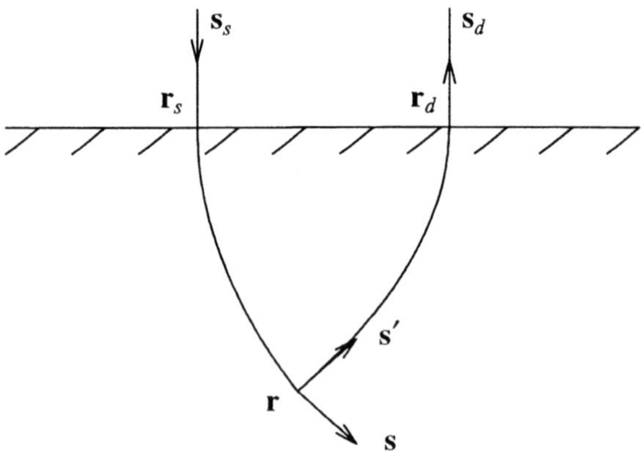

Fig. 1. Schematic diagram of the proposed model.

An ultrashort laser pulse, which is located at $\mathbf{r}_s$ with incident direction $\mathbf{s}_s$, is launched at time $t=0$ into a highly scattering medium. Photons are received by a detector, located at $\mathbf{r}_d$ with receiving direction $\mathbf{s}_d$, at time t. The medium is characterized by the scattering coefficient $\mu_s(\mathbf{r})=\rho(\mathbf{r})\sigma_s(\mathbf{r})c$, with ρ the density of scatterers, σ_s the scattering cross section, and c the light speed in the medium, the absorption coefficient $\mu_a(\mathbf{r})=\rho(\mathbf{r})\sigma_a(\mathbf{r})c$ with σ_a the absorption cross section, and the phase function $P(\mathbf{s}',\mathbf{s},\mathbf{r})$ (angular distribution of the scattering cross-section) normalized to $\int ds' P(\mathbf{s}',\mathbf{s})=1$.

The Boltzmann kinetic equation for the photon distribution function $I(\mathbf{r},\mathbf{s},t)$ as a function of time t, position $\mathbf{r}$ and direction $\mathbf{s}$ is

$$\partial I(\mathbf{r},\mathbf{s},t)/\partial t + c\mathbf{s}\cdot\nabla_{\mathbf{r}} I(\mathbf{r},\mathbf{s},t) + \mu_a(\mathbf{r}) I(\mathbf{r},\mathbf{s},t)$$

$$= \mu_s(\mathbf{r}) \int P(\mathbf{s},\mathbf{s}',\mathbf{r})[I(\mathbf{r},\mathbf{s}',t) - I(\mathbf{r},\mathbf{s},t)]\, ds'$$

$$+ S_0 \delta(\mathbf{r}-\mathbf{r}_s)\delta(\mathbf{s}-\mathbf{s}_s)\delta(t-0) \tag{1}$$

A perturbative method is used, that takes a multiple small-angle scattering in a uniform background medium as the zero-order approximation. It takes a single large-angle scattering combining with two multiple small-angle scatterings, shown in Fig. 1, as the first-order perturbation. The phase function is divided into two part: $P(\mathbf{s}',\mathbf{s},\mathbf{r}) = P_1(\mathbf{s}',\mathbf{s}) + P_2(\mathbf{s}',\mathbf{s},\mathbf{r})$. Here $P_1(\mathbf{s}',\mathbf{s})$ includes only the forward peak of the phase function (small-angle scattering), and is taken as position independent. The remaining part of phase function is $P_2(\mathbf{s}',\mathbf{s},\mathbf{r})$. Let $\int ds' P_1(\mathbf{s}',\mathbf{s}) = \alpha$. A uniform background is assumed, with parameters μ_s^0 and μ_a^0. The zero-order Boltzmann equation is written as

$$\partial I^{(0)}(\mathbf{r},\mathbf{s},t)/\partial t + c\mathbf{s}\cdot\nabla_{\mathbf{r}} I^{(0)}(\mathbf{r},\mathbf{s},t) + \mu_a^0 I^{(0)}(\mathbf{r},\mathbf{s},t)$$

$$= \mu_s^0 \int P_1(\mathbf{s},\mathbf{s}')[I^{(0)}(\mathbf{r},\mathbf{s}',t) - I^{(0)}(\mathbf{r},\mathbf{s},t)]\, ds'$$

$$+ S_0 \delta(\mathbf{r}-\mathbf{r}_s)\delta(\mathbf{s}-\mathbf{s}_s)\delta(t-0) \tag{2}$$

This equation describes the photon transport motion through small-angle scattering only in a uniform medium. The first order perturbative contribution, shown in Fig. 1, to the normalized scattered light strength is given by

$$I^{(1)}(\mathbf{r}_d,\mathbf{s}_d,t) = \int dt' \int d\mathbf{r} \int ds'\, I^{(0)}(\mathbf{r}_d,\mathbf{s}_d,t-t' \mid \mathbf{r},\mathbf{s}')$$

$$\times \mu_s(\mathbf{r}) \{ \int P_2(\mathbf{s}',\mathbf{s},\mathbf{r}) I^{(0)}(\mathbf{r},\mathbf{s},t' \mid \mathbf{r}_s,\mathbf{s}_s)\, ds$$

$$- (1-\alpha) I^{(0)}(\mathbf{r},\mathbf{s}',t' \mid \mathbf{r}_s,\mathbf{s}_s) \} \tag{3}$$

In Eq. (3), $I^{(0)}(\mathbf{r}_2,\mathbf{s}_2,t \mid \mathbf{r}_1,\mathbf{s}_1)$ is a solution of Eq. (2), that is the photon distribution function at time t, position $\mathbf{r}_2$, and direction $\mathbf{s}_2$, from a source $\delta(\mathbf{r}-\mathbf{r}_1)\delta(\mathbf{s}-\mathbf{s}_1)\delta(t-0)$, obtained through only multiple small-angle scattering in a uniform background medium. The correction due to inhomogeneity of absorption coefficients is not included at this moment.

The optical reciprocity theorem[5,6] is used in form

$$I^{(0)}(\mathbf{r}_2,\mathbf{s}_2,t \mid \mathbf{r}_1,\mathbf{s}_1) = I^{(0)}(\mathbf{r}_1,-\mathbf{s}_1,t \mid \mathbf{r}_2,-\mathbf{s}_2)\,, \tag{4}$$

Eq. (3) can be rewritten as

$$I^{(1)}(\mathbf{r}_d,\mathbf{s}_d,t) = \int dt' \int d\mathbf{r} \int ds'\, I^{(0)}(\mathbf{r},\mathbf{s}',t-t' \mid \mathbf{r}_d,-\mathbf{s}_d)$$

$$\times \{\mu_s(\mathbf{r}) \int P_2(-\mathbf{s}',\mathbf{s},\mathbf{r}) I^{(0)}(\mathbf{r},\mathbf{s},t' \mid \mathbf{r}_s,\mathbf{s}_s)\, ds$$

$$- (1-\alpha) I^{(0)}(\mathbf{r},-\mathbf{s}',t' \mid \mathbf{r}_s,\mathbf{s}_s) \} \tag{5}$$

The solution of Eq. (2) in a small-angle scattering approximation has been studied by Polishchuk et al[1], where the distribution of photon density is given by

$$N(\mathbf{r}_2, t \mid \mathbf{r}_1 = 0, \mathbf{s}_1) = \int d\mathbf{s}_2 I^{(0)}(\mathbf{r}_2, \mathbf{s}_2, t \mid \mathbf{r}_1 = 0, \mathbf{s}_1)$$

$$= \frac{1}{(4\pi D_L(t)ct)^{1/2}} \exp\left[-\frac{(\mathbf{r}_L - \mathbf{r}_c)^2}{4D_L(t)ct}\right]$$

$$\times \frac{1}{(4\pi D_T(t)ct)} \exp\left[-\frac{(\mathbf{r}_T)^2}{4D_T(t)ct}\right] \exp(-\mu_a^0 t), \qquad (6)$$

where $\mathbf{r}_L$ is the component of $\mathbf{r}_2$ in the $\mathbf{s}_1$ direction, and $\mathbf{r}_T$ is the component of $\mathbf{r}_2$ that perpendicular to $\mathbf{s}_1$. Eq. (6) describes a nonisotropic diffusion starting from a moving center $\mathbf{r}_c$,

$$\mathbf{r}_c = \mathbf{s}_1 \bar{l}_t [1 - \exp(-ct/\bar{l}_t)], \qquad (7)$$

The time-dependent nonisotropic diffusion coefficients described by $D_T(t)$ and $D_L(t)$ in Eq. (6) is given by

$$D_T(t) = \frac{\bar{l}_t^2}{2ct}\left[\frac{2}{3}\bar{\tau} - f_1 + \frac{1}{9} f_3\right] \qquad (8.1)$$

and

$$D_L(t) = \frac{\bar{l}_t^2}{2ct}\left[\frac{2}{3}\bar{\tau} - \frac{2}{9} f_3 - f_1^2\right]. \qquad (8.2)$$

Here $\bar{\tau} = ct/\bar{l}_t$ and $f_n = 1 - \exp(-n\bar{\tau})$. In Eq. (7), (8.1) and (8.2), the average transport mean free path of small-angle scattering $\bar{l}_t = 1/\langle\rho\sigma\langle\theta^2\rangle_{P_1}\rangle_{\mathbf{r}}$. The inside $\langle \ \ \rangle_{P_1}$ means integration over $\mathbf{s}$ with the phase function P_1 as the weight function and the outside $\langle \ \ \rangle_{\mathbf{r}}$ means an average over spatial coordinates $\mathbf{r}$. Fig. 2 shows $\mathbf{r}_c$, D_L, and D_T as functions of time.

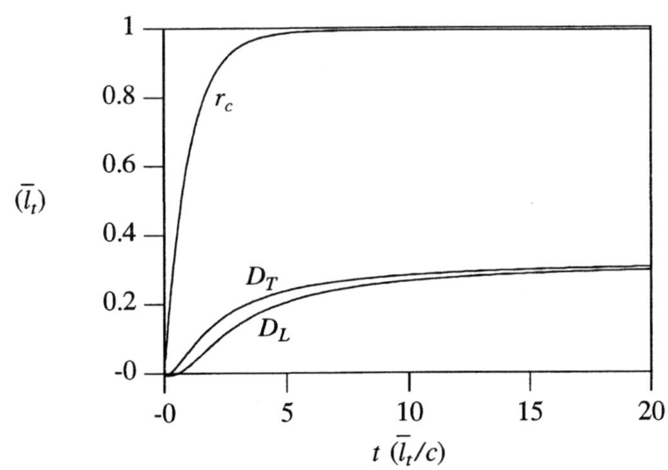

Fig. 2 shows the process of photon migration, form the ballistic mode to the snake mode, then to the diffusion mode. At small times photon's motion is nearly ballistic, described by motion of the center $\mathbf{r}_c(t)$. With increase of time the motion of the center becomes slower, while nonisotropic diffusion gradually increases (snake mode). At large times the center stops moving and Eq. (6) gradually tends to the solution of the conventional diffusion equation with the transport mean free path $\bar{l}_t$.

The average angular distribution is given by[1]

$$F(\mathbf{s}_2, t \mid \mathbf{s}_1) = \int d\mathbf{r}_2 I^{(0)}(\mathbf{r}_2, \mathbf{s}_2, t \mid \mathbf{r}_1, \mathbf{s}_1)$$

$$= \sum_{l=0}^{\infty} \frac{2l+1}{4\pi} \exp\left[-\frac{c}{2\bar{l}_t} l(l+1)t\right] P_l[\cos(\mathbf{s}_2\mathbf{s}_1)], \qquad (9)$$

where P_l are Legendre polynomials. For small times $t \to 0$, $F(\mathbf{s}_2, t \mid \mathbf{s}_1)$ reduces to $\delta(\mathbf{s}_2 - \mathbf{s}_1)$. For large times it reduces to an isotropic distribution $1/4\pi$.

A complete expression for $I^{(0)}(\mathbf{r}_2, \mathbf{s}_2, t \mid \mathbf{r}_1, \mathbf{s}_1)$ is not available in Ref. 1. The following approximate expression will be used in our calculation,

$$I^{(0)}(\mathbf{r}_2, \mathbf{s}_2, t \mid \mathbf{r}_1, \mathbf{s}_1) = N(\mathbf{r}_2, t \mid \mathbf{r}_1, \mathbf{s}_1) F(\mathbf{s}_2, t \mid \mathbf{s}_1)$$

$$- \frac{3}{4\pi} H(ct - \mid \mathbf{r}_2 - \mathbf{r}_1 \mid)$$

$$\times (D_L(t)\nabla_L + D_T(t)\nabla_T) N(\mathbf{r}_2, t \mid \mathbf{r}_1, \mathbf{s}_1) \cdot \mathbf{s}_2. \qquad (10)$$

The first term in Eq. (10) is an accurate distribution at the center $\mathbf{r}_c$, where the second term is equal to zero. At early times, the photons concentrate near $\mathbf{r}_c$, the first term dominates. At large times, the second term produces the required diffusion current, while the first term becomes isotropic and does not contribute to Eq. (5). H in Eq. (10) is the Heaviside function, which is equal to 1 for positive value of the argument and zero otherwise. This H function is needed to prevent propagation faster than speed of light induced by the diffusion Green's function.

Eq. (5) together with Eq. (10) are the main formulae of our model. A Legendre expansion of the known phase function is used in Eq. (5):

$$P_2(\mathbf{s}, -\mathbf{s}', \mathbf{r}) = \sum_n C_n(\mathbf{r}) P_n(\cos(\mathbf{s}\mathbf{s}')) \qquad (11)$$

Let angle $(\mathbf{s}\mathbf{s}') = \beta$, $P(\mathbf{s}, -\mathbf{s}', \mathbf{r})$ is $P(\pi - \beta, \mathbf{r})$, because the optical reciprocity theorem is used in Eq. (5). Denoting $\mathbf{s} = (\theta, \phi)$, $\mathbf{s}' = (\theta', \phi')$, and using the well known addition formula

$$P_n(\cos\beta) = \sum_{m=0}^{\infty} \eta_m \frac{(n-m)!}{(n+m)!}$$

$$\times P_n^m(\cos\theta) P_n^m(\cos\theta') \cos[m(\phi - \phi')], \qquad (12)$$

where $\eta_0 = 1$, $\eta_m = 2$, $(m = 1, 2, \ldots)$, and P_n^m are the associated Legendre functions, we have the orthogonality relation

$$\int ds_2 P_n(\cos(s_1 s_2)) P_{n'}(\cos(s_2 s_3))$$

$$= \frac{4\pi}{2n+1}\delta_{nn'} P_n(\cos(s_1 s_3)) \quad (13)$$

Therefore, integration over s and s′ in Eq. (5) can be analytically performed. Using this model in optical backscattering tomography, computational time is substantially reduced.

The effect of large-angle multiple scattering can be partly introduced by renormalization of $\bar{l}_t$. This concept is similar to the "mass renormalization" or "charge renormalization" in the quantum field theory. A large-angle scattering can be approximately described as a result of several small-angle scatterings. Therefore, The "bare" Green's function $I^{(0)}$, which represents only the small-angle multiple scattering, can be "dressed" through replacing $\bar{l}_t$ by a smaller l_t, that partly includes effect of large-angle multiple scattering. The value of l_t can be determined by fitting experimental data. If $l_t = 1/\langle \rho\sigma \langle 1-\cos\theta \rangle_P \rangle_r$ is chosen, here the inside $\langle \;\; \rangle_P$ means integration over s with the whole phase function P as the weight function, then the dressed Green's function $I^{(0)}$ at large time tends to the conventional diffusion Green's function. This model works both for objects the near surface and for deeper objects, hence, it is convenient for latter inversion efforts.

Results and Discussion

This time-resolved backscattering model is applied to a simulated semi-infinite uniform medium problem. Spherical scatterers, with radius 1.0 μm and index of refraction of 1.59, are uniformly distributed in a background medium with index of refraction of 1.33. The phase function $P(\mathbf{s}, \mathbf{s}', \mathbf{r})$ in Eq. (1) is calculated using Mie theory,[7] with the wavelength of incident light assumed to be 1.25 μm. The cut-off for small angle, P_1, is taken to be 33°, corresponding to α=0.89. The density of scatterers is chosen to correspond a transport mean free path l_t= 2 mm. The background medium has no absorption is assumed, $\mu_a^0 = 0$.

The source-detector setup is illustrated in Fig. 1. Light emitted from the source and light received by the detector are perpendicular to the surface of medium. A quantitative evaluation, according to Eqs. (5) and (10), has been performed using our approach. The time-dependent profiles of the normalized received intensity with different source-detector separations, 1 mm, 2 mm, 4 mm, and 10 mm are plotted in Fig. 3 (a), (b), (c), and (d).

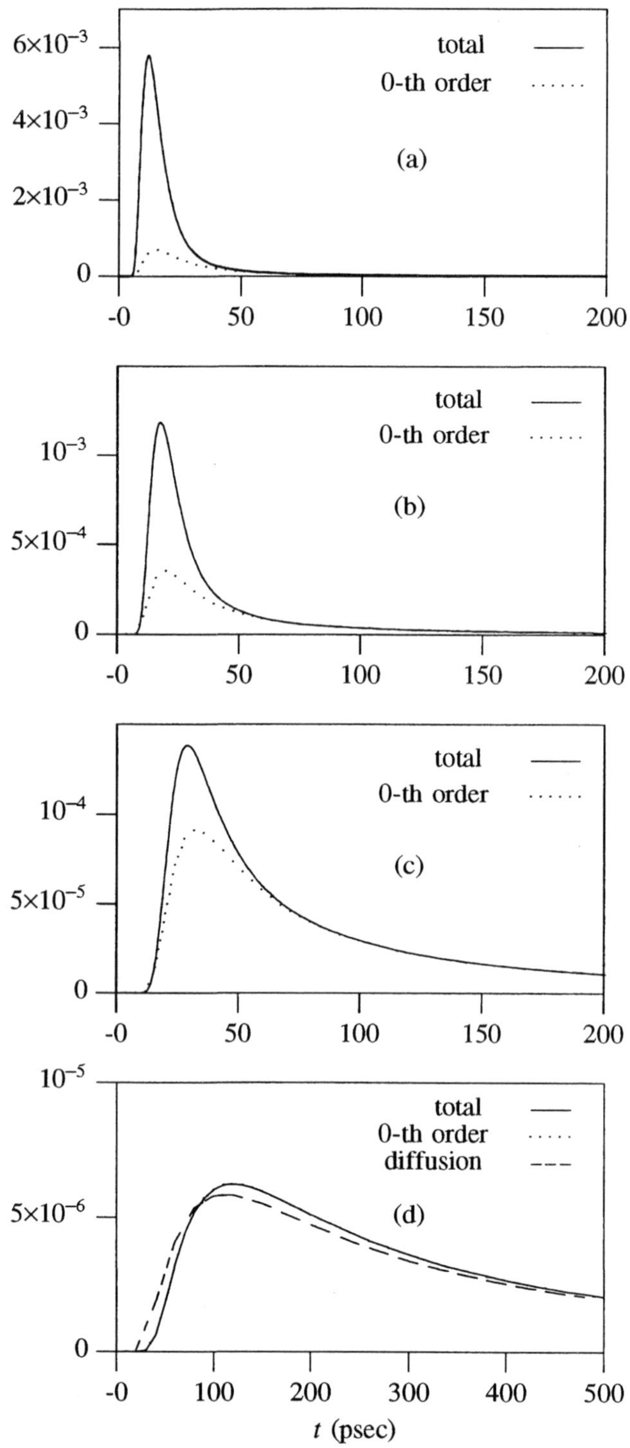

Fig. 3 Simulated temporal profile of backscattering of a half-infinite uniform medium with $l_t = 2$ mm, and different source detector separations x_d: (a) $x_d = 0.5\, l_t$, (b) $x_d = 1.0\, l_t$, (c) $x_d = 2.0\, l_t$, and (d) $x_d = 5.0\, l_t$. Note the different scales.

In Fig. 3, dotted curves represent the contribution from the zero-order term, namely, through only multiple small-angle scattering. The solid curves represent the total intensity, including the first-order contribution, a large angle scattering process shown in Fig. 1, and the zero-order contribution. The dashed curve in Fig. 3(d) is the solution of the conventional diffusion equation. When the distance between source and detector, x_d, is small, the contribution from the first-order term dominates. With increase of x_d, the zero-order contribution increases. When $x_d > 5\,l_t$, the zero-order contribution dominates, and the result is similar to that obtained from the conventional diffusion approximation, if a suitable renormalized l_t is chosen.

Inhomogeneity of the medium is included through **r** dependence of $\mu_s(\mathbf{r})$ and $P_2(-\mathbf{s}', \mathbf{s}, \mathbf{r})$ in Eq. (5). These factors reflect inhomogeneity of density and structure (for example, size of sphere) of scatterers, which are important parameters for image reconstruction of a highly scattering turbid medium. The spatial dependence of the absorption coefficient $\mu_a(\mathbf{r}) - \mu_a^0$ can also be included, which we will consider in future work.

The zero-order Green's function, $I^{(0)}(\mathbf{r}_2, \mathbf{s}_2, t \mid \mathbf{r}_1, \mathbf{s}_1)$, used in this paper is that for an infinite spatial geometry.[1]. Error due to this approximation is not as serious as when the same approximation is used in the conventional diffusion approach. Since the direction of source light is pointed toward into medium and small angle scattering dominates, the Green's function is highly anisotropic and number of photons in a region above the semi-finite medium could be negligible in a short time period. For the case of large separation of source-detector and long time delay, the boundary effect for $I^{(0)}(\mathbf{r}_2, \mathbf{s}_2, t \mid \mathbf{r}_1, \mathbf{s}_1)$ should be considered, which is a subject we will study.

Summary

In summary, we have developed a time-resolved backscattering model, combining multiple small angle scattering with single large angle scattering. The anisotropic character of light scattering at early time has been properly treated. The temporal profile of intensity of scattered light received by a detector located near source is calculated at different source-detector separations. Inhomogeneity can be included in our model, so that this model could be served as a forward model for image reconstruction of highly scattering media in a sub-surface area.

We thank A. Ya. Polishchuk for useful discussions. This work is supported in part by NASA, DOE, ARO, Mediscience Technology Corp., New York State Science and Technology Foundation.

[1]A. Ya Polishchuk, R. R. Alfano, Opt. Lett. Vol 21, p916-918 (1996).

[2]L. T. Perelman, J. Wu, I. Itzkan, and M. S. Feld, Phys. Rev. Lett. 72, p1341-1344 (1966).

[3]L. S. Dolin and V. A. Saveliev, Izv. Akad. Nauk SSSR Ser. Fiz. Atmos. Okeana 7, p505-510 (1971).

[4]I. L. Katsev, E. P. Zege, A. S. Prikhach, I. N. Polonsky, J. Opt. Soc. Am. A, Vol. 14 p1338-1346 (1997).

[5]B. Devison, *Neutron Transfer Theory*, (Clarendon, Oxford) 1958.

[6]E. P. Zege, A. P. Ivanov, and I. L. Katsev, *Image Transfer through a Scattering Medium* (Springer-Verlag, Heidelberg) 1991.

[7]C. F. Bohren and D. R. Huffman, *Absorption and Scattering of light by small particles* (John Wiley & Sons, New York) 1983.

Optical characterization of tissue and other scattering media

Study of Large Scale Fluctuations in Turbid Media by Photon-Density-Wave Fluctuation Correlation Spectroscopy

V. Toronov, M. Filiaci, S. Fantini, and E. Gratton

Laboratory for Fluorescence Dynamics
Department of Physics
University of Illinois at Urbana-Champaign
1110 West Green Street, Urbana, Illinois 61801-3080

Abstract

We propose photon-density-wave fluctuation correlation spectroscopy as a method to study the macroscopic dynamics of tissue optical properties. Phantom measurements have shown that our frequency-domain instrument is appropriate to explore intensity, modulation and phase fluctuations in the frequency band from 0 to 125 Hz. Preliminary *in vivo* studies have revealed a reach optical dynamics in human tissues.

Keywords

(170.5270) Photon density waves; (170.6510) Spectroscopy, tissue diagnostics

1. Introduction

Many physiological processes (such as metabolism, hemodynamics and neuron activity) cause changes of tissue optical properties at the macroscopic level. Characteristic time and space scales of such changes can be about 10 ms and 1mm, respectively. It was shown that the propagation of intensity-modulated light in tissues can be described in terms of photon-density waves [1]. A photon-density wave at a given point is characterized by its intensity, modulation amplitude and phase. A change in these parameters may be related to changes in the tissue absorption and scattering properties. We propose photon-density-wave fluctuation correlation spectroscopy (PDWFCS) as a method to study the dynamics of tissue optical properties. One should distinguish photon density wave fluctuation correlation spectroscopy from diffusing-wave spectroscopy (DWS) [2]. The latter relates fluctuations in the scattered electromagnetic field to the microscopic motion of light scatterers. The time and spatial scales of these field fluctuations are smaller than 0.1 ms and the light wavelength, respectively. Different time and spatial scales probed by PDWFCS and DWS require different experimental approaches. Furthermore, PDWFCS measures intensity (DC), modulation amplitude (AC) and phase fluctuations, while in a DWS experiment the acquired signal is the light intensity only.

2. Instrumentation and method

We use a frequency-domain instrument in which 750 or 830 nm light emitted by laser diodes (~2 mW average power) is guided to the medium through a multimode silica optical fiber having a core diameter of 600 μm and a numerical aperture of 0.39. A glass fiber bundle (internal diameter: 3.2 mm, numerical aperture: 0.56) collects the scattered light and conducts it to the photomultiplier tube detector (PMT). The laser current and PMT voltage are modulated by synchronized synthesizers at 110 and 110.005 MHz, respectively. The signal from the PMT at the difference frequency (cross-correlation frequency) is applied to the input of an interface card for an IBM-PC computer, where the data acquisition and processing is performed to obtain DC, AC and phase values. The time series of these values are processed to calculate their fluctuation power spectra. The bandwidth and the frequency resolution of the spectra are determined by the sampling rate and length of the time sweep. For the measurements reported in this work the acquisition time per point was 8 ms corresponding to 40 periods of the cross-correlation wave. All results were verified to be in good quantitative agreement with those obtained at different acquisition times (4 and 16 ms). An acquisition time of 4 (16) ms corresponds to a frequency bandwidth of 125 (31.25) Hz in the power spectrum. The total time of a sweep was 8.2 s, which corresponds to a spectral resolution of 0.125 Hz.

3. Phantom studies

To test the method and to characterize the instrumentation capabilities, we performed a series of phantom experiments in which we constructed a dynamical phantom. The central part of our experimental setup consisted of a container (a 1-L beaker) filled with an aqueous Liposyn suspension, in which black rubber particles underwent turbulent motion caused by a magnetic-stick stirrer. The average velocity of particle motion is controlled by adjustment of the stirrer rotation frequency. In each experiment we used particles of approximately equal size, either 1 mm or 1.6 mm effective diameter.

Figure 1 shows the DC and phase time traces obtained with a sampling rate of 8 ms. The left part of the plot corresponds to the condition when the mixer is off, so that the particles rest on the bottom of the beaker, and the right part (starting at t~2.5 sec) corresponds to the fluctuation regime, when the stirrer is on. One can see that when the mixer is on, both the phase and the DC exhibit fluctuations over some average value, which is different from that when the mixer is off.

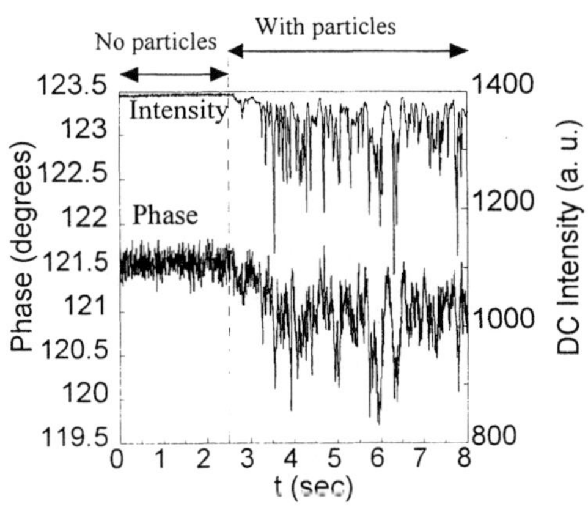

Figure 1. Time traces of the DC and phase signals before and after the stirrer is turned on. The data are obtained under the following experimental conditions: particle effective radius and concentration are respectively 0.5 mm and 0.3 cm^{-3}, magnet rotation frequency is 4 Hz, μ_a =0.043 cm^{-1} and μ_s' =9.98 cm^{-1}.

The Fourier transforms of the time traces were used to obtain power spectra. The dot symbols in Figure 2 show typical experimental power spectra of the DC and phase fluctuations obtained by averaging 100 modulus - squared Fourier transforms. One can see that both DC and phase spectra exhibit narrow resonance peaks (at the stirrer rotation frequency, its harmonics and sub-harmonics) rising over the bell-shaped background. While the peak frequencies are equal for both the DC and phase spectra, the background width of the phase spectrum is significantly smaller than that of the DC. This fact demonstrates the difference between the DC and phase information and originates from the difference of the space regions probed by the DC and phase.

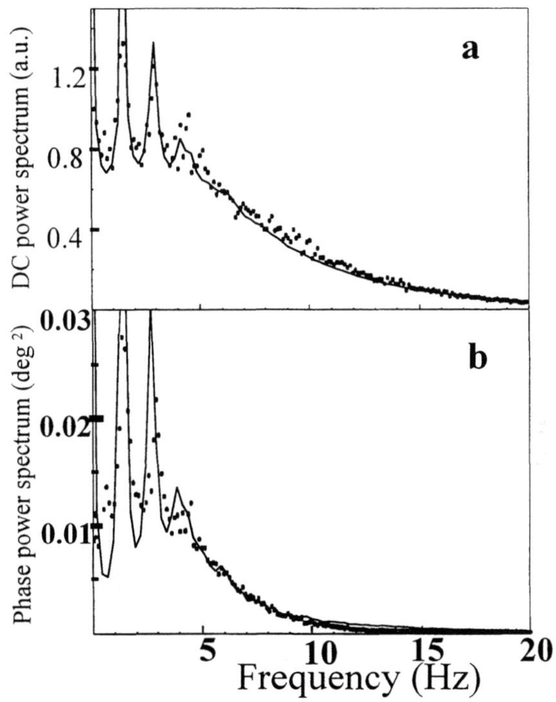

Figure 2. (a) DC and (b) phase power spectra. The points represent the experimental data obtained at the same conditions as in Fig.1. The lines are the best fits provided by model equations. In the theoretical curves the effective particle rotation frequency is 1.5 Hz. Although the frequency band investigated extends up to 125 Hz, the frequency axis is limited to 20 Hz because the particle motion does not contribute to the power spectrum at larger frequencies

We elaborated a theoretical model predicting the DC and phase power spectra. This model is based on the following assumptions:

1) The motion of each absorbing particle is independent of the motion of others and is a superposition of a Brownian stochastic process characterized by the diffusion constants D_x, D_y and D_z, and a deterministic rotation with a given frequency;

2) Due to the small size of absorbing particles and their low concentration, the total response of both DC and phase to the ensemble of particles can be expressed as an algebraic sum over single particle contributions of the form

$$\delta u(\xi,\eta) \approx -S\frac{(3\mu_s')^2}{c}\frac{a^2}{2+3a\mu_s'}\frac{\exp(-\sqrt{3\mu_a\mu_s'}(\xi+\eta))}{4\pi\xi\eta} \quad (1)$$

for the DC and

$$\delta\phi(\xi,\eta) \approx R\frac{3\mu_s' a^2}{2+3a\mu_s'}\mathrm{Im}[\frac{\exp(ik(\xi+\eta-R))}{\xi\eta}] \quad (2)$$

for the phase, where ξ and η are the distances form the object to the light source and the detector respectively, μ_a and μ_s' are the medium absorption and reduced scattering coefficients, c is the speed of light, S is the source power factor, R is the source-detector separation and k is the photon density wave complex wave number. Note that according to Eq. (1), the effective region of the DC sensitivity is localized immediately around the source-detector line, while Eq. (2) shows that the phase perturbation is zero on the source-detector line and maximal when the object is situated within a ring area surrounding this line. The solid lines in Fig. 2 show the power spectra fitted to the experimental data. The fitting parameters are the particle diffusion constants D_x, D_y and D_z. The best fit occurs at $D_x = D_y = D_z = 6$ cm²/s, which is consistent with the observed average speed of particles (about 5-10 cm/s).

4. *In vivo* studies

We have performed a number of *in vivo* measurements of optical fluctuations in tissues including measurements on muscles, the forehead and the brain cortex. Figure 2 shows the power spectra obtained on a human forearm for the cases of normal blood circulation and ischemia caused by a pneumatic cuff inflated at 240 mmHg.

Except for the peaks at the heart beat frequency and its harmonics in the normal circulation spectrum, both curves nearly coincide. The power spectrum acquired from a silicone pad at the same instrument adjustments (the lowest curve in Fig.3) represents the instrument noise. Comparison of the *in vivo* power spectra with the one obtained on the pad shows that the frequency band of the physiological fluctuations ranges from 0 to 5 Hz

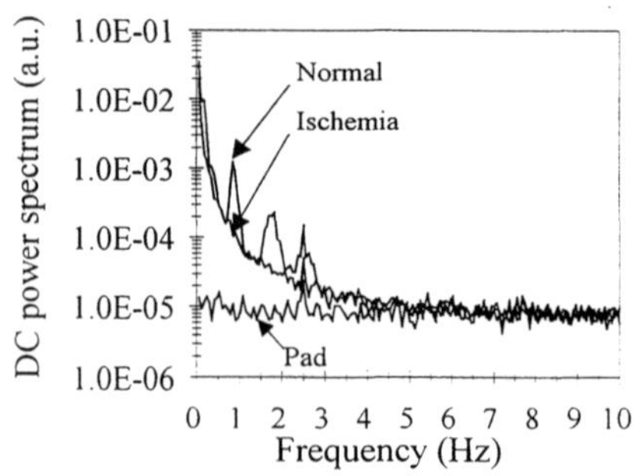

Figure 3. DC power spectra obtained *in vivo* from a human forearm at normal blood concentration and ischemia. In the case of ischemia the pressure of the cuff is 240 mmHg. The lowest curve, obtained from the silicon pad, represents the instrument noise.

5. Conclusion

Phantom and *in vivo* studies confirm that photon-density-wave fluctuation correlation spectroscopy is a promising method to study physiological dynamics. Information on spatial coherence can be obtained by calculating cross-correlation spectra of signals in different locations [3]. Measurements show that our instrument is appropriate to investigate optical fluctuations in the frequency band of 0-125 Hz (processes on a time scale of 8 msec or slower), and over a spatial size as small as 1 mm.

Our future work will be focused on the better localization of the probed region and the study of different physiological processes including blood circulation and brain activity.

Acknowledgment
This research is supported by NIH grants RR03155 and CA57032.

References

[1] J.B. Fishkin and E. Gratton, J. Opt. Soc. Am. A **10,** 127 (1993)

[2] G.Maret and P.E.Wolf, Z. Phys. B **65**, 407 (1987)

[3] M.Filiaci, V. Toronov, S. Fantini, and E. Gratton, "Optical probe and frequency-domain instrumentation to study spatial and temporal correlations of fluctuations in tissues", in this issue

Time-Sliced Transillumination Imaging of Normal and Cancerous Breast Tissues

S. K. Gayen, M. E. Zevallos, B. B. Das and R. R. Alfano

Institute for Ultrafast Spectroscopy and Lasers, New York State Center for Advanced Technology for Ultrafast Photonic Materials and Applications, Department of Physics, The City College of the City University of New York, 138th Street at Convent Avenue, New York, NY 10031

Abstract

Transillumination images of *in vitro* human female breast tissue specimens were recorded using an ultrafast electronic gated imaging camera system. Breast tissue specimens comprising normal and cancerous tissues were illuminated with 800 nm, approximately 130 fs, 1 kHz repetition-rate light pulses from a self-mode-locked Ti:sapphire laser and amplifier system. The gated imaging system formed a sequence of two-dimensional images using approximately 80-ps time slices of the transmitted light pulse. Images recorded with different time slices of transmitted light highlighted different types of tissues, and allowed distinguishing between normal and cancerous tissues. Transillumination imaging as a function of wavelength over the 1225-1300 nm range using a Cr:forsterite laser demonstrated that the optical absorption band of adipose tissues around 1203 nm could be used as a 'spectral finger print' to obtain images of adipose tissues distinct from that of the fibrous tissues.

Key Words

(170.6920) time-resolved imaging; (170.6510) spectroscopy, tissue diagnostics; (290.7050) scattering, turbid media; (999.9999) breast cancer.

Introduction

Development of optical modalities for screening of breast cancer, a major and increasing cause of death that accounted for 4.4% of all US female deaths in 1990,[1] is an intensely pursued area of contemporary research. The impetus for this pursuit stems from a desire to overcome some of the limitations of the current 'gold standard', x-ray mammography, as well as to exploit the spectroscopic advantage and noninvasive nature of optical methods. Over the last ten years a number of time-resolved, frequency-domain, and steady-state laser-spectroscopic techniques and inverse reconstruction methods have evolved in an attempt to image interior structure of tissues.[2-9]

In this paper, we present the initial results of a time-sliced transillumination imaging approach to image and distinguish between normal and cancerous regions of an *in vitro* breast tissue sample. Results of a complementary near-infrared (NIR) spectroscopic imaging study over the 1225-1300 nm range using a Cr:forsterite laser demonstrate that a spectroscopic difference may be used to highlight different tissue constituents.

Methods and Materials

The constituents of normal female human breast include adipose, glandular, and fibrous tissues, ducts, blood vessels, and water. Breast lesions include malignant and benign tumors, cysts and mastitis. The differential transmission of light through different types of breast tissues, such as normal and cancerous or adipose and fibrous, provides the physical basis for formation of shadow image in transmitted light. Optical imaging of breast is complicated by the extremely low ratio of image-bearing photons to the image-degrading multiple-scattered photons, and paucity of high contrast targets that are indicative of the presence of cancer and suitable for imaging. Commonly used time-gating methods attempt to reduce the image-blurring effect of strong light scattering in tissues by sorting out early-arriving image-bearing photons. The time-sliced two-dimensional imaging approach used in this work is an extension of the time-gated methods. It uses the optimal picosecond-scale slices of the transmitted light pulse to form images, and demonstrates that in specific cases later-arriving light may be as useful in distinguishing between normal and cancerous tissues.

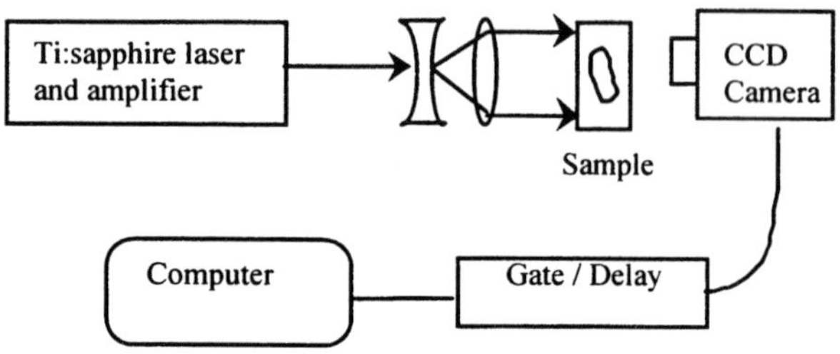

Figure 1. A schematic diagram of the experimental arrangement for time-sliced imaging measurements.

The experimental arrangement for time-sliced transillumination imaging, displayed schematically in Fig. 1, used 800-nm, approximately 130-fs duration, 1 kHz repetition-rate pulses from a self-modelocked Ti:sapphire laser system consisting of an oscillator (Spectra Physics Tsunami), a regenerative amplifier and a multipass power amplifier system (Quantronix Titan).[10]. The average beam power used in the experiment was approximately 400 mW, reduced from 5W available from the laser system. The beam was expanded and a 3-cm diameter central part of it was selected out using an aperture to illuminate the sample. Time-slicing was accomplished using an ultrafast electronic gated imaging camera system comprising a compact time-gated image intensifier unit fiber-optically coupled to a charge-coupled device (CCD) camera. The transmitted light from the sample was collected by a camera lens and directed to the image intensifier. The system provided an electronic gate pulse whose full-width-at-half-maximum duration could be adjusted to a minimum of approximately 80 ps. The gate position could be varied in steps of 25 ps over a 20 ns range. The signal recorded by the system at a particular gate position was a convolution of the transmitted light pulse with the gate pulse centered on the gate position. The time-sliced image was recorded by the CCD camera and displayed on a personal computer.

The experimental arrangement for NIR spectroscopic imaging used the continuous-wave mode-locked output of a Cr^{4+}:forsterite laser pumped by a Nd:YAG laser. The linearly polarized laser output was tuned from 1225 to 1300 nm using an intracavity birefringent plate. The average beam power was maintained at approximately 35 mW for all the wavelengths using appropriate neutral density filters. A Fourier space gate[11] and a polarization gate[12] were used to sort out a fraction of image-bearing photons and discriminate against image-blurring multiple-scattered photons. A 50 mm focal-length camera lens collected and collimated the low-spatial-frequency light filtered by the Fourier gate and directed it to the 128x128 pixels sensing element of an InGaAs NIR area camera (Sensors Unlimited SU 128-1.7RT).

Two different breast tissue samples were used. The first sample obtained from the left breast of a 61 year old female patient consisted of a cancerous piece placed between two normal pieces of 5 mm nominal thickness and lateral dimensions between 8-14 mm. Available literature values[13] of transport length l_t for normal and cancerous breast tissues at 789 nm are approximately 1.7 mm and 1.0 mm, respectively. It was used for time-sliced imaging experiments. The second sample was derived from the left breast of a 28 year old female patient following reduction mammoplasty. A 35 mm X 14 mm X 5 mm section of it comprising adipose tissue in the middle and broad strands of firm fibrous tissue on the two sides was used in the spectroscopic imaging experiment. Available literature values[13] of transport length l_t for normal adipose and fibrous breast tissues at 789 nm are approximately 1.3 mm and 0.7 mm, respectively. Both the tissue specimens were made available to us by National Disease Research Interchange under an IRB approval at the City College of New York. Each of the samples were placed between two glass plates and compressed in an attempt to provide uniform thickness and good physical contact between the adjacent pieces of tissue.

Results

Time-sliced Imaging

Time-sliced transillumination images of the first sample for gate positions at 0 and 200 ps are displayed in Figs. 2(a) and 2(b), respectively. The zero position was taken to be the time of arrival of the light pulse through a 5-mm thick glass cell filled with water. The spatial distribution of transmitted light intensity along a horizontal area covering the normal-cancerous-normal (NCN) regions in the tissue sample is displayed in Fig. 2(c). The profile allowed a comparison of the relative magnitude of light intensity transmitted through the normal and the cancerous tissues at different delay times.

The cancerous and normal regions are clearly distinct in both the 0-ps and the 200-ps images. The spatial profile of the 0-ps image, displayed by the thick line in Fig. 2(c), shows a marked dip indicating much lower light transmission through the cancerous region. The width of the dip corresponds to the spatial extent of the cancerous region. Interestingly, the situation is completely reversed in the spatial intensity profile of the 200-ps image of Fig. 2(b). The profile, displayed by the thin line in Fig. 2(c), exhibits a peak indicative of higher light transmission through the cancerous region at later time.

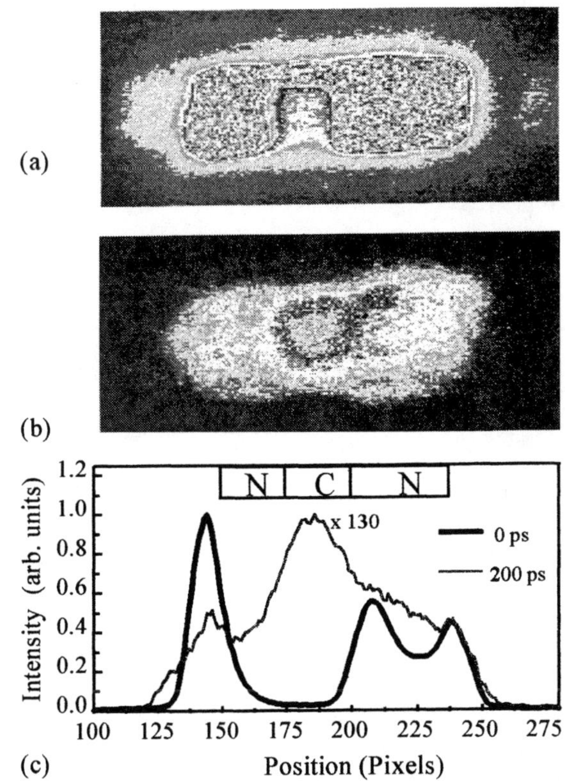

Figure 2. Time-sliced transillumination image of the first breast tissue sample described in the text for gate delays of (a) zero, and (b) 200 ps. (c) Normalized spatial profiles of the 0-ps image (thick line) and the 200-ps image (thin line) integrated over a horizontal area of 20 pixel vertical width.

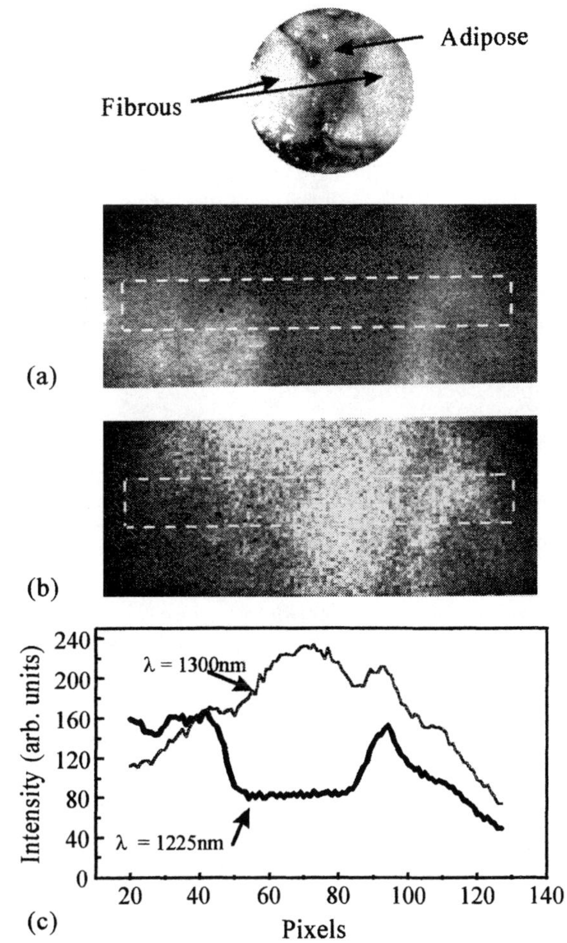

Figure 3. Two-dimensional transillumination images of a 5 mm thick human breast tissue sample comprising adipose and fibrous tissues recorded using light of wavelength (a) 1225 nm, and (b) 1300 nm. (c) Spatial intensity profiles of the images integrated over the horizontal area highlighted by the white dashed lines in Figs. 3(a) and 3(b). The inset at the top is a photograph of the exit surface of the sample.

We attribute the difference in the relative light transmission between the normal and the cancerous human breast tissues at the two different times to the higher scattering of light by the cancerous tissue. The observation is in qualitative agreement with the literature values of transport length of 1.7 mm and 1 mm at 789 nm for normal and cancerous breast tissues, respectively.[13] Photons transiting through the cancerous region scattered more and came out later compared to those transiting through the normal region. The zero and 200 ps time slices accentuated this difference and provided the observed contrast in the transillumination images. The results demonstrate the importance of time-domain imaging approach.

Spectroscopic Imaging

Spectroscopic imaging holds the promise for developing a biomedical imaging modality with genuine diagnostic ability based on the spectroscopic differences between normal and diseased tissues. To test if a spectroscopic difference between two tissues in a specimen provides a distinguishable signature in the transillumination image, we obtained images of the second sample using light of several wavelengths in the 1225-1300 nm range. Some of the wavelengths were near-resonant with the adipose optical absorption band around 1203 nm,[14] while others were far removed from the resonance peak.

Figures 3(a) and 3(b) show a 'near-resonant image' recorded using 1225-nm light, and a 'nonresonant image' recorded using 1300-nm light, respectively. Optical transmittance of a 3 mm thick human breast tissue sample was reported to be 0.23 at 1225 nm and 0.32 at 1300 nm.[14] Scattering is responsible for the low value of transmission of light even at wavelengths that are not absorbed by breast tissues. Spatial profiles of the images in Figs. 3(a) and 3(b) integrated over the horizontal area outlined by the dashed lines are shown in Fig. 3(c). Since 1225-nm light was selectively absorbed by adipose tissues in the middle of the specimen, the corresponding region in the 'resonant image' appeared much darker and distinct from the fibrous regions on the two sides. The region appeared as a deep trough in the corresponding spatial intensity distribution. As the laser output was tuned out of the absorption band, the contrast between the adipose and fibrous regions in the images decreased. The depth of the trough in the spatial intensity profile decreased and the region appeared as a crest in the profile of the 'nonresonant image.' These results clearly demonstrate that an appreciable spectroscopic difference may significantly enhance the contrast between different types of breast tissues in a transillumination image. The ratio of and/or the difference between images recorded using resonant (or near-resonant) and nonresonant light may enhance the image contrast even further.

Summary

The time-sliced imaging experiment demonstrates that optical imaging using optimal time slices is a promising method for imaging biomedical media. Results of the NIR spectroscopic imaging experiments demonstrate the diagnostic potential of optical imaging. Both the methods show selectivity of tissue types based on differences in photon transport and spectroscopic properties. Time-sliced method provides much more information about tissue interior than that available from steady-state and frequency-domain optical methods. Combined with spectroscopic imaging it has the potential to provide much more information than even the x-ray techniques.

Acknowledgements

We acknowledge M. Alrubaiee and J. Evans for technical help. The work is supported by the New York State Science and Technology Foundation, NASA IRA Program, DOE through the Center of Excellence Program, the HEAT program of the City University of New York, and Mediscience Technology Corp.

References

1. G. Harper and B. Englisbe, "Prevention and screening for breast cancer," Cancer Prevention and Detection 17, p. 551, 1993.
2. For a brief review of optical imaging techniques see S. K. Gayen and R. R. Alfano, “Emerging optical biomedical imaging techniques”, Opt. Phot.News 7(3) 17-22 (1996).
3. K.M. Yoo and R. R. Alfano, “Time-resolved coherent and incoherent components of forward light scattering in random media”, Opt. Lett. 15 320-322 (1990).
4. L. Wang, P. P. Ho, G. Liu, G. Zhang, R. R. Alfano, “Ballistic 2-D imaging through scattering walls using an ultrafast optical Kerr Gate,” Science 253 769-771 (1991).
5. E. Gratton, W. W. Mantulin, M. J. vande Ven, J. B. Fishkin, M. B. Maris, and B. Chance, "A novel approach to laser tomography," Bioimaging, 1 40-46 (1993).
6. M. S. Patterson, B. Chance, and B. C. Wilson, "Time resolved reflectance and transmittance for the noninvasive measurement of tissue optical properties," Appl. Opt. 28 2331-2336 (1989).
7. M. Brezinski, G. Tearny, S. A. Boppart, B. Bouma, and J. G. Fujimoto, "High resolution intra-arterial imaging with optical coherence tomography". In *Advances in Optical Imaging and Photon Migration*, 1996 Technical Digest (Optical Society of America, Washington DC 1996), pp. 21-23.
8. J. Hebden, R. Kruger, and K. Wong, "Time-resolved imaging through a highly scattering medium," Appl. Opt. 30 788-794 (1991).
9. M. A.O'Leary, D. A. Boas, B. Chance, and A. G. Yodh, "Experimental images of heterogeneous turbid media by frequency-domain diffusing-photon tomography," Opt. Lett. 20 426-428 (1995) and references therein.
10. Q. Fu, F. Seier, S. K. Gayen, and R. R. Alfano, "High-average-power kilohertz-repetition-rate sub-100-fs Ti:sapphire amplifier system," Opt. Lett. 22 712-714 (1997).
11. J. Dolne, K. M. Yoo, F. Liu, and R. R. Alfano, "IR Fourier space gate and absorption imaging through random media," Lasers Life Sci. 6 131-141 (1994).
12. S. G. Demos and R. R. Alfano "Temporal gating in highly scattering media by the degree of optical polarization," Opt. Lett. 21 161-163 (1996).
13. T. Troy, D. Page, and E. Sevick-Muraca, "Optical properties of normal and diseased breast tissues: prognosis for mammography," J. Biomed. Opt. 1 342-355 (1996). Tabulated values of optical transport properties vary widely. We chose values corresponding to those specimens in the table that closely matched the characteristics of our sample, such as, relative distribution of fibrous and adipose tissues in the specimen, and age of the patient.
14. F. A. Marks, "Optical determination of the hemoglobin oxygenation state of breast biopsies and human breast cancer xenografts in nude mice," in *Proceedings of Physiological Monitoring and Early Detection Diagnostic Methods*, Thomas S. Mang and Abraham Katzir (eds.), SPIE vol.1641, Bellingham, Washington, 1992, pp. 227-237.

Analytic Expressions for Determining the Concentrations of Absorber in Turbid Media by Time-Gating Measurements

Yutaka Tsuchiya, Yukio Ueda, Hedong Zhang *, Yutaka Yamashita, Motoki Oda and Tsuneyuki Urakami

Central Research Laboratory, Hamamatsu Photonics K.K., 5000 Hirakuchi Hamakita 434-8601, Japan
**On leave from State Key Laboratory of Modern Optical Instrumentation, Zhejiang University, Hangzhou, 310027, P.R. China*

Abstract

We propose analytic expressions newly developed based on the microscopic Beer-Lambert law to determine the concentrations of absorber in turbid media by time-gating measurements. The changes in the concentration and the absolute concentrations of an absorber are simply determined by three observable values; the time-gated reemission, newly defined mean pathlength, and extinction coefficient of the absorber. The validity of the expressions has been confirmed by Monte Carlo simulations and experiments at various absorption levels with different time-gates. Our expressions thus can be used for quantitative imaging and tomography of turbid media.

Keywords: Photon migration, Quantitation of absorber concentration, Tissue imaging, Optical computed tomography.

1. Introduction

Optical transmission imaging and tomography of turbid media such as human tissues are currently an active area of research [1]. However, the overwhelming predominance of scatter makes the image blurry and the spatial resolution poor. To improve the spatial resolution, it is essential to detect photons which suffer less scattering, so-called early arriving and quasi-straightforward-propagating photons. These photons have been extracted from the transmitted emission by various time-gates [2-6], collimated photon detection [7], and polarization-modulation method [8]. These methods are however currently restricted to small media since the straightforward photon decreases rapidly with the increase of pathlength. Therefore, it is important in practical application to use not only very early but also relatively early-arriving photons which provide larger signals. In these cases, analytic descriptions of photon migration in the medium are of particular interest: analytic expressions give a general framework for determining the optical properties of turbid media from measured values and provide a foundation for computer-based inversion methods. However, none of reports has clearly addressed the analytic expressions for time-gated reemission.

At present, the photon diffusion approximation is the tool of choice in developing algorithms for the optical tomography [1]. The diffusion approximation with the exact boundary conditions is quite accurate for relatively large systems and allows us to determine the absolute transport scattering and absorption coefficients from the measured reemission. Mismatched boundary conditions are incorporated in the analysis to take specular reflections due to the refractive-index mismatch at the boundary into account. However, analytic solutions with exact boundary conditions are only available for the media that have simple shapes. In addition, these approaches yield somewhat intricate expressions in which the reemission is described by explicit functions for source-detector separation and boundary conditions. Thus, the analytic expressions for the time-gated reemission may not be simply derived from those approaches.

In contrast, the analytic expressions that we have derived based on the microscopic Beer-Lambert law are simple and intuitive. They do not include the source-detector separation, refractive-index-mismatched boundary conditions and exterior shape of the medium as explicit variables [9-13]. The purpose of this paper is to extend our above idea to arbitrarily time-gated reemission maintaining the above mentioned advantages.

2. Theory

Although the human tissue has complex microstructures, for many applications it is sufficient to consider a small volume element to have mesoscopic optical properties. In such case, we can describe the tissue with non-uniform distribution of the mesoscopic properties. At the same time we can specify the average scattering and absorption coefficients for such a medium. Consequently, the non-uniform distribution of absorber in the medium can be described with deviations of the mesoscopic absorption coefficient from the average value. In addition, during the quantitation of absorber concentration, it makes the problem simple that the scattering properties are regarded

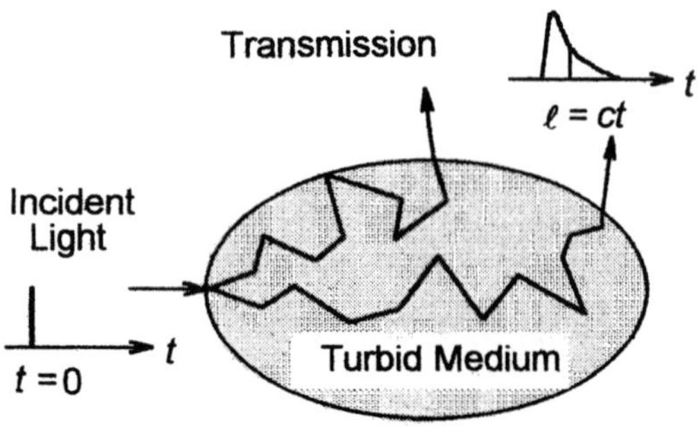

Figure 1. Measurement for determining the concentration of absorbing substances in a turbid medium which has a non-reentrant surface. The medium may have various exterior shapes.

as homogeneous. Very recently, a tomographic image of tissue-like phantom (80 mm in diameter) was successfully obtained with the above method [14]. In the present study, we assume a homogeneous medium having a non-reentrant surface and specify the medium by the average (macroscopic) properties; the refractive index n, the anisotropic scattering coefficient μ_s, the average cosine of scattering angle g, and the absorption coefficient μ_a. We also assume that photons in the medium can be treated as neutral particles traveling at a velocity c determined by n and that wave phenomena can be disregarded in the medium. In practice, absorbing substances may affect n and thus c, but these effects are small enough to be considered insignificant.

During the measurement, photons injected into the medium undergo multiple scattering. Some of the photons then reach the surface and escape (see Fig.1). The escaping photons are detected as time-resolved transmissions for example. Consider a single photon incident on a medium and $h(t)$ to be the probability density function for the escape of the single photon at time t from a certain point on the surface. The temporal behavior of an ensemble of many photons, i.e. reemission, also conforms to $h(t)$. Thus, the impulse response of the medium is expressed by $h(t)$. Now we assume that an optical impulse is incident at time $t = 0$ on a homogeneous medium having a non-reentrant surface and that the microscopic Beer-Lambert law holds in the medium. If we regard the parameter μ_a as a variable, the impulse response $h(t)$, which is a time-causal function, is obtained as [10-13]

$$h(t) \equiv h(\mu_s, \mu_a, t) = s(\mu_s, t)\exp(-\mu_a ct) \tag{1a}$$

$$\frac{\partial}{\partial \mu_a} \ln h(\mu_s, \mu_a, t) = -ct = -\ell. \tag{1b}$$

The term $s(\mu_s, t)$ represents the probability of the scattered photons that reach the detector without absorption, and is specified by $h(t)$ when $\mu_a = 0$. Thus, $s(\mu_s, t)$ is independent of absorption but is affected by the scattering properties, the source-detector separation, the boundary conditions, and the exterior shape of the medium. The term $\exp(-\mu_a ct)$ represents the survival probability of the photons at time t for the absorption process and is independent of scattering. The zigzag photon pathlength ℓ is given by $\ell = ct$ for the photons detected at time t. Here we should note that Eqs. (1a) and (1b) are valid for homogeneous turbid media that have arbitrary non-reentrant and refractive-index mismatched boundary surfaces.

The time-gated reemission $I_T(\mu_s, \mu_a)$ for an arbitrary gate-period $[t_1, t_2]$ $(t_2 > t_1 > 0)$ is given by definite temporal integration of $h(t)$. From Eq.(1) we thus obtain

$$\frac{\partial}{\partial \mu_a} \ln I_T(\mu_s, \mu_a) = -\frac{c\int_{t_1}^{t_2} t\, s(\mu_s, t)\exp(-\mu_a ct)dt}{\int_{t_1}^{t_2} s(\mu_s, t)\exp(-\mu_a ct)dt}$$

$$= -c\langle t\rangle \equiv -L_T(\mu_s, \mu_a). \tag{2}$$

Here we defined the mean pathlength $L_T(\mu_s, \mu_a)$, which is a function of μ_s, μ_a and the time-gate, and is affected by the boundary conditions and the exterior shape of the medium. The mean pathlength $L_T(\mu_s, \mu_a)$ is equivalent to the weighted mean value $c\langle t\rangle$ for the photons detected with the time-gate, and can be calculated from the time course of $h(t)$ in the period $[t_1, t_2]$. From Eq. (2) we obtain

$$\ln I_T(\mu_s, \mu_a) = -\int_0^{\mu_a} L_T(\mu_s, \mu_a)d\mu_a + \ln\int_{t_1}^{t_2} s(\mu_s, t)dt. \tag{3}$$

The last term is the constant of the integration and is obtained as $I_T(\mu_s, \mu_a)$ when $\mu_a = 0$. To simplify the description, we will use $L_T(\mu_a)$ instead of $L_T(\mu_s, \mu_a)$ as long as μ_s remains constant. This rule will be also adapted to other functions.

First, assume that a turbid medium contains an absorber and that the absorption coefficient μ_a changes from μ_{a1} to μ_{a2} but that the scattering coefficient μ_s remains constant during two successive measurements. From Eq. (3) we obtain

$$\ln\frac{I_T(\mu_{a1})}{I_T(\mu_{a2})} = \int_{\mu_{a1}}^{\mu_{a2}} L_T(\mu_a)d\mu_a = (\mu_{a2} - \mu_{a1})L_T(\mu_x), \tag{4}$$

where $\min(\mu_{a1}, \mu_{a2}) \le \mu_x \le \max(\mu_{a1}, \mu_{a2})$. Since $L_T(\mu_a)$ is a monotonic function with μ_a, the mean pathlength $L_T(\mu_x)$ is expressed with a factor α $(0 \le \alpha \le 1)$ as $L_T(\mu_x) = \alpha L_T(\mu_{a1}) + (1-\alpha)L_T(\mu_{a2})$. The linear approximation predicts $\alpha = 0.5$. The change in the concentration of the absorber ΔC during the two successive measurements is obtained from the Beer-Lambert law as $\Delta C = (\mu_{a2} - \mu_{a1})/\varepsilon$, where ε is the extinction coefficient of the absorber. The extinction coefficient ε can be determined with a spectrometer for example. Therefore, we can determine changes in the concentration of the absorber ΔC by successive time-gating measurements.

Second, consider dual-wavelength spectroscopy of a turbid medium with an absorber. We assume that the scattering property of the medium is independent of wavelength in the range of interest and that the wavelengths of incident light are λ_1 and λ_2, where the absorption coefficients are μ_{a1} and μ_{a2}, respectively. It is obvious that Eq. (4) holds for the time-gated reemission obtained by the dual-wavelength spectroscopy. As before, $L_T(\mu_x)$ is expressed with a factor α, and $L_T(\mu_{a1})$ and $L_T(\mu_{a2})$ at the respective wavelengths. The linear approximation predicts $\alpha = 0.5$. The concentration of the absorber C is expressed using its extinction coefficients ε_1 and ε_2 at the respective wavelengths: $C = (\mu_{a2} - \mu_{a1})/(\varepsilon_2 - \varepsilon_1)$. The extinction coefficients ε_1 and ε_2 can be determined with a spectrometer for example. Therefore, we can determine the absolute concentrations of the absorber C by time-gating dual-wavelength spectroscopy. If the number of injected photons passing through a surface differs at the two wavelengths, we can introduce a factor b denoting the ratio of the injected photons at λ_1 and λ_2 to compensate the ratio $I_T(\mu_{a1})/I_T(\mu_{a2})$ in Eq. (4). It is obvious from the above discussion that we can determine the concentrations of m-type absorbers by multi-wavelength (m+1) spectroscopy.

3. Monte Carlo simulation

The validity of our prediction was evaluated by Monte Carlo (MC) simulation, which is similar to that of van der Zee *et al* [15]. Figure 2 shows the geometry for the simulation. The medium examined is an infinitely wide, 20 mm thick slab for which optical properties are; $n = 1.33$, $\mu_s = 3$ mm^{-1}, $g = 0.85$, and the boundary conditions are refractive index matched and mismatched. The absorption coefficient was changed as follows: (a) $\mu_a = 0.005$, (b) 0.01, (c) 0.02, and (d) 0.03 mm^{-1}. Time-resolved reflections (R_1, R_2, and R_3) and transmissions (T_0, T_1, T_2, and T_3) of injection of 2×10^8 photons into the medium are detected by ideal photo-detectors (1 mm in diameter) arranged each with a 5 mm separation. We thus obtained 7 types of time-resolved reemission (MC data) at 4 different absorption levels with both boundary conditions. One pair of MC data for 2 different values of μ_a, but at the same detection point, are sufficient to determine changes in the concentration of

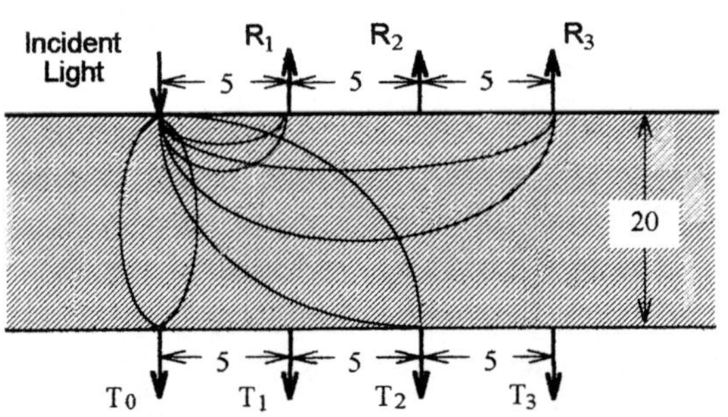

Figure 2. Geometry for Monte Carlo simulation. The boundaries are refractive index matched and mismatched.

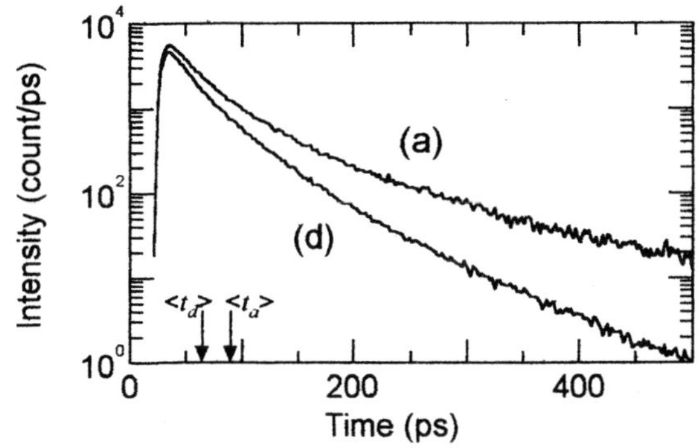

Figure 3. An example of a pair of MC data. Reflection R_1 for (a) $\mu_a = 0.005$ and (d) $\mu_a = 0.03$ mm^{-1}. The symbols $\langle t_a\rangle$ and $\langle t_d\rangle$ stand for the weighted mean time for the curves (a) and (d), respectively.

absorber ΔC. Figure 3 shows an example of a pair of MC data for R_1 (the source-detector separation is 5 mm and the boundary is refractive index matched; (a) $\mu_a = 0.005$, and (d) 0.03 mm^{-1}. The symbols $\langle t_a\rangle$ and $\langle t_d\rangle$ stand for the weighted mean time for the curves (a) and (d), respectively.

We determined the differences $\Delta\mu_a$ using Eq. (4). Since each reemission (R_1, R_2, R_3, T_0, T_1, T_2, and T_3) was calculated for 4 different values of μ_a, we evaluated the errors in $\Delta\mu_a$ for 6 pairs of MC data at each detection point. Table I shows an example of a series of results for errors in $\Delta\mu_a$ determined by 6 pairs of MC data for T_2 with various time-gates. The boundary is matched. We applied 4 types of time-gates; (i) the gate-period [0, 1 ns], (ii) [0, $\langle t\rangle_1$], (iii) [0, t_{p1}], and (iv) [t_{p1}, $\langle t\rangle_1$]. The symbols $\langle t\rangle_1$ and t_{p1} denote the weighted mean time and the peak time respectively, and were specified for the MC data obtained with the absorption coefficient μ_a which is smaller than that of the other MC data in the pair. In Table I, μ_{a1} indicates the smaller μ_a.

Table I. An example of errors in $\Delta\mu_a$ determined by 6 pairs of MC data for T_2 with refractive-index matched boundary by various time-gates (i)-(iv); (i) [0, 1 ns], (ii) [0, $\langle t_1\rangle$], (iii) [0, t_{p1}], and (iv) [t_{p1}, $\langle t_2\rangle$].

Absorption (mm^{-1}) For MC data T_1			Errors ($\times10^{-3}$ mm^{-1}) in $\Delta\mu_a$			
μ_{a2}	μ_{a1}	$\Delta\mu_a$	(i)	(ii)	(iii)	(iv)
0.01	0.005	0.005	0.0	-0.2	-0.6	-0.1
0.02	0.01	0.01	-0.2	-0.1	-0.1	0.0
0.03	0.02	0.01	0.1	0.3	0.0	0.6
0.02	0.005	0.015	-0.3	-0.3	-0.3	-0.3
0.03	0.01	0.02	-0.3	0.2	0.2	0.3
0.03	0.005	0.025	-0.8	-0.2	-0.7	0.0

For all pairs of MC data with both boundary conditions, the differences $\Delta\mu_a$ were determined with small fluctuations in which errors appeared at random in the range of $-1.3\times10^{-3} \sim 2.2\times10^{-3}$ mm^{-1}, but almost no correlation was seen among the errors and the MC parameters. The errors may be caused by the signal-to-noise ratio of the MC data. We thus believe that the validity of our prediction has been numerically well demonstrated.

4. Experiment

4.1 Quantitation of absorber concentration

To demonstrate the validity of the dual-wavelength time-gated spectroscopy (TGS) method, we measured a tissue like phantom with transmission geometry (see Fig. 4). The experimental setup has been described in detail elsewhere [16]. Briefly, two picosecond light pulsers (PLP, Hamamatsu), which have a repetition rate of 5 MHz, peak power of ≈100 mW, pulse width of ≈50 ps (FWHM) and peak wavelengths of λ_1 = 831 nm and λ_2 = 782 nm, were used as light sources. A graded index optical fiber of 200 µm in diameter was used to deliver light to the tissue like phantom. A 5-mm-diameter optical bundle fiber collected the photons re-emitted from the opposite surface of the phantom (source-detector separation is 40 mm) and fed the light to a time-correlated single photon counting system (TCSPC). The signals from the TCSPC system were then digitized by an A/D converter (Oxford/Tennelec, PCA III), accumulated in the memory of a personal computer, and finally the temporal profiles of re-emission were obtained.

Measurements were conducted in 1% aqueous suspension of Intralipid-20% (Pharmacia AB, Stockholm), which acted as a scattering medium. The suspension of 420 ml was filled into an acrylic resin box (120 mm × 120 mm × 40 mm) of which inner parts were sprayed black. Greenish brown ink (Chugai Kasei) was used as a pure absorber and added into the Intralipid suspension in steps

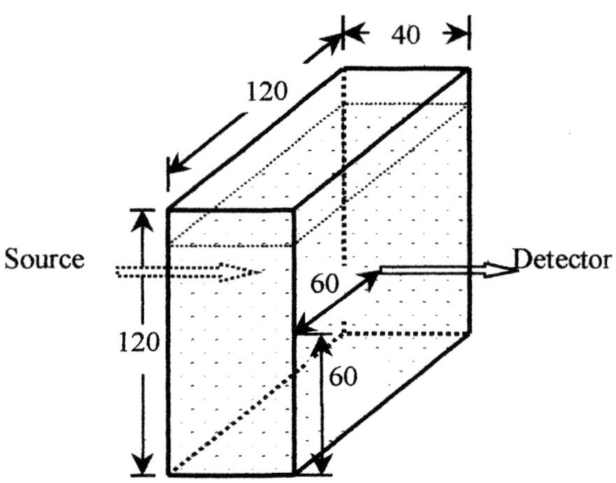

Figure 4. Structure of a tissue like liquid phantom.

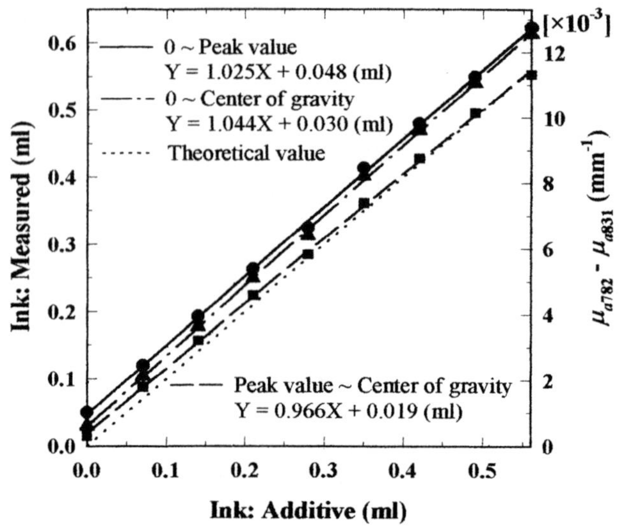

Figure 5. Results of the dual wavelength spectroscopy of tissue like phantom with three types of time-gate; $[0, t_p]$, $[0, \langle t\rangle]$ and $[t_p, \langle t\rangle]$. The values t_p and $\langle t\rangle$ were specified for the time course of time resolved reemission obtained without ink additive.

of 0.07 ml until the total volume of the ink reached 0.56 ml. The absorption coefficients of the ink were determined by a spectrometer at various concentrations. The values, along with the absorption of water (measured by a spectrometer as 2.45×10^{-3} mm^{-1} at 782 nm and 3.10×10^{-3} mm^{-1} at 831 nm), yielded absorption coefficients of the phantom in a range from 2.45×10^{-3} to 2.94×10^{-2} mm^{-1} at 782 nm and from 3.10×10^{-3} to 1.85×10^{-2} mm^{-1} at 831 nm.

The absolute concentrations of the absorber (ink) were determined by the dual-wavelength TGS method. The results obtained by the three types of gate were shown in Fig. 5; gate-period $[0, t_p]$, $[0, \langle t\rangle]$ and $[t_p, \langle t\rangle]$. The symbols $\langle t\rangle$ and t_p respectively referred to the weighted mean time (the center of gravity) and the peak time specified for the time course of time resolved reemission obtained without ink additive.

We see good agreement between measured and added ink volumes, especially in the slopes. The value of the y-intercept indicates the offset due to the systematic errors. The offset may result from the inaccuracy in the determination of zero point on the time axis.

4.2 Imaging with time-gate method

We studied the feasibility of TGS method for imaging a turbid medium using a tissue-simulated solid phantom. The body of the solid phantom is made of epoxy including silica and dyes as the respective scattering and absorbing substances and has a size of 140 mm × 180 mm × 70 mm as shown in Fig. 6. The phantom contains four cylindrical absorbers of which absorption coefficient differs from that of the body; absorption coefficients of the body and the

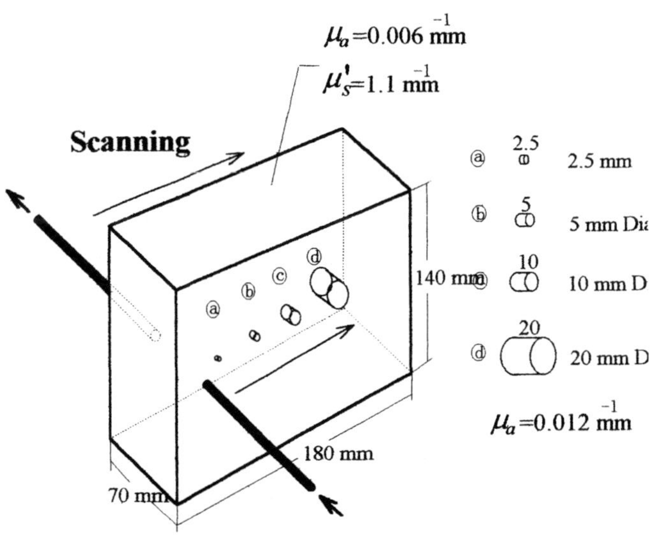

Figure 6. Imaging of tissue-simulated solid phantom contains four cylindrical absorbers. The phantom was scanned along the horizontal line, joining the four absorbers, across the interface by moving the source-detector pair.

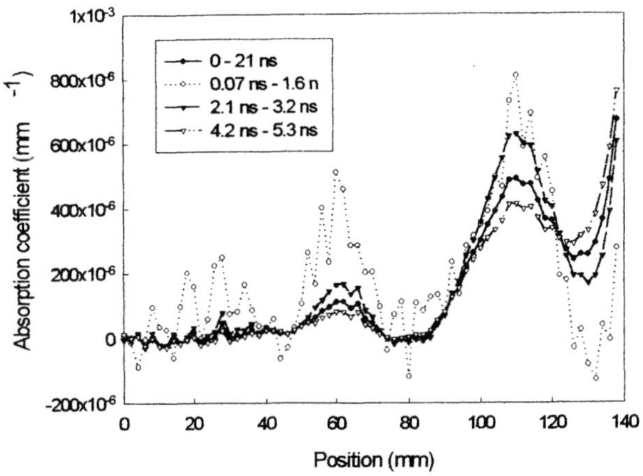

Figure 7. Profile of absorption coefficient along the scan. The absorber (a), (b), (c) and (d) are located at positions ≈ 10 mm, ≈ 30 mm, ≈ 60 mm and ≈ 110 mm. We can see that the modulation in profile is improved ≈ 30% by time-gate [0.07 ns, 1.6 ns], which demonstrates a strong point of the use of early arriving photons.

absorbers are 0.006 mm^{-1} and 0.012 mm^{-1}, respectively. The transport scattering coefficients designed for both the body and absorbers are 1.1 mm^{-1}. The sizes of the four cylindrical absorbers are; (a) 2.5 mm, (b) 5 mm, (c) 10 mm and (d) 20 mm in height and diameter.

The phantom was scanned along the horizontal line, joining the four absorbers, across the solid interface by moving the light incident and detection points, simultaneously. We can thus measure the distribution of absorbing substance in terms of relative change of absorption coefficient. Figure 7 shows the change in absorption coefficient along the scan position, which was obtained by four types of time-gate; (1) [0, 21 ns], (2) [0.07 ns, 1.6 ns], (3) [2.1 ns, 3.2 ns] and (4) [4.2 ns, 5.3 ns]. The changes in the absorption coefficient were calculated with Eq. (4).

We can clearly see that the modulation is improved by time-gate [0.07 ns, 1.6 ns]. This demonstrates a strong point of the use of early arriving photons. A potential application of the above technique may be in mammography to diagnose breast cancer. Another application may be in measurement of temporal changes in the absorber concentration in highly scattering media.

5. Summary

In summary, we have derived analytic expressions for various time-gated re-emissions to determine the concentrations of absorber in homogeneous medium having non-reentrant surfaces. Validity of the expression has been confirmed by Monte Carlo simulation and experiments at various absorption levels with different time-gates. We also experimentally showed the advantage of the use of early arriving photons. Our expressions hold for arbitrary timing and duration of the time-gate and do not include the source-detector separation and exterior shape of the medium as variables. It is well known that the distribution of path of detected photons can be varied by changing the timing and duration of the gate in transmission and reflection measurements. Our results therefore have significant implications for quantitative imaging of absorptive objects hidden in turbid media, for example; (i) optical tomography in which the non-uniform distribution of absorber is described by deviations of the mesoscopic absorption coefficient from the average value [14]; (ii) optical transmission and reflection imaging using Eq. (4) in which the timing and duration of the gate are controlled and the sample is scanned by moving the source-detector pair, and (iii) optical topography [17] using Eq. (4) in which the timing and duration of the gate are controlled.

References

1) See, for example, R. R. Alfano and J. G. Fujimoto, eds: *OSA TOPS on Advances in Optical Imaging and Photon Migration*, Vol. **2** (1996).
2) S. Andersson-Engels, R. Berg, S. Svanberg and O. Jarlmann: "Time-resolved transillumination for medical diagnostics," Opt. Lett. **15**, 1179-1181 (1990).
3) J. A. Moon and J. Reintjes: "Image resolution by use of multiply scattered light," Opt. Lett. **19**, 521-523 (1994).
4) F. Liu, K. M. Yoo and R. R. Alfano: "Transmitted

photon intensity through biological tissues within various time windows," Opt. Lett. **19**, 740-742 (1994).

5) J. C. Hebden and A. H. Gandjbakhche: "Experimental validation of an elementary formula for estimating spatial resolution for optical transillumination imaging," Med. Phys. **22**, 1271-1272 (1995).

6) A. J. Joblin: "Method of calculating the image resolution of a near-infrared time-of-flight tissue-imaging system," Appl. Opt. **35**, 752-757 (1996).

7) Y. Yamashita and M. Kaneko: "Visible and infrared diaphanoscopy for medical diagnosis," in *Medical Optical Tomography: Functional Imaging and Monitoring*, G. J. Müller, *et al.* ed: SPIE IS**11**, 283-316 (1993).

8) H. Horinaka, K. Hashimoto, K. Wada and Y. Cho: "Extraction of quasi-straightforward-propagating photons from diffused light transmitting through a scattering medium by polarization modulation," Opt. Lett. **20**, 1501-1503 (1995).

9) M. Oda, Y. Yamashita, G. Nishimura and M. Tamura: "A simple and novel algorithm for time-resolved multiwavelength oximetry," Phys. Med. Biol. **41**, 551-562 (1996).

10) Y. Tsuchiya and T. Urakami: "Photon migration model for turbid biological medium having various shapes," Jpn. J. Appl. Phys. **34**, L79-81 (1995).

11) Y. Tsuchiya and T. Urakami: "Frequency domain analysis of photon migration based on the microscopic Beer-Lambert law," Jpn. J. Appl. Phys. **35**, 4848-4851 (1996).

12) Y. Tsuchiya and T. Urakami: "Quantitation of absorbing substances in turbid media such as human tissues based on the microscopic Beer-Lambert law," Opt. Commun. **144**, 269-280 (1997).

13) Y. Tsuchiya and T. Urakami: "Optical quantitation of absorbers in variously shaped turbid media based on the microscopic Beer-Lambert law: A new approach to optical computerized tomography," in *Advances in Optical Biopsy and Optical Mammography*, R. R. Alfano, ed. Ann. N.Y. Acad. Sci. **838**, 75-94 (1998).

14) Y. Ueda, K. Ohta, M. Oda, M. Miwa, Y. Yamashita and Y. Tsuchiya: "Optical imaging reconstruction using the average value as the reference," Proc. SPIE **2979**, 795-806 (1997).

15) P. van der Zee and D. T. Delpy: Simulation of the point spread function for light in tissue by a Monte Carlo method," Adv. Exp. Med. Biol. **215**, 179-191 (1987).

16) H. Zhang, M. Miwa, T. Urakami, Y. Yamashita and Y. Tsuchiya: "Simple subtraction method for determining the mean path length traveled by photons in turbid media," Jpn. J. Appl. Phys. **37**, 700-704 (1998).

17) A. Maki, Y. Yamashita, Y. Ito, E, Watanabe and H. Koizumi: "Spatial and temporal analysis of human motor activity using noninvasive optical topography," in *OSA TOPS on Advances in Optical Imaging and Photon Migration*, R. R. Alfano and J. G. Fujimoto, eds. Vol. **2**, 357-362 (1996).

Diffusely Backscattered Polarized Light Mueller Matrix Imaging of a Turbid Medium

Brent D. Cameron

Biomedical Engineering Program, Texas A&M University College Station, Texas 77843-3120

M.J. Raković

Center for Theoretical Physics, Department of Physics, Texas A&M University, College Station, Texas 77843-4242
And Institute of Physics, University of Belgrade, P.O. Box 57, 11001 Belgrade, Yugoslavia

Mehrübe Mehrübeoğlu

Department of Electrical Engineering, Texas A&M University, College Station, Texas 77843-3128

George W. Kattawar

Center for Theoretical Physics, Department of Physics, Texas A&M University, College Station, Texas 77843-4242

Sohi Rastegar, Lihong V. Wang, Gerard L. Coté

Biomedical Engineering Program, Texas A&M University, College Station, Texas 77843-3120

Abstract

The diffusely backscattered Mueller matrix of a turbid medium consisting of 2.02 μm polystyrene spheres has been measured experimentally and simulated with Monte Carlo analysis.

Key Words

(290.1350) Backscattering, (290.7050) Turbid media, (290.4020) Mie theory, (260.5430) Polarization, Monte Carlo

Introduction

Recently, a few groups have investigated the spatially dependent diffusely backscattered intensity patterns which arise when a polarized laser beam is incident on a turbid media[1,2,3,4]. Some of the reported applications include the measurement of the optical properties of the medium including the average particle size, the scattering coefficient, and the anisotropy factor[2], as well as, potential biomedical diagnostic applications such as cancer detection[4].

In our study, we further these investigations by using a Monte Carlo based approach to independently reproduce the elements of the two-dimensional backscattered Mueller matrix, we further compare these results with experimental measurements.

Monte Carlo Method

A flow chart of the implemented Monte Carlo method is presented in Figure 1[5].

Experimental Materials and Methods

The turbid medium utilized in this study consists of a suspension of 2.02 μm polystyrene spheres. The sample specifically consists of 0.075 ml of a 10 weight percent suspension of spheres in 15 ml of deionized water. The relative index of refraction is 1.192 with a scattering coefficient and anisotropy factor of 11.88 cm^{-1} and 0.912, respectively at a wavelength of 632.8 nm.

Figure 2 depicts the experimental setup used for the collection of the diffuse backscattered images. The light source utilized is a 10 mW HeNe laser emitting a wavelength of 632.8 nm. In order to avoid saturation of the CCD camera array, the initial power of the laser is attenuated by a 10% transmittance neutral density filter. The beam exiting the filter is then passed through a set of polarization optics, which consists of various linear and circular polarizers that are interchanged throughout the course of the experiment to obtain the desired input polarization. The beam is then focused by a lens with a focal length of 10 cm onto the sample. The focused beam is passed through an approximate 2 mm diameter hole in a mirror mounted at 45 degrees with respect to the sample. The reflected image of the diffusely backscattered light is then imaged though a polarization analyzer and lens system onto a 12-bit cooled CCD camera array. The polarization analyzer consists of another set of polarization

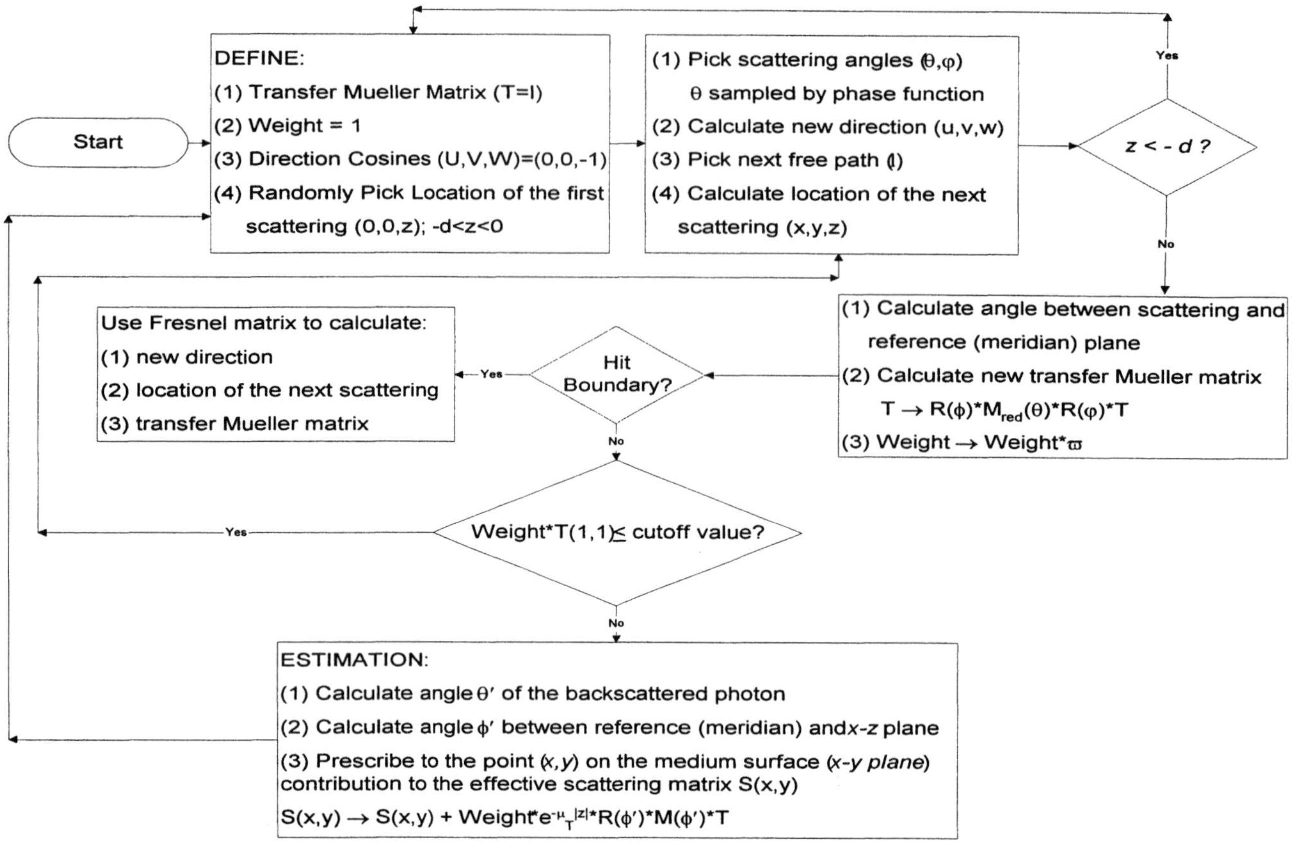

Figure 1. Flow chart of the Monte Carlo method. Notation: φ: rotation in *x-y* plane on surface, θ: scattering angle, ϕ: rotation between photon trajectory planes, ϖ: single-scattering albedo, M: single-scattering Mueller matrix, and R: rotation matrix

optics which are interchanged throughout the course of the experiment to analyze a certain type of polarized light.

In order to fully experimentally measure the 16 elements of the Mueller matrix of a sample, a Stokes vector-Mueller matrix approach to light scattering is used[4]. This approach requires a total of 49 individual images to be taken at various combinations of input and output analyzer polarization states. Presented in Figure 3 is a subset of 4 (M_{12}, M_{22},M_{23}, M_{24}) of the 16 Mueller matrix elements experimentally measured and simulated with the Monte Carlo based method in our study for the 2.02 μm polystyrene sphere phantom. The small opaque circle in the center of all the experimental images is due to the hole in the mirror. As can be seen, there is agreement in the number of lobes, their orientations, and size between the experimental and simulated images. The total Mueller matrix of the sample completely characterizes the sample's optical polarization properties and information such as the average particle size, scattering coefficient, etc. may be extracted through analysis of the elements.

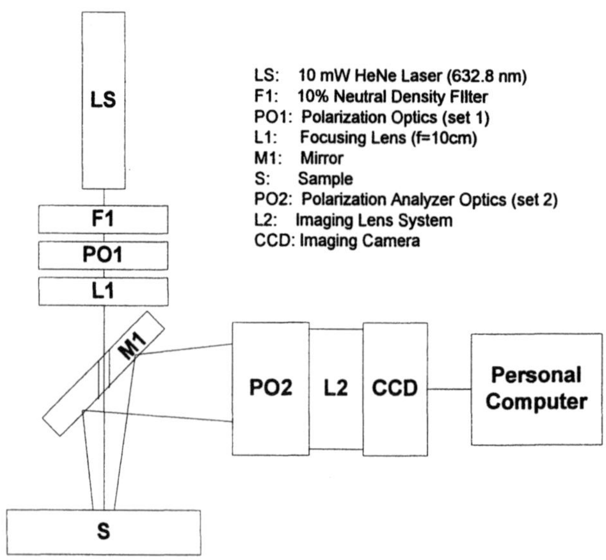

Figure 2. Block diagram of diffuse backscattering Mueller matrix imaging experimental setup.

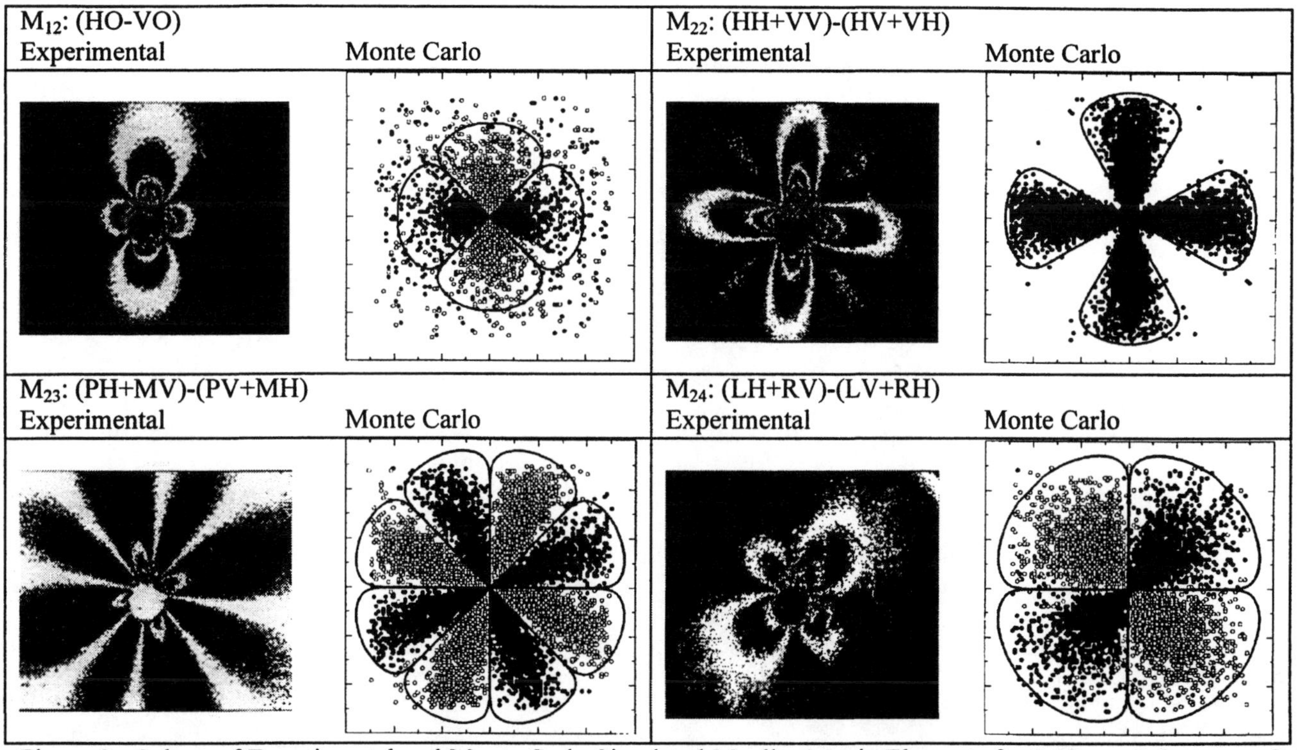

Figure 3. Subset of Experimental and Monte Carlo Simulated Mueller Matrix Elements for 2.02 μm Polystyrene Sphere Suspension. For the experimental elements, the individual elements are obtained by adding and subtracting a series of images corresponding to differing input and output analyzer polarizations. The individual images are denoted by a two letter combination which denotes the input polarization and output analyzer orientation (i.e. HV would denote horizontal input polarized light and a vertical polarization analyzer). The corresponding symbols denoting polarization are: V: vertical, H: horizontal, P: +45° , M: -45° , R: right, L: left, O:open or no polarization optics.

Discussion

We have shown in this investigation the ability to simulate the two-dimensional backscattered Mueller matrix of a turbid medium using a Monte Carlo based approach and obtain agreement through experimental measurements. This ability offers several fascinating and potential applications for the eventual use of this method in the biomedical diagnostic field such as in the detection of cancerous lesions. Hopefully, the ability to simulate the diffusely backscattered Mueller matrix of a given turbid medium through the use of Monte Carlo methods will help facilitate the eventual design and implementation of such a diagnostic tool.

References

1. M. Dogariu and T. Asakura, J. Opt., **24**, 271 (1993).
2. A.H. Hielscher, J.R. Mourant, and I.J. Bigio, Appl. Opt., **36**, 125 (1997).
3. S.L. Jacques, M. Ostemeyer, L. Wang, and D. Stephens, Proc. SPIE, **2671**, 199 (1996).
4. A.H. Hielscher, A.A. Eick, J.R. Mourant, and I.J. Bigio, Proc. SPIE, **2976**, 298 (1997).
5. B.D. Cameron, M.J. Raković, *et.al.,* Optics Letters, **23(7)**, (1998)

Effect of chicken breast tissue anisotropy on absorption and scattering spectra

Guillermo Marquez and Lihong V. Wang
Optical Imaging Laboratory, Biomedical Eng. Program
Texas A&M University
College Station, TX 77843-3120, USA
Phone: (409) 847-9040, Fax: (409) 845-4450, E-mail: LWang@tamu.edu

Sharon L. Thomsen, Ph.D.
Laser Biology Research Laboratory, Box 17
University of Texas M. D. Anderson Cancer Center
1515 Holcombe Blvd., Houston, TX 77030, USA

Abstract

We used oblique incidence reflectometry technique to deduce absorption and reduced scattering spectra of chicken breast tissue from wavelength-resolved measurements of the relative diffuse reflectance profile of white light as a function of source-detector distance with the probe oriented at 0 and 90 degrees relative to the muscle fibers.[1, 2] We found that the deduced optical properties varied with the probe orientation due to the tissue's structural anisotropy.

Key Words

(290.5820) Scattering measurements; (300.6170) Spectra

Theoretical

The fundamental parameter of oblique incidence reflectometry is a shift in the positions of the point sources in the x direction [Fig. 1]. This shift, Δx, is

$$\Delta x = \frac{\sin(\alpha_t)}{0.35\mu_a + \mu_s'} \tag{1}$$

where μ_a is the absorption coefficient, μ_s' the reduced scattering coefficient, and α_t is the angle of refraction, which may be calculated from the relative index of refraction n and the angle of incidence α_i.

The modified dipole source diffusion theory model gives diffuse reflectance[3]

$$R(x) = \frac{\Delta z(1+\mu_{eff}\rho_1)\exp(-\mu_{eff}\rho_1)}{4\pi\rho_1^3} + \frac{(\Delta z + 2z_b)(1+\mu_{eff}\rho_2)\exp(-\mu_{eff}\rho_2)}{4\pi\rho_2^3} \tag{2}$$

where ρ_1 and ρ_2 are the distances from the two point sources to the point of interest.

We first measured the diffuse reflectance profile with our fiber optic probe. The light collected by each detection fiber was input to the detection system for spectral resolution. We have thus measured the spatial distribution of diffuse reflectance at many wavelengths simultaneously. The reflectance profile at any wavelength can be extracted and analyzed to deduce the optical properties at that wavelength. Next, we find the position of the center of far diffuse reflectance relative to the light entry point (Δx) and calculate the diffusion coefficient.

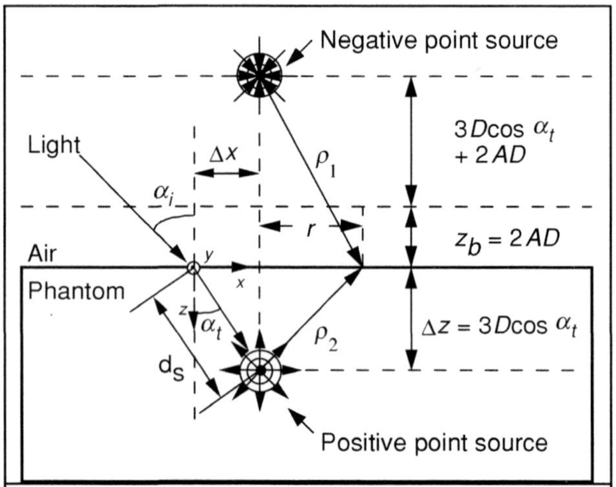

Figure 1. Obliquely incident light. Note the shift in the center of diffuse reflectance, Δx.

Then a non-linear least squares fit to Eq. (2) with the Levenberg-Marquardt method yielded the effective attenuation coefficient, μ_{eff}.[4] The absorption coefficient, μ_a, and the reduced scattering coefficient, μ_s', were computed with the following equations

$$\mu_a = \frac{\mu_{eff}^2 \Delta x}{3\sin(\alpha_t)} \tag{3}$$

and

$$\mu_s' = \frac{\sin(\alpha_t)}{\Delta x} - 0.35\mu_a \tag{4}$$

Experimental

The experimental system is drawn schematically in Fig. 2. White light (Dolan-Jenner Industries, Fiber-Lite® High Intensity Illuminator 170-D) was delivered, and the diffuse reflectance was collected with a fiber optic probe made from black delrin and 600 μm diameter, low-loss optical fibers. The source fiber was oriented at a 45° angle of incidence, and the nine collection fibers, arranged in a linear array, collected the diffuse reflectance. The detection system was an optical multi-channel analyzer (OMA) that utilized a spectrograph (Instruments S.A., HR320), an intensified CCD camera (Princeton Instrument Inc., TRY-700G/RAR/R), and a Macintosh computer to record the spectrum of collected white light.

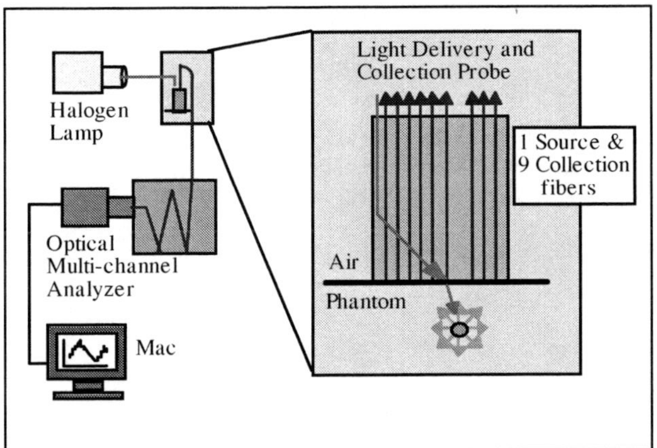

Figure 2. Experiment setup.

Frozen, raw, chicken breasts were allowed to thaw at room temperature for 20 minutes. The skin and fat were removed, and the chicken breasts were cut so as to have only one muscle fiber orientation per piece. The pieces were stacked in a small black container with the muscle fibers aligned [Fig. 3a]. An exposed x-ray film was placed on top of the chicken tissue to approximate a matched boundary condition for diffusion theory. Four measurements each were performed with the probe oriented at 0° (parallel) and 90° (perpendicular) relative to the muscle fibers [Fig. 3b]. Absorption and reduced scattering spectra were evaluated for each diffuse reflectance measurement, and the results for each probe orientation were averaged.

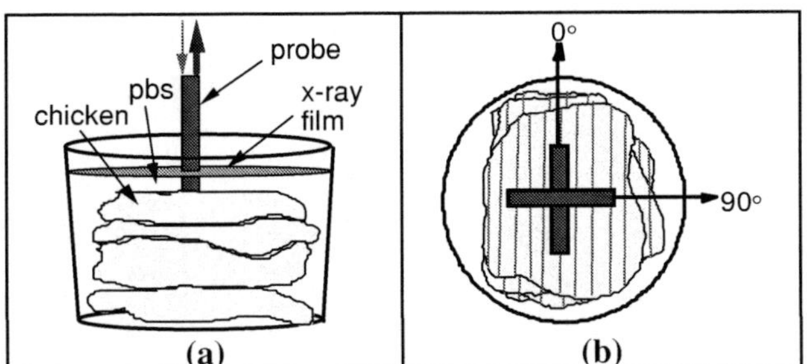

Figure 3. Chicken breast tissue setup. (a) The container: 10cm deep and 10cm in diameter. (b) Probe orientation relative to the muscle fibers: 0° is parallel and 90° perpendicular.

After the measurements, all the chicken breasts were blended using a blender until the purée appeared homogeneous by visual inspection. We placed the purée back in the black container making sure there were no air or oil pockets created by the blending. Then six measurements at each probe orientation were performed again.

Results

As shown in Fig. 4 we could measure the full visible spectrum of the average absorption coefficient (μ_a) and reduced scattering coefficient (μ_s') for both probe orientations. In Fig. 4a, we also included the scaled absorption coefficient of deoxy-hemoglobin for comparison. Two peaks at approximately 430 and 550 nm were observed in all three absorption curves. The absorption coefficient was consistently lower at 0° than that at 90°. The reduced scattering coefficient curves peaked at approximately 450 nm then decreased smoothly with increasing wavelength [Fig. 4b]. The 90° curve had a lower μ_s' than the 0° curve below the crossover wavelength at about 585 nm. The relative difference in the measured absorption coefficients between the 0° and 90° probe orientations was as large as 50% and in the measured reduced scattering coefficients was up to 10%.

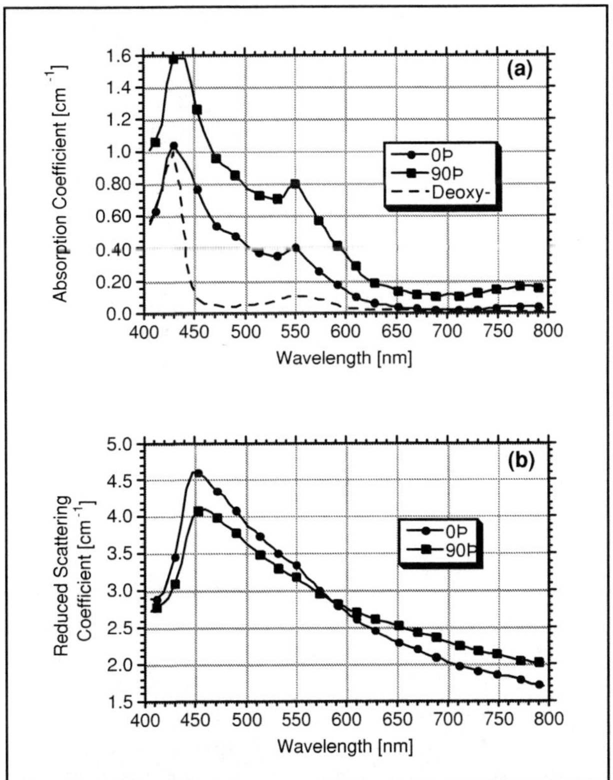

Figure 4. Probe aligned at 0° (parallel) and 90° (perpendicular) with respect to the muscle fiber orientation. (a) Absorption spectra. The relative absorption spectrum of deoxy-hemoglobin is also plotted. (b) Reduced scattering spectra.

The average absorption and reduced scattering coefficients for the homogenized chicken tissue are plotted in Fig. 5. Figure. 5a shows two absorption peaks at 430 and 540 nm, and Fig. 5b shows a reduced scattering coefficient peak at 450 nm wavelength. The absorption coefficient of the homogenized chicken tissue was approximately equal to the average of the absorption coefficients for 0° and 90° probe orientations. We also found that the reduced scattering coefficient for homogenized tissue was higher than that for either 0° or 90° probe orientation.

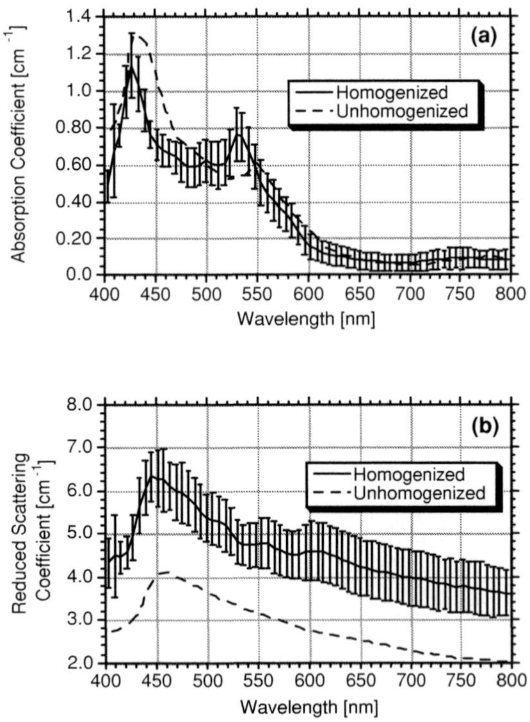

Figure 5. Homogenized chicken breast tissue: (a) The absorption coefficient spectrum. (b) The reduced scattering coefficient spectrum.

References

1. S. P. Lin, "Oblique-incidence fiber-optic reflectometry for measuring absorption and scattering in turbid media," M.S. thesis (Rice University, Houston, Texas, 1996).
2. G. Marquez, L.-H. Wang, S.-P. Lin, J. A. Schwartz, and S. L. Thomsen, "Anisotropy in the absorption and scattering spectra of chicken breast tissue," Applied Optics 37, 798-805 (1998).
3. T. J. Farrell, M. S. Patterson, and B. C. Wilson, "A diffusion theory model of spatially resolved, steady-state diffuse reflectance for the non-invasive determination of tissue optical properties *in vivo*," Med. Phys. 19, 879-888 (1992).
4. W. H. Press, B. P. Flannery, S. A. Teukolsky, and W. T. Veterlin, *Numerical Recipes in C*, 2nd ed. (Cambridge U. Press, Cambridge, UK, 1992), Section15.5.

Effective Optical Properties of Two-Layered Turbid Media Using the Frequency-Domain Multi-Distance Method

Lelia Adelina Paunescu, Maria Angela Franceschini, Sergio Fantini, Albert E. Cerussi and Enrico Gratton

Laboratory for Fluorescence Dynamics, Department of Physics, University of Illinois at Urbana-Champaign 1110 West Green Street, Urbana, Illinois 61801-3080

Abstract

We measured the effective absorption and reduced scattering coefficients on the surface of two-layered turbid media using a frequency-domain multi-distance method. We measured the amplitude and the phase of the modulated intensity at source-detector distances ranging from 1.5 to 3.0 cm. We considered several combinations of one layer (L) (thickness: 0.1-1.6 cm) placed on top of one block (B) (thickness ~ 5 cm). We found that for layers less than about 0.40 cm thick the frequency-domain multi-distance method yields the optical properties of the underlying block. We also found that for layers more than 1.3 cm thick the frequency-domain multi-distance method yields a reduced scattering coefficient which reproduces that of the superficial layer. For a layer thickness in the range of 1.3-1.6 cm, the effective absorption coefficient approaches the layer absorption coefficient but does not match it to within experimental errors.

Keywords:

(300.6500) Time-Resolved Spectroscopy, (300.1030) Absorption, (290.7050) Turbid Media, (290.4210) Multiple Scattering.

1. Introduction

The quantitative determination of the optical properties of tissues has a valuable potential in noninvasive diagnostics. In recent years, newly developed techniques have achieved accurate measurements of the absorption and reduced scattering coefficients in homogeneous turbid media [1-3]. While these techniques are accurate in homogeneous media, there are still questions about their validity *in vivo*, where the presence of layers and inhomogeneities may significantly affect optical measurements. For this reason, it is of interest to investigate the photon migration in layered turbid media [4, 5]. The purpose of this article is to investigate the meaning of the effective optical properties measured in two-layered turbid media with a multi-distance protocol based on the semi-infinite geometry and assuming homogeneous media [6].

2. Methods

The data were collected using a frequency-domain (110 MHz) spectrometer we have previously described [7] (ISS, Inc. tissue oximeter Model 96208). The light sources were 4 laser diodes emitting at 750 nm, and 4 at 840 nm. Four source-detector separations were used (between 1.5 and 3 cm) (Fig. 1(a)). The samples (underlying blocks and superficial layers) were made of gelatin. Different quantities of black India ink and white paint were added in order to have various absorption and reduced scattering coefficients. From 5 batches of gelatin (each having different optical properties), we made 5 blocks (labeled 1 to 5) and 5 sets of 4 superficial layers with different thicknesses (labeled a to d) (Fig. 1(b)). The measurements were performed on all the combinations of one underlying block and one superficial layer. From the raw data, using a frequency-domain multi-distance approach for the semi-infinite geometry [3] we recovered the optical properties of each block ($\mu_a^{(B)}$ and $\mu_s'^{(B)}$) and the effective optical coefficients of all the combinations of one superficial layer on top of one underlying block ($\mu_a^{(eff)}$, $\mu_s'^{(eff)}$).

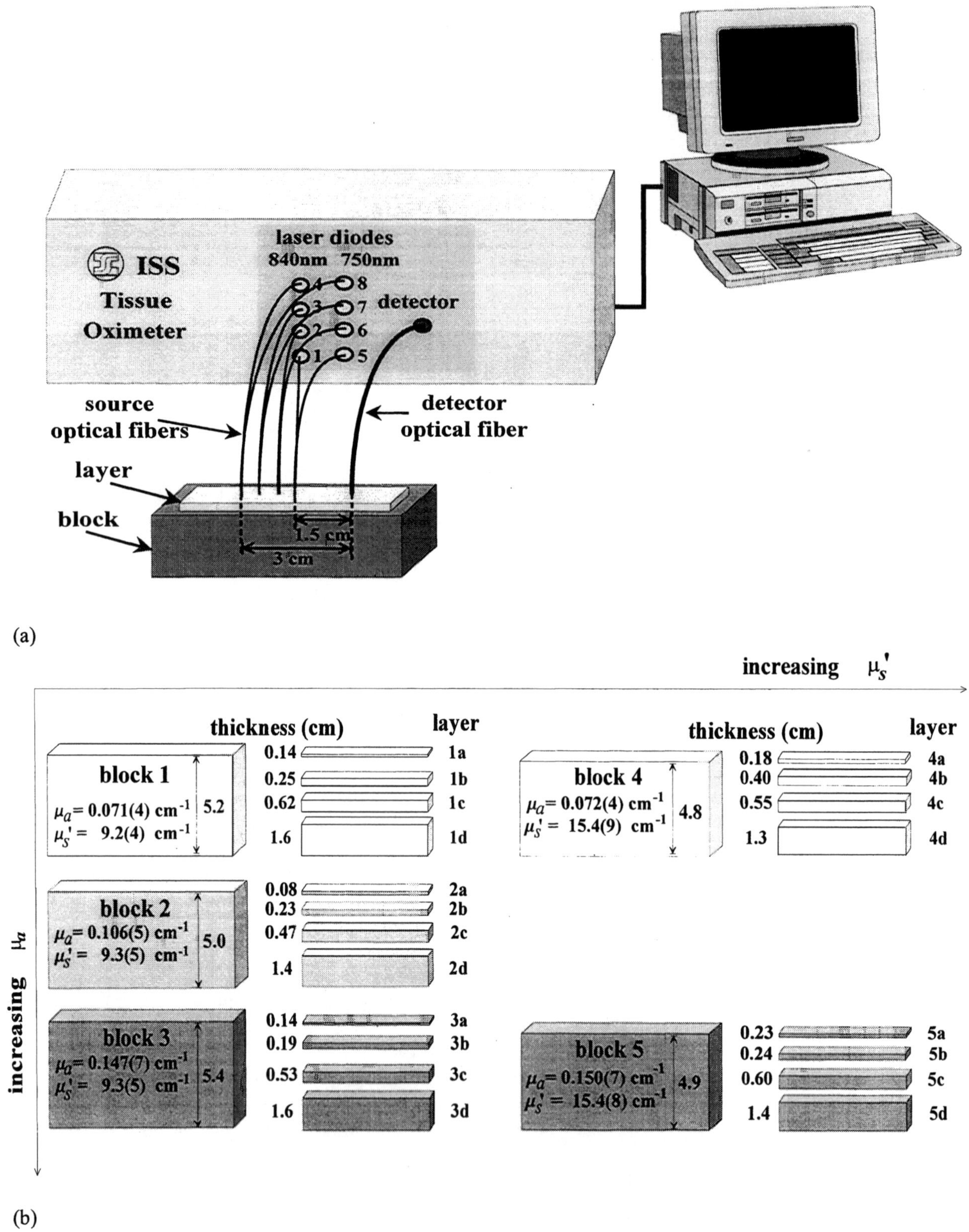

Figure 1. (a) Schematic diagram of the experimental apparatus. (b) Thickness, absorption (μ_a), and reduced scattering (μ_s') coefficients at 750 nm of the blocks and the layers employed in this study.

3. Results and Discussion

When the underlying block and the superficial layer had the same optical properties, we found that, in most cases, $\mu_a^{(eff)}$ and $\mu_s'^{(eff)}$ recovered the common values of μ_a and μ_s' to within the experimental error of 10% and 20%, respectively. This means that there was no significant inhomogeneity introduced by the contact surface between the block and the layer.

Figure 2(a) shows the recovered optical coefficients ($\mu_a^{(eff)}$ and $\mu_s'^{(eff)}$) for different thicknesses of one superficial layer having the same scattering and different absorption than the underlying block. For all thicknesses, the recovered reduced scattering coefficient was close to the common value of 9.2 cm^{-1} (maximum deviation 13%). At thicknesses up to 0.62 cm, the recovered absorption coefficient was close to that of the underlying block (maximum deviation 3%), while at a layer thickness of 1.6 cm, the recovered absorption coefficient was closer to the layer value.

(a)

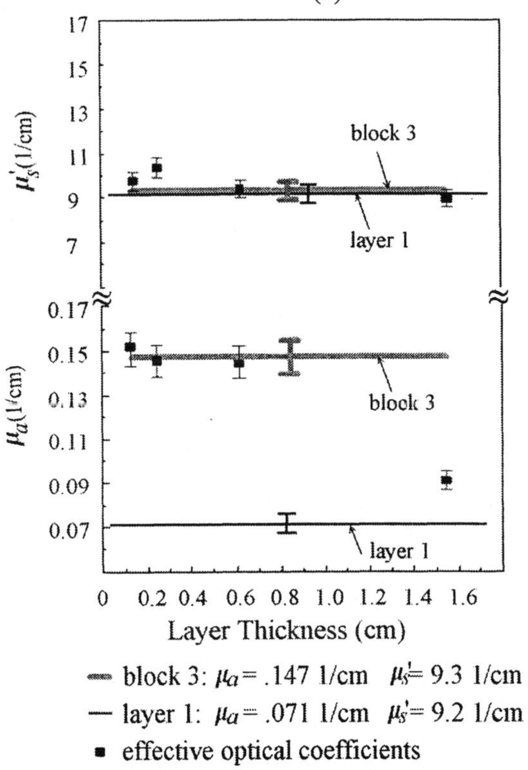

(b)

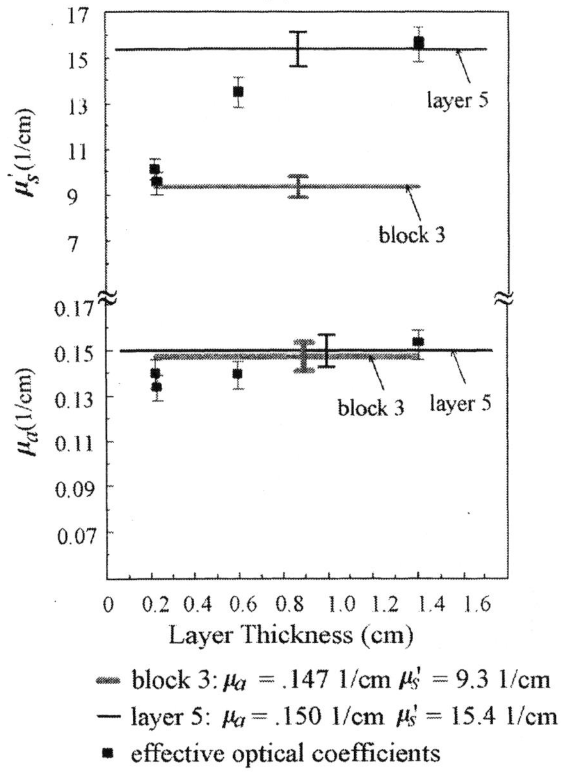

(c)

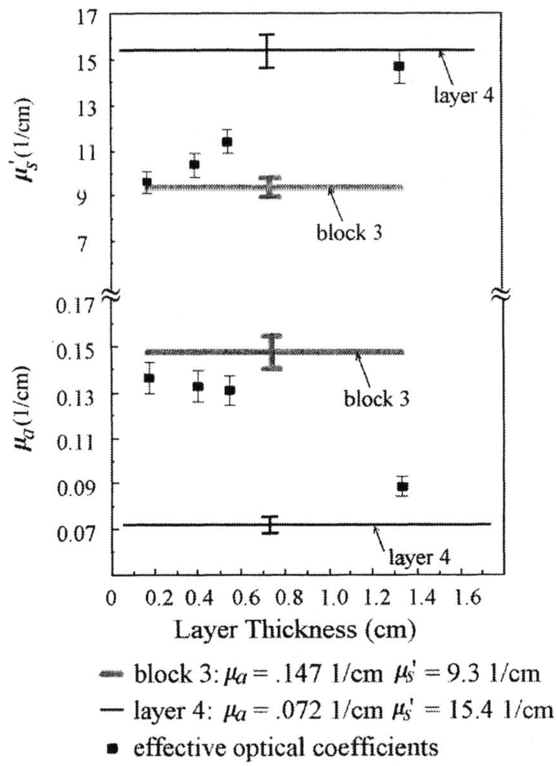

Fig. 2. Effective optical properties versus layer thickness in three different cases:
(a) layer and block with different absorption and same reduced scattering coefficients;
(b) layer and block with the same absorption and different reduced scattering coefficients;
(c) layer and block with different absorption and different reduced scattering coefficients.

Figure 2(b) reports the results for different thicknesses of one superficial layer having the same absorption and different scattering than the underlying block. For all thicknesses, the recovered absorption coefficient was close to the common value of 0.147 cm^{-1} (maximum deviation 11%). At thicknesses up to 0.24 cm, the reduced scattering coefficient was close to that of the underlying block (maximum deviation 8%), while at a layer thickness of 1.3 cm, the recovered scattering coefficient was to within 2% of that of the superficial layer. For the intermediate layer thickness of 0.60 cm the recovered absorption and scattering coefficients are in between the block and layer values.

A similar trend is found in the cases where the layer and the block differ in both absorption and scattering. Figure 2(c) shows the case for different thicknesses of layer 4 on top of block 3. In the case of layer thickness less than 0.40 cm, the recovered absorption and reduced scattering coefficients coincide with those of the underlying block to within 10%. In the case of the thickest layer (1.6 cm), the recovered reduced scattering coefficient coincide with that of the upper layer to within 5%, while the absorption coefficient is does not reproduce the upper layer. For the intermediate layer thickness of 0.55 cm the recovered absorption and reduced scattering coefficients are as shown in figure 2(c).

Figure 3 reports a summary of results obtained on the layer/block combinations where $\mu_a^{(L)}$ and/or $\mu_s'^{(L)}$ differ from $\mu_a^{(B)}$ and $\mu_s'^{(B)}$, respectively. This figure reports a normalized deviation between $\mu_a^{(eff)}$ ($\mu_s'^{(eff)}$) and $\mu_a^{(B)}$ ($\mu_s'^{(B)}$). This normalized deviation is defined as:

$$\frac{\mu^{(eff)} - \mu^{(B)}}{\mu^{(L)} - \mu^{(B)}}$$

where μ indicates either μ_a or μ_s'.

This normalized deviation takes a value of 0 when $\mu^{(eff)}=\mu^{(B)}$ and a value of 1 when $\mu^{(eff)}=\mu^{(L)}$.

(a) Absorption coefficient

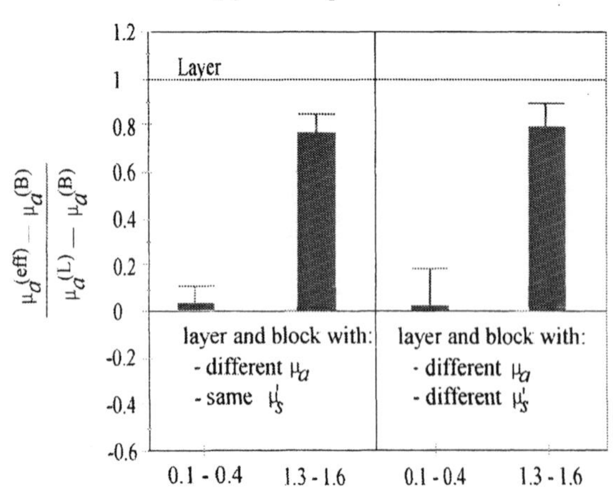

(b) Reduced scattering coefficient

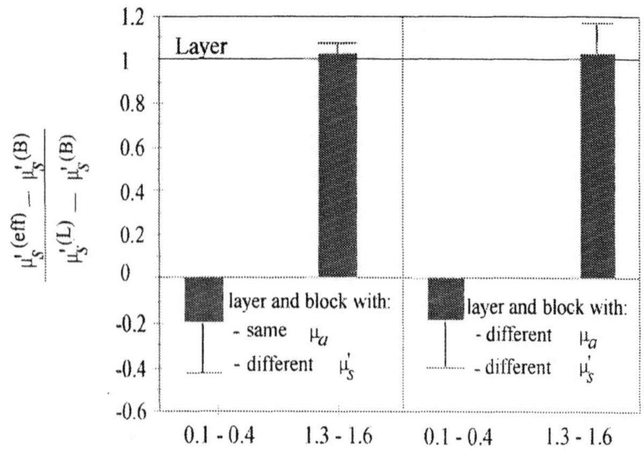

Fig. 3. Normalized deviations between the effective optical coefficients ((a) absorption; (b) reduced scattering) and the corresponding coefficients of the block. The various thicknesses, and the relationship between the layer and block optical properties are indicated in the legends.

We can divide all the cases considered into three categories:

(1) Layer and block with same μ_a and different μ_s' - ($\mu_a^{(L)}=\mu_a^{(B)}$; $\mu_s'^{(L)}\neq\mu_s'^{(B)}$)

In this case, $\mu_a^{(eff)}$ differed from the value $\mu_a^{(L)}=\mu_a^{(B)}$ by less than 10% for all of the layer thicknesses (except for 2 cases: layer 1c on top of block 4 and layer 3c on top of block 5 for which the difference was about 30%). The difference between $\mu_s'^{(eff)}$ and $\mu_s'^{(B)}$ was less than 20% for layer thicknesses in the range 0.08-0.5 cm. For layer thicknesses 1.3-1.6 cm $\mu_s'^{(eff)}$ coincided with $\mu_s'^{(L)}$. For thicknesses in the range of 0.5-0.62 cm, the values of the optical coefficients were in between the block and layer values.

(2) Layer and block with different μ_a and same μ_s' - ($\mu_a^{(L)}\neq\mu_a^{(B)}$; $\mu_s'^{(L)}=\mu_s'^{(B)}$)

The difference between $\mu_a^{(eff)}$ and $\mu_a^{(B)}$ was less than 10% for layer thicknesses in the range 0.08-0.62 cm. For layer thickness in the range of 1.3-1.6 cm, $\mu_a^{(eff)}$ approached $\mu_a^{(L)}$, from which, however it usually differed by more than the experimental error. The difference between $\mu_s'^{(eff)}$ and the common value $\mu_s'^{(L)}=\mu_s'^{(B)}$ was less than 20% for all the layer thicknesses.

(3) Layer and block with different μ_a and different μ_s' - ($\mu_a^{(L)}\neq\mu_a^{(B)}$; $\mu_s'^{(L)}\neq\mu_s'^{(B)}$)

In this case, $\mu_a^{(eff)}$ differed from the value $\mu_a^{(B)}$ by less than 10% for layer thicknesses in the range of 0.08-0.40 cm. For layer thickness in the range of 1.3-1.6 cm $\mu_a^{(eff)}$ approached $\mu_a^{(L)}$. For layer thicknesses in the range of 0.40-0.62 cm the values are in between the $\mu_a^{(B)}$ and $\mu_a^{(L)}$.

The difference between $\mu_s'^{(eff)}$ and $\mu_s'^{(B)}$ was less than 20% for layer thicknesses in the range 0.08-0.55 cm. For layer thicknesses in the range of 1.3-1.6 cm, $\mu_s'^{(B)}$ coincided with $\mu_s'^{(L)}$. For thicknesses 0.55-0.62 cm, the values of the optical coefficients are in between the block and the layer values.

5. Conclusions

From our data taken in the range of distances 1.5-3 cm, we conclude that in the presence of superficial layers less than about 0.40 cm thick, the effective optical coefficients measured with the frequency-domain multi-distance method are representative of the optical coefficients of the underlying block.

For layers having thickness between 1.3-1.6 cm, the recovered value $\mu_s'^{(eff)}$ reproduced the reduced scattering coefficient of the top layer, whereas $\mu_a^{(eff)}$ approached $\mu_a^{(L)}$ but it does not match it to within experimental errors.

For layer thickness of 0.5-0.6 cm the recovered effective optical coefficients are in between the layer and the block values.

6. Acknowledgments

This research is supported by National Institutes of Health grant CA57032 and by Whitaker-NIH grant RR10966.

7. References

[1] M. S. Patterson, B. Chance, and B. C. Wilson, "Time Resolved Reflectance and Transmittance for the Non-Invasive Measurement of Optical Properties," Appl. Opt. **28**, 2331-2336 (1989).

[2] S. Fantini, M. A. Franceschini, J. B. Fishkin, B. Barbieri, and E. Gratton, "Quantitative Determination of the Absorption Spectra of Chromophores in Strongly Scattering Media: a Light-Emitting-Diode Based Technique," Appl. Opt. **33**, 5204-5213 (1994).

[3] S. J. Madsen, E. R. Anderson, R. C. Haskell, and B. J. Tromberg, "Portable, high-bandwidth frequency-domain photon migration instrument for tissue spectroscopy", Opt. Lett. **19**, 1934-1936 (1994).

[4] A. H. Hielscher, H. Liu, L. Wang, F. K. Tittel, B. Chance, and S. L. Jacques, "Determination of Blood Oxygenation in the Brain by Time Resolved Reflectance Spectroscopy (I): Influence of the Skin, Skull, and Meninges," Proc. SPIE **2136**, 15-25 (1994).

[5] A. Kienle, M. S. Patterson, N. Dögnitz, R. Bays, G. Wagnières, and H. van der Bergh, "Noninvasive Determination of the Optical Properties of Two-Layered Turbid Media," Appl. Opt. **37**, 779-791 (1998).

[6] S. Fantini, M. A. Franceschini and E. Gratton, "Semi-infinite-geometry Boundary Problem for Light Migration in Highly Scattering Media: a Frequency-domain Study in the Diffusion Approximation," J. Opt. Soc. Am. B **11**, 2128-2138 (1994).

[7] M. A. Franceschini, D. Wallace, B. Barbieri, S. Fantini, W. W. Mantulin, S. Pratesi, G. P. Donzelli, and E. Gratton, "Optical Study of the Skeletal Muscle During Exercise with a Second Generation Frequency-Domain Tissue Oximeter," Proc. SPIE **2979**, 807-814 (1997).

Tissue Structure Characterization by Measuring the Specific Correlation Scales of the Scattered Light Fluctuations

D. A. Zimnyakov[1], V.V. Tuchin[1], A. G. Yodh[2]
[1] *Optics Department, Saratov State University, Astrakhanskaya 83, Saratov, 410026, Russia*
[2] *Department of Physics, University of Pennsylvania, Philadelphia, PA19104, USA*

Abstract

Possibilities of tissue structure characterization by measuring the scale parameter such as exponential factor and specific correlation times of speckle intensity fluctuations are discussed. These fluctuations are induced due to coherent beam scattering by non-stationary tissue structures. For "steady-structure" tissues (for example, *in-vitro* tissue samples) speckle dynamics can be provided by sample scanning. Polarization and correlation measurements of the scattered optical fields are used for evaluation of specific correlation time and corresponding spatial scales of depolarization and decorrelation of the probe light field. Some experimental results obtained with phantom objects such as water suspensions of polystyrene spheres as well as with optically cleared native human sclera are presented.

Key words: (180.0180) Microscopy, Scattering, Depolarization, Autocorrelation.

Introduction

One of the important applications of laser sources in the biology and medicine is tissue structure diagnostics by using the analysis of statistics and correlation properties of the dynamic speckles induced by laser light scattering in the tissue volume. Non-stationary character of the analyzed speckle patterns is caused by the certain dynamics of the scattering structures; biospeckles as the manifestation of blood microcirculation are the classical example of such dynamic speckle patterns. These speckles induced by quasi-stochastic movements of the scattering particles are characterized by certain spatial-temporal dynamics which is usually determined as "boiling" dynamics (for more details of the description of boiling motions of speckles see, e.g., Ref.1). By measuring autocorrelation functions or power spectra of intensity fluctuations of such boiling speckles we can in certain cases obtain parameters of blood microcirculation in the upper layers of tissue.

Non-stationary speckles induced by regular movements of the "steady-structure" scattering media illuminated by coherent light are characterized by more complicated form of the spatial-temporal dynamics which can be considered in the common case as the combination of two types of speckle motions - "translation" and "boiling". In the dependence on illumination and detection conditions as well as on scattering properties of the regularly moving medium one type of speckle dynamics can predominate and we will observe more expressed boiling or more expressed translation motion of speckles.

Correlation analysis of the speckle intensity fluctuations in the fixed detection point can also be used for tissue structure diagnostics; such approach based on the usage of the sample scanning by the focused coherent beam can be determined as the "spatial speckle correlometry" [2]. In this work we discuss some scale properties as well as relations between polarization and correlation characteristics of the scattered optical fields observed in the speckle-correlometric experiments with tissues demonstrating transition from multiple scattering to single scattering mode. Optically cleared human sclera has been chosen as the similar object with controlled optical properties; optical clearing was artificially stimulated by using the special matching agent. Phantom objects such as water suspensions of polystyrene spheres have also been used to study the relations between correlation and polarization properties of the multiply scattered optical fields.

Scale properties of speckle intensity fluctuations

Transition between multiple and single scattering mode of light propagation in tissue samples is manifested in the characteristical changes of temporal autocorrelation functions of detected intensity fluctuations. As it can be shown [3], for single scattering objects field and intensity autocorrelation functions of the far-zone speckles in the fixed detection point do not depend on the structure parameters of the scattering object. In this case form and halfwidth of the field autocorrelation function $G_1(\tau) = \langle U(0)U^*(\tau) \rangle$ and intensity autocorrelation function $G_2(\tau) = \langle I(0)I(\tau) \rangle$ are determined only by field distribution of the illuminating beam in the scattering zone and velocity of the scanned object movement. On the contrary, with an increase of tissue scattering coefficient and, correspondingly, rising of the sample turbidity field and intensity autocorrelation functions demonstrate the diminishing of correlation time and characteristical changes of their form. Thus, for single scattering mode and Gaussian illuminating beam we have form of $G_1(\tau)$ and $G_2(\tau)$ close to the Gaussian one, but for multiple scattering regime rather "exponential" form is characteristical.

Good example of similar changes of the statistical properties of the dynamic speckle patterns induced by coherent beam scanning of the tissue under study are the results of correlation experiments with optically cleared human sclera [2,4-7]. In these experiments such optical clearing was stimulated by administration of special immersion agent - 60% water solution of Trazograph. Interpretation of the optical clearing due to diffusion of the "matching" agent into the tissue volume has been discussed in details in the cited references. Matching of the refractive indices of two basic components of sclera (collagen fibrils and "ground" substance - water solution of polysaccharides) leads to the dramatic changes of path distribution for photons travelling through sclera layer even in the case of partially cleared sclera; thus, ballistic (unscattered) and "snake" (low-step scattered) photons significantly dominate in comparison with long-path photons.

Experiments have been carried out with samples of human sclera using technique described in Ref.2. Fig.1,a illustrates the evolution of the form of normalized autocorrelation function $\tilde{g}_2(\tau) = G_2(\tau)/ \langle I^2 \rangle - 1$ with the progress of optical clearing of human sclera caused by matching agent application, and Fig.1,b shows the dependence of exponential factor of speckle intensity fluctuations on time elapsed after matching agent application. $\tilde{g}_2(\tau)$ is the normalized autocorrelation function of the speckle intensity fluctuations and exponential factor ν is determined by the scale properties of corresponding autocorrelation and structure functions for small time delays [8]. ν can be obtained as the slope of logarithmically scaled structure function in the vicinity of zero value of its argument. It should be noted that semi-qualitative analysis of the scale properties of temporal fluctuations of optical fields induced by multiple scattering carried out in terms of random walk approach [9], gives the value of ν equal to 1 for the case of regular movement of the "steady-structure" scattering media. In other words, in this case induced speckle intensity fluctuations can be considered as classical one-dimensional Brownian process [10].

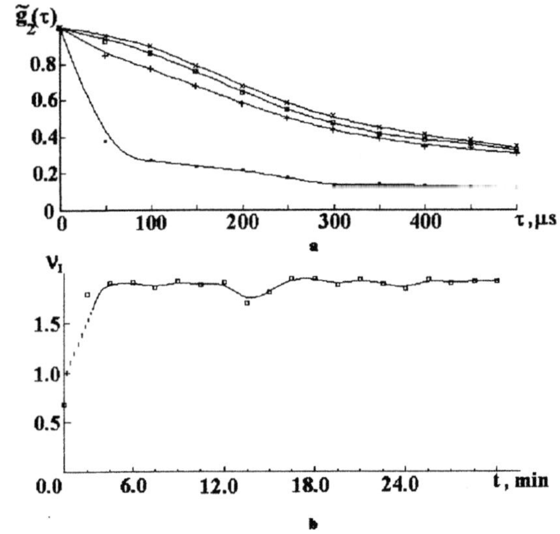

Figure1. a - evolution of $\tilde{g}_2(\tau)$ form with the progress of optical clearing of the human sclera; time after matching agent application: • - 100 s, + - 200 s, □ - 300s, × - 400 s. b - dependence of exponential factor of speckle intensity fluctuations for human sclera (thickness 0.6 mm) on time elapsed after matching agent (Trazograph) application.

This behavior allows us to propose exponential factor of the speckle intensity fluctuations as the generalized parameter for characterization of the tissue scattering properties. If single-scattering events become predominating in the scattered field formation and multiple-scattering events are significantly suppressed exponential factor of the detected intensity fluctuations will be close to 2. On the other hand, with an increase of contribution of multiple-scattering processes in the speckle field formation ν decreases, asymptotically reaching the value equal to 1.

One of the prospective applications of such correlation analysis of the speckle intensity fluctuations is the imaging of the scattering tissues by using the local estimations of exponential factor as visualization parameter. In particular, possibility of monitoring of the progress of human skin diseases (for example, such as psoriasis) has been demonstrated by the formation of ν-images of in-vitro samples of the diseased human epidermis [2,11].

Specific correlation time as generalized parameter of multiply scattered optical fields

Significant interest to the investigations of polarization phenomena accompanying propagation of the polarized coherent light through optically dense media is caused by the possible applicability of these effects for tissue characterization and imaging in biology and medicine (see, e.g., [12,13]). If the probe beam with the linear polarization is used in the light scattering experiments such generalized parameter of the total scattered field as its polarization degree can be introduced like:

$$P = \frac{< I_{II} > - < I_{\perp} >}{< I_{II} > + < I_{\perp} >}, \quad (1)$$

where $< I_{II} >$ and $< I_{\perp} >$ are the mean intensities of the co-polarized and cross-polarized components obtained by an ensemble averaging across the all possible realizations of the scattering structure. For non-stationary scattering media such as ensembles of Brownian particles averaging can be carried out over the time series of the intensity fluctuations of both components detected in the fixed observation point. For «steady-structure» scattering media and fixed point observation conditions detected intensity fluctuations can be induced by the media transverse movement with respect to the probe beam and detector.

As it was mentioned above, normalized autocorrelation function of the scattered field $g_1(\tau)$ and autocorrelation function of intensity fluctuations $g_2(\tau)$ are traditionally used for characterization of coherent light interaction with multiple scattering media and quantification of the scattering ensembles of moving particles [14,15]. For scattered fields with Gaussian statistics of amplitude induced by statistically homogeneous and ergodic scattering systems $g_1(\tau)$ and $g_2(\tau)$ are related with each other through Siegert relation [16].

In this work we have also studied the relations between correlation and polarization characteristics of the scattered fields for non-single scattering media in the region of the transition from multiple-scattering mode of light propagation to single-scattering mode. Water suspensions of polystyrene spheres and the human sclera with artificially controlled optical parameters have been used as studied objects in our experiments.

In the general form scattering properties of the turbid media can be described by using the probability density function $\rho(s)$, which describes the distribution of propagation paths of partial scattered waves forming the total scattered field. This function can be introduced in terms of random walk model of light-media interaction [see, e.g., Ref.9]. For optically dense scattering media, when diffusion mode of light transport through medium takes place, $\rho(s)$ can be interpreted as the probability density of path distribution of the diffusing photons. In terms of propagation path distribution all the statistical and correlation moments of scattered field, which describe in the general form its polarization and correlation properties, can be obtained as the integral transforms of $\rho(s)$ with corresponding weighting functions. In particular, field autocorrelation function for multiple scattering ensembles can be expressed as Laplas transform of $\rho(s)$ [17]:

$$g_1(\tau) \sim \int_0^{\infty} \exp\left[-sk_0^2 W(\tau)/3l^*\right]\rho(s)ds, \quad (2)$$

where k_0 is the scattering vector module, l^* is mean transport length for light propagation through a scattering medium and $W(\tau)$ is the mean-square particle displacement as a function of time delay. $\rho(s)$ depends on boundary conditions which describe the scattering object geometry as well as on the illumination and detection conditions.

On the other hand, polarization degree of light scattered by multiple scattering media can be expressed as [18]:

$$P \sim \int_0^{\infty} \exp(-s/\xi)\rho(s)ds, \quad (3)$$

where ξ is characteristic spatial scale determining propagation path required to destroy the initial polarization state of the probe beam. Comparing expressions (2) and (3), we can see that for exponential form of weighting function $\varphi_p(s)$ equality of polarization degree and field autocorrelation function for certain fixed value of time delay τ_{dp} should take place. In particular, for scattering systems consisting of non-interacting spherical Brownian particles this specific correlation time is equal to $\tau_{dp} = l^* / 2\xi D_s k_0^{\,2}$, where D_s is diffusion coefficient. Taking into account that ratio l^* / ξ for given scattering spheres does not depend on their concentration ϕ and is determined only by scattering properties of the single particle we should expect constant value of τ_{dp} for the wide range of concentration of scattering particles.

To verify such simple relations between characteristic scales of polarization and correlation decay correlation experiments with phantom objects such as water suspensions of polystyrene spheres were performed. Experiments have been carried out with forward scattering slab geometry with the slab thickness equal to 10 mm; single-mode Ar ion laser operating at 514 nm wavelength was used as an illumination source; scattered light was collected by single-mode fiber. Used diameters of polystyrene spheres (0.460 μm and 1.07 μm) and experimental conditions correspond to Mie scattering regime for single particle. For small volume concentrations of suspensions used in our experiments (less than 0.01) scattering particles can be considered as the uncorrelated spheres. Photon-counting photomultiplier tube (Hamamatsu HC-120) and digital autocorrelator BIC-9000 (Brookhaven Instruments Corp., USA) were used for the detection of intensity fluctuations and processing of the detected signal.

Obtained results have shown that, despite the dramatic decrease of «conventional» correlation time estimated as the halfwidth of the autocorrelation peak with an increase of ϕ, variations of τ_{dp} parameter in the analyzed region of scattering concentration are enough small to confirm the validity of the discussed relation between characteristic scales of polarization and correlation decay. τ_{dp} changes very small (maximal variation is less than 15 per cents) in comparison with «conventional» correlation time.

Estimations of ratios ξ/l^* made by using the obtained values of τ_{dp} ($(\xi/l^*)_{1.07\mu m} \approx 11.9$ and $(\xi/l^*)_{0.46\mu m} \approx 3.4$) are in a satisfactory agreement with experimental and simulation values presented in Ref.18 (for example, $(\xi/l^*)_{0.46\mu m} \approx 3.84$ obtained by estimations from data presented by D.Bicout et al.) and can be considered as verification of the discussed approach.

Coherent light scattering experiments with optically cleared human sclera [2,4-7] also demonstrate simultaneous changes of correlation and polarization characteristics of scattered light caused by evolution of $\rho(s)$ with a decrease of tissue scattering coefficient due to application of special matching agent. We should note the qualitative difference between above discussed case of dilution of the water suspensions of polystyrene spheres and optical clearing of sclera. In the later case with an increase of l^* we will have the simultaneous changes of ξ/l^* ratio due to the matching of refractive indices of collagen fibrils and ground substance. Thus, specific correlation time will depend on time elapsed after matching agent administration and this dependence can be used for analysis of "slow" dynamics of scattering structure during the optical clearing.

Dependencies of the "conventional" correlation time $\Delta\tau_{0.5}$ for co-polarized and cross-polarized components on polarization degree are presented in Fig.2 for an arbitrary selected sclera sample. Peculiarity in behavior of $\Delta\tau_{0.5}$ for co-polarized component (less values in comparison with the values for cross-polarized component as well as slow increase with an increase of polarization degree) in the case of cleared sclera can be explained by significant contribution of the specular (unscattered) component for highly cleared tissue fragments appeared along the scanning trace. Such contribution should strongly distort the form of the autocorrelation peak. Thus, in the presence of specular component separate analysis of cross-polarized and co-polarized components can be used in polarization-correlation "forward-scattering" measurements for selection of long-paths part and, respectively, short-path part of s distributions.

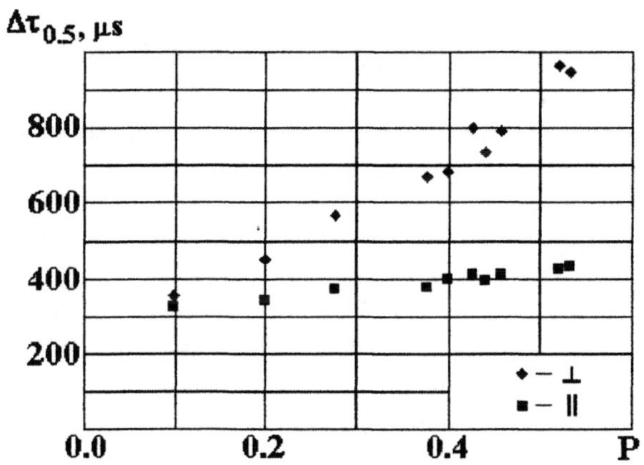

Figure 2. Conventional correlation time $\Delta\tau_{0.5}$ vs polarization degree for optically cleared human sclera. Scanning velocity is equal to 4mm/s.

Conclusion

Thus, specific correlation time τ_{dp} can be considered as one of the multiple scattering media that describes relation between decorrelation and depolarization of coherent light propagating through media. Being evaluated from simultaneous polarization and correlation measurements this parameter seems an attractive tool for analysis of averaged optical characteristics of individual scatterers in the non-stationary scattering ensemble as well as for analysis of peculiarities of ensemble dynamics. In this turn usage of the controlling technique on the basis of solute-induced changes of tissue optical properties (see, e.g., Refs.5,19) in combination with polarization and correlation measurements can be recommended for tissue structure diagnostics. Object of further investigations could be the relation between correlation and polarization properties of scattered light in the intermediate region between single-scattering and multiple-scattering modes of the light propagation through tissue (when a significant contribution of specular (unscattered) and low-step scattered components in the optical paths statistics takes place).

Acknowledgment

The research described in this publication was made possible in part by Award N RB1-230 of the U.S. Civilian Research & Development Foundation for the Independent States of the Former Soviet Union (CRDF).

References and Notes

1. T. Yoshimura, "Statistical properties of dynamic speckles," J. Opt. Soc. Am. A 3, 1032-1054 (1986).
2. D.A. Zimnyakov, V.V. Tuchin, A.A. Mishin, "Spatial speckle correlometry in applications to tissue structure monitoring," Appl. Opt. 36, 5594-5607 (1997).
3. S.M. Rhytov, U.A. Kravtsov, V.I. Tatarsky, Introduction to statistical radiophysics, P.2, Random Fields, (Nauka Publishers, Moscow, 1978).
4. D.A. Zimnyakov, V.V. Tuchin, A.A. Mishin, I.L. Kon, A.N. Serov "In-vitro human sclera structure analysis using tissue optical immersion effect", in SPIE Proceedings, Vol.2673, pp.233-242 (1996).
5. V.V. Tuchin, I.L. Maksimova, D.A. Zimnyakov, I.L. Kon, A. Kh. Mavlutov, A.A. Mishin, "Light propagation in tissues with controlled optical properties," J. Biomedical Optics 2, 401-417 (1997).
6. V.V. Tuchin, I.L. Maksimova, V.I. Kochubey, I.L. Kon, A. Kh. Mavlutov, A.A. Mishin, S.V. Tuchin, D.A. Zimnyakov,"Optical and osmotic properties of human sclera," in SPIE Proceedings, Vol.2979, pp.658-675 (1997).
7. D.A. Zimnyakov, V.V. Tuchin, K.V. Larin, A.A. Mishin, "Speckle pattern polarization analysis as an approach to turbid tissue structure monitoring", in SPIE Proceedings, Vol.2981, pp.172-180 (1997).
8. D.A. Zimnyakov, V.V. Tuchin, "Fractality of speckle intensity fluctuations", Appl.Opt. 35, 3325-3333 (1996).
9. F.C.MacKintosh, S.John, "Diffusing-wave spectroscopy and multiple scattering of light in correlated random media", Phys.Rev.B 40, 2383-2406 (1989).
10. J. Feder, Fractals, (Plenum Press, New York and London, 1988).
11. D.A. Zimnyakov, V.V. Tuchin, S.R. Utts, "A study of statistical properties of partially developed speckle fields as applied to the diagnostics of structural changes in human skin, " Opt. Spectrosc. (USSR) 76, 838-844 (1994).
12. C.Salzman, S.B.Singham, R.G.Johnston, C.F.Bohren, "Light scattering and cytometry", in Flow Cytometry and Sorting, (Wiley-Liss, Inc, New York, 1990), pp.81-107.
13. J. M. Schmitt, A. H. Gandjbakhche, R. F. Bonner, "Use of polarized light to discriminate short-path photons in a multiply scattering medium", Appl. Opt. 31, 6535-6546 (1992).
14. M.J.Stephen, "Temporal fluctuations in wave propagation in random media", Phys.Rev.B 37, 1-5 (1988).
15. E.O.Schultz-DuBois, ed., Photon Correlation Technique in Fluid Mechanics, (Springer-Verlag, Berlin, 1983).
16. See, for example, A.Ishimaru, Wave Propagation and Scattering in Random Media, (Academic Press, New York, 1978).
17. A.G.Yodh, D.J.Pine, P.D.Kaplan, M.H.Kao, N.Georgiades, "Speckle-fluctuations and their use as probes of dense random media", Mol. Cryst. Liq. Cryst. Sci. Technol. - Sec. B: Nonlinear Optics 3, 149-160 (1992).
18. D.Bicout, C.Brosseau, A.S.Martinez, J.M.Schmitt, "Depolarization of multiply scattering waves by spherical diffusers : Influence of the size parameter", Phys. Rev. E 49,1767-1770 (1994).
19. H. Lui, B. Beauvoit, M. Kimura, B. Chance, "Dependence of tissue optical properties on solute-induced changes in refractive index and osmolarity", J. Biomed. Optics 1, 200-211 (1996).

An Non-contact Instrument for Measuring Motions in Thick Tissues

Cecil Cheung, Anatoly Chaussky, and A. G. Yodh
Department of Physics and Astronomy, University of Pennsylvania
Philadelphia, Pennsylvania 19104
Office: 215-898-8422, fax: 215-898-2010

D. A. Boas
Electro-Optics Technology Center, Tufts University
Medford, Massachusetts 02155
Office: 617-627-3136, fax: 617-627-3135

Light diffusing in a turbid medium generally experiences small changes in its frequency due to the Doppler shift induced by moving scatterers. In mammal's tissues, the main source of this Doppler shift is from the blood flow. By measuring the autocorrelation function of the intensity of the diffusing light, one can measure the flow speed or the diffusion coefficient if the blood is stagnant[1]. Importantly, with diffusing light, we can probe deeper (i.e. centimeters) than that by optical Doppler tomography (ODT)[2] which relies on single scattering events.

In previous work, we have had some successes on probing skin burns[3, 5], and probing blood flow in the arm during an ischemia[4, 5]. However, we also encountered the following problems: (1) physical contacts with the skin caused changes in the near surface blood flow and the optical properties, which affected the accuracy of the diagnosis; (2) there were measurement errors due to the relative motions of the probe to the skin; (3) since both the polarized and the de-polarized components of the diffusing light were collected, the analyses were more complicated than it was necessary and the signal-to-noise was degraded; (4) the optical properties of the tissue were not measured concurrently. In this work, we introduce an instrument that tackled all the above problems.

We use a lens to image the surface of the tissue onto the probe which is situated behind the lens. This arrangement makes the measurement contact-free. The depth-of-field of the lens substantially reduces the measurement errors due to the axial motion of the tissue. With a camera zoom lens, we found that the measurements yield the same results when the tissue sample is placed within 5 mm of the object plane. An added advantage of using a lens is that by changing the magnification of the lens, one can probe a wide range of penetration depths. Compared with previous versions of the instrument, the signal-to-noise ratio is greatly improved by collecting only the de-polarized component of the diffusing light.

Figure 1 shows a schematic diagram of the instrument. There are two sets of optical fibers in the probe. The first set is for the correlation measurements. The laser source is a high power (0.5W) single-mode solid state diode laser with a very narrow linewidth ($\sim$ 5 MHz). The wavelength is continuously tunable from 770 nm to 800 nm. Polarized laser light is delivered from

Reprinted from Advances in Optical Imaging & Photon Migration, 1998 Technical Digest,

a polarization-maintaining optical fiber, through the lens, onto the sample. The diffused light coming out from the sample is gathered by the same lens and focused on a linear array of singe-mode fibers. The polarization of the collected light is controlled by a polarizer in front of the detecting fibers. This array of single-mode fibers provides a range of source-detector separations, which translates into a range of penetration depths. Then the light intensities are measured by photomultipliers. A newly developed multi-channel correlator is used to compute the autocorrelation functions. This correlator is a cost-effective way to speed up the whole measurement as the autocorrelation functions are computed simultaneously.

To interpret the measurements from the correlation functions more quantitatively, we need to know the optical properties of the tissue samples, which can be measured independently using diffuse photon density waves[6]. The measurement is facilitated by a second set of optical fibers in the probe. One of the many source fibers deliver an intensity modulated laser light onto the sample. Laser diodes are used. With a detection optical fiber connected to an avalanche photodiode and then to a lock-in amplifier, the amplitude and the phase (with respect to the source) of the diffusing light are measured. With multiple sources and hence multiple source-detector separations, the reduced scattering coefficient and the absorption coefficient can be found. Oxy-hemoglobin and deoxy-hemoglobin have different absorption coefficients in the range of the wavelengths that we use. By repeating the measurement with different wavelengths, the relative blood oxygenation and the blood volume in the tissue may be estimated.

References

[1] Boas, Campbell, and Yodh, Phys. Rev. Lett. **75**, 1855 (1995).

[2] Chen *et al.*, Opt. Lett. **22**, 1119 (1997); Huang *et al.*, Science **254**, 1178 (1991); Izatt, Hee, Owen, Swanson, and Fujimoto, Opt. Lett. **19**, 590 (1994).

[3] Boas, Yodh, Schomacker, and Nishioka, Proceedings of Conference on Lasers and Electro-Optics 96 (1996).

[4] Boas, Meglinsky, Zemany, Campbell, Chance, and Yodh, Coherence Domain Methods in Biomedical Optics SPIE **2732** (1996).

[5] Boas, Ph.D. thesis, University of Pennsylvania (1996).

[6] Yodh and Chance, Phys. Today **34** (March 1995).

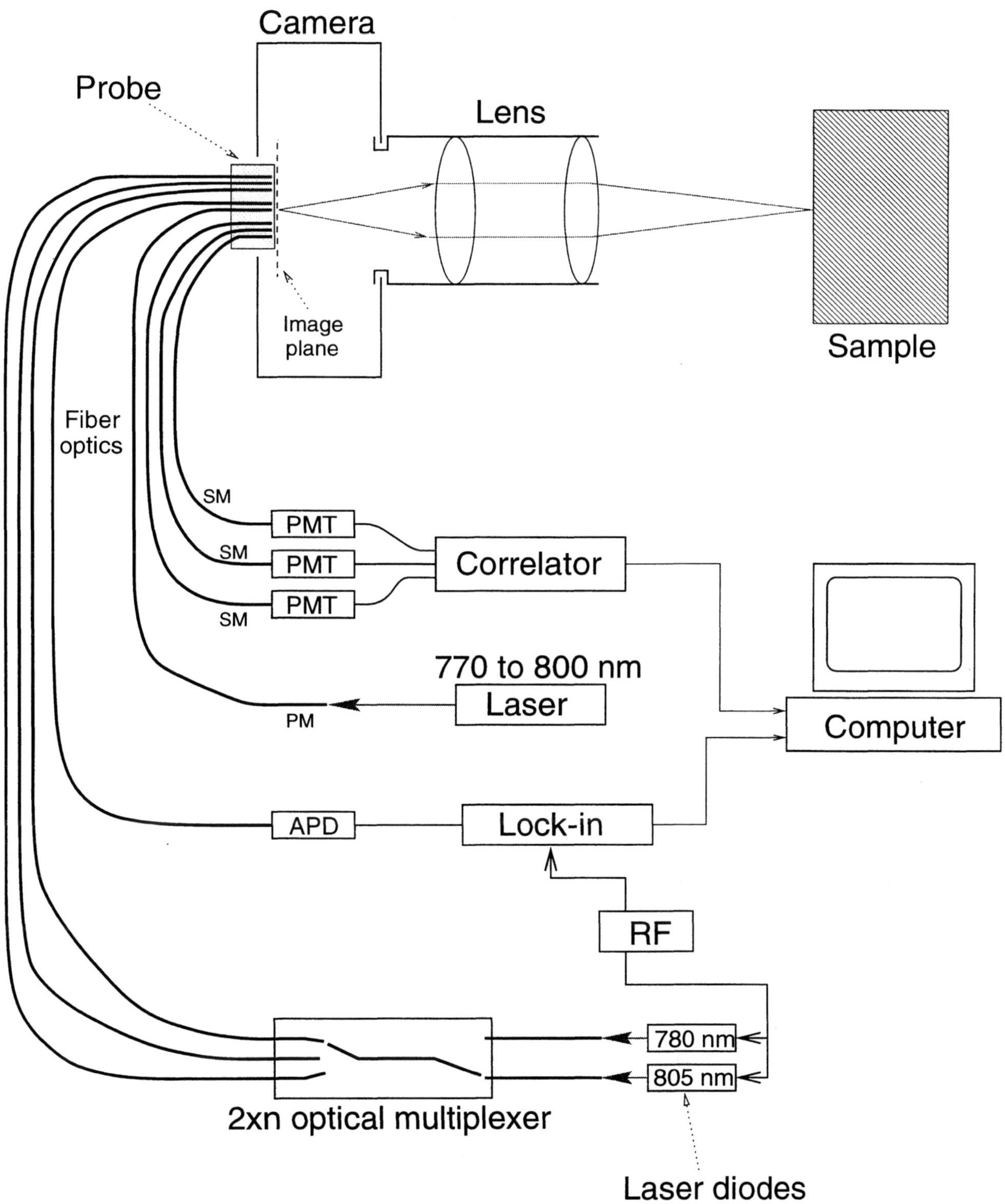

Figure 1: A schematic of the instrument. Key: PMT: photomultiplier, APD: avalanche photodiode, RF: radio frequency generator, PM: polarization-maintaining fiber, SM: single-mode fiber.

Determination of Particle Size Distribution in Concentrated Suspensions: Photon Migration Approach

Huabei Jiang, Guillermo Marquez*, Lihong Wang*

Department of Physics & Astronomy, Clemson University, Clemson, SC 29634-1911
Tel: (864)656-5307 Fax: (864)656-0805 Email: hjiang@clemson.edu
**Biomedical Engineering Program, Texas A&M University, College Station, TX 77843-3120*

Abstract

We demonstrate a new approach for determining particle size distribution in concentrated suspensions from spectral measurement of isotropic scattering coefficients using steady-state, continuous-wave photon migration techniques. Successful recovery of particle size distribution for TiO_2 suspensions in the form of a lognormal function is achieved through a regularized inverse algorithm, in which a synthesized scheme of Marquardt and Tikhonov regularizations has been incorporated. Our results for dense TiO_2 suspensions with three different particle concentrations are in excellent agreement with the size distribution measured with x-ray sedimentation.

Key Words: (290.4210) Multiple scattering, (290.1990) Diffusion, (290.5850) Scattering, particles

Particle sizing in non-dilute suspensions is important in the areas such as environmental remote sensing and on-line control of chemical- and pharmaceutical-based industrial processes. Among the existing particle sizing techniques, optical methods such as dynamic light scattering and optical turbidity or extinction spectrum measurements have been particularly attractive for portability and the potential for rapid measurement.[1-2] While these optical techniques have been used to provide size information noninvasively about scattering particles, they have been restricted to diluted suspensions and often require extensive calibrations on the suspensions they are intended to monitor.

To overcome the problems associated with the present optical techniques for particle sizing, a new optical method has been proposed recently, based on frequency-domain measurements of photon migration in scattering suspensions.[3] Since this technique depends upon multiply scattered light, it is particularly suitable for non-dilute suspensions and has great potential for on-line process monitoring. In addition, the technique does not require calibrations on the suspensions.

However, the measurements we conducted have been based on a frequency-domain Ti:Sapphire laser system not able to provide light beam in a wide range of spectrum. Further, the procedures for optical data collection were relatively slow since measurements at multi-wavelength were needed. In this paper, we experimentally demonstrate particle sizing in concentrated TiO_2 suspensions for the first time, using continuous-wave based photon migration techniques. Compared with the frequency-domain techniques, this method allows faster and simpler optical measurements with the same accuracy as the frequency-domain techniques on the measurements of optical properties in a scattering medium.[4] In addition, the cost for a continuous-wave based system is lower than its frequency-domain counterpart. Using a regularized inverse algorithm, we have reconstructed particle size distributions of TiO_2 suspensions with three different concentrations. The results are in excellent agreement with the x-ray sedimentation size distribution measurements.

Particle size distributions are obtained with a regularized inverse algorithm under Mie scattering theory as reported in Ref. 3. Our approach casts particle sizing as an optimization problem in which the optimization parameters are coefficients in a probability function such as lognormal or Gaussian distribution function known as *a priori*. The details of our inverse algorithms can be found in Ref. 3; we present a brief overview here for context.

It is known that particle size distribution in a multiple scattering medium is correlated with the isotropic scattering coefficient of the medium by the following relationship:[5-6]

$$\mu_s'(\lambda)=\int_0^{\infty}\frac{3Q_{scat}(x,n,\lambda)[1-g(x,n,\lambda)]}{2x}\phi f(x)dx \tag{1}$$

where $(1-g)\mu_s$ is the isotropic scattering coefficient, g is the mean cosine of the scattering angle from a single particle, Q_{scat} is the scattering efficiency, n is the refractive index of the medium and λ is the wavelength of light; f(x) (x is the diameter of the particles) is the particle volume size distribution and ϕ is the total volume fraction of the particles. Both quantities of g and Q_{scat} can be computed using Mie theory.[6]

For concentrated suspensions, we usually have *a priori* knowledge about the number of modes and their size distributions. Generally, particulate suspensions can be characterized by Gaussian or log-normal distributions. We use lognormal distribution in this study for TiO_2 suspensions:[7] $f(x)=\exp[-(\ln x-a)^2/2b^2]/\sqrt{2\pi}\,bx$, where parameters a and b denote the mean and variance of the distribution, respectively. Using our regularized optimization methods, we obtain the following system of equations leading to inverse problem solution:[3]

$$(\Im^T\Im+\alpha I)\Delta\zeta=\Im^T[(\mu_s')^o-(\mu_s')^c] \tag{2}$$

where the Jacobian matrix $\Im$ represents the sensitivity of isotropic scattering coefficients measured at multiple wavelengths, the elements of which are numerically computed using Eq. (1). I is the identity matrix and α may be a scalar or a diagonal matrix. The vectors, $(\mu_s')^o$ and $(\mu_s')^c$ contain the observed and computed values of isotropic scattering coefficients at the wavelengths of interest. $\Delta\zeta$ is the vector updating the three parameters, a, b and c: $\Delta\zeta=(\delta a,\delta b,\delta c)^T$ (we choose ϕ as the third parameter). Thus, the basic idea for determining a, b and c is to make measurements of the isotropic scattering coefficient of the suspensions at multiple wavelengths. Then the particle sizing task is to make estimates [which are updated and improved through solution of Eq. (2)] of parameters a, b and c that are required for the measured scattering coefficients to be sustained. Inasmuch as the matrix $\Im\Im^T$ is known to be ill-conditioned,[2-3] a way to regularize or stabilize the decomposition of $\Im\Im^T$ is needed. We used a hybrid technique that synthesizes standard Marquardt and Tikhonov regularization schemes.[3]

The experimental system is an arc lamp/CCD camera combined spectral system, which was based on oblique incident reflectometry.[4] In this system, white light (Oriel, 75 W Xenon Arc Lamp) was delivered, and the diffuse reflectance was collected with a fiber optic probe made from black delrin and 600 μm diameter, low-loss optical fibers. The source fiber was oriented at a 45° angle of incidence, and the nine collection fibers, arranged in a linear array, collected the diffuse reflectance. To correct for slight variations in collection efficiency from one detection fiber to the next, a correction factor was calculated for each fiber based on a calibration procedure using standard phantoms. The detection system was composed of an imaging spectrograph (Oriel, Multispec 257), a CCD camera (Princeton Instrument, 1530P), and a personal computer to automatically record the spectrum of the collected light through the wavelength range of 400 nm–800 nm. The spectral resolution of the measurements was 0.5 nm.

The fiber probe was placed on the surface of the concentrated suspensions. An exposed x-ray film was attached to the top of the suspensions to approximate a matched boundary condition for diffusion theory. Four measurements were performed with the probe oriented at 0°, 30°, 60°, and 90° with respect to an arbitrary reference orientation. The isotropic scattering spectra was evaluated for each diffuse reflectance measurement, and the results were averaged. The concentrated suspensions is a mixture of TiO_2, India ink and polyacrylamide gel,[8] which produced TiO_2 concentration values of 8.8×10^{-3}, 11.7×10^{-3}, and 14.6×10^{-3} by weight, respectively. The mixture was poured into a plastic container with 8 cm in diameter and 6 cm in height. Figure 1 presents the isotropic scattering spectra measured for the TiO_2 suspensions with three different concentrations.

Using our reconstruction algorithm, particle size distributions for all three TiO_2 suspensions have been recovered, as shown in Figure 2. The symbolized lines denote the reconstructed values from the photon migration measurements while the dashed line represents the size distribution obtained from the sedimentation measurement for comparison. As can be seen, excellent agreement between our results and that by the sedimentation has been reached. Since we have used the concentrated suspensions (whereas dilute solution was used in the sedimentation measurement) and a lognormal distribution in our reconstruction, the errors between the results from photon migration and sedimentation may be expected. To reconstruct a typical particle size distribution shown here, only a few seconds were needed with a SunSparc 5 workstation. We found that the inversion results were not very sensitive to the choices of the initial estimates of the reconstructed parameters (a, b and c), which were up to 100% greater than the true parameters, depending of course on the optical property measurements. It should be noted that the inverse solution procedure was quite stable indicating the high accuracy of our optical property measurements, whereas an unstable solution procedure has been experienced using frequency-domain photon migration measurements.[3] It is also interesting to note that, while the shape of particles in TiO_2 suspensions is not perfectly spherical, the spherical particulate-based Mie theory still provides us accurate results from our inverse algorithm. This could be due to the use of the hybrid regularization in our iterative optimization procedure, which provides a compensation mechanism for the geometrical discrepancy occurred in the theory.

In summary, we have used continuous-wave photon migration measurements of isotropic scattering for particle sizing in concentrated suspensions. Our study shows that fast, simple continuous-wave photon migration measurements in a wide range of wavelength, coupled with an efficient inverse algorithm, can provide particle size distribution of concentrated suspensions in a matter of seconds. This indicates its potential for on-line process monitoring.

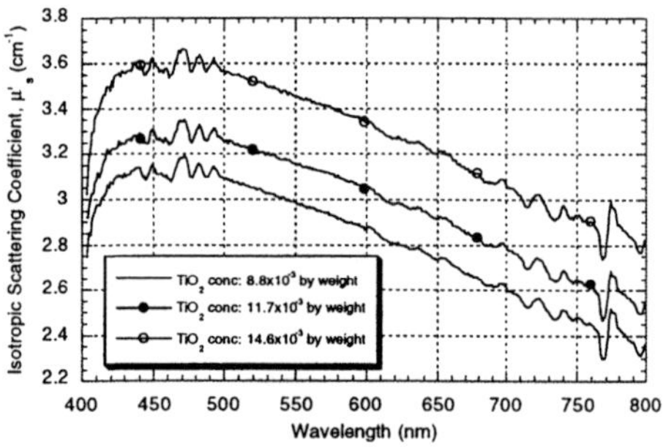

Figure 1. Isotropic scattering spectra [μ_s' (cm^{-1}) versus wavelength, λ (nm)], which was determined by the method of oblique incident reflectometry.

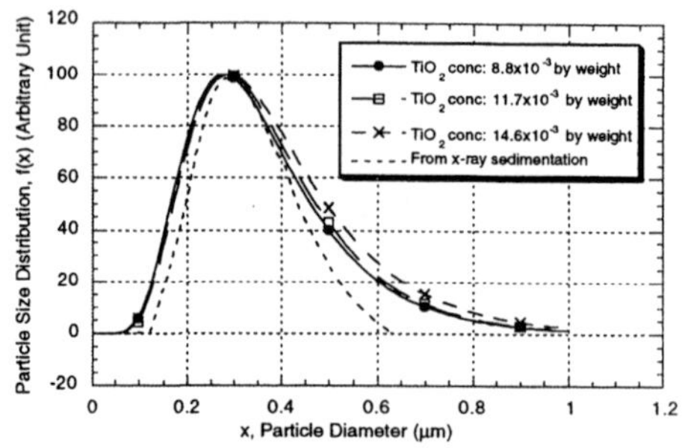

Figure 2. The particle size distribution, f(x) (arbitrary unit), as a function of diameter, x, (μm) for the TiO_2 samples as inverted from the diffuse reflectance measurements (symbolized lines) and as measured from the sedimentation (dashed line).

References

1. T. Allen, *Particle Size Measurement* (Chapman and Hall, New York, 1990).
2. J. Wang, and F.R. Hallet, Appl. Opt. **35**, 193(1996).
3. H. Jiang, J. Pierce, J. Kao, E. Sevick-Muraca, Appl. Opt. **36**, 3310(1997).
4. S.-P. Lin, L.-H. Wang, S. L. Jacques, F. K. Tittel, Appl. Opt. **36**, 136(1997).
5. H. Van de Hulst, *Light scattering by small particles* (Dover Publications, New York, 1983).
6. G. Bohren, D. Hoffman, *Absorption and Scattering of Light by Small Particles* (John Wiley, New York, 1983).
7. E. L. Crow, K. Shimizu, *Lognormal Distributions* (Marcel Dekker, Inc., New York, 1988).
8. M. Lee, *New gel phantoms simulating optical properties of biological tissue*, M. S. Thesis (Texas A&M University, College Station, 1997).

On-line Monitoring of Changes in Size Distribution of Cell-Debris Suspension Through Frequency Domain Measurements of Photon Migration

Ganesh Balgi and Eva M. Sevick-Muraca
School of Chemical Engineering, Purdue University, West Lafayette, IN 47907, USA
http://photon.ecn.purdue.edu/~chepmi/ppml.html
Phone: (765) 496-2377, Fax: (765) 494-0805, e-mail: sevick@ecn.purdue.edu

Abstract

Dual wavelength measurements of frequency domain photon migration (FDPM) are used to monitor the changes in size distribution of a cell-debris suspension due to appearance of large granules in the effluent stream of a centrifugation process. The ratio of scattering coefficients at two distinct wavelengths is used as an indicator of changes in size distribution of this process stream.

Key Words

Photon migration; Light scattering; Particle size distribution; Cell-debris suspension.

Introduction

In most of the chemical, pharmaceutical, and biotechnological industries characteristics of particulate dispersions in the intermittent process stream usually govern the properties of the finished product, or are indicators of the success of a particular unit operation process. Implementation of an effective control strategy in these processes is often limited by the on-line measurement technique used for assessing important characteristics such as size distribution, solids volume fraction, or the structure set up due to intermolecular interactions. Optical techniques are suitable for these measurements because they are fast, robust, and non-invasive. However, most intensity-based techniques such as turbidity measurements need sample dilution to prevent multiple scattering. These techniques also require frequent external standard calibrations to account for any changes in absorption properties of the process stream. As a result, these optical techniques are not suitable for on-line use. On the other hand, non-optical techniques such as electro-acoustic measurements have to resort to external calibration to obtain sizing information from concentrated dispersions [1]. Frequency domain photon migration (FDPM) technique provides unique advantages over other optical and non-optical measurement techniques for on-line assessment of particulate dispersions. The FDPM technique does not require sample dilution because measurements are based on multiple scattering of light. For very high solids concentrations the analysis of FDPM measurements can account for intermolecular interactions via the static structure factor $S(q)$ [2]. In addition, FDPM measurements allow independent estimation of the two optical coefficients: the isotropic scattering coefficient, μ'_s, and the absorption coefficient, μ_a. Hence, FDPM measurements do not require any external standard calibration. The purpose of this communication is to use the FDPM technique for an on-line application from biotechnological industry.

Background

Many important pharmaceutical proteins are obtained in the form of granules by harvesting appropriately mutated *E-coli* bacterial cells [3]. The mean size of the cell-debris generated upon homogenization of the cellular suspension is usually about 0.2 μm. The protein granules of 1~2 μm in size are separated from this cell debris using a multi-stage centrifugation process. Unforeseen fluctuations in the upstream process parameters can cause undesirable build-up of granules inside the centrifuge. Unanticipated build-up can result in a substantial loss of granules in the effluent stream containing cell-debris, causing inefficient operation. The focus of this investigation is to use the FDPM measurements on the effluent stream containing the cell-debris suspension for on-line monitoring of any changes in size distribution due to appearance of protein granules. In other words, we look to identify the appearance of 1~2 μm sized granule bodies within a cell-debris suspension of 0.2 μm mean size.

Theory

In the FDPM technique the intensity of the incident light of wavelength, λ, is modulated sinusoidally at frequency, ω, ranging from KHz to MHz. This multiply scattered "photon density wave" is attenuated and phase-shifted relative to the incident light. The measurements of phase-shift, θ, and modulation ratio, M, at various values of modulation frequency can be analyzed using solutions of diffusion approximation to the radiative transfer equation describing the transport of photon in multiply scattering sample. Details of this technique have been described elsewhere [4-6]. The experimental measurements of frequency domain photon migration (FDPM) yield two optical parameters characterizing the particulate process. The isotropic scattering coefficient, μ_s', is a function of solids volume fraction, ϕ, the particle size distribution, f(x) based on volume, and colloidal interactions *via* static structure factor, S(q). Here $q = \left(4\pi n/\lambda\right) Sin(\varphi/2)$ for the scattering angle, φ, and n is the refractive index of the particle of size x. For spherical particles of relative refractive index η,

$$\mu_s'(\lambda) = \phi \int_0^\infty \frac{3}{2x}(1-g)\, Q_{scat}\, f(x)\, dx\,, \qquad (1)$$

where the scattering efficiency, $Q_{scat}(x,\lambda,\eta)$, and the anisotropy factor, $g(x,\lambda,\eta)$, can be computed using appropriate scattering theory, e.g. Mie theory [7], coupled with a suitable model for S(q) [2]. The absorption coefficient μ_a is a function of the concentration of all chemical species in the sample which absorb the incident radiation.

Any change in the isotropic scattering coefficient of the effluent stream containing cell-debris suspension determined from single wavelength measurements of FDPM would be due to change in the quantity ϕ f(x). In order to monitor change in the size distribution, f(x), independent of changes in the solids volume fraction, ϕ, the FDPM measurements must be conducted at two distinct wavelengths λ_1 and λ_2. The ratio of isotropic scattering coefficients at these two wavelengths can be expressed as

$$^{rel}\mu_s' = \frac{\mu_s'(\lambda_1)}{\mu_s'(\lambda_2)} = \frac{\displaystyle\int_0^\infty \frac{3}{2x}(1-g(\lambda_1))Q_{scat}(\lambda_1)f(x)dx}{\displaystyle\int_0^\infty \frac{3}{2x}(1-g(\lambda_2))Q_{scat}(\lambda_2)f(x)dx}\,. \qquad (2)$$

Equation 2 states that any change in the ratio $^{rel}\mu_s'$ for the effluent stream is indicative of changes in its size distribution alone. The underlying assumptions for this conclusion are that, the changes in the scattering coefficient due to wavelength dependence of the refractive index of scattering particles is negligible, and S(q) equals unity due to negligible intermolecular interactions. For the cell-debris suspensions investigated in this study the above assumptions remain valid.

Experiments

The cell-debris suspension samples from the effluent stream were kindly provided by Eli Lilly and Company. The cell-debris suspension was assumed to be a reference sample devoid of any granules. Freeze-dried granules were added to the reference sample to prepare suspensions with approximately 1% and 5% protein granules by volume. The two wavelengths selected for this investigations were 670 nm using laser diode, and 820 nm using pico-second pulsed Titanium-Sapphire laser. The details of the instrumentation have been described elsewhere [4-6]. The light was delivered in a 1000 ml sample through an optical fiber of 1 mm diameter. The scattered light was collected at distance r from the source with a 1 mm diameter detector optical fiber placed parallel to the source fiber in the sample. The sample geometry satisfied the infinite media boundary condition for the diffusive transport of light. All samples were refrigerated at 5 degree C, and their pH was adjusted to 8.4 prior to the FDPM measurements. The samples were gently stirred during the measurements. The phase shifts were measured at four different source-detector separations of 1.0 cm, 1.5 cm, 2.0 cm, and 2.5 cm respectively for varying modulation frequencies. These measurements were used to compute the relative phase shift, $\Delta\theta$, as a function of modulation frequency, ω, corresponding to relative detector-detector separation, Δr, of 0.5 cm, 1.0 cm, and 1.5 cm.

Results and Discussion

Figure 1 shows the experimentally measured relative phase shift data for the reference cell-debris sample at 670 nm wavelength of incident light. From the solution of the diffusion approximation to the radiative transfer equation the relative phase shift can be expressed as

$$\Delta\theta = \Delta r \sqrt{\frac{3\mu_a\left(\mu_s^{/}+\mu_a\right)}{2}\sqrt{\left[1+\left(\frac{\omega}{c\mu_a}\right)^2\right]}-1}\ , \qquad (3)$$

where c is the speed of incident light in the continuous medium [8]. Equation 3 is regressed to the measured relative phase shift data from Fig. 1 to determine the values of the isotropic scattering coefficient, $\mu^{/}_s$, and the absorption coefficient, μ_a, for the reference cell-debris suspension at 670 nm. The isotropic scattering coefficients of all samples at 670 nm and 820 nm wavelengths of incident light are presented in Fig. 2. These plots indicate that for

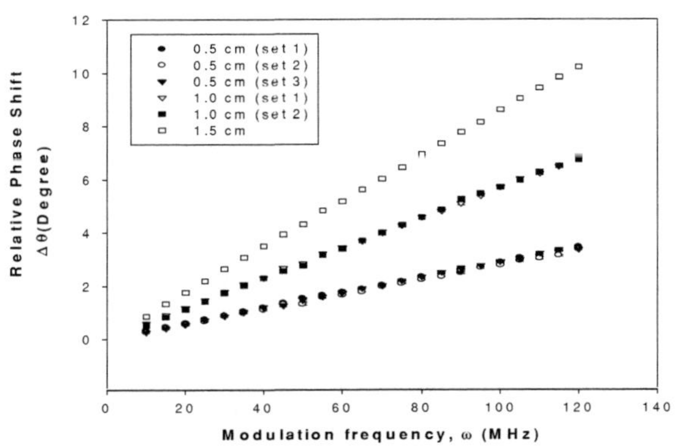

Fig. 1. Experimentally measured relative phase shift at varying modulation frequencies for the reference cell-debris suspension (0 % added granules) at 670 nm. The legends indicate the relative detector-detector separation, Δr, for the corresponding data set.

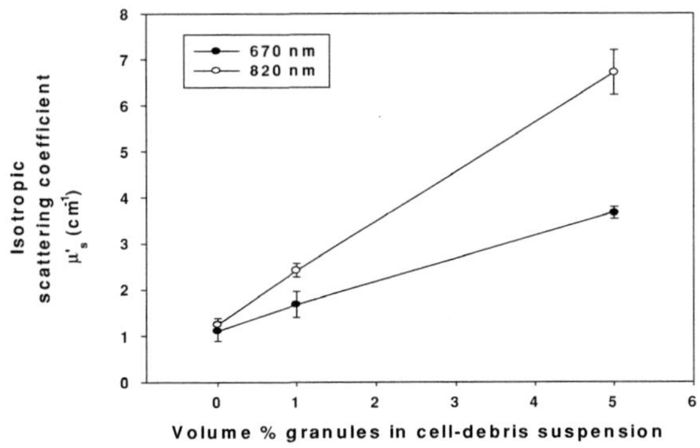

Fig. 2. The isotropic scattering coefficient of cell-debris suspension as a function of volume % of added granules.

measurements conducted at a single wavelength, the scattering coefficient increases as the volume % of added granules in the cell-debris suspension is increased. The ratio of the scattering coefficients measured at two different wavelengths is plotted in Fig. 3 as a function of the amount of added granules in the suspension. The result shows that the ratio, ${}^{rel}\mu^{/}_s$, is sensitive to changes in the size distribution caused by addition of granules in the cell-debris suspension.

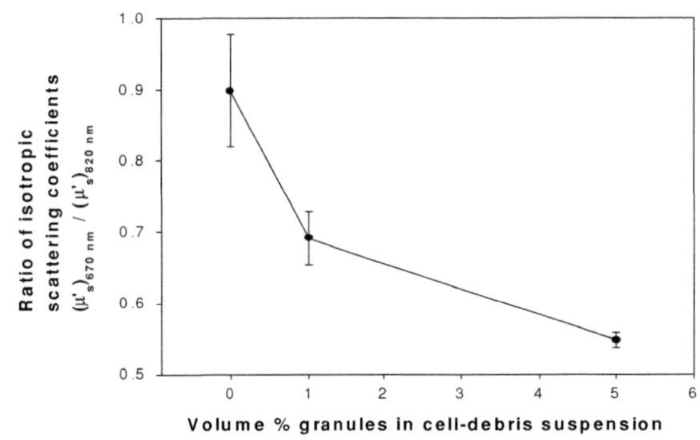

Fig. 3. The ratio of isotropic scattering coefficient of cell-debris suspension as function of volume % of added granules from dual wavelength FDPM measurements.

The work presented in this communication exhibits the potential of the FDPM technique for on-line process monitoring of particulate streams. In the case of the effluent stream containing cell-debris suspension, FDPM measurements of scattering coefficient at dual wavelengths in near infrared regime (670 nm and 820 nm) can be used to monitor changes in size distribution due to appearance of larger granules in the stream. In order to obtain quantitative information about the amount of granules from the measurement of ${}^{rel}\mu^{/}_s$, future investigations must be conducted with reference cell-debris samples devoid of any granules. The two wavelengths for FDPM measurements must be selected judiciously, so as to obtain maximum sensitivity of ${}^{rel}\mu^{/}_s$ to a small amount of granules appearing in the effluent stream.

Acknowledgments

The authors would like to acknowledge Jeffery Reynolds and Ralf Mayer of Purdue University Photon Migration Laboratory for their assistance in developing the data acquisition system. We would also like to thank Eli Lilly and Company for providing the cell-debris suspensions. This work was supported in part by Eli Lilly and Company, DuPont, and NSF Young Investigator Award.

References

1. E. M. Sevick-Muraca, "A review on coupling particle size measurements for control and monitoring of particulate processes," IFPRI report SAR, 12-18 (1997).
2. R. Shinde, G. Balgi, S. Richter, S. Banerjee, J. Reynolds, J. Pierce, and E. M. Sevick-Muraca, "Investigation of static structure factor in dense suspensions using multiply scattered light," submitted to Appl. Opt. (1998).
3. C. S. Shin, M. S. Hong, C. S. Bae, and J. Lee, "Enhanced production of human mini-proinsulin in fed-batch cultures at high cell density of Escherichia coli BL21(DE3)[pET-3aT2M2]," Biotech. Prog. 13 249-257 (1997).
4. S. M. Richter, R. R. Shinde, G. V. Balgi, and E. M. Sevick-Muraca, "Particle sizing using frequency domain photon migration," Part. & Part. Sys. Char. 15 9-15 (1998).
5. E. M. Sevick-Muraca, J. Pierce, H. Jiang, and J. Kao, "Photon migration measurements of latex size distribution in concentrated suspensions," AIChE J. 43 655-664 (1997).
6. H. Jiang, J. Pierce, J. Kao, and E. M. Sevick-Muraca, "Measurement of particle size distribution and volume fraction in concentrated suspensions using photon migration techniques," Appl. Opt. 36 3310-3318 (1997).
7. C. Bohren, and D. Hoffman, Absorption and scattering of light by small particles. John Wiley & Sons, New York 1983.
8. R. C. Haskell, L. O. Svaasand, T.-T. Tsay, T.-C. Feng, M. C. McAdams, and B. J. Tromberg, "Boundary conditions for the diffusion equation in radiative transfer," J. Opt. Soc. Am. A 11 2727-2741 (1994).

Optics of the Human Sclera: Photon Migration, Imaging and Spectroscopy

Valery V. Tuchin
Saratov State University, 83 Astrakhanskaya, Saratov, 410026 Russia, Ph: (8452) 72-34-61, FAX: (8452) 24-04-46, e-mail: tuchin@sgu.ssu.runnet.ru

Abstract

Optics, thermodynamics and structure of the human eye sclera impregnated by osmotically active chemical agents are studied. Photon migration analysis, imaging and spectroscopy of treated scleral tissue are presented.

Key Words: (300.6340) Spectroscopy, infrared; (180.0180) Microscopy

Introduction

In vivo tissue optical properties control is very important for many medical applications. For example, the human eye compression technique allows to perform transscleral laser coagulation of the ciliary body and retina/choroid [1]. The possibility of selective translucence of the upper tissue layers should be very useful for the developing of the functional tomography and spectroscopy optical technologies. For example, a number of results on noninvasive glucose concentration monitoring within human body using NIR light scattering techniques based on tissue components refractive indices matching were reported [2,3], the results on the human sclera optical properties controlled by tissue impregnation by osmotically active chemicals, such as *trazograph, glucose and polyethylene glycol* (PEG) were recently reported as well [4].

In this paper further investigation of the treated scleral tissue optics is presented. All experiments were performed *in vitro* using samples of the human and bovine sclera cut from the autopsy eyes. The optical transmission and reflection spectra in the visible and NIR as well as dynamical response of sclera transmittance at the selected wavelengths via period of a chemical agent administration were measured. The statistical parameters of the far-zone speckle intensity fluctuations caused by the sample scanning perpendicular to the focused laser beam were recorded and analyzed from the point of view of their application for tissue imaging in a process of its controlled transition from multiple to single scattering and back. Thermodynamics and histology of the treated scleral tissue samples were studied. The nonlinear effects of tissue interaction with chemical agents were revealed.

Methods and materials

The human sclera is a turbid, nontransparent medium cover about 80% of the eye ball and serving as a protective membrane. It is a strong fibrous tissue which is mainly consists of conjunctive collagen fibers packed in lamellar bundles which are immersed within amorphous ground (interstitial) substance containing glycosaminoglycans, proteins and proteinpolysaccharides complexes [1]. The fibers in the scleral stroma have a diameter ranging from 30 to 300 nm with an average diameter of about 100 nm, the mean distance between fibrils centers is about 285 nm. These fibrils are arranged in the individual bundles in a parallel fashion but more randomly than that in the cornea. Collagen bundles show a wide range of widths (1-50 μm) and thickness (0.5-6 μm). Such ribbonlike structures are multiply crosslinked, their length can be a few millimeters. They cross each other in all directions but remain parallel to the scleral surface. The thickness of the sclera is variable in the range 0.3-1.8 mm. Hydration of the human sclera can be estimated as 68%. About 75% of its dry weight is due to collagen as well as 10% is due to other proteins, and 1% - to mucopolysaccharides. The refractive indices of the scleral fibrils (hydrated collagen) and interstitial medium at λ=600 nm can be estimated as n_c=1.474 and n_I=1.345, correspondingly [1,4].

The collimated light transmission by a tissue layer of thickness l is defined as

$$T_c = I/I_0 = \exp(-\mu_t l), \qquad (1)$$

where I_0 and I are the intensities of the collimated incident and detected light, respectively; $\mu_t=\mu_a+\mu_s$ is the extinction coefficient, μ_a and μ_s are the absorption and scattering coefficients, respectively. For normal scleral tissue $\mu_a<<\mu_s$. In its turn, the ratio of the scattering coefficients for different degree of refractive indices matching is estimated as [4]

$$\mu_{s2} \cong \mu_{s1}[(m_2-1)/(m_1-1)]^2. \qquad (2)$$

For example, for $m_1=n_C/n_{I1}=1.1$ and $m_2=n_C/n_{I2}=1.001$, $\mu_{s2} \sim 10^{-4}\mu_{s1}$.

Because the scleral tissue shows a large variability of the scattering properties due to refractive index matching we should include in consideration the coherent effects. Method of the spatial-domain correlation analysis of the scattered light seems to be of great interest for tissue imaging [4,5]. The space-time transfer of intensity fluctuations can be provided by the scanning of the probe beam. The corresponding intensity autocorrelation function can be expressed as

$$g_2(\tau) = < I_s(t) I_s(t+\tau) > / < I_s^2(t) >. \qquad (3)$$

Due to fibrous structure of the sclera it is possible to present it as a permeable membrane, then for the large external volume when concentration of chemical agent c_1 can be considered as a constant we have an exponential dependence for this agent concentration inside tissue [4]

$$c_2 \cong c_1\{1-\exp(-t/\tau)\}, \quad \tau = l^2/D. \qquad (4)$$

where D is the diffusion coefficient for chemical agent in tissue, l is the tissue thickness.

The spectrophotometric measurements in the wavelength range 400-800 nm were provided by spectrophotometer Cary-2415 allowing to measure collimated transmittance, total transmittance and diffuse reflectance using integrating sphere. Individual spectrum was registered during 85 s.

Results and Discussion

The human sclera collimated and total transmission spectra $T_c(\lambda)$, $T(\lambda)$ and diffusion reflection spectra $R_d(\lambda)$ measured for the different time intervals of *trazograph* (60%) administration are presented in Fig.1. It is easily seen that the native sclera is a poor transparent tissue for the visible and NIR light, *trazograph* impregnation makes this tissue highly transparent, up to 75% at 800 nm for the sample kept in solution during Δt=7-10 min, as well as its spectral reflectivity goes down from more than 30% to slightly less than 15% in this wavelength range.

The Monte Carlo simulation of the sclera transmission and reflection spectra was provided using the probability function for the free photon path l_{ph} [4]. The results of such modeling for scleral sample of 1 mm thickness, with collagenous fibrils of mean diameter equal to 120 nm and mean separation between fibrils centers equal to 285 nm, refractive index of colagen n_C = 1.47 and of interstitial medium n_I = 1.35-1.45 presented in Figs 2-4. It is well seen that designed model describes well all important features of immersed tissue.

The comparison of calculated and experimental spectra shows that refractive indices matching is the main mechanism responsible for tissue enhanced translucence. The developed Monte Carlo simulation technique allows to describe the transition from a completely diffusion to a partially-coherent transmission of the sclera caused by refractive indices matching. Such transition is well illustrated by the histograms in Figs 3 and 4. The numbers of back and forward scattered photons collected by integrating spheres was calculated These histograms show that for sclera with unmatched or slightly matched refractive indices there is approximately uniform distribution of the number of scattering events which forward traveling photon undergo. For fairly matched refractive indices there is appear ballistic photons (the coherent part of transmitted light). For matched refractive indices the unscattered and low-step scattered photons dominate in both directions - forward and back. In the NIR the optical clearing of tissue and transformation of scattering mode from multiple to low-step or even single begins much earlier than for the visible light.

For finding of optical parameters of sclera in a process of its translucence the following model was used. Collimated laser beam with the finite diameter and with the flat distribution across its cross-section falls on the five layered medium, transmitted and reflected light was detected using the integrating spheres. The computer model took into account the real geometry of the integrating sphere – finite diameter and geometry of the sphere windows. The five layered model contained cuvette glass walls, solution of osmotically active chemical and scleral sample itself. We suggested that in the directions perpendicular to the laser beam incidence direction medium is infinite. The calculated spectra for optical parameters of the human sclera are presented in Fig.5.

The time-dependent collimated transmittance of the sclera sample measured at the selected wavelength concurrently with chemical agent impregnation can be used for estimation of diffusion coefficients of the interacting fluids: water and *trazograph*, water and *glucose*, etc. (see, Eqs (1), (2) and (4)).For *trazograph* solution (60%) $D_T=(1.46\pm0.19)\cdot10^{-5}$ - $(4.00\pm1.15)\cdot10^{-6}$ cm^2/s (4 samples of the human sclera), for *glucose* (45%) the averaged for 5 samples of bovine sclera $D_G=(2.13\pm1.48)\cdot10^{-4} cm^2/s$, for PEG solution (80%) the averaged for 4 samples of bovine sclera $D_{PEG}=(1.91\pm1.24)\cdot10^{-4} cm^2/s$.

During the turbidity decay process samples under study were scanned by the focused probe beam using spatial speckle correlometer [4,5]. The trace length was usually equal to 8 mm. Visually the characteristical changes of the far-zone speckle structure are observed clearly after the chemicals administration. Firstly there were no any visually observed speckles in the far diffraction zone, after a few minutes (1-3) isotropic speckle pattern with small-sized speckles appears in the detection plane, then the speckle pattern becomes inhomogeneous with the large-sized speckles in the central part, finally the main part of the transmitted light is concentrating around the incident beam axis.

Besides, the low-amplitude quasi-periodic oscillations of the averaged transmitted intensity and the correlation peak halfwidth with the characteristical time scales about 1.5-2.0 min were seen. Such quasi-periodic oscillations can be induced by the temporal-spatial inhomogeneity of osmolyte diffusion within the tissue volume. The oscillating character of tissue response can be explained as a multi-step origin of fluid diffusion [4].

Histological and polarimetric investigations of the treated by *trazograph* human sclera showed that for its concentration of 60% and duration of impregnation less than 30 min no any serious unreversible changes of tissue structure took place.

Static speckle patterns were recorded by using Electrim digital video recorder without any imaging optical system between CCD array and sclera sample. Slightly diverging beam of He-Ne laser was used for the illumination of the sclera sample. Beam diameter was approximately equal to 0.5 mm. Distance between sclera sample and CCD array was equal to 350 mm. After the recording data image files were converted to 8-bit bitmap format for further processing.

Fig.6 displays the changes of recorded speckle patterns with the progress of translucence process.. Evolution of form of the empirical probability density functions of speckle intensity with progress of sclera translucence is illustrated by Fig.7. For early stages of translucence speckle statistics is close to the one typical for so-called fully developed speckle fields with zero value of mean amplitude. Such form of probability density functions of intensity distributions is approximated by the negative exponential function (non-zero minimal value of intensity for presented histograms is caused by non-zero bias used for video recording). For later stages of translucence we can see qualitative changes of intensity statistical distributions manifested in the appearing of bimodality of the intensity probability density functions. This effect is typical for single scattering objects causing only phase modulation of the probe beam and characterizing by relatively large values of standard deviation and correlation length of the boundary field phase fluctuations (see distributions d,e of Fig.7). Similar behavior can be explained in terms of appearing of specular component of scattered field and suppressing of the diffuse component. Thus, analysis of speckle intensity statistical distributions allows to estimate time elapsed after *trazograph* administration, when phase modulation of the probe beam becomes adequately weak.

Specific changes of the first- and second-order statistics of the speckle intensity distributions with the transition from multiple scattering mode to the single scattering mode are also manifested in the evolution of the form of spatial autocorrelation functions: for early stages of translucence their form is more "exponential", as for later stages this form is close to "Gaussian". Such behavior is related with the changes of temporal autocorrelation function in the case of sclera scanning by the focused coherent beam and fixed point detection of the scattered light intensity fluctuations (see, e.g. [5]). Thus, increase of exponential factor estimated for temporal as well as for spatial intensity fluctuations up to the values close to 2 shows that single scattering in tissue volume is predominate in the formation of scattered field.

During the translucence far-zone speckles observed at the middle and later stages of process demonstrate the slow dynamic behavior which can be interpreted as speckle "boiling" with very low frequencies. This dynamics is caused by the spatial-temporal inhomogeneities of the *trazograph* concentration in the tissue volume.

Conclusion

This study shows that impregnation of the sclera by osmotically active chemical agents affects the reversible refractive indices matching of the collagen fibrils and interstitial media what leads to dramatic reducing of the tissue scattering ability.

Dynamics of tissue enhanced translucence caused by osmolytes is defined by the time response about 3-10 min. This is in a good agreement with results obtained by Rol for mechanical stress or laser beam induced local heating [1].

The speckle-correlometry technique should be a good method for tissue structure imaging in a process of enhanced translucence, especially to investigate the space-time nonlinear oscillations of a chemical agent diffusion.

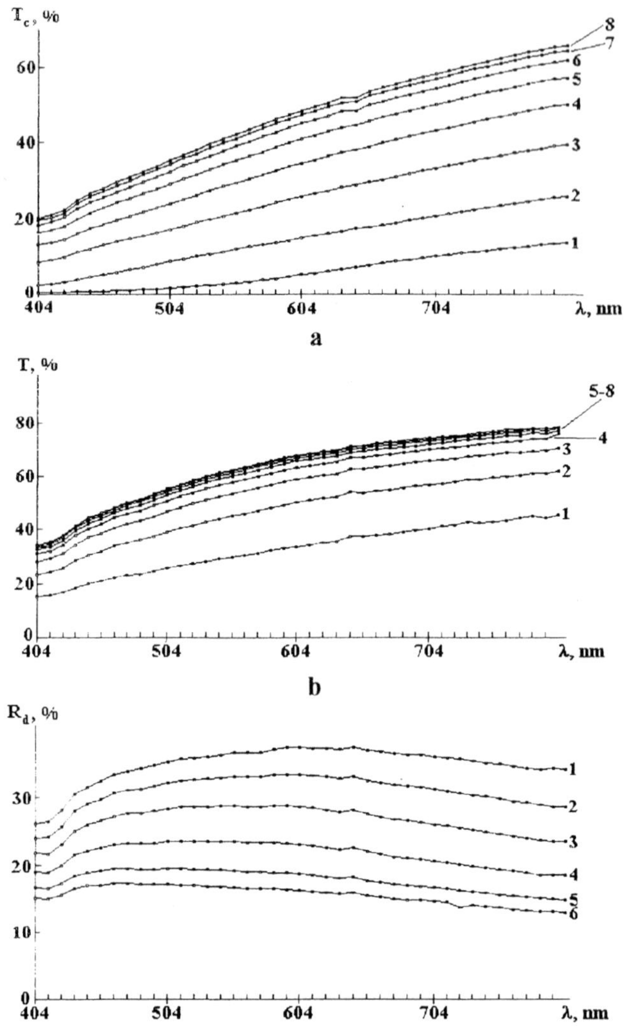

Fig.1. Typical experimental spectra of the human sclera samples measured for different time intervals of 60% trazograph solution administration. Spectra numbered 1 were measured in 1 min after sample immersing into solution, spectra 2-8 were measured subsequently at 2 min intervals. An individual spectrum measuring time scanning from higher to lower wavelengths was about 85 sec. a - the collimated transmittance, T_c, sample thickness - 0.6 mm; b - the total transmittance, T, sample thickness - 0.7 mm; c - the diffusion reflection R_d, sample thickness - 0.7 mm (heavy pigmented tissue)

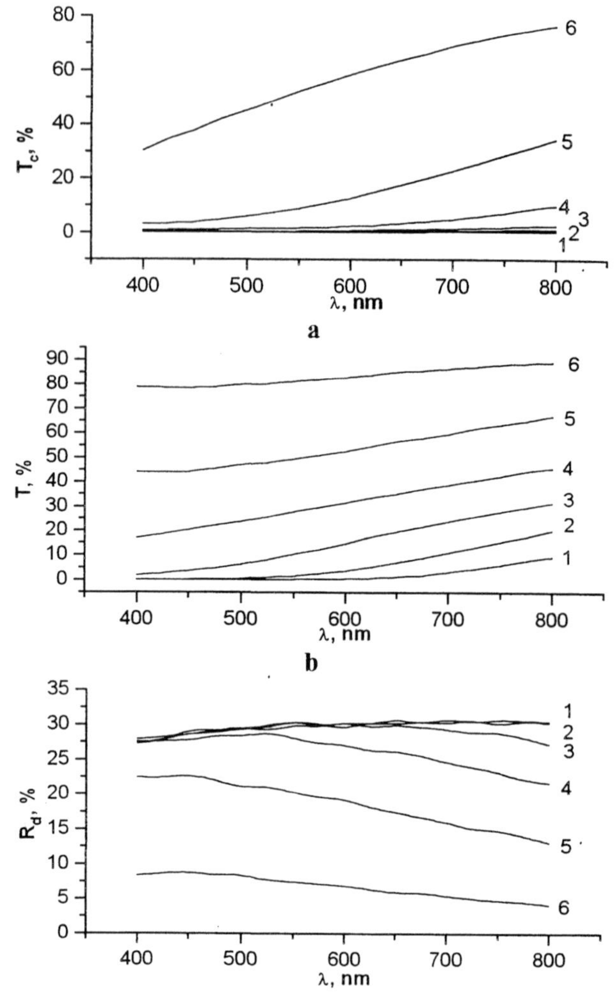

Fig.2. The collimated (a) and total (b) transmittance spectra as well as diffuse reflectance (c) spectra calculated for different refractive indices matching for geometry very close to experimental one ; n_C = 1.47 and n_I = 1.35(1), 1.37 (2), 1.39 (3), 1.41(4), 1.43 (5), 1.45 (6).

Acknowledgments

Author greatly appreciate help of his colleagues Dr. D.A.Zimnyakov, Dr. V.I.Kochubey, and A.A.Mishin in experimental work, MD T. N. Semenova in providing of scleral samples, Dr. I.L.Maksimova and student A. H. Mavlutov in calculations.

The research described in this publication was made possible in part by grant N RB1-230 of the U S Civilian Research & Development Foundation for the Independent States of the Former Soviet Union (CRDF) and grant of the Russian Foundation of the Basic Research "Leading Scientific Schools" N 96-15-96389.

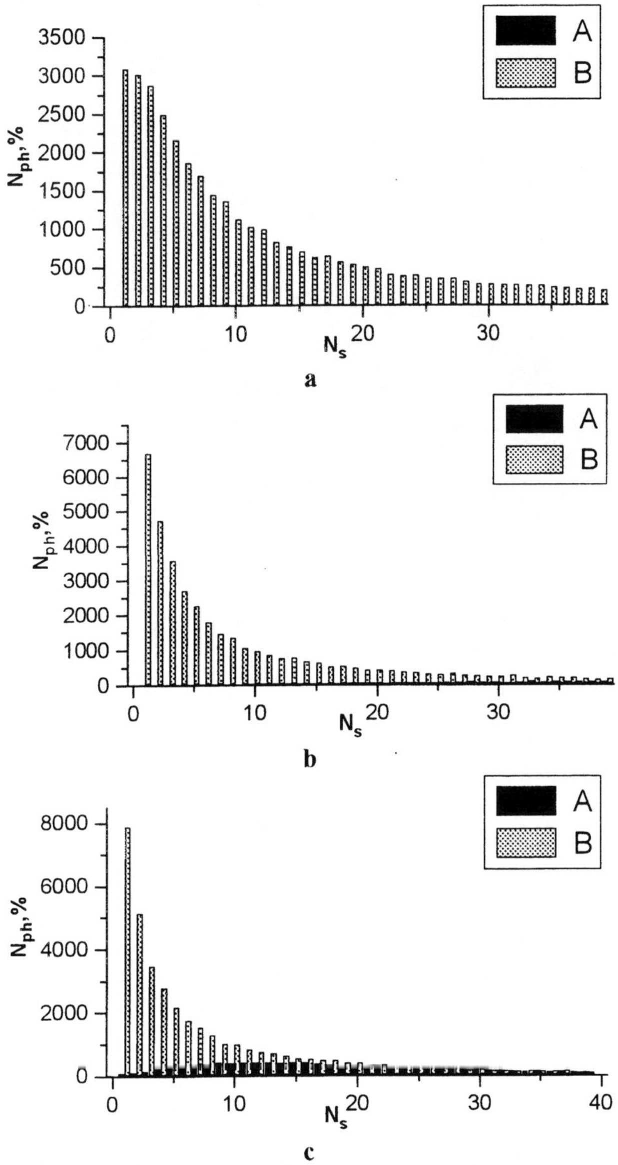

Fig.3. Calculated distributions of the number N_{ph} of forward (A) and back scattered (B) photons via number of collisions N_s (two integrating sphere geometry) for slightly matched refractive indices of the human scleral components (n_C = 1.47, n_I = 1.37): a - λ = 400 nm; b - λ = 600 nm; c - λ = 800 nm.

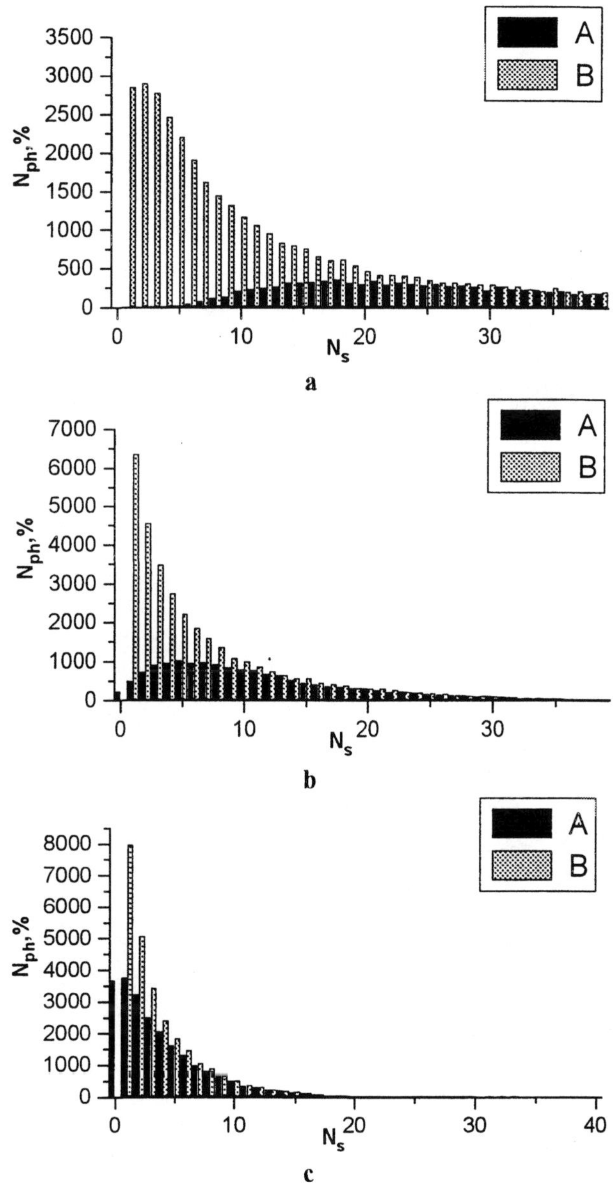

Fig.4. Calculated distributions of the number N_{ph} of forward (A) and back scattered (B) photons via number of collisions N_s (two integrating sphere geometry) for partly matched (mid level) refractive indices of the human scleral components (n_C = 1.47, n_I = 1.41): a - λ = 400 nm; b - λ = 600 nm; c - λ = 800 nm.

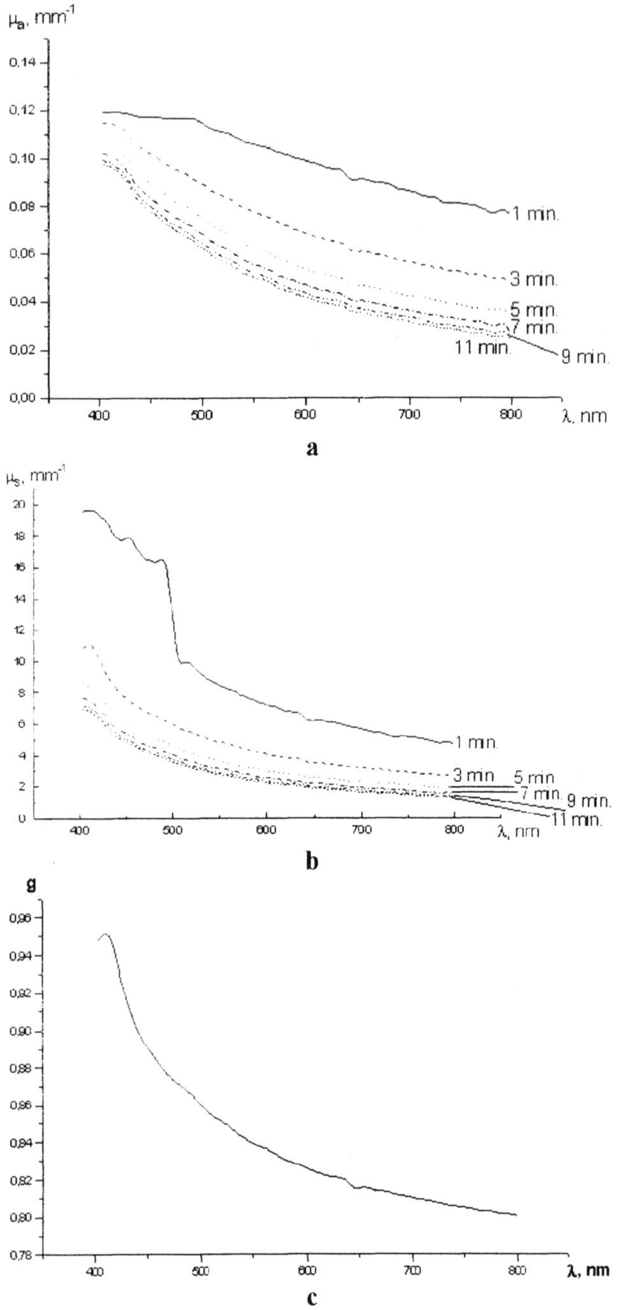

Fig.5. Spectra of optical parameters of the human sclera calculated using experimental spectra presented in Fig.1.

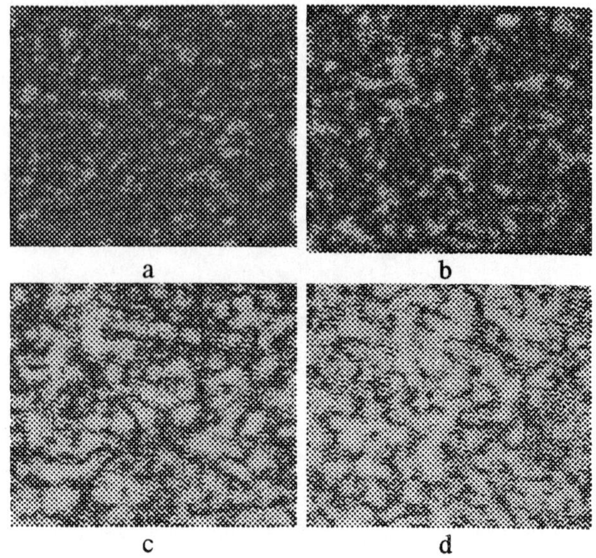

Fig.6. Fragments of the images of speckle patterns induced by He-Ne laser beam scattering by translusent human sclera. Time elapsed after *trazograph* administration: a - 1 min; b - 2 min; c - 4 min; d - 5 min.

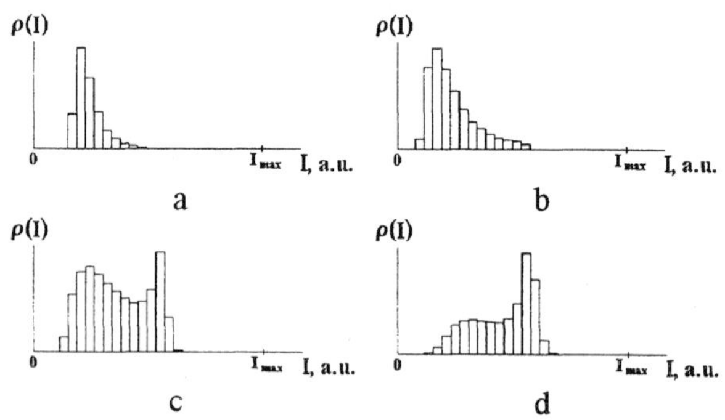

Fig.7. Evolution of empirical probability density functions of the speckle intensity with the progress of translucence of human sclera; time elapsed after *trazograph* administration: a - 1 min; b - 2 min; c - 3min; d - 4 min; e - 5 min.

References

1. P. O. Rol, *Optics for Transscleral Laser Applications,* Dissertation for the degree of Doctor of Natural Sciences, Inst. Biomedical Engineering, Zurich, Switzerland (1992).
2. S. Maier, S. A. Walker, S. Fantini, M. A. Franceschini, E. Gratton, "Possible correlation between blood glucose concentration and the reduced scattering coefficient of tissues in the near infrared", *Opt. Lett.* **19**, 2062-2064 (1994)
3. Kohl, M. Cope, M. Essenpreis, D. Bocker, "Influence of glucose concentration on light scattering in tissue-simulating phantoms", *Opt. Lett.* **19**, 2170-2172 (1994).
4. V.V.Tuchin, I.L.Maksimova, D.A.Zimnykov et al., "Light propagation in tissues with controlled optical properties", *J. Biomedical Optics* **2**(4),401-417 (1997).
5. A. Zimnykov, V. V. Tuchin, A. A. Mishin, "Spatial tissue structure monitoring", *Applied Optics* **36**, 5594-5607 (1997).

Laser Backscattering Polarization Patterns from Turbid Media: Theory and Experiment

George W. Kattawar and Milun J. Raković
Department of Physics, Texas A&M University, College Station, Texas 77843-4242

Brent D. Cameron
Biomedical Engineering Program, Texas A&M University, College Station, Texas 77843-3120

Abstract

We present both experimental measurements and Monte Carlo based simulations of the diffusely backscattered intensity patterns which arise from illuminating a turbid medium with a polarized laser beam. A new numerical method that allows us to calculate all 16 elements of the two-dimensional Mueller matrix is used. It is shown that only seven of these matrix elements are independent. To validate our method we have compared simulations to measurements from a turbid medium which consisted of $2.02\mu m$ - diameter polysterene spheres in deionized water. Two-dimensional images corresponding to each element of the diffusely backscattering Mueller matrix were obtained. The experimental and numerical results are in good agreement.

Key Words

260.5430 Polarization; 290.1350 Backscattering; 290.4210 Multiple scattering; 290.7050 Turbid media.

Introduction

A few recent studies have demonstrated that information on the properties of a turbid medium can be measured by shining a polarized laser beam onto a sample and then analyzing the state of polarization of the diffusely backscattered light. The investigated applications of this technique include the measurements of the average particle size, scattering coefficients and anisotropy factor of particle suspensions [1], cloud diagnostics [2,3], the study of biological materials [4-6], and the measurements of the average photon pathlength [7].

In order to achieve full experimental characterization of the optical properties of the sample under investigation, Hielcsher et al [6] used a Stokes vector/Mueller matrix approach to polarized light scattering. They generalized the concept of an effective Mueller matrix [8] and measured the two-dimensional Mueller matrix of backscattered light. In recent theoretical work, Ambirajan and Look [9] used a Monte Carlo technique in studying the backscattering of a polarized laser beam from a plane-parallel medium. They studied the degree of polarization of the diffuse light when the incident beam was right circularly polarized.

In this study we present a numerical method that allows us to calculate all 16 elements of the two-dimensional Mueller matrix. We show that there are only seven independent elements in this matrix and that the remaining elements are obtained from these by simple rotations or equality. To validate our method, we compare numerical simulations to measurements on polysterene sphere suspensions.

Mueller matrix/Stokes vector formulation

Let $\mathbf{k}$ be the wave vector which defines the propagation of the plane monochromatic light (electromagnetic) wave. The plane wave is completely described by the two complex transverse electric field components. To define them, let us choose as a plane of reference any plane containing the vector $\mathbf{k}$ and then by $\hat{\mathbf{n}}$ denote the unit vector perpendicilar to this plane and by $\hat{\mathbf{p}}$ the unit vector in this plane perpendicular to $\mathbf{k}$ so that $\hat{\mathbf{p}} \times \hat{\mathbf{n}} = \mathbf{k}/k$. The electric field is then $\mathbf{E} = Re[E_p\hat{\mathbf{p}} + E_n\hat{\mathbf{n}}]$, where the transverse components E_p and E_n are complex oscillatory functions.

Consider now the case of single scattering by a particle located at the origin (of the coordinate frame). In the far zone, the transverse components of the scattered light are [10]

$$\begin{pmatrix} E_p^s \\ E_n^s \end{pmatrix} = \frac{\exp(ikr - i\mathbf{k}\cdot\mathbf{r})}{ikr} \begin{pmatrix} S_2 & S_3 \\ S_4 & S_1 \end{pmatrix} \begin{pmatrix} E_p^i \\ E_n^i \end{pmatrix} \tag{1}$$

where the scattering amplitude matrix $\mathbf{S} = \begin{pmatrix} S_2 & S_3 \\ S_4 & S_1 \end{pmatrix}$ depends on the scattering angles θ, ϕ and the superscripts i and s correspond to incident and scattered beams respectively. In general each element of $\mathbf{S}$ is complex and therefore has both an amplitude and a phase; however, there are only a total of seven independent parameters since only the relative phase is important. In the case of spherical scatterers, the scattering process possesses axial symmetry with respect to direction of the incident wave and the matrix $\mathbf{S}$ depends only on the angle θ between the directions of the incident and scattered light. Moreover, it has only two nonzero elements, namely S_1 and S_2. Note that in the above equation it is assumed that the reference plane of both incident and scattered wave coincides with the scattering plane.

Instead of complex components and matrices it is convenient to use an equivalent description in terms of real variables. Four components of a real vector, known as a Stokes vector $\mathbf{I}^t = (I, Q, U, V)$, which completely describe the state of the radiation field are defined as follows

$$I = E_p E_p^* + E_n E_n^*, \quad Q = E_p E_p^* - E_n E_n^*,$$
$$U = E_p E_n^* + E_n E_p^*, \quad V = i(E_p E_n^* - E_n E_p^*), \tag{2}$$

where the asterisk denotes complex conjugation and the superscript t denotes transpose. Note that the first component is proportional to the radiance of the light and

$$I^2 = Q^2 + U^2 + V^2. \tag{3}$$

The single scattering process can now be described by a 4×4 real matrix, known as the Mueller matrix $\mathbf{M}$, which connects the Stokes vector of the incident and scattered wave:

$$\mathbf{I}^s = \mathbf{M}\mathbf{I}^i. \tag{4}$$

Of course, the 16 real elements of the Mueller matrix are completely determined by the four complex elements of the matrix $\mathbf{S}$ [10]. There are only seven independent elements for a single scattering Mueller matrix and nine relations which connect the others together. When Mueller matrices from an ensemble of particles differing in size, orientation , morphology, or optical properties are added incoherently, six of these eqaulities become inequalities [11].

In the simplest, and very important case, of light scattering by a homogeneous sphere, the Mueller matrix has been calculated by solving Maxwells equations with appropriate boundary conditions (Mie theory). In this case the Mueller matrix takes the relatively simple form [10]:

$$\mathbf{M}(\theta) = \begin{pmatrix} a(\theta) & b(\theta) & 0 & 0 \\ b(\theta) & a(\theta) & 0 & 0 \\ 0 & 0 & d(\theta) & -e(\theta) \\ 0 & 0 & e(\theta) & d(\theta) \end{pmatrix}, \tag{5}$$

where the four independent elements, a, b, d and e, expressed in terms of series of Bessel functions, depend on the scattering angle θ, the refractive index and size parameter of the scattering sphere.

Suppose now that we have to change the reference plane of a given Stokes vector $\mathbf{I}$. Let the first reference plane coincide with the second one after a counterclockwise rotation by the angle ϕ around the direction of the wave propagation. Then the two Stokes vectors are connected by the rotational Mueller matrix

$$\mathbf{R}(\phi) = \begin{pmatrix} 1 & 0 & 0 & 0 \\ 0 & \cos 2\phi & \sin 2\phi & 0 \\ 0 & -\sin 2\phi & \cos 2\phi & 0 \\ 0 & 0 & 0 & 1 \end{pmatrix}. \tag{6}$$

So far we have defined Stokes vectors for the case of a monochromatic plane wave, and the Mueller matrix for single scattering. These concepts are generalized to more complicated situations. The Stokes vectors are defined in cases of a quasimonochromatic wave [12], then in the case of partially polarized light, instead of Eq.(3) one has the inequality [10]

$$I^2 \geq Q^2 + U^2 + V^2. \tag{7}$$

In general, eg., in the case of a radiative (incoherent) light field in a highly scattering medium, the complete description of the light is given by specifying Stokes vectors for each direction of propagation at each point of the medium, so that the first component of the vector I coincides with the radiance (intensity) of the light [13].

Backscattering Mueller matrix and symmetry relations

We assume that the scattering of light is incoherent. The incoming narrow light beam propagates downward along the z-axis and is injected into an optically deep conservative scattering medium which is assumed to occupy the lower half-space, $z \leq 0$ (see Fig.1).

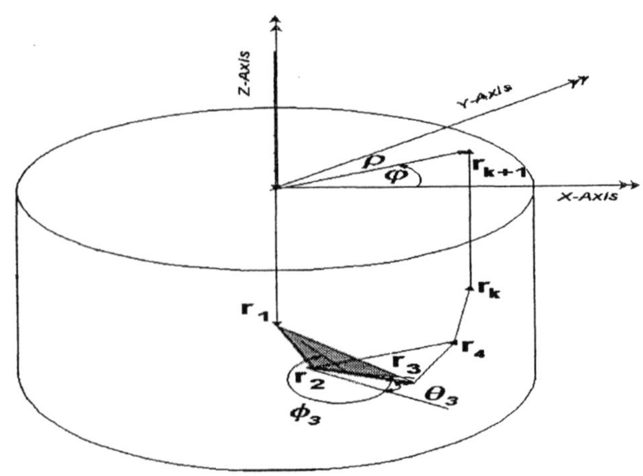

Figure.1 Geometry of the multiple scattering.

Let $\mathbf{F_0}$ be the Stokes vector corresponding to the irradiance of the incident laser beam with respect to the $x-z$ reference plane. We assume that the light crosses a small surface element ds_0, so that the vector describing the incident total light power is $\mathbf{P_0} = \mathbf{F_0}ds_0$. After being scattered by particle scatterers, light leaves the medium in all possible directions. Let $\mathbf{I}^{bs}(\rho,\phi)$ be the Stokes vector describing the radiance at the detector, i.e. at the point $(\rho,\phi,z=0)$ on the surface of the scattering medium, of the light which has been scattered exactly backwards. For the reference plane we choose again the $x-z$ plane. Then

$$\mathbf{I}^{bs}(\rho,\phi) = \mu_s^2\mathbf{S}(\rho,\phi;\mu_s,\mu_T)\mathbf{P_0} \tag{8}$$

where μ_s and μ_T are scattering and extinction coefficients respectively [13] and by definition $\mathbf{S}$ is the effective backscattering Mueller matrix. This matrix is obtained by solving a radiative transfer equation [13] with appropriate boundary conditions. Later we shall numerically calculate the matrix $\mathbf{S}$ using the Monte Carlo tecnique. Here we discuss important general relations satisfied by the elements of $\mathbf{S}$.

If the scattering medium is homogeneous, the system possesses axial symmetry with respect to the z-axis. To see how this affects the form of the backscattering matrix, assume as above that the incoming light beam propagates downward along the z-axis and that the backscattered light leaves the medium at the pont $(\rho,\phi,z=0)$. However, let the plane of reference be (instead of the fixed $x-z$ plane) the plane containing the incoming and backscattered light beam. Then

$$\mathbf{I'}^{bs}(\rho,\phi) = \mu_s^2\mathbf{S}'(\rho,\phi;\mu_s,\mu_T)\mathbf{P}'_0, \tag{9}$$

where primes correspond to a different reference plane. Now, axial symmetry of the system is expressed by the fact that the matrix $\mathbf{S}'$ does not depend on ϕ, i.e..

$$\mathbf{S}'(\rho,\phi;\mu_s,\mu_T) \equiv \tilde{\mathbf{S}}(\rho;\mu_s,\mu_T), \tag{10}$$

where $\tilde{\mathbf{S}}$ is the reduced backscattering matrix. From this it follows that the effective Mueller matrix takes the form

$$\mathbf{S}(\rho,\phi;\mu_s,\mu_T) = \mathbf{R}(-\phi)\tilde{\mathbf{S}}(\rho;\mu_s,\mu_T)\mathbf{R}(-\phi), \tag{11}$$

where $\mathbf{R}$ is given in Eq.(6). We see that the radial and azimuthal dependence of the backscattering matrix are factorized. Setting $\phi = 0$ in Eq.(11) one obtains

$$\tilde{\mathbf{S}}(\rho;\mu_s,\mu_T) = \mathbf{S}(\rho,\phi=0;\mu_s,\mu_T). \tag{12}$$

Furthermore, $\tilde{\mathbf{S}}$ can be represented as a sum over all orders of the multiply scattered light, namely,

$$\tilde{\mathbf{S}}(\rho;\mu_s,\mu_T) = \sum_{n=2}^{\infty}\tilde{\mathbf{S}}_n(\rho;\mu_s,\mu_T) \tag{13}$$

where $\tilde{\mathbf{S}}_n$ denotes the contribution from the light that has been scattered exactly n times by the scattering particles. It can be shown that $\tilde{\mathbf{S}}$ depends on ρ, μ_s and μ_T only through two scaled variables

$$\tilde{\mathbf{S}}(\rho;\mu_s,\mu_T) \equiv \tilde{\mathbf{S}}(\rho_s;\varpi), \tag{14}$$

where $\rho_s = \mu_T\rho$ and $\varpi = \mu_s/\mu_T$ is the single scattering albedo. In scaled variables the above expansion formula becomes

$$\tilde{\mathbf{S}}(\rho;\mu_s,\mu_T) = \sum_{n=2}^{\infty}\varpi^{n-2}\mathbf{L}_n(\rho_s) \tag{15}$$

Now we shall show that not all elements of the backscattering Mueller matrix, Eq.(8), are independent. First define the projection matrix $\mathbf{P}$

$$\mathbf{P} = diag(1,1,-1,1), \quad \mathbf{P}^2 = \mathbf{E}, \tag{16}$$

where $\mathbf{E}$ is the identity marix. Let us now assume that the single scattering Mueller matrix satisfies the relation

$$\mathbf{M}^t = \mathbf{PMP}. \tag{17}$$

This is true, e.g., for spherical scatterers, see Eq.(5). Then, it can be shown [14] that the same relation is satisfied by the corresponding reduced backscattering matrix, i.e.:

$$\tilde{\mathbf{S}}^t(\rho;\mu_s,\mu_T) = \mathbf{P}\tilde{\mathbf{S}}(\rho;\mu_s,\mu_T)\mathbf{P}. \tag{18}$$

The last equation defines the most general form that the matrix $\tilde{\mathbf{S}}$ can take:

$$\tilde{\mathbf{S}} = \begin{pmatrix} \tilde{S}_{11} & \tilde{S}_{12} & \tilde{S}_{13} & \tilde{S}_{14} \\ \tilde{S}_{12} & \tilde{S}_{22} & \tilde{S}_{23} & \tilde{S}_{24} \\ -\tilde{S}_{13} & -\tilde{S}_{23} & \tilde{S}_{33} & \tilde{S}_{34} \\ \tilde{S}_{14} & \tilde{S}_{24} & -\tilde{S}_{34} & \tilde{S}_{44} \end{pmatrix}. \tag{19}$$

Eqs.(11) and (19) now define the most general forms for the matrix elements of $\mathbf{S}$. From these general forms it follows that there are only seven independent elements, $S_{11}(\rho)$, $S_{12}(\rho,\phi)$, $S_{14}(\rho)$, $S_{22}(\rho,\phi)$, $S_{23}(\rho,\phi)$, $S_{24}(\rho,\phi)$ and $S_{44}(\rho)$, and the other nine,

$$S_{13}(\rho,\phi) = S_{12}(\rho,\phi+\pi/4), \quad S_{21}(\rho,\phi) = S_{12}(\rho,\phi),$$

$$S_{31}(\rho,\phi) = S_{12}(\rho,\phi-\pi/4), \quad S_{32}(\rho,\phi) = S_{23}(\rho,\phi\pm\pi/4),$$

$$S_{33}(\rho,\phi) = -S_{22}(\rho,\phi\pm\pi/4), \quad S_{34}(\rho,\phi) = S_{24}(\rho,\phi-\pi/4),$$

$$S_{41}(\rho,\phi) = S_{14}(\rho), \quad S_{42}(\rho,\phi) = S_{24}(\rho,\phi),$$

$$S_{43}(\rho,\phi) = S_{24}(\rho,\phi+\pi/4), \tag{20}$$

can be obtained by simple rotations at most.

We conclude this subsection by noting that the most general form of the single scattering Mueller matrix $\mathbf{M}$ for which Eqs.(19) and (20) hold is the same as the one for $\tilde{\mathbf{S}}$, Eq.(19), since the only condition imposed on $\mathbf{M}$ was

Eq.(17). (Of course the matrix elements M_{ij} are functions of the scattering angle θ.) This form coincides with the single scattering matrix for optically active spherical scatterers. In addition, it describes the effective single scattering matrix which corresponds to an ensemble of particles of any kind provided that for each particle there is a particle in the reciprocal position. It is easy to show that Eqs.(19) and (20) also hold when the medium which contains scattering particles is optically active. Finally, for the symmetry relation to hold, it is not necessary that the scattering medium fill the entire halfspace, it is sufficient for it to have rotational symmetry around the initial laser beam, e.g. it can be a cylinder of finite size whose axis coincides with the direction of the initial laser beam.

Monte Carlo calculations

We used the Monte Carlo method to calculate the effective backscattering Mueller matrix for the case of spherical scatterers. Essentially, we represent S, Eqs.(8) and (11), as the sum over a large number of randomly chosen "photon trajectories" which all start at the origin $\mathbf{r}_0 = 0$. Each trajectory is defined by the collection of points, $\mathbf{r}_i$, $i = 1, 2, ..., n$, where the light scatterings take place, and possible additional points of reflection from the surface of the medium.

The trajectories are generated in the following way (we describe only the trajectories without surface reflection, the generalization is straightforward). Since the light enters the medium propagating downward along the z-axis, the first scattering point $\mathbf{r}_1 = (0, 0, z_1 < 0)$ is determined by choosing the length $l_1 = |z_1|$. This, as well as all subsequent "free" path lengths $l_{i+1} = |\mathbf{r}_{i+1} - \mathbf{r}_i|$, $i = 1, 2, ...$, are chosen using a random number generator so that their probability distribution is $w(l) = \mu_T \exp(-\mu_T l)$. Suppose now that the points $\mathbf{r}_i$, $i = 1, 2, ..., k$ and l_{k+1} are already defined, the next point $\mathbf{r}_{k+1}$ is determined by choosing the unit vector $|\mathbf{r}_{k+1} - \mathbf{r}_k|/l_{k+1}$. This is defined by the scattering angle θ_k, which is the angle between vectors $\mathbf{r}_{k+1} - \mathbf{r}_k$ and $\mathbf{r}_k - \mathbf{r}_{k-1}$, and angle ϕ_k between vectors $(\mathbf{r}_{k+1} - \mathbf{r}_k) \times (\mathbf{r}_k - \mathbf{r}_{k-1})$ and $(\mathbf{r}_k - \mathbf{r}_{k-1}) \times (\mathbf{r}_{k-1} - \mathbf{r}_{k-2})$. Azimuth angles ϕ are chosen according to the uniform distribution $w(\phi) = 1/2\pi$, while the scattering angles θ are chosen according to the distribution defined by an element of the single scattering matrix, $w(\theta) = 2\pi a(\theta) \sin\theta$, see Eq.(5).

To each point $\mathbf{r}_i$ we associate a matrix $\mathbf{M}_i$ and weight w_i using the following procedure. To $\mathbf{r}_0$ we associate $\mathbf{M}_0 = \mathbf{1}$ and weight $w_0 = 1$. Suppose that $\mathbf{M}_i$, w_i $i = 1, 2, ..., k$ are already defined. Then, $\mathbf{M}_{k+1} = \mathbf{M}(\theta_k)\mathbf{R}(\phi_k)\mathbf{M}_k/a(\theta_k)$, $w_{k+1} = \varpi w_k$. If between $\mathbf{r}_k$ and $\mathbf{r}_{k+1}$ there is a point of reflection from the surface (boundary) of the medium, then the matrix $\mathbf{M}_{k+1}$ is in addition multiplied by the appropriate Fresnel matrix. The trajectory is not infinite, it is truncated when, either the product $(\mathbf{M}_k)_{11} w_k$ is below some predetermined cutoff value, or when the optical depth $\mu_T |z_k|$ is greater than a certain predetermined value.

Each triplet $\mathbf{r}_i, \mathbf{M}_i, w_i$ gives a contribution to the backscattering matrix S in the following way. Effectively, the resulting matrix S is given as a matrix function defined on a discrete set of points which define a certain grid on the $z = 0$ plane. Now, the ith triplet gives an additive contribution to the value of S associated with the grid point which is closest to the point x_i, y_i, the contribution being

$$w_i \exp(-\mu_T |z_i|)\mathbf{R}(\phi_i')\mathbf{M}(\theta_i')\mathbf{M}_i \tag{21}$$

where θ_i' is the angle between $\mathbf{r}_i - \mathbf{r}_{i-1}$ and the positive z-axis, while ϕ_i' is the angle between the plane $(\mathbf{r}_{i-1}, \mathbf{r}_i, \mathbf{r'}_i)$ and $x - z$ plane, where $\mathbf{r'}_i = (x_i, y_i, 0)$. We see that each trajectory effectively carries contributions of as many trajectories as there are scattering points along the trajectory.

The procedure described above uses the Monte Carlo method to calculate the matrix S, while we have seen that, due to Eq.(11), it is sufficient to calculate the matrix $\tilde{\mathbf{S}}$. From Eq.(11) it follows that, to calculate this matrix, it is sufficient to sandwich each contribution, Eq.(21), between the two matrices $\mathbf{R}(\phi_i)$ and to associate the result to the point ρ_i, where ρ_i, ϕ_i, z_i are the cylindrical coordinates of $\mathbf{r}_i$. We note that this last transformation improves our statistics enormously.

Comparison with experiment

A schematic view of the experimental apparatus used for collection of the diffuse backscattered images is shown in Fig.2.

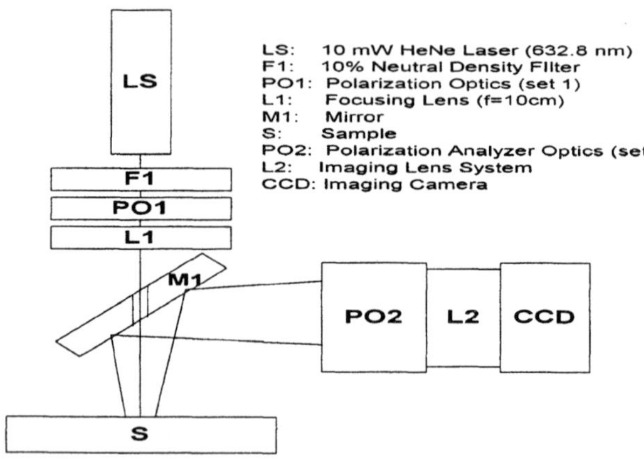

Figure 2. Schematic diagram of the experimental setup.

A collimated light beam emitted by a 10 mW HeNe laser at a wavelength of 632.8 nm passes through a 10% transmittance neutral density filter (FI). This beam is then polarized via various polarization optics (PO1) (i.e. linear and circular polarizers) to obtain the desired input polarization. This polarized beam is then focused with the lens (L1, $f = 10$ cm) through a hole ($d \approx 2$ mm) in a mirror (M1), mounted at 45 degrees, onto the sample. The diffusely backscattered light from the sample is then imaged through a polarization analyzer (PO2) using a cooled 12-bit CCD camera (SpectraSource, Westlake Village, CA). The polarization analyzer consists of a variety of optics which were interchanged in order to analyze a specific type of polarized light (i.e. vertical, horizontal, ±45° linear, and left, right circular polarized light) for a respective image used to reconstruct the Mueller matrix. The CCD array size is 336x243 pixels with a pixel size of 10x10 μm^2. Each image was collected using an exposure time of 1.7 seconds.

The phantom utilized in this study was comprised of a suspension of 2.02 μm polysterene spheres (Ernest F. Fullam, Latham, NY). The sample was created by diluting 0.075 ml of a 10 weight percent suspension of polysterene spheres with 15 ml of deionized water. The index of refraction of the spheres is 1.59 and the sample has a relative index of refraction n/n_w of 1.192 where n_w is the index of refraction for water. The sample has a scattering coefficient (μ_s) and anisotropy factor (g) of 11.88 cm^{-1} and 0.912, respectively at 632.8 nm, and has a negligible absorption coefficient $\mu \approx 0$ ($\varpi \approx 1$).

In Fig. 3 we show a comparison of the measurement of the Mueller matrix elements with the Monte Carlo calculations. To determine each of the 16 experimental Matrix elements, a total of 49 images are taken at various combinations of input and output analyzer polarization states. The approximate size of each image is 1.6cm × 1.6cm. Each of the 16 experimental elements is calculated by adding or subtracting a series of images. The individual images are represented by a two letter combination which denotes the input polarization and output analyzer orientation (i.e. HV would denote horizontal input polarized light and a vertical polarization analyzer). The corresponding symbols denoting polarization are: V vertical, H horizontal, P +45°, M -45°, R right, L left and O open or no polarization optics. For the Monte Carlo simulations, the average number of collisions per trajectory is 10. We see that, in the case of multiple scattering of the incident radiation from a suspension of polysterene spheres, the numerical results for intensity patterns of the elements of the two-dimensional backscatter Mueller matrix are in good agreement with the experimental measurements, especially the azimuthal dependence. The small discrepancy observed in elements S_{34}, S_{42} and S_{43} comes from the influence of the mirror M1 in the experimental setup which distorts these elements.

For the suspension used in our study, the transport mean free path was $mfp' = 1/(\mu_a+\mu_s(1-g)) = 0.967cm$. It appears that, for distances that exceed two transport mean free paths, the azimuthal dependence of the patterns becomes less pronounced because multiple scatterings tend to randomize the polarization state of the light. Although the laser beam in the experiment is coherent, the light scattering treated in the Monte Carlo simulations is incoherent. Because the correlation time of the laser-induced speckles is of the order of 10 ms, the experimental images with the exposure time of 1.7 seconds have the speckle effect averaged out. Therefore, the coherence effect is not important in our images.

Figure 3. Experimental and Monte Carlo backscattered Mueller matrix.

Acknowledgments

This research was partially supported by the Office of Naval Research under contract N00014-95-1-0275, the Office of Vice President for Research at Texas A&M University, and the Whitaker Foundation.

References

1. A.H. Hielscher, J.R. Mourant and I.J. Bigio, "Influence of particle size and concentration on the diffuse backscattering of polarized light from tissue phantoms and biological cell suspensions", Appl. Opt. **36**, 125-135 (1997).
2. A. I. Carswell and S. R. Pal, "Polarization anisotropy in lidar multiple scattering from clouds", Appl. Opt. **19**, 4123-4126 (1980).
3. S. R. Pal and A. I. Carswell, "Polarization anisotropy in lidar multiple scattering from atmospheric clouds", Appl. Opt. **24**, 3464-3471 (1985).
4. S.L. Jacques, M. Ostemeyer, L. Wang and D Stephens, "Polarized light transmission through skin using video reflectometry: toward optical tomography of superficial tissue layers", Proc. SPIE, **2671**, 199-220 (1996).
5. S.G. Demos and R.R. Alfano, "Optical polarization imaging", Appl. Opt. **36**, 150-155 (1997).
6. A.H. Hielscher, A.A. Eick, J.R. Mourant and I.J. Bigio, "Biomedical Diagnostic with Diffusely Backscattered Linearly and Circularly Polarized Light", Proc. SPIE, **2976**, 298-305 (1997).
7. M. Dogariu and T. Asakura, "Photon pathlength distribution from polarized backscattering in random media", Opt. Eng. **35**, 2234-2239 (1996).
8. W.S. Bickel and W.M. Bailey, "Stokes vectors, Mueller matrices, and polarized scattered light", Am. J. Phys. **53**, 468-478 (1985).
9. A. Ambirajan and D.C. Look, "A Backward Monte Carlo study of the multiple scattering of a polarized laser beam", J. Quant. Spectrosc. Radiat. Transfer **58**, 171-192 (1997).
10. H. C. van de Hulst, *Light scattering by small particles* (Dover, New York, 1981).
11. Edward S. Fry and George W. Kattawar, "Relationships between elements of the Stokes matrix", Appl. Opt. **20**, 2811-2814 (1981).
12. Max Born and Emil Wolf, *Principles of Optics* (Pergamon Press, Oxford, 1975).
13. S.Chandrasekhar, *Radiative transfer* (Oxford University Press, London, 1950).
14. Brent D. Cameron, M. J. Raković, Mehrube Mehrubeoglu, George Kattawar, Sohi Rastegar, Lihong V. Wang and Gerard Coté, "Measurement and Calculation of the Two-Dimensional Backscattering Mueller Matrix of a Turbid Medium", Opt. Lett. **23**, 485-487 (1998)).

Relationship between tissue redox potentials and hemoglobin oxygenation based on fluorescence of NADH / FAD and absorption spectroscopy of oxy / deoxy hemoglobin

Nimmi Ramanujam, Yanping Li, Britton Chance
University of Pennsylvania, Johnson Research Foundation, Philadelphia, PA, 19104
Phone: (215) 898-4342, Fax: (215) 898-1806
Douglas Heintzelman, Rebecca Richards-Kortum
University of Texas, Biomedical Engineering, Austin, TX, 78712
Phone: (512) 471-2104, Fax: (512) 471-0616
Sharon Thomsen
MD Anderson Cancer Center, Dept. Pathology, Houston, TX, 77030
Phone: (713) 792-7962, Fax: (713) 792-3995
Anita Mahadevan-Jansen
Vanderbilt University, Biomedical Engineering, Nashville, TN, 37235
Phone: (615) 343-4787, Fax: (615) 343-7919

Neoplastic changes in tissues are reflected by biochemical (tissue metabolism, blood oxygenation, blood volume) and morphological changes. These changes can be monitored non invasively using fluorescence / reflectance spectroscopy in the ultraviolet / visible spectral regions. Biochemical interpretation of spectral data obtained in situ has been limited by the fact that it is modulated by both the absorption and scattering properties of the medium. A technique that can provide a biochemical assay of the spectra measured in situ could provide insight into the biochemical basis for the differences between neoplastic and non-neoplastic tissues measured spectroscopically in situ. As a first step towards addressing this issue, the goal of this study was to perform a biochemical assay and characterize the relationship of two parameters involved in the neoplastic process, metabolic activity and hemoglobin oxygenation from animal and human tissues using a low temperature redox (reduction-oxidation) scanner.

The low temperature redox scanner is a spectrometer that can be used to measure cellular metabolism and hemoglobin oxygenation at a spatial resolution of 50 μm from the surface of the metabolically arrested specimens at -196 °C. In particular, it measures the fluorescence of metabolic components, pyridine nucleotides (Pn) (340 nm excitation, 450 nm emission) and flavoproteins (Fp) (460 nm excitation, 540 nm emission) which reflect tissue metabolism. The ratio of the fluorescence: Fp/(Fp+Pn) provides the redox potential; an increase in the redox ratio relates to a decrease in metabolic activity. Hemoglobin oxygenation is obtained by measuring the diffuse reflectance at two wavelengths straddling the isosbestic point, 557 and 578 nm. The ratio of the difference and sum of the reflectance at the two wavelengths provides a relative measure of hemoglobin saturation; an increase in this ratio corresponds to an increase in hemoglobin oxygenation independent of changes in blood volume.

To establish the sensitivity and dynamic range of the redox potential and hemoglobin oxygenation measurements, tissue samples from the rat liver, in various metabolic states were studied first. The animals were segregated into three groups: the first group inhaled carbogen to induce hyperoxia, the second or normoxic group inhaled air and the third group underwent nitrogen inhalation to induce hypoxia. The redox potential and hemoglobin oxygenation were measured from metabolically arrested liver samples from each animal in all three groups and characterized. In the second part of the study, redox potentials and hemoglobin oxygenation were measured from cross-sections of metabolically arrested human cervical biopsies in various histological states (normal, inflammation, dysplasia). The relationship between the redox potential and hemoglobin oxygenation were compared for the histologically different tissues.

Figure 1 displays the relationship between the redox potential and hemoglobin oxygenation for the rat liver acini in various metabolic states. The hyperoxic and normoxic liver samples display a redox potential which is independent of the hemoglobin saturation. This can be explained by the fact that under normal conditions, the oxygen tension of the mitochondria is atleast an order of magnitude less than that in the vascular bed, and the supply of oxygen in the blood far exceeds the cellular demands. However, when hypoxia is induced, there is a decrease in the hemoglobin saturation as expected. Furthermore, the redox potential is now correlated to hemoglobin saturation displaying a gradient coherence. This is reflective of the fact that under oxygen deprivation, the cellular metabolic demands become equivalent to or begin to exceed the supply of oxygen in the vascular bed. This sigmoidal curve demonstrates the oxygen gradients in tissue for hemoglobin saturations that span the hyperoxic to hypoxic range.

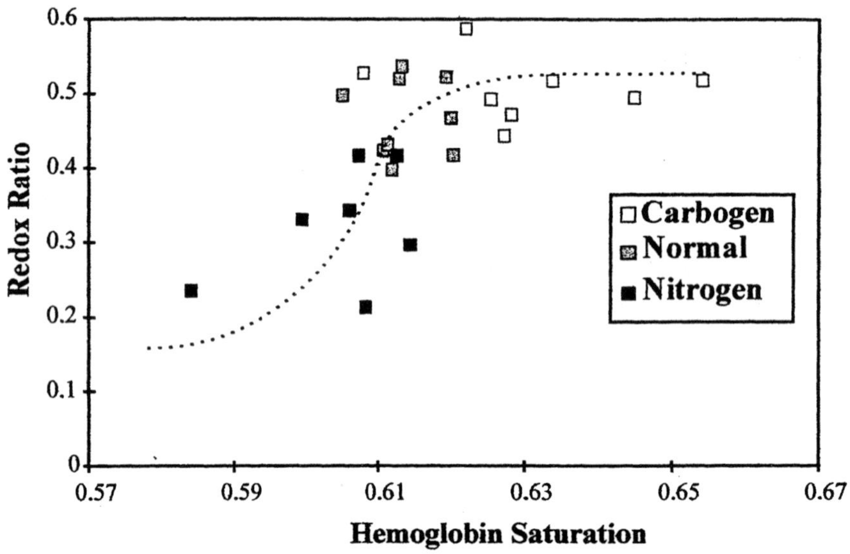

Figure 1. Oxygen gradients in the rat liver acini

Figure 2 displays the oxygen gradient for human cervical biopsies in various histological states. Normal cervical tissues display an oxygen gradient curve similar to that obtained from the liver samples in various metabolic states. An evaluation of the samples with inflammation indicate that the hemoglobin saturation range is similar to that for normal tissues, but the metabolic rate is increased as reflected by the lower redox potential. Finally, the dysplastic tissues, like the inflammatory tissues display a similar hemoglobin saturation range as normal tissues; however, a strong correlation is observed between the redox potential and hemoglobin saturation indicating a strong gradient coherence which reflects the increased metabolic demand of these tissues relative to their normal counterparts.

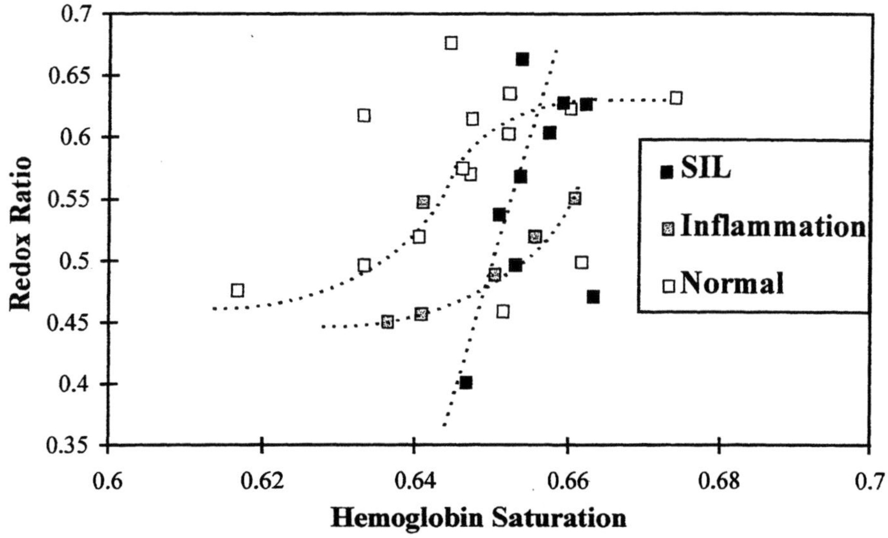

Figure 2. Oxygen gradients in human cervical tissue.

In summary, the relationship between tissue metabolism and hemoglobin oxygenation (oxygen gradient) provides a direct indication of oxygen gradients in tissue. The data from the liver study indicate that when the rate limiting parameter is hemoglobin oxygenation, there is a sigmoidal relationship between tissue metabolsim and hemoglobin oxygenation. However, when the metabolic demands of tissue increase as would be the case with neoplastic growth, there is a strong gradient coherence between redox potential and hemoglobin oxygenation even under normoxic hemoglobin saturations. Finally, these results indicate that it may be important to probe tissue metabolism and hemoglobin oxygenation simultaneously in situ in order to obtain greater selectivity and specificity of tissue characterization in situ.

The Characterization of a Two Wavelength Time-resolved System for the Determination of Hemoglobin Saturation through Comparison with Continuous Wave and Frequency Domain Systems

Robert J. Hunter, Michael S. Patterson, Robert A. Weersink, Jody T. Bruulsema and Joseph E. Hayward

Hamilton Regional Cancer Centre and McMaster University, 699 Concession Street, Hamilton, Ontario L8V 5C2 Canada

Abstract

The absorption coefficient of hemoglobin is dependent upon its oxygenation status and on the wavelength of interest. The known absorption differences can be used to monitor noninvasively hemoglobin saturation, which is indicative of the oxygenation state of the tissue volume being probed. Time-resolved measurements combined with diffusion theory provide a means of quantifying the optical properties of a turbid medium, like tissue. To characterize the ability of our time domain system to measure optical properties, its performance was compared to a continuous wave and a frequency domain system in our laboratory. The response to changes in either optical property was determined. Following these investigations, preliminary *in vivo* measurements were performed. The time domain system proved sensitive enough to detect induced oxygenation changes in the hand of a volunteer.

Key Words: (170.6510) Spectroscopy, tissue diagnostics, (170.1470) Blood/tissue constituent monitoring and (170.3660) Light propagation in tissue

Introduction

Due to an increased interest in optical techniques for use in both therapeutic and diagnostic medicine, the measurement of the optical properties of tissue has become an important field of research. Since it is desirable to determine these properties noninvasively, it is necessary to use the information retained in the light that is scattered out of the tissue to estimate values for the scattering and absorption coefficients of the tissue. It is possible to use these properties to study many physiological functions, including hemoglobin saturation and tissue metabolism [1, 2].

Methods for the indirect measurement of tissue optical properties fall into three categories: steady-state, frequency domain and time domain techniques. Regardless of which technique is used to measure the reflectance, a suitable model of light transport is needed to provide the best estimates of the optical properties of the medium investigated. The diffusion equation has been shown to have good agreement with both phantom measurements and Monte Carlo simulations [3, 4].

We have designed a two wavelength system which operates in time-correlated single photon counting mode to measure noninvasively the optical properties of tissue. By exploiting the known differences in absorption between oxyhemoglobin and deoxyhemoglobin at 750 nm and 810 nm, we hope to use this system to monitor hemoglobin saturation in a method similar to that described by Chance *et al.* [1]. The main differences between our system and existing systems is the use of pulsed diode lasers in combination with an inexpensive miniature photomultiplier tube. However, prior to performing oxygenation measurements, the performance of the system was compared to a continuous wave (cw) and a frequency domain system in our lab. Following investigation of the system's ability to predict variations in optical properties, *in vivo* measurements were undertaken. This report will briefly outline the instrumentation of the time-resolved system, the results of a series of phantom measurements performed on all three systems, and preliminary *in vivo* results.

Experimental

The experimental set up for both the frequency domain and the steady-state systems are described in detail elsewhere [5]. The steady-state technique gave results over a range of wavelengths but the wavelength of interest for this study was 810 nm which corresponds to one of the wavelengths used by the two other systems. The dual wavelength apparatus used in the time domain system is shown in Figure 1; the phantom experiments described here were

Reprinted from OSA TOPS Vol. 22, Biomedical Optical Spectroscopy and Diagnostics / Therapeutic Laser Applications, Eva Sevick-Muraca, Joseph A. Izatt, and Marwood N. Ediger (eds.)

performed at only one wavelength. The outputs from two pulsed diode lasers (PLP-01, LDH082 and LDH075 Hamamatsu Photonics K.K.) were coupled together with a 400 μm fiber optic combiner (Fibersense and Signals, Concord, ON). The signal was then split into three fibers, the first fiber received most of the power and delivered light to the sample, the second fiber was used as a reference pulse to account for instrumental time drifts and the third portion was routed to a fast photodiode to provide a stop pulse for the time-to-amplitude converter (TAC). This optical trigger was found to provide better stability than the electronic trigger provided by the laser drivers. The remainder of the set up is typical of a single photon counting experiment and has been described previously [6]. The principle difference is the choice of detector; the detector in this case is an inexpensive miniature photomultiplier tube (Metal Package PMT R5600U Series, Hamamatsu Photonics K.K.). Although this detector does not have the time resolution of more expensive microchannel plate PMTs, the FWHM is about 250 ps, it has proven to be adequate for our purposes.

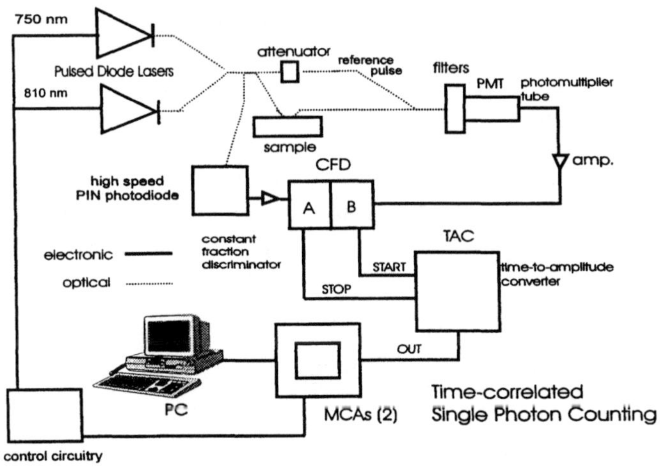

Figure 1: Experimental apparatus for the time-correlated single photon counting system: the dotted lines represent optical signals, the solid lines show electronic signals.

Theory

The reflectance data were fit to diffusion theory using a non-linear least squares fitting routine. The instrument response was measured and accounted for in the fitting procedure. Three parameters were varied in the fits, μ_a, μ_s' and a scaling factor. The diffusion model used the modified extrapolated boundary condition described by Haskell *et al.* [7]. The integral for the time-resolved reflectance involves the fluence rate, $\Phi(\rho,z,t)$ and the flux, $R_f(\rho,t)$ which are given by Kienle[4] as:

$$\Phi(\rho,z,t)=\frac{c}{(4\pi Dct)^{3/2}}\exp(-\mu_a ct)\times \left\{\exp\left[-\frac{(z-z_0)^2+\rho^2}{4Dct}\right]-\exp\left[-\frac{(z+z_0+2z_b)^2+\rho^2}{4Dct}\right]\right\} \tag{1}$$

and

$$R_f(\rho,t)=\frac{t^{-5/2}}{2(4\pi\ Dc)^{3/2}}\exp(-\mu_a ct)\times \left\{z_0\exp\left[-\frac{r_1^2}{4Dct}\right]+(z_0+2z_b)\exp\left[-\frac{r_2^2}{4Dct}\right]\right\} \tag{2}$$

where $D = 1/[3(\mu_a + \mu_s')]$ is the diffusion constant, $z_o = 1/(\mu_a+\mu_s')$, $r_1^2 = z_o^2 + \rho^2$, $r_2^2 = (z_o + 2z_b)^2 + \rho^2$ and the value of z_b is given by equation 3:

$$z_b=\frac{1+R_{eff}}{1-R_{eff}}2D \tag{3}$$

where R_{eff} is the fraction of photons that are internally diffusely reflected at the boundary [4,7].
The time-resolved reflectance is given by equation 4:

$$R(\rho,t)=\frac{1}{4\pi}\int_{2\pi}\left[\Phi(\rho,z,t)+3D\frac{\delta\Phi(\rho,z,t)}{\delta z}\cos\theta\right]\times[1-R_{fres}(\theta)]\cos\theta\, d\Omega \tag{4}$$

where $R_{fres}(\theta)$ is the Fresnel reflection coefficient. It was assumed that the boundary conditions were matched i.e. n=1 and in this situation, equation (4) reduces to:

$$R(\rho,t)\ =\ 0.25\,\Phi(\rho,z=0,t)\ +0.5\,R_f(\rho,t) \tag{5}$$

The cw method fit relative intensity to the corresponding steady-state equations for the range of 1 to10 mm, while the frequency domain approach fit relative phase and intensity from 5 to 21 mm to the appropriate diffusion theory expressions.

Results and Discussion

Phantom Measurements

Two experiments were performed in which either the absorption or the scattering coefficients were varied. The use of polystyrene microspheres of known diameter suspended in water and Mie theory allowed for the theoretical calculation of the scattering coefficient. The absorption coefficient and the index of refraction were assumed to correspond to water at 810 nm. Spheres of two different sizes, 0.653 μm and 1.53 μm, were used in varying concentrations to provide a range of scattering coefficients. Time domain measurements were performed at a source detector separation of 10 mm only. Since there did not appear to be significant differences between the results obtained with the two sizes of spheres, only results from the larger diameter spheres are shown in Figure 2 below. Figure 2a shows the measured scattering versus expected scattering coefficient. All three systems produced linear increases in the scattering coefficients estimated. The time domain results overestimated the scattering by about 25%, the steady-state system underestimated the value by about 15% , while the frequency domain results appear to be in good agreement with theory, within 5%. Investigations in fitting Monte Carlo data indicated that some errors are to be expected in the optical properties due to limitations of the diffusion model at a source detector separation of 10 mm. However, the errors should be less than 10% in μ_s'. Although the sources of the discrepancies between the three methods and from the theoretical predictions are not clear, many factors which could influence the results have been investigated. Remaining possibilities include: inaccuracies in diffusion theory, directionality of source and detector, mismatches in the boundaries or the presence of a scattering time-delay [8].

In this experiment, the absorption coefficient remained constant at 0.002 mm^{-1}, the absorption of water at 810 nm. The measured absorption coefficients from the three measurement methods are shown in Figure 2b. The frequency domain method did not recover the absorption coefficient successfully at higher values of μ_s'. The other two techniques did not provide accurate estimates for μ_a when the scattering was low. However, in the time domain measurements, for the last five cases, the accuracy of μ_a increased.

The second experiment used polystyrene spheres to provide a known scattering coefficient but indocyanine green (ICG) was added to the solution to vary the absorption coefficient. Absorbance measurements were performed on dilute concentrations of ICG in order to obtain the relationship between the concentration and absorption coefficient. The results from this investigation are displayed in Figure 3. Due to large errors in the previous experiment at small source detector separations, measurements were performed from 10 to 40 mm, in steps of 10 mm. The measured absorption coefficients are shown in Figure 3a. Frequency domain and time domain measurements performed at 10 mm both deviate from the expected results. Results for larger ρ values for the time-resolved system and the cw system show better agreement with the expected results. Only 30 mm results in the time domain are shown, but similar accuracy was obtained at 20 and 40 mm.

The results for estimating the scattering are shown in Figure 3b. The scattering coefficient was held constant in this experiment at 1.093 mm^{-1}. Spatial and frequency domain estimates of μ_s' were similar to those obtained in the previous experiment. The time domain results at 10 mm again overestimated the expected value by about 25%. Improvements were observed at 20 mm, where the error was about 20% and at 30 and 40 mm where the errors in μ_s' were less than 10%. Again, only the 10 and 30 mm time domain results are displayed. Under optimal conditions

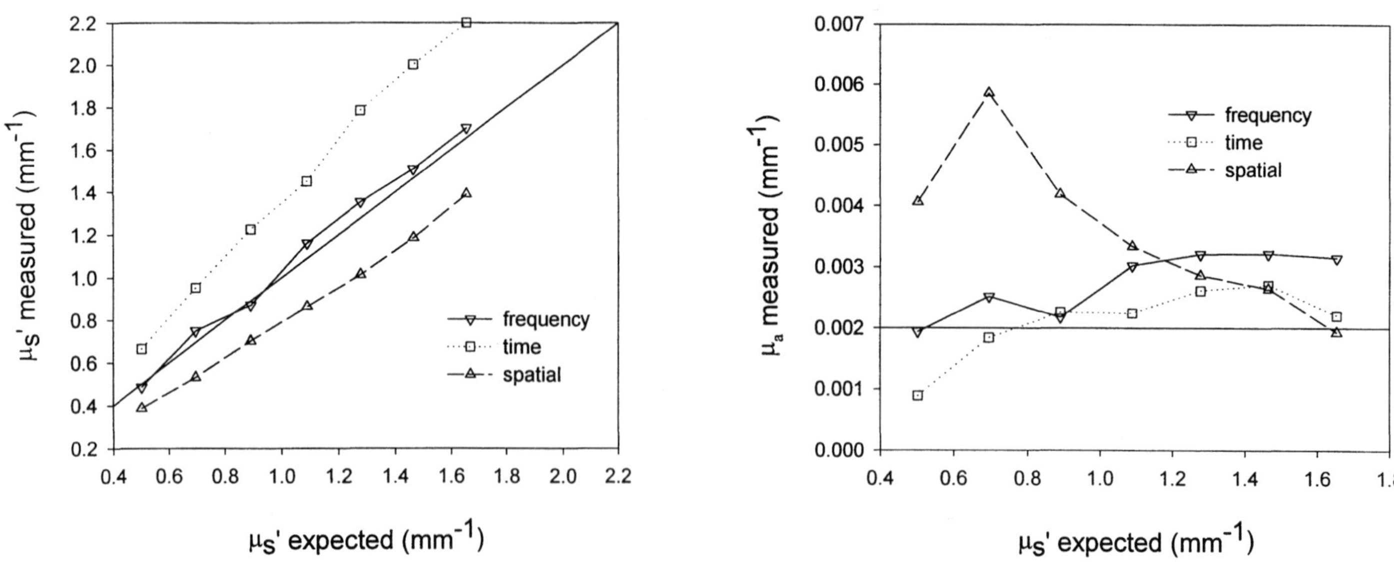

Figure 2: (a) Measured μ_s' values plotted against expected μ_s' values for polystyrene spheres of diameter 1.53 μm in water at 810 nm.

(b) Measured μ_a values plotted against expected μ_s' values for polystyrene spheres at 810 nm. Time domain results are for ρ = 10 mm.

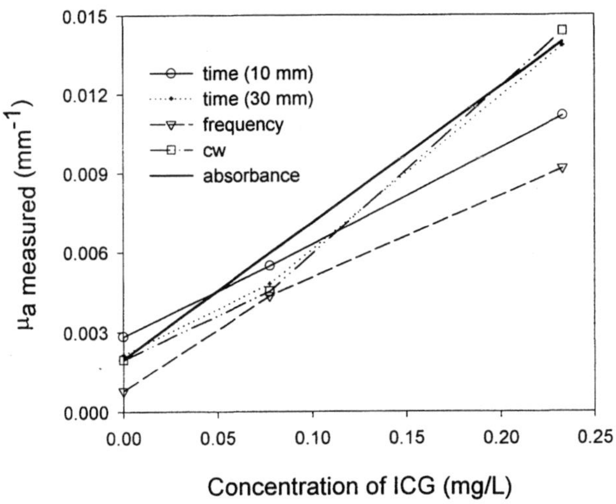

Figure 3: (a) Measured absorption coefficient versus concentration of ICG in polystyrene and water at 810 nm.

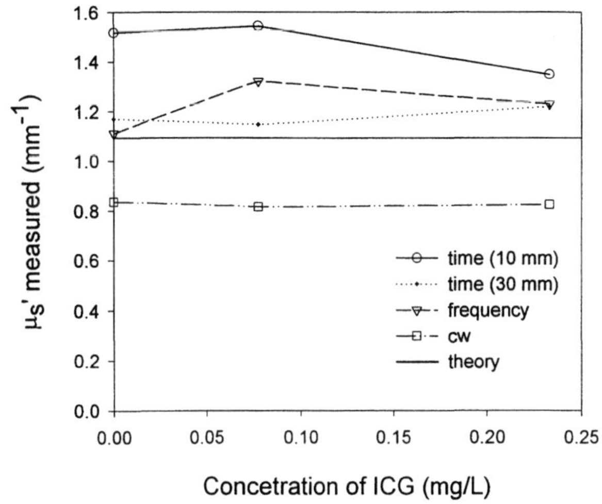

(b) Measured scattering coefficient versus concentration of ICG. Theoretical μ_s' was 1.093 mm^{-1}.

for ρ values greater than 20 mm and an absorption coefficient greater than pure water, errors in μ_s' were about 10% and μ_a errors were about 2%. Since the system had been shown to estimate optical properties with acceptable accuracy, especially the absorption coefficient, *in vivo* measurements were undertaken.

Measurements Performed *in vivo*

To determine if the system was sensitive enough to monitor induced changes in oxygenation, measurements were performed on the palm of the hand of a volunteer. From the absorption coefficients estimated at two wavelengths, it is possible to obtain the hemoglobin saturation of the tissue volume probed. The hemoglobin saturation is the fraction of hemoglobin molecules bound to oxygen and is indicative of the tissue oxygenation. The concentrations of oxy- and deoxyhemoglobin can be calculated from the following equations:

$$\mu_a(750)=\varepsilon_{Hb}(750)[Hb]+\varepsilon_{HbO2}(750)[HbO_2]+\mu_{a(back,750)} \quad (6a)$$

$$\mu_a(810)=\varepsilon_{Hb}(810)[Hb]+\varepsilon_{HbO2}(810)[HbO_2]+\mu_{a(back,810)} \quad (6b)$$

where $\varepsilon_i(x)$ is the specific absorption of molecule i at wavelength x and $\mu_{a(back,x)}$ is the background tissue absorption at wavelength x. The background absorption is estimated to be 70% the absorption of water at the wavelength of interest. Using this technique, the hemoglobin concentrations and the hemoglobin saturation were measured on the palm, under resting conditions. A blood pressure cuff at a pressure of 240 mm Hg was used to occlude blood flow to the hand. Three successive measurements were performed on the same location. The cuff was then released and a final measurement was made. A typical fit from the *in vivo* data is shown in Figure 4. This measurement was performed under normal conditions, at a wavelength of 750 nm and a source detector separation of 10 mm. There is good agreement over the entire fitting range. The χ^2 value is reasonable at 1.62 and the optical properties are similar to other results obtained in our laboratory. The results are summarized in Figure 5 which displays the temporal variation of four oxygenation parameters. The hemoglobin saturation decreased following occlusion as did the concentration of HbO_2, as expected. The deoxyhemoglobin concentration increased and the total concentration remained fairly constant, again as expected.

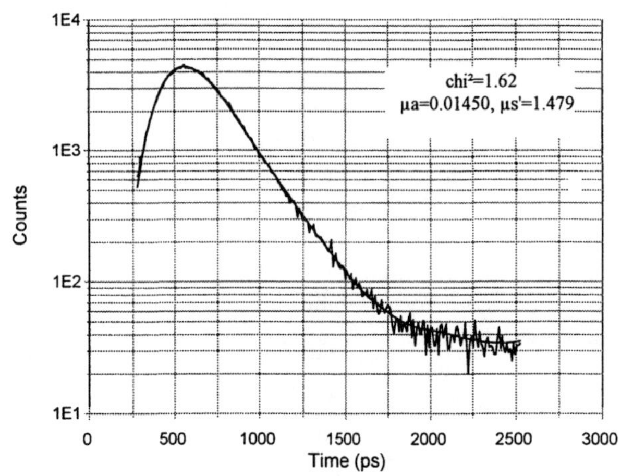

Figure 4: Data and fit from *in vivo* measurement on the palm of the hand at 750 nm and ρ=10mm.

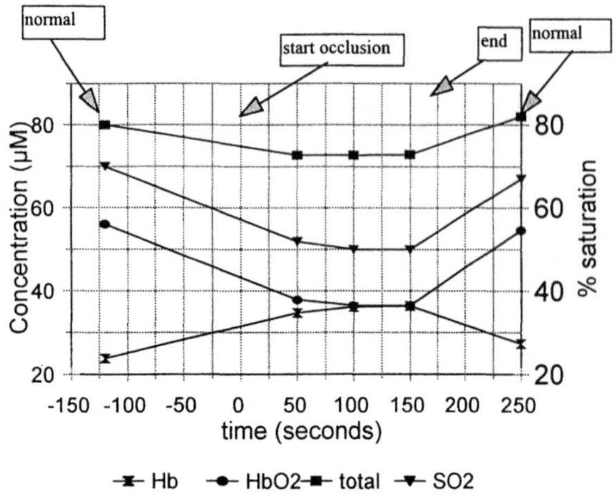

Figure 5: Variation of oxygenation parameters, as determined by time-resolved measurements, with time during occlusion experiment.

There is no gold standard to which these can be compared. However, the parameters changed as expected and values are similar to results from other researchers. For example, Fantini found that hemoglobin saturation decreased from 80% to about 45% following occlusion of blood flow in the arm [9], although large variations could be expected depending on the experimental protocol and the subject. Furthermore, the hemoglobin concentration of 80 μM is within the range of total hemoglobin concentrations encountered clinically, between 20 μM and 100 μM [10]. It is necessary to perform further measurements on a system in which the oxygenation status can be controlled, in order to quantify the accuracy of the system in monitoring hemoglobin saturation.

Conclusions

From fitting Monte Carlo data, it is clear that there will be errors associated with modeling the reflectance with diffusion theory. The large errors in μ_s' are due, in part, to the ρ value used; more accurate results are possible at larger ρ values. Another possible explanation is the presence of a scattering time delay not accounted for by diffusion theory. However, the accuracy with which the absorption coefficient was estimated by the time domain system was encouraging. Experiments performed *in vivo* to determine hemoglobin saturation showed that the system is sensitive enough to monitor induced oxygenation changes in tissue. Results obtained for various oxygenation parameters seem reasonable when compared to work done by other researchers. To perform a thorough test of the accuracy of the system, a phantom in which the oxygenation state can be monitored is needed. If both the optical properties and the oxygenation parameters of a phantom could be calculated or measured directly, it would be possible to explore the full potential of this system for the noninvasive monitoring of hemoglobin saturation.

Acknowledgements
This work was supported by Photonics Research Ontario.

References

1. B. Chance, J.S. Leigh, H. Miyake, D.S. Smith, S. Nioka, R. Greenfeld, M. Finander, K. Kaufmann, W. Levy, M. Young, H. Yoshioka and R. Boretsky, "Comparison of time-resolved and -unresolved measurements of deoxygenation in brain," Proc. Natl. Acad. Sci. **85**, 4971-4975 (1988).
2. M. Cope and D.T. Delpy, "System for long-term measurement of cerebral blood and tissue oxygenation on newborn infants by near infrared transillumination," Med. Biol. Eng. Comput. **26**, 289-294 (1988).
3. M.S. Patterson, B. Chance and B.C. Wilson, "Time-resolved reflectance and transmittance for the non-invasive measurement of tissue optical properties," Appl. Opt. **28**, 2331-2336 (1989).
4. A. Kienle and M.S. Patterson, "Improved solutions of the steady-state and time-resolved diffusion equations for reflectance from a semi-infinite turbid medium," J. Opt. Soc. Am. A **14**, 246-254 (1997).
5. M.S. Patterson, "Noninvasive Measurements of Tissue Optical Properties: Current Status and Future Prospects," Comments Mol. Cell. Biophys. **8**, 387-417 (1995).
6. S.J. Madsen, B.C. Wilson, M.S. Patterson, Y.D. Park, S.L. Jacques and Y. Hefetz, "Experimental tests of a simple diffusion model for the estimation of scattering 7and absorption coefficients of turbid media from time-resolved diffuse reflectance measurements," Appl. Opt. **31**, 3509-3517 (1992).
7. R.C. Haskell, L.O. Svaasand, T. Tsay, T. Feng, M.S. McAdams and B.J. Tromberg, "Boundary conditions for the diffusion equation in radiative transfer," J.Opt. Soc. Am. A **11**, 2727-2741 (1994).
8. I.V. Yaroslavsky, A.N. Yaroslavsky, V.V. Tuchin and H.-J. Schwarzmaier, "Effect of the scattering delay on time-dependent photon migration in turbid media," Appl. Opt. **36**, 6529-6538 (1997).
9. S. Fantini, M.A. Franceschini-Fantini, J.S. Maier, S.A. Walker, B. Barbieri and E. Gratton, "Frequency-domain multichannel optical detector for noninvasive tissue spectroscopy and oximetry," Opt. Eng. **34**, 32-42 (1995).
10. S.J. Matcher and C.E. Cooper, "Absolute quantification of deoxyhaemoglobin concentration in tissue near infrared spectroscopy," Phys. Med. Biol. **39**, 1295-1312 (1994).

Diffuse optical imaging: Theory, simulation, and experiment

The development of a 32-channel time-resolved optical tomography system

Florian E. W. Schmidt, Martin E. Fry, Jeremy C. Hebden and David T. Delpy

University College London,
Department of Medical Physics & Bioengineering,
11-20 Capper Street, London WC1E 6JA, UK

Phone: +44-171 209 6262, Fax: +44-171 209 6269
Email: florian@medphys.ucl.ac.uk; mefry @medphys.ucl.ac.uk;
jem@medphys.ucl.ac.uk; ddelpy@medphys.ucl.ac.uk;

1. Introduction

Preterm infants have an increased risk of cerebral haemorrhages or hypoxic-ischaemic injuries. If undetected these can lead to severe neurodevelopmental disorders, in particular motor disorders. Cerebral damage in the term infant is mostly a result of birth asphyxia, which frequently gives rise to learning disabilities. There is currently no device capable of continuously monitoring cerebral blood volume and oxygenation at the bedside. In common with a number of other research groups[1,2,3] it is our objective to develop such an instrument using near-infrared (NIR) laser light to image the neonatal brain.

Unlike x-rays, NIR photons do not travel straight paths through biological tissue but undergo multiple scattering events[4]. This severely degrades the spatial resolution and contrast achievable. One approach is to overcome these limitations by using *pulsed* laser light and extracting the image information not from the *intensity distribution* of light emerging from the tissue, but from the *temporal distribution* of individual photons exiting the tissue surface[5]. Statistical parameters, such as the mean photon flight time, are extracted from a finite set of these **T**emporal **P**oint **S**pread **F**unctions (TPSFs), each corresponding to a particular source/detector pair, and fed into a reconstruction algorithm to produce an image.

A multichannel time-resolved imaging system, MONSTIR (**M**ulti-Channel **O**pto-electronic **N**ear-infrared **S**ystem for **T**ime-resolved **I**mage **R**econstruction), is currently being developed at University College London (UCL) suitable for optical tomography of the neonatal brain and the human breast. Previously, we have discussed the theoretical considerations which led to our choice of design[6]. This paper describes the current state of the hardware employed in the imaging system.

2. Imaging system

A block diagram of the imaging system is shown in Figure 1. Light from a pulsed high-power picosecond laser source is switched *sequentially* into one of 32 fibres which are attached to the tissue surface (e.g. head of a neonate). The photons undergo multiple scattering events in the tissue which causes a broadening of the pulse to several nanoseconds. It is the task of the detector system to record the temporal distribution (the TPSF) of light exiting the tissue at certain positions around the circumference of the head with a temporal resolution of better than 100ps. This is accomplished by utilising 32 fully simplex ultrafast photon counting detectors.

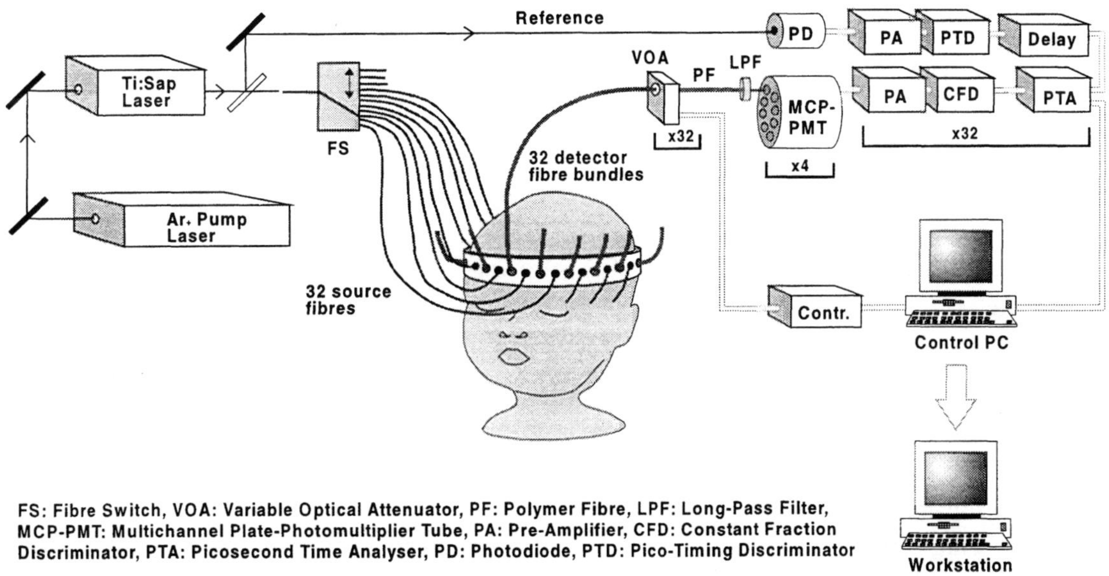

Figure 1 Schematic diagram of the MONSTIR imaging system.

Photons exiting the tissue surface are collected by 32 low-dispersion, large diameter (2.5 mm) fibre bundles which are coupled to 32 stepper motor driven **V**ariable **O**ptical **A**ttenuators (VOAs). Because of the large dynamic range of light intensities around the circumference of the head, the VOAs are required to ensure that the detectors are not saturated or damaged and that the system operates within the single photon counting regime. If more than one photon per duty cycle is incident on the detector, only the first photon would be recorded. Subsequent photons are lost due to the system dead time, which is equivalent to a distortion of the resulting histogram of photon flight times.

Light transmitted via the VOA is collected by a short 3.0 mm diameter single polymer fibre and relayed to the photocathodes of four ultrafast 8-anode Multichannel Plate-Photomultiplier Tube (MCP-PMT) detectors via a visible light blocking filter. The resulting electronic pulse is preamplified, converted into a logic pulse by a discriminator unit, and a histogram of photon flight times recorded and transferred to the control computer. The control software (written in Borland Delphi 2 under Windows95) will enable the system to be fully automated. Its functions include: instrumentation monitoring to ensure safe and optimal operation of the system, control of various parts of the hardware (including the fibre switch and the VOAs), transfer of acquired TPSFs from the local memory of the Picosecond Time Analyser units to the host PC, data processing and display.

A dedicated image reconstruction software package, TOAST (**T**ime-resolved **O**ptical **A**bsorption and **S**cattering **T**omography[7,8]), is used to reconstruct tomographic images of the absorption and scattering profiles.

3. Conclusion

The system was designed to acquire 32 channels of time-resolved data simultaneously, with sub-100ps temporal resolution, and at count rates of up to a few 100k photons per second per channel.

Preliminary tests have shown that the overall time resolution of the system is ~80ps (FWHM), which should provide a high resolution system for investigating and developing near-infrared imaging methodology.

Development of the system hardware and software is continuing, while phantom studies are progressing simultaneously. We envisage that during 1998 the system will be fully automated and ready for preliminary clinical studies at the UCL neonatal intensive care unit. The latest results will be presented at the conference.

References

1. Y. Hoshi and M. Tamura, "Dynamic multichannel near-infrared optical imaging of human brain activity", J. Appl. Physiol. **75,** 1842-1846, 1993
2. P. Van Houten, D. A. Benaron, S. Spilman and D. K. Stevenson, "Imaging Brain Injury Using Time-Resolved Near Infrared Light Scanning", Pediatric Research, **39**, 1996
3. Q. Luo, S. Nioka, B. Chance, "Functional near-infrared imager", SPIE Proc., **2979**, 84-93, 1997
4. M. S. Patterson, B. C. Wilson and D. R. Wyman, "The Propagation of Optical Radiation in Tissue I. Models of Radiation Transport and their Application", Lasers Med Sci **6**, 155-168, 1991
5. J. C. Hebden, S. R. Arridge and D. T. Delpy, "Optical imaging in medicine: I. Experimental techniques", Phys. Med. Biol. **42**, 825-840, 1997
6. K. Wells, J. C. Hebden, F. E. W. Schmidt and D. T. Delpy, "The UCL multichannel time-resolved system for optical tomography", Proc. SPIE **2979**, 599-607, 1997
7. S. R. Arridge, "Inverse Methods for Optical Tomography", in: Information Processing in Medical Imaging '93, edited by H. H. Barrett and A. F. Gmitro (Springer, Berlin, 1993), 259-277
8. TOAST Website: http://www.medphys.ucl.ac.uk/toast/index.htm

Detection of Objects Buried in Diffusive Media in the Presence of Interfaces. Contrast Effects on Measurement

J. Ripoll* and M. Nieto-Vesperinas

Instituto de Ciencia de Materiales de Madrid, Consejo Superior de Investigaciones Científicas, Campus de Cantoblanco, 28049 Madrid. Spain.

(April 1, 1998)

The surface integral formalism is used here to derive the integral equations for the scattering of diffusive waves, which are numerically solved without approximations. The Extinction Theorem for diffusive waves is here introduced to obtain the boundary values of both the wave photon density and the current photon density. We apply this theory to the simulation of diffusive objects buried in diffusive media in the presence of interfaces, and study the effect caused by the variation of the contrast between the two semi-infinite media.

I. INTRODUCTION

Recently there has been much interest on the scattering and propagation of diffusive waves, due to its potential applications in medical diagnosis (see for example refs. [1], [2] and references therein). Light traveling in turbid media in the strong scattering regime can be described by the diffusion approximation [3]. Such is the case of brain and tumor tissues illuminated with light of wavelength in the range of 650-1300nm (the so called *spectral window*, see ref. [2]). Several studies have been made to determine the optical properties of diffusive media both in the frequency and time domains, (see for example [4–7]), and for the detection and characterization of objects in diffusive media [8–13], but none of them, to our knowledge, solves exactly the propagation equations for multiply connected scattering domains taking into account the surface contributions and the presence of contrast across them.

In this paper we present a frequency-domain theoretical study of the diffusion equation in the case of multiply connected homogeneous diffusive media. We present the scattering integral equations for homogeneous diffusive media and the extinction theorem for scattered diffusive scalar waves. We deal with a configuration consisting of two diffusive media, separated by a plane interface, with a diffusive object immersed in one of them. This may represent a reasonable model to *in vivo* experiments for detecting, for example, hematomae in brain tissue covered by a diffusive solution with known scattering and absorption coefficients. In order to study what medium would be the most appropriate to place outside, we vary the contrast between both semi-infinite media. For the sake of computing time and memory, we illustrate numerical results in 2D configurations without loss of generality.

II. SCATTERING EQUATIONS FOR DIFFUSIVE WAVES

On taking the Boltzmann transport equation and considering a dense medium in which there is: quasi-isotropic angular photon flux, an isotropic photon source, and the variations in time of the photon current density $\mathbf{J}$ occur on a time scale much larger that the time between photon collision with the scattering particles of the medium, one obtains the so-called diffusion approximation for the total photon density $\phi(\mathbf{r},t)$ [3]:

$$\frac{\partial\phi(\mathbf{r},t)}{\partial t} + \nabla\cdot\mathbf{J}(\mathbf{r},t) + c\mu_a(\mathbf{r})\phi(\mathbf{r},t) = q_0(\mathbf{r},t). \qquad (1)$$

In Eq.(1) c is the speed of light in the medium, q_0 is the photon source term in photons per second, μ_a is the absorption coefficient, and $\mathbf{J}(\mathbf{r},t)$ is the photon current density, given by Fick's Law, $\mathbf{J}(\mathbf{r},t) = -cD(\mathbf{r})\nabla\phi(\mathbf{r},t)$, where $D(\mathbf{r})$ is the diffusion constant. In this study we will consider μ_a and D time independent, and c constant and equal throughout the media. Assuming that the photon source is modulated by a frequency ω, i.e. $q_0(\mathbf{r},t) = c\cdot S_0(\mathbf{r})\cdot[1+A\exp(-i\omega t)]$, where A is the modulation of the source, and omitting from now on the DC contribution ($\omega = 0$), we can express the diffusion equation in the frequency regime as:

$$\nabla^2\phi(\mathbf{r}) + \frac{\nabla D(\mathbf{r})}{D(\mathbf{r})}\cdot\nabla\phi(\mathbf{r}) + \kappa^2(\mathbf{r})\phi(\mathbf{r}) = -\frac{S_0(\mathbf{r})}{D(\mathbf{r})}, \qquad (2)$$

where $\kappa^2(\mathbf{r}) = -\frac{\mu_a(\mathbf{r})}{D(\mathbf{r})} + i\frac{\omega}{D(\mathbf{r})v}$. In Eq.(2), the product $\nabla D(\mathbf{r})\cdot\nabla\phi(\mathbf{r})$ arises from introducing the expression for the photon current density given by Fick's Law into Eq.(1). From now on we will place this term as a source term.

Let us consider a diffusive homogeneous medium of volume V defined by constant parameters κ_1, D_1 and μ_{a_1}, embedded in an isotropic and homogeneous diffusive medium of volume $\tilde{V}$ in which the parameters are κ_0, D_0 and μ_{a_0} constant. Defining by S the surface that separates both media, $\nabla D(\mathbf{r})$ has a non zero value only at the volume boundaries, i.e. at the surface S. It is more convenient to present the scattering integral equations as

*E mail: jripoll@everest.icmm.csic.es

OSA TOPS Vol. 21 Advances in Optical Imaging and Photon Migration, James G. Fujimoto and Michael S. Patterson (eds.)

a combination of two sets with their corresponding Green funcions: one defining the diffusive wave at the outside medium (ϕ^+) and the other defining the diffusive wave inside the object (ϕ^-).

- Outside Medium:

$$\nabla^2\phi^+(\mathbf{r}) + \kappa_0^2\phi^+(\mathbf{r}) = -\frac{S_0(\mathbf{r})}{D_0},$$
$$\nabla^2 G(\kappa_0|\mathbf{r}-\mathbf{r}'|) + \kappa_0^2 G(\kappa_0|\mathbf{r}-\mathbf{r}'|) = -4\pi\delta(\mathbf{r}-\mathbf{r}'). \quad (3)$$

- Inside Medium:

$$\nabla^2\phi^-(\mathbf{r}) + \kappa_1^2\phi^-(\mathbf{r}) = 0,$$
$$\nabla^2 G(\kappa_1|\mathbf{r}-\mathbf{r}'|) + \kappa_1^2 G(\kappa_1|\mathbf{r}-\mathbf{r}'|) = -4\pi\delta(\mathbf{r}-\mathbf{r}'). \quad (4)$$

These sets of equations must satisfy the saltus conditions:

$$\phi^+(\mathbf{r})|_{S(+)} = \phi^-(\mathbf{r})|_{S(-)}$$
$$D_0\left[\frac{\partial\phi^+(\mathbf{r})}{\partial\mathbf{n}}\right]_{S(+)} = D_1\left[\frac{\partial\phi^-(\mathbf{r})}{\partial\mathbf{n}}\right]_{S(-)}, \quad (5)$$

that correspond to photon density and photon density current continuity, respectively. In Eq.(5) $S^{(+,-)}$ represents surface S when approached from volumes $\tilde{V}$ and V, respectively. On applying Green's theorem, $\int_v (U\nabla^2 G - G\nabla^2 U)d^3r = \int_s (U\nabla_\mathbf{r} G - G\nabla_\mathbf{r} U)\cdot\mathbf{ds}$ to $\phi(\mathbf{r})$ and $G(\mathbf{r},\mathbf{r}')$ on Eqs.(3) and (4), considering the possible locations of $\mathbf{r}$ and $\mathbf{r}'$ either in V or $\tilde{V}$, these vectors being denoted by either $\mathbf{r}_>$ or $\mathbf{r}_<$ according to whether they belong to $\tilde{V}$ or V, one obtains the following integral scattering equations for diffusive media:

- $\mathbf{r}',\mathbf{r} \in \tilde{V}$

$$\phi^+(\mathbf{r}_>) = \phi^{(inc)}(\mathbf{r}_>) + \frac{1}{4\pi}\sum\nolimits_{S(+)}(\mathbf{r}_>) + \frac{1}{4\pi}\sum\nolimits^{(\infty)}(\mathbf{r}_>) \quad (6)$$

- $\mathbf{r}' \in \tilde{V}, \mathbf{r} \in V$

$$0 = \phi^{(inc)}(\mathbf{r}_<) + \frac{1}{4\pi}\sum\nolimits_{S(+)}(\mathbf{r}_<) + \frac{1}{4\pi}\sum\nolimits^{(\infty)}(\mathbf{r}_<) \quad (7)$$

- $\mathbf{r}',\mathbf{r} \in V$

$$\phi^-(\mathbf{r}_<) = -\frac{1}{4\pi}\sum\nolimits_{S(-)}(\mathbf{r}_<). \quad (8)$$

- $\mathbf{r}' \in V, \mathbf{r} \in \tilde{V}$

$$0 = -\frac{1}{4\pi}\sum\nolimits_{S(-)}(\mathbf{r}_>). \quad (9)$$

where $\partial/\partial\mathbf{n} = \hat{\mathbf{n}}(\mathbf{r})\cdot\nabla_\mathbf{r}$ is the normal derivative. In Eqs.(6)-(9) we have written the volume integral over the photon source, i.e. $\int_{\tilde{V}} G(\kappa_0|\mathbf{r}-\mathbf{r}'|)\frac{S_0(\mathbf{r}')}{D_0}\mathbf{d}^3\mathbf{r}'$, as $\phi^{(inc)}$, on placing the photon injection source in volume $\tilde{V}$. Notice that in Eqs.(8) and (9) the minus sign of the surface integral comes form the scalar product $\nabla\phi\cdot\hat{\mathbf{n}}$ which is negative when defining surface S from inside. If we look at Eq.(7) we will see the *Extinction Theorem for Diffusive Waves.* In Eqs.(6)-(9), the surface integrals $\sum_{S(+,-)}$ and $\sum^{(\infty)}$ are:

$$\sum\nolimits_{S(+)}(\mathbf{r}) = \int_S \left\{\phi^+(\mathbf{r}')\frac{\partial G(\kappa_0|\mathbf{r}-\mathbf{r}'|)}{\partial\mathbf{n}'} - G(\kappa_0|\mathbf{r}-\mathbf{r}'|)\frac{\partial\phi^+(\mathbf{r}')}{\partial\mathbf{n}'}\right\}dS',$$
$$\sum\nolimits_{S(-)}(\mathbf{r}) = \int_S \left\{\phi^-(\mathbf{r}')\frac{\partial G(\kappa_1|\mathbf{r}-\mathbf{r}'|)}{\partial\mathbf{n}'} - G(\kappa_1|\mathbf{r}-\mathbf{r}'|)\frac{\partial\phi^-(\mathbf{r}')}{\partial\mathbf{n}'}\right\}dS'$$
$$\sum\nolimits^{(\infty)}(\mathbf{r}) = \int_{S(\infty)} \left\{\phi^+(\mathbf{r}')\frac{\partial G(\kappa_0|\mathbf{r}-\mathbf{r}'|)}{\partial\mathbf{n}'} - G(\kappa_0|\mathbf{r}-\mathbf{r}'|)\frac{\partial\phi^+(\mathbf{r}')}{\partial\mathbf{n}'}\right\}dS'. \quad (10)$$

where $S^{(\infty)}$ is the surface of a sphere at infinite enclosing the infinite volume $\tilde{V}$. We assume from now on that $\mu_{a_0} > 0$, and therefore $\sum^{(\infty)} = 0$. Rearranging Eqs.(6)-(9), by using the relation between the normal derivatives at the surface as expressed in Eqs. (5), the integral equations for a constant homogeneous diffusive medium of volume V, delimited by surface S, embedded in a constant infinite homogeneous diffusive medium with a known density of photon sources, with the saltus conditions defined by Eq.(5), are:

- $\mathbf{r}',\mathbf{r} \in \tilde{V}$

$$\phi^+(\mathbf{r}_>) = \phi^{(inc)}(\mathbf{r}_>) + \frac{1}{4\pi}\int_{S(+)}\left\{\phi^+(\mathbf{r}')\frac{\partial G(\kappa_0|\mathbf{r}_> - \mathbf{r}'|)}{\partial\mathbf{n}'} - G(\kappa_0|\mathbf{r}_> - \mathbf{r}'|)\frac{\partial\phi^+(\mathbf{r}')}{\partial\mathbf{n}'}\right\}dS'. \quad (11)$$

- $\mathbf{r}' \in \tilde{V}, \mathbf{r} \in V$

$$0 = \phi^{(inc)}(\mathbf{r}_<) + \frac{1}{4\pi}\int_{S(+)}\left\{\phi^+(\mathbf{r}')\frac{\partial G(\kappa_0|\mathbf{r}_< - \mathbf{r}'|)}{\partial\mathbf{n}'} \quad G(\kappa_0|\mathbf{r}_< - \mathbf{r}'|)\frac{\partial\phi^+(\mathbf{r}')}{\partial\mathbf{n}'}\right\}dS'. \quad (12)$$

- $\mathbf{r}',\mathbf{r} \in V$

$$\phi^-(\mathbf{r}_<) = -\frac{1}{4\pi}\int_{S(-)}\left\{\phi^+(\mathbf{r}')\frac{\partial G(\kappa_1|\mathbf{r}_< - \mathbf{r}'|)}{\partial\mathbf{n}'} - \left[\frac{D_0}{D_1}\right] G(\kappa_1|\mathbf{r}_< - \mathbf{r}'|)\frac{\partial\phi^+(\mathbf{r}')}{\partial\mathbf{n}'}\right\}dS'. \quad (13)$$

• $\mathbf{r}' \in V, \mathbf{r} \in \tilde{V}$

$$0 = -\frac{1}{4\pi} \int_{S^{(-)}} \left\{ \phi^{+}(\mathbf{r}') \frac{\partial G(\kappa_1 |\mathbf{r}_{>} - \mathbf{r}'|)}{\partial \mathbf{n}'} - \left[\frac{D_0}{D_1} \right] G(\kappa_1 |\mathbf{r}_{>} - \mathbf{r}'|) \frac{\partial \phi^{+}(\mathbf{r}')}{\partial \mathbf{n}'} \right\} dS'. \quad (14)$$

As seen in Eqs. (11)-(14), there appears a $\frac{D_0}{D_1}$ factor inside the surface integral due to the contrast ΔD of the diffusive constant of both media which appears when inserting the saltus conditions expressed in Eq.(5). We wish to emphasize here the importance of this factor for situations in which there is contrast in D.

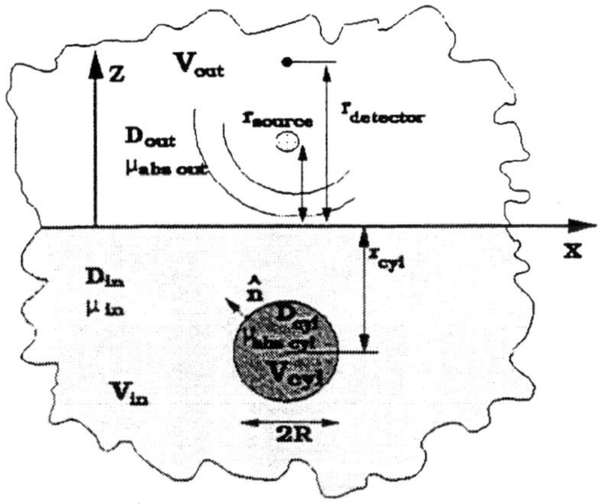

FIG. 1. Scattering Geometry

III. OBJECT DETECTION IN THE PRESENCE OF AN INTERFACE

In order to study the propagation of diffusive waves in inhomogeneous media we have taken into consideration a two dimensional configuration as shown in Fig. 1. It consists of a flat interface at $z = z_0$ that separates a semi-infinite diffusive medium $V_{out}[z > z_0]$, characterized by a linear, spatially uniform and isotropic diffusion coefficient D_{out}, and absorption coefficient $\mu_{a_{out}}$, from a semi-infinite diffusive medium $V_{in}[z < z_0]$ with a linear, spatially uniform and isotropic diffusion coefficient D_{in}, and absorption coefficient $\mu_{a_{in}}$. A cylinder of radius R, diffusion coefficient D_{cyl}, absorption coefficient $\mu_{a_{cyl}}$ and axis along OY is immersed in V_{in}. A point source is placed in the upper medium $z > z_0$ at point $\mathbf{r}_0$, its intensity being sinusoidally modulated at frequency ω. Throughout these three diffusive media, we impose flux continuity and conservation of the total photon current across their common boundaries. After proceeding as in Section II but for three constant diffusive media, we obtain a closed set of integral equations for the scattering geometry represented in Fig. 1. These integral equations and their derivation can be found in reference [15]. In the present configuration, our incident field will be of the form $\phi^{(inc)}(\mathbf{r} - \mathbf{r}_0) = \frac{S_0}{4\pi D_{out}} G(\kappa_{out}|\mathbf{r} - \mathbf{r}_0|)$ which in 2D represents a damped cylindrical wave with its origin at $\mathbf{r}_0$. The set of equations obtained for the three connected diffusive media enable us to obtain the solution for the scattering of a diffusive wave, which we shall define as: $\phi^{(total)} = \phi^{(inc)} + \phi^{(sc)}$, where $\phi^{(inc)}$ and $\phi^{(sc)}$ are the incident and scattered photon density waves, respectively. We shall study separately the relative amplitude $|\phi^{(total)}(\mathbf{r})|/|\phi^{(total)}_{no-object}(\mathbf{r})|$ and the relative scattered phase $\Delta(\mathbf{r}) - \Delta_{no-object}(\mathbf{r})$, i.e. $\phi^{(total)}(\mathbf{r}) = |\phi^{(total)}(\mathbf{r})| \exp i\Delta(\mathbf{r})$. In order to obtain the numerical solution for this set of equations we have employed a numerical procedure as in references [15]- [18].

IV. NUMERICAL RESULTS

Following experimental procedures (see for example [19] and [20]), we have considered a source with a modulation frequency $\omega = 200MHz$, of wavelength 780 nm. Since we are going to vary the diffusion constant of the outside medium, the parameters chosen for the three media V_{out}, V_{in} and V_{cyl}, are: Medium inside, brain tissue, as defined in [2] i.e. $\mu_{a_{in}} = 0.04cm^{-1}$ and $D_{in} = 0.0190042cm$. For the cylinder, we have used the parameters of a simulated brain hematoma, as described in [21], $\mu_{a_{cyl}} = 1cm^{-1}$ and $D_{cyl} = 0.019608cm$. For the outside media we have used $\mu_{a_{out}} = 0.03cm^{-1}$ and we have varied the diffusion constant so that it is the same, twice, and half the value of D_{in}, i.e. $D_{out} = D_{in}$, $D_{out} = 2D_{in}$, and $D_{out} = D_{in}/2$, respectively. In all cases the speed of light in the media was $v = 2.25 \times 10^{10} cm/s$. In order to reach numerical convergence, we have used a discretization $dx = 0.015cm$ for an interface of length $8cm$ and $ds = 0.004cm$ for the cylinder.

As can be seen in Fig. 2(a), where we represent the relative variation in amplitude in the presence of the object, for different diffusive contrast between the outside and the inside media, we can separate two different cases: when measuring in reflection and when measuring in transmission. Representing amplitude in reflection (see inset of Fig. 2(a)), we can see that the perturbation caused by the object, is greater when $D_{out} > D_{in}$ and that a maximum can be observed. This maximum is caused by the constructive interference of the wave reflected from the surface and the wave that has interacted with the object. If we look at the inset of Fig. 2(a) we see that we can still find this maximum even when matching both diffusive constants. This is because, even though the diffusion constants are the same, i.e. $D_{out} = D_{in}$, the absorption coefficients are different and therefore the complex diffusion wave number is also different. On the other hand, when measuring the total amplitude in transmission (see Fig. 2(a)), the greater the fraction D_{in}/D_{out}

is, the bigger the perturbation in amplitude caused by the presence of the object. Of course, there is a limit value for D_{out}, because, the lower the value of D_{out}, the greater the real and imaginary values of κ_{out} and therefore the greater the absorption of the outside media.

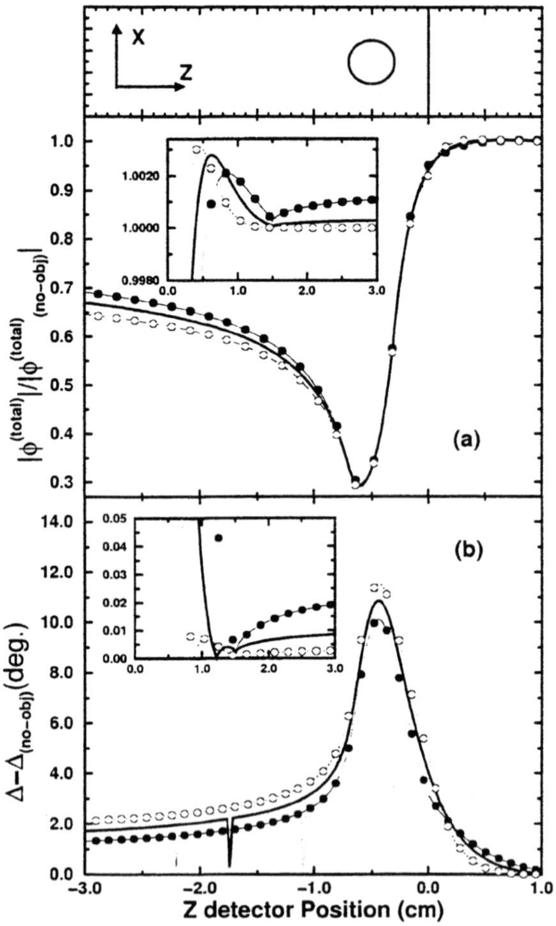

FIG. 2. **(a):** Relative amplitude, $|\phi^{(total)}|/|\phi^{(total)}_{no-object}|$ in the presence of a flat interface at $z_0 = 0$ for $\mathbf{r_{source}} = (0, 15.0mm)$, $\mathbf{r_{detector}} = (0.0, z)$. The distance cylinder-surface is $d = 5mm$, and radius of the object is $R = 2mm$. The media parameters are (see text): For Outside medium: $\mu_{a_{out}} = 0.03cm^{-1}$, $D_{out} = D_{in}$ (thick solid line), $D_{out} = 2D_{in}$ (thin solid line [•]) and $D_{out} = D_{in}/2$ (thin dotted line [○]). Brain tissue for Inside medium. Hematoma tissue for Object. **(b):** Same as (a) but for relative phase $\Delta - \Delta_{no-obj}$, i.e. $\phi = |\phi| \exp i\Delta$.

This effect can also be appreciated when measuring the relative phase, as shown in Fig. 2(b), where the greater changes in phase are when $D_{out} = D_{in}/2$. The sudden drops that can be appreciated in the relative phase measurements are the points where the photon density wave with the object and the photon density wave without the object are in phase. It must be stated that, since we are representing in Fig. 2 the relative values, these may be sometimes confusing. This is because, as we measure further away from sources, the measured amplitudes tend exponetially to zero, but when representing the relative values, this effect cancels out. It is interesting to see that when we measure the relative phase in reflection (see inset of Fig. 2(b)) we obtain a greater change in phase with the lowest diffusion contrast, as we saw in the relative reflected amplitude (see inset of Fig. 2(a)), but this effect varies with the position of the source. In order to see this, we have represented different results for the same media with different source positions in Fig. 3.

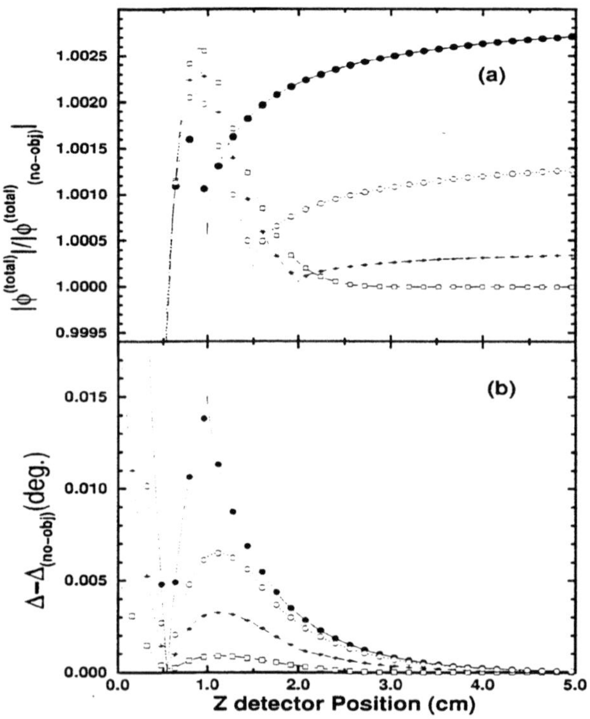

FIG. 3. **(a):** Relative amplitude, $|\phi^{(total)}|/|\phi^{(total)}_{no-object}|$ in the presence of a flat interface at $z_0 = 0$ for $\mathbf{r_{detector}} = (0.0, z)$. The distance cylinder-surface is $d = 5mm$, and radius of the object is $R = 2mm$. The media parameters are (see text): For Outside medium: $\mu_{a_{out}} = 0.03cm^{-1}$, $D_{out} = 2D_{in}$. Brain tissue for Inside medium. Hematoma tissue for Object. Source positions: $\mathbf{r_{source}} = (0.0, 10mm)$ (solid line [•]), $\mathbf{r_{source}} = (0.0, 15mm)$ (dotted line [○]), $\mathbf{r_{source}} = (0.0, 20mm)$ (dashed line [⋆]), $\mathbf{r_{source}} = (0.0, 30mm)$ (dotted line [□]). **(b):** Same as (a) but for relative phase $\Delta - \Delta_{no-obj}$, i.e. $\phi = |\phi| \exp i\Delta$.

As can be appreciated from Figs. 2 and 3, placing the source at smaller distances from the interface has a similar effect as incrementing the diffusion constant of the outside medium, when placing the detector behind the source.

V. CONCLUSIONS

We have developed a method to solve numerically the diffusion equation in the presence of bodies of arbitrary shape without any approximation whatsoever (although, of course within the diffusion approximation context, that is). We have presented the scattering integral equations for diffusive media, which are the equivalent to the scattering integral equations for both scalar and electromagnetic fields in potential scattering, and the *Extinction Theorem for Diffusive Waves* which constitutes a boundary condition for finding the sources, and have solved such equations for multiply connected volumes. We have found that by changing the contrast between the medium outside and the medium inside, it is possible to make detection of small objects (in this case hematomae in brain tissue) more feasible, either in reflection or in transmission. We would like to emphasize that so far we have not seen any experiments which model this situation, that is, an object buried in a medium in the presence of an interface, and it is not yet possible to make comparisons with experimental data.

ACKNOWLEDGMENTS

We would like to thank R.Carminati and A.Madrazo for relevant discussions. This research has been supported by Comisión Interministerial de Ciencia y Tecnología of Spain under grant PB95-0061 and by the Fundación Ramón Areces. J.Ripoll acknowledges a scholarship from Ministerio de Educación y Cultura of Spain.

---

[1] E. B. de Haller, "Time-Resolved Transillumination and Optical Tomography", J. Biomed. Opt. **1**,7-17 (1996)

[2] A. Yodh and B. Chance, "Spectroscopy and Imaging with diffusing light", Phys. Today March 1995

[3] A. Ishimaru, "Wave Propagation and Scattering in Random Media", Academic, New York, 1978, Vol. I.

[4] K.Michelsen, H. de Raedt, J. Przeslawski and N.García, "Determination of tissue optical properties by time-resolved transillumination and reflection experiments", *Submitted*

[5] M.S.Patterson,B.Chance, and B.C.Wilson, "Time resolved reflectance and transmittance for the non invasive measurement of tissue optical properties", Appl. Opt. **28**, 2331-2336 (1989)

[6] S. Fantini, M.A. Franceschini and E. Gratton, "Semi-infinite-geometry boundary problem for light migration in highly scattering media: a frequency-domain study in the diffusion approximation", J.Opt.Soc.Am. B, **11**, 2128-2138 (1994)

[7] J.B.Frishkin, O.Coquoz, E.R.Anderson, M.Brenner and B.J.Tromberg, "Frequency-domain photon migration measurements of normal and malignant tissue optical properties in a human subject", Appl. Opt. **36**, 11-20 (1997)

[8] P.N. den Outer and T.M.Niewenhuizen, A.Lagendijk, "Location of Objects in multiple-scattering media", Opt. Soc. Am. A, **10**, 1209-1218 (1993)

[9] X.D.Li, T. Durduran, and A.G. Yodh, B.Chance, D.N.Pattanayak, "Diffraction tomography for biochemical imaging with diffuse-photon density waves", Opt. Lett. **22**, 573-576 (1997)

[10] S.B.Colak, D.G. Papaioannou, G.W.'t Hooft, M.B. van der Mark, H.Schomberg, J.C.J. Paasschens, J.B.M. Melissen and N.A.A.J. van Asten, "Tomographic image reconstruction from optical projections in light-diffusing media", Appl. Opt. **36**, 181-213 (1997)

[11] C.L.Matson, N.Clark, L- McMackin and J.S.Fender, "Three-dimensional tumor localization in thick tissue with the use of diffuse photon-density waves", Appl. Opt. **36** 215-219 (1997)

[12] S.C. Feng, F. Zeng and B.Chance, "Photon migrations in the presence of a single defect",Appl. Opt. **34** 3826-3837 (1995)

[13] D.A. Boas, M.A. O'Leary, B. Chance, and A.G. Yodh, "Detection and characterization of optical inhomogeneities with diffuse photon density waves: a signal-to-noise analysis", Appl. Opt. **36**, 75-92 (1997)

[14] M.Nieto-Vesperinas *Scattering and Diffraction in Physical Optics*, Wiley-Interscience, 1991.

[15] J.Ripoll and M.Nieto-Vesperinas, "Scattering Integral Equations for Diffusive Waves. Detection of Objects Buried in Diffusive Media in the Presence of Interfaces", *Submitted to JOSA A*

[16] J.A.Sánchez-Gil and M.Nieto-Vesperinas, "Light scattering from random rough dielectric surfaces", J. Opt. Soc. Am. A **8**, 1270-1286 (1991)

[17] A.Madrazo and M.Nieto-Vesperinas, "Scattering of light and other electromagnetic waves from a body buried beneath a highly rough random surface", J. Opt. Soc. Am A **14**, 1859-1866(1997)

[18] J.Ripoll, A.Madrazo and M.Nieto-Vesperinas, "Scattering of electromagnetic waves from a body over a random rough surface", Opt. Comm. **142**, 173-178 (1997)

[19] D.A.Boas, M.A.O'Leary, B.Chance and A.G.Yodh, "Scattering of diffusive photon density waves by spherical inhomogeneities within turbid media: Analytic solution and applications", Proc. Natl. Acad. Sci. USA **19**, 4887-4891 (1994)

[20] M.A. O'Leary, D.A. Boas, B. Chance and A.G. Yodh, "Refraction of Diffusive Photon Density Waves", Phys. Rev. Lett **69**, 2658-2661 (1992)

[21] F.F.Jöbsis, "Noninvasive, infrared monitoring of cerebral and myocardial oxygen sufficiency and circulatory parameters", Science **198** 1264-1267 (1977)

Suppression of scattering effect using spatially dependent point spread function

Yoichi ONODERA, Yuji KATO, and Koichi SHIMIZU

Graduate School of Engineering, Hokkaido University,
North 13 West 8, Sapporo, 060-8628 Japan

Tel:+81-11-706-7229, Fax:+81-11-706-7219
E-mail:ono@bme.eng.hokudai.ac.jp

Abstract

A technique was developed to suppress the scattering effect in a reconstructed image of an optical CT. We have incorporated the process of the deconvolution using depth-dependent point spread function (PSF) in the reconstruction algorithm. Through a numerical simulation and a phantom experiment, the effect of scattering suppression in the CT imaging was demonstrated and the fundamental feasibility of this technique was verified.

Keywords
(170.0170) Medical optics and biotechnology; (170.3010) Image reconstruction techniques; (170.6960) Tomography; (100.1830) Deconvolution

Introduction

Near-infrared light has relatively high transmission through mammalian tissue. However, the transillumination image is blurred due to the strong scattering in the tissue[1-3]. This image blurring can be considered as the convolution of a point spread function (PSF) of the scattering medium. Then, we can suppress the scattering effect by the deconvolution with the PSF. This principle has been applied to the optical CT, but in these applications the PSF was defined according to the thickness of the scattering medium[4,5]. Since the degree of blurring is dependent on the depth in the scattering medium, the PSF has to be depth-dependent[6,7]. In this study, we proposed a technique to suppress the scattering effect using the depth-dependent PSF in the process of reconstruction of an optical CT.

Point spread functions in scattering media

First, we evaluated how much the scattering effect is dependent on the depth. The experimental system is illustrated in Figure 1. A beam of incident light (He-Ne laser, λ = 632.8 nm) and a detector were collimated. The collimated pair was scanned along the surface of the scattering medium to obtain a profile of the projection data. As a scattering medium, Intralipid suspension (μ_s' = 0.5 /mm) was filled in a cell with 50 mm inner distance. We defined the depth as the length from the incident side to the detection side. A metal rod (3 mm in diameter, painted black) was placed in the cell as an absorber. The measured intensity profiles are shown in Figure 2. When

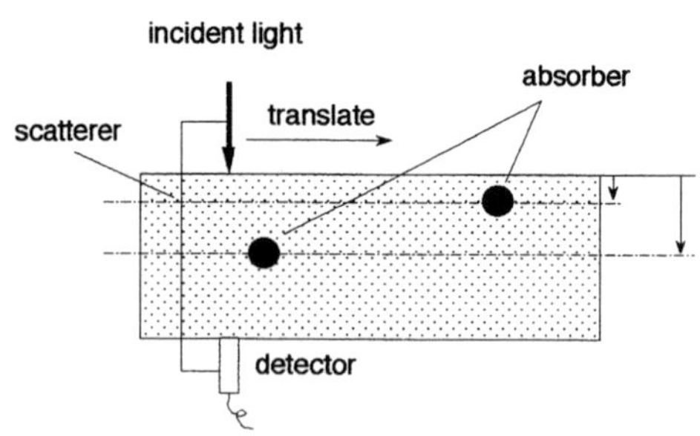

Figure 1. Measurement of depth-dependent profiles.

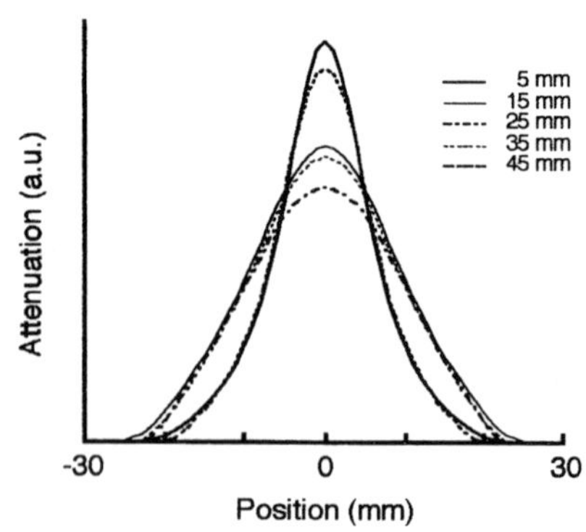

Figure 2. Measured profiles of 3 mmϕ rod at different depths.

the rod was placed at 5 mm depth, the scattering effect was weak and a sharp profile was obtained. As the depth increased toward the middle (25 mm depth), the scattering effect became strong and the profile became flat. As the depth increased further toward the detection side, the scattering effect became weak again. Almost the same profiles were obtained at the symmetric depth with respect to the center of the cell. In this way, the depth-dependence of the projections was verified and its fundamental characteristics were found. Without the scattering effect, the projection of a rod with 3 mm diameter should be the rectangular shape with 3 mm width. So, PSF's were obtained deconvolving the blurred projection by 3 mm width rectangular function. In this way, we estimated the PSF at each depth in the Intralipid suspension of 50 mm thickness. The estimated depth-dependent PSF's were illustrated in Figure 3.

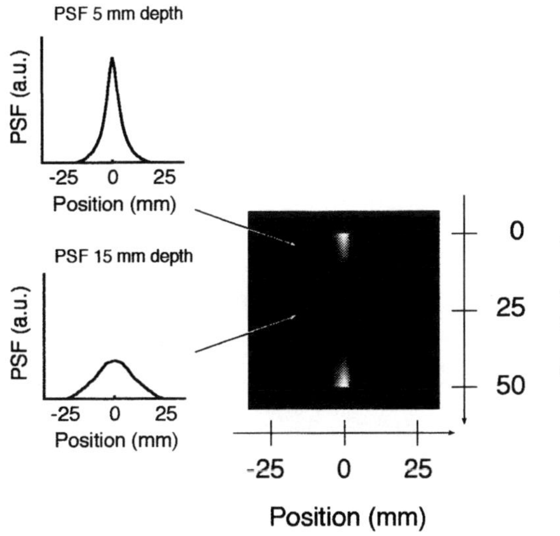

Figure 3. Estimated PSF's in each depths.

Generally, in biological tissue the absorption coefficient is much smaller than the scattering coefficient and the shape of the PSF substantially depends on the scattering coefficient. The problem is how to reconstruct the distribution of the absorption coefficient in the scattering medium with an almost homogeneous scattering coefficient. Figure 4 illustrates how the depth-dependent PSF's are convolved into a profile of an optical projection. The observed projection of an absorber in scattering medium is blurred due to the scattering effect. We consider this blurred projection is the integral of the absorber-profiles convolved with the depth-dependent PSF at each depth. When we use these blurred projections for the CT reconstruction, the computed image is blurred further. If we can deconvolve the profile with an appropriate PSF at each depth, the CT image should be improved. Based on this concept, we developed a technique to improve the CT image using the blurred projection data. Figure 5 illustrates the principle of the proposed technique. The process of this technique is as follows. First, we obtain the PSF's at each depth either by a measurement or by a computer simulation. Then:

(1) A CT image is obtained using the measured projection data that are blurred due to the scattering effect. Here, a filtered backprojection algorithm is used for the CT reconstruction.

(2) The blurred image is sampled at each depth, and the profile of absorption distribution is deconvolved with the PSF at the depth.

(3) The corrected profiles are integrated in a direction of projection to make a corrected projection data.

(4) Using the corrected projection data, a new CT image is reconstructed in which the effect of scattering is suppressed.

Numerical simulation

To verify the fundamental feasibility of this technique in an optical CT, a computer simulation was carried out. We used a numerical phantom illustrated in Figure 6(a). The circle in the figure means the valid field of view, and its diameter is 50 mm. Two disks were supposed to be the cross sections of cylindrical absorbers (10 mm in diameter). The depth dependent PSF's in a vertical direction were shown in Figure 6(b). With these PSF's, the original absorption profiles were blurred at each depth, and blurred projections were obtained in every direction. The

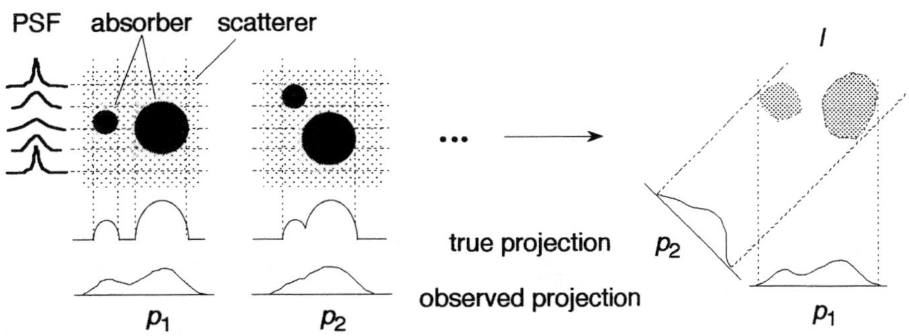

Figure 4. Convolution of scattering effect in CT image reconstruction.

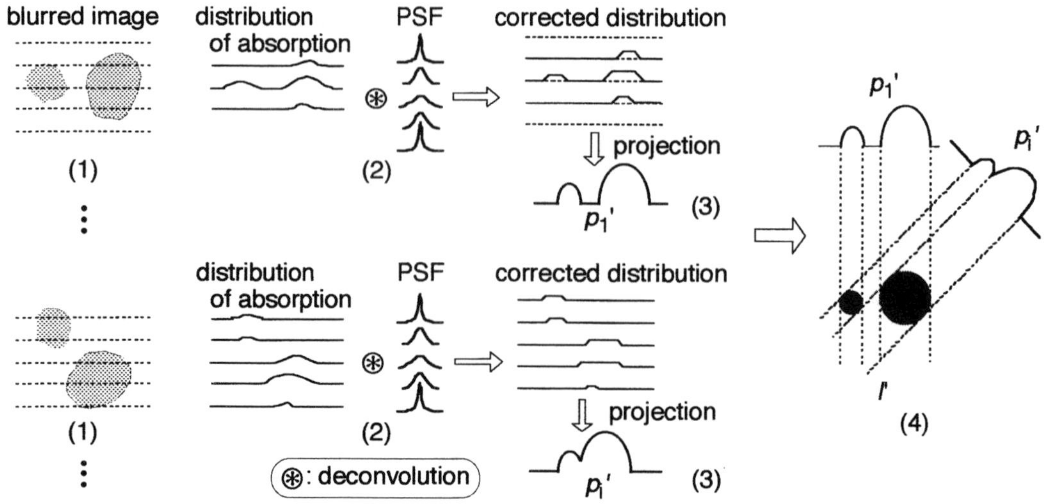

Figure 5. Principle of proposed technique. Numbers correspond to the order of reconstruction process. Detailed explanation is in the text.

blurred image (Figure 6(c)) was reconstructed using the blurred projection data. Then, the proposed technique was applied to this blurred image and we obtained a corrected image shown in Figure 6(d). Although some artifacts appeared in the peripheral region, the image was improved considerably. This artifact seems to be caused by the irreversibility of the computer calculation of the deconvolution using the fast Fourier transform.

Application to phantom experiment

To verify the effect of the proposed technique in the practical CT imaging, a phantom experiment was carried out. The depth-dependent PSF's obtained in Chapter 2 were used for the deconvolution. The projection data for image reconstruction were obtained in the same experimental system shown in Chapter 2. The structure of an experimental phantom was illustrated in Figure 7. Three black metal posts were placed in a scattering medium, or Intralipid suspension ($\mu_s' = 0.5$ /mm). The cross-sections of the posts were circle, triangle and square with a typical dimension of 10 mm. The projection data were collected in the translate/rotate method, and a CT image was reconstructed by the filtered backprojection algorithm. The projection data was measured at 128 points with a 0.5 mm step in a translate scan. This was repeated in 18 orientations with a 10 degree step. At first, the effect of this technique was examined in the projection data (Figure 8). The broken line and the thick solid line are the projection data before and after the deconvolution process, respectively. The thin solid line is the projection measured in clear water. The two peaks were well separated by the deconvolution technique.

Second, the effectiveness of this technique was examined in the CT imaging with an experimental phantom. CT images were reconstructed using the projection data such as those shown in Figure 8, i.e., before deconvolution (Figure 9(a)) and after the deconvolution

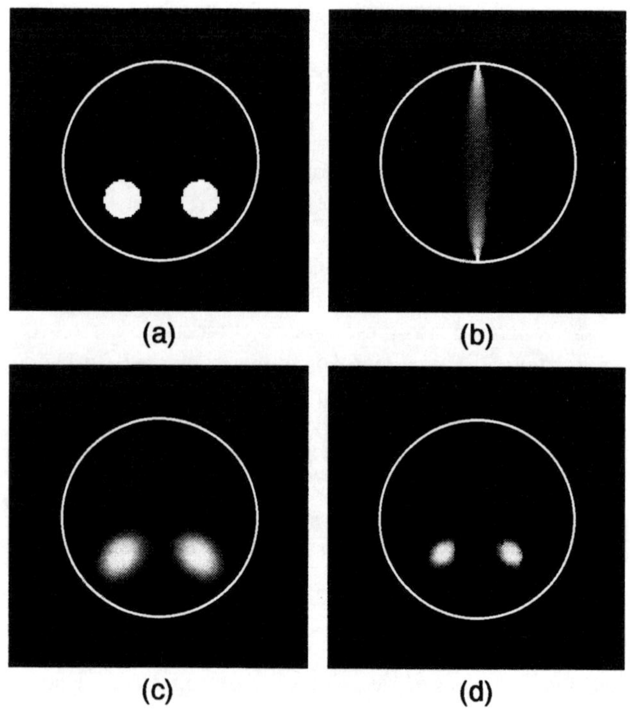

Figure 6. Result of computer simulation,
(a) numerical phantom,
(b) depth dependent PSF's,
(c) blurred image with conventional method,
(d) corrected image with developed technique.

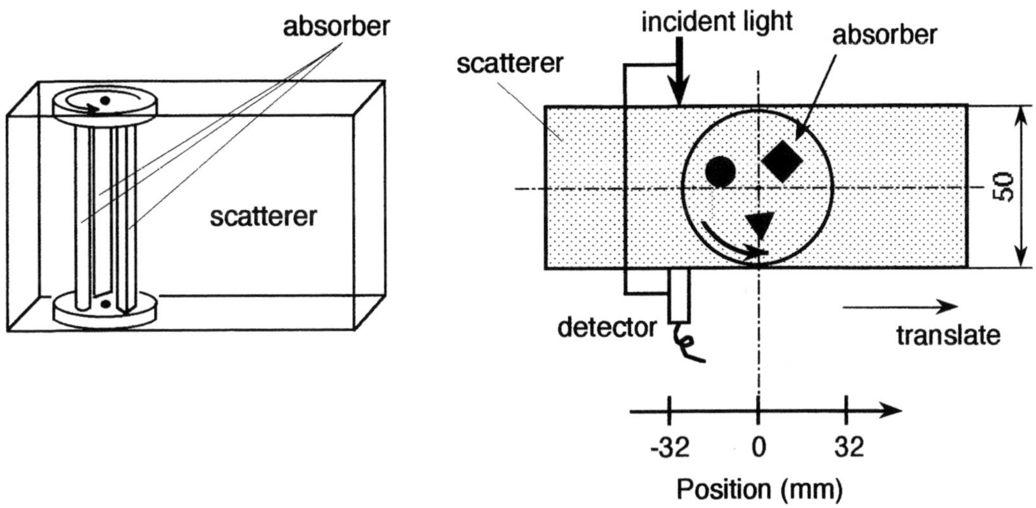

Figure 7. Structure of phantom used in experiment.

(Figure 9(b)). The conditions of the image presentation were set to be common among these two images. In Figure 9(a), the effect of scattering was overwhelming, and the internal structure could be hardly seen. In Figure 9(b), the scattering effect was suppressed to some extent, and we could recognize the existence of the three posts. However, the shape of the cross-sectional image was not still clear.

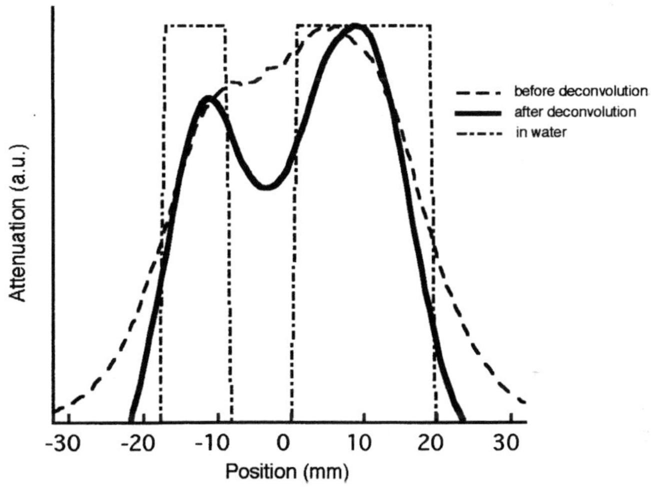

Figure 8. Example of projection : before and after deconvolution.

Conclusions

In order to suppress the scattering effects, a new deconvolution technique was developed for the reconstruction of an optical CT. It became possible to use the depth-dependent PSF in the deconvolution technique. The process of the deconvolution could be incorporated in the reconstruction of the filtered backprojection principle. Through the computer simulation and the phantom experiment, the fundamental feasibility of the proposed technique was verified to suppress the scattering effect in an optical CT. This technique seems to provide a useful tool in the practical application of an optical CT.

The experimental verification with living tissue is the next subject of this research.

This research is supported in part by the Ministry of Education, Science, Sports and Culture of Japan.

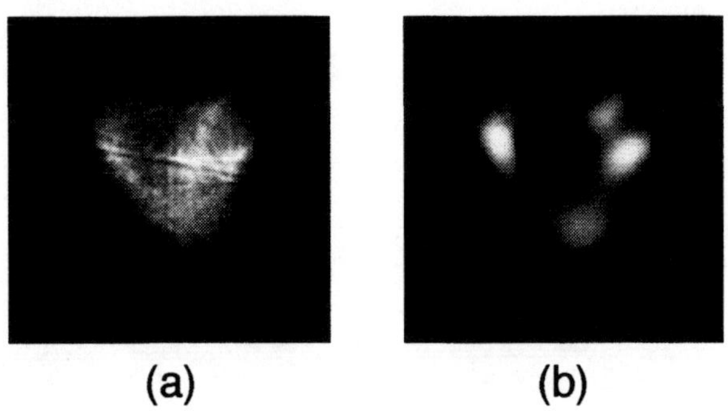

Figure 9. Reconstructed images,
(a) before deconvolution,
(b) after deconvolution.

References

1. B. Chance ed. : Time-resolved spectroscopy and imaging of tissues, Proceedings of SPIE, 1431 (1991)
2. G. J. Müller ed. : Medical Optical tomography - functional imaging and monitoring, SPIE Press, Bellingham (1993)
3. B. Chance and R. R. Alfano ed. : Optical tomography, photon migration, and spectroscopy of tissue and model

media -theory, human studies and instrumentation-, Proceedings of SPIE, 2389 (1995)

4. R. Araki and I. Nashimoto : Near-infrared imaging in vivo. Imaging of Hb oxygenation in living tissues, Proceedings of SPIE, 1431, 321/332 (1991)
5. P. van der Zee and D. T. Delpy : Computed point spread functions for light in tissue using a measured volume scattering function. Advances of Experimental Medicine and Biology, 222, 191/197 (1988)
6. S. B. Corlak, D. G. Papaioannou, G. W. 't Hooft, M. B. van der Mark, H. Schomberg, J. C. J. Paasschens, J. B. M. Melissen, and N. A. A. J. van Asten : Tomographic image reconstruction from optical projections in light-diffusing media, Applied Optics, 36, 1, 180/213 (1997)
7. Y. Onodera, Y. Kato, and K. Shimizu : Suppression of scattering effect of optical CT image using point spread function. Japanese Journal of Medical Electronics and Biological Engineering, 35, 4, 400/406 (1997)

Imaging of absorption distribution in diffuse medium using backscattered light

Aki AWATA, Yuji KATO, and Koichi SHIMIZU

Graduate School of Engineering, Hokkaido University,
North 13 West 8, Sapporo, 060-8628, Japan

Tel: +81-11-706-7229, Fax: +81-11-706-7219
E-mail: awata@bme.eng.hokudai.ac.jp

Abstract

A technique was developed to reconstruct the cross-sectional image of the absorption distribution in a diffuse medium using backscattered light. In this technique, we illuminate an object with an ultra-short pulse, and measure the time-resolved pulse shape of the light backscattered from the object. The absorption distribution of the scattering object can be estimated using the path-length distribution of photons at each detection time and the intensity of backscattered light at the corresponding detection time. The feasibility of this technique was verified through a simulation and a measurement with a biological model.

Keywords

(170.0170) Medical optics and biotechnology; (290.1350) Backscattering; (170.3010) Image reconstruction techniques; (170.6920) Time-resolved imaging; (170.6960) Tomography

Introduction

In the optical CT[1-4], two types of light-detection schemes have been used. One detects the transmitted or forwardly scattered light, and another the backwardly scattered light (here we call it as backscattered light). Using the transmitted light, we can have the whole cross-sectional image of the subject. But the thickness of the subject has to be thin enough to get the transmitted light. Using the backscattered light, we can obtain the cross-sectional image of only a certain part near the surface of the subject. However, the applicability of this technique is not restricted by the thickness of the subject. In the combination with optical fibers, it is particularly useful in the clinical applications, such as the imaging of inner walls of digestive tracts, blood vessels and heart chambers.

We have developed a new technique to reconstruct the cross-sectional image of diffuse medium using the backscattered light. This paper presents the basic idea of this technique, some results of a computer simulation and a measurement to examine the capability of this technique.

Principle of inverse problem

Figure 1 shows the arrangement of the light incidence and the detection of backscattered light. In practice, they are made with the optical fibers in close contact with the body tissue to illuminate a light and to guide a backscattered light to a photodetector.

In this technique, the cross-sectional image of a subject is reconstructed in two processes. In the first process, we obtain the absorption distribution in the depth-direction of the subject using the principle described below. In the second process, we scan the pair of two optical fibers along the surface of the subject to obtain the horizontal variation. By combining these two processes, we can reconstruct the cross-sectional image of the subject in the same way as the B-mode scan in an ultrasonic imaging.

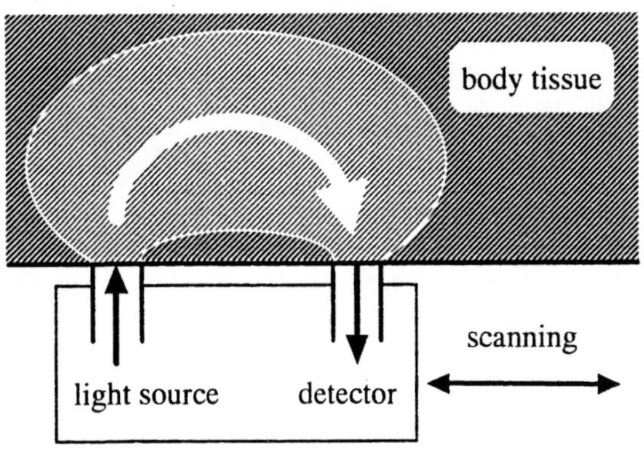

Figure 1. Method of measurement.

Figure 2 illustrates the principle to obtain the depth-distribution of the absorbance. An ultra-short light pulse enters the medium from an optical fiber and travel through some layers to reach another fiber. The light backscattered from the medium is guided to a detector (such as a streak camera) and a time-resolved pulse shape is measured. Photons propagated through shallow layers are detected in earlier times than those reached to deeper layers. Thus, if the scattering coefficient of the object is homogeneous, the backscattered light intensities at different times have some correlation with the geometrical distribution of the absorption coefficient. In this technique, we try to obtain this distribution from the measured temporal pulse shape of the backscattered light.

In the explanation of this technique, we follow the general manner of the Monte Carlo method[5, 6] to simulate the light scattering and absorption in random media. The word "photon" here represents a group of photons that propagate in the medium as if they are a particle with a certain optical power to be attenuated in the propagation. We assume that the scattering coefficient of the medium is homogeneous.

In Fig. 2, injected k-th photon with a unit power reaches to a detector in the time t_j after some collisions against scatterers. Then, the optical power reached in the detector is given by,

$$i_k(t_j) = \exp\left\{-\left(A_1 d_{k1} + A_2 d_{k2} + \cdots + A_n d_{kn}\right)\right\} \quad (1),$$
$$t_j = \left(d_{k1} + d_{k2} + \cdots + d_{kn}\right)/c$$

where A_1, A_2, ..., A_n are the absorption coefficients of the 1st, 2nd, ... n-th layers, d_{k1}, d_{k2}, ..., d_{kn} are the distances in which the k-th photon propagates in the 1st, 2nd, ... n-th layers respectively, and c is the velocity of light in the medium. Since the incident impulse of light is considered as a collection of many photons, the total received intensity of transmission delay t_j is expressed as the sum of Eq. (1),

$$I(t_j) = \sum_{k=1}^{N(t_j)} i_k(t_j) \quad (2),$$

where $N(t_j)$ is the number of photons detected at the detection time t_j. Equation (2) is nonlinear with respect to the absorption coefficients and too complex to resolve the set of $\{A_m\}$. So we apply a linear approximation to Eq. (1) at a certain constant α, and get

$$i_k(t_j) \approx \exp(-\alpha)\left(1 + \alpha - \sum_{m=1}^{n} A_m d_{km}\right) \quad (3),$$
$$t_j = \left(d_{k1} + d_{k2} + \cdots + d_{kn}\right)/c$$

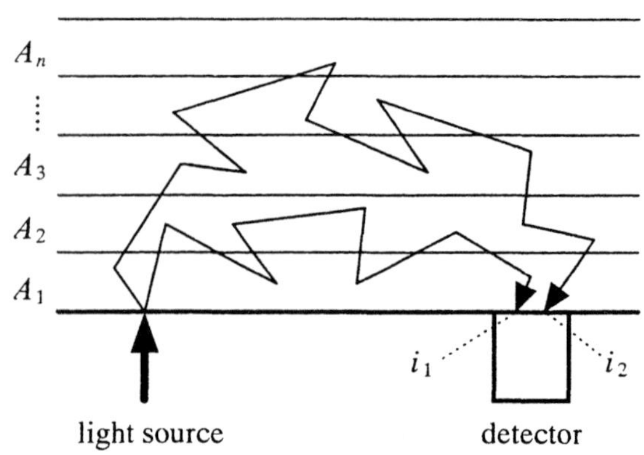

Figure 2. Principle of measurement.

where $\alpha - \sum_{m=1}^{n} A_m d_{km} << 1$.

Strictly, α of each photon should be determined independently. But photons detected at the same time are considered to have traveled through similar paths, and the attenuation of these photons should be in the same level. Thus, we consider α to be constant over the k-th photons (k=1, 2, ..., $N(t_j)$). Then the total detected intensity in the detection time t_j is expressed as,

$$I(t_j) \approx \exp(-\alpha)\left\{N(t_j)(1+\alpha) - \sum_{m=1}^{n} A_m D_m(t_j)\right\} \quad (4),$$

where $D_m(t_j)$ is the total path-length in the m-th layer at the detection time t_j, or

$$D_m(t_j) = \sum_{k=1}^{N(t_j)} d_{km}, \quad \sum_{m=1}^{n} d_{km} = ct_j \quad (5).$$

Equation (5) is considered to be the probability of the photon propagation in the m-th layer.

In Eq. (4), the constant α for the linear approximation is determined as below using the detection time t_j. A reasonable choice of α was found to be the average of light attenuation in the detection time t_j, i. e.,

$$\alpha \approx \sum_{m=1}^{n} A_m d_{km} \approx A_{av} \sum_{m=1}^{n} d_{km} = A_{av} ct_j \quad (6),$$

where A_{av} is the average of the absorption coefficients which the photons detected at time t_j propagated through. Then the received intensity of a single photon in the detection time t_j is given by,

$$i_k(t_j) \approx \exp\left(-A_{av}ct_j\right) \quad (7).$$

Since Eq. (7) is independent of the subscript k, the total received intensity in the detection time t_j is expressed as,

$$I(t_j) = \sum_{k=1}^{N(t_j)} i_k(t_j) \approx N(t_j)\exp\left(-A_{av}ct_j\right) \quad (8)$$

$$\text{or} \quad A_{av} = \frac{1}{ct_j}\ln\frac{N(t_j)}{I(t_j)} \quad (9).$$

Using Eqs. (4), (6) and (9), we finally get the linear equation of the time-resolved transmitted intensity, i. e.,

$$N(t_j)\ln\frac{N(t_j)}{I(t_j)} = \sum_{m=1}^{n} A_m D_m(t_j) \quad (10).$$

The set of distances $\{D_m(t_j)\}$ depends on the scattering property of the tissue but is almost independent of the absorption property. Therefore, if we know the $\{D_m(t_j)\}$ a priori, we can obtain the set of unknowns $\{A_m\}$ from the measured pulse shape $I(t_j)$.

In practice, this principle is conducted in the following procedure. First, we obtain $N(t_j)$ and $\{D_m(t_j)\}$ in a computer simulation or in the measurement using a model phantom. Then we illuminate a short pulse of light on the body tissue, and measure the received pulse shape $I(t_j)$. In the reconstruction process, we sample the pulse shape $I(t_j)$ along the time axis at n different positions. We divide the object in some imaginary layers to get the absorption distribution in the depth-direction. The number n corresponds to the number of these layers. In this way, we can have the same number of simultaneous equations as the number of unknowns. By solving the linear simultaneous equations, we can obtain the set of unknowns $\{A_m\}$.

Computer simulation of layered structure

In order to examine the applicability of this technique to a diffuse medium, a computer simulation based on the Monte Carlo method was conducted. Figure 3 shows the model for the simulation. A detector which has a round cross-section of 2mm diameter is located on the medium surface, separated 10mm from the light incidence point. The light incidence area is assumed as an ideal point. The diffuse medium is assumed to be divided as 10 layers, and each layer have a reduced scattering coefficient of μ_s'=1.0mm^{-1} and an absorption coefficient of μ_a=0.002 to 0.03mm^{-1}. Each layer is 0.5mm thickness. About 10^9 photons were injected to get a time-resolved pulse shape.

Figure 4 shows the comparison between the given

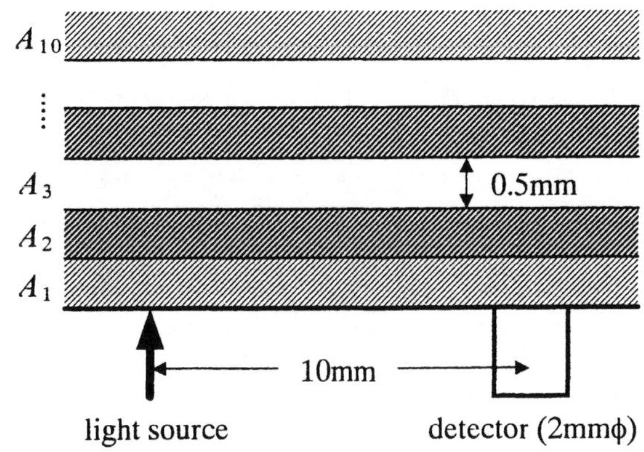

Figure 3. Numerical phantom for simulation.

distribution and the reconstructed distribution. When the variation of the absorption coefficient was small among layers, the reconstructed distribution agreed well with the given distribution (Fig. 4 (a)). However, with large variation, the error became large particularly in deep layers (Fig. 4 (b)). These errors seemed to be due to the error in the linear approximation.

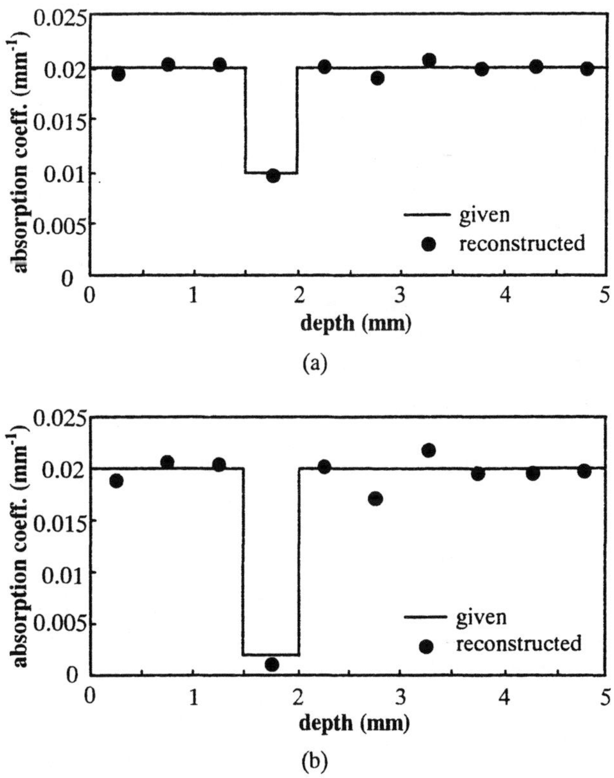

Figure 4. Results of reconstruction; (a) small absorption variance; (b) large absorption variance.

Measurement with two-layered phantoms

To verify the fundamental feasibility of this technique, we attempted to reconstruct the absorption distribution of two-layered phantoms from the measured data. Figure 5 shows the measurement system. Ultra-short light pulses with a pulse width of 30ps at FWHM, λ=752nm were irradiated to the phantom. The light emitted from the phantom was caught by the detection fiber to measure time-resolved pulse shape using a streak camera system. The distance between light injection and detection was 10mm.

The phantom was made of gelatin. The scattering and absorption properties of the phantom was controlled by the amount of Intralipid[7] and ICG[8] solution mixed in the gelatin. Surface layer was 5mm thick and the second layer was much thicker than the fiber separation, or 10mm. The scattering coefficients of two layers were common, i. e. μ_s'=1.0mm^{-1}, and the absorption coefficients were varied from μ_a=0.012 to 0.036mm^{-1}. $N(t_j)$ and $D_m(t_j)$ were obtained in the Monte Carlo simulation beforehand. The absorption coefficients A_1 and A_2 were obtained from the measured time-resolved data $I(t_j)$ using the Eq. (10).

Figure 6 shows the comparison between the given distribution and the reconstructed distribution. Generally, the estimated absorption coefficients agreed reasonably well with the given values. The estimation error increased when the variance of absorption became large (Fig. 6 (b)) and when the surface layer became absorbing (Fig. 6 (c)). The latter seemed to be due to the significant decrease in the probing light intensity in the deep layer.

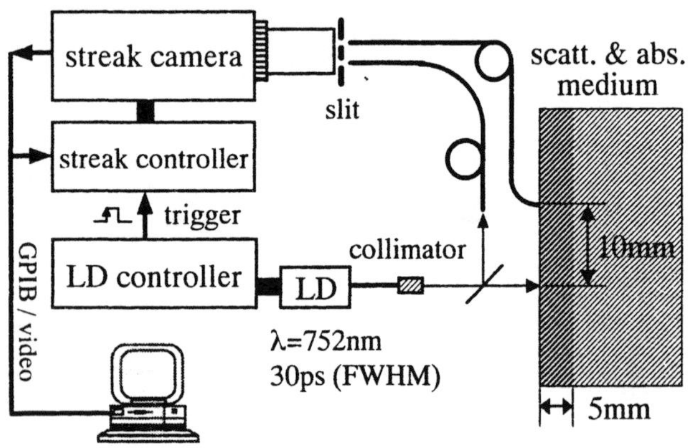

Figure 5. Measurement system.

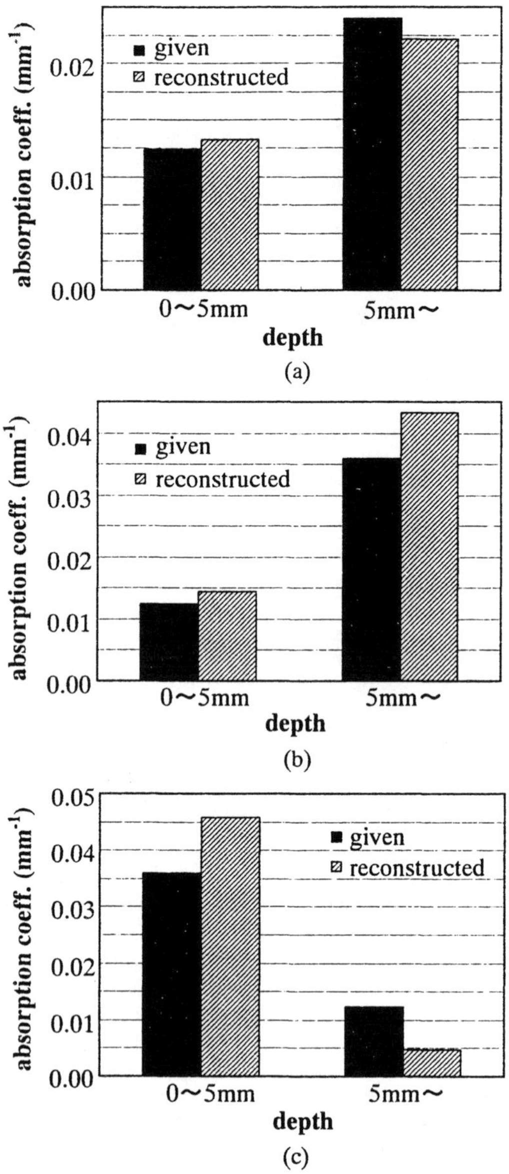

Figure 6. Result of measurements with two-layered phantoms.

Discussion

The principle of this technique is based on the solution of the simultaneous equation, or Eq. (10). The accuracy of this technique is largely dependent on the independence of the term $D_m(t_j)$ with respect to the layer number m. When t_j is small, the received light had passed only shallow layers. When t_j becomes large, the propagation path of the received light extends to large area. However, the spread of the propagation path forms a well-known banana-shape, and its axis can be centered in a certain layer. This means that the light intensity received at a certain detection time represents well the absorption coefficient of a certain layer. Therefore, we can expect to improve the accuracy of the

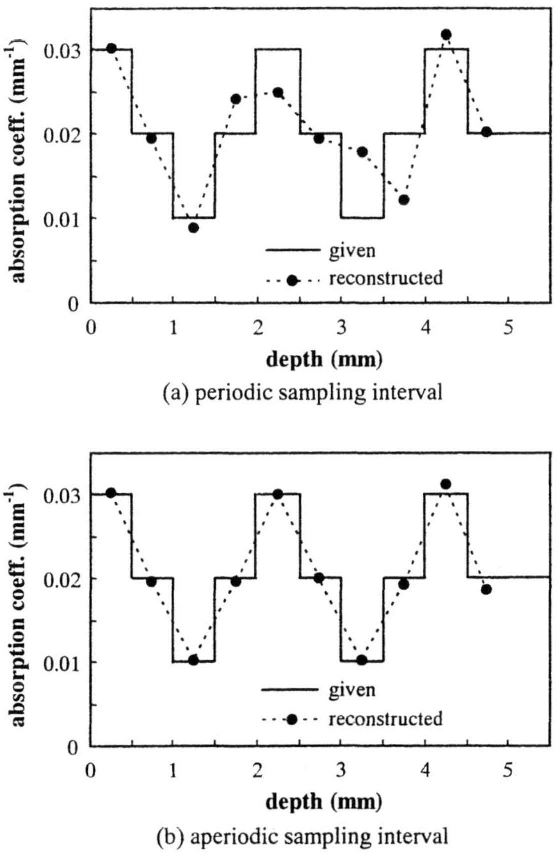

Figure 7. Improvement in reconstruction of absorption distribution.

reconstruction by an appropriate choice of the detection time.

We assumed the axis of the banana-shape distribution as an arc of the circle which passed the given three points. They are the points of light incidence, the center of the light detector and the center of the layer which is in the same distance to the above two points. The detection time was determined as the time of the light propagation along this arc. This gives an aperiodic sampling intervals in the time-resolved pulse shape.

Figure 7 shows the reconstructed distribution using this aperiodic sampling in comparison with that using previous periodic sampling. The simulation model was the same as that shown in Fig. 3. This result shows that the reconstruction error in the case of large absorption variance (corresponds to the case of Fig. 4 (b)) was improved considerably with the proposed aperiodic sampling.

Conclusions

To realize the optical CT using backscattered light, a technique was newly developed to reconstruct the absorption distribution in diffuse media. Through a computer simulation and a real measurement, the fundamental effectiveness of this technique was verified. In order to apply this technique to a more complex structure, experimental study for more than three-layered phantoms is requested.

The accuracy of this technique can be increased by the improvement in the linear approximation of the fundamental equation (Eq. (3)), and in the appropriate choice of sampling times of the detected pulse shape.

This research is supported in part by the Ministry of Education, Science, Sports and Culture of Japan.

References

1. S. R. Arridge, "The forward and inverse problems in time resolved infra-red imaging," *Medical Optical Tomography*, SPIE Press, pp. 35-64 (1993).
2. H. L. Graber, J. Chang, R. Aronson and R. L. Barbour, "A perturbation approach for optical diffusion tomography using continuous-wave and time-resolved data," *Medical Optical Tomography*, SPIE Press, pp. 87-143 (1993).
3. Special Issue, "Optical properties of mammalian tissue," Appl. Opt., Vol. 28, No. 12, pp. 2207-2357 (1989).
4. I. Oda, H. Eda, Y. Tsunazawa, M. Takeda and Y. Yamada, "Optical tomography by the temporally extrapolated absorbance method," Appl. Opt., Vol. 35, No. 01, pp. 169-175 (1996).
5. B. C. Wilson and G. Adam, "A Monte Carlo model for the absorption and flux distributions of light in tissue," Med. Phys., Vol. 10, pp. 824-830 (1983).
6. P. van der Zee and D. T. Delpy, "Simulation of the point spread function for light in tissue by a Monte Carlo method," Adv. Exp. Med. Biol., Vol. 215, pp. 179-191 (1986).
7. S. T. Flock, S. L. Jaques, B. C. Wilson, W. M. Star and M. J. C. van Gemertet, "Optical properties of Intralipid: A phantom medium for light propagation studies," Lasers Surg. Med., Vol. 12, pp. 510-519 (1992).
8. K. J. Baker, "Binding sulfobromophthalein (BSP) sodium and indocyanine green (ICG) by plasma a_1 lipoprotains," Proc. Soc. Exp. Biol. Med., Vol. 122, No. 4, pp. 957-963 (1966).

Inverse reconstruction of three-dimensional tomographic images of objects in turbid media from time-sliced two-dimensional transmission measurements

W. Cai, S. K. Gayen, M. Xu, M. Lax and R. R. Alfano

Institute for Ultrafast Spectroscopy and Lasers
New York State Center of Advanced Technology for Ultrafast Photonic Materials and Applications
Department of Physics
The City College and Graduate Center of City University of New York
New York, NY 10031

Abstract

Time-sliced two-dimensional intensity distribution of transmitted light measured using a time-gated imaging system is used to reconstruct three-dimensional images of an object inside a turbid medium. The inversion algorithm combines a two-dimensional matrix inversion with a one-dimensional Fourier transform inversion in cylindrical coordinates for a fast three-dimensional image reconstruction. A L-curve method is used for choosing suitable regularization parameters. The method was used to reconstruct images of an absorbing object located inside a highly scattering turbid medium using both simulated and experimental data.

Key Words

(170.6920) Time-resolved imaging; Time-sliced image; (170.3010) Image reconstruction techniques; (290.7050) Scattering, turbid media; (170.6090) Tomography.

Introduction

Near-infrared optical imaging of the internal structure of biological tissues, such as a tumor in a breast, is an active area of contemporary photonic and biomedical research.[1-5] Multiple scattering of light by biological tissues is a major problem that impedes direct imaging using transillumination shadowgram approach. Inverse reconstruction of images of objects inside turbid media using time-resolved light intensity distribution is an appealing approach for following reasons. First, it may make use of both less-scattered and multiple-scattered light. So even for thick tissues adequate signal is available, while the paucity of image-bearing light may be a limiting factor for transillumination imaging. Second, use of time-resolved data for inverse reconstruction may provide three-dimensional (3D) images, while transillumination imaging or even x-ray mammography may provide only 2D images.

In this article, we report on an inverse reconstruction algorithm and a time-sliced data acquisition approach that allow fast data acquisition and image reconstruction. The algorithm uses a 2D matrix inversion with a 1D Fourier transform inversion based on symmetry in cylindrical coordinates that greatly reduces computation time compared to direct 3D inversion. The data is collected in a transillumination geometry using an ultrafast electronic gated imaging system consisting of a gated image intensifier coupled to a charge-coupled device (CCD) camera. It provides a sequence of picosecond time-gated two-dimensional spatial intensity distributions at different times, $I(x, y, t_i)$, suitable for 3D-image reconstruction. Since the inversion problem is ill-posed, we use the 'L-curve method' to select optimal regularization parameters.[6] We present 3D images of an absorbing object hidden inside a turbid medium reconstructed using the above-mentioned algorithm and both simulated and real experimental data.

Algorithm

The experimental arrangement to test the inverse image reconstruction algorithm is displayed in Fig. 1.

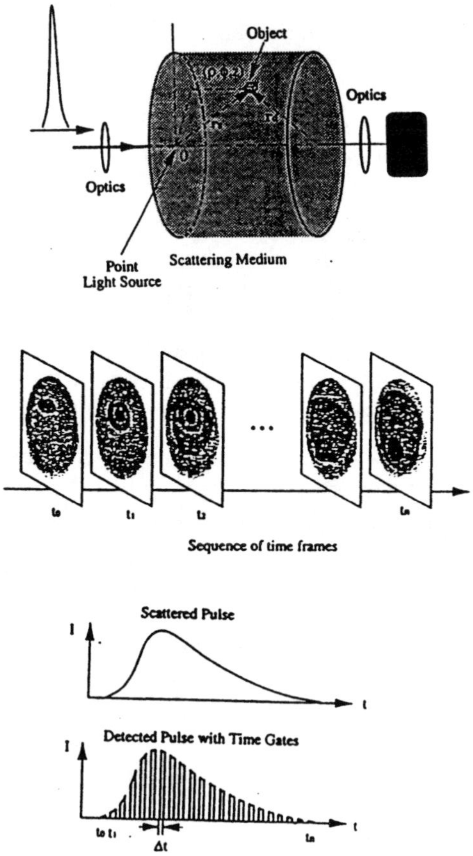

Figure 1. A schematic of the experimental geometry.

The inverse algorithm is based on the diffusion approximation of the radiative transport theory for photon migration in a scattering medium.[3] The setting is on a slab geometry $(-z_0, z_0)$ formed by cylindrical coordinates (ρ, ϕ, z). A point source (or a ϕ symmetric finite source) is located (or is centered) at $(0, 0, z_0)$. The detector system is located on the $-z_0$ plane, and is mapped to a CCD camera through a lens. The ρ boundary is assumed to be at infinity. However, it can be replaced by a finite boundary using a more complex Green's function. The reference system is assumed to be uniform.

The forward problem is represented as a matrix of form:

$$Y = WX, \tag{1}$$

where Y is an array containing measurements of change in intensity due to hidden objects, and X is an array of optical parameter changes (absorption coefficient and scattering coefficient) due to hidden objects. W is called the weight function,[3] which, in the linear inversion case, is related only to the reference but not to hidden objects. Under these conditions and assumptions, W satisfies ϕ-rotation invariance, namely, it is a function of $\phi-\phi_d$, where ϕ_d is the angle-coordinate of a point on the detector plane, ϕ is the angle-coordinate of a voxel. For a time-resolved absorption tomography, W is written as:

$$W_{3D}(\rho_d, \rho, z, \phi-\phi_d, t) = \frac{\Delta V}{G^0_\rho(\rho_d, t)G^0_z(-z_0, z_0, t)} \times\int_0^t d\tau \frac{1}{4\pi Dc(t-\tau)}\exp\left[\frac{-\rho^2-\rho_d^2+2\rho\rho_d\cos(\phi-\phi_d)}{4Dc(t-\tau)}\right] \times G^0_z(z, -z_0, t-\tau)\, G^0_\rho(\rho, \tau)\, G^0_z(z, z_0, \tau) \tag{2}$$

where $G^0_z(z, z_0, t)$ is a 1D slab Green's function for reference system with source at z_0. $G^0_\rho(\rho, t)$ is a 2D Green's function with the source located (or centered) at $\rho=0$. For a point source, it is of the form:

$$G^0_\rho(\rho, t) = \frac{1}{4\pi Dct}\exp\left[\frac{-\rho^2}{4Dct}\right], \tag{3}$$

and for a finite size spot source, it is written as

$$G^0_\rho(\rho, t) = \frac{1}{2Dct}\int f(\rho_s)\exp\left[-\frac{\rho^2+\rho_s^2}{4Dct}\right] I_0\left[\frac{2\rho\rho_s}{4Dct}\right]\rho_s d\rho_s, \tag{4}$$

where I_0 is the zeroth-order Modified Bessel function, and $f(\rho_s)$ is the intensity distribution of the incident light spot.

After making a one-dimensional Fourier transform over $\phi-\phi_d$, we obtain K independent 2D matrices, $W_{2D}(k)$, parameterized by k, with K the number of grid-points in the Fourier k-space. $W_{2D}(k)$ is a $M\times N$ matrix, with $M=$(number of ρ_d division)×(number of time slices), and $N=$(number of ρ division)×(number of z division). The indices of $\phi-\phi_d$ in the 3D real space are transferred to indices of k in the Fourier space. The normal form, $W^T_{2D}(k)W_{2D}(k)$, is a $N\times N$ matrix. We separately calculate the K inverse matrices $[W_{2D}(k)^T W_{2D}(k)+\Lambda(k)]^{-1}$, $k=1,2,\ldots K$, where $\Lambda(k)$ is a matrix for regularization. These inverse matrices are stored as a database for later reconstruction of images in different hidden object cases. The computational complexity now is K times that of inverting a W_{2D} matrix, which is much less than that of inverting a W_{3D} matrix.

Having experimental data $Y(\rho_d, \phi_d, t)$, we make a Fourier transform over ϕ_d to obtain $Y(\rho_d, t, k)$, $k=1, 2, \ldots, K$. The technique for fast discrete Fourier convolution is used. We then use simple matrix multiplication to obtain the image in the k space:

$$X(k) = Y(k)^T W_{2D}(k) [W_{2D}(k)^T W_{2D}(k) + \Lambda(k)]^{-1}. \quad (5)$$

After $X(k), k = 1, 2, \ldots, K$, is obtained, the inverse 1D Fourier transform over k gives the 3D distribution of the change of absorption coefficients or scattering coefficients, $X(\rho, \phi, z)$.

Experimental Arrangement and Procedure

A schematic diagram of the experimental setting is shown in Fig. 1. The object to be imaged was a black $3\times3\times10\,\text{mm}^3$ rectangular parallelepiped suspended inside Intralipid-10% suspension in water contained in a 60 mm long and 200 mm diameter cylindrical Plexiglas cell. It was placed on-axis at a distance of 15 mm from the end face of the cylinder closer to the detector. It was illuminated through the opposite end face of the cylinder with 800-nm, 150-fs, 1 kHz repetition-rate pulses of near-infrared light from a Ti:sapphire laser and amplifier system.[7] The 3-mm diameter collimated laser beam with an average power of approximately 100 mW was incident along the cylinder axis. The concentration of Intralipid-10% suspension in water was adjusted to provide an estimated[8] reduced scattering coefficient μ_s' of 0.4 mm^{-1} and absorption coefficient μ_a of 0.02 mm^{-1} at 800 nm.

The intensity distribution of light emergent from the detector-end face was collected by a camera lens and recorded by an ultrafast electronic gated imaging system (UEGIS). The UEGIS is a compact gated image intensifier unit fiber-optically coupled to a charge-coupled device (CCD) camera. It provides an electronic time gate pulse whose full-width-at-half maximum (FWHM) duration could be adjusted from 2 ns to a minimum of approximately 80 ps. The gate position could be varied in steps of 25 ps over a 20 ns range. The signal recorded by the system at a particular gate position is a convolution of the transmitted light pulse with the gate pulse centered on the gate position, and provided a two-dimensional intensity distribution at a particular instant of time, $I(x, y, t_i)$. This time-sliced intensity distribution was recorded by the CCD camera and displayed on a personal computer.

In the experiment reported here, the gate width was adjusted to 80 ps and time-sliced transmitted intensity distributions, $I(x, y, t_i)$, were recorded over a 5 ns range by varying the gate pulse position in steps of 100 ps. The zero time was taken to be the instant when the incident light pulse entered the front surface of the cell. $I(x, y, t_i)$ provided the experimental data for inverse image reconstruction.

Results

a) Image from simulated data

In the simulation, the thickness of the slab was taken to be $2z_0 = 60$ mm. A point light source was considered. The detector system was taken as a circle with a radius $R_d = 30$ mm. To choose grids of equal area, the angle 2π was uniformly divided by 36, and the radius R_d was divided by 30 in a ratio of interval $\Delta\rho_d \sim \rho_d^{-1/2}$. In each simulated temporal profile, intensities at 30 time slices uniformly distributed from 450 ps to 1900 ps were taken. The number of sampling data was 30×36×30. Data both with and without hidden objects were computed to produce $Y = -\log(I/I_0)$, after adding uniformly distributed noise in the ±5% range, where I and I_0 are intensity with and without the hidden object, respectively. The volume of sample was divided into 10×36×20 voxels, with 20 divisions along the z direction, and 10 divisions along ρ direction in a 30 mm range, in a way similar to that used for detector division.

Elements of W were calculated using Eq. (2) with a background transport mean free path $l_t^0 = 2.5\,\text{mm}$ and absorption length $l_a^0 = 500\,\text{mm}$. The parameters of hidden objects were $l_a = 2.5\,\text{mm}$ and $l_t = 2.5\,\text{mm}$. The regularization parameter $\lambda = 1.0$ was taken based on the L-curve analysis. Image of the absorption object obtained using Eq. (5) and 1D Fourier transform is shown in Fig. 2. The 20 consecutive circles in Fig. 2, counted by column first (from left to right) then by row (from top to bottom), represent the images at different positions along the cylinder axis (z-axis), from the source side to the detector side. The sample of hidden object was set in a voxel located at 15th division along z-axis, which was about 15 mm away from the detector, shown as a darkest area in the 15th circle of Fig. 2.

The image of a hidden object with size of about 3 mm in length can be clearly seen, with the resolution at (ρ, ϕ) plane is about 6-7 mm. The resolution along z direction is about 10 - 15 mm. With increasing distance between the hidden object and the detector, the image resolution becomes worse. In this case, reversing the z-positions the source-detector pair would improve resolution.

b) Image from experimental data

In the inversion using experimental data, we approximated the laser beam to be a point source located at center of $z_0 = 30$ mm plane. The ρ boundary of the container was assumed to be infinity. The time-resolved 384×286 pixels image from the CCD camera, with and without a hidden object, were transformed to provide 30×36×30 of Y data, as was done in simulation. The plot of $\|X_\lambda\|$ vs. $\|WX_\lambda - Y\|$ for computing the regularization parameter, λ, showed only a small kink on the curve instead of a clear L-shape. The norm of the residual vector $\|WX_\lambda - Y\|$ at kink position was about 2 orders-of-magnitude larger than

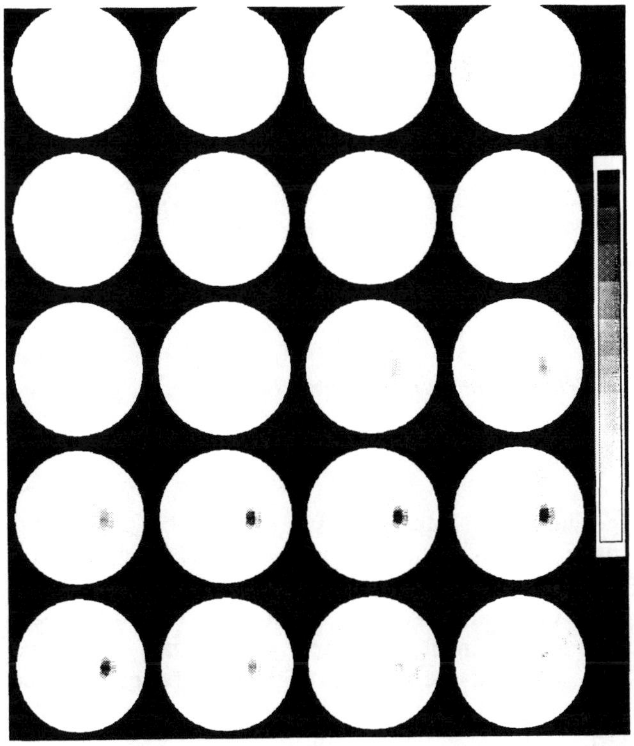

Figure 2. Images from simulated data

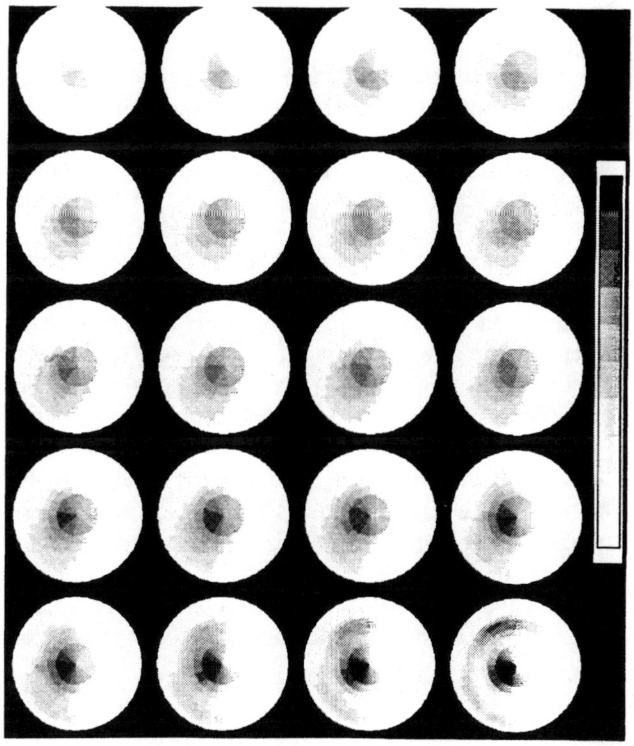

Figure 3. Images from experimental data

that in the simulated case. This indicates existence of deviation between theoretical forward model and experiment and experimental noise. Fig. 3 shows the 3D image of the hidden object using experimental data. The lateral position of the object is correctly presented in the image map. However, the resolution in z-direction is limited. The position of the object in the image is much more shifted towards the detector plane with low resolution. Our computation shows a 3D image of $36 \times 20 \times 10$ voxels can be obtained with a running time of less than 1 minute on a Silicon Graphic Origin 2000 computer.

In summary, our initial results demonstrate that time-sliced transmission measurements using an electronic time-gated camera and the diffusion tomographic inverse algorithm show promise of obtaining 3D images of objects in a turbid medium in a short time.

This work is supported in partly by NASA, DOE, ARO, Mediscience Technology Corp. and New York State Science and Technology Foundation.

Reference

[1]For a brief review see S. K. Gayen and R. R. Alfano, "Emerging optical biomedical imaging techniques," Opt. Photon. News **7(3)** 16 (1996).

[2]B. Chance, K. Kang, L. He, J. Weng, and E. Sevick, "Highly sensitive object location in tissue models with linear in-phase and anti-phase multi-element optical arrays in one and two dimensions," Proc. Natl. Acad. Sci. USA **90** 3423-3427(1993).

[3]S. R. Arridge, "The forward and inverse problems in time resolved infra-red imaging" Medical optical tomography: functional imaging and monitoring SPIE Institutes, Vol. IS11, G. Muller (ed.) 31-64, (1993).

[4]W. Cai, B. B. Das, F. Liu, M. Zevallos, M. Lax, R. R. Alfano, "Time resolved optical diffusion tomographic image reconstruction in highly scattering turbid media", Proc. Natl. Acad. Sci. USA **9** 1561-1564 (1996).

[5]W. Cai, B. B. Das, F. Liu, Fan An Zeng, M. Lax, R. R. Alfano, "Three dimensional image reconstruction in highly scattering turbid media", SPIE **2979** Proceedings of Optical Tomography and Spectroscopy of Tissue: Theory, Instrumentation, Model, and Human Studies II, 241-248 (1997).

[6]Per Christian Hansen, "Analysis of Discrete Ill-Posed Problems by Means of the L-Curve", SIAM **34** 561-580, (1992).

[7]Q. Fu, F. Seier, S. K. Gayen, and R. R. Alfano, "High-averagepower kilohertz-repetition-rate sub-100-fs Ti:sapphire amplifier system", Opt. Lett. **22** 712 (1997).

[8]H. J. van Staveren, C. J. M. Moes, J. van Marle, S. A. Prahl, and M. J. van Gemert, "Light scattering in Intralipid-10% in the wavelength range of 400-1100 nm," Appl. Opt. **31** 4507-4514 (1991).

Near-infrared imaging and optical properties of tissue phantoms having curved boundaries and clear layers

Yong Wang [1,3], Robert M. Danen [1, 2], William S. Thayer [1], Britton Chance[3] and Arjun G. Yodh [2]

[1] *NIM Incorporated, 3624 Market Street, Suite 508, Philadelphia, PA 19104*

[2] *Dept. of Physics and Astronomy, University of Pennsylvania, Philadelphia, PA 19104*

[3] *Dept. of Biochemistry and Biophysics, University of Pennsylvania, Philadelphia, PA 19104*

Abstract

A multi-source multi-detector near-infrared frequency-domain instrument was used to characterize the optical properties of tissue phantoms having curved boundaries and a clear layer to simulate the brain and cerebrospinal fluid. Cylindrical models with varying diameters and a three-layer model were studied. As radius of curvature of the model was decreased, both μ_a and μ_s' decreased compared with a flat surface model. The presence of a clear layer dramatically decreased μ_s', but had relatively little effect on μ_a. When a black object was located beneath the clear layer, an image of the object could still be obtained from the surface.

Key words

(110.3080) Infrared imaging; Brain Model; Cylindrical model; Three-layer model.

Introduction

The determination of optical properties provides a basis for new diagnostic applications in medicine. Infinite or semi-infinite boundary conditions and homogeneous medium are often applied for calculation of optical properties and imaging. In the NIM Regional Imager, we have used semi-infinite boundary conditions to obtain a near-infrared image of the brain [1]. However, cylindrical and layered structures more closely resemble human heads. For a cylindrical model, the absorption coefficient as derived from measurement along the Z axis of the cylinder is weakly dependent on curvature when considered according to a semi-infinite geometry [2]. Measurements of model systems and theoretical calculations show that thin clear layers (as a simulation of the cerebrospinal fluid (CSF) between the skull and brain) have a significant effect upon the light distribution in the medium [3]. In this study we investigate the effect of cylindrical boundaries for measurements along the azimuthal direction and the effect of a clear layer in a three-layered structure on images and the optical properties.

Experimental Design

Instrument

We measured the average optical properties of the cylinder model using our infrared Regional Imager [1]. This instrument is a frequency-domain, multi-wavelength system with multiple source and detector positions (Figure 1; see also Danen et al., these proceedings) Light is sent to and from the sample via optical fibers, which are positioned against the model by an optical pad. The pad is a semi-flexible, sturdy rubber sheet that holds 13 source fibers and 4 detector fiber bundles arranged in a 9.6 cm x 4.8cm rectangular area. This configuration gives separations (on a flat surface) between sources and detectors ranging from 2 to 6.8 cm. We measure the amplitude and phase of the transmitted diffusive wave. We then plot ln(ρ^2· Amplitude) and phase as functions of the source-detector separation ρ for all source-detector combinations. The slopes of lines fit to these plots give the real and imaginary components of the wavevector, k_r and k_i. These quantities determine the absorption coefficient μ_a and scattering coefficient μ_s' of the medium .

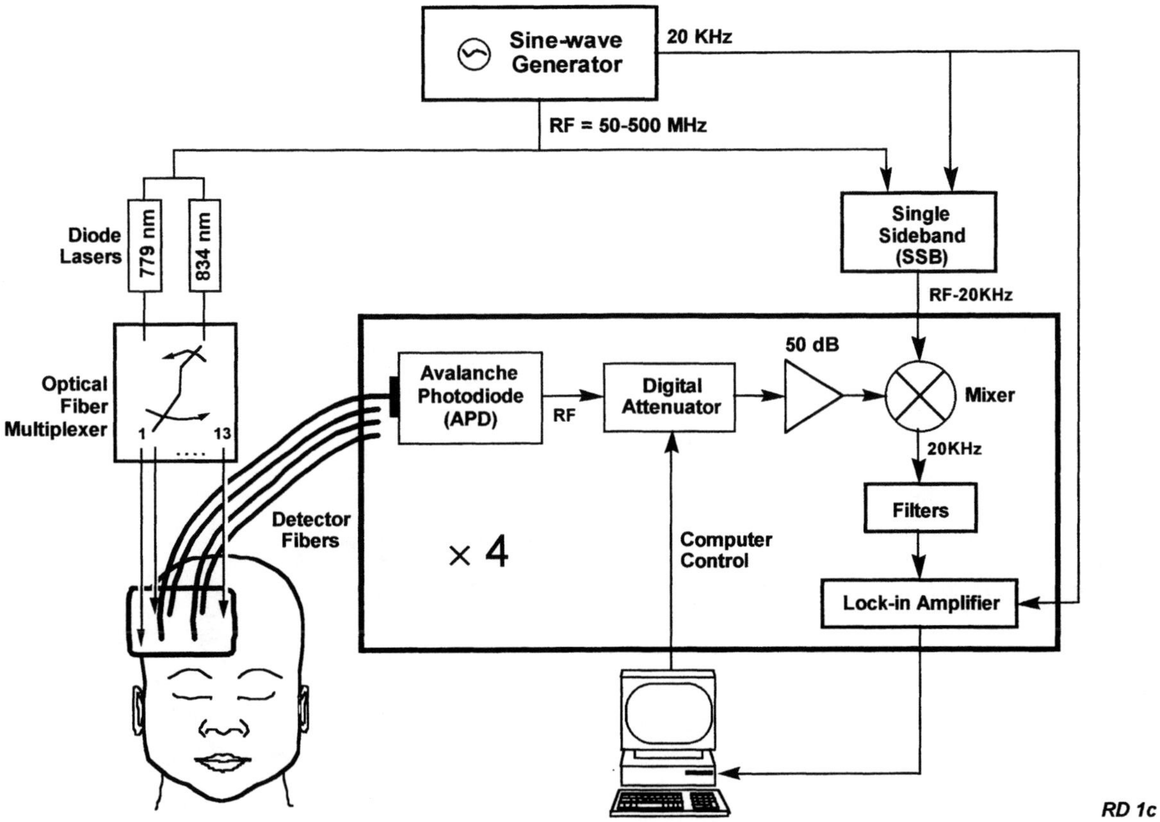

Figure 1. Electronic configuration of the NIM Regional Imager. The Regional Imager contains four identical detection circuits that include an avalanche photodiode, computer-controlled amplification electronics, mixer, filters and lock-in amplifier

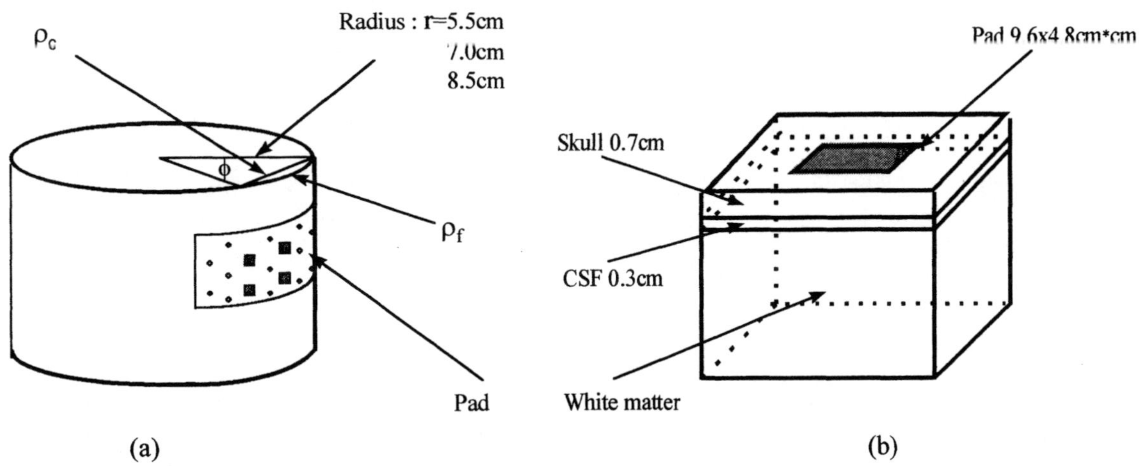

Figure 2. (a) The cylindrical phantom model, with a diagram of the optical pad. On the pad, squares represent detectors and circles represent sources. (b) The three-layer phantom model simulating various anatomical layers within the head.

Model Construction

(a) Cylindrical Models

In order to test the effects of a curved surface on absorption and scattering parameters under semi-infinite boundary condition, we prepared gel phantoms, made from agarose, TiO_2, and black India ink, cast into cylinders of 3 different diameters: 11cm, 14cm, 17cm as well as a flat surface phantom. The composition of the phantoms was identical; only the diameters were different. The optical pad was placed against the model oriented with the long dimension of the pad around the circumference (see Fig. 2a).

(b) Three-layer Model

To investigate the effects of the thin layer of CSF between the brain and skull, we constructed a three-layer model (Figure 2b). The model included a top layer (representing skull and scalp) of 0.7 cm thickness, a clear layer (simulating the CSF) of 0.3 cm thickness, and a bottom layer (simulating brain tissue) of several cm thickness. The clear layer was composed only of agarose, whereas the top and bottom layers contained the same amounts of TiO_2, which scatters light, but different amounts of India Ink in order to correspond approximately to the optical properties of the relevant layers of the head.

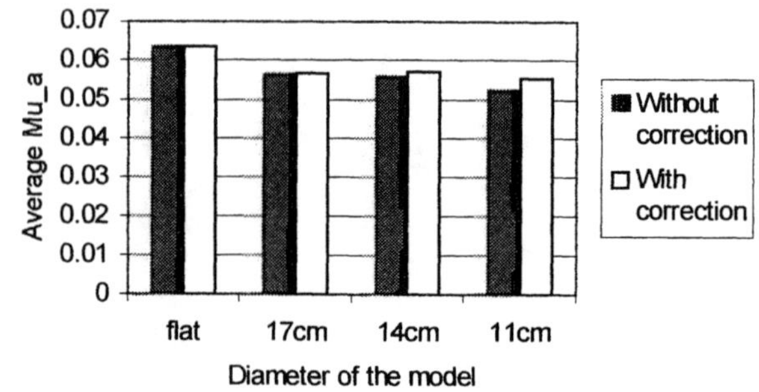

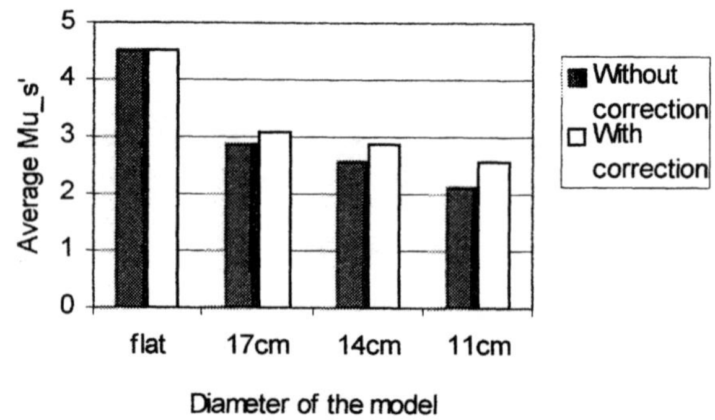

Figure 3. Average μ_a and μ_s' vs the diameter of the model.

Results

(a) Cylindrical Models

For our model studies, the laser diode wavelength was 834 nm and the modulation frequency was 200 MHz. The gray shading in Fig. 3 shows the variation of μ_a and μ_s' along the circumferential direction vs diameter of the cylinder. In comparison to the values of a flat model, both μ_a and μ_s' decreased with decreasing diameter in the cylindrical models. However, the effect on μ_s' was more substantial than that on μ_a. One reason for this finding is that the separation distance between a particular source and detector pair on a curved surface is shorter than that on a flat surface because of the physical geometry. The white shading of Fig. 3 shows that the effect of curvature is less pronounced when we use $\rho_c=2r\sin(\rho_f/r)$ instead of ρ_f in fitting the data for determination of optical parameters. This procedure may be useful as a correction in analyzing data from the baby head.

Table 1. Average μ_a and μ_s' values in the three-layer phantom model.

	Top layer only (cm^{-1})	Bottom layer only (cm^{-1})	Three-layer model [from top] (cm^{-1})
μ_a	0.109	0.080	0.103
μ_s'	4.7	4.9	1.2

(b) Three-layer Model

Table 1 shows optical properties of the individual top and bottom layers used in the three layer model and the overall optical properties measured on the three layer model itself. The measured absorption coefficient of the three-layer model remained approximately the same as the absorption coefficients of the top and bottom layers. This result is consistent with the findings of Hielscher et al. [4]. The measured scattering coefficient, however, changed dramatically with the addition of a clear layer.

In order to obtain further information about the possible effect of a CSF layer on imaging the brain, we obtained μ_a images of a black object at different depths within the bottom layer of the three-layer model. We used the back projection reconstruction algorithm discussed in [1]. The object was a black object having a 6.5 mm diameter and 20 mm length. We put the object at depths of 1.5 and 2.5 cm below the pad. Fig.4 shows images obtained with the clear layer (Figs. 4a and 4b), and without a clear layer (Figs. 4c and 4d).

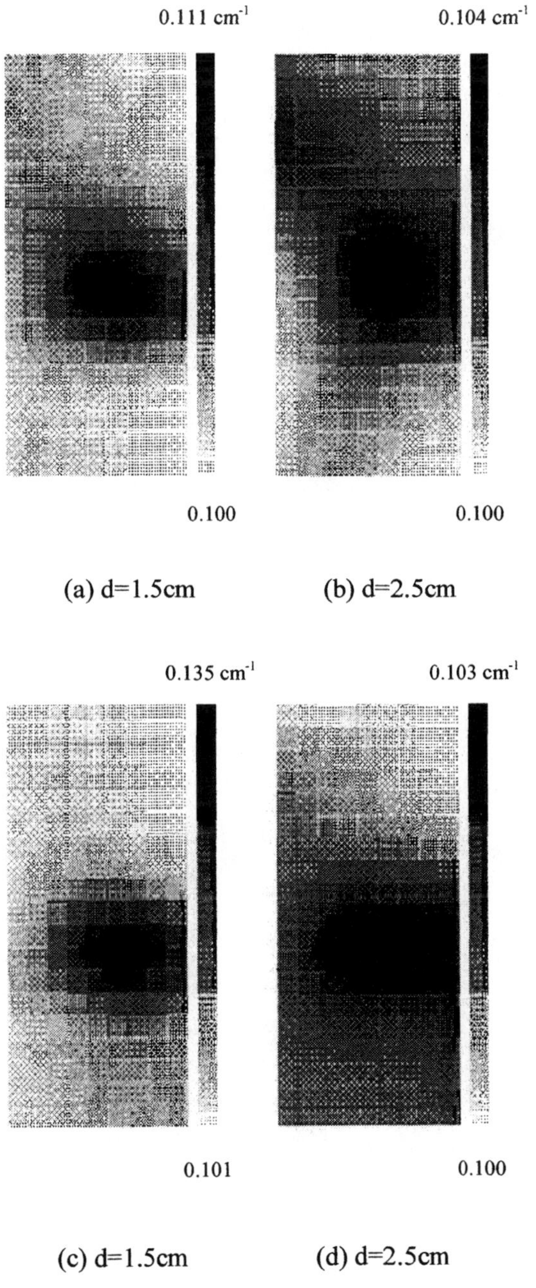

Figure 4. Images of μ_a with the object located at different depths: (a) and (b) with a clear layer, (c) and (d) without a clear layer.

The results show that the object could be detected even when a clear layer was between the object and the optical pad. However, the range of values for the optical parameters was altered slightly by the presence of the clear layer. The amplitude signals were larger when the clear layer was present than when it was not. In addition, this effect was more noticeable at longer separation distances between source and detector than at short separation distances.

Conclusion

Our results show that both curvature of a head and the presence of the thin layer of CSF need to be taken into account when imaging or measuring the optical properties of the brain. Smaller radius of curvature leads to greater underestimation of the true values of μ_a and μ_s'. The presence of a CSF-like layer changes μ_s' dramatically, whereas the effect on μ_a is much less significant. In spite of these effects, images can still be obtained.

References

1. R.M. Danen, Yong Wang, X.D. Li, W.S. Thayer, and A.G. Yodh, "Regional Imager for low-resolution functional imaging of the brain with diffusing Near-infrared light," Photochemistry and Photobiology 67(1), pp. 33-40 (1998).
2. Albert Cerussi, John Maier, Sergio Fantini, Maria Angela Franceschini and Enrico Gratton, "The frequency domain multi-distance method in the presence of curved boundaries," in Technical Digest of Conference on Advances in Optical Imaging and Photon Migration (Optical Society of America, Orlando FL), paper BWB3-1, pp. 24-26 (1996).
3. Michael Firbank, Martin Schweiger, David T Delpy, "Investigation of 'light piping' through clear regions of scattering objects" Proc. SPIE-Int. Soc. Opt. Eng. 2389, pp167-173 (1995).
4. Andreas H. Hielscher Hanli Liu, Britton Chance, Frank K.Tittel and Steven L. Jacques "Phase Resolved reflectance spectroscopy on layered turbid media" Proc. SPIE-Int. Soc. Opt. Eng. 2389, pp248-256 (1995).

Non-invasive characterization of breast lesion optical and structural properties using multi-wavelength, multi-frequency photon density waves co-registered with ultrasound imaging

X.D. Li, A.G. Yodh [a]
S.N. Huang, B. Chance [b]
[a]Department of Physics and Astronomy
[b]Department of Biochemistry and Biophysics
University of Pennsylvania

J.B. Fishkin, N. Shah, B.J. Tromberg [c]
D. Kidney [d]
J. Butler [e]
[c] Laser Microbeam and Medical Program
Beckman Laser Institute
[d] Department of Radiology
[e] Chao Cancer Center
University of California-Irvine

We present the preliminary results of quantitative characterization of normal and malignant breast tissues in a human subject. The tissue was probed non-invasively by multi-wavelength (674, 786, 803, 849, and 956 nm) multi-frequency (50MHz-1GHz) diffuse photon density waves (DPDW). The optical characterization of the lesion was co-registered with Ultrasound. Ultrasound images provide us information about the lesion size and position which were used for optical characterization of localized inhomogeneity (lesion). Multi-frequency allows us to extract the optical properties (the absorption and reduced scattering coefficients) at a single source-detector separation. We assign the optical properties of the lesion by fitting the data to analytic solutions of diffuse photon density waves for a semi-infinite heterogeneous medium. We find that the lesion and normal tissues are clearly resolvable. We also find that the optical contrast is elevated by analyzing the data with the sophisticated *heterogenous* model (as opposed to a *homogeneous* model). Multi-wavelength spectroscopic measurements indicate that the lesion might have higher water concentration.

Reprinted from Advances in Optical Imaging & Photon Migration, 1998 Technical Digest,

Sensitivity Study on Scattering Media with MRI-Guided Spatial Heterogeneity

Y. Pei[1], H. L. Graber[2], F.-B. Lin[1], and R. L. Barbour[2,3,4]

[1]Department of Civil Engineering, Polytechnic University, 6 Metrotech Center, Brooklyn, NY 11201

[2] Program in Physiology and Biophysics, SUNY Health Science Center at Brooklyn, 450 Clarkson Ave., Brooklyn, NY 11203

[3] Department of Pathology, SUNY Health Science Center at Brooklyn, 450 Clarkson Ave., Brooklyn, NY 11203

[4] Department of Electrical Engineering, Polytechnic University, 6 Metrotech Center, Brooklyn, NY 11201

Abstract
The sensitivity of NIR optical measurements of the breast, as a function of parameters associated with data collection and of target properties, have been computed from analysis of MRI data. Specific parameters explored include: source condition (DC, 200MHz), view angle (coronal vs. sagittal view), optical contrast of background tissues, contrast of embedded tumor, tumor to breast cross-sectional area ratio, and breast diameter. Results obtained demonstrate the importance of background heterogeneity and previously undocumented an improvement in sensitivity to embedded objects with increasing target diameter. In addition, the sensitivity dependence on the explored parameter space was not found to scale directly with target size.

Key Words
Scattering media, Photon migration, Light propagation in tissue.

Introduction

NIR technologies afford potentially significant advantages over established imaging methods for investigations of the breast, owing to their low cost, compact instrumentation and high intrinsic sensitivity. Presently, numerous laboratories worldwide are engaged in the development of practical systems [1]. Key to these efforts is the need to adopt methodologies that identify the expected system performance under *defined* realistic conditions. Such results provide a guide to rational instrument development, serve as a basis of comparison to clinical studies and aid in test validation. An approach we have taken in this venue has been to evaluate anatomically accurate models of human tissue as defined by MRI images [2,3]. In the work reported on here, we have extended these studies and undertaken a systematic survey of the parameter space tied to measurement sensitivity. This includes parameters associated with data collection and target properties. The former comprises the source condition (*e.g.*, DC or AC illumination) and view angle (*e.g.*, coronal and sagittal) while the latter includes the geometry and composition of the breast and embedded objects simulating tumors. The survey conducted was extensive, and limitations on space restrict the scope of reported findings here. Results are restricted to coronal views using DC source illumination. More complete details of the study will be described elsewhere.

Methods

MRI Acquisitions: T1 weighted MRI images of the breast were acquired using the GE Signa system fitted with a breast coil. A total of 24 slices were acquired and segmented according to tissue type (mainly fat and parenchyma) using both coronal and sagittal views. Reported here are results restricted to studies on coronal views. For selected slices, a centrally located inclusion, simulating a tumor, was introduced.

FEM Model Development: Figure 1(a) shows an image of a representative MRI slice (displayed at low resolution) used in these studies. Figure 1(b) shows the finite element model derived from the MRI map. The two maps differ because the external boundary of the breast was extended to conform to a circular geometry. This was adopted to reflect the expected measurement geometry achieved using a newly constructed imager (*i.e.*, IRIS-OPTIScanner) described elsewhere in this volume. Three different tissue types are identified: fat (dark region in MRI map), parenchyma (light region in MRI map) and a centrally positioned tumor shown in the FEM map. The extended region was assigned coefficients corresponding to fat. A semiautomatic image segmentation code, provided by C. R. Johnson, University of Utah, was used to segment the map to produce boundary regions required for FEM mesh generation. The map shown in Figure 1(b) was used for all breast diameters considered. The number of nodes and mesh points used in the different models varied depending on breast size. For small diameters, the number of nodes and mesh points were on the order of 1500 and 3000, respectively. For large diameters, these values were increased by a factor of 15. To improve on the efficiency of the FEM calculation, a multigrid solver (KASKADE) was used that produced solutions for successively finer densities of the nonuniform mesh, increasing the mesh density by a constant multiplicative factor each time.

Assignment of Optical Coefficients and Target Geometries: Optical absorption and scattering coefficients of the background tissues were assigned according to Table 1. Table 2 defines the optical coefficients assigned to an embedded tumor. Table 3 lists the range of diameters of breasts and tumors explored.

Forward Model: Solutions to the 2-D diffusion equation were computed by the FEM method using Dirichlet boundary conditions. Depending on the breast geometry, sources and detectors were located 1-2 mean free paths lengths below the surface. The arrow in Figure 1(b) illustrates the source location used for the reported studies and is designated as zero angle. Angular increments in steps of 10° proceed in the clockwise direction.

Parameter Space Explored: The complete matrix described by Table 3 (35 cases) was computed for each of the nine different contrast values of the embedded tumor for the case 3 background medium (Table 1). A similar calculation was performed for each of the eight different background cases for the case 3 tumor contrast (Table 2). A total of 595 cases were explored for each view and source condition.

Results and Discussion

I. Contrast Studies

A. Effect of Tumor Contrast. The results in Figure 2 show the computed relative intensity change caused by a change in the tumor's absorption coefficient as a function of the view angle for a specific background medium, breast size and ratio of cross-sectional area of tumor to breast size. The absorption coefficient of the tumor was varied over the range 0.08-0.4 cm^{-1}. As expected, for the target geometries (breast and tumor size) and background media explored, the relative change increases with increasing angle from source and with increasing μ_a.

B. Threshold Contrast Study. The parameter titration summarized in Figure 2, for each of the 35 breast-tumor geometry cases explored, was

Table 1: Optical properties of tumor tissue

Cases	μ_a (cm^{-1})	μ_s' (cm^{-1})
1	0.20	40
2	0.20	20
3	0.20	10
4	0.20	5
5	0.08	40
6	0.08	20
7	0.08	10
8	0.08	5
9	0.40	10

Table 2: Optical properties of background tissue

Cases	Fat		Parenchyma	
	μ_a (cm^{-1})	μ_s' (cm^{-1})	μ_a (cm^{-1})	μ_s' (cm^{-1})
1	0.04	10	0.04	10
2	0.02	10	0.08	7
3	0.04	10	0.08	7
4	0.08	10	0.08	7
5	0.04	10	0.08	15
6	0.04	10	0.08	25
7	0.04	15	0.08	7
8	0.04	25	0.08	7

Table 3: Diameters of breast and embedded tumor size

Cases	Breast Diameter (cm)	Tumor Diameter (cm) (Ratio of cross-sectional area of tumor to breast size)				
		(0.0625 %)	(0.25%)	(1%)	(2.5%)	(4%)
I	16	0.4	0.8	1.6	2.4	3.2
II	12	0.3	0.6	1.2	1.8	2.4
III	10	0.25	0.5	1.0	1.5	2.0
IV	8	0.2	0.4	0.8	1.2	1.6
V	6	0.15	0.3	0.6	0.9	1.2
VI	4	0.1	0.2	0.4	0.6	0.8
VII	2	0.05	0. 1	0.2	0.3	0.4

used to estimate the contrast required to produce a 1% relative intensity change for a specific background medium (case 3) and view angle (180°). As shown in Figure 3, we observe that: (a) increasing tumor size, for a fixed breast size, reduces the contrast required to produce a 1% relative change; (b) unexpectedly, *increasing* breast size for a fixed ratio of cross-sectional area of tumor to breast size *reduces* the tumor contrast required to meet the specified threshold; (c) the dependence of this response differs by a constant ratio for different tumor sizes; (d) the required tumor contrast increases rapidly with smaller breast sizes.

II. Influence of Breast Size. The dependence of relative intensity change on breast size with a fixed tumor contrast and cross-sectional area ratio is shown for two different backgrounds in Figures 4(a) and 4(b). Breast diameters were varied between 2 and 16 cm to include the expected range of diameters in the vicinity of the nipple and the chest wall. In all cases we observe that: (a) the maximum relative intensity change increases with increasing breast size; (b) the full width at half maximum (FWHM) decreases with increasing breast size, indicating improved available resolution; and (c) a biphasic sensitivity response is observed at intermediate angles as a function of breast size.

III. Influence of Background Heterogeneity.
A. Effect of increased scattering. Comparison of results shown in Figures 4(a) and 4(b), illustrates the effect that an increase in the difference between background scattering coefficients of fat and parenchyma tissue has on computed sensitivity. Most notable is a shift in the angle where maximum sensitivity is observed, from 180 to 200 degrees. Also observed is a marked narrowing of the FWHM response for larger diameter breasts.

B. Effect of target-object size. The results in Figures 5(a) and 5(b) show the relative intensity change as a function of breast and tumor diameter for a specific background medium (case 3) and view angle (180°). We observe that: (a) increasing the cross-sectional area ratio of an embedded tumor for a fixed breast size increases sensitivity; (b) increasing breast size with a fixed cross-sectional area ratio also increases sensitivity. Interestingly, opposite responses are observed when breast and tumor sizes are considered together; *i.e.*, the maximum rate of change in sensitivity occurs in large breasts having small tumors and small breasts having proportionally large tumors. These findings indicate that sensitivity responses do not simply scale with target size.

C. Comparison to homogeneous media. Data in Figures 6(a) and 6(b) compare the ratio of the relative intensity change of the case 5 background medium to a homogeneous background medium (case 1) as a function of target and object size for a specific view angle (180°) and tumor contrast (case 3). The following results were obtained. The dependence of object sensitivity on the background heterogeneity examined varies principally with breast size. In the case of small breast diameters, enhanced sensitivity is observed, while the opposite occurs for larger diameters. In addition, whereas these differences are mainly independent of tumor size for small breast diameters, they vary at larger breast sizes, but only with larger tumors. These findings reinforce the indication that sensitivity responses do not simply scale with target size.

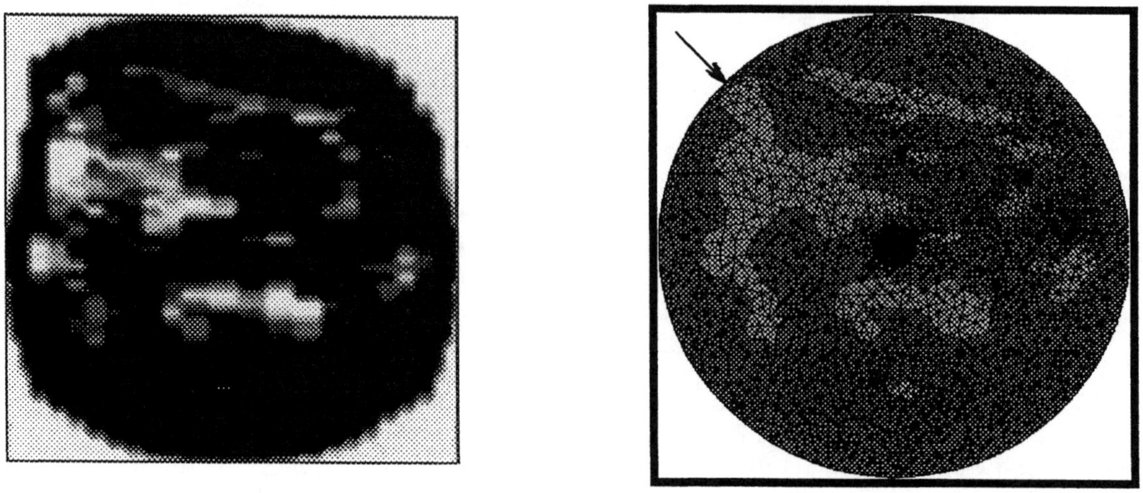

(a) (b)

Figure 1: (a) Image of a coronal slice of a breast derived from MRI maps; (b) FEM model for computation.

D. Importance of view angle. The results in Figure 7 show the dependence of sensitivity on view angle (180° vs. 200°) as a function of the cross-sectional area ratio and the breast diameter for the case 6 background medium. The dependence of the sensitivity response on the angle for this case was previously shown in Figure 4(b). Recall that increasing the scattering of the parenchyma tissue produces a shift in the angle where a maximal response occurs. This comparison is made to model the influence that restricted view measurements in a transmission mode (*e.g.*, planar compression [4]) can have on expected sensitivity. We observe that whereas little loss of sensitivity is observed for small diameter breasts, at larger diameters this increases to more than 50% and is greatest for cases involving smaller size tumors.

Conclusions

1. Unexpectedly, improved sensitivity to embedded tumors is seen in breasts having proportionally larger sizes.

2. Variations in background contrast, in particular due to increased scattering, can significantly influence the angle dependence of measured sensitivity. This finding may complicate the expected utility of measurements involving simple planar compression.

3. Parameter dependence of sensitivity does not scale directly with breast size.

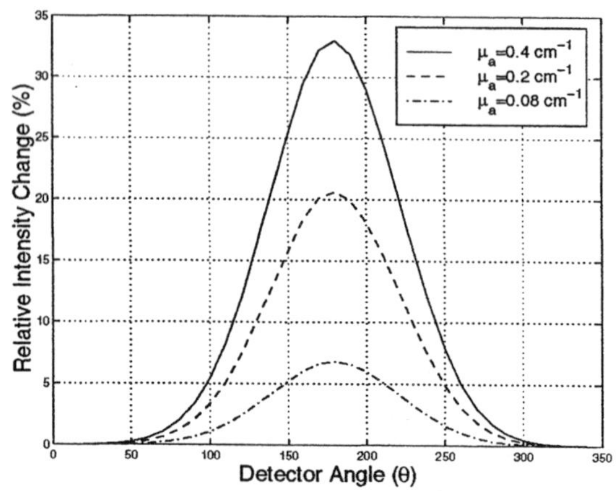

Figure 2: Relative intensity change versus view angle, for different absorption coefficients of a central tumor whose cross-sectional area is 1% that of the breast and $\mu_s' = 10\ \text{cm}^{-1}$, embedded in type 3 background medium with breast diameter of 8 cm .

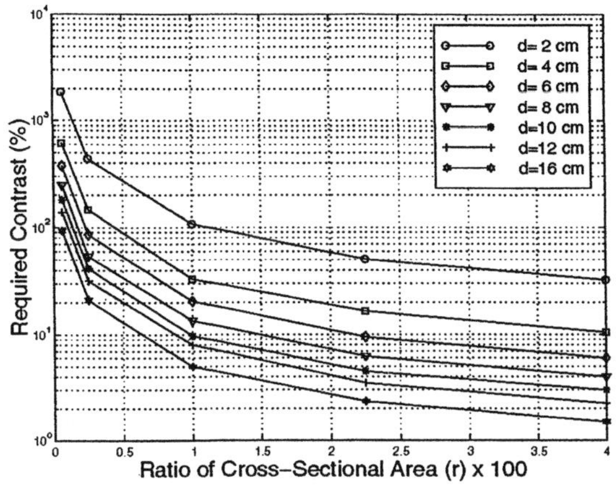

(a)

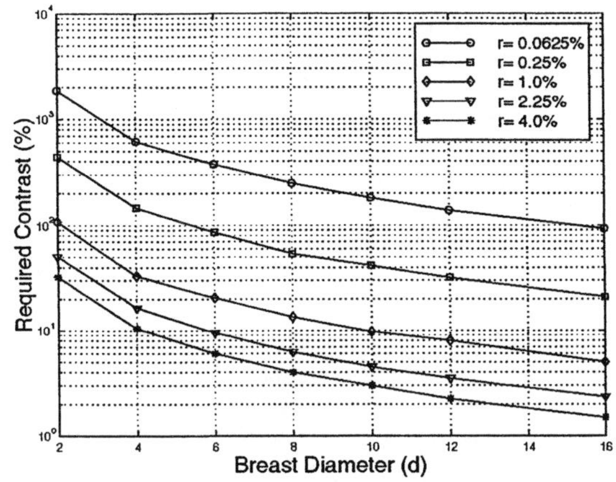

(b)

Figure 3: Tumor contrast required to produce 1% relative intensity change versus (a) the tumor-to-breast cross-sectional area ratio for different breast sizes and (b) breast diameter for different cross-sectional area ratios with case 3 background medium at 180° view angle.

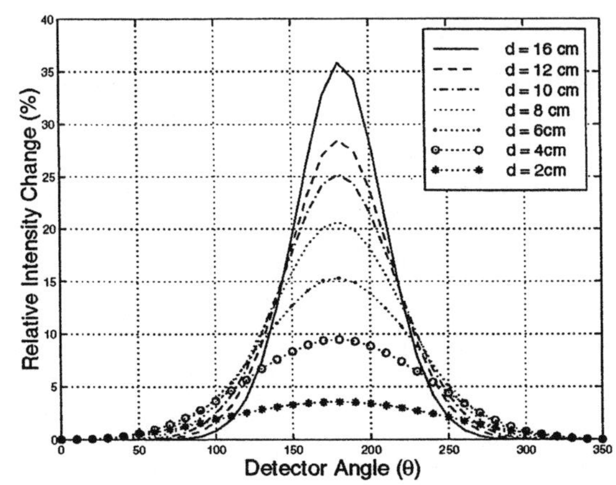

(a)

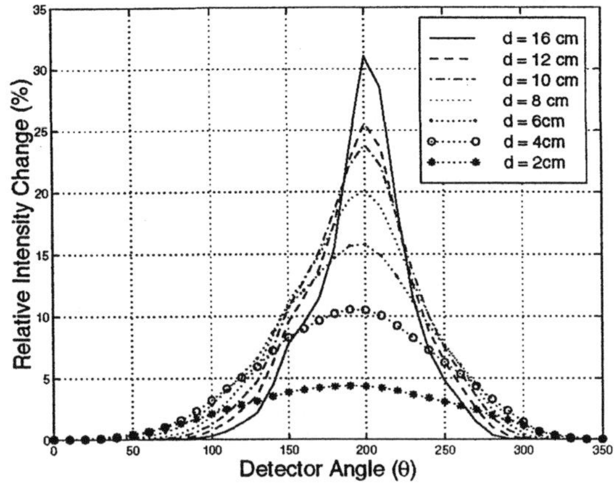

(b)

Figure 4: Percent change in relative intensity versus view angle for different breast size, with case 3 tumor (1% of total cross-sectional area), embedded in (a) case 3 and (b) case 6 background medium.

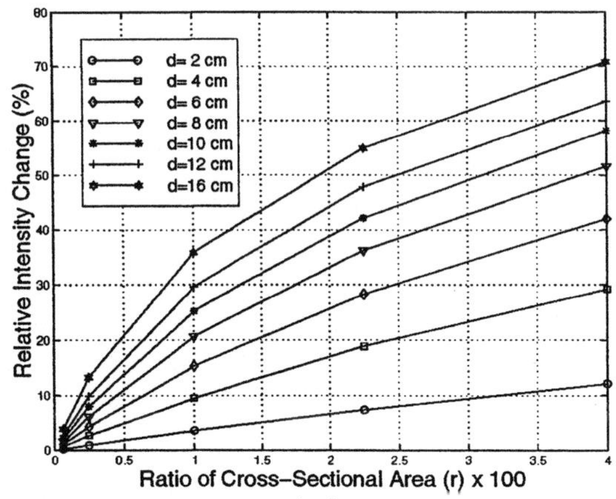

(a)

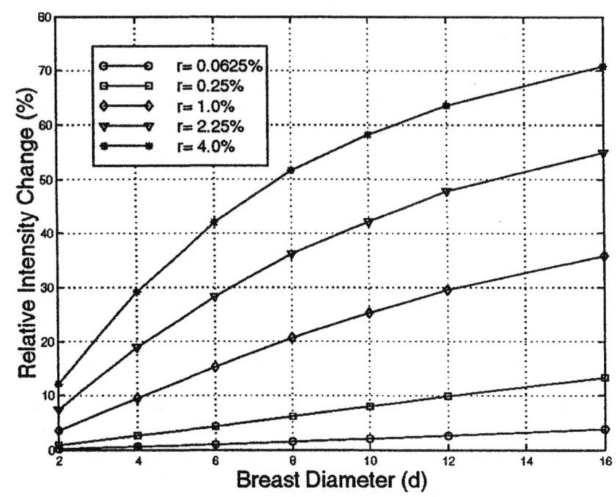

(b)

Figure 5: Percent change in relative intensity, versus (a) the cross-sectional area ratio for different breast diameters and (b) the breast diameter for different cross-sectional area ratio with case 3 background medium at 180° view angle.

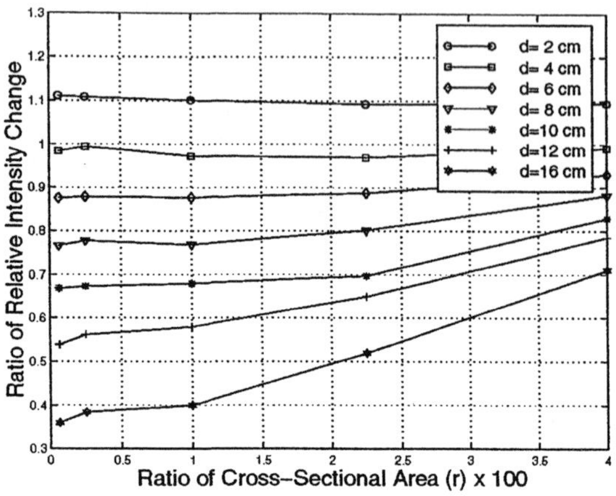

(a)

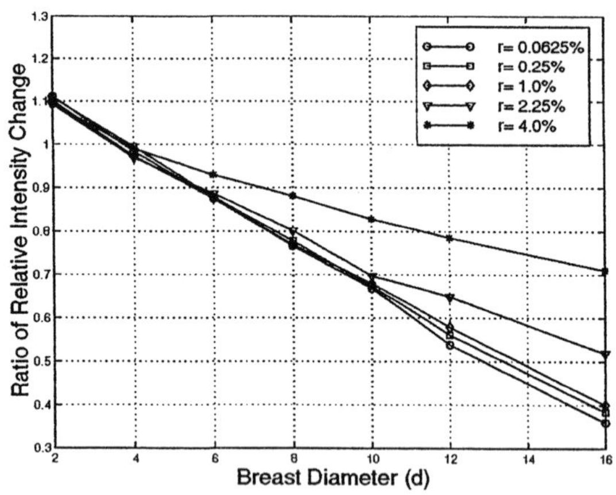

(b)

Figure 6: Ratio of relative intensity change of case 5 background medium to homogeneous background medium (case 1) versus (a) the cross-sectional area ratio for different breast diameters and (b) the breast diameter for different cross-sectional area ratio with case 3 tumor contrast at 180° view angle.

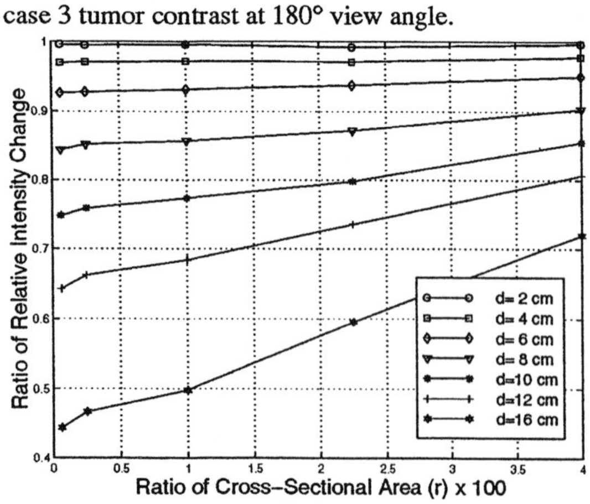

(a)

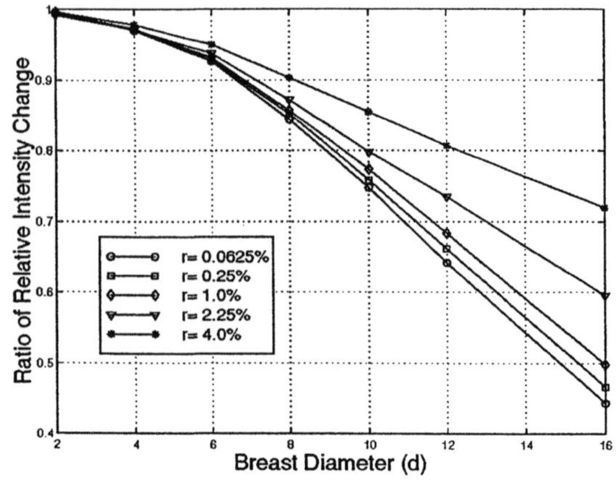

(b)

Figure 7: Ratio of relative intensity change at 180° view angle to that at 200° versus (a) the cross-sectional area ratio for different breast diameters and (b) the breast diameter for different cross-sectional area ratio with case 5 background tissue and case 3 tumor contrast.

Acknowledgements These studies were supported in part by NIH grant no. CA66184-A02.

References

1. See other papers in this volume.
2. R. L. Barbour, H. L. Graber, J. Chang, S.-L. Barbour, P. C. Koo, and R. Aronson, "MRI-Guided Optical Tomography: Prospects and Computation for a New Imaging Method", *IEEE Computational Science & Engineering*, **2**(4), 63–76, (1995).
3. J. Chang, H. L. Graber, P. C. Koo, R. Aronson, S.-L. Barbour and R. L. Barbour, "Optical Imaging of anatomical maps derived from magnetic resonance images using tim-independent optical sources", *IEEE Trans. Med. Imag.*, **9**, 68–77, (1997).
4. S. Fantini, S. A. Walker, M. A. Franceschini, M. Kaschke, P. M. Schlag and K. T. Moesta, "Assessment of the size, position, and optical properties of breast tumors *in vivo* by noninvasive optical methods", *Applied Optics* **37**, 1982–1989 (1998).

Reconstructed images for oxy and deoxy hemoglobin for a phantom with varied local oxygenation using optical CT

Ichiro Oda, Hideo Eda, Yasunobu Ito,Yukio Oikawa, Yukihisa Wada and Yoshio Tsunazawa
Technology Research Laboratory, Shimadzu Corporation
380-1, Horiyamasita, Hadano, Kanagawa 259-13, Japan
phone:+81-463-88-8731, fax:+81-463-88-8763

Motoki Oda, Kazuyoshi Ohta, Yutaka Yamashita, Yutaka Tsuchiya
Central Research Laboratory, Hamamatsu Photonics K.K.
5000, Hirakuchi, Hamakita, Shizuoka 434, Japan
phone:+81-53-586-7111, fax:+81-53-586-6180.

Introduction

Many investigators have reported the reconstructed images for phantoms with black absorbers in a scattering media or absorbers with some fixed (finite) absorption coefficient. However the final target of the optical CT should be the imaging of the oxygenation value of human subject or animals. For this approach, the authors constructed a phantom with variable oxygenation and obtained the images of the local distribution of oxy and deoxy hemoglobin from the μa images of the two or three wavelength using a newly built optical CT system.

Measurement system and image reconstruction

The optical tomography system has three wavelength pulse lasers; 761,791 and 835 nm and 64 photo multiplier tubes, 64 Time to Amplitude Converters, 64 Constant Fraction Discriminators. It can measure 64 temporal profiles simultaneously. [1] This time 16x15 continuous wave data were used by integrating 16x15 temporal profiles. Images with these CW data were reconstructed by TOAST developed by UCL[2,3]

Phantom with variable oxygenation

The body of the phantom was made of resin and the blood of different oxygenation circulated through it. The figure 1 shows the top view of the phantom whose body was cylindrical, 80 mm in diameter. It was made of silicone resin which contained scattering material so as to make the reduced scattering coefficient be 1 mm^{-1}. It had two holes, 10 mm in diameter, along the axis through which the oxygenation variable solution was circulated by the peristaltic pump. Red blood cell-milk solutions were used for this purpose. The concentration of the red blood

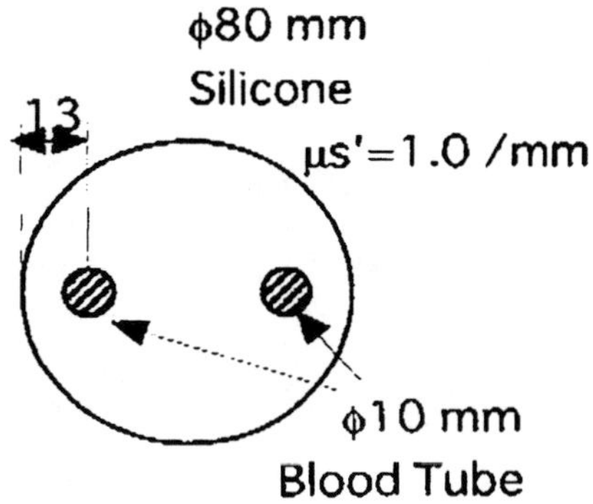

Figure 1 Top view of the phantom

Reprinted from Advances in Optical Imaging & Photon Migration, 1998 Technical Digest,

cells in the phosphate buffer was 1.5% or 3.0%. Milk was added to the solution to make the reduced scattering coefficient about 1 mm^{-1}. A nitrogen-oxygen gas system adjusted the oxygenation of the solution. Authors report the case of complete (100%) oxygenation and the case of complete deoxygenation in this experiment.

Results and discussion

1) image of absorption coefficient

Primary μa images were reconstructed for three wavelengths 761nm, 791nm and 805nm. Figure 4 shows four reconstructed images of the phantom with 1.5% erythrocyte solutions as the combination of different wavelengths (761nm or 835nm) and oxygenation (0% or 100% saturation). The corresponding wavelength and the oxygenation appear in each figure.

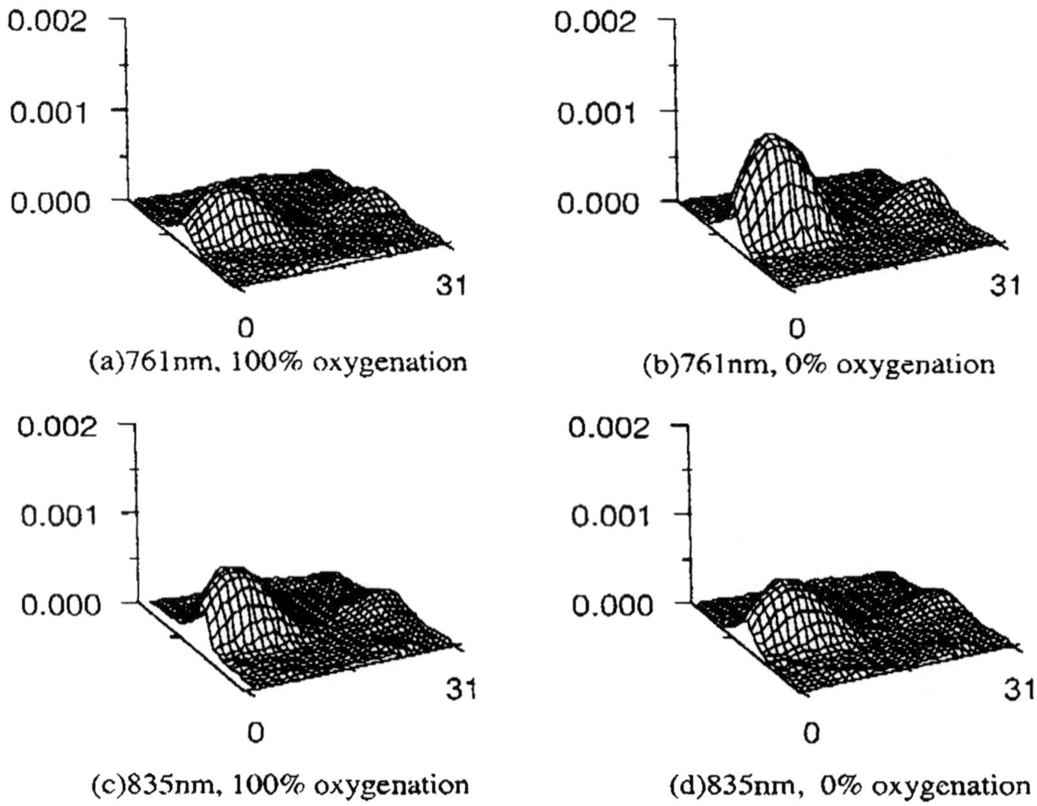

Figure 2 Images of absorption coefficient

2) image of oxy-deoxy hemoglobin

In order to convert a set of μ_a images to the images of hemoglobin, the following equations were used.

$$[\text{oxyHb}] = -2.791\mu_a(761nm) + 5.977\mu_a(835nm)$$

$$[\text{deoxyHb}] = 3.707\mu_a(761nm) - 2.349\mu_a(835nm)$$

Final images of oxy and deoxy hemoglobins were thus calculated and are shown in the figure 3 and 4 respectively. In the figure 3, as was expected, the oxy-hemoglobin image became almost flat in case with 0% oxygenation. In the same way the image of deoxy Hb was found nearly flat for 100% oxygenation.

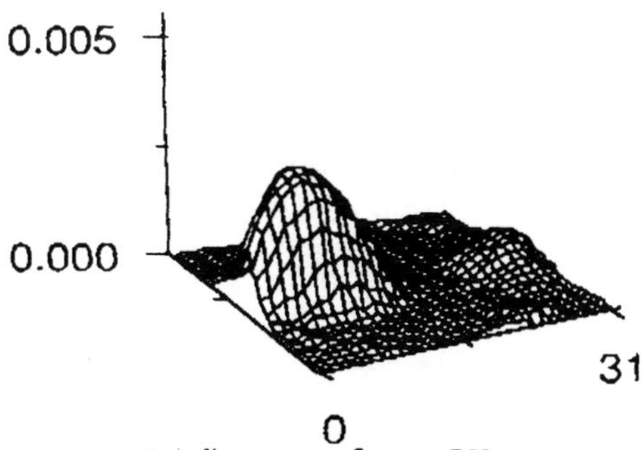

(a) Image of oxy Hb
for 100% oxygenation

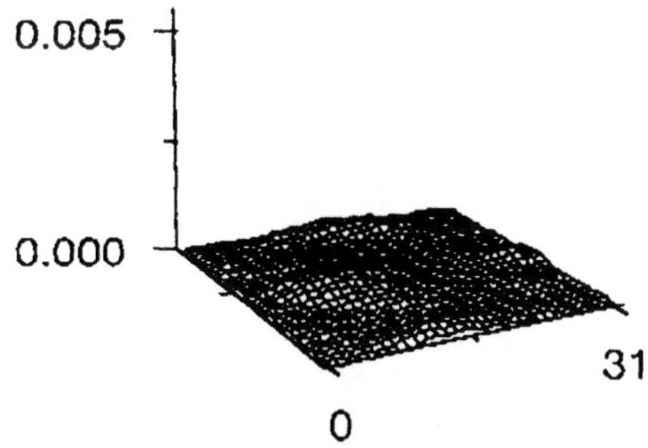

(b) Image of oxy Hb
for 0% oxygenation

Figure 3 Image of oxy hemoglobin

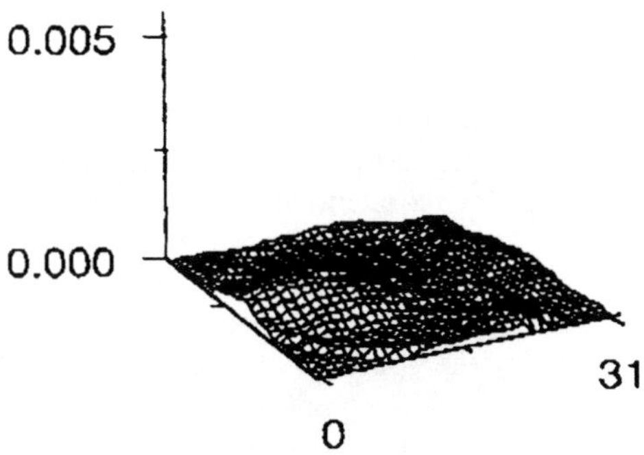

(a) Image of deoxy Hb
for 100% oxygenation

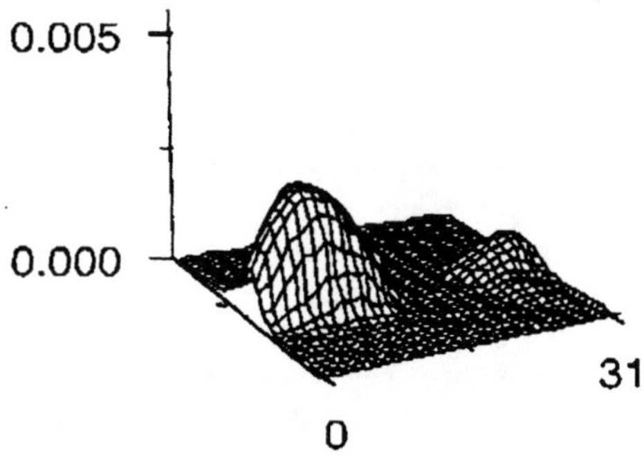

(b) Image of deoxy Hb
for 0% oxygenation

Figure 4 Image of deoxy hemoglobin

References

1) M.Oda, Y.Yamashita, H.Kan, H.Miyajima, A.Sawaki, T.Nakano, S.Suzuki, A.Suzuki, K.Shimizu, S.Muramatsu, N.Sugiura, K.Ohta and Y.Tsuchiya, "Advanced Devices for Near-infrared Time-resolved Spectroscopy and Optical Computed Tomography: High Sensitive/Fast PMT, High Power PLP, Miniaturized CFD/TAC Module and High Speed Multi-Channel Signal Acquisition Unit", SPIE Vol. 2979, pp765-773, 1997

2) M. Schweiger, S. R. Arridge, M. Hiraoka and D. T. Delpy, "The Finite Element Model for the Propagation of Light in Scattering Media: Boundary and Source Conditions", Med. Phys. 22(11), 1779-1792 (1995).

3) S. R. Arridge and M. Schweiger, "Reconstruction in Optical Tomography Using MRI Based Prior Knowledge", in Information Processing in Medical Imaging '95 edited by Y. Bizais, C. Barillot and R. di Paola" (Springer, Berlin, 77-88 (1995).

Acknowledgment

This work was supported by the National Research and Development Program for Medical and Welfare Apparatus sponsored by NEDO (New Energy and Industrial technology Development Organization).

Tomographic Imaging of BiologicalTissue by Time-Resolved, Model-Based, Iterative Image Reconstruction

Andreas H. Hielscher[1,*], Alexander Klose[1,2,*], David Catarious Jr.[1], and Kenneth M. Hanson[2]

Los Alamos National Laboratory
[1]Bioscience and Biotechnology, CST-4
[2]Image Analysis, DX-3
MS E535, Los Alamos, New Mexico 87545,

Abstract

Currently available tomographic image reconstruction schemes for photon migration tomography (PMT) are mostly based on the limiting assumptions of small perturbations and *a priori* knowledge of the optical properties of a reference medium. In this work a model-based iterative image reconstruction (MOBIIR) method is presented, which does not require the knowledge of a reference medium or that the encountered heterogeneities are small perturbations. After a description of the major code structure and a review of the mathematical background, the clinical relevant examples of brain imaging and breast cancer detection are discussed. It is shown that ventricular bleedings in the brain can be detected and that cysts and tumors in the breast can be distinguished using the MOBIIR technique.

Keywords

(100.6950) Tomographic image processing; (170.3010) Image reconstruction techniques; (170.3880) Medical and biological imaging; (170.6920) Time-resolved imaging; (290.7050) Turbid media

1. Introduction

In recent years researchers have invested considerable efforts towards tomographic imaging systems that use near-infrared (NIR) light. [1-4] Besides Photon Migration Tomography (PMT) various other names are commonly used for this novel imaging modality, such as Medical Optical Tomography, Diffusion Tomography, or Transillumination Imaging, to name a few. Based on measurements of transmitted and/or reflected light intensities on the surface of the medium, a reconstruction is attempted that provides the spatial distribution of the optical properties, e.g. absorption coefficient, $\mu_a(\mathbf{r})$, and scattering coefficient, $\mu_s(\mathbf{r})$, inside the medium. In initial clinical trials, performed by various groups around the world, PMT has shown great promise for brain-blood-oxygenation monitoring in preterm infants, hematoma detection and location, cognition analysis, and breast cancer diagnostics. [5-9]

The technology for making light-transmission measurement on human subjects is nowadays readily available. [5,6,10-12] However, a major challenge remains the development of computer algorithms that transform these measurements into useful images of the interior of large organs. Other than x-rays, the near-infrared photons used in PMT do not cross the medium on a straight line from the source to the detector. Light is scattered and absorbed throughout the system. Hence, standard backprojection method have only limited success, [13,14] and other analytical reconstruction methods are only available for highly restrictive problems. [15-17] In general, either the optical properties of a reference medium have to be known, or/and it has to be assumed that the heterogeneities constitute only a small perturbation in a homogeneous background medium. To overcome this limitation research has increasingly turned to model-based iterative reconstruction (MOBIIR) algorithm. [18-21]

The goal of these algorithms is to reconstruct an image of the optical properties (absorption, $\mu_a(\mathbf{r})$, and scattering, $\mu_s(\mathbf{r})$, coefficients; $\mathbf{r}$ is spatial location) inside a medium, from a given set $\mathbf{Y}$ of measurements on the circumference of the medium. These iterative reconstruction schemes in general consist of three components (see Fig. 1): **(1)** ***Forward Model***: This model is a theory or algorithm that predicts a set of measured signals, $\mathbf{U}$, based on the position of the light source and the spatial distribution, $\zeta = (\mu_a(\mathbf{r}), \mu_s(\mathbf{r}))$, of optical properties; **(2)** ***Analysis Scheme:*** Here an objective function, ϕ, is defined, which describes the difference between the measured and predicted data. A simple example is the least-square error norm $\phi(\zeta) \sim (\mathbf{Y} - \mathbf{U}(\zeta))^2$. Since the problem is highly ill-posed, a regularization term, **R**, is usually added to the objective function. **(3) *Updating Scheme:*** Once the objective function is defined, the task becomes to minimize ϕ by successively modifying the distribution ζ. This means that the parameters of the forward model have to be updated to generate a second, third, etc., prediction **U**, which can be compared to the actual measurement data **Y**. The reconstruction is complete once a distribution ζ is found for which $\phi(\zeta)$ is smallest.

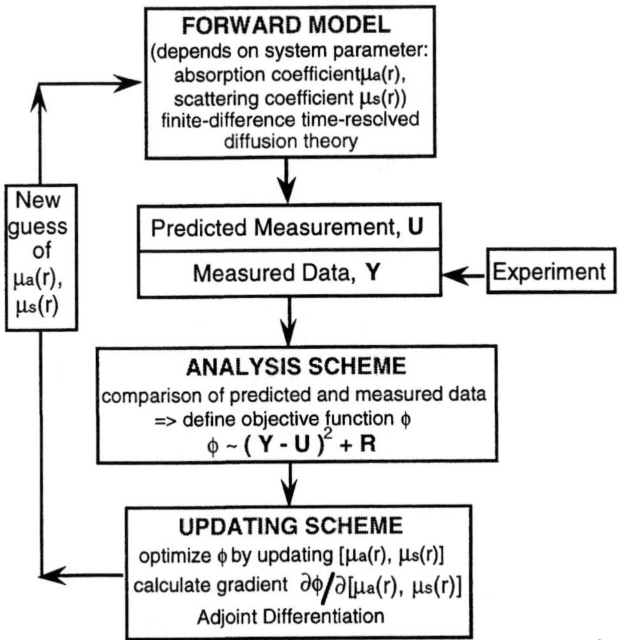

Fig. 1. Flow diagram of model-based iterative image reconstruction scheme.

In this work I will first review some details of the MOBIIR code, which was recently developed at Los Alamos National Laboratory. [20,21] The three components, (1) forward model, (2) analysis scheme, and (3) updating scheme, used in this particular algorithms are outlined and discussed. This is followed by several simulated reconstruction examples, which illustrate the performance of this code. Besides simple reconstructions of well defined objects in a homogeneous background, reconstructions of breast and brain tissue are presented.

2. Mathematical Background

2.1. Forward Model

As a forward model that describes the photon propagation in the turbid medium and predicts the measurements at the detector position $\mathbf{r_d}$, we use in this study a 2-dimensional, time-resolved diffusion equation:

$$\frac{\partial U}{\partial t} = \frac{\partial}{\partial x_p}\left(D\frac{\partial U}{\partial x_p}\right) - c\mu_a U + S, \quad p \in \{1,2\} \qquad (1)$$

where $U(x,y,t)$ is the diffuse intensity and $S(x,y,t)$ is the source strength at position (x,y) and time t. The position-dependent absorption and diffusion coefficients are denoted by μ_a and $D = c/(3[\mu_a+(1-g)\mu_s])$ respectively. Here c = 22 cm/ns is the speed of light in the medium and g is the scattering anisotropy value, equal to the average value of the cosine of the angle through which photons are scattered. $\mu_s' = (1-g)\mu_s$ is known as the effective or transport scattering coefficient.

The equation (1) is solved by replacing the temporal and spatial derivatives by their finite-difference approximations as follows:

$$\frac{\partial U}{\partial t} \approx \frac{U^n - U^{n-1}}{\Delta t} \qquad (2)$$

$$\frac{\partial}{\partial x_p}\left(D\frac{\partial U}{\partial x_p}\right) \approx \delta_{x_p}(U_{i,j}, D_{i,j}) \qquad (3)$$

with

$$\delta_{x_1}(U_{i,j,k}) \equiv [D_{i+1/2,j,k}(U_{i+1,j,k} - U_{i,j,k}) - D_{i-1/2,j,k}(U_{i,j,k} - U_{i-1,j,k})] / \Delta x^2 \qquad (3a)$$

$$\delta_{x_2}(U_{i,j,k}) \equiv [D_{i,j+1/2,k}(U_{i,j+1,k} - U_{i,j,k}) - D_{i,j-1/2,k}(U_{i,j,k} - U_{i,j-1,k})] / \Delta x^2 \qquad (3b)$$

By substituting the finite-difference approximations (2, 3) in the diffusion equation (1), we obtain a difference equation that needs to be solved forward in time. The finite-difference approximations to the spatial derivatives (3) can be evaluated at time index n+1 or n, when we are solving the difference equation for U^{n+1}. Methods that evaluate the spatial derivatives (3) at the past time instance (n) are called explicit, while methods that evaluate the spatial derivatives (3) at the present time instance (n+1) are called implicit. Here we use a technique known as "alternating directions implicit" (ADI) method. [22-24] In this method, the computation of U^{n+1} from U^n is broken up in two time steps. In the first 1/2 time step, only the spatial derivative in one direction is evaluated at the present time instance (implicit) and the other spatial derivative is evaluated at the previous time instance (explicit). In the next 1/2 time step, the implicit and explicit directions are switched, etc..

We can compactly write the ADI method as

$$AU^{n+1/2} = BU^n + S^{n+1/4} \qquad (4a)$$

or

$$U^{n+1/2} = A^{-1}BU^n + A^{-1}S^{n+1/4} \qquad (4b)$$

The structures of the matrices A and B remain the same in all two 1/2-time steps but the absolute values of the matrix elements differ. The advantage of the ADI method is that by correct ordering the matrix A can always be made triadiagonal, e.g., for the first 1/2 step:

$$A_{ii} = 1 + c\mu_a\frac{\Delta t}{2} + (D_{i+1/2,j} + D_{i-1/2,j})\frac{\Delta t}{2\Delta x} \qquad (5a)$$

$$A_{i,i\pm1} = -D_{i\pm1/2,j}\frac{\Delta t}{2\Delta x} \qquad (5b)$$

The matrix A needs to be inverted to compute $U^{n+1/2}$ from U^n (Eq. 4b). Since A is always tri-diagonal the inversion can be easily done in O(N) computation. Furthermore, the ADI method is unconditionally stable for any value of Δt. From now on, we do not distinguish between the two 1/2-time steps for notational simplicity. For more detailed information on explicit, implicit and ADI methods see refs. [20-24].

2.2. Analysis Scheme

The forward model is used to calculate detector responses at a set of detector position M. Let Y denote the measurements of the diffuse intensity U for all $s \in M$. We will assume in this paper for simplicity that the measurements are corrupted by uncorrelated Gaussian noise. However, note that the method we propose is not limited to this choice, and we could as well have chosen a more complex model such as poisson noise. In this work the objective function is defined as:

$$\Phi(\zeta) = \sum_{s \in M} 1/2\sigma(s)^2 \sum_{n} (Y_s^n - U_s^n(\zeta))^2 + R(\zeta) \tag{6}$$

Here $R(\zeta)$ describes any additional regularization term, which in general depends on the spatial distribution of optical properties $\zeta = [c\mu_a(r), D(r)]$. For example, assume that we know the spatial distribution of optical properties, ζ_H, of a healthy breast without any tumors or cysts. We could use this information for the reconstruction by defining

$$R(\zeta) = \lambda \sum_r \left| \frac{\zeta_H(r) - \zeta(r)}{\sigma(r)} \right|^p . \tag{7}$$

The parameter λ is a weighting factor for the regularization term, which determines the influence of the regularizer on the final solution. The regularization term compares ζ at position **r**, with ζ_H at position **r**. Differences between ζ_H and ζ are more or less penalized depending on the choice of p. In this work we used p = 2.0. Note that the noise variance $\sigma(s)$ is space varying.

The goal of the reconstruction algorithm is now to minimize this objective function, Φ, by appropriately choosing the system parameters $D(\mathbf{r})$ and $c\mu_a(\mathbf{r})$.

2.3. Gradient Calculation

The effective solution to optimization problems involving many variables (here the spatial distribution of optical properties) relies on knowing the gradient of the objective function with respect to the variable, $d\phi/d\zeta$. In this work we employ the method of reverse-adjoint differentiation. To outline this method we assume for simplicity that no regularization term is used, i.e. $R(\zeta) = 0$. Furthermore we chose the following notation: $p,q,r,d \in \Omega$ are grid points; $d \in \Omega_d \subset \Omega$ are detector positions on the boundary $\partial\Omega$ and M denote measurments on the boundary $\partial\Omega$. The derivative of the objective function (Eq. 6) with respect to the optical properties ζ is given by:

$$\frac{d\Phi}{d\zeta_r} = \sum_{n \in T} \sum_{p \in \Omega} \frac{d\Phi}{dU_p^n} \frac{\partial U_p^n}{\partial \zeta_r} \tag{8}$$

Here we simply applied the chain rule, and the first term in the sum is the outer derivative, while the second term is the inner derivative of Eq. 6. The second term is easily obtained by differentiating Eq. 4a at time n+1/2 with respect to the optical parameters, which yields:

$$\frac{dA}{d\zeta_r} U^{n+1/2} + A \frac{\partial U^{n+1/2}}{\partial \zeta_r} = \frac{dB}{d\zeta_r} U^n$$

$$\Rightarrow \quad \frac{\partial U^{n+1/2}}{\partial \zeta_r} = A^{-1} \left(\frac{dB}{d\zeta_r} U^n - \frac{dA}{d\zeta_r} U^{n+1/2} \right) \tag{9}$$

Equation 9 can be easily calculated by means of the known intensity U and the matricies A and B of the forward model.

More involved is the calculation of derivative of the objectiove function Φ with respect to the intensities U^n, which is the first term in Eq. (8). We obtain this term recursively by applying the chain rule and stepping backward (reverse) in time from n+1/2 to n,

$$\frac{d\Phi}{dU_p^n} = \sum_{q \in \Omega} \frac{d\Phi}{dU_q^{n+1/2}} \frac{dU_q^{n+1/2}}{dU_p^n} + \frac{\partial \Phi}{\partial U_p^n} . \tag{10a}$$

with

$$\frac{d\Phi}{dU_q^N} = \frac{\partial \Phi}{\partial U_q^N} \text{ for } N = \max(n \in T). \tag{10b}$$

Here, $\partial\Phi / \partial U_p^n$ denotes the change in ϕ when only U_p^n is varied keeping all other variables constant, while $d\phi/dU_p^n$ denotes the total change in ϕ when U_p^n is varied along with all variables that depend on U_p^n. Partially differentiating Eq. 6 with respect to U_p^n, we obtain

$$\frac{\partial \Phi}{\partial U_p^n} = \begin{cases} \frac{1}{\sigma^2}(M_p^n - U_p^n) & p = d \in \Omega_d \\ 0 & otherwise \end{cases} \tag{11}$$

The term $dU_p^{n+1/2}/dU_p^n$ can again be calculated from the forward finite difference Eq. 4b and we obtain:

$$\frac{dU_q^{n+1/2}}{dU_p^n} = A^{-1} B I_p \text{ with } I_p = \begin{cases} 1 & q = p \\ 0 & q \neq p \end{cases} \tag{12}$$

where I_p is a column vector that is zero everywhere except at the spatial point p, where it is unity.

The method of reverse-adjoint differentiation provides an efficient means to calculate the needed gradients for a complex sequence of calculations, such as the forward simulation of the migration of photons described above. Its power lies in its ability to calculate the derivatives with respect to all the variables in a CPU time that is comparable to the forward calculation.

3. Results

3.1. Problem Setup

The MOBIIR algorithm was tested by simulating time-resolved measurements on heterogeneous media and using this simulated measurements as input for the reconstruction code. The measurements were simulated with the finite-difference, time-resolved diffusion code. To mimic real-world measurements, gaussian noise was added to each measurement at a signal to noise (SNR) level of 30 db. This is a typical value for time-resolved single photon counting measurement. [25,26] When single photon counting is used, smooth data can be obtained over 2-3 orders of magnitude. Signals that are approximately 1000 times smaller than the maximum are increasingly effected by statistical noise.

Once the detector readings are simulated, they are used as the actual measurement data, Y, in the reconstruction code. To start the reconstruction program an initial guess of the optical properties is necessary. If not stated otherwise a medium with a constant diffusion coefficient was used as a first guess. Based on this guess the forward code calculates the detector responses, U, and compares these predicted measurements with the measurement data Y, by calculating the objective function, ϕ (Eq. 6). The derivatives of ϕ with respect to ζ at all grid points are calculated and a new spatial distribution of D and $c\mu_a$ is put in place for the next forward calculation. All calculations were done on a HP-9000/770 workstation.

3.2. Simple Systems

First we consider a simple heterogeneous system (Fig. 2a), which consists of a background medium with D = 0.9 cm^2ns^{-1} and two objects with D = 0.5 cm^2ns^{-1} and 1.46 cm^2ns^{-1} respectively. To simulate an 8x8cm medium a 40x40 x-y grid with a spatial resolution of 0.2 cm is used. The medium is surrounded by 16 source detector positions. Moving the source around the medium results in 16x15 = 240 detector readings. Each detector reading consists of 50 time points, which are equally space by $\Delta t = 0.1$ns. In the reconstruction code the derivatives of each time-point with respect to each grid point is calculated, which results in (240x50)x(40x40) = 19.200.000 derivative calculations in each iteration step. Figure 2b shows the reconstructed D-image, which was obtained after 15 iterations (20 minutes). The initial guess was a homogeneous medium with D=1.0 cm^2ns^{-1}. The locations of the two objects and absolute values of D are reconstructed with high accuracy.

Fig. 2c shows the original $c\mu_a$ distribution for the same medium. The reconstructed image is shown in Fig. 2d. The Again the locations of the inhomogeneities are determined with high accuracy. However the absolute values of $c\mu_a$ are not as accurate. For example, the small object with $c\mu_a = 9.0$ ns^{-1} appears as a larger object with $c\mu_a = 3.2$ ns^{-1}. Similar results have been observed by other researchers using different diffusion based image algorithms. Further studies are necessary to fully explore the reasons for this behavior

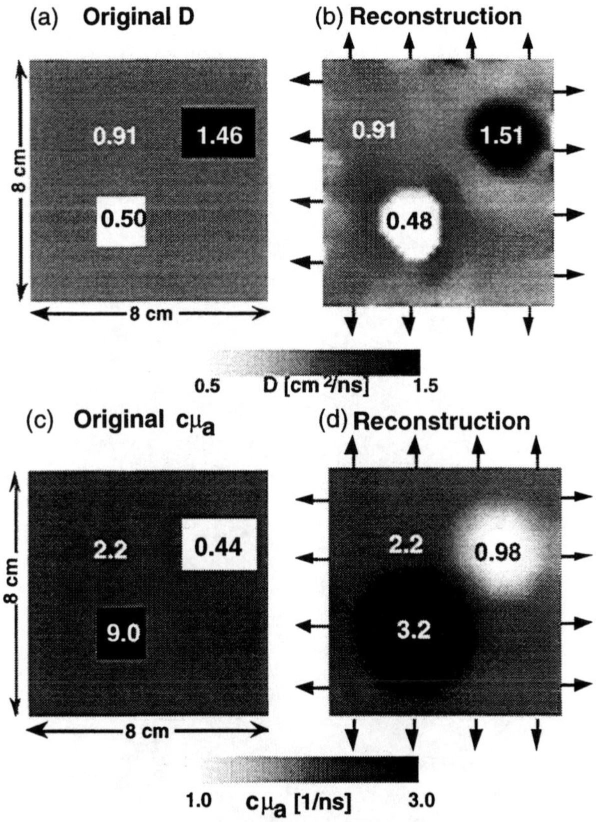

Fig. 2. (a,c) The original medium that contains two inhomogeneities, which differ in the the diffusion and absorption coefficient. (b,d) Here the reconstructed medium after 15 iterations (20 minutes) is shown for both optical parameters. The arrows indicate the 16 source/detector positions surrounding the medium.

3.3. Brain Imaging

To test the algorithms in a more realistic situation, MRI density maps were used to generated optical property maps of the brain. MRI imaging techniques allow to distinguish between skin, skull, white matter, gray matter, blood and cerebrospinal-fluid-filled spaces in the head. These different tissues appear in MRI scans with different densities. From this an optical property map ($c\mu_a(\mathbf{r})$, $D(\mathbf{r})$) is obtained by assigning different optical properties to different density values. Fig. 3a shows such a segmented scan for a slice through the head with the variations in D. Outstanding features are the cerebro-spinal-fluid-filled ventricle in the center of the brain and a hematoma near the forehead. Furthermore, the brain is surrounded by cerebrospinal fluid as can be seen in the light areas in Fig. 3a. Figure 3b shows the reconstruction based on measurements from 12 source-detector positions. In this case 60 iterations were performed which took about 70 minutes. While fine structures are not resolved, clearly visible in the reconstruction are the ventricles and other areas filled with cerebrospinal fluid. Also the hematoma at the forehead can be seen. Figure 3c shows the same reconstruction, only this time the ventricles in the center of the brain were filled with blood. Ventricular bleeding often occurs in preterm infants and leads to server brain damage or death of the infant. The reconstructions shown in 3 b and c, clearly show that photon migration imaging can detect ventricular bleeding and can possibly be used as an early warning system, which allows physicians to react in time.

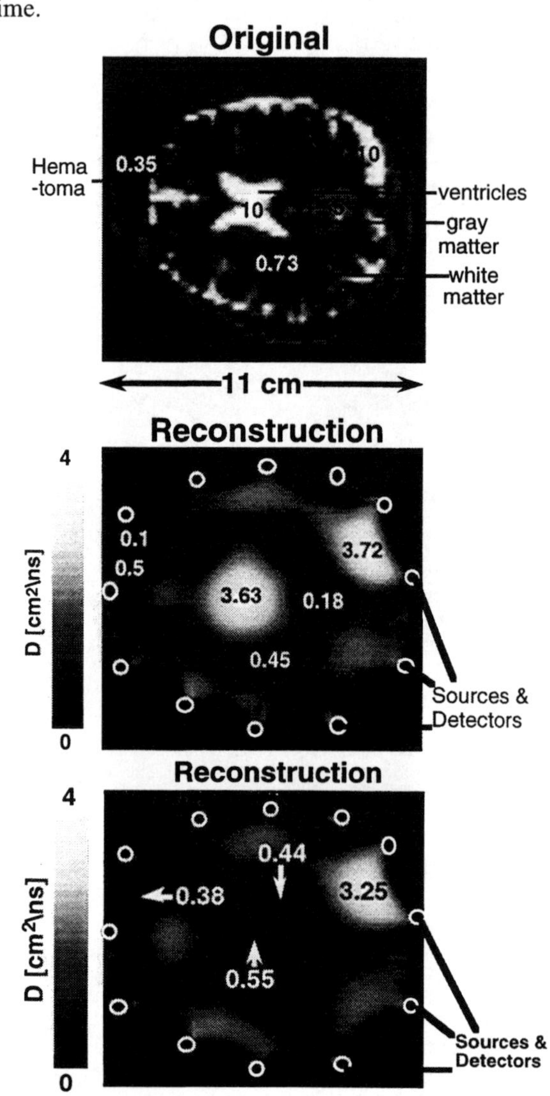

Fig. 3: Reconstruction example of the brain. (a) Original distribution of the diffusion coefficient D [cm^2ns^{-1}]. The spatial resolution is 0.2 cm. (b) Reconstruction results after 60 iterations. The circles indicate source and detector positions. (c) Reconstruction results after 60 iteration for the case that ventricles in center of image are filled with blood.

3.4. Breast Imaging

Another field of possible application of photon migration tomography is breast cancer detection. To simulate breast image reconstructions, an MRI-density map of the breast was converted into a map of optical properties. Fig. 4a shows such a segmented scan for a sagittal slice, which contains the nipple and chestwall. Tumors consist of dense tissues that are highly vascularized, therefore D is lower. On the otherhand, cysts are filled with fluids that scatter less than the surrounding tissue. Therefore, D is higher. It was assumed that the breast is immersed in a matching fluid with optical properties similar to the one of the breast. This is a technique currently proposed by several researchers in the field.

The challenge in this type of problem is to detect the diseased tissue (here tumor [a] and cysts [b]) in a healthy heterogeneous background. For this example the spatial resolution was chosen to $\Delta x = \Delta y = 0.28$ cm, resulting in a 32x56 grid for the finite-differencing scheme. 15 source/detector positions were assigned on three accessible sides of the breast.

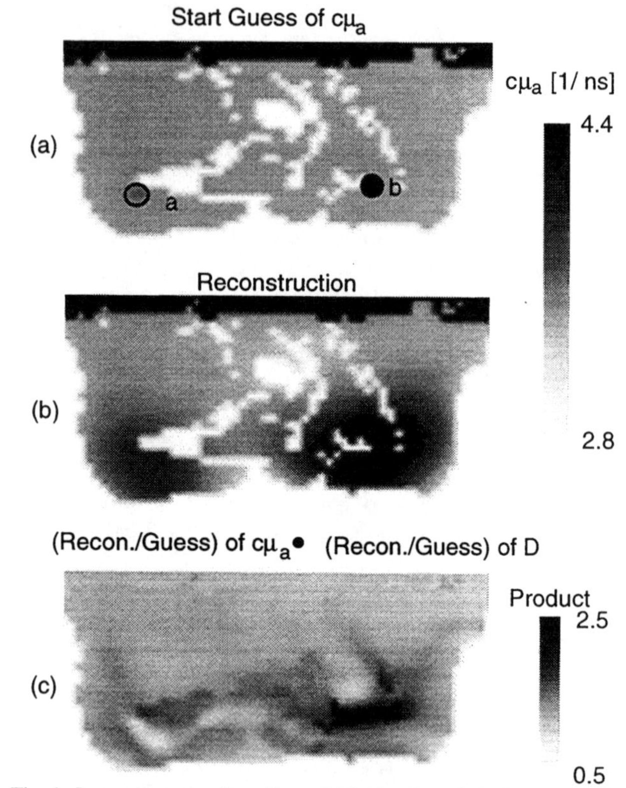

Fig. 4. Image reconstruction of a sagittal slice through the breast. (a) The initial guess for the reconstruction is the healthy breast without tumor, a, or cyst, b. (b) Reconstruction result after 30 iterations for $c\mu_a$. (c) Product of D and $c\mu_a$. reconstructions divided by initial guesses D and $c\mu_a$ distributions.

In this example we used as initial guess and regularization term the information about the healthy breast (Fig. 4a; see also Eq. 7). This approach may be useful in cases where MRI data is available, and one needs to obtain additional information about a suspicious mass in the breast. Fig. 4b shows the result of the reconstruction for the absorption coefficient after 30 iterations (~35min). The tumor and cyst appear as darker areas. However, from this reconstrction alone it is not clear if one has just two tumors of different size or if one of the objects is a cyst. The simultanious reconstruction of D allows for additional analysis. Fig. 4c shows the results of a multiplication of the D and μ_a reconstruction, devided by the initial guesses for D and μ_a. We found that this procedure best identifies cysts (dark area) in the breast and allows to distinguish cysts with high μ_a and D from tumors with high μ_a and low D.

4. Summary

We have developed a model-based iterative reconstruction algorithm for time-resolved photon migration tomography. This algorithm is not limited to small perturbation nor does it require knowledge of a reference medium. The algorithm consists of three components: (1) a forward model that is used to predict the measurements assuming a certain distribution of optical properties. The forward model is currently based on a finite-difference formulation of the time-resolved diffusion equation; (2) an analysis scheme in which an objective function is defined, which needs to be minimized; (3) an updating scheme, which allows to subsequent guess the optical properties of the medium based on the gradient of the objective function.

The code was successfully used to reconstruct simple heterogeneous media, which contain well defined object in a homogeneous background. The code was also tested on MRI generated optical property maps of the brain and the breast. The initial results are encouraging as tumors, cysts, and other inhomogeneities appear to be reconstructable even in an highly heterogeneous background.

Acknowledgments

This work was in part supported by a fellowship from the director's office at Los Alamos National Laboratory.

References

1. B. Chance and R. R. Alfano, eds., Optical Tomography and Spectroscopy of Tissue: Theory, Instrumentation, Model, and Human Studies, Proc. 2979, SPIE-The International Society for Optical Engineering, Bellingham, WA 1997.
2. R.R. Alfano and J.G. Fujimoto, eds., Advances in Optical Imaging and Photon Migrations, OSA Trends in Optics and Photonics Series, vol. 2, Optical Society of America, Washington, DC, 1996.
3. B. Chance and R. R. Alfano, eds., Optical Tomography , Photon Migration and Spectroscopy of Tissue and Model Media: Theory, Human Studies, and Instrumentation, Proc. 2389, Part 1 and 2, SPIE-The Internat. Soc. for Optical Engineering, Bellingham, WA 1995.
4. A. Yodh , B. Chance, "Spectroscopy and Imaging with diffusing light," Physics Today 48, 34-40 (1995).
5. H. Jess, H. Erdl, K.T. Moesta, S. Fantini, M.A. Franceschini, E. Gratton, and M. Kaschke, " Intensity Modulated Breast Imaging: Technology and Clinical Pilot Study Results," OSA Trends in Optics and Photonics on Advances in Optical Imaging and Photon Migration, Vol. 2, R.R. Alfano and J.G. Fujimoto, eds., Optical Society of America, Washington, DC, pp. 126-129 (1996).
6. J.P. Vanhouten, D.A. Benaron, S. Spilman, and D.K. Stevenson, "Imaging brain injury using time-resolved near-infrared light scanning," Pediatric Research 39, pp. 470-476, 1996.
7. S.P. Gopinath, C.S. Robertson, C.F. Contant, R.K. Narayan, R.G. Grossman, and B. Chance, "Early detection of delayed traumatic intracranial hematomas using near-infrared spectroscopy," J. Neurosurgery 83, 438-444 (1995).
8. J. R. Mourant, T. M. Johnson, and D. S. Jack, "Physiological monitoring of brain function with a broadband, multifiber, continous-wave optical system," in Biomedical Sensing, Imaging, and Trackin Technologies II, Proc. SPIE 2976, Bellingham, WA, 1997, pp. 32-39.
9. G. Gratton, M. Fabiani, D. Friedman, M.A. Franceschini, S. Fantini, P. Cobrallis, and E. Gratton, "Rapid changes of optical parameters in the human brain during a tapping task," J. Cognitive Neuroscience 7, 446-456 (1995) .
10. K. Wells, J.C. Hebden, F.E.W. Schmidt, and D.T. Delpy, " The UCL multichannel time-resolved system for optical tomography", in Optical Tomograpgy and Spectroscopy of Tissue, B. Chance and R.R. Alfano, eds., Proc. SPIE 2979, pp. 599-607, 1997.
11. M. Miwa and Y. Ueda, "Development of time-resolved spectroscopy system for quantitative noninvasive tissue measurement," in Optical Tomography, Photon Migration, and Spectroscopy of Tissue and Model Media, B. Chance and R.R. Alfano, eds., Proc. SPIE 2389, pp.142-149, 1995.
12. S. Fantini, M.A. Franceschini, J.S. Maier, S.A. Walker, B. Barbieri, and E. Gratton, "Frequency domain multichannel optical detector for noninvasive tissue spectroscopy and oximetry," Opt. Eng. 34, 32-42 (1995).
13. S.A. Walker, S. Fantini, and E. Gratton, "Image recon-struction by backprojection from frequency domain optical measurements in highly scattering media," Appl. Opt. 36, 170-179 (1997).
14. S. B. Colak, D. G. Papaioannou, G. W. 't Hooft, M. B. van der Mark, H. Schomberg, J. C. J. Paasschens, J. B. M. Melissen, and N. A. A. J. van Asten, "Tomographic image reconstruction from optical projections in light-diffusing media," Appl. Opt. 36, 180-213 (1997).
15. Y. Q. Yao, Y. Wang, Y. L. Pei, W. W. Zhu, R. L. Barbour, " Frequency-domain optical imaging of absorption and scattering distrinbutions by Born iterative method," J. Opt. Soc. Am. A 14, 325-342 (1997).
16. R. L. Barbour, H. L. Graber, J. W. Chang, S. L. S. Barbour, P. C. Koo, R. Aronson, "MRI-guided optical tomography: Prospects and computation for a new imaging method," IEEE Computational Science & Engineering 2, 63-77 (1995).
17. M. A. Oleary, D. A. Boas, A. G. Yodh, "Experimental images of heterogeneous turbid media by frequency-domain diffusing photon tomography," Opt. Lett. 20, 426-428 (1995).
18. M. Schweiger and S.R. Arridge, " A system for solving the forward and Inverse Problems in optical spectroscopy and imaging," in Advances in Optical Imaging and Photon Migrations, OSA Trends in Optics and Photonics Series, vol. 2, R.R. Alfano and J.G. Fujimoto, eds., Optical Society of America, Washington, DC, 1996.
19. H. B. Jiang, K. D. Paulsen, U. L. Osterberg, B. W. Pogue, and M. S. Patterson, "Optical image recon-struction using frequency domain data: Simulations and experiments," J. Opt. Soc. Am. A 13, 253-266 (1996).
20. S. S. Saquib, K. M. Hanson, and G. S. Cunningham, "Model-based image reconstruction from time-resolved diffusion data", in Medical Imaging: Image Processing, Proc. SPIE 3034, pp. 369-380, 1997.
21. A.H. Hielscher, "Model-Based Iterative Image Recon-struction for Photon Migration Tomography," in Methods for Solving Ill-Posed Inverse Imaging Problems: Medical and Nonmedical Applications, R.L. Barbour, M.J. Carvlin, and M.A. Fiddy, eds., SPIE- Proc. 3171, pp.106-117, 1997.
22. W.H. Press, S. A. Teukolsky, W.T. Vetterling, and B.P. Flannery, Numerical Recipes in C, Cambridge University Press, New York, NY, 1992.
23. G. Birkhoff and R.E. Lynch, Numerical Solution of Elliptical Problems, SIAM, Philadelphia, PA, 1984, pp. 173-178.
24. W.F. Ames, Numerical Methods for Partial Differential Equations, Academic Press, New York, NY, 1977, pp. 148-157.
25. D.V. O'Connor and D. Phillips, Time-correlated single photon counting, Academic Press, Orlando, FL, 1984.
26. H. Liu, M. Miwa, B. Beauvoit, N.G. Wang,a nd B. Chance, "Characterization of absorption and scatetring properties of small volume biological samples using time-resolved spectroscopy, Anal. Biochem. 213, 378-385 (1993).

* As of April 1, 1998, Andreas H. Hielscher and Alexander Klose are with the State University of New York, Health Science Center at Brooklyn, Department of Pathology, Box 25, 450 Clarkson Avenue, Brooklyn, NY 11203. e-mail: hielsch@challenge.path.hscbklyn.edu.

Investigation of Alternative Data Types for Time-Resolved Optical Tomography

Jeremy C. Hebden
University College London, Department of Medical Physics, 11-20 Capper Street, London WC1E 6JA

Simon R. Arridge
University College London, Department of Computer Science, Gower Street, London WC1E 6BT

Martin Schweiger
University College London, Department of Medical Physics, 11-20 Capper Street, London WC1E 6JA

Abstract

The technique known as optical tomography involves reconstructing a two- or three-dimensional map of the internal distribution of optical properties of an object from a series of measurements made of the light transmitted between pairs of points on the surface. In this paper we consider the form in which time-resolved measurements should be provided in order to achieve optimum simultaneous reconstruction of scatter and absorption. Various data types are proposed, and an examination is reported of the effect of stochastic noise on their intrinsic uncertainty, and on the ability to uniquely specify the global internal optical properties using pairs of different data types.

Key Words

(170.3010) Image reconstruction techniques; (100.3190) Inverse problems; (170.6920) Time-resolved imaging.

Introduction

Optical tomography is based on an assumption that, given a set of measurements of transmitted light between pairs of points on the surface of an object, there exists a unique three-dimensional distribution of internal scatterers and absorbers which would yield that set. Thus imaging becomes a task of solving an inverse problem using an appropriate model of photon transport [1]. The success of the reconstruction process will ultimately depend on the ability of the algorithm to identify the correct solution despite the presence of noise in the data. Meanwhile, the influence of noise on the reconstruction, both stochastic and systematic, will depend on the choice of the type of data to be employed [2,3].

We consider the optimal data type to be one which facilitates simultaneous reconstruction of both the internal absorbing and scattering properties with the greatest spatial resolution and contrast. The ideal data type therefore has a sensitivity to the precise location of a change in scatter and/or absorption at large depths below the surface, and a sensitivity to the change at large depths which is not negligible compared to the sensitivity to the change near the surface. It is self-evident that measurements of one data type at one source-detector combination results in a single number which contains insufficient information to recover two parameters. Instead we need two or more numbers. But whether multiple data types or multiple source-detector positions are better is not completely resolved. Following a study by Schweiger and Arridge [2] of image reconstruction using a single data type and multiple measurement positions, it was argued that multiple data types are probably essential. Indeed, if only total intensity is used, the non-uniqueness of the inverse problem has been rigorously proven [4]. Total intensity alone is insufficient to distinguish effects of absorption and scatter, since an absorption distribution can always be found that reproduces the set of intensity measurements produced by an arbitrary distribution in scatter, and vice versa.

Because of the relative simplicity of the instrumentation required to measure it, much of the groundwork in theoretical and experimental optical tomography has been performed using total intensity [5,6]. However, intensity measurements are particularly susceptible to common types of systematic noise (variation in source intensity or detector efficiency) and are overwhelmingly sensitive to absorption occurring at or very near the object surface. As a consequence intensity may be far less effective and easy to measure in practice than is often supposed. For the provision of further, more useful data types there has been considerable interest in measurements of one or more characteristics of the temporal distribution of transmitted light [7-10], or their equivalent in the frequency domain [11,12]. The temporal distribution of transmitted light is conventionally known as the temporal point spread function (TPSF).

The sensitivity of a data type to the precise location of a change in scatter or absorption is related to spatial resolution. In optical tomography, the achievable spatial resolution is also inextricably linked with the ability of the reconstruction method to yield a unique solution. Although the inverse problem is severely ill-posed, it is improved to some extent by using a large set of source-detector measurements acquired over as much of the object surface as possible, and by imposing appropriate *a-priori* information. Naturally, this improvement is also critically influenced by the degree of uncertainty due to noise in the available data. As part of our effort to identify the optimal data type(s) for optical tomography, we have examined the effect of stochastic noise on the intrinsic uncertainty of a given measurement, and on the ability to uniquely specify the scattering and absorbing properties of a (homogenous) medium from pairs of different data types.

Data types

All the data types we have examined may be expressed in terms of a normalized integral transform, where the TPSF weighted by a filter function is divided by the total integrated intensity. In this way, we remove the overwhelming sensitivity of the data to surface absorption, and to variation in source intensity and detector efficiency. Furthermore, for several data types direct methods can be found to calculate them, resulting in a substantial improvement in the speed of analysis [9,10]. The temporal moments, or Mellin transforms, are defined by:

$$M_n = E^{-1} \int_0^\infty t^n \, \Gamma(t) \, dt \quad , \qquad (1)$$

where Γ(t) is the temporal distribution of transmitted light, or TPSF, and E is the total intensity obtained by integrating Γ(t) over all time. The central moments are defined by:

$$C_n = E^{-1} \int_0^\infty (t-\langle t\rangle)^n \, \Gamma(t) \, dt \quad , \qquad (2)$$

where <t> is the mean flight time, equivalent to M_1. The specific moments examined here include the mean flight time M_1, and the second and third central moments C_2 and C_3, which we will refer to as the variance and skew.

Another class of data type, recently examined by Schweiger and Arridge [10], is the normalized Laplace transform of the time-resolved intensity:

$$L(s) = E^{-1} \int_0^\infty e^{-st} \, \Gamma(t) \, dt \quad . \qquad (3)$$

This data type can be considered to be equivalent to time-gating using an exponentially-weighted time window with a $1/e$ temporal width equal to the reciprocal of the parameter s. Since shorter flight-time photons sample a narrower volume surrounding the line-of-sight between the source and detector, time-gating can enhance the inherent spatial sensitivity of the data. However, there is obviously a trade-off between the higher resolution image information resulting from a decrease in time window and the greater dominance of noise in the evaluation of L(s). Previous time-domain imaging experiments using tissue-equivalent phantoms have indicated that for tissue thicknesses of five or six centimeters, very little intensity is detectable over the first nanosecond after source photons first enter the medium [13]. Thus for the investigation presented here we have chosen a parameter value of s = 0.001 ps^{-1}, which we anticipate being approximately the largest which can employed in practice. Our fourth class of data type we call the Mellin-Laplace transform, defined as follows:

$$ML_n(s) = E^{-1} \int_0^\infty t^n \, e^{-st} \, \Gamma(t) \, dt \quad , \qquad (4)$$

which also represents a form of time-gating, but with a window that is now constrained to zero weight at zero time. Some control over the shape and width of the effective time window is achieved through appropriate selection of the parameters n and s. Values of n equal to 1, 2, and 3 were again investigated.

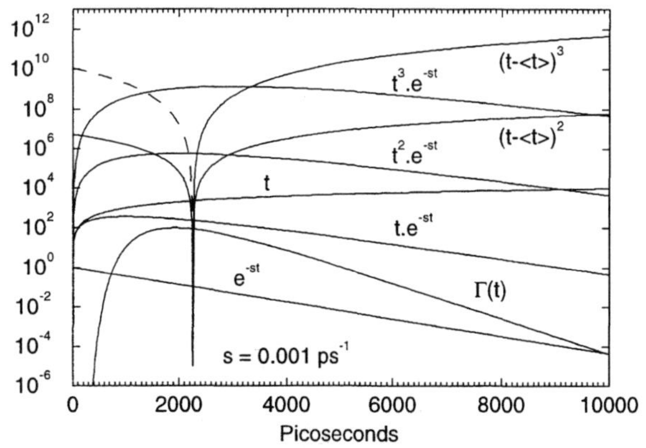

Figure 1. Filter functions for data types and Green's function Γ(t) for selected optical properties. Dashed line represents negative weighting.

The various filter functions employed by our selected data types are represented graphically in figure 1. A value of 2250 ps was selected for the mean flight time <t>, in accordance with the model parameters described in the following section. It is evident that the temporal and central moments give increasing weight to longer flight-time photons, whereas the Laplace and Mellin-Laplace data types are able to give the greatest weight to the shortest or intermediate flight-time photons. Note that the amplitude of the skew weighting function is negative for t less than <t>.

Data type standard error

In order to evaluate the effects of stochastic noise on the various data types, we chose to represent the ideal TPSF by a Green's function solution to the time-dependent diffusion equation for an infinite, homogenous medium and a fixed source-detector separation d [14]. This has the form:

$$\Gamma(t) = A.\,(\mu_a + \mu_s')^{-\frac{3}{2}}.\, t^{-\frac{5}{2}}.\, \exp(-\mu_a c t).\, \exp\left(-\frac{3d^2}{4ct}(\mu_a+\mu_s')\right) \quad (5)$$

where μ_a is the absorption coefficient, μ_s' is the transport scatter coefficient, A is an arbitrary amplitude term and c is the speed of light in the medium. For this investigation, we elected to evaluate the function for a single set of optical parameters which we anticipate being representative of the typical properties which would be encountered for clinical applications: $\mu_s' = 1.0\ \mathrm{mm}^{-1}$, $\mu_a = 0.01\ \mathrm{mm}^{-1}$, and $d = 55$ mm. The Green's function $\Gamma(t)$ calculated for these specific values is also shown in figure 1. In practice, our investigation was concerned primarily with the relative influence of data types and their combinations, and the selected values of the optical properties were not highly significant, providing $\mu_s' >> \mu_a$ and $\mu_s'd >> 1$.

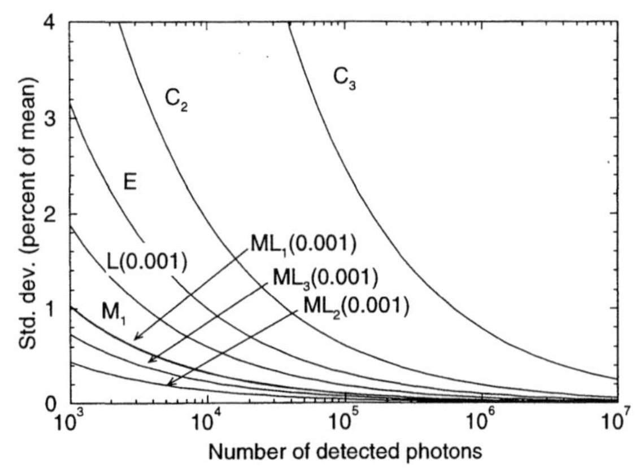

Figure 2. Standard deviation expressed as a percentage of the mean for different data types as a function of the number of detected photons.

Intrinsic statistical uncertainty in the evaluation of various data types has been investigated previously by Arridge *et al.* [15] and Schweiger and Arridge [2]. Their analysis showed that the standard error in the temporal moment measurement for N detected photons (where N is reasonably large) is given by:

$$\sigma_{M_n} = \sqrt{\frac{M_{2n} - M_n^2}{N}} \quad . \quad (6)$$

And for the standard error in the central moment we obtain:

$$\sigma_{C_n} = \sqrt{\frac{C_{2n} - C_n^2}{N}} \quad . \quad (7)$$

Similarly, the standard error in the evaluation of the Laplace measurement is given by:

$$\sigma_{L(s)} = \sqrt{\frac{L(2s) - L(s)^2}{N}} \quad , \quad (8)$$

and the standard error for Mellin-Laplace is given by:

$$\sigma_{ML_n(s)} = \sqrt{\frac{ML_{2n}(2s) - ML_n(s)^2}{N}} \quad . \quad (9)$$

The intrinsic uncertainties of all our measurement types as expressed by the above equations are compared in figure 2, using the optical parameters listed above and a Laplace parameter of $s = 0.001\ \mathrm{ps}^{-1}$. The number of detected photons N was selected over the range from 10^3 to 10^7, which covers the range we expect to encounter for clinical measurements. To enable a straightforward comparison to be made between data types, the standard errors are expressed as a percentage of the mean measurement value. It is particularly interesting to note that the relative uncertainty associated with the evaluation of the mean flight time is significantly lower than that of the total intensity, but the relative uncertainties for variance and skew are very much greater. For example, only 10^3 detected photons are required in order to estimate the mean flight time with a one percent accuracy, but almost 10^6 detected photons are required to obtain a similar accuracy in the estimation of the skew. Of course, these values are dependent on the thickness and optical properties of the medium, but the general behaviour of the measurement types exhibited here are found to be typical over a fairly broad range of physical conditions.

Recovery of μ_a and μ_s' from noisy TPSFs

The absorption and scattering coefficients of a homogeneous medium can be estimated from an experimental measurement of the TPSF by performing a least-squares fit between the data and a suitable analytical model, such as a Green's function. Indeed, this has become a relatively common method of evaluating the optical properties of large thicknesses of biological tissues and phantom materials [16]. In practice, the accuracy of the derived values of μ_a and μ_s' will depend on the noise in the acquired measurement. In order to study this dependency, the Green's function model described by equation (4) was employed to generate a series of noise-limited TPSF profiles using our standard optical parameters. The detection of a fixed number N of transmitted photons was simulated by adding Poisson noise to the analytically derived function $\Gamma(t)$ and then resampling [15]. One thousand noisy Green's functions were generated for each value of N selected in the range from 10^3 to 10^7. The same model was then fitted using a least-squares routine to each of the simulated TPSFs, and values of μ_a and μ_s' were obtained.

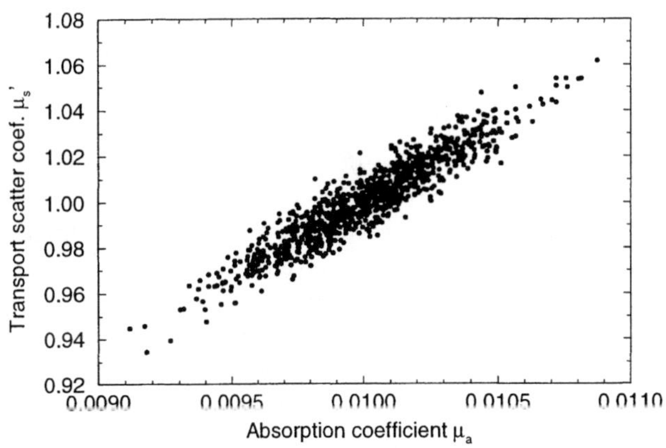

Figure 3. Values of μ_a and μ_s' derived from model fits to simulated TPSFs with photon noise corresponding to detection of 10^4 photons.

Figure 3 shows the distribution of recovered values for 10^4 detected photons. Note that the two coefficients are strongly positively correlated. Despite using all the time information, the resulting variance is quite elongated, indicating that ambiguity in distinguishing μ_a and μ_s' is still prevalent. For larger numbers of photons, the distribution becomes more tightly concentrated around the expected (i.e. correct) values. The spread of values was quantified in terms of the standard deviation as a percentage of the mean for each value of N and the results are shown in figure 4.

Recovery of μ_a and μ_s' from data types

Image reconstruction procedures employed in optical tomography involve recovering the internal optical properties

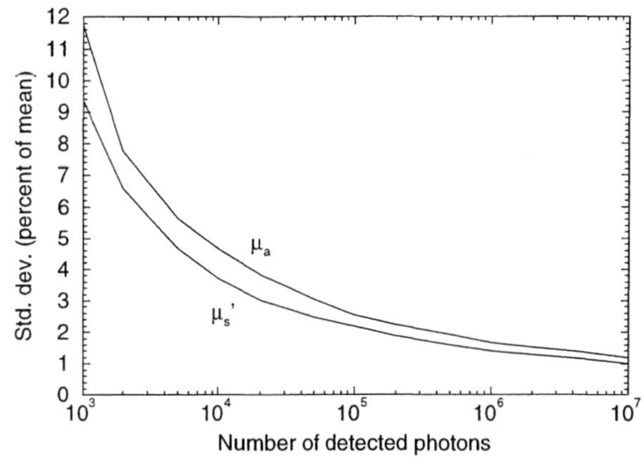

Figure 4. Standard deviation expressed as a percentage of the mean of μ_a and μ_s' derived from model fits to simulated TPSFs with photon noise.

of a medium from a series of external measurements. Implicit in this process is an assumption that the measurements, expressed as data type values, contain sufficient and necessary information. The principal motivation for this investigation has been to test this assumption by determining how accurately the global optical properties of a medium can be recovered from pairs of data types in the presence of noise. Starting with the Green's function given in equation (4), analytic expressions for each class of data type were calculated, and their derivatives were obtained with respect to the coefficients μ_a and μ_s'. These were then employed in a least-squares fitting routine to derive the two coefficients from pairs of data types evaluated from the various sets of noisy TPSFs generated for the investigation described in section 4. The results reveal how accurately certain combinations of data types can predict the true optical properties of the medium as a function of the number of detected photons N. A broad variety of combinations were investigated. Figure 5 shows the standard deviation in the values of the two coefficients for four combinations of data types: total intensity E and mean time M_1, mean time M_1 and variance C_2, mean time M_1 and skew C_3, and variance C_2 and skew C_3. The error in the absorption coefficient is represented by the solid line, and the error in the scatter coefficient by the dashed line. It was immediately apparent that all combinations involving total intensity produced relatively low uncertainty, very similar to the E,M_1 combination in figure 5. Most of the rest of the combinations of data types which were investigated generally fell into one of two categories. The relatively low error produced by mean time and variance was typical of many other combinations shown collectively in figure 6, while mean time and skew produced somewhat greater error characteristic of those combinations shown in figure 7. One exception was a combination involving the variance and skew which produced a strikingly poor convergence, resulting in very high uncertainty in the derived values of the coefficients, as shown in figure 5. In order to compare the relative magnitudes of the error, the vertical scale in figures 5, 6, and 7 is the same.

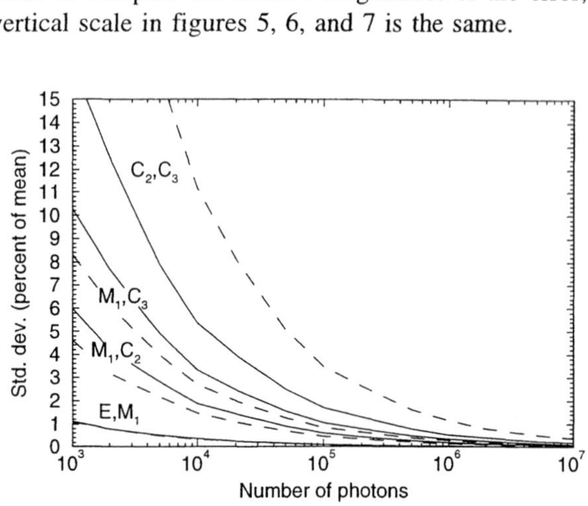

Figure 5. Standard deviation as a percentage of the mean of μ_a (solid line) and μ_s' (dashed line) derived from pairs of data types as shown.

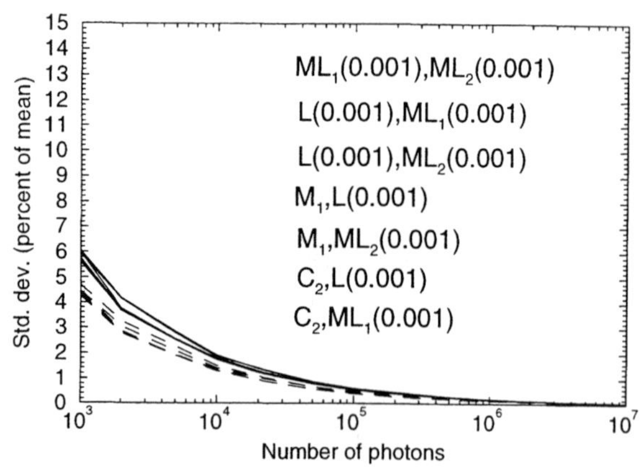

Figure 6. Standard deviation as a percentage of the mean of μ_a (solid line) and μ_s' (dashed line) derived from listed pairs of data types.

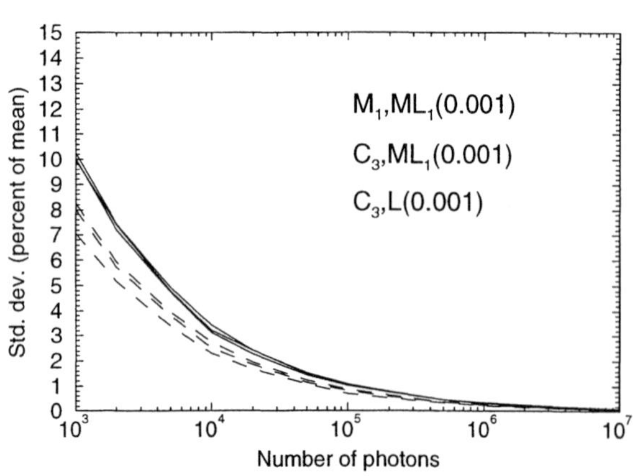

Figure 7. Standard deviation as a percentage of the mean of μ_a (solid line) and μ_s' (dashed line) derived from listed pairs of data types.

Discussion

The ability to reconstruct the absorption and scatter coefficients of a perturbing region from different data types has been investigated earlier by Schweiger and Arridge [2]. They employed an FEM-based forward model, and considered the simple problem of a homogenous circle and a single embedded perturbation region (with properties μ_a and μ_s'), and calculated an error norm $\phi(\mu_a,\mu_s')$ representing the difference between the homogenous reference data and perturbation data summed over all measurements, weighted by the known standard deviation in the measurement. Maps were generated of $\phi(\mu_a,\mu_s')$ as a function of the two coefficients. All maps consisted of a narrow elongated valley indicating that simultaneous recovery of both coefficients is ambiguous for any single data type. Although a reconstruction algorithm can easily converge towards some point within the valley from an arbitrary pair of starting values, the very small gradient along the valley will result in an inability to converge any further towards the true solution. Under these circumstances the reconstruction is said to be ill-posed. However, the orientation of the valley with respect to the two axes was shown to differ between data types, suggesting that an appropriate combination of data types can enable the true solution to be identified. In theory, the ideal combination is one where the valleys are orthogonal, so that the sums of the error norms exhibit a clearly identified minimum. The distinctive orientation of the valley characteristic of the total intensity data type E was such that it could produce a clear minimum when combined with any of the other data types. The valleys for intensity and mean time, for example, were virtually perpendicular. The effectiveness of combinations involving intensity is echoed by the results presented here. The poor convergence exhibited by the combination of variance and skew in figure 5 is similarly supported by the very strong similarity of the error maps for the data types generated by Schweiger and Arridge [2]. This earlier work suggested that among the most promising combinations not involving total intensity were skew and mean time, and skew and L(0.01). Although the *s* parameter used here was lower than that employed in the earlier work, neither of these pairs exhibited dramatically superior convergence in the study presented here.

One critical aspect of the selection of data type for image reconstruction which has not yet been fully addressed is the relative susceptibility of each to variability or noise in the measuring system. The inherent difficulties associated with the use of total intensity due to variation in source power and detector sensitivity has already been mentioned, but other data types can also be severely corrupted due to measurement errors or systematic noise. For example, temporal nonlinearity (an unfortunate characteristic of streak camera systems) will severely distort all time-resolved data types. Work is currently in progress to characterize the influence of systematic effects on various data types using the UCL multichannel imaging system [17], although fortunately temporal nonlinearity is not a significant contributing factor.

In conclusion, we find that only combinations involving total intensity yielded outstanding recovery of the optical coefficients, and most other combinations produced similar results. However, almost all combinations yielded more accurate recovery than direct evaluation of coefficients from a least-squares fit of the Green's function. Although we have not yet managed to identify an obvious ideal combination of data types, there are potentially many more which can be investigated in future.

Acknowledgments

This work has been supported by the Wellcome Trust and Action Research.

References

1. S. R. Arridge and J. C. Hebden, "Optical imaging in medicine II: modelling and reconstruction," Phys. Med. Biol. **42**, 841-854 (1997).
2. M. Schweiger and S. R. Arridge, "Optimal data types in optical tomography," IPMI 1997, Lecture Notes In Computer Science **1230**, 71-84 (1997).
3. D. Grosenick, H. Wabnitz, and H. Rinneberg, "Time-resolved imaging of solid phantoms for optical mammography," Appl. Opt. **36**, 221-231 (1997).
4. S. R. Arridge and W. R. B. Lionheart, "Non-uniqueness in diffusion-based optical tomography," Opt. Lett. (In press).
5. H. L. Graber, J. Chang, J. Lubowsky, R. Aronson, and R. L. Barbour, "Near-infrared absorption imaging of dense scattering media by steady-state diffusion tomography," Proc. SPIE **1888**, 372-386 (1993).
6. J. R. Singer, F. A. Grünbaum, P. D. Kohn, and J. P. Zubelli, "Image reconstruction of the interior of bodies that diffuse radiation," Science, **248**, 990-993 (1990).
7. M. Schweiger, S. R. Arridge, and D. T. Delpy, "Application of the finite-element method for the forward and inverse models in optical tomography," J. Math. Imaging & Vision **3**, 263-285 (1993).
8. I. Oda, H. Eda, Y. Tsunazawa, M. Takada, Y. Yamada, G. Nishimura, and M. Tamura, "Optical tomography by the temporally extrapolated absorbance method," Appl. Opt. **35**, 169-175 (1996).
9. S. R. Arridge and M. Schweiger, "Direct calculation of the moments of the distribution of photon time of flight in tissue with a finite-element method," Appl. Opt. **34**, 2683-2687 (1995).
10. M. Schweiger and S. R. Arridge, "Direct calculation with a finite-element method of the Laplace transform of the distribution of photon time of flight in tissue," Appl. Opt. **36**, 9042-9049 (1997).
11. H. Jiang, K. D. Paulsen, U. L. Osterberg, B. W. Pogue, and M. S. Patterson, "Simultaneous reconstruction of optical absorption and scattering maps in turbid media from near-infrared frequency-domain data," Opt. Lett. **20**, 2128-2130 (1995).
12. M. A. O'Leary, D. A. Boas, B. Chance, and A. G. Yodh, "Experimental images of heterogeneous turbid media by frequency domain diffusing photon tomography," Opt. Lett. **20**, 426-428 (1995).
13. J. C. Hebden, D. J. Hall, M. Firbank, and D. T. Delpy, "Time resolved optical imaging of a solid tissue-equivalent phantom," Appl. Opt. **34**, 8038-8047 (1995).
14. S. R. Arridge, M. Cope, and D. T. Delpy, "Theoretical basis for the determination of optical pathlengths in tissue : temporal and frequency analysis," Phys. Med. Biol. **37**, 1531-1560 (1992).
15. S. R. Arridge, M. Hiraoka, and M. Schweiger, "Statistical basis for the determination of optical path-length in tissue," Phys.Med.Biol. **40**, 1539-1558 (1995).
16. H. Heusmann, J. Kölzer, and G. Mitic, "Characterization of female breasts in vivo by time resolved and spectroscopic measurements in NIR spectroscopy," J. Biomed. Opt. **1**, 425-434 (1996).
17. K. Wells, J. Hebden, F. Schmidt, and D. Delpy, "The UCL multichannel time-resolved system for optical tomography," Proc. SPIE **2979**, 599-607 (1997).

Three-Dimensional Optical Image Reconstruction: Finite Element Approach

Huabei Jiang

Department of Physics & Astronomy, Clemson University, Clemson, SC 29634-1911, U.S.A.
Tel: (864)656-5307 Fax: (864)656-0805 Email: hjiang@clemson.edu

Abstract

A finite element-based three-dimensional optical image reconstruction algorithm has been implemented. Successful 3D image reconstruction has been achieved using simulated experimental data.

Key Words: (170.6960) Tomography, (170.3010) Image reconstruction techniques, (290.7050) Turbid media

To date, studies on optical image reconstruction in multiple scattering media have been limited to two-dimensional problems both theoretically and experimentally.[1-10] While promising two-dimensional results have been achieved using tissue phantoms in the laboratories,[1-10] difficulties are anticipated when we are to conduct 2D imaging in a clinical sitting due to the essential 3D nature of light propagation in tissue. This necessitates our consideration of optical image reconstruction in a 3D framework.

We report here on our initial studies on 3D optical image reconstruction using simulated data. Our 3D reconstruction algorithm is a natural extension of our 2D algorithm that has been based on powerful finite element methods. While details on our 2D algorithms can be found in literature,[4-10] we present here a brief description of the algorithm under 3D framework.

Given a time-independent incident beam, light propagation in tissue can be described by the well-known diffusion equation:

$$\nabla \cdot D(x,y,z)\nabla\Phi(x,y,z) - \mu_a(x,y,z)\Phi(x,y,z) = -S(x,y,z) \qquad (1)$$

where Φ is the photon density, D is the diffusion coefficient, μ_a is the absorption coefficient, and S is the source term. The diffusion coefficient D can be written as $D = 1/3[\mu_s' + \mu_a]$, where μ_s' is the reduced scattering coefficient (since $\mu_s' >> \mu_a$ in turbid medium, we work directly with D and μ_a during reconstruction). In this study, a point source, $S = S_o\delta(x - x_o, y - y_o, z - z_o)$, and Type III boundary conditions (BCs), $\alpha D\nabla\Phi \cdot \hat{n} = \Phi$, are used where S_o expresses the strength of the source, and α is related to the internal reflection at the boundary.

Making use of a finite element discretization, we can obtain a discrete matrix equation for (1) and realize other derived matrix relations (through differentiation) which lead to a set of equations capable of inverse problem solution[4-10]

$$[A]\{\Phi\} = \{b\} \qquad (2)$$

$$[A]\{\partial\Phi / \partial\chi\} = \{\partial b / \partial\chi\} - [\partial A / \partial\chi]\{\Phi\} \qquad (3)$$

$$(\Im^T\Im + \lambda I)\Delta\chi = \Im^T(\Phi^o - \Phi^c) \qquad (4)$$

where the elements of matrix [A] are $a_{ij} = \langle -D\nabla\phi_j \cdot \nabla\phi_i - \mu_a\phi_j\phi_i \rangle$ with < > symbolizing integration over the 3D problem domain; ϕ_i, ϕ_j are locally spatially-varying Lagrangian basis function; χ expresses D or μ_a; $\Im$ is the Jacobian matrix which should be formed from $\partial\Phi / \partial\chi$ at the boundary measurement sites; $\Delta\chi = (\Delta D_1, \Delta D_2, \ldots \Delta D_N, \Delta\mu_{a,1}, \Delta\mu_{a,2}, \ldots \Delta\mu_{a,N})^T$ is the update vector for the optical property profiles where N is the total number of nodes in the finite element mesh used; $\Phi^o = (\Phi_1^o, \Phi_2^o, \ldots \Phi_M^o)^T$ and $\Phi^c = (\Phi_1^c, \Phi_2^c, \ldots \Phi_M^c)^T$, where Φ_i^o and Φ_i^c are observed and calculated data for i=1, 2, ... M boundary locations. Note that in order to estimate D and μ_a spatially, we expand these quantities in a similar fashion to Φ as a finite sum of unknown coefficients multiplied by the locally defined Lagrangian basis functions. In optical imaging, D and μ_a are unknown and Φ is also generally not known except possibly at a finite number of measurement sites. The basic idea for determining D and μ_a is to make measurements of diffusive light around the perimeter of the problem domain for a set of known optical excitation positions. Then the image formation task is to make estimates (which are updated and improved through solution of (4)) of the D and μ_a distributions that are required to sustain the measured boundary quantities. Since the matrix $\Im^T\Im$ is known to be ill-conditioned[1-10], a way to regularize or stabilize the decomposition of $\Im^T\Im$ is needed. We have used a hybrid technique in this study which synthesizes standard Marquardt and Tikhonov regularization schemes.

Our initial 3D reconstruction simulation has been based on the geometry shown in Figure 1, where an offset target cylinder (1 cm high, 1 cm in diameter) is embedded in the background cylinder (5 cm high, 43 mm in diameter). The simulated background optical properties used are μ_s'=1.0/mm and μ_a=0.005/mm. At this time, we have just performed the reconstruction using an absorption heterogeneity only with 2:1 contrast with the background, i.e., the target has μ_a=0.01/mm and μ_s'=1.0/mm. The 3D finite element mesh used has 2169 nodes and 9984 tetrahedral elements. The images obtained are the results of only 10 iterations of our 3D algorithm with about one hour per iteration in a Sun workstation. Figure 2 presents the results from this 3D image reconstruction, where Figures 2(a)-(b) show the image at the cut-planes Z=0 and X=0 (see Figure 1), respectively.

We note that the image quality obtained has already been promising, although in our algorithms we haven't applied any of the image enhancement schemes developed in our 2D reconstructions. Yet the images shown have been the first of their kind to demonstrate the ability of 3D image reconstructions. We believe the image quality of our 3D reconstructions can be further improved/enhanced considerably after we optimize our algorithms and apply our novel image enhancement schemes, as has been evidenced in our 2D image reconstructions.[7,8,10] The computational costs for 3D reconstructions can also be reduced by incorporating the dual mesh scheme as we have experienced in 2D reconstruction[7,8] and by taking advantage of the massively parallel structure of these types of image reconstruction algorithms.[11]

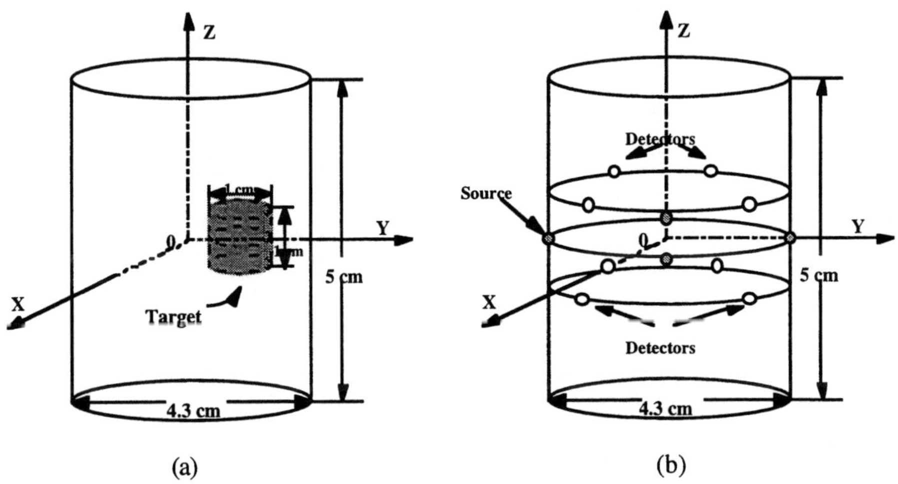

Figure 1. *The test geometry for our 3D image reconstructions. (a): The target configuration; (b) The source/detector locations (Demonstration purpose only; we have actually used 16 (sources)x 64 (detectors) in this study.*

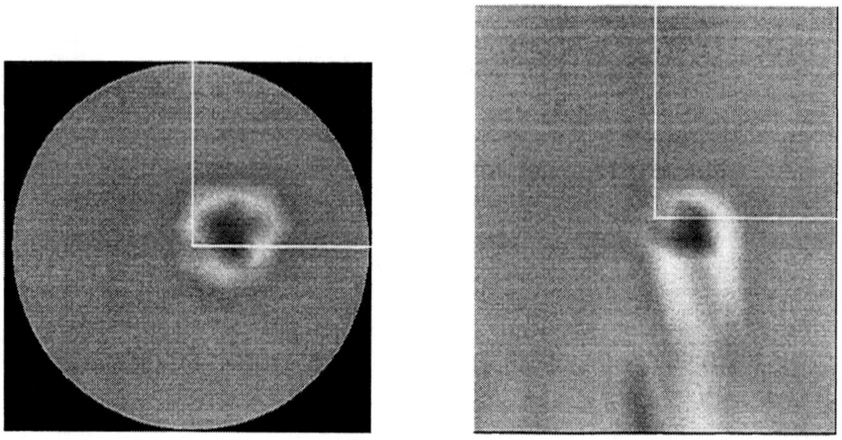

(a) (b)

Figure 2. *(a) Reconstructed absorption image displayed at the cut-plane Z=0; (b) Reconstructed same image displayed at the cut-plane X=0. The darker region is the recovered target.*

This work was supported in part by grants from the Greenville Hospital System/Clemson University Biomedical Cooperative and the Clemson University Research Grant Committee.

References

1. R.L. Barbour, H. Graber, Y. Wang, J. Chang and R. Aronson, "Perturbation approach for optical diffusion tomography using continuous-wave and time-resolved data", *Medical Optical Tomography*: Functional Imaging and Monitoring, G. Muller (Eds), SPIE Press, Bellingham, WA, pp.87-120, 1993.
2. S.R. Arridge, M. Schweiger , D.T. Delpy, "Iterative reconstruction of near infra-red absorption images", *Proc SPIE*, **1767**: 372-383, 1992.
3. M. A. O'Leary, D. A. Boas, B. Chance and A. G. Yodh, "Experimental images of heterogeneous turbid media by frequency-domain diffusing-photon tomography", *Opt. Lett.* **20**, 426-428(1995).
4. H. Jiang, K. D. Paulsen, U. L. Österberg, B. W. Pogue and M. S. Patterson, "Simultaneous reconstruction of absorption and scattering maps in turbid media from near-infrared frequency-domain data", *Optics Letters* **20**, 2128-2130(1995).
5. H. Jiang, K. D. Paulsen and U. L. Österberg, B. W. Pogue and M. S. Patterson, "Optical image reconstruction using frequency-domain data: simulations and experiments", *J. of Optical Society of America A* **13**, 253-266(1996).
6. H. Jiang, K. D. Paulsen and U. L. Österberg, "Optical image reconstruction using DC data: simulations and experiments", *Physics in Medicine and Biology* **41**, 1483-1498(1996).
7. H. Jiang, K. D. Paulsen, U. L. Österberg and M. S. Patterson, "Frequency-domain optical image reconstruction in heterogeneous media: an experimental study of single-target detectability", *Applied Optics* Feature Issue: Diffusing Photons in Turbid Media **36**, 52-63(1997).
8. H. Jiang, K. D. Paulsen, U. L. Österberg and M. S. Patterson, "Frequency-domain optical image reconstruction for breast imaging: initial evaluation in multi-target tissue-like phantoms", *Medical Physics*(in Press).
9. K. D. Paulsen and H. Jiang, "Spatially-varying optical property reconstruction using a finite element diffusion equation approximation", *Medical Physics* **22**, 691-702(1995).
10. K. D. Paulsen and H. Jiang, "Enhanced frequency-domain optical image reconstruction in tissues through total variation minimization", *Applied Optics* Feature Issue: Optical Technologies in Medicine and Biology **35**, 3447-3458(1996).
11. Z. Chen, P. Fci, H. Zheng, "A parallel quasi newton algorithm for unconstrained optimization", *Computing* **55**, 125-133(1995).

Experimental Validation of an Algorithm for 3-Dimensional Tumor Localization in Turbid Media with a 2-Dimensional Frequency-Domain Measurement

Hanli Liu, Kelly Lau, and Charles L. Matson*
Biomedical Engineering, University of Texas at Arlington, Arlington, TX 76019
voice: (817) 272-2054, fax: (817) 272-2251, email: hanli@utarlg.uta.edu

*Optical Sensing Division, Air Force Phillips Laboratory, Kirtland AFB, NM 87117-6008
voice: (505) 846-2049, fax: (505) 846-2045, email: matson@plk.af.mil

1. Introduction

Over the last few years, common geometries used for near infrared imaging with diffuse photon density waves (DPDW's) are source-detector pairs scanned 1) horizontally or 2) circularly.[1,2] In both cases, mechanical movement is required, and only two-dimensional images can be obtained. Recently, Matson et al proposed a new approach to 3-dimensional (3-D) tumor localization in turbid media with the use of frequency-domain (FD) measurements in a single plane.[3] The theoretical basis of this novel reconstruction algorithm stems from Fourier Optics: we apply a Fourier Transform (FT) to the diffusion equation that DPDW's obey, and then extend the solution of Fourier Optics to DPDW's. The FT reconstruction algorithm has several advantages over other approaches: 1) the basic method does not require any mechanical movement for the measurements (although, for our initial measurements, we do use a single detector in a scanning configuration), 2) arbitrary light source geometries can be accommodated, and 3) the processing time is very fast. The theory needs to be verified by experiments because experiments permit studying boundary effects on the reconstructed images by using finite sized tissue phantoms. In addition, we can determine optimal system parameters by performing experiments with various scan dimensions, pixel sizes, and modulation frequencies. In this abstract, we will first introduce the theoretical foundation for the FT reconstruction algorithm, followed by a brief description of our experimental setup, materials and methods. Then we will show experimental results and reconstructed images, and finally we will draw conclusions from this study.

2. Theoretical Background and Reconstruction Algorithm

It has shown[3] that the following transfer function, $H_{\Delta z}(f_x, f_y)$, describes how a DPDW propagates from a plane defined by $z=z_1$ to a plane defined by $z=z_0$ in a homogeneous medium, where $z_1<z_0$

$$H_{\Delta z}(f_x,f_y)=\frac{A_{z_0}(f_x,f_y)}{A_{z_1}(f_x,f_y)}=\exp\left\{i\Delta z\sqrt{\left(\frac{-v\mu_a+i\omega}{D}\right)-(2\pi f_x)^2-(2\pi f_y)^2}\right\}. \qquad (1)$$

where μ_a and D are optical properties, v is the speed of light traveling inside the turbid medium, ω corresponds to the modulation frequency, $\Delta z=z_1-z_0$ is the distance between the two planes, and f_x, f_y are Fourier frequency variables. Since Eq. (1) describes the change in the Fourier transform of the photon density wave as it propagates in the forward direction, Eq. (1) can be used to invert the propagation and to reconstruct the photon density wave behind the detection plane.

To reconstruct the location of a tumor in an otherwise homogeneous medium, we first measure the amplitude and phase of a DPDW in a plane after it passes through the tissue phantom containing the tumor. For background subtraction, we also measure the amplitude and phase of a DPDW in a plane after it passes through a homogeneous sample. After the amplitudes and phases are measured as a function of position in a xy plane at $z=z_0$, we accomplish the following procedures to reconstruct 3-D images: 1) based on the measured amplitudes and phases, calculate the complex photon density waves $U_h(x,y,z_0)$ and $U_{inh}(x,y,z_0)$, respectively, for both homogeneous sample and the sample containing objects; 2) subtract the homogeneous wave from the inhomogeneous wave, $U(x,y,z_0)=U_{in}(x,y,z_0)-U_h(x,y,z_0)$; 3) FT $U(x,y,z_0)$ with

Reprinted from Advances in Optical Imaging & Photon Migration, 1998 Technical Digest, © 1998 Optical Society of America

respect to x and y to obtain its Fourier transform $A(f_x, f_y, z_0)$; 4) divide $A(f_x, f_y, z_0)$ by the transfer function to execute the reconstruction process at z_l; 5) multiply $A(f_x, f_y, z_l)$ by a low-pass filter to regularize the reconstruction; 6) inverse FT $A(f_x, f_y, z_l)$ to obtain the reconstruction in the $z=z_l$ plane; 7) repeat steps 4) to 6) for different values of z_l to reconstruct the entire 3-D volume; 8) determine the x, y coordinates of the hidden object at the detection plane by finding the peak amplitude in the detection plane; 9) at the corresponding x and y coordinates, plot the amplitude of the reconstructed wave versus z to localize the tumor in z direction.

3. Experimental Setup

The components of our FD system include two radio frequency (RF) signal generators (9 kHz to 1.04 GHz), two 50% power splitters, two RF amplifiers, a NIR laser diode able to be modulated up to 100 MHz at 780 nm and 680 nm with laser powers of 3 mW and 9 mW, an Avalanche Photo Diode (APD), two frequency mixers, two low pass filters with the corner frequency at 1 MHz, a lock-in amplifier, and a computer-controlled, two-dimensional positioning stage.

The frequencies of the two signal generators are chosen from 10 to 100 MHz with an offset Δf=25 kHz for the reference signal with respect to the measurement signal in order to achieve a heterodyne system. The intensity-modulated, manufacture-collimated laser light illuminates a tissue phantom and then is detected by a movable APD to convert the optical signal to an electrical signal. The output of APD is amplified before the electrical signal was mixed with the reference signal. By utilizing two low pass filters, only the signal with frequency at 25 kHz passes to the lock-in amplifier. The corresponding amplitudes and phases are recorded automatically by a computer; in turn, feedback from the computer controls sensitivity and time constant settings for the lock-in amplifier based on the readings. In addition, through a GPIB interface, the computer sets the RF frequencies for the two signal generators and controls a XY positioning stage, which held the APD for 2-D imaging scans.

4. Materials and Methods

Tissue phantoms are made out of plastic resin that is hardened in molds with catalyst. The resin is combined with TiO_2 powder for scattering properties and sometimes with NIR dye for absorption properties.[4] The optical properties of homogeneous phantoms are determined using the slope algorithm[5] in transmission geometry. Since the algorithm requires the information on the homogeneous background medium that contains an object/objects, multiple samples are made as a set at the same time to ensure that the samples of the set have the same background optical properties. The phantom sizes are typically 14cm x 14cm x 5cm, and the absorption and scattering coefficients of the background media are within a range of 0.01 to 0.1 cm^{-1} and 3 to 17 cm^{-1}, respectively. The objects used as inhomogeneities are plastic black, white, and clear beads with various diameters and shapes.

All of the scans used are performed on areas of 8 cm x 8 cm having 32x32 pixels. The modulation frequencies are chosen to be 20 and 40 MHz. The experiments are performed in transmission geometry with the light source fixed at one side of the sample and the movable detector located on the other side of the sample. The software package Interactive Data Language (IDL) is used to process the measured data and to accomplish the reconstructions.

4. Experimental Results and Image Reconstruction

Figure 1 is a set of examples showing amplitude images of DPDW's measured at the detection plane with 780 nm laser scanned on a phantom containing two black objects. The phantom has a thickness of 4.2 cm; the two hidden black beads have a diameter of 9 mm and are imbedded at the center of the phantom in Z direction and with a separation of 4 cm. Figures 1(a) and 1(b) are amplitude images measured from the homogeneous and inhomogeneous phantoms, and the amplitude of the difference of the complex fields from the two samples is shown in Figure 1(c). These three images are obtained at the detection plane and show clearly two hidden objects after the simple subtraction. Furthermore, if the FT reconstruction algorithm is applied, we obtain the reconstructed amplitude images of DPDW's at 1.05 cm (Figure 1(d)) and 2.1 cm (Figure 1(e)) behind the detection plane. These two figures demonstrate that reconstructed XY images provide correct tumor locations in XY plane, and that the images reconstructed near the hidden objects have better spatial resolutions.

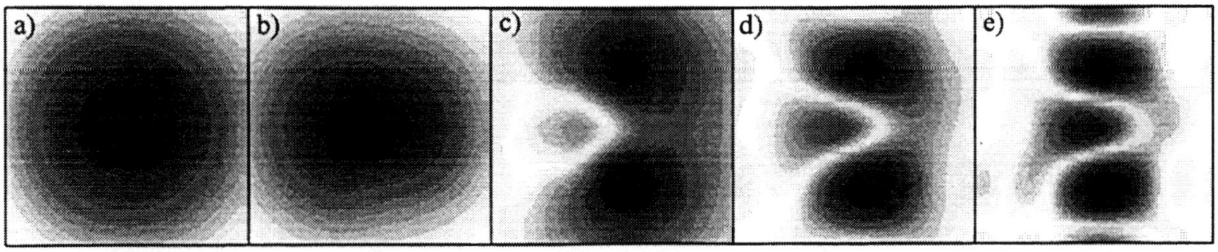

Figure 1. Amplitude images of the DPDW's measured from a phantom with two high-absorbing objects, scanned at 20 MHz with an array size of 32 x 32. a) taken from a homogeneous phantom, b) from an inhomogeneous phantom, c) inhomogeneous image –homogeneous image, d) reconstructed image at 1.05 cm behind the detection plane, and e) reconstrcted image at 2.1 cm behind the detection plane. For d) and e), filter used is 2.

To obtain the object location in z direction, we first determine the x and y coordinates of the detected object(s) based on XY images, such as those shown in Figures 1(c) to 1(e), and then plot the amplitude versus Z to obtain the peak position. The z coordinate at the peak of the amplitude will be the location of the hidden object in z direction. To show this strategy, we plot a z plot in Figure 2 based on Figure 1(c) at X_{pixel}=20, $Y1_{pixel}$=8, $Y2_{pixel}$=24, which are the pixel coordinates of the observed, two dark spots for the hidden black beads. Figure 2 displays a peak at z=2.1 cm behind the detection plane in both cases, implying that both of the two imbedded objects are located 2.1 cm behind the detection plane.

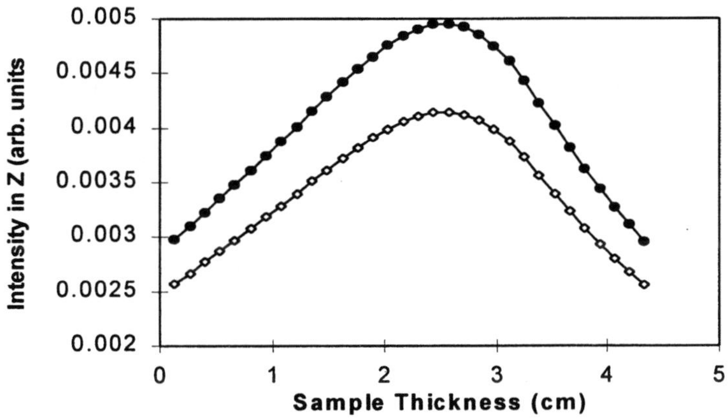

Figure 2. Z plot along the line perpendicular to the detection plane and containing the black beads. Solid circles are plotted at Y1=8 pixels (for bottom spot), open diamonds are plotted at Y2=24 pixels (for top spot).

5. Summary

We perform experiments on tissue phantoms using a frequency-domain photon migration system. The system includes an amplitude-modulated laser diode, two RF signal generators, an APD, and a lock-in amplifier. The tissue phantoms are made with inhomogeneities imbedded and are then scanned with the system at 20 and 40 MHz. The data from the amplitude and phase measurements are analyzed using the FT reconstruction algorithm. The reconstructed data provides clear validation of the FT algorithm's ability to localize inhomogeneities three-dimensionally from a single 2-D measurement.

6. References:

1. M. A. O'Leary, D. A. Boas, B. Chance, and A. G. Yodh, SPIE, **2389**, p.328-339 (1995).
2. Brain W. Pogue, Michael S. Patterson, and Tom J. Farrell, SPIE, **2389**, p.328-339 (1995).
3. C. L. Matson, N. Clark, L. McMackin, and J. S. Fender, Appl. Opt. **36**, 214-220 (1997).
4. M. Firbank, M. Hiraoka, and D. T. Delpy, SPIE, **1888**, 264-274 (1993).
5. J. B. Fishkin and E. Gratton, J. Opt. Soc. Am. A **10**, 127-140 (1993).

Analysis of a 3-D Tumor Localization Algorithm from a Single 2-D Measurement Using Diffraction Tomography

Charles L. Matson
Optical Sensing Division, Air Force Phillips Laboratory, Kirtland AFB, NM 87117-6008

Hanli Liu
Department of Biomedical Engineering, University of Texas at Arlington, Arlington, TX 76019

Abstract

Diffraction tomography theory describes both the forward and inverse problems in media where electromagnetic radiation propagation must be modeled in terms of its wave nature. In general, this theory assumes media in which the radiation is not attenuated. We have recently extended this theory to include the attenuating media case. Here, we apply this newly-developed theory to analyze a tumor localization algorithm which permits three-dimensional tumor localization from a single two-dimensional measurement. We demonstrate that this algorithm can be viewed as a backpropagation algorithm. With this insight, show how tumor localization occurs.

Key Words

(170.3880) Medical and biological imaging; (100.3020) Image reconstruction/restoration; (170.3660) Light propagation in tissue.

Introduction

The use of optical radiation for imaging human tissue has many desirable advantages over other imaging methodologies[1]. One promising category of approaches, called frequency-domain methods,[2,3] is being explored for use in localizing and characterizing inhomogeneous structures imbedded in turbid media. Recently, we have developed an inversion algorithm that permits three-dimensional localization of inhomogeneities from a single two-dimensional measurement when using a diffuse photon density wave (DPDW).[4] A heuristic justification for the success of this algorithm was provided in that paper. In this paper, we develop an analytical solution to the inverse problem for small inhomogeneous structures using our newly-developed diffraction tomographic theory.[5] This analytical solution provides a theoretical explanation for the success of our algorithm.

The forward problem formulation

The physical geometry of our observation scenario is shown in Figure 1, where the object is imbedded in a homogeneous and infinite turbid medium. The illumination plane is located at $z=z_1$, the detection plane is located at $z=z_o$, where $z_o>z_1$, and the object is located between the two planes. We assume plane wave illumination, where the illumination plane is defined as the location where the plane wave has unit amplitude and zero phase. In this section, we present a general expression for the Fourier transform of the detected wave under the Born approximation and then specialize it to the case of small inhomogeneities.

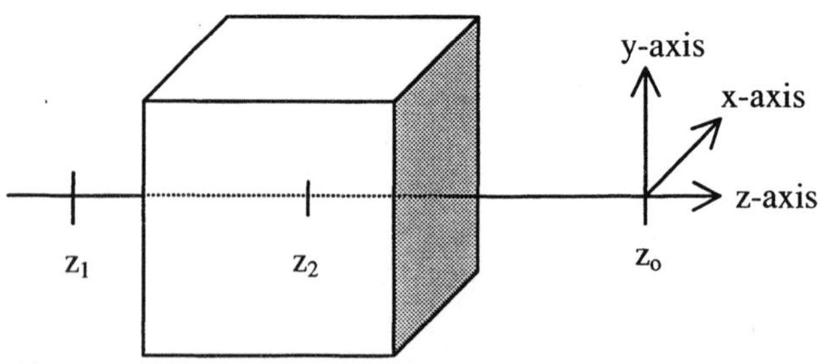

Figure 1. Conceptual diagram showing the geometry for the diffraction tomography development. The inhomogeneity is represented by the cube, which is assumed to be imbedded in an infinite homogeneous turbid medium. The illumination plane is located at z_1, the center of the inhomogeneity is located at $x=0$, $y=0$, $z=z_2$, and the detection plane is located at z_o. The illumination and detection apparatus are not shown.

The general expression for the two-dimensional Fourier transform of the detected wave, denoted by $U_B(\omega_x,\omega_y;z_o)$, is given by

$$\begin{aligned} U_B(\omega_x,\omega_y;z_o) &= \frac{1}{2\gamma_\omega}\exp[-(z_o-z_1)k_i] \\ &\times\exp(-iz_1k_r - iz_o\gamma_{\omega i})\iiint[o_a(x',y',z') \\ &-ik\gamma_\omega o_s(x',y',z')]\exp[-(z_o-z') \\ &\times(\gamma_{\omega r}-k_i)]\exp[-ix'\omega_x - iy'\omega_y \\ &-iz'(-\gamma_{\omega i}-k_r)]\,dx'dy'dz' \end{aligned} \tag{1}$$

where $o_a(x,y,z)$ and $o_s(x,y,z)$ are related to the absorptive and scattering structures of the inhomogeneity,[6] respectively, k is the complex wavenumber of the homogeneous medium, and where

$$\begin{aligned} \gamma_\omega &\equiv \gamma_{\omega r} + i\gamma_{\omega i} \\ &= \mathrm{Re}\left(\sqrt{\omega_x^2+\omega_y^2-k^2}\right) + i\,\mathrm{Im}\left(\sqrt{\omega_x^2+\omega_y^2-k^2}\right) \end{aligned} \tag{2}$$

where Re() and Im() denote the real and imaginary parts of a complex number, respectively. To continue, we substitute previously-derived expressions[6] for $o_a(x,y,z)$ and $o_s(x,y,z)$ into Eq.(1). We assume that the inhomogeneity structure is a cube of linear dimension a, centered at x=0, y=0, and $z=z_2$, where $z_1<z_2<z_o$. We also assume that a is very small, so the terms in the integrand are approximately constant in the support of the structure. This is equivalent to assuming that our object is a point object. Substituting these expressions for the object and making the required simplifications gives

$$\begin{aligned} U_B(\omega_x,\omega_y;z_o) &= \frac{a^3}{2}\exp\{-(z_o-z_1)k_i \\ &-(z_o-z_2)(\gamma_{\omega r}-k_i) \\ &-i[z_1k_r + z_o\gamma_{\omega i} + z_2(\gamma_{\omega i}+k_r)]\} \\ &\times\left(\frac{3\mu_s'\delta\mu_a}{\gamma_\omega} - \frac{ik}{\mu_s'}\delta\mu_s'\right) \end{aligned} \tag{3}$$

In the next section, we will use Eq.(3) as the starting point for our inversion algorithm.

The inverse problem

In this section we develop a backpropagation algorithm for reconstructing the three-dimensional location of an inhomogeneity from our planar measurement. Our approach to developing a backpropagation algorithm is to extend the standard diffraction tomographic backpropagation algorithm[7] to attenuating media. As done in standard diffraction tomography, we develop the backpropagation algorithm in terms of the homogeneous background. Our methodology is to reconstruct the three-dimensional inhomogeneity structure by reconstructing planes of the structure which are parallel to the detection plane. Consider reconstructing one such plane at a location z_3 where $z_1<z_3<z_o$; that is, our reconstruction plane lies between the illumination plane and the detection plane. As previously reported,[4] we accomplish the backpropagation operation by dividing the detected scattered wave, $U_B(\omega_x,\omega_y;z_o)$, by the transfer function $H(\omega_x,\omega_y;z_o-z_3)$ which corresponds to forward propagating from the $z=z_3$ plane to the $z=z_o$ plane. This transfer function is given by

$$H(\omega_x,\omega_y,z_o-z_3) = \exp[-(z_o-z_3)\gamma_\omega] \tag{4}$$

and the Fourier transform of our reconstructed plane, $U_{recon}(\omega_x,\omega_y;z_3)$, is found by dividing Eq.(3) by Eq.(4), which gives

$$\begin{aligned} U_{recon}(\omega_x,\omega_y;z_3) &= F(\omega_x,\omega_y)\frac{a^3}{2}\exp\{-(z_3-z_1)k_i \\ &-(z_3-z_2)(\gamma_{\omega r}-k_i) \\ &-i[z_1k_r + z_3\gamma_{\omega i} + z_2(\gamma_{\omega i}+k_r)]\} \\ &\times\left(\frac{3\mu_s'\delta\mu_a}{\gamma_\omega} + i\frac{k}{\mu_s'}\delta\mu_s'\right) \end{aligned} \tag{5}$$

where $F(\omega_x,\omega_y)$ is a regularizing filter. Notice that Eq.(5) is equal to Eq.(3) (ignoring the regularizing filter) except that z_o is replaced with z_3. We interpret this to mean that the reconstruction process effectively moves the detection plane backwards. It can be shown that this concentrates the scattered wave's energy for $z_3>z_2$. When the reconstruction plane is beyond the location of the inhomogeneity, it can be shown that the reconstruction algorithm spreads the energy back out again in a cosine type of structure. By defining the location of the inhomogeneity as the place where the amplitude of the reconstructed wave has its maximum value, we can localize the inhomogeneity.

To illustrate the previous discussion, we use Eq.(5) to create the amplitude of the reconstructed scattered wave for the case of an 8cm x 8cm detection plane, with the illumination plane 8cm in front of the detection plane. The background values for μ_a and μ_s' are 0.1 cm^{-1} and 12 cm^{-1}, the DPDW modulation frequency is 200 MHz, $a^3\delta\mu_a = 0.11$, and $\delta\mu_s' = 0$. The inhomogeneity is located at z_2=4cm, and we display the amplitude of the reconstructed scattered wave in the y=0 plane. The result is shown in Figure 2. We see that the inhomogeneity is localized quite accurately. As expected, we see that the amplitude of the scattered wave becomes more concentrated as the distance from the detection plane increases, up to the location of the inhomogeneity. Beyond this point, the amplitude expands again in a cosine fashion.

Conclusions

We have developed an analytical solution for the backpropagation algorithm used to reconstruct the location of an inhomogeneity in a turbid medium. This algorithm successfully reconstructs the three-dimensional location of an inhomogeneity from a single two-dimensional measurement. The solution accurately predicts the results we have previously obtained using computer-simulated data.

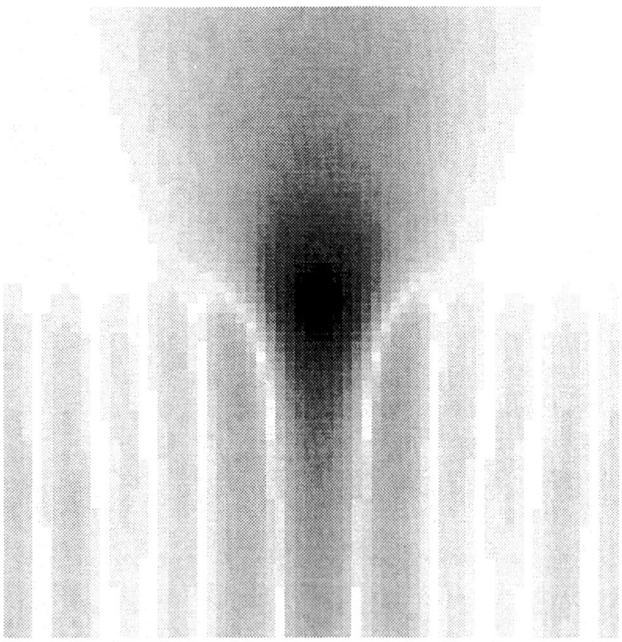

Figure 2. Amplitude of the reconstructed scattered wave, using Eq.(5), in the y=0 plane. The detection plane is at the top of and perpendicular to the figure.

References

[1] *Medical Optical Tomography: Functional Imaging and Monitoring*, R.F. Potter, ed., SPIE Optical Engineering Press **IS11** (1993)

[2] M.A. O'Leary, D.A. Boas, B. Chance, and A.G. Yodh, "Experimental images of heterogeneous turbid media by frequency-domain diffusing-photon tomography," *Optics Letters* **20**, 426-428 (1995)

[3] H.B. Jiang, K.D. Paulsen, U.L. Osterberg, M.S. Patterson, "Frequency-domain optical-image reconstruction in turbid media – an experimental study of single-target detectability," *Applied Optics* **36**, 52-63 (1997)

[4] C.L. Matson, N. Clark, L. McMackin, and J.S. Fender, "Three-dimensional tumor localization in thick tissue with the use of diffuse photon-density waves," *Applied Optics* **36**, 214-220 (1997)

[5] C.L. Matson, "A diffraction tomographic model of the forward problem using diffuse photon density waves," *Optics Express* **1**, 6-11 (1997)

[6] M.A. O'Leary, D.A. Boas, B. Chance, and A.G. Yodh, "Simultaneous scattering and absorption images of heterogeneous media using diffuse waves within the Rytov approximation," in *Optical Tomography, Photon Migration, and Spectroscopy of Tissue and Model Media: Theory, Human Studies, and Instrumentation*, Proceedings SPIE **2389**, 320-327 (1995)

[7] A.J. Devaney, "Reconstructive tomography with diffracting wavefields," *Inverse Problems* **2**, 161-183 (1986)

Sensitivity of the Elliptic Systems Imaging Algorithm to Prior Knowledge of the Background Medium

Michael V. Klibanov and Thomas R. Lucas
Department of Mathematics
University of North Carolina at Charlotte
Charlotte, NC 28223
Tel. (704) 547-2645, e-mail MKLIBANV@NEWMAIL.UNCC.EDU

Abstract

The Elliptic Systems Method reduces the original inverse problem of Optical Tomography (OT) to a coupled system of elliptic Partial Differential Equations. In this manuscript the sensitivity of the ESM to a prior estimate of the background is tested for a set of experimental data for Intralipid Solution. The robustness of the algorithm is demonstrated.

Key words: Background sensitivity, optical tomography

1 Introduction

While a number of imaging algorithms in Optical Tomography (OT) have been developed (c.f. [1-5]), all of them require many locations for the light source, for each region of interest, which leads to an increase of the acquisition time as well as to some technical inconveniences in the hardware design. In addition, since the original problem of OT is ill-posed, a regularization parameter α must be chosen for its solution. However, the choice of α is usually troublesome. Indeed, the generic theory of ill-posed problems [6] assures that α depends on at least such factors as the background medium, geometry, source-detectors configuration and noise level. Furthermore, small fluctuations of these factors usually lead to large corrections in α. In the practical scenario of optical medical imaging this implies that while α can be successfully chosen for one patient, another patient would require a new choice for α. Generally, the choice of α requires solution of many forward/inverse problems for the diffusion equation, which is time consuming. On the other hand, as soon as α is successfully chosen, the rest of the image reconstruction usually can be performed reasonably rapidly.

The authors have recently developed a new approach to OT, the Elliptic Systems Method (ESM) [7, 8]. The ESM is based on a novel mathematical algorithm, in which the original inverse problem is reduced to a coupled system of elliptic Partial Differential Equations (PDEs). This system can be solved by use of the Finite Element Method (FEM); the resulting matrix system is sparse and can be solved rapidly using either direct [9] or iterative [10] methods. The choice of the regularization parameter is not a problem for the ESM: the number N of this elliptic system is such a parameter. We discovered that $N = 3$ or 4; independently of both the background media, and the level of noise; both values $N = 3$ & $N = 4$ provide good quality images (see an explanation of this fact below). In addition, the ESM requires only a *single* source location for a region of interest, which should potentially reduce the acquisition time.

In this paper we discuss the sensitivity of the ESM to a prior estimate of the background medium using experimental data for Intralipid Solution as a representative example.

2 Statement of the inverse problem

Let Ω be a bounded 2 or 3-dimensional domain of interest, with boundary $\partial\Omega$. Let $x = (x_1, x_2, x_3)$, t denotes time, the function $u(x,t)$ be the light intensity, and $x_0 \in \partial\Omega$ be the position of the light source. Then the function $u(x,t)$ satisfies the diffusion equation in Ω with the Robin boundary conditions of the 3rd kind, which are equivalent to the well-known extrapolated boundary conditions,

$$u_t = \text{div}\,(\mathcal{D}(x)\nabla u) - a(x)\,u,\ x \in \Omega,\ t \in (0,T) \tag{2.1}$$

$$u(x,0) = \delta(x - x_0) \tag{2.2}$$

$$\frac{\partial u}{\partial n} + \gamma u = 0, \text{ for } x \in \partial\Omega,\ t \in (0,T). \tag{2.3}$$

Where $\delta(x - x_0)$ is the delta function, n is the unit outward vector on $\partial\Omega$, $\gamma = \text{const.} = 1/(z\, l_t)$, where l_t is the transport mean free path in the medium, $z \cdot l_t$ is the distance between $\partial\Omega$ and the extrapolated boundary, z is "responsible" for the mismatch of refractive indexes between Ω and the air. In our computations below we took $z = 1.1$. In a low absorbing medium the diffusion coefficient $\mathcal{D}(x) = c\ l_t(x)/3$ and the absorption term $a(x) = c/l_a(x)$, where c is the speed of light in the medium and $l_a(x)$ is the mean absorption length [11].

We will consider the following **Inverse Problem**: *Let the function $\mathcal{D}(x)$ be known and $a(x) = a_0(x) + h(x)$, where the function $a_0(x)$ is given and $h(x)$ is a small unknown perturbation of $a_0(x)$. Further, let $\{x_i\}_{i=1}^m$ be a set of detectors on $\partial\Omega$. Determine the perturbation term $h(x)$ given measurements of the light intensity $\varphi(x_i,t) = u(x_i,t)$ for $t \in (T_0,T)$, where T_0 is a positive constant.*

Interpolating the detector readings $\varphi(x_i,t)$ over the entire boundary $\partial\Omega$, we obtain the function $\varphi(x,t)$,

$$u(x,t) = \varphi(x,t), \text{ for } x \in \partial\Omega,\ t \in (T_0,T). \tag{2.4}$$

Because of (2.3), the flux of photons is also known on the boundary. That is,

$$\frac{\partial u}{\partial n}|_{\partial\Omega} = -\gamma\varphi(x,t)\ , \text{ for } x \in \partial\Omega, t \in (T_0,T). \tag{2.5}$$

3 Description of the Experiment

An experimental data set was delivered to us by the group led by Professor R.R. Alfano (City College of CUNY, New York). The turbid medium used was uniform colloid, the so-called Intralipid Solution with $l_t \approx 2.7$ mm, $l_a \approx 500$ mm and $c \approx 0.225$ $\frac{mm}{ps}$. The Intralipid Solution was in a container with transparent walls. The container was a rectangular prism whose horizontal cross-section was a 60 mm $\times$ 60 mm square with a height of 90 mm. The light source and detectors were located on the boundary of the horizontal mid-plane. The inclusion was a black absorbing rod of diameter 4.8 mm and height 90 mm. The abnormality was consecutively placed in three (3) different positions A, B, and C. The center of each inclusion was 10 mm off the center of the square, see Fig. 1. We will refer to these inclusions as A, B, and C.

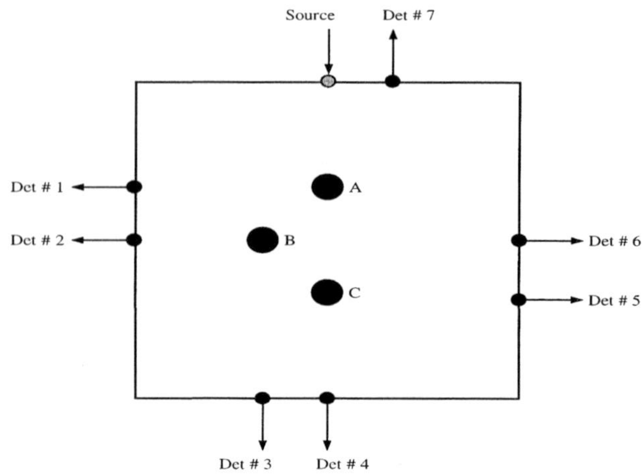

Figure 1. Source/detectors/inclusions configuration for the experiment.

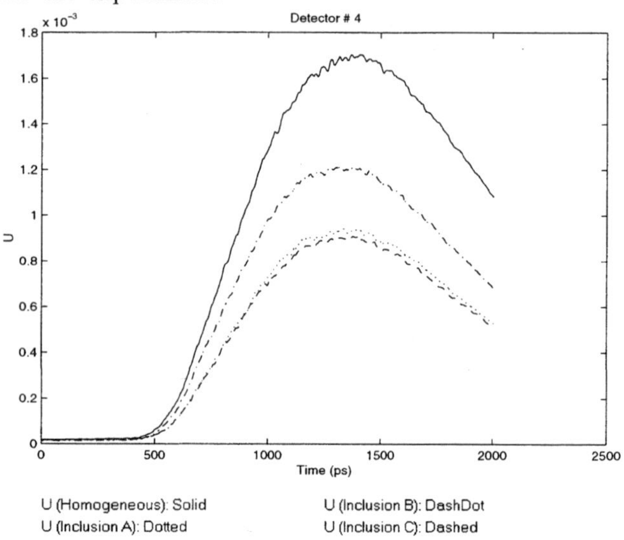

Figure 2. Experimental data for detector #4 (opposite the light source).

The light source was an ultrafast laser with pulses of 100 femtoseconds duration at a wavelength of 625 nanometers. On each detector x_i the time resolved output light intensity was collected by optical fibers using a streak camera on the time interval $t \in (d_i, 2000)ps$, where $d_i = |x_i - x_0|/c$. The time between two consecutive readings was 1 ps. The output light was measured for the following scenarios: (i) for the case when the inclusion was absent, i.e. the response of the homogeneous reference medium, and (ii) separately, for the cases of the inclusions A, B, and C. As an example of the measured output light intensities, Fig. 2 displays the time resolved curves on detector #4, which was opposite the source. The solid curve represents the response of the reference medium; the dotted, dash-dotted, and dashed curves represent the light intensities of the cases A, B, and C respectively.

Note that since the inclusion was a black absorber of non-zero diameter with $l_t = l_a = 0$ inside, then within this absorber

$$h(x) = \infty. \tag{3.1}$$

Images from the same experimental data were computed earlier by the CUNY group using a different image reconstruction algorithm [3]. However, unlike ourselves, this group had treated the data for the A, B, and C cases as coming from the same inclusion using three (3) different source positions and seven (7) detectors per source, which were obtained by three consecutive rotations by 90° of the entire medium.

4 Weak and Strong Absorbers

By the perturbation approach, one assumes that the function $h(x)$ is much smaller than the background function $a_0(x)$, leading to a linearization with respect to $h(x)$ on each consecutive iteration of the ESM. As an example, consider the case of a single inclusion of a circular shape with a radius r. For the sake of simplicity, assume that

$$h(x) = \begin{cases} h_0(x) = \text{const.} > 0, & \text{within inclusion} \\ 0, & \text{otherwise} \end{cases}.$$

Also, assume that $a_0(x) = \text{const.} = a_0 > 0$. Then in the case of *weak* absorbers

$$\frac{\|h\|_{L_2(\Omega)}}{\|a_0\|_{L_2(\Omega)}} = \frac{\pi r^2 h_0}{A(\Omega)|a_0|} = d << 1, \tag{4.1}$$

where $A(\Omega)$ is the area of the domain Ω (in the $2\mathcal{D}$ case). If, however, $d \approx 1$, or $d > 1$, then we call such inclusions *strong* absorbers. For example, in the case of the above described experiment, each of the inclusions A, B, and C is a strong absorber because of (3.1). Thus, the convergence of a perturbation approach for such strong absorbers need not occur. One might hope, however, that the first iteration of this approach would provide a satisfactory image. This is exactly the case of our studies: we obtain good locations of all three inclusions A, B, and C on the first iterative step of the ESM. However, we sometimes observed divergence of the later iterations for this set of experimental data (because of (3.1). The convergence of the perturbation approach should be expected in the case of weak absorbers, as was demonstrated in [7].

5 The Elliptic Systems Method (ESM)

In this section we briefly outline the ESM referring to [7] for more details. The ESM is a version of the perturbation approach. Thus a linearized problem (LP) is solved on each iteration of this method, and this solution is used for the update. The most difficult step is the solution of the LP; updates are easier [7]. Thus, we describe here only the solution of the LP, which amounts to the first iteration of the ESM.

Let the function $u_0(x,t)$ be the solution of the boundary value problem (2.1)-(2.3) for the case of the background coefficient $u(x) = a_0(x)$. That is,

$$\begin{aligned} u_{0t} &= div\,(D(x)\nabla u_0) - a_0(x)\,u_0, \quad (5.1)\\ x &\in \Omega,\ t \in (0,T),\\ u_0(x,0) &= \delta(x - x_0),\\ \frac{\partial u}{\partial n} + \gamma u &= 0, \text{ for } x \in \partial\Omega, t \in (0,T). \end{aligned}$$

We consider the normalized solution $H(x,t) = (u/u_0 - 1)(x,t)$. It follows from (2.1) and (5.1) that the equation for the function H is

$$H_t = div\,(D\nabla H) + 2D\frac{\nabla u_0}{u_0}\nabla H - h(x)(1 + H). \tag{5.2}$$

Linearization of (5.2) with respect to h entails dropping the product hH. Hence the equation for H, linearized with respect to the perturbation term $h(x)$, is

$$H_t = div\,(D\nabla H) + 2D\frac{\nabla u_0}{u_0}\nabla H - h(x). \tag{5.3}$$

In addition, it was proven in [7] that $H(x,0) = 0$. The idea now is to eliminate the perturbation term $h(x)$ from the equation (5.3). We achieve this by differentiating (5.3) with respect to t using the fact that $\frac{\partial}{\partial t}h(x) \equiv 0$. Let $p(x,t) = \frac{\partial H}{\partial t}$. Then

$$H(x,t) = \int_0^t p(x,\tau)\,d\tau. \tag{5.4}$$

Thus, (5.3) and (5.4) lead to an integro-differential equation for the function p:

$$div\,(D\nabla p) + 2D\frac{\partial}{\partial t}\left[\frac{\nabla u_0}{u_0}\int_0^t \nabla p(x,\tau)\,d\tau\right] - p_t = 0. \tag{5.5}$$

There are two boundary conditions for the function $p(x,t)$, but no initial conditions. To derive these boundary conditions, we use (2.4) and (2.5). By (2.4) and (2.5), for $x \in \partial\Omega, t \in (T_0, T)$

$$p = \frac{\partial}{\partial t}\left(\frac{\varphi(x,t)}{u_0(x,t)}\right) = f(x,t); \quad (5.6)$$

$$\frac{\partial p}{\partial n} = -\frac{\gamma u}{u_0} + \frac{\gamma u}{u_0} = 0.$$

Once the function p is computed, then the perturbation term $h(x)$ can be reconstructed from (5.3) by

$$h(x) = \frac{1}{T - T_0}\int_{T_0}^{T}\left[div(D\nabla H) + 2D\frac{\nabla u_0}{u_0}\nabla H - H_t\right]dt \quad (5.7)$$

where the function H is determined by (5.4). We take the average value in (5.7) because in practical computations the integrand has a t dependence. We will now focus on our approach to solving the boundary observation problem (5.5), (5.6), which represents the major computational effort of the ESM.

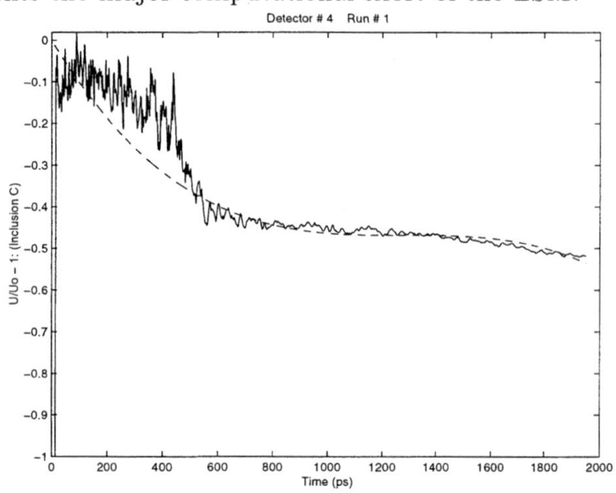

Figure 3. Function $H(x_4, t)$ (solid line) and its fit (5.9) (dashed line) for detector #4, case C.

Let $\{a_k(t)\}_{k=1}^{\infty}$ be an orthonormal basis in $L_2(T_0, T)$, such that all functions $a_k(t)$ are real valued and analytic as functions of the real variable t for $t > 0$. In our numerical work we have chosen the Legendre polynomials normalized over (T_0, T), for this basis. Other options can also be explored. Let $N \geq 1$ be an integer. We assume the following expansion of the function $p(x,t)$ in a generalized Fourier series with respect to $\{a_k(t)\}$

$$p(x,t) \approx \sum_{k=1}^{N} a_k(t)\, Q_k(x), for x \in \Omega,\ t \in (0,T), \quad (5.8)$$

To estimate the number N, consider Fig. 3, which displays the function $H(x_4, t)$ for detector #4 and the inclusion C (Fig. 1). That is, this curve represents the ratio of the upper and lowest curves of Fig. 2 (1 is subtracted from this ratio). This is a *typical* time dependent behavior of the function $H(x,t)$ at a detector: we have repeatedly observed very similar behavior in our numerical studies. One can see that this curve is quite noisy for $t < 500$ ps. On the other hand, it looks almost like a straight line for $t > 600$ ps. The latter implies that the function $H(x_4, t)$ can be interpolated by a low order polynomial for $t \in (600, 2000)$ ps. That is why we typically choose $N = 3$ or 4 in our Legendre expansion (5.8). Thus, the choice of the regularization parameter N is not a costly problem for the ESM.

Next, we perform a least squares polynomial fit for H with respect to t to obtain the generalized Fourier coefficients $Q_k(x)$ required in the function $p(x,t)$ of (5.8). This is our approach to the noise smoothing, which has proven to be very successful. Those generalized Fourier coefficients are found fitting with

$$\widehat{a}_k(t) = \int_0^t a_k(\tau)\, d\tau,\ 1 \leq k \leq N, \quad (5.9)$$

the integrals of $a_k(t)$, using least squares, giving $H(x,t) = \sum_{k=1}^{N} \widehat{a}_k(t)\, Q_n(x)$. Thus in practice f will be well approximated by (5.8). Now we want to obtain a coupled elliptic system with respect to the generalized Fourier coefficients $Q_k(x)$. Introduce the vector valued functions $Q(x) = (Q_1(x), ..., Q_N(x))$ and $\alpha(x) = (\alpha_1(x), ..., \alpha_N(x))$. Multiply both sides of (5.5) by $a_k(t)$ for $k = 1, ..., N$, and integrate with respect to t over the interval (T_0, T). Since the functions $\{a_k(t)\}$ are orthonormal on (T_0, T), by using (5.8), (5.9), we obtain the following elliptic boundary value problem

$$A(Q) := div(D\nabla Q) - \sum_{j=1}^{2} B_j(x)\frac{\partial Q}{\partial x_j} - CQ = 0 \quad (5.10a)$$

$$Q|_{\partial\Omega} = \alpha(x),\ \frac{\partial Q}{\partial n}|_{\partial\Omega} = 0. \quad (5.10b)$$

The $N \times N$ matrices B_j depend on the function u_0, and are given by

$$(B_1, B_2)_{ks} = -2D\int_{T_0}^{T} a_k(t)\frac{\partial}{\partial t}\left[\frac{\nabla u_0}{u_0}\int_0^t a_s(\tau)\, d\tau\right]dt$$

$$= -2Da_k(t)\frac{\nabla u_0}{u_0}\int_0^t a_s(\tau)\, d\tau|_{T_0}^{T}$$

$$+2D \int_{T_0}^{T} a_k'(t) \frac{\nabla u_0}{u_0} \int_0^t a_s(\tau)\, d\tau dt$$

using integration by parts. The elements of the $N \times N$ matrix C are given by

$$c_{ks} = \int_{T_0}^{T} a_s'(t)\, a_k(t)\, dt; k, s = 1, ..., N.$$

Equation (5.10a) is an overdetermined elliptic system because of the two boundary conditions (5.10b), rather than just one. This leads to solving an associated 4th order elliptic system. Let A^* be the operator formally adjoint to the operator A, $A^*(Q) := div(D\nabla Q) + \sum_{j=1}^{n} \frac{\partial}{\partial x_j}\left(B_j^T(x)\, Q\right) - C^T Q$, where the superscript "T" denotes a matrix transpose, and $n = 2$ or 3, depending on the dimension of the space K. We consider the boundary value problem for the following 4th order elliptic system:

$$(A^* A)(P) = 0, \tag{5.11a}$$

$$P|_{\partial\Omega} = \alpha(x), \frac{\partial P}{\partial n}|_{\partial\Omega} = 0. \tag{5.11b}$$

The biggest advantage of the resulting system (5.11) is its *differential* form. Therefore, the solution of this system by use of the FEM leads to the factorization of a sparse matrix which can be done rapidly [9, 10]. We note that the problem (5.12) is equivalent to the problem of the minimization of the cost functions $J(Q)$, subject to the boundary conditions (5.11b)

$$J(Q) = \int_{\Omega} |AQ|^2\, dx \to \min.$$

6 RESULTS

We have observed in the numerical experiments of this paper that the first iteration of the ESM provides inclusion location values, which we evaluate as good, given such a limited number of detectors and only a single source. However, the subsequent Born-like iterations diverged in some cases (not always). On the other hand, we have commonly observed the convergence of iterations for weak absorbers, for computationally simulated data. Thus, we attribute these divergent cases to (3.1), which implies the violation of the natural condition (4.1) of the convergence. For this reason, we decided to use only the first iteration of the ESM when working with this experimental data. Likewise we took $(T_0, T) = (600, 2000)$ ps and $N = 4$ in all numerical experiments of this section.

To test the sensitivity of these images to prior knowledge of the background medium, we have varied the diffusion coefficient D taking

$$\widetilde{D} = (1 + r)\, D, \tag{6.1}$$

where D is the value of the diffusion coefficient used in the experiment, $D = 0.2025 \frac{\text{mm}^2}{\text{ps}}$. In (6.1) the parameter r characterizes the deviation of the prior guess of the value of the diffusion coefficient from its correct value. The function u_0 in (5.1) was computed using $\widetilde{D}$. On the other hand, the boundary conditions (5.11b) do not depend on $\widetilde{D}$ since only the experimental data is used for values of u_0 at the detectors. Thus, the deviation in D affected only the coefficients of the operator A in (5.11a). We have consistently taken $r \geq 0$ in (6.1), because we observed that better quality images were obtained when $r \geq 0$, rather than for $r < 0$, i.e. for $\widetilde{D} \geq D$. That is, it is better to over estimate the diffusion coefficient, rather than to underestimate it.

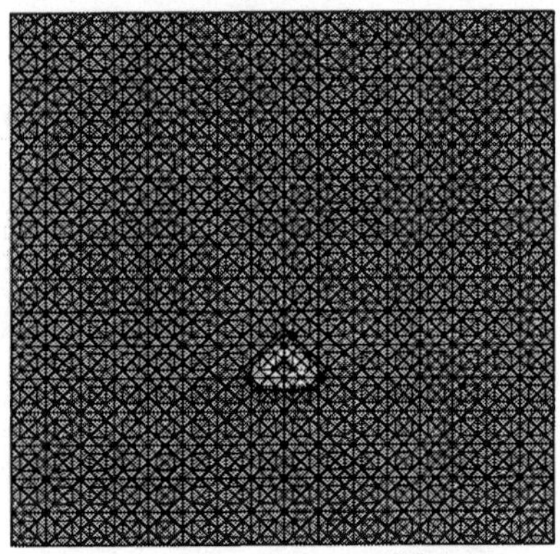

Figure 4. Correct image for the inclusion C.

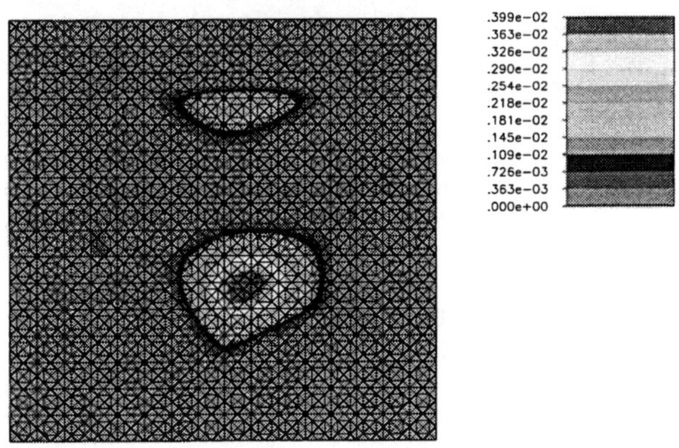

Figure 5. Computed image of C with $r = 0.5$.

We have constructed images using r values of 0; 0.1; 0.3; and 0.5, which implies errors in D of 0% to 50%, over inclusions A, B, and C. A typical image is presented on Fig. 4, for the inclusion C with $r = 0.5$

in (5.13). Fig. 5 displays the correct image of C. The location of the inclusion C is imaged with a good accuracy.

In all images consider ellipses within which the value of the computed function $h(x)$ is not less than 90% of its maximal value. In most cases the small axis of these ellipses was about 3.7 mm, and the big axis was about 4.5 mm. This is close to the correct diameter of each of the inclusions, which was 4.8 mm. Imaging results of the experimental data are summarized in Table 1.

Table 1. Location (in mm) of the point of the maximal value of the computed function $h(x)$ for different values of $r \cdot 100\%$.

Inclusion	0%, 10%	30%, 50%
A $(0, 10)$	$(0.6, 13.1)$	$(0.6, 13.1)$
B $(-10, 0)$	$(-8.1, 1.8)$	$(-9.3, -4.3)$
C $(0, -10)$	$(1.8, -8.1)$	$(1.8, -8.1)$

One can conclude from this table that the obtained images are quite stable with respect to prior knowledge of the diffusion coefficient for values of up to an error of 50%. The results at the error level of 100% (not shown) are poor for A and B but fair for C.

7 Discussion

The performance of the ESM for the case of experimental data was demonstrated and its sensitivity to prior estimates of the background medium was studied. One can conclude from Table 1 that the obtained images are quite stable with respect to the uncertainties of up to 50% in the prior estimate of the diffusion coefficient.

The robustness of the ESM is further demonstrated by its ability to work with only a single location for the light source and a very limited number of detectors, which potentially could lead to a significant decrease of the acquisition time and a simplification/lower price of the hardware design.

In addition, the choice of the regularization parameter N is not a time consuming problem for the ESM, unlike many competing techniques. We have given a good explanation that typically $N = 3$ or 4. The total CPU time for each of the above images was 2.5 minutes on a Silicon Graphics Indigo 2: (i) 114 seconds for the solution of the forward problem (5.1) and (ii) 36 seconds for the solution of the resulting elliptic system (5.11) by the FEM using 1024 quadratic finite elements and with the number of coupled elliptic partial differential equations $N = 4$.

8 Acknowledgments

The authors express their appreciation to the experimental group led by Prof. Robert R. Alfano from City College of CUNY, New York for the use of their experimental data. This research was supported by NSF grant DMS-9704923.

9 References

1. S.R. Arridge, M. Schweiger, and D.T. Delphy, "Iterative reconstruction of near infrared absorption images", Proc. SPIE, Vol. 1767, 372-383, (1992).

2. R.L. Barbour, H.L. Graber, J. Chang, S.-L. S. Barbour, P.C. Koo, and R. Aronson, "MRI-guided optical tomography", IEEE Computational Science and Engineering, Vol. 2, 63-77, (1995).

3. W. Cai, B.B. Das, F. Liu, M. Zevallos, M. Lax, and R.R. Alfano, "Time-resolved optical diffusion tomographic image reconstruction in highly scattering turbid media", Proc. Natl. Acad. Sci. USA, Vol. 93, 13561-13564, (1996).

4. A.H. Hielscher, "Model-based iterative image reconstruction for photon migration tomography", Proc. SPIE, Vol. 3171, 106-117, (1997).

5. H. Jiang, K.D. Paulsen, and U.L. Österberg, "Optical image reconstruction using DC data: simulations and experiments", Physics in Medicine and Biology, Vol. 41, 1483-1498, (1996).

6. V. Isakov, "Inverse Problems for Partial Differential Equations" (Springer, Applied Mathematical Sciences, Vol. 127, New York, 1998).

7. M.V. Klibanov, T.R. Lucas, and R.M. Frank, "A fast and accurate imaging algorithm in optical/diffusion tomography", Inverse Problems, Vol. 13, 1341-1361 (1997).

8. M.V. Klibanov and T.R. Lucas, "Method and apparatus for detecting abnormality with a host medium", United States patent pending; patent's assignee is University of North Carolina at Charlotte.

9. A George and J. Lin, Computer Solution of Large Sparse Positive Definite Systems (Prentice Hall, Englewood Cliffs, NJ, 1981).

10. O. Axelsson and N. Munksgaard, "A class of preconditioned conjugate gradient methods for the solution of a mixed finite element discretization of the biharmonic operator", International Journal Of Numerical Methods in Engineering, Vol. 14, 1001-1019 (1979).

11. B.B. Das, F. Lin, and R.R. Alfano, "Time resolved fluorescence and photon migration studies in biomedical and model random media", Rep. Prog. Phys., Vol. 60, 227-292, (1997).

A fundamental limitation of linearized algorithms for diffuse optical tomography

D. A. Boas
Electro-Optics Technology Center, Tufts University, Medford, MA 02155
phone: (617)627-3136 fax: (617)627-3151 email: dboas@tufts.edu

When light is incident on a strongly scattering medium like tissue or clouds, the ballistic light is quickly attenuated through scattering, and imaging capabilities are generally assumed to be lost. Recently, there have been significant research efforts to model the transport of light through strongly scattering media in order to characterize the optical properties of spatially uniform media and to image spatial variations in optical properties [1-4]. Imaging methods have been developed using continuous-wave light (CW domain) [5], pulsed lasers (time domain) [6,7], and intensity modulated light (frequency domain) [8-10]. Many imaging algorithms rely on assumptions which linearize the relationship between the optical contrast of the inhomogeneity and measurements of the scattered light. In this paper, we show that this linearization results in incorrect moments and thus accurate quantitative imaging is not possible. We also discuss the implications that this result has on Diffuse Optical Tomography (DOT); primarily that quantitative DOT requires non-linear reconstruction algorithms which correctly model the moments. Although the results presented here are based on an analysis in the frequency domain, the results also apply to the time domain (as the two are related by a Fourier transform) and the CW domain (which is the zero frequency limit of the frequency domain).

Common examples of linearized image reconstruction algorithms for DOT are based on the Born approximation [4,8,10] (and Rytov approximation [8], not detailed here). The Born approximation assumes that the DPDW is given by $\Phi(\mathbf{r_s},\mathbf{r_d}) = \Phi_{inc}(\mathbf{r_s},\mathbf{r_d}) + \Phi_{sc}(\mathbf{r_s},\mathbf{r_d})$ while the Rytov approximation assumes that the DPDW is given by $\Phi(\mathbf{r_s},\mathbf{r_d}) = \Phi_{inc}(\mathbf{r_s},\mathbf{r_d}) \exp(\Phi_{sc}(\mathbf{r_s},\mathbf{r_d}))$. $\Phi_{inc}(\mathbf{r_s},\mathbf{r_d})$ is the DPDW measured without any inhomogeneities in the optical properties for a light source at position $\mathbf{r_s}$ and a detector at position $\mathbf{r_d}$. The scattered DPDW is given by $\Phi_{sc}(\mathbf{r_s},\mathbf{r_d})$. For the Born approximation,

$$\Phi_{sc}(\mathbf{r}_s,\mathbf{r}_d) = -\int \Phi_{inc}(\mathbf{r}_s,\mathbf{r}_d)\mathcal{L}(\mathbf{r})G(\mathbf{r}_s,\mathbf{r}_d)d\mathbf{r} \quad . \qquad (1)$$

$\mathcal{L}(\mathbf{r})$ is an operator which depends on the spatially varying absorption coefficient and the spatially varying reduced scattering coefficient [8]. Images of the spatially varying optical properties are reconstructed given measurements of $\Phi_{sc}(\mathbf{r_s},\mathbf{r_d})$, calculations of $\Phi_{inc}(\mathbf{r_s},\mathbf{r_d})$ and $G(\mathbf{r_s},\mathbf{r_d})$ from the known background optical properties, and inverting eq. (1) to find $\mathcal{L}(\mathbf{r})$. This inversion generally requires discretizing the integral equation to obtain a matrix equation which is then solved by matrix inversion or algebraic reconstruction techniques.

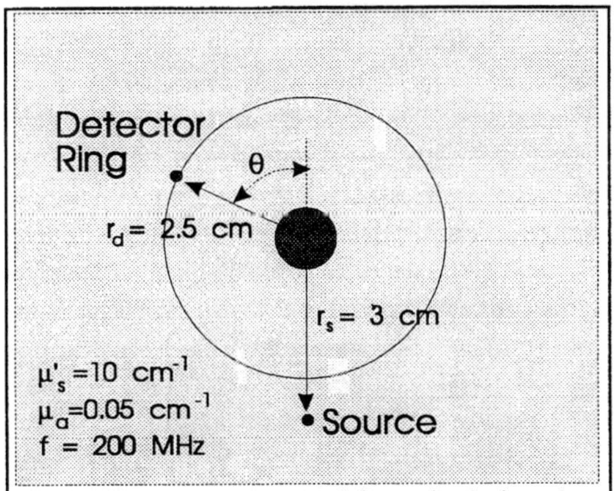

Fig. 1 The geometry used for calculating the scattered DPDW (see text).

Quantitative imaging is possible only if the linearized solution, eq. (1), produces scattered waves which agree well with exact solutions. The fact that sensitivity to the moments of the scattered wave is critical to quantitative characterization of optical inhomogeneities, as described by Boas *et al.* [11], suggests that we should analyze the moments of the linear solutions used for image reconstruction algorithms. To do this, we calculated eq. (1) for a spherical object embedded in an otherwise uniform, infinite medium, and compared the numerical results with the analytic solution for DPDW's scattering from spherical objects [12,13]. The geometry used for the calculations is indicated in fig. 1. The optical properties of the infinite medium were μ_s'=10 cm^{-1} and μ_a=0.05 cm^{-1}. The source was positioned 3 cm along the z axis from the center of the object, which was placed at the origin. The source had a modulation frequency of 200 MHz (or angular modulation frequency of 2π 200 MHz). Detectors were placed every 2° in a circle 2.5 cm from the center of the object. The spherical object had a fixed reduced scattering coefficient, μ_s'=10 cm^{-1}, and a varying absorption coefficient and radius ($0.1 \le \mu_a \le 1.0$ cm^{-1} and $0.1 \le a \le 1.0$ cm). The integrals in eq. (1) were evaluated by dividing the bounding cube of the spherical object into 10x10x10 voxels and summing the contribution from each voxel. The moments were calculated from the projection of the detected complex scattered wave onto spherical harmonics $Y_{l,m}(\theta)$ for m=0 (no dependence on azimuthal angle). We do not explore the dependence on the azimuthal index m as our geometry was chosen for the moments with $m \neq 0$ in order to have zero amplitude.

Reprinted from Advances in Optical Imaging & Photon Migration, 1998 Technical Digest, © 1998 Optical Society of America

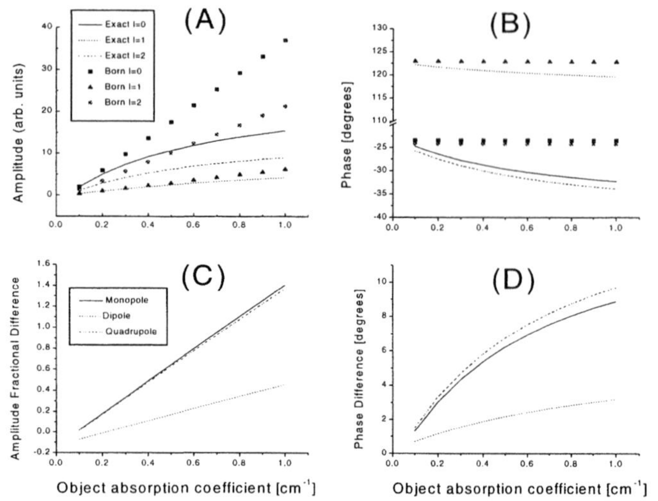

Fig. 2 Comparison of the first order Born approximation with the exact solution for a 1 cm diameter absorbing object with different absorption coefficients.

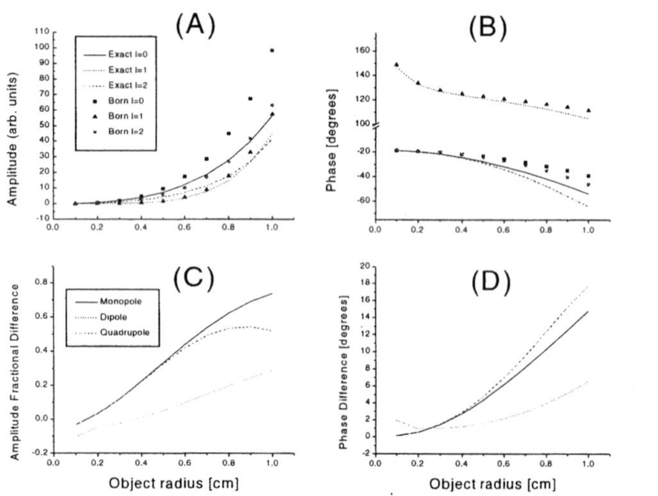

Fig. 3 Comparison of the first order Born approximation with the exact solution for an absorbing object with an absorption coefficient of 0.3 cm^{-1} and different radii.

In fig. 2 we plot the relative amplitudes and phases of the monopole, dipole, and quadrupole moments of the scattered wave for the Born approximation and the exact solution, to show their differences. Results are graphed versus the absorption coefficient of the object for a 1.0 cm diameter object. When the absorption coefficient of the object is 0.10 cm^{-1}, the amplitude (phase) of the exact and Born moments agree within a few percent (1 degree). As the object absorption coefficient increases, the amplitudes disagree by more than 20% and the phases disagree by more than a few degrees.

In fig. 3 we plot the relative amplitudes and phases of the monopole, dipole, and quadrupole moments of the scattered wave versus radius for an object with an absorption coefficient of 0.3 cm^{-1}. Similar discrepancies are observed. The results in fig. 2 and fig. 3 clearly demonstrate that the first order Born approximation does not accurately predict the amplitude and phase of the scattered DPDW. Thus, image reconstruction algorithms based on the first order Born approximation will not accurately quantify the optical properties of the medium.

The trend observed in fig. 2 for the Born moments is interesting. As the object absorption coefficient increases, the amplitudes of the moments increase proportionally and the phases of the moments remain fixed. This result is expected. Within the Born approximation, a voxel of the discretized integral produces a monopole scattered wave (given a voxel with an absorption coefficient different than the background). The strength of this monopole wave is linearly proportional to the absorption difference between the voxel and background. Doubling the absorption difference will double the monopole moment. An object comprised of a single absorbing voxel will not produce higher order moments (e.g. dipole and quadrupole). It is only from a group of absorbing voxels that these higher moments are produced. Furthermore, as the existence of these moments depends only on the spatial distribution of voxels, increasing the delta absorption coefficient by a multiplicative factor X will increase the amplitudes of the moments by a factor X. Thus, the relative amplitude of the moments is not dependent on the absorption coefficient (contrary to that observed with the exact solution). Furthermore, the phase remains constant because of this linear dependence on the absorption difference.

It is useful to consider what effect this discrepancy between the Born approximation and the exact solution has on image reconstructions. Reconstruction algorithms typically find the image which gives the best fit to the experimental data by means of minimizing the least-squares difference between the experimental measurement (in this case the exact solution) and the model used by the algorithm (in this case the first Born approximation). This is not strictly true as some algorithms use regularization methods that also minimize a norm of the image [6]. Consider a 1.0 cm diameter spherical absorber with an absorption coefficient of 0.3 cm^{-1}. From fig. 2 and fig. 3 we see that minimizing the least-squares difference of the amplitude can be accomplished by decreasing the size and/or absorption coefficient of the object. Or the size can be overly decreased and compensated by an increase in the absorption coefficient, or vice-versa. To minimize the least-squares difference in the phase, the size of the object must increase, as changing the absorption coefficient has no effect. This suggests that the reconstruction algorithm would reconstruct an object which has a larger diameter and smaller absorption coefficient than the real object parameters. More insight on specific algorithms operating on particular clinical problems could be obtained by correlating the performance of the algorithm with observations like those described in this paragraph.

In summary, we have shown that linearized algorithms for diffuse optical tomography do not accurately model the amplitude and phase of scattered DPDW's. Quantitative imaging is thus not feasible as it relies on the use of accurate models for predicting the amplitude and phase of the scattered DPDW. The most common linearized algorithms are based on the first order Born approximation and the Rytov approximation. Each approximation poorly models the scattered DPDW. Although not shown here, the Rytov approximation calculates the monopole and quadrupole moments scattered from an absorbing object more accurately than the Born approximation, but the Born approximation is much more accurate for the dipole moment. The main consequence of the results presented here is that non-linear imaging algorithms must be used to achieve quantitative imaging. Elsewhere, we have shown that higher order Born approximations can more accurately calculate the scattered wave [14]. When the absorption coefficient is less than 0.4 cm^{-1} and the diameter is less than 8 mm, the fifth order Born approximation generally provides sufficient accuracy. However, implementing an Nth order Born approximation in a non-linear imaging algorithm is much more computationally expensive than using multi-grid finite-difference or finite-element methods.

1. B. Chance, *Photon Migration in Tissues* (Plenum Press, New York, 1988).

2. A. J. Welch and M. J. C. van Gemert, *Optical-Thermal Response of Laser-Irradiated Tissue* (Plenum Press, New York, 1995).

3. A. Yodh and B. Chance, "Spectroscopy and imaging with diffusing light," Physics Today **48:**34-40 (1995).

4. S. R. Arridge and J. C. Hebden, "Optical Imaging in Medicine: II. Modelling and reconstruction," Phys. Med. Biol. **42:**841-854 (1997).

5. R. L. Barbour, H. L. Graber, Y. Wang, J. Chang and R. Aronson, "Perturbation approach for optical diffusion tomography using continuous-wave and time-resolved data," in *Medical Optical Tomography: Functional Imaging and Monitoring*, ed. G. Muller, B. Chance, R. Alfano, S. Arridge, J. Beuthan, E. Gratton, M. Kaschke, B. Masters, S. Svanberg and P. van der Zee, Proc. SPIE **IS11**:87-120 (1993).

6. S. R. Arridge, "Forward and inverse problems in time-resolved infrared imaging," in *Medical Optical Tomography: Functional Imaging and Monitoring*, ed. G. Muller, B. Chance, R. Alfano, S. Arridge, J. Beuthan, E. Gratton, M. Kaschke, B. Masters, S. Svanberg and P. van der Zee, Proc. SPIE **IS11**:35-64 (1993).

7. D. A. Benaron, D. C. Ho, S. Spilman, J. P. Van Houten and D. K. Stevenson, "Tomographic time-of-flight optical imaging device," Adv. Exp. Med. Biol. **361:**609-617 (1994).

8. M. A. O'Leary, D. A. Boas, B. Chance and A. G. Yodh, "Experimental images of heterogeneous turbid media by frequency-domain diffusing-photon tomography," Opt. Lett. **20:**426-428 (1995).

9. H. Jiang, K. D. Paulsen, U. L. Osterberg, B. W. Pogue and M. S. Patterson, "Optical image reconstruction using frequency-domain data: Simulations and experiments," J. Opt. Soc. Am. A **13:**253-266 (1996).

10. Y. Yao, Y. Wang, Y. Pei, W. Zhu and R. L. Barbour, "Frequency-domain optical imaging of absorption and scattering distributions using a born iterative method," J. Opt. Soc. Am. A **14:**325-342 (1997).

11. D. A. Boas, M. A. O'Leary, B. Chance and A. G. Yodh, "Detection and characterization of optical inhomogeneities with diffuse photon density waves: a signal-to-noise analysis," Appl. Opt. **36:**75-92 (1997).

12. P. N. den Outer, T. M. Nieuwenhuizen and A. Lagendijk, "Location of objects in multiple-scattering media," J. Opt. Soc. Am. A **10:**1209-1218 (1993).

13. D. A. Boas, M. A. O'Leary, B. Chance and A. G. Yodh, "Scattering of diffuse photon density waves by spherical inhomogeneties within turbid media: analytic solution and applications," Proc. Natl. Acad. Sci. USA **91:**4887-4891 (1994).

14. D. A. Boas, "A fundamental limitation of linearized algorithms for diffuse optical tomography," Optics Express **submitted:**(1997).

Characterization of small inhomogeneities

Jean Virmont and Guy Ledanois
Laboratoire P.M.I., Ecole Polytechnique, 91128 Palaiseau cedex, France
tel: +33 1 6933 3269, fax: +33 1 6933 3023, e-mail: virmont@lpmi.polytechnique.fr

Introduction and methods

X-ray mammography, yet very useful, has limitations. Even with considerable training, it produces false negatives and false positives. Statistics show that the median size of breast tumours at detection is ~1.75 cm.[1] So alternative methods which would (help to) reliably detect 0.5 cm or even 1 cm tumours would be most welcome. Can IR light do that? While light can be detected through breast, scattering induces highly blurred transmission images. As a result, using simple transmission IR imaging proved insufficient. During the last 15 years, efforts to improve the resolution, particularly using picosecond pulsed lasers, yielded limited success. Numerical solutions of the inverse problem (*i.e.* the numerical reconstruction of the absorption and/or scattering maps inside the organ from information obtained with a set of external light sources and detectors) were proposed, however with large computation times.

More recently faster (but approximate) inversion techniques have been proposed.[2,3] It has also been recognized that tumor detection is not only a matter of resolution, but also one of contrast, and that both the absorption and scattering coefficients may be affected in benign and malign tissue modifications. While measuring both may provide more information and specificity, it also makes the problem more difficult. Several new prototypes of optical mammographer have been recently built and are being tested.[2,4,5] Other imaging prototypes have been built in neonatology,[6] to detect brain hematoma,[7] and to measure brain activity.[8] This shows the continuing clinical interest for the optical imaging of large organs.

'Detection' and 'characterization' of suspicious areas should be distinguished.[9] The 'burried object' can be detected if it produces a modification of the measured signal (transmitted light) larger than the 'noise level'. If the other conditions are specified (noise, geometry, background properties), this yields a minimum object size: the detection threshold. Characterizing the object means evaluating with a reasonable accuracy its volume and its optical properties (absorption and scattering coefficients). This yields another minimum size, larger than the previous one: the characterization threshold. These thresholds have been evaluated by D. Boas *et al.* for a purely absorbing object and for a purely scattering one.[9]

In this paper, we consider the characterization of more complex (more realistic) objects, with both absorption and scattering contrasts. We use the Bayesian approach to study the error bars on the parameters to be reconstructed, and on combinations of them. We show that CW illumination cannot characterize absorbing and scattering objects except for very large sizes. We demonstrate that a significant reduction of the characterization threshold can be obtained if complete characterization is replaced by a partial one, giving information only on the nature (absorbing versus scattering contrast) of the burried object.

We consider a standard model problem of optical imaging: a slab of 'background' uniform medium with absorption and scattering coefficients μ_{a0} and μ'_{s0}, a 'burried sphere' in the center with volume V and optical coefficients $\mu_a = \mu_{a0} + \Delta\mu_a$ and $\mu'_s = \mu'_{s0} + \Delta\mu'_s$. Source and detector are scanned in tandem along opposite faces. We consider either CW or modulated (frequency-domain) illumination. We assume uncorrelated noise with levels ±0.1% on amplitude and ±0.1° on phase. We consider values typical of mammography: slab thickness T = 7 cm, $\mu_{a0} = 0.05$ cm^{-1} and $\mu'_{s0} = 10$ cm^{-1} We choose a modulation frequency of 200 MHz, small enough not to reduce too much the transmitted amplitude and large enough to provide phase information. An accuracy of ±30% on unknowns is considered as reasonable. An alternative choice T = 5 cm and $\mu_a = 0.025$ cm^{-1} reduces our thresholds by ~20%.

In the Bayesian approach,[10] the measurements (transmitted intensities for a set of source/detector positions) yield a data set **M** which is considered as a gaussian random vector, with average $\mathbf{M_0}$ and covariance matrix **B**: $B_{ij} = < \delta M_i \, \delta M_j >$ where $\delta M_i = M_i - M_{i0}$ is the fluctuation on M_i. The unknowns (V, μ_a, μ_s') are also considered as a random gaussian vector **P**. The method provides for **P** a mean value $\mathbf{P_0}$ and a covariance matrix **C**. So, in addition to most probable values, standard deviations (S.D.) are obtained for the unknowns P_i , and also for any combination of them. The Bayesian approach also can handle prior information on the unknowns, *e.g.* that scattering is small or that the size is approximately known.

Results and discussion

We first consider continuous (CW) illumination: we demonstrate that inhomogeneities with both absorption *and* scattering contrasts are much more difficult to characterize than only absorbing or only scattering ones. Fig.1 shows that the sphere diameter must be larger than 2.1 cm in the former case. In the latter ones a minimum diameter of ~1.2 cm is obtained (figure not shown). A diameter of 2.1 cm is unacceptably large for early cancer identification. On the other hand, if the illumination is modulated at 200 MHz (frequency-domain), the minimum size is ~1.2 cm in all cases, which is marginally acceptable.

These results can be put as follows. For a size of 1.2 cm for which the gaussian-like CW image can provide information on 2 unknowns, adding the phase image provides information on 3 with comparable accuracy. For the CW image to provide information of 3 unknowns, a larger size is needed: the image shape must deviate from the gaussian, and the signal-to-noise ratio must be large enough for the details of the shape to be significant.

Then we study the errors as a function of the contrast and nature of the inhomogeneity. In the plane (μ_a, μ_s'), let O correspond to the origin, B to the background, and I to the inhomogeneity. The contrast of the inhomogeneity can be defined by C = |BI| / |OB|, and its nature by the angle ß between **BI** and the axis μ_a (from 0° to 360°). The inhomogeneity considered in Fig.1 corresponds to C = 1 and ß = 45°. We find that the characterization threshold does not depend much on C (for reasonable values C ~ 0.5-1), nor on ß (except for the purely absorbing case ß ~ 0°, which is worse for CW illumination). So Fig.1 is representative of most inhomogeneities expected in tissues.

Is it possible to say 'something' about the inhomogeneity when its size is smaller than the characterization threshold? It is known from the perturbation approach of the forward problem that the image profile mainly depends on two composite quantities: $V\Delta\mu_a$ and $V\Delta\mu_s$'. We can also consider their ratio, or more precisely the angle ß defined previously, since tg ß = $V\Delta\mu_s' / V\Delta\mu_a$. As shown by Fig.2 for CW illumination, the angle ß remains accurate within ±20° down to a diameter of 1.2 cm. For modulated illumination (200 MHz), the corresponding value is 0.8 cm (figure not shown). So information on the 'nature' of the inhomogeneity remains available significantly below the characterization threshold.

We enforce that by the 'nature' of the inhomogeneity we mean: scattering versus absorbing contrast, and also positive versus negative contrasts (*e.g.* $\mu_a < \mu_{a0}$ vs. $\mu_a > \mu_{a0}$). Determining only the angle ß can be called a 'partial characterization' of the sphere. The good accuracy obtained on ß can be understood as follows. For the sizes of interest, the image width depends more on the nature of the inhomogeneity than on its size: the half-width is increased by ~0.21 cm from the purely scattering case to the purely absorbing one, and only by ~0.05 cm when the diameter is varied from 0 to 1.2 cm.

Concerning the role of prior information, we find that an inhomogeneity can be considered as purely absorbing (with the corresponding threshold reduction) only if the scattering contrast is smaller than a few percent, which seems difficult to meet (even using two wavelengths). Information on the size would be useful in the CW case, where knowing the size within ±30% reduces the threshold for partial characterization from 1.2 to 0.8 cm.

Little reliable information is available on contrasts to be expected inside the human breast between fatty tissue, glandular tissue, tumours, cysts, ...[11] Recent measurements should help to estimate the clinical value of our results.

References

1. Osteen R.T. et al., *J. Nat. Cancer Instit. Monographs*, **16**,43-46 (1994).
2. Colak S.B. et al., *Applied Optics*, **36**, 180-213 (1997).
3. Walker S.A., Fantini S., Gratton E., *Applied Optics*, **36**, 170-179 (1997).
4. Grable R.J., "Optical tomography breast imaging", *SPIE Proceedings*, **2979** (1997).
5. Jess H. et al.,*Trends in Optics and Photonics*, **2**, 126-129 (1996).
6. Benaron D.A. et al., *SPIE Proceedings*, **2389**, 582-596 (1995).
7. Gopinath S.P. et al., *J. Neurosurg.*, **79**, 43-47 (1993).
8. Maki A. et al., *Med. Phys.*, **22**, 1997-2005 (1995).
9. Boas D.A., O'Leary M.A., Chance B., Yodh A.G., *Applied Optics*, **36**, 75-92 (1997).
10. Jaynes E.T., *Papers on probabilitiy, statistics and statistical physics*, R.D. Rosencrantz Ed., Dordrecht (1983).
11. Troy T.L., Page D.L., Sevick-Muraca E.M., *J. Biomed. Optics*, **1**, 342-355 (1996).

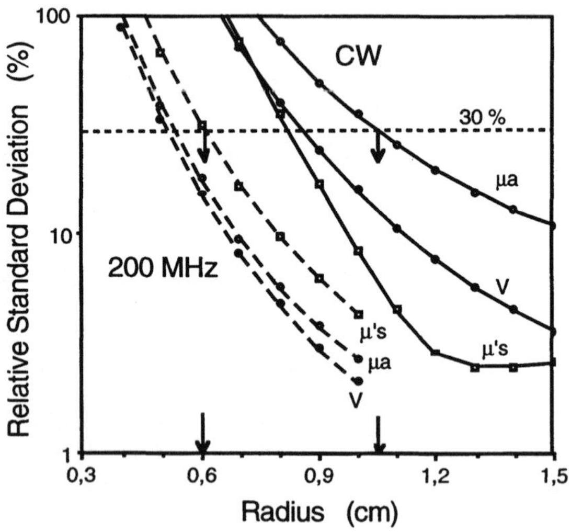

Fig. 1 - Error bars on the unknowns V, μa and μs, versus the radius of the sphere (inhomogeneity). Comparison of CW illumination (solid lines) and modulated illumination (dotted lines).

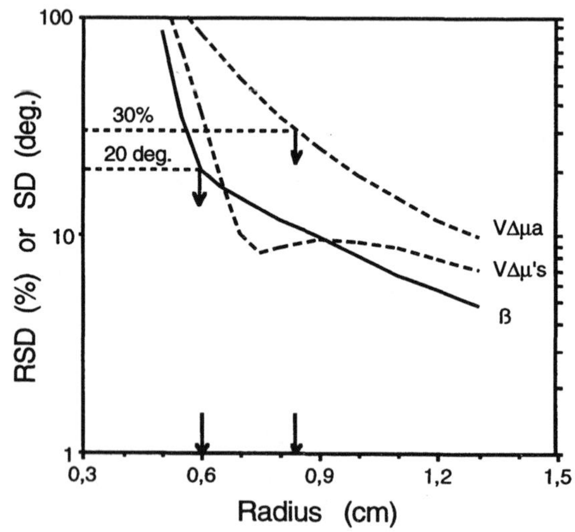

Fig. 2 - Error bars on composite quantities: Vμa, Vμ's and angle ß, versus sphere radius, for CW illumination.

Location and characterization of objects in scattering media using arrays of vertical cavity surface emitting lasers (VCSELs)

Jean-Pierre Marie, Bernard Le Jeune, Jack Cariou, Jean Lotrian

Université de Bretagne Occidentale, Laboratoire de Spectrométrie et Optique Laser
6, avenue Le Gorgeu, B.P. 809, 29285 Brest Cedex, France

Abstract

This paper experimentally shows that using an array of Vertical Cavity Surface Emitting Lasers to probe a scattering medium in which an inhomogeneity is embedded enables one to localize and characterize this inhomogeneity. It reports on an optical experiment based on the study of the beams transmitted through the medium. Two types of experiments, the first one without optical elements and the second one with a polarimetric analyzer, are presented. Finally, the advantages of the technique developed hereabove are discussed.

Key Words

(140.2010) Diode laser arrays; (290.5820) Scattering measurements; (260.5430) Polarization; (999.9999) Inhomogeneities.

Introduction

The use of arrays of vertical cavity surface emitting lasers (VCSELs) in the near-infrared constitutes an interesting approach for the location and characterization of objects in scattering media [1-4]. Their small size, array lay-out and optical power along with the fact that they are mutually incoherent confer them interesting properties and allow their integration into systems devoted to the detection of inhomogeneities.

Our purpose was to localize an object embedded in a scattering medium from an intensity- and polarimetry-characterization.

Principle of experiment

The method, described here and illustrated in figure 1, enabled us to localize an object in a medium by analyzing the shades produced in the detector plane.

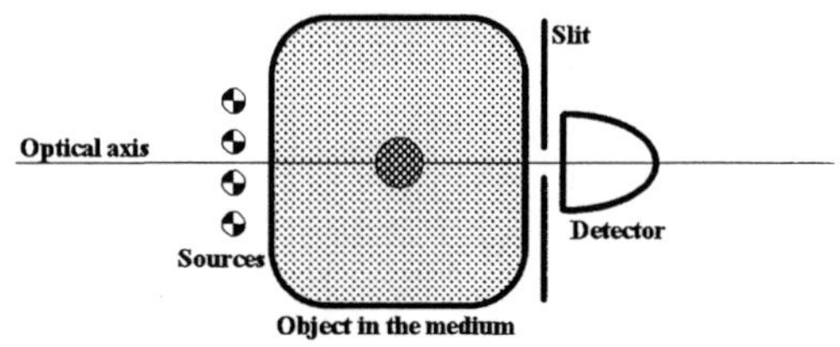

Figure 1. Schematic view of the scattering system with an embedded object.

The system was composed of the light source, constituted by an array of 4 x 4 VCSELs of 10 µm in diameter and 125 µm in spacing presented in figure 2, and the detector, a silicon photodiode behind a narrow slit. These elements were set in a fixed position. The medium studied was constituted by a mixture of milk and water in a glass tank (25 mm thick) containing an object. The sources emitted 850 nm circular Gaussian beams.

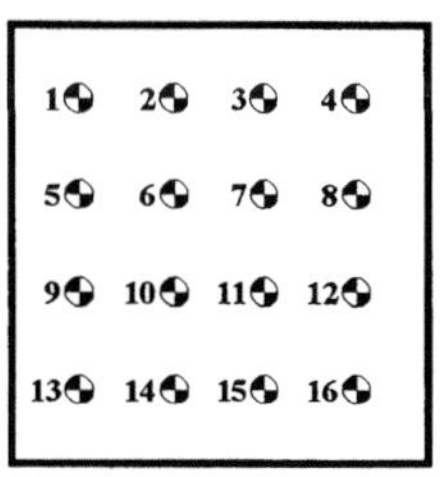

Figure 2. Spatial configuration of the sixteen VCSELs.

The different sources were amplitude-modulated at various frequencies. Shifting the tank in a direction perpendicular to the system optical axis allowed us to precisely determine the object position in the container and its size. The medium was thus moved inside the system. We studied the evolution of a beam issued from an off-line source and hitting the detector. The determination of the intensity just behind the slit gave us precisely the distance between the sources and the object.

Description of experiment

Figure 3 represents a planar view of the experiment for 2 sources situated symmetrically on either side of the optical axis and spaced by the distance $2a$. The interval between the sources and the detector is D, and the width of the slit is e. We stated x_0 the position of the object shadow in the detector plane, and X_0 the position of the object in the medium along the axis perpendicular to the optical axis. The unknown parameters were d the distance between the sources and the object, and b the size of the object. We finally named b' the shadow of the object in the detector plane.

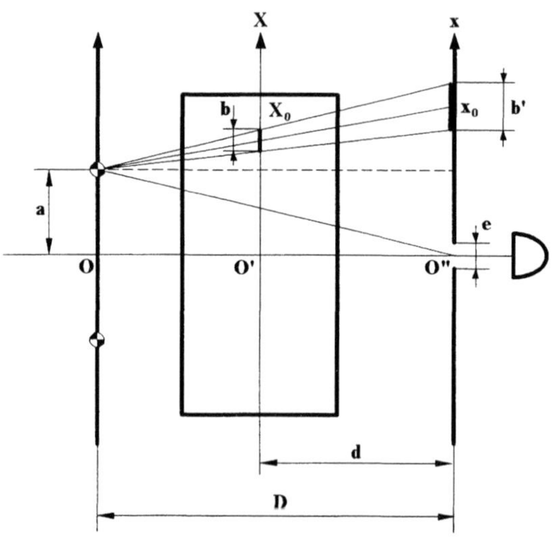

Figure 3. Experimental description.

Calculus focused on the detection of the 2 edges of an object located in a plane perpendicular to the optical axis. We first considered the object alone.

Thus, the relation between b and b' without scattering medium is:

$$b' = b\frac{D}{D-d}$$

which gives immediately:

$$\frac{X_0 - a}{D-d} = \frac{x_0 - a}{D}$$

So, we can express x_0 as a function of X_0 by:

$$x_0 = a + \frac{D}{D-d}(X_0 - a)$$

The flux $I(x_0)$ detected at the abscissa x_0 in the detector plane can be expressed in first approximation by:

$$I(x_0) = I_0 \int_{-\infty}^{+\infty} \Pi_e(x)\left[1 - \Pi_{b'}(x - x_0)\right]dx$$

where

$$\Pi_e(x) = \begin{cases} 1 \ for & -\frac{e}{2} \le x \le \frac{e}{2} \\ 0 \ elsewhere & \end{cases}$$

If we define A_1, A_2, A_3, and A_4 by the following relations:

$$A_1 = \frac{1}{2D}\left[-bD + 2ad - e(D-d)\right]$$

$$A_2 = \frac{1}{2D}\left[-bD + 2ad + e(D-d)\right]$$

$$A_3 = \frac{1}{2D}\left[bD + 2ad - e(D-d)\right]$$

$$A_4 = \frac{1}{2D}\left[bD + 2ad + e(D-d)\right]$$

we then obtain the expression of the normalized flux expressed in the object plane and issued from the source at the abscissa a:

$$I(X_0) = \frac{D}{e(D-d)}\left[-X_0 + \frac{1}{2D}\left[2ad - bD + e(D-d)\right]\right]$$

$$\text{for } A_1 \le X_0 \le A_2$$

$$I(X_0) = 0$$

$$\text{for } A_2 \le X_0 \le A_3$$

$$I(X_0) = \frac{D}{e(D-d)}\left[X_0 + \frac{1}{2D}\left[-2ad - bD + e(D-d)\right]\right]$$

$$\text{for } A_3 \le X_0 \le A_4$$

and

$$I(X_0) = 1 \quad elsewhere$$

The same expressions can be established for the source at the abscissa $-a$. The coefficients A'_1, A'_2, A'_3 and A'_4 are deduced from A_1, A_2, A_3, and A_4 by changing a into $-a$. We thus obtain the curves presented in figure 4.

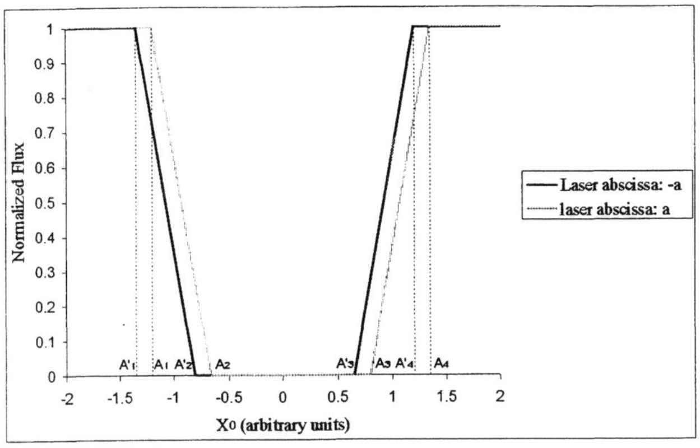

Figure 4. Theoretical profile for a rectangular object.

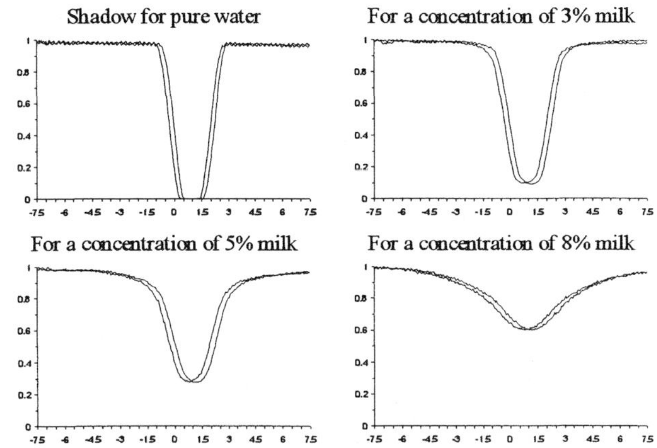

Figure 5. Experimental results for a rectangular bar in the middle of the medium, with $D = 30$ mm, and $e = 1$ mm. The curves represent the normalized flux as a function of X_0 and give $d \approx 15$ mm.

We measured the following parameters:

$$\alpha = \overline{A_1A_2} = \overline{A'_1A'_2} = \overline{A_3A_4} = \overline{A'_3A'_4} = e\left(1 - \frac{d}{D}\right)$$

$$\beta = \overline{A_2A_3} = \overline{A'_2A'_3} = b - e\left(1 - \frac{d}{D}\right)$$

$$\gamma = \overline{A'_1A_1} = \overline{A'_2A_2} = \overline{A'_3A_3} = \overline{A'_4A_4} = 2a\frac{d}{D}$$

Then:

$$b = \overline{A_1A_3} = \alpha + \beta$$

gave us the size of the object, and:

$$d = \frac{\gamma}{2a}D$$

the distance between the object and the detector as a function of the distance D.

Experimental results

The results obtained with two sources set along the same line of the array, *e.g.* lasers 1 and 4, are illustrated in figure 5 for a rectangular bar, and in figure 6 for a glass cylindrical object. The 2 lasers were spaced of 375 μm. The distance between the curves gave us the distance d with a good approximation.

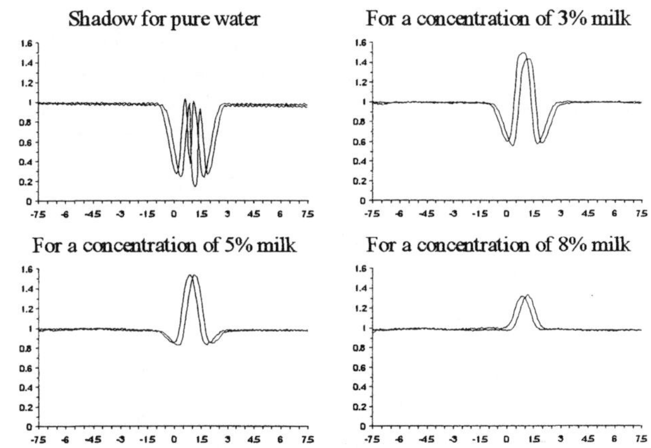

Figure 6. Experimental results for a polished glass cylindrical object under the same conditions as above.

Figure 6 shows the profile changes due to the transmission through the cylindrical glass which acts as a lens. The same results, with similar profiles, were obtained for a scattering cylindrical object. This case will be detailed in the polarimetric analysis experiment.

Polarimetric characterization

As the localization of an object in a scattering medium remains a difficult task, the knowledge of how beams are transmitted through the medium is interesting and studying the alterations in the polarization of incident beams can thus be a good solution to solve this problem. The different sources constituting the array are polarized in the same direction along one of array axes. Knowing the polarization state behind the system enabled us to determine interesting properties of the considered object [5]. It is why we perfected the system described in figure 7.

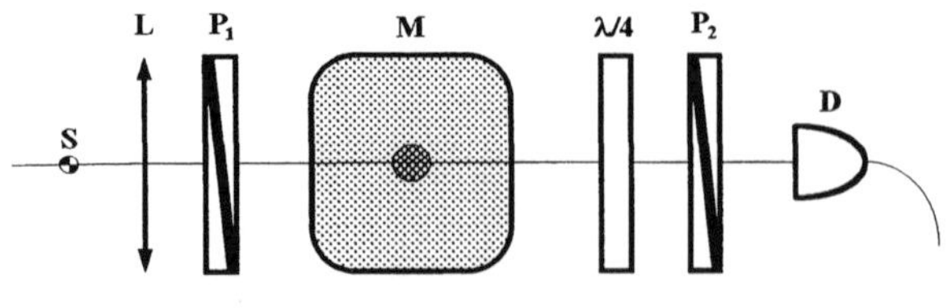

Figure 7. Schematic view of the polarization analyzer.

We used a VCSEL as the source *S*, a lens *L* to parallelize the beam, and a polarizer P_1 to set the initial polarization. The polarizer P_2 and the quarter-wave plate constituted the analyzer system. The detector *D* was a photomultiplier. Polarization was set before beams entered the medium. The total intensity reaching the detector can be written as a function of the angles between the reference axis and one of the quarter-wave plate axis. We determined the Stokes vector behind the system from four different positions. We thus exhibited the way an object in the medium modified the polarization state.

This configuration is quite interesting for objects that are not totally opaque. We studied the case of a scattering cylindrical object with 2 initial polarization states, the first one being in the direction of the object axis, and the second one in a 45° direction. The Stokes parameters S_0, S_1, S_2, S_3, the polarization degree D_P, and the Azimuth of the polarization ellipse θ are represented in figures 8 and 9. The concentration of milk in the mixture was 8%.

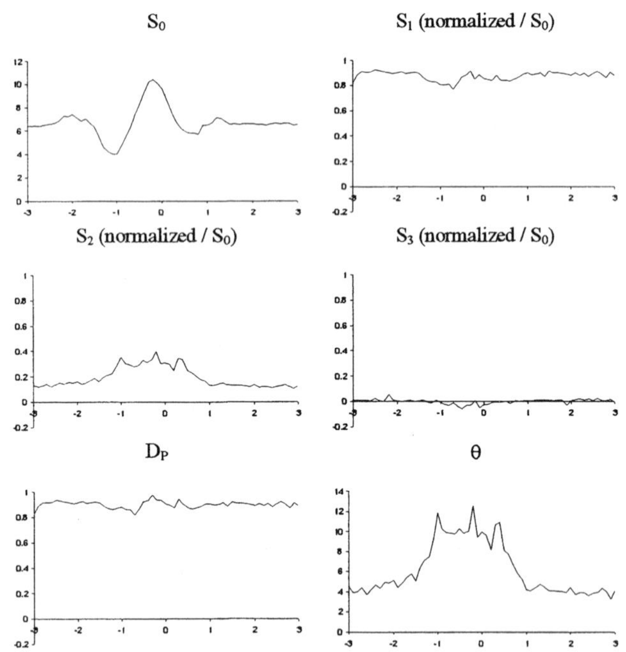

Figure 8. Parameters for an initial polarization state parallel to the object symmetry axis.

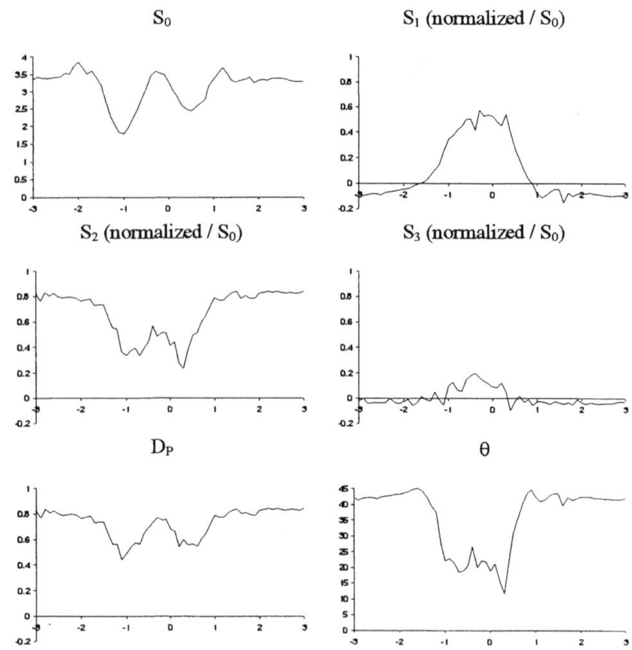

Figure 9. Parameters for an initial polarization state at 45° of the object symmetry axis.

These figures evidence the presence of one object with its depolarization properties. Therefore, a scattering object in a scattering medium entails depolarization and polarization transformation. The 45° polarization permits here to detect the dichroïsm phenomenon produced by the vertical structure of the object.

Conclusion

In summary, we experimentally showed the possibility of localizing an object with an array of vertical cavity surface emitting lasers. Its use was justified by the fact that object shapes were not necessarily regular. Its configuration makes possible the localization of the object in a scattering medium. Moreover, polarization provides one with a better spatial definition of this object and information on its depolarization properties. To characterize more accurately the object, another quarter-wave plate can be placed before the medium. The use of the Stokes-Mueller formalism gives thus the Mueller matrix corresponding to the studied object [6].

References

1. P.N. den Outer, T. M. Nieuwenhuizen, "Location of objects in multiple-scattering media," J. Opt. Soc. Am. A 10 1209-1218 (1993).
2. S. B. Colak, D. G. Papaioannou, G. W. 't Hooft, M. B. van der Mark, H. Schomberg, J. C. J. Paasschens, J. B. M. Melissen, N. A. A. J. van Asten, "Tomographic image reconstruction from optical

projections in light-diffusing media," Appl. Opt. 36 180-213 (1997).

3. M. A. O'Leary, "Imaging with diffuse photon density waves," Ph.D. dissertation (U. Pennsylvania, 1996).
4. A. Knüttel, J. M. Schmitt, R. Barnes, J. R. Knutson, "Spatial localization using interfering photon-density waves: Contrast enhancement and limitations," SPIE 1888 322-333 (1993).
5. S. P. Morgan, M. P. Khong, M. G. Somekh, "Effects of polarization state and scatterer concentration on optical imaging through scattering media," Appl. Opt. 36 1560-1565 (1997).
6. P. Eliès, B. Le Jeune, J.-P. Marie, J. Cariou, J. Lotrian, "Surface rugosity and polarimetric analysis," Proceedings of. SPIE 2782 719-729 (1996).

An investigation of the effects of layered tissue on NIR imaging of deep heterogeneous structure

Xuefeng Cheng*, Miljan R. Stankovic**, Phillip G. Stubblefield**, David Boas*
*Electro-Optics Technology Center, Tufts University, 4 Colby St., Medford, MA 02155
** Boston University School of Medicine, 1BMC Place, Boston MA 02118

Introduction

Near-infrared spectroscopy/imaging has been used for non-invasive monitoring of changes in hemoglobin oxygenation and volume in the human brain caused by a variety of physiological and pathophysiological conditions. The distribution of photons inside the brain determines the volume imaged and the spatial resolution. Due to the layered structure of the head including the scalp, the skull, and the cerebrospinal fluid (CSF) surrounding the brain tissue, the behavior of photon migration is very different from that in a semi-infinite homogeneous model. In this study, we began to examine the effect of these layers on the sensitivity that near infrared spectroscopy offers for detecting and imaging absorption changes caused by variations in hemoglobin concentration. We used phantoms with three layers to stimulate the structure of the head. The third layer (representing the brain) contains absorbers to simulate localized changes in total hemoglobin concentration. By using our portable CW and phase modulated instruments we imaged deep heterogeneous structures and systematically studied the effects of the layered structures on the amplitude of the back reflected light, the average path length, and the penetration depth of light. Here, we mainly discuss the effects of the CSF.

System description

The block diagram of our CW imager and phase-modulated system are shown in Fig.1a and 1b. For the CW system, laser diodes are used as sources and are modulated at 1 kHz to enable lock-in detection to reduce electronic noise and reject room lights. All the sources and detectors are mounted in a central control box and controlled by a computer. Source light is coupled into the fibers and delivered to the sample. The back reflected light is collected by fibers and delivered to the detectors. In this way we can easily configure multiple source and detector arrays by arranging the fiber ends on our probe. Fig. 1a shows our nine-by-four source-detector array probe. Our phase-modulated system is shown in Fig. 1b. The ultimate system will be able to scan the modulation frequency from 50 MHz to 800 MHz. Fig.2 shows our three-layer model of the adult head. The phantoms were made of a water solution of titanium dioxide powder and carbon powder, used as scatters and absorbers respectively. The CSF layer is a piece of transparent glass with a thickness of 2 mm. For the scalp, scull and brain tissue, we used a reduced scattering coefficient of 15 cm^{-1}, and absorption coefficient of 0.1 cm^{-1}. The thickness of the scalp and skull is 15 mm and that of the gray and white matter is taken as infinite.

Figure1a CW system

Figure1b Phase modulated system

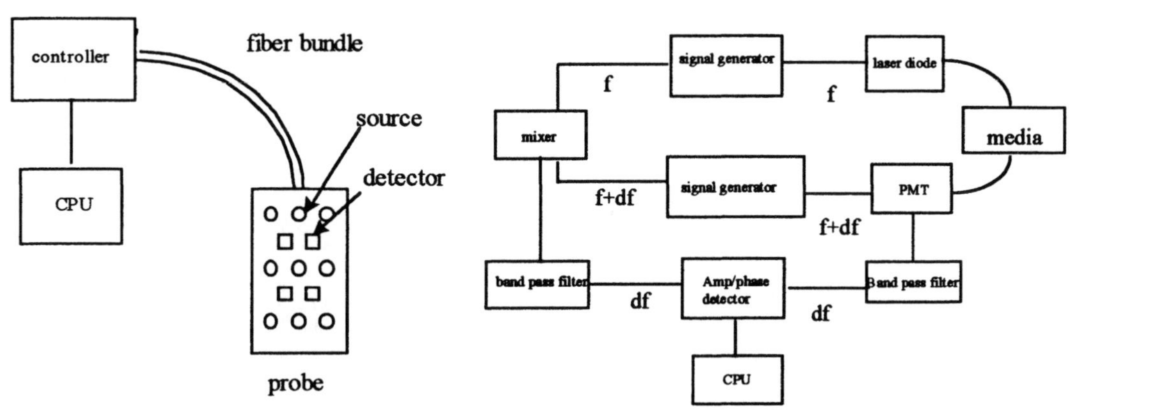

Experiment and Discussions

Fig. 3a shows the effect of the CSF layer based on the back reflectance measurement on a semi-infinite model. We see that the transparent CSF layer reduces the effective attenuation coefficient of the media for source-detector separations larger than the depth of the CSF layer. This is most likely a result of the CSF guiding light away from the source. When the CSF layer is deeper, this effect is reduced. Fig. 3b shows the results of the experiment depicted in fig .2, where the source-detector separation is fixed and the depth of the absorber is changed. We observed that with the CSF layer, there is a decreased sensitivity to the depth of the absorber. Figure 4 shows the effect of the CSF layer on imaging objects with different sizes. Two highly absorbing objects with diameters of 1.5 cm and 0.8 cm were buried in the medium at a depth of 3 cm. The probe shown in Fig. 1a was used to image the absorbers. We can see that the image of the smaller ball fades away with the presence of the CSF layer. The intensity plot along the line connecting the centers of the balls again shows that the CSF layer has the effect of reducing the sensitivity of the back reflectance to the absorbers.

Fig2. set-up for measurement local refelctance Vs depth of absorber

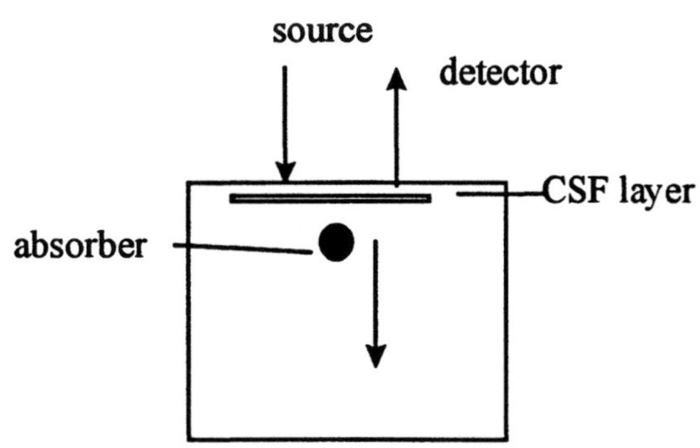

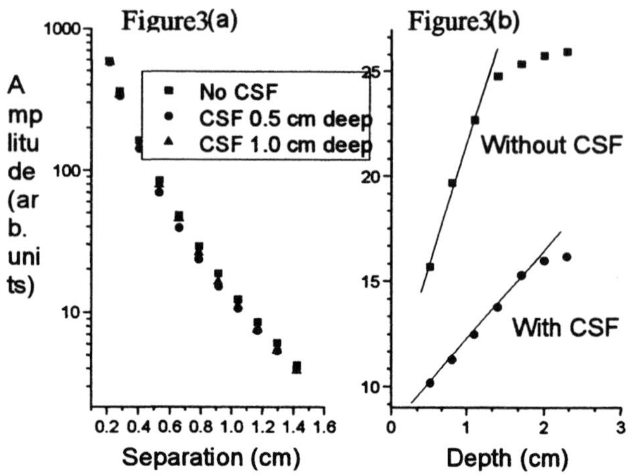

Figrue4. Imaging the absorbers with and without CSF

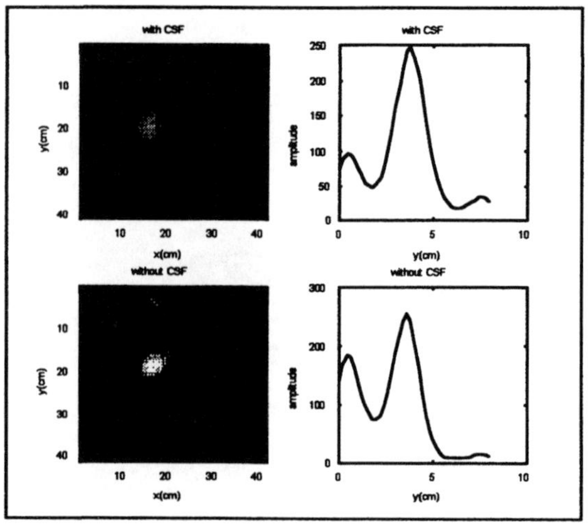

Conclusion

In this study the effects of the CSF layer of the adult head on NIR image has been investigated with a CW imager. The transparent CSF highly affects the photon transportation inside the head by acting like a wave-guide. This effect reduces the penetration depth of the light, thus decreasing the sensitivity to localized changes in cerebral hemoglobin oxygenation and volume. Note that the index mismatching of our phantom CSF layer and the surrounding layers may have amplified this result. An investigation of the index mismatch will be presented. Furthermore, a more systematic study of these effects on imaging will be presented. In addition, a phase-modulated system will be used to study the path length changes due to the multiple layered structure of the head.

References

1. C.C.Piantadosi, T.M.Hemstreet and F.F. Jobsis, "Near infrared spectrophotometric monitoring of oxygen distribution to intact brain and skeletal muscle tissue, "Critical Care Med. **14**:698-706, 1986.

2.W.F.Cheong, S.A.Prahl and A.J. Welch," A review of the optical properties of biological tissues," IEEE J. Quantum Electron., **26**:2166-2158, 1990.

3.W.Dirbank, M.Schweifer and D.Tdelpy, "Investigation of 'light piping' through clear regions of scatter objects," Proc.SPIE **2389**:167-173, 1995.

4.W.Cui, L.E.Ostrander, "The relationship of surface reflectance measurements to optical properties of layered biological media,"IEEE Trans. Biomed. Eng. **39**: 194-201, 1992.

Solution of the Perturbation Equation in Optical Tomography Using Weight Functions as a Transform Basis

Erh-Ya Lin*, Yao Wang*, Yaling Pei*, and Randall L. Barbour#
* *Polytechnic University, Brooklyn, NY 11201 and* #SUNY Health Science Center, Brooklyn, NY 11203

Abstract

This paper describes a new inverse solver for optical tomography. As with prior studies, we employ an iterative perturbation approach, which at each iteration requires the solution of a forward problem and an inverse problem. The inverse problem involves the solution of a linear perturbation equation, which is often severely underdetermined. To overcome this problem, we propose to represent the unknown image of optical properties by a set of linearly independent basis functions, with the number of basis functions being equal to or less than the number of independent detector readings. The accuracy of the solution depends on the choice of the basis. We have explored the use of the weight functions associated with different source and detector pairs (i.e. the rows in the weight matrix of the perturbation equation) as the basis functions. By choosing those source and detector pairs which have uncorrelated weight functions, the inverse problem is transformed into a well-posed, uniquely determined problem. The system matrix in the transformed representation has a dimension significantly smaller than the original matrix, so that it is feasible to perform the inversion using singular value decomposition (SVD). This new method has been integrated with a previously reported forward solver, and applied to data generated from numerical simulations using diffusion approximation. Compared to the Conjugate Gradient Descent (CGD) method used in previously reported studies, the new method takes substantially less computation time, while providing equal, if not better, image reconstruction quality at similar noise levels.

Keywords (100.3010) Image reconstruction techniques; (100.3080) Infra-red imaging.

Introduction

The imaging problem in optical tomography deals with the reconstruction of the absorption and the scattering coefficients of a heterogeneous scattering medium from the measurements of multiple scattered light signals on the medium surface. The problem is inherently difficult because the scattered field is non-linearly related to the optical properties of the underlying medium. To overcome this difficulty, we have developed an iterative perturbation approach [1,2]. Each iteration involves the solution of a forward problem to set up a new perturbation equation, and the solution of the perturbation equation, which constitutes the inverse problem. This approach is equivalent to the Gauss-Newton method (also known as Newton method). When Tikhonov regularization is employed in the least squares solution, it leads to the Levenberg-Marquardt method. The same overall approach has been taken by research groups of Arridge [3] and Paulsen [4,5], although our approaches to solving the forward and inverse problems differ from theirs.

One challenging problem in solving the perturbation equation,

$$\mathbf{Wx}=\mathbf{y}, \tag{1}$$

is that the affordable number of source and detector (SD) pairs, M, is usually substantially smaller than the number of unknowns, N. That is, the system is *severely underdetermined.* Among many possible solutions, we are interested in obtaining the *minimum norm solution*, which is equivalent to finding the minimal perturbation from the previous reconstruction to satisfy the measurement data. Theoretically, one can obtain this solution using singular value decomposition (SVD) of the weight matrix **W** [6]. However, the very large dimension of **W** can make the computation of SVD unfeasible in terms of both computation time and memory requirement. To overcome this problem, we have adopted the Conjugate Gradient Descent (CGD) method in our prior studies [1], which obtains the solution by minimizing the error norm $\|\mathbf{Wx}-\mathbf{y}\|^2$ through a conjugate gradient descent scheme. A problem with this approach is that the resulting solution depends on the initial solution. There is no easy way to guarantee the solution is minimum norm. In our previous studies, the initial solution is set to zero, which usually gives good solutions.

Paulsen, *et al.* uses Gauss elimination to solve the normal equation associated with the perturbation equation. The problem with this approach is that it gives a meaningful (least squares) solution only when the system is overdetermined. When the problem is severely underdetermined, the solution can be far from the desirable one. For this reason, in their earlier work, a coarse mesh is used to represent the target medium so that the number of

OSA TOPS Vol. 21 Advances in Optical Imaging and Photon Migration, James G. Fujimoto and Michael S. Patterson (eds.)

elements is roughly the same as the number of SD pairs. However, using a coarse mesh hinders the accuracy of the forward solver, and consequently the accuracy of the reconstructed image. Sufficiently accurate results can be obtained only when the object size is large relative to the medium size [4]. In order to overcome the underdetermined problem, they proposed a dual mesh technique that uses a fine mesh for the forward solution, but a coarse mesh for the inverse solution. Very good reconstruction results have been obtained from experimental data for a test medium containing a single target [5].

In this paper, we propose a new method for solving the perturbation equation which is significantly faster than the CGD method and the direct SVD method, when the number of SD pairs is an order of magnitude smaller than the number of unknowns.

The WTSVD Method

As described in Introduction, the perturbation equation given in Eq. (1) is usually severely underdetermined. In addition, depending on the SD configurations, many rows in the weight matrix (the weight functions) may not be linearly independent, which causes the weight matrix to be rank deficient. Rather than solving this ill-posed linear equation directly, we first perform a preprocessing on the original perturbation equation to delete rows that are known to be correlated *a priori*, we then apply a linear transformation to the unknowns before using SVD to solve the transformed equation.

Preprocessing of the Perturbation Equation

In general, with an arbitrary SD configuation, one needs to apply some sophisticated operation to derive independent rows in the weight matrix. However, in the most common SD configuration, in which the sources and detectors share the same set of locations, there is a simpler way. Because of the symmetry of $\mathbf{r}_s$ and $\mathbf{r}_d$ in the integral formula for calculating the weight matrix, exchanging source and detector positions gives us the same weight functions, and therefore the rows in the weight matrix that corresponding to these two pairs of SD are correlated. Instead of solving the original equation, we replace every two rows in the augmented matrix $[\mathbf{W}\ \mathbf{y}]$ that correspond to the two SD pairs with exchanged positions by their average. The SD pairs that do not have counterparts are those where the source and detectors are co-located. Because the detector readings at exactly the source locations are not very informative, we choose not to use these SD pairs. It is easy to show that the number of rows after such a reduction is $K=L(L-1)/2$, where L represents the number of SD locations. To make the solution of the equation more stable numerically, we further normalize the row vectors in the resulting weight matrix. We denote the resulting equation as

$$\tilde{\mathbf{W}}\mathbf{x}=\tilde{\mathbf{y}} \qquad (2)$$

Transformed Representation Using Weight Functions

With the transformed representation, the unknown $\mathbf{x}$ is spanned by the row vectors $\tilde{\mathbf{w}}_k$ in the weight matrix as

$$\mathbf{x}=\sum_k t_k \tilde{\mathbf{w}}_k = \tilde{\mathbf{W}}^T \mathbf{t} \qquad (3)$$

where $\mathbf{t}=[t_1, t_2, \ldots, t_K]^T$. Substituting Eq. (3) into Eq. (2) leads to

$$\mathbf{At}=\tilde{\mathbf{y}} \quad \text{with} \quad \mathbf{A}=\tilde{\mathbf{W}}\tilde{\mathbf{W}}^T. \qquad (4)$$

Once $\mathbf{t}$ is determined by solving the above transformed equation, one can derive the desired solution by using Eq. (3), which can be considered as the inverse transform. The overall solution can be represented as

$$\mathbf{x}=\tilde{\mathbf{W}}^T(\tilde{\mathbf{W}}\hat{\mathbf{W}}^T)^{-1}\tilde{\mathbf{y}}. \qquad (5)$$

With the weight functions as the basis functions, as given in Eq. (3), the contribution of each SD pair to the final solution is in the form of the weight function associated with it. *Since the weight functions tell us where we can see the object more clearly given a source and detector pair, by using this transformation basis, we only try to solve the part of the object that can be seen by the source and detector pairs used, thus reducing the ambiguity of the inverse problem.* The reconstruction of the original unknowns $\mathbf{x}$ from the transformed vector $\mathbf{t}$, i.e., the inverse transform in Eq. (3), resembles the well known back-projection process. It is easy to recognize that the final solution in Eq. (5) is in fact the minimum-norm solution for an underdetermined system with rank equal to the number of rows. Note that the *ij*-th element in the transformed matrix $\mathbf{A}$ is $a_{ij}=(\tilde{\mathbf{w}}_i, \tilde{\mathbf{w}}_j)$, the inner product of the normalized weight functions of two SD pairs. This ensures that the diagonal elements in $\mathbf{A}$ is equal to or larger than the off-diagonal elements. Although we still need to solve a linear equation, the properties of $\mathbf{A}$ are very different from that of $\tilde{\mathbf{W}}$. A is a square symmetric matrix with a dimension equal to the number of uncorrelated SD pairs, and it is diagonally dominant in that all diagonal elements are one and off-diagonal terms are less than one in magnitude. The decay of the off-diagonal elements, however, is not very rapid.

Solution of the Transformed Equation

Theoretically, as long as the K normalized weight functions are independent, then the square matrix $\mathbf{A}$ is guaranteed to be non-singular, and the transformed equation (4) can be solved reliably using the simplest inversion method --- the Gauss elimination method. In practice, however, some rows in the weight matrix could still be correlated, and the Gauss elimination method may lead to unstable solutions. In order to circumvent this problem, we solve the transformed equation using the SVD method with Tikhonov regularization. This is feasible because the dimension of $\mathbf{A}$, K, is usually on the order of hundreds, for a practically feasible number of SD pairs. For example, if $L=18$, than

$K=153$. The incorporation of regularization is to suppress the noise effect. Let α_k represent the singular value of **A**, arranged in a descending order, and $\mathbf{u}_k$ and $\mathbf{v}_k$ the corresponding left and right singular vectors, the regularized SVD solution of Eq. (4) is

$$\mathbf{t}=\sum_k \left(\frac{\alpha_k}{\alpha_k^2+\lambda^2}\right)(\mathbf{u}_k,\tilde{\mathbf{y}})\,\mathbf{v}_k \quad . \tag{6}$$

The regularization parameter λ is chosen based on the Miller's L-curve criterion [7].

Simulation Results

The proposed method has been used to accomplish simultaneous reconstruction of absorption and diffusion coefficients from numerically calculated data for several test media. In all the simulations, a frequency modulated source is employed with a modulation frequency of 200 MHz. The detector measures the photon density on the surface, which is a complex variable. All the test media have a circular boundary with a radius of 4 cm. The medium is represented with an FEM mesh with 1800 nodes so that the number of unknowns is N=3600 (2 unknowns for each node). This spatial resolution is used for both the forward and inverse calculations. The sources and detectors are located uniformly on the circular boundary sharing the same set of locations. Two SD configurations are tested, with 10x10 ($L=10,K=45$) and 18x18 ($L=18,K=153$) SD pairs, respectively. The forward calculation is based on the diffusion approximation using an FEM method, using the Dirchlet boundary condition [8]. For the inverse problem, we compared the WTSVD method and the CGD method. With the CGD method, the original perturbation equation is solved directly. The number of iterations at each perturbation step varies but is limited to at most 1000 iterations. With both methods, the perturbation iteration stops when the ratio of the energy (sum of squares) of the current perturbation solution to the energy of the total perturbation until this iteration is less than 10^{-3}, or when the number of perturbations reaches 10. In the following, we describe the reconstruction results for two test media. The properties of these test media are summarized in Table 1. In either case, the initial estimate is set to the same as the background medium.

Test Medium I Figure 1 shows the reconstructed images under different test conditions. In the noise free case, the results obtained using WTSVD and CGD are very similar when 10x10 SD pairs are used, with the reconstructed objects having a size noticeably larger than the true object size. With 18x18 SD pairs, the shape of the reconstructed object by the WTSVD method is very close to the true object, better than the CGD result. In the noise-added case (10% or 20dB), the reconstructed object by WTSVD is blurred compared to the noise-free case, but the object structure is clearly visible. The CGD method has produced a similar reconstruction result. Figure 2 plots the root mean square error (RMSE) between the reconstructed $\Delta\mu_a$ and ΔD and the true $\Delta\mu_a$ and ΔD. This plot clearly shows that the WTSVD method converges faster than the CGD method.

Test Medium II Figure 3 shows reconstruction results under different test conditions, all obtained with 18x18 SD pairs. The reconstructed ΔD images have a severe ringing artifact initially, which gradually goes away with more iterations. It can be seen that images obtained with WTSVD is much sharper than with CGD. With both WTSVD and CGD method, the true $\Delta\mu_a$ values are underestimated, whereas the recovered ΔD values are quantitatively accurate. With noise-added data (3% or 30 dB), the object is blurred, but is reconstructed at the right location. At higher noise levels, reconstructed images after first one or two iteration usually were correct in terms of location and contrast of the objects, but the reconstruction results after subsequent iterations become unacceptable. We have found that the reconstruction of a medium with two objects is more sensitive to noise than a medium with a single object. Also, reconstruction from 18x18 SD set-up is more sensitive to noise than from 10x10 SD set-up. Figure 4 compares the RMSEs of the reconstructed $\Delta\mu_a$ and ΔD after different iterations, using the WTSVD and CGD methods, respectively. In this case, the WTSVD method yielded lower RMSE for both $\Delta\mu_a$ and ΔD after one iteration, than CGD. But upon convergence, the WTSVD has a higher RMSE for $\Delta\mu_a$, but lower RMSE for ΔD. Note that with the CGD method, the RMSE for $\Delta\mu_a$ continues to decrease and eventually may become lower than that of WTSVD, but the RMSE for ΔD either stays flat or even increase with more iterations. Visually, the reconstructed ΔD image by the CGD method tends to become noisier on the boundary as the number of perturbations increases. The rational behind this also needs further investigation.

Comparison of Computational Complexity The results presented before showed that the WTSVD method and the CGD method achieves about the same quality, with the WTSVD method giving slightly sharper images with fewer boundary artifacts. The more noteworthy advantage of the WTSVD method is its reduced computation time. In our study, all the computations were done using a SUN Ultra SPARCstation with 290 Mbytes of memory. The forward part was implemented using Fortran. The CGD method was integrated with the forward code in a single Fortran code, while the WTSVD method was implemented using Matlab 4.2 platform. Table 2 summarizes the required computation time for different parts of the reconstruction algorithm for 10x10 and 18x18 SD cases. It is clear that the WTSVD method requires significantly shorter time in each iteration. For example, with 18x18 SD pairs, the computation time for each iteration (including both forward and inverse solutions) is 13 minute with the WTSVD, while the CGD method requires 60 minutes. As described previously, the WTSVD method takes a smaller number of iterations to converge than the CGD method in the simulation results. Roughly, five iterations are sufficient with the WTSVD method, whereas ten or more iterations are needed for the CGD method. From Table 2, in the 18x18 SD case, 5 iterations using WTSVD will take 65 minutes, whereas 10 iterations using CGD would take 600 minutes, a saving factor of 10. Note that the computation time required by the WTSVD method is likely to be reduced further once it is implemented outside the Matlab environment.

Discussion

The proposed WTSVD algorithm has been shown to require significantly shorter computation time than the previously reported CGD method, while providing similar reconstruction quality. It not only requires less time to solve the perturbation equation in each iteration, but also requires fewer iterations. The total saving factor is on the order of 10. Note that the computational advantage of the WTSVD method decreases when the number of SD pairs increases. Specifically, using an analysis of the flops required by different methods, we have found that the CGD algorithm will become more efficient, when the number of SD pairs exceeds 68x68.

The WTSVD method represents the unknown image using a basis consisting of normalized, independent weight functions. There may be other bases which can represent the unknown image better. Although using the weight functions enables one to attempt to recover the part of the image that can be "seen" by the SD pairs, they are not good basis functions in terms of their spatial-frequency support regions. Fourier or Wavelet basis functions may be more efficient in that with the same number of basis functions, these bases may represent an image more accurately. As a future research, we plan to investigate how to adapt standard wavelet bases to fit with node distributions in the finite element mesh representation of an arbitrarily shaped medium and explore the potential of using such bases for representing the unknown image.

The focus of this work is to examine the effectiveness of the WTSVD algorithm in solving a given perturbation equation and how does it compare with the CGD method in terms of accuracy, robustness, and computational complexity. Therefore, we have only tested for a fixed type of source and detector configuration and for the case when the initial estimate is the same as the background medium. When all such conditions are equal, we have indeed show that the WTSVD method can provide significant computation savings over the CGD method. It is not our intention to examine the capability of the overall perturbation approach in terms of minimally detectable target contrast and size, and the influence of the SD configurations on the reconstruction performance. Nor did we try to investigate the effect of the mismatch between the actual background medium and the assumed one on the convergence behavior of the overall reconstruction scheme. These issues will depend not only on the inverse algorithm, but also the forward algorithm. A thorough evaluation in this direction requires further investigation.

Acknowledgments

This work is supported in part by the NIH under grant 1-RO1-CA66184-01A2.

References

1. R. L. Barbour, H. L. Graber, Y. Wang, J.-H. Chang, and R. Aronson, "A Perturbation Approach for Optical Diffusion Tomography Using Continuous--Wave and Time--Resolved Data," In *Medical Optical Tomography --- Functional Imaging and Monitoring*, G. Muller ed., SPIE Inst. IS11, pp. 87--120, 1993.
2. Y. Yao, Y. Pei, Y. Wang, and R. L. Barbour, "A Born Type Iterative Method for Imaging of Heterogeneous Scattering Media and Its Application to Simulated Breast Tissue," in *Optical Tomography and Spectroscopy of Tissue: Theory, Instrumentation, Model, and Human Studies II*, Proc. SPIE, Vol. 2979, pp. 232-240, Feb. 1997.
3. S. R. Arridge, "The forward and inverse problems in time resolved infra-red imaging", In *Medical Optical Tomography --- Functional Imaging and Monitoring*, G. Muller ed., SPIE Inst. IS11, pp. 35-64, 1993.
4. H. Jiang, K. D. Paulsen, and U. L. Osterberg, "Optical image reconstruction using frequency-domain data: simulations and experiments," *J. Opt Soc. Am. A*, 13(2):253-266 (1996).
5. H. Jiang, K. D. Paulsen, U. L. Osterberg, and M. Patterson, "Frequency-domain optical image reconstruction in turbid media: an experimental study of single target detectability," *Applied Optics*, 36(1): 52-63 (1997).
6. H. Golub and C.F. Van Loan, *Matrix Computations*, Johns Hopkins Univ. Press. 2nd ed.,1991.
7. K. Miller, "Least squares methods for ill-posed problems with a prescribed bound," *SIAM J. Math. Anal.* 1: 52-74, (1970)
8. Y. Pei, Y. Yao, F.-B. Lin, and R. L. Barbour, "Frequency domain fluorescence optical imaging using a finite element method," in *Optical Tomography and Spectroscopy of Tissue: Theory, Instrumentation, Model, and Human Studies II*, Proc. SPIE, Vol. 2979, pp. 164-175, Feb. 1997.

Table 1 Optical Properties of Test Media

Test Cases	Background media			Target geometry		Target 1		Target 2	
	Radius	ua(cm^{-1})	us(cm^{-1})	Radius	Separation	ua(cm^{-1})	us(cm^{-1})	ua(cm^{-1})	us(cm^{-1})
Case I	4 cm	0.02	5	0.75 cm		0.04	10	N/A	N/A
Case II	4 cm	0.05	10	0.35 cm	1.5 cm	0.1	20	0.1	20

Table 2 Computer Time (in minutes) Required by Different Methods

Methods	One Iter.			5 Iter.	10 Iter.
	Forw.	Inv.	Total		
CGD10 (1000 iterations)	2.0	20.0	22	110	220
WTSVD10	2.5	1.5	4	20	40
CGD18 (1000 iterations)	6.0	54.0	60	300	600
WTSVD18	6.5	6.5	13	65	130

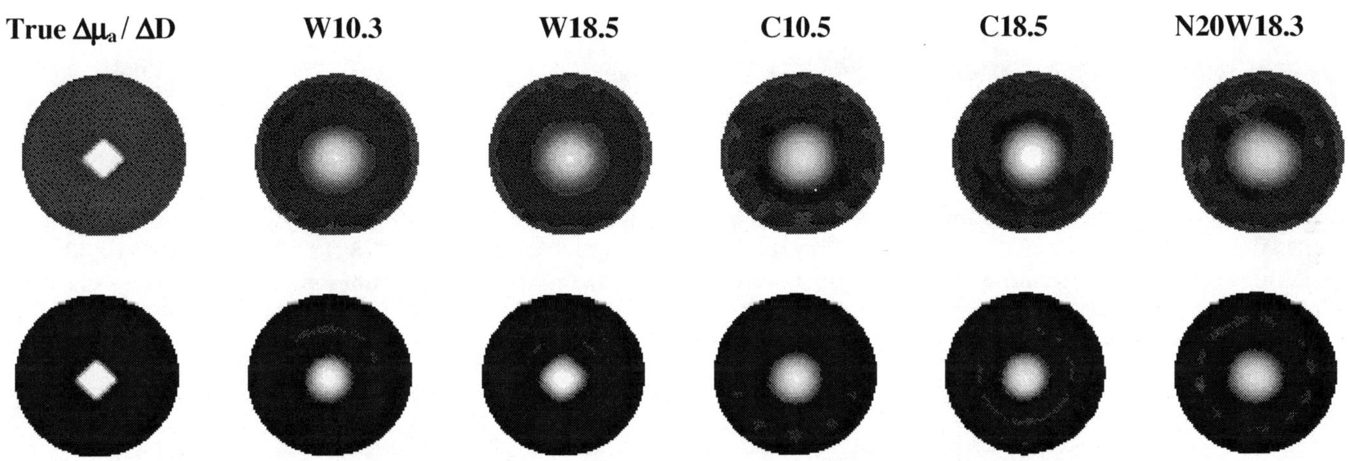

Figure 1 Reconstruction results for Test Medium I. Top row is for $\Delta\mu_a$ and bottom row for ΔD. From left to right are: 1) the true image; 2) and 3) reconstructed using WTSVD from noise-free data of 10x10 and 18x18 SD pairs, respectively; 4) and 5) reconstructed using CGD from noise-free data of 10x10 and 18x18 SD pairs, respectively; 6) reconstructed using WTSVD from noise-added data of 18x18 SD pairs, noise level is 10% (SNR=20dB). The last number in the label of each image indicates the number of iterations used.

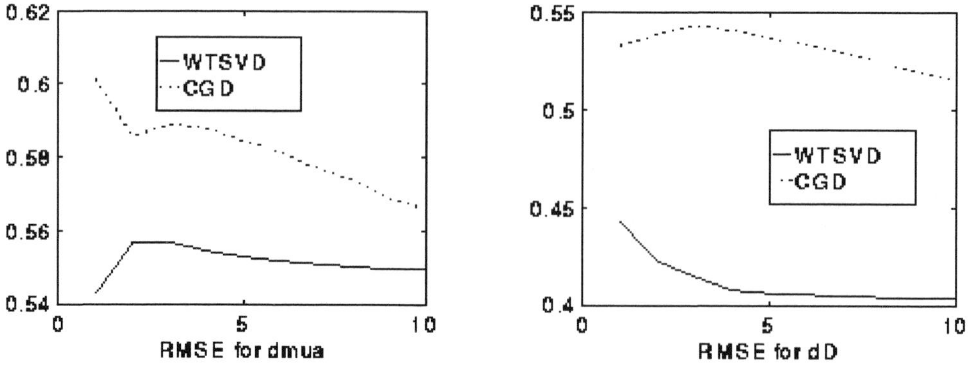

Figure 2 RMSEs of reconstructed images after successive iterations for Test Medium I.

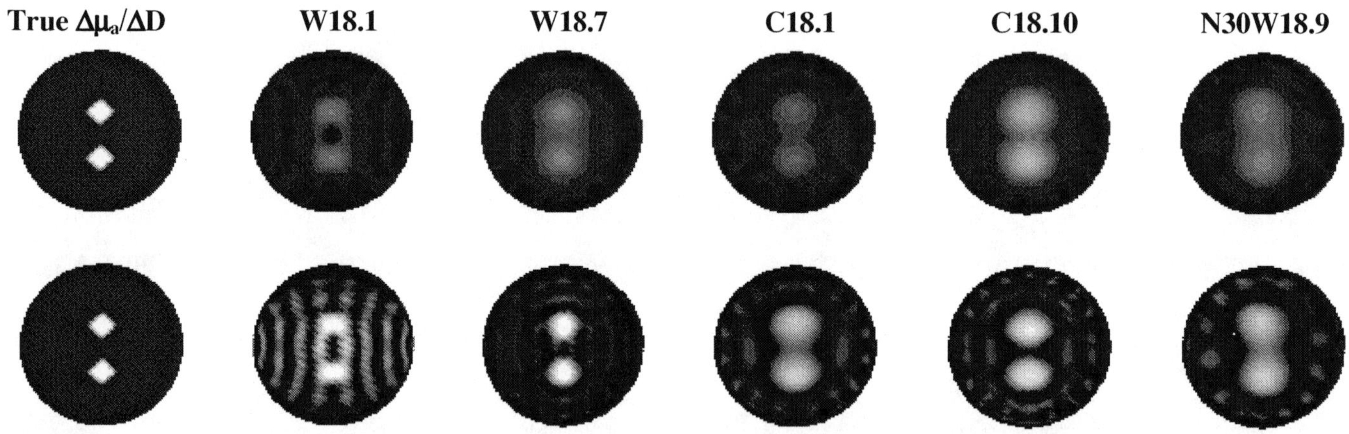

Figure 3 Reconstruction results for Test Medium II. Top row is for $\Delta\mu_a$ and bottom row for ΔD. From left to right are: 1) the true image; 2) and 3) reconstructed using WTSVD from noise-free data of 18x18 SD pairs, after 1 and 7 (reaching convergence) iterations, respectively; 4) and 5) reconstructed using CGD from noise-free data of 18x18 SD pairs, after 1 and 10 iterations respectively; 6) reconstructed using WTSVD from noise-added data of 18x18 SD pairs, noise level is 3% (SNR=30dB).

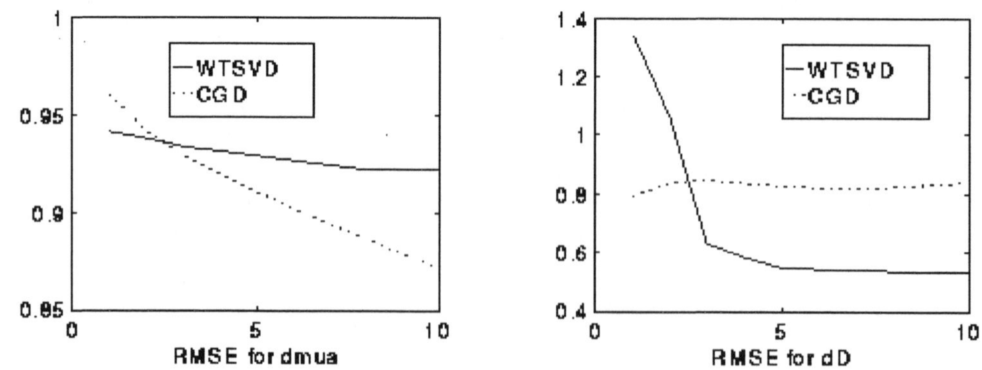

Figure 4 RMSEs of reconstructed images after successive iterations for Test Medium II.

Heuristic Determination of Spatial Resolution from Viewpoint of Data Fitting in Transmission Optical Computed Tomography Based on Coherent Detection Imaging

Tetsuya YUASA, Balasigamani DEVARAJ[1], Yuuki WATANABE[1], Takao AKATSUKA, and Humio INABA[2]

Yamagata University, 4-3-16 Jonan, Yonezawa, Yamagata 992, Japan.
[1]Biophotonics Information Laboratories, 2-2-1 Matsuei, Yamagata, Yamagata 990, Japan.
[2]Tohoku Institute of Technology, 35-1 Yagiyama-Kasumicho, Sendai, Miyagi 982, Japan.
Tel: 81-238-26-3323, Fax: 81-238-26-2082, E-Mail: yuasa@eat72.yz.yamagata-u.ac.jp

Abstract

A new evaluation method for analysis of the spatial resolution of the coherent detection imaging (CDI) system is proposed with a new formulation of the computed tomography based on the algebraic equation system. The spatial resolution is defined as the optimal square grid size in the sense of data-fitting in the algebraic reconstruction, and is determined with the minimum description length (MDL) principle devised by Rissanen. Our method can simultaneously consider the effects brought about by the numbers of projections and bins, the beam width, the observation noise, the truncated projections, and so on. Moreover, a method to generate the weight matrix faithfully representing data-acquisition processes by the Gaussian beams is proposed to consider the profile of the incident beams. We demonstrate the effectiveness with simulated data.

Key Words: *Optical imaging, Spatial resolution, Computed tomography, Coherent detection imaging, Minimum data length principle*

1. Introduction

Optical imaging in biomedicine is increasingly receiving attention as futuristic diagnostic modalities that would complement or provide an alternative to existing method such as X-ray CT, MRI (Magnetic Resonance Imaging), PET (Positron Emission Tomography) and ultrasonography. However, the complex and highly scattering nature of tissues presents a formidable problem for the detection of the emerging photons and subsequent image reconstructions based on existing knowledge. Modified versions of the well known filtered backprojection (FBP) [1] and algebraic reconstruction techniques (ART) [2] were used to reconstruct image using off-axis emerging photons.

Inaba and co-workers earlier proposed and demonstrated the coherent detection imaging (CDI) system based on optical heterodyne detection technique, that has the ability to discriminate and select the high directional forward scattered photons emerging from tissues [3]. The CDI method was used to obtain transillumination laser CT images of biological specimens *in vivo* and *in vitro* [3-5]. Image reconstruction in the CDI method is carried out by the FBP.

The pioneers of CT have studied the fundamental properties with respect to the FBP in detail, mainly, for the X-ray CT [6-8]. However, the results can not be straightforwardly applied to the optical CT, because optical radiation is so different from X-ray in the physical features. For example, coherence of the incident beam excessively decreases on the object surface. In addition, from a viewpoint of laser imaging, the beam size is not negligible compared to the object, and the beam profile is Gaussian, and so on. Therefore, the conventional analysis based on the rigid mathematics can not treat these effects.

In this paper, we propose an evaluation method for

analysis of the spatial resolution of the CDI-based imaging system. We define the spatial resolution as the optimal square grid in the CT reconstruction based on the algebraic equation system, used in early days of X-ray transmission CT [9]. The square is not spatially isotropic, but defining the square grid will be practically useful, as the reconstructed images are displayed as pixels arranged squarely on a monitor. Furthermore, the effects brought about by the beam profile, the observation noise, the limitted nimbers of projections and bins, the truncated projections can be easily incorporated into the matrix equation in algebraic reconstructions. We investigate the effectiveness of our method with simulation data.

2. Theory

We view how the CT reconstruction based on the algebraic equation system obtains solutions from an algebraic equation system representing measurement processes. On superimposing grids to the object as shown in Fig. 1, and let f_j ($1 \leq j \leq N$) and g_i ($1 \leq i \leq M$) be, respectively, the jth pixel value and the ith projection data, where N and M are the numbers of pixels and rays, respectively. In addition, n_i represents the error component when measuring the ith projection data. We obtain

$$g_i = \sum_{j=1}^{N} h_{ij} f_j + n_i \qquad (1 \leq i \leq M), \tag{1}$$

where h_{ij} is the line segment length when the ith ray traverses

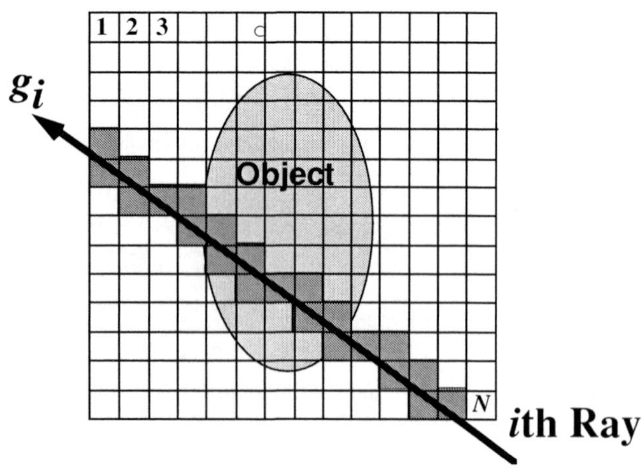

Fig. 1 Formalization of algebraic reconstruction

the jth pixel. Each equation of Eq. (1) is equivalent to discretized representation of the corresponding Radon transform. Representation of Eq. (1) by matrix and vectors yields

$$\boldsymbol{g} = \boldsymbol{H}\boldsymbol{f} + \boldsymbol{n}, \tag{2}$$

where

$$\boldsymbol{g} = \begin{pmatrix} g_1 \\ g_2 \\ \vdots \\ g_M \end{pmatrix}, \boldsymbol{H} = \begin{pmatrix} h_{11} & h_{12} & \cdots & h_{1N} \\ h_{21} & h_{22} & & \\ \vdots & & \ddots & \vdots \\ h_{M1} & & & h_{MN} \end{pmatrix}, \boldsymbol{f} = \begin{pmatrix} f_1 \\ f_2 \\ \vdots \\ f_N \end{pmatrix}, \boldsymbol{n} = \begin{pmatrix} n_1 \\ n_2 \\ \vdots \\ n_M \end{pmatrix} \tag{3}$$

are the projection vector, the weight matrix, the parameter vector to be estimated and the random vector, respectively. Accordingly, the algebraic CT reconstruction leads to solving the linear equation system (2) using the least squares method.

We have previously reported that selection of grids to be estimated is important in CT reconstruction from truncated noisy projections [10]. That is, the pixels to be estimated should be restricted to only the pixels covering the object. This operation will be effective also to the usual CT reconstruction based on the algebraic equation system, because the excessive variables which need not to be estimated should be removed for more precise estimation in the least squares problem. In addition, we can not only, thereby, do without unnecessary computer memories, but also save the computational cost [10].

Based on this fact, we reformalize the CT reconstruction based on the algebraic equation system. Let $f(x, y)$ and k be the cross-section of the object and the number of grids, *i.e.*, the pixels, required for covering the object, respectively, where x and y are the spatial coordinates (Fig. 2). The reconstructed image is represented as $\bar{f}(x, y)$. Here, we define the linear operator, $\boldsymbol{P}_i$, for the ith ray to acquire the projection data as

$$\boldsymbol{P}_i \bar{f}(x, y) = g_i \quad (1 \leq i \leq M), \tag{4}$$

where g_i is the projection data for the ith ray, and $\boldsymbol{P}_i$ is not

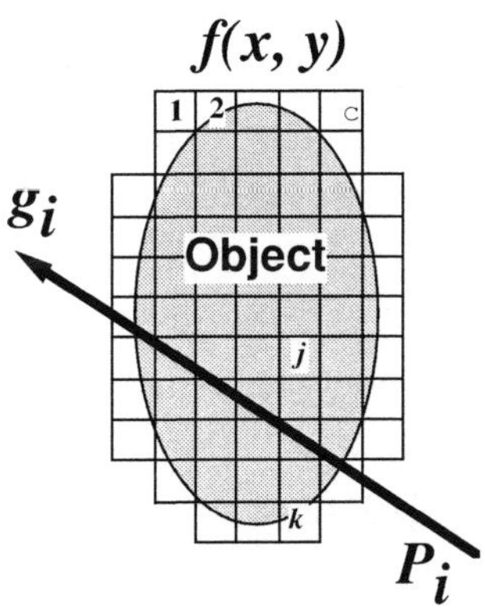

Fig. 2 Reformalization of algebraic reconstruction

always a line integral. We introduce the orthonormal bases

$$\phi_j(x, y) = \begin{cases} 1/A & (x, y) \in grid\ j \\ 0 & otherwise \end{cases}, \tag{5}$$

where A is the edge length of each grid. Therefore, $\bar{f}(x, y)$ can be represented as

$$\bar{f}(x, y) = \sum_{j=1}^{k} f_j\, \phi_j(x, y)\ . \tag{6}$$

Using $\boldsymbol{P}_i\, \phi_j(x, y) = h_{i\,j}$, we can obtain the equation system,

$$\boldsymbol{P}_i \bar{f}(x, y) = \boldsymbol{P}_i \left| \sum_{j=1}^{k} f_j\, \phi_j(x, y) \right| = \sum_{j=1}^{k} f_j\, h_{i\,j} = g_i \quad (1 \le i \le M)\ . \tag{7}$$

From the above equations, the undetermined parameters are the grid size of the orthonormal bases, *i.e.*, A, and the data-acquisition processes, *i.e.*, h_{ij}.

We determine the parameters, *i.e.*, A and h_{ij}, based on the minimum description length (MDL) devised by Rissanen [11]. It is represented as

$$MDL = \frac{1}{2\,\sigma^2} \left\| \boldsymbol{g} - \boldsymbol{H}\,\hat{\boldsymbol{f}} \right\|^2 + \frac{k}{2} \log M, \tag{8}$$

when the observation noise obeys Gaussian distribution, whose mean and standard deviation are 0 and σ, respectively. The observed data vector, the weight matrix, and the least squares estimator of the pixel values are $\boldsymbol{g}$, $\boldsymbol{H}$, and $\hat{\boldsymbol{f}}$, respectively [10]. MDL is the amount of information required to describe the observed data according to a model, and the model taking the minimum MDL value must be selected as the optimal model from among prepared models.

3. Simulation Results and Discussion

3.1 Spatial Resolution by Line Beams

First, we investigate if the proposed spatial resolution reflects its intuitive understanding, using simulated projections obtained by line integrals. A Shepp-Logan's numerical phantom was used (Fig. 3) [12]. An edge length of the image in Fig. 3 was set to 2.0. Each projection was analytically generated from the phantom. We selected the optimal model by substituting the least squares estimator $\hat{f}$ obtained from Eq. (7) using the conjugate gradient method (CGM) into Eq. (8). The translational step, L, of the incident ray was 2.0/64, and the size of grid was increased by an incremental step of 1/11 L. Fig. 4 shows the behavior of MDL value against the grid

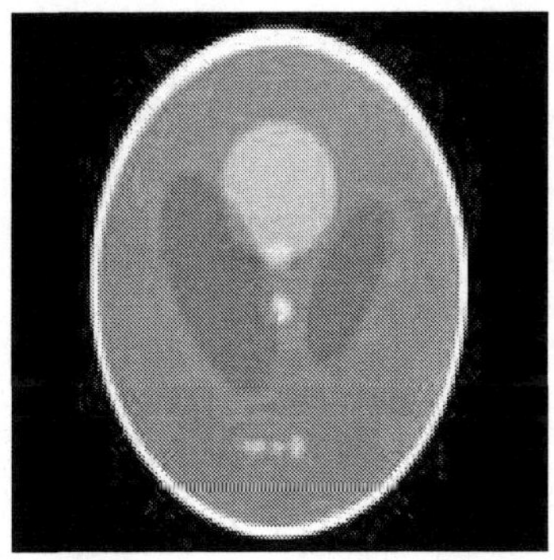

Fig. 3 Numerical phantom

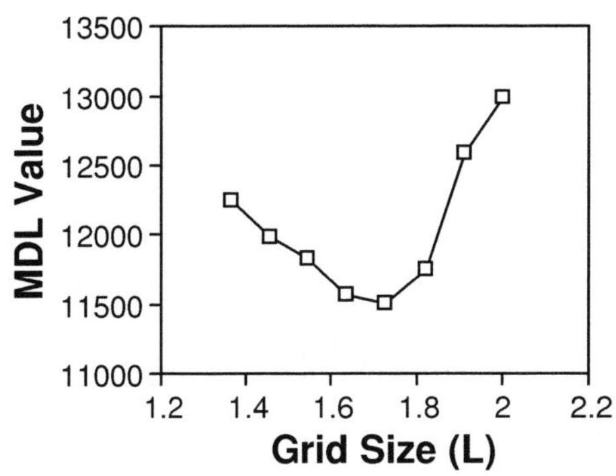

Fig. 4 MDL vs. grid size

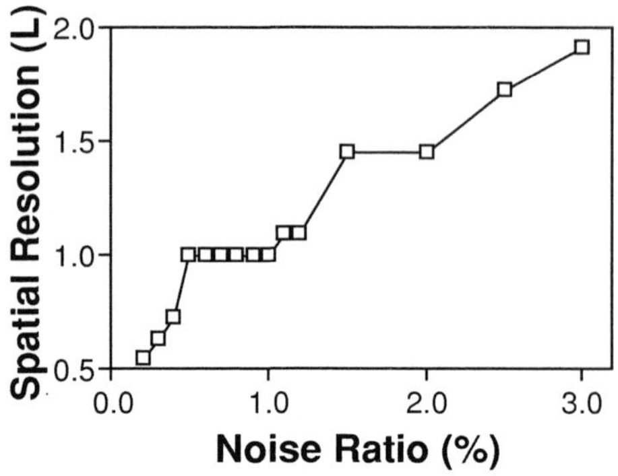

Fig. 5 Dependence of spatial resolution on noise ratio

Table 1 Dependence of spatial resolution on numbers of projections

A \ B	0.3%	1.0%
180	0.636	1.0
90	0.818	1.182
60	0.818	1.182
45	0.818	1.455
36	0.909	1.636

A = Number of Projections
B = Noise Ratio

size, where the number of projections was 180, and the Gaussian noise, whose mean and standard deviation's magnitude are 0 and 2.5% was added to each projection data. From Fig. 4, it can be seen that the spatial resolution is 1.727*L*.

Fig. 5 shows the dependence of spatial resolution on the added noise, where the vertical and the horizontal axes are the spatial resolution and the ratio of the standard deviation to the maximum projection data, respectively. The number of projections was 180. The result corresponds to the intuitive understanding that the spatial resolution deteriorates with increasing noise.

Table 1 shows the dependence of the spatial resolution on the number of projections when 0.3% and 1.0% noise were added. The results also correspond to the intuitive understanding that the spatial resolution deteriorates with decreasing projection number.

To determine the influence of truncated projections, we set two data points at the both sides in each projection to zero. The number of the projections and the noise ratio are 60 and 1.0%, respectively. The spatial resolution was 1.727*L*, and degrades similarly to that shown in Table 1. Also in this case, the result corresponds to the intuitive understanding that the resolution deteriorates on image reconstruction from truncated projections.

3.2 Spatial Resolution by Gaussian Beams

We have used the FBP by regarding the data-acquisition process as the line integral. Actually, the beam size is not negligible compared to the object. In addition, the laser beam takes on the Gaussian shape. These factors influence the spatial resolution.

As it is difficult to calculate the weight matrix analytically, we propose a method to generate the weight matrix based on the Monte Carlo method. That is, for a bin in each projection, Gaussian random numbers are generated around the bin, and then the line corresponding to a random number is determined. The weight matrix elements are calculated using the same method as by a line integral. The individual sums of weight matrix elements, which are obtained in each procedure, converge to those acquired by a Gaussian beam, as these procedures are repeated (Fig. 6).

We compare the result obtained by the weight matrix by pencil beams to that by the Gaussian beams whose standard deviation is *L*, where the projection data are generated by the Gaussian beams with the standard deviation of *L*, the translational step *L*=2.0/64, and the projection number and

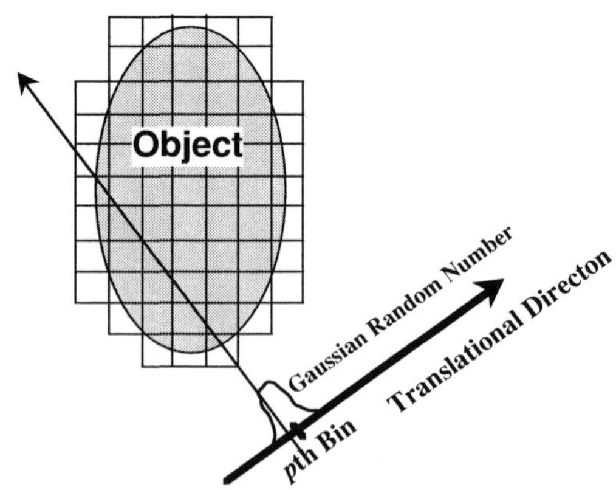

Fig. 6 Schematic diagram of generation of the weight matrix by Gaussian beams

the noise ratio are, respectively, 60 and 1.0%. The results by the pencil beams and by the Gaussian beams are 1.545*L* (MDL = 7201) and 2.273*L* (MDL = 4310), respectively. The results appear to suggest that the weight matrix by the pencil beams should be adopted, since the spatial resolution by the line integrals is superior to that by the Gaussian beams. However, as the spatial resolution was originally defined from the viewpoint of data fitting, the weight matrix by Gaussian beams should be adopted, since the corresponding MDL is less than that of the weight matrix by the pencil beams. As a result, we conclude that the spatial resolution is 2.273*L*, and the weight matrix faithfully representing the data-acquisition processes should be adopted as the weight matrix. The simulation shows that MDL can determine both the grid size of the orthonormal bases $\phi_j(x, y)$ $(1 \leq j \leq k)$ and the data-acquisition processes h_{ij}.

Reconstructed images are shown in Figs. 7 and 8. Fig. 7 is reconstructed using the weight matrix by the pencil beam with the grid size of 1.545*L*. Fig.8 (a), (b), and (c) are reconstructed using the weight matrices by Gaussian beams with the grid sizes of 2.182*L*, 2.273*L*, and 2.364*L*, respectively. In all images the iteration of optimization was halted, when the following condition held at the first time:

$$\frac{\sum_{j=1}^{k} \left| f_j^n - f_j^{n-1} \right|}{\sum_{j=1}^{k} \left| f_j^n \right|} < 0.08, \tag{9}$$

where f_j^n is the *j*th pixel value in the *n*th iteration. It is difficult objectively to evaluate and compare the qualities of these images, but Fig. 8 (b), which took on the smallest MDL value, appears to be the best result from a subjective evaluation.

4. Conclusion

We defined the spatial resolution as the optimal size of the grids in algebraic scheme using the MDL-principle, and investigated its dependence on the observation noise, the profile of incident beam, the number of projections, and the truncated projections from the viewpoint of data fitting. Our result shows that the factors which influence the spatial resolution can be easily incorporated and calculated by our proposed method.

In this research, we employed the conjugate gradient

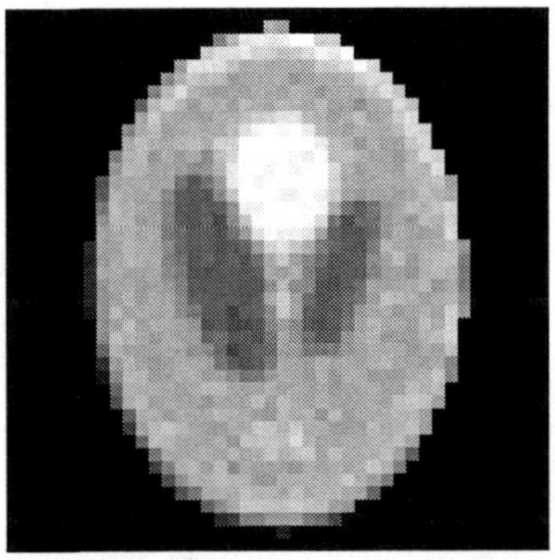

Fig.7 Image reconstructed with weight matrix by pencil beam

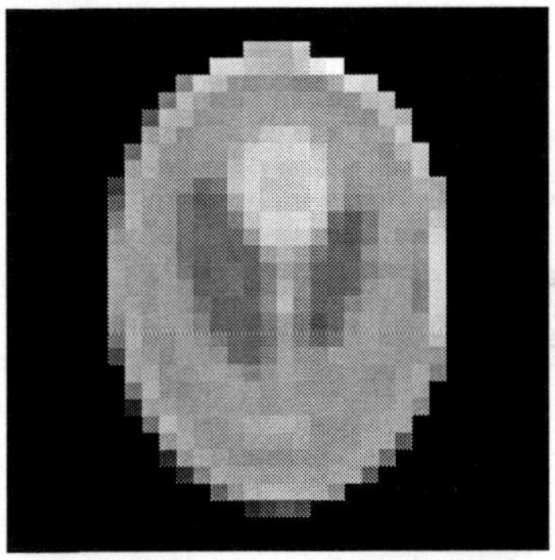

(a) 2.182*L*

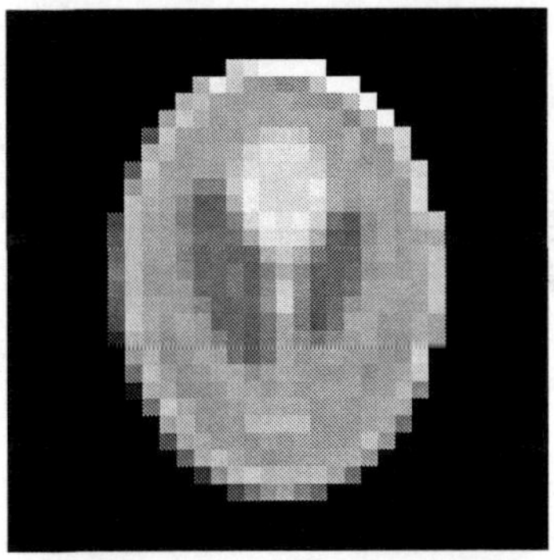

(b) 2.273*L*

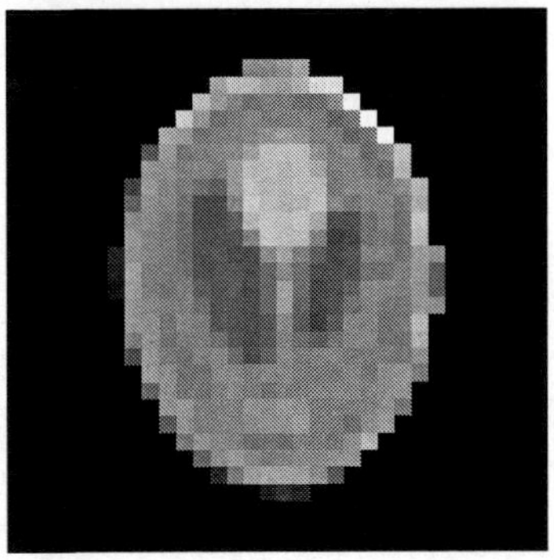

(c) 2.364*L*

Fig.8 Images reconstructed with weight matrices by Gaussian beams method, whose convergence rate is relatively high, as an optimization algorithm to process a great deal of data. But, the convergence represented a slight instability, or the results more or less depended on the stopping rule, which is generally characteristic of the iterative algorithms for an inverse problem [13, 14]. In order to obtain more reliable results, we must take a more stable algorithm, or develop a more reliable stopping rule.

In addition, we did not consider the surface effect of the object, because the properties have not yet been clarified. However, if they will be restricted, our method can easily incorporate the effect.

If these problems are resolved, the proposed method will not only specify the practical resolution of our imaging system, but also derive a data-acquisition scheme more suitable to the existing system.

References

1. S.B. Colak, H. Schomberg, G.W.'t Hooft, and M.B. van der Mark, "Optical Backprojection Tomography in Heterogeneous Diffusive Media," *TOPS on Advances in Optical Imaging and Photon Migration*, Vol. 2, pp. 282-289, 1996.
2. S.R. Arridge, "The Forward and Inverse Problems in Time Resolved Infra-Red Imaging," in *Medical Optical Tomography: Functional Imaging and Monitoring*, G. Muller, *et al.* (eds.), SPIE Institutes for Advanced Optical Technologies Series, Vol. IS11, pp. 35-64, SPIE Opt. Eng. Press, Bellingham, WA, 1993.
3. H. Inaba, "Coherent Detection for Medical Laser Tomography," in *Medical Optical Tomography: Functional Imaging and Monitoring*, G. Muller, *et al.* (eds.), SPIE Institutes for Advanced Optical Technologies Series, Vol. IS11, pp. 317-347, SPIE Opt. Eng. Press, Bellingham, WA, 1993.
4. B. Devaraj, M. Takeda, M. Kobayashi, M. Usa, P. Chan, Y. Watanabe, T. Yuasa, T. Akatsuka, M. Yamada, and H. Inaba, "In Vivo Laser Computed Tomographic Imaging of Human Fingers by Coherent Detection Imaging Method Using Different Wavelengths in Near Infrared Region," *Appl. Phys. Lett.*, Vol. 69, pp. 3671-3673, 1996.
5. B. Devaraj, M. Usa, K.P. Chan, T. Akatsuka, and H. Inaba, "Recent Advances in Coherent Detection Imaging (CDI) in Biomedicine: Laser Tomography of Human Tissues In Vivo and In Vitro," *IEEE J. Selected Topics in Quantum Electron.*, Vol. 2, pp. 1008-1016, 1996.
6. L.A. Shepp, and B.F. Logan, "The Fourier Reconstruction of a Head Section," *IEEE Trans. Nucl. Sci.*, Vol. NS-21, pp. 21-43, 1974.
7. D.A. Chesler, S.J. Riederer, and N.J. Pelc, "Noise Due to Photon Counting Statistics in Computed X-Ray Tomography," *J. Computer Assisted Tomography*, 1(1), pp. 64-74, 1977.
8. K.M. Hanson, "Detectability in the Presence of Computed Tomographic Reconstruction," *Proc. SPIE*, Vol. 127, pp. 304-312, 1977.
9. F. Natterer, *The Mathematics of Computerized Tomography*. John Wiley, 1986.
10. T. Yuasa, B. Devaraj, Y. Watanabe, T. Sato, Y. Sasaki, A. Hoshino, H. Inaba, and T. Akatsuka, "Minimum-Description-Length Based Approach to CT Reconstruction Using Truncated Projections from Objects," *Proc. SPIE*, Vol. 3167, pp. 186-197, 1997.
11. J. Rissanen, "Modeling by Shortest Data Description," *Automatica*, Vol. 14, pp. 465-471, 1978.
12. A. Rosenfeld, and A.C. Kak, *Digital Picture Processing*. Academic Press, 1982.
13. A.V. Lakshminarayanan, and A. Lent, "Methods of Least Squares and SIRT in Reconstruction," *J. Theor. Biol.*, 76, pp. 267-295, 1979.
14. E. Levitan, and G.T. Herman, "A Maximum A Posteriori Probability Expectation Maximization Algorithm for Image Reconstruction in Emission Tomography," *IEEE Trans. Med. Imag.*, Vol. MI-6, No. 3, pp. 185-192, 1987.

A new algorithm based on time-dependent contrast functions, used to evaluate optical characteristics of an abnormality hidden within a tissue-like phantom

Victor Chernomordik[1], Jeremy C. Hebden[2], Ralph Nossal[1], and Amir H. Gandjbakhche[1]

[1]Laboratory of Integrative and Medical Biophysics, National Institute of Child Health and Human Development, National Institutes of Health, Bethesda, MD 20892, USA
[2]Department of Medical Physics, University College London, United Kingdom WC1E 6JA

Abstract

We have developed a methodology that can be used in reconstruction algorithms to quantify the optical coefficients and the geometrical cross section of a weakly abnormal optical target embedded in an otherwise homogeneous medium. This novel procedure uses different time-dependent point spread functions to analyze the diffusive and absorptive contrasts obtained from time-of-flight measurements. To test the robustness of our reconstruction algorithm, we analyzed two time-resolved data sets obtained in different experimental geometries for a tissue-like phantom. We have shown that results for the two cases are consistent and agree with known values.

1. Introduction

Quantification of the optical characteristics of a small weakly abnormal inclusion embedded in thick highly scattering tissue is one of the most challenging tasks in Biomedical Imaging. We recently have suggested a new approach to address this problem, which uses time-dependent contrast functions [1].

The main idea of our methodology is to use varying time delay for time-gated photon detection to separate the influence of scattering and absorption perturbations on the observed image contrast. We take advantage of their different relative contributions to the observed contrast for different time delays, Δt, determined by the behavior of the point spread function (PSF) of the imaging system. In Section 2 we present the outline of our theoretical model. Details of our model have been presented elsewhere [1].

In Section 3 the experimental set-up used to measure temporal light distributions for a real tissue-like phantom is briefly described. The suggested procedure for data processing is presented in Section 4.

To test the robustness and consistency of our analysis we now consider (Section 5) two available data sets, obtained for the same phantom but different experimental geometries ("fan" and transillumination, see Fig. 1). The previous application of our algorithm was to the former case, with a stationary laser source and scanning detector (streak camera) [1]. Results of our analysis are presented in Section 6.

2. Theoretical model

Our model is based on the theory of lattice random walks. A tissue continuum is replaced by a simple cubic lattice, and photons are represented by random walkers permitted to move only between lattice points [2]. In this picture scattering events are represented by steps taken by a random walker moving from one lattice site to another. The tissue structure is considered to be homogeneous, except for a compact region containing an abnormal site at which the corrected scattering and the absorption coefficients (μ_{sc}' , μ_a) are both different from those of the background.

Assuming a time-of-flight (TOF) measurement, the intensity $I(x,\Delta t)$, is recorded at position x for different Δt. We choose point x_0 of the analyzed scan with the maximal detected intensity. Then we use $I(x_0,\Delta t)$ as a reference to calculate time-dependent contrast functions from experimental data:

$$C(x,\Delta t) = \frac{I(x,\Delta t) - I(x_0,\Delta t)}{I(x_0,\Delta t)}. \qquad (1)$$

OSA TOPS Vol. 21 Advances in Optical Imaging and Photon Migration, James G. Fujimoto and Michael S. Patterson (eds.)

For a small (size $d_i << L$, L - slab thickness) abnormally absorbing inclusion, consisting of N_a^3 independent absorbers, located on a cubic lattice with step $d_s \sim 2^{1/2}/\mu_{sc}^{(0)}$, the contrast function is [3]:

$$C_a(x,\Delta t) \approx \sum \{\eta_{eff} \frac{W(x,\Delta t)}{p(\Delta t)}\}. \qquad (2)$$

Here, $p(\Delta t)$ is the probability that in the absence of the inclusion an injected photon will reach the detector with time delay Δt. $W(x,\Delta t)$ is the probability that a photon, having visited the inclusion, will reach the detector at Δt. $\eta_{eff} = (\mu_a - \mu_a^{(0)})/\mu_{sc}^{(0)\prime}$ is the effective absorptivity of an elementary purely absorber . We define $\Delta\mu_a = (\mu_a - \mu_a^{(0)})$.

Increased scattering inside an inclusion results in an additional time delay for photons [1]:

$$\Delta\tau \approx \frac{d_i^2(\mu_{sc}' - \mu_{sc}^{(0)\prime})}{2c} \approx \frac{d_i^2 \Delta\mu_{sc}'}{2c}, \qquad (3)$$

so, for a purely scattering inclusion, the contrast is given by:

$$C_{sc}(x,\Delta t) \approx \frac{1}{p(\Delta t)} \frac{dW(x,\Delta t)}{dt} \Delta t. \qquad (4)$$

We have found that $C_a(x,\Delta t)$ is practically constant for relatively large Δt, but steeply increasing for small Δt. $C_{sc}(x,\Delta t)$ depends on the *slope* of PSF (i.e., time derivative of $W(x,\Delta t)$). Hence, the largest $C_{sc}(x,\Delta t)$ corresponds to the smallest available time delays Δt. For $\Delta t > \Delta t_{max}$ (Δt_{max} corresponds to the maximum of the PSF), the contribution from $C_{sc}(x,\Delta t)$ is negligible (this part of the data is used to estimate the perturbation due to extra absorption). That is, we use the different dependencies on time delay to separate scattering from absorptive perturbations.

3. Experimental Data

We have used two sets of experimental data provided by investigators working at University College, London. These researchers studied a phantom both in the "fan" geometry (stationary laser source aligned with the center of the target and the detector moving in 1 mm steps) and in transillumination (the source and the detector move in tandem in 1 mm steps on the opposite sides of the slab). In both geometries scanning is realized along a single line (**x**-axis).

The phantom was made of a rectangular slab, of thickness D = 55mm, with transport corrected scattering $\mu_{sc}^{(0)\prime}$ = 0.7mm^{-1} , and absorption coefficient $\mu_a^{(0)}$ = 0.006mm^{-1}, containing a single perturbing cylinder having diameter 5mm and length 5mm. The cylinder, whose absorption and corrected scattering coefficients are twice those of the surrounding medium, was located midway between the two faces of the slab. Temporal pathlengths distributions, $I(x, \Delta t)$, were experimentally obtained for each position of the detector. However, due to low intensity, no photon data were provided for Δt < 500ps.

4. Outline of Data Analysis

In summary, the procedure for the data processing based on the described above algorithm is the following:

- Perform preliminary analysis of the time-gated intensity scans, corrected for the experimental geometry, to find the scan position x_0 corresponding to the maximum detected photon intensity $I(x_0,\Delta t)$ ("quasi-undistorted") that can be considered as a "reference" point for the background;
- Estimate "undistorted" parameters (scattering and absorption coefficients) of the tissue surrounding the inclusion;
- Determine geometrical parameters of the inclusion, i.e., the center position inside the tissue slab and effective dimensions. To do so, use contrast scans for the shortest available time delays and curve fit the observed contrast assuming a cubic inclusion represented by $N_a \times N_a \times N_a$ independent, identical elementary lattice points (N_a is an input parameter, estimated physical dimensions of the inclusion do not change significantly with its variation);

- Perform similar curve fitting for a chosen set of time delays Δt;
- Analyze the behavior of η_{eff} (Δt) (it is expected that for relatively large time delays, i.e. $\Delta t>\Delta t_{max}$, $\eta_{eff}(\Delta t)$ is constant);
- Estimate the parameter K ,which is defined as: $K = (\Delta\mu_a/\mu_{sc}^{(0)\prime})(\mu_{sc}{}' / \mu_{sc}^{(0)\prime}) \approx \eta_{eff}$ characterizing optical properties of the inclusion from these large-Δt data;
- Use short Δt data to evaluate the "scattering" contrast component and the perturbation of the scattering coefficient due to the inclusion;
- Combine the results of the two previous steps and estimate the perturbation of the absorption coefficient due to the inclusion.

5. Results

For both geometries, unperturbed values of scattering and absorption coefficients estimated by curve fitting from the relatively undistorted photon pathlength distributions (characterized by the largest intensity (or smallest observed contrast)) proved to be quite consistent:

1) from transillumination data:

$\mu_{sc}^{(0)\prime} = 0.79\text{mm}^{-1}$ and $\mu_a^{(0)} = 0.0053\text{mm}^{-1}$;

2) from "fan" data:

$\mu_{sc}^{(0)\prime} = 0.77\text{mm}^{-1}$ and $\mu_a^{(0)} = 0.0056\text{mm}^{-1}$.

The fitting parameters chosen for the curve fitting are the effective absorptivity of elementary absorber η_{eff}, offset (vertical displacement) of the contrast ΔC, the position of the center of the inclusion x_i, and the distance between two lattice sites of the target d_s . To illustrate that both sets of data are consistent, contrast scans, for Δt = 900ps, are presented in Fig. 1.

Analysis of the contrasts were done for Δt = 800, 900, 1200, 1500ps. Two fitted "fan" contrast curves for Δt = 750, 1500ps are presented for illustration in Figs. (2a-b). Parameters evaluated from both data sets are very close. For example, the absorption and the scattering coefficients of the inclusion obtained from contrast scans for Δt = 900 and 1200ps are :

1) from transillumination data:

$\mu_{sc}{}' = 1.64\text{mm}^{-1}$ and $\mu_a = 0.0112\text{mm}^{-1}$;

2) from "fan" data:

$\mu_{sc}{}' = 1.60\text{mm}^{-1}$, $\mu_a = 0.011\text{mm}^{-1}$ and the size $d_\perp = 5.4\text{mm}$.

These values compare favorably with the stated values 1.4mm^{-1} and 0.012mm^{-1}. The position of the inclusion (along the axis of scanning) from both data sets is also estimated with good accuracy. Our analysis shows that the determined values of the inclusion characteristics do not depend significantly on the particular time delays chosen for the data fitting, Δt_1 and Δt_2, as long as $\Delta t_2 \geq \Delta t_{max}$ and Δt_1 is considerably smaller than Δt_{max} (the smaller the Δt_1 that are available, the higher is the accuracy of the estimate).

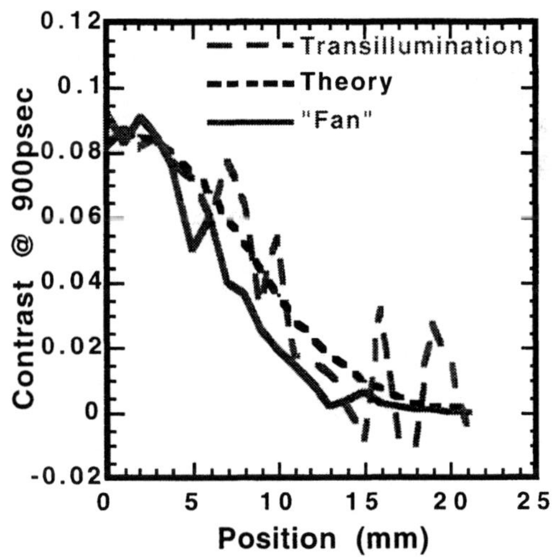

Fig. 1. Experimental contrasts obtained from "fan" and transillumination for Δt = 900ps. The fitted theory is also shown .

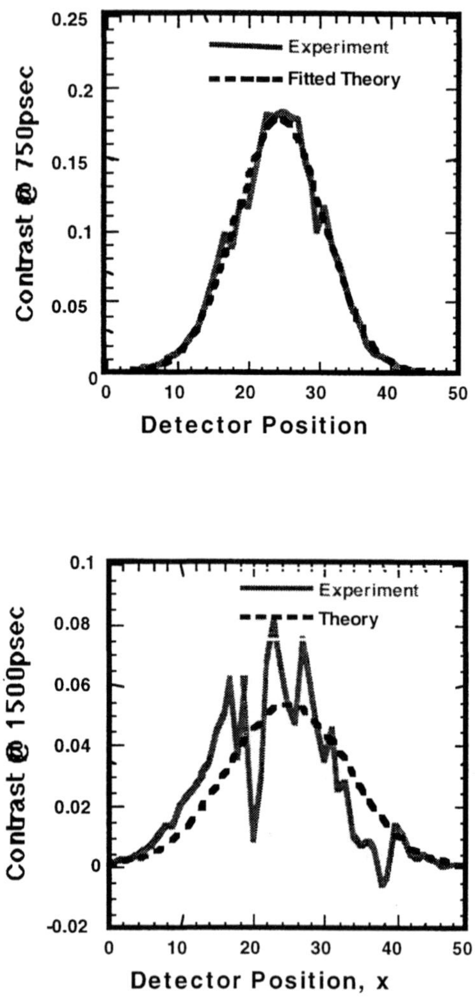

Fig.2. Experimental contrasts (solid lines) and the corresponding theoretical fits (dashed lines) obtained from "fan" geometry for a) Δt = 750ps, and b) Δt = 1500ps.

6. Summary

We have tested an algorithm, which uses available time-of-flight measurements, to reconstruct the optical properties of an abnormality embedded in otherwise homogeneous, optically turbid background. This algorithm, based on analysis of corresponding contrast functions, can discriminate between absorbing and scattering contributions to the total detected contrast. We have applied our method to two independent sets of time-of-flight measurements on a tissue-like phantom. For both data sets our approach gives almost the same optical parameters of the inclusion in question, and provides reasonable estimates of its position.

References

1. A. H. Gandjbakhche, V. Chernomordik, J. C. Hebden and R. Nossal, "Use of time-dependent contrast functions for quantitative imaging in time-resolved transillumination experiments", Appl. Opt., **37**, 1973-1981, 1998

2. A. H. Gandjbakhche and G.H. Weiss, "Random walk and diffusion-like model of photon migration in turbid media", Progress in Optics, Ed. E. Wolf, **34**, 333-402, 1995

3. A.H. Gandjbakhche, R.F. Bonner, R. Nossal, and G.H. Weiss, "Absorptivity contrast in transillumination imaging of tissue abnormalities", Appl. Opt., **35**, 1767-1774, 1996.

Localization of Fluorescent Objects Imbedded In Turbid Media In the Presence of Background Fluorescence

L. T. Perelman, K. Chen, R. R. Dasari, and M. S. Feld

George R. Harrison Spectroscopy Laboratory
Massachusetts Institute of Technology
Cambridge, Massachusetts 02139

We are developing a fluorescence-based photon migration imaging system for use in detecting small tumors in the turbid biological tissue of the breast, brain and other organs. Our scheme observes tumors in which exogenous fluorescent dye has selectively accumulated. The major difficulty with fluorescent imaging is the background fluorescence, which comes from the dye absorbed by the surrounding normal tissue. This signal may overwhelm the useful fluorescence from the tumor. However, our study shows that this difficulty can be overcome by detecting the early arriving photons. Fluorescence provides contrast, and early-time detection provides both resolution and suppression of background fluorescence.

The key point is that only the dye molecules accessible to the excitation photons contribute to the detected signal. The formation and collection of the fluorescence can be viewed as a three step process: (1) an excitation photon from the ps laser propagates through the scattering medium to a point (in either the object or the background) where it is absorbed; (2) it is then converted into a fluorescent photon, which (3) subsequently propagates through the scattering medium to the detector. Significant time (several ns) is required for the propagating photons to fully diffuse throughout the medium. Thus, at early detection times the effective volume of background fluorophores is small, much smaller than at later times at which diffusion has set in. This physical picture indicates that interference from background fluorescence will be smallest at the rising edge of the total fluorescence signal.

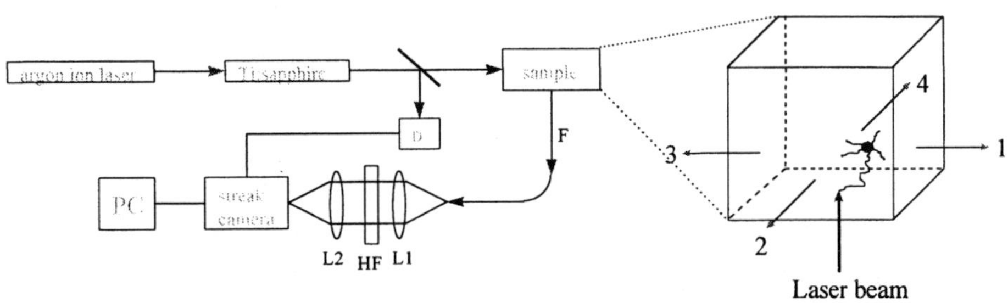

Figure 1. Experimental setup. The turbid medium is placed in a cubic glass container 63.5 mm on a side. Fibers 1 and 3 are oriented along the x direction, and fibers 2 and 4 along the y direction. In the current experiments the fluorescent object is placed on the plane of the collection fibers.

Our experiments employ a tomographic set-up in which ps pulses of excitation light are delivered to a glass container containing a turbid medium ($\mu_s' = 7.84cm^{-1}, \mu_a = 0.04cm^{-1}$) with a small imbedded fluorescent object, which mimics a dye-containing tumor. The same dye is added to the medium to achieve 10:1 object-background contrast. The volume of the dye-containing cell is $0.065ml$; the volume of the scattering medium is $256ml$. Therefore, the background medium contains 400 times more dye than the object. The returning fluorescence is collected by an array of fibers surrounding the container and detected by a streak camera. Localization of the imbedded object is accomplished by analyzing the early portion (few 100 ps) of the time-dependent fluorescence signal.

Typical data displaying the time-dependent fluorescence signals from the two collection fibers positioned at opposite sides of the container along the *x*-axis are shown in the right-hand panel of Fig. 2. Note that for this data the fluorophore was not positioned at the center of the cell. As expected, the rising portions of these two signals are distinct (early-arriving photon regime), but the tails coalesce (diffusion regime). Similar results were obtained from the *y*-directed collection fibers.

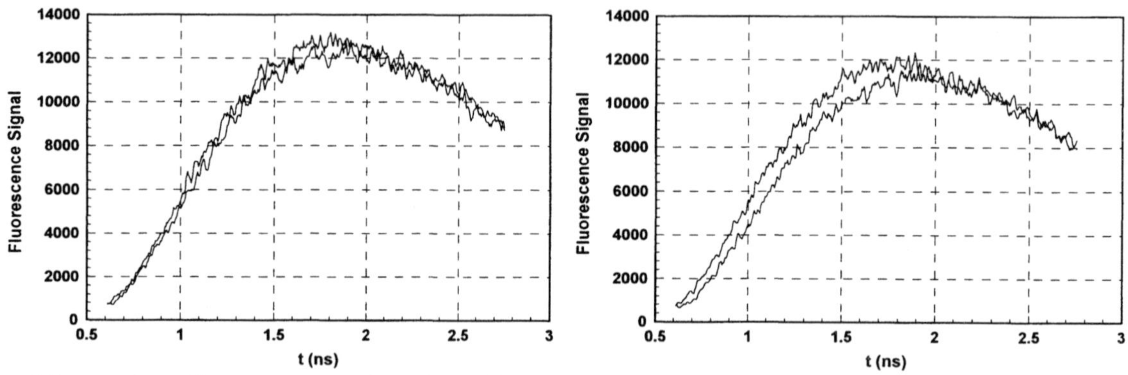

Figure 2. Fluorescence signals collected by the two fibers (1 and 3) along the *x* direction. The fluorophore concentration of the turbid medium is 10% of that of the fluorescent object. The left panel shows the fluorescence signals in the absence of the cell. These signals are entirely due to the background dye; the small difference is due to misalignment of the laser beam. The right panel shows the signals with the fluorescent cell inside the medium. The cell is positioned 1.9 cm from channel 1 and 4.5 cm from channel 3. Differences are evident in the rising edges of the two signals, whereas at long times ($t > 2ns$) the signals converge.

A model based on diffusion approximation has been developed to compute the position of the cell inside the turbid medium from the fluorescence ratio signals. The model takes into account the propagation of the excitation and fluorescence photons in the turbid medium, and can invert data even in the presence of background fluorescence. We derive the following approximate formula for the ratio of the signals from two opposite channels:

$$\frac{f_1 + B_1}{f_3 + B_3} = \beta + 2\varepsilon\left[\sinh\left(\frac{L}{2Dct}x_f\right) - \frac{x_f}{L}\cosh\left(\frac{L}{2Dct}x_f\right)\right]\exp\left[-\frac{1}{4Dct}\left(x_f^2 - (L - z_f)^2\right)\right]$$

Here, $f_i(t)$ and $B_i(t)$ are the contributions to the observed fluorescence signal from the imbedded object and the background, respectively, detected by fiber i; x_f and z_f are the coordinates of the embedded cell, with the origin at the center of the container; D is diffusion coefficient of the turbid medium; $2L$ is the length of the side of the container; ε (<<1) describes the effective cell-to-background fluorophore amount; and β is a calibration factor of order unity which accounts for the lack of centering of the excitation laser beam and small differences in collection between the two fibers. The predicted and the actual positions of the object are shown in Fig. 3. The object was shifted from position 1 to position 2 and then to position 3.

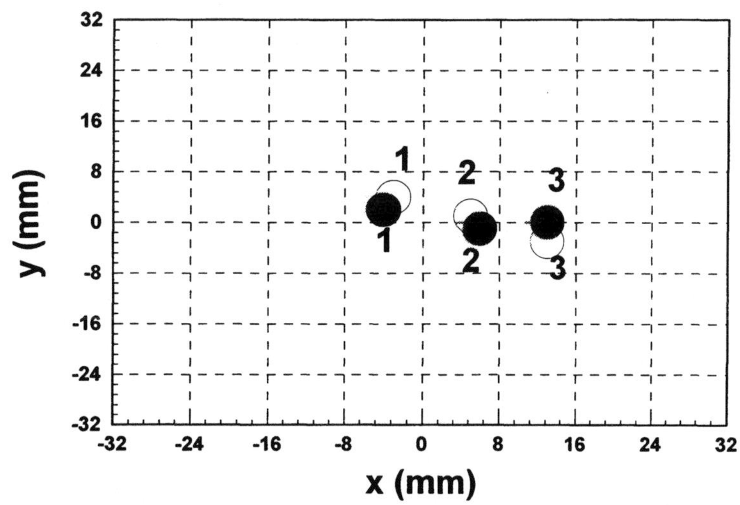

Figure 3. Object positions obtained by inverting the fluorescence ratio data (open circles).

In conclusion, we have shown that early arriving fluorescent photons can be used to localize an embedded fluorescent object under realistic conditions with accuracy of 2-3 mm. It appears that with improved algorithms, this accuracy can be further improved.

Diffuse optical imaging and functional assessment: Clinical and *in vivo* studies

Rapid and Sensitive Optical Imaging of Tissue Functional Activity, and Breast

Chance, B., Endla Anday*, E. Conant**, S. Nioka, S., Zhou, H. Long,
Department of Biochemistry and Biophysics, Johnson Research Foundation,
***Department of Radiology,*
School of Medicine, University of Pennsylvania, Philadelphia, PA 19104
**Allegheny UniversityHealth Sciences, Philadelphia, 19131*

Abstract

The better understanding of photon migration dynamics coupled with improved electronics, particularly frequency domain technologies has brought new possibilities for not only more precise oxygenation measurements, but most important localization of responeses either of functional activity or pathological heterogeneities introduced, for example, by functional activity for aberrant metabolic characteristics of tumors with blood volume increases, stroke, hematomas, aneurysms, etc. Furthermore, outreach to populations of neonates, pre-term and full-term are now in hand, particularly for functional imaging of parietal and occipital regions. Eventual studies of internal organs of the body, particularly the fetus inutero in normal and abnormal states will be milestones of future progress towards a goal where optical tomography can image with reasonable resolution most of the larger organs of the body and eventually the smaller ones. While NMR has similar possibilities, the safety, affordability and patient friendliness make this a preferred technology.

Key Words

Photon migration; frequency domain; functional activity; blood volume; functional imaging; fetus inutero.

Introduction

The normal brain, the diseased brain, the neonate brain and cancer bearing breast are presented herewith. Arjun Yodh will later tell you how low frequency photon migration waves are generated by high frequency modulation (100 MHz) of laser diodes (1).

The wavelength is about 10 cm and we observe the usual high attenuation of photon patterns. We find that the resolution of imaging with the usual source and detector combinations is not really enough to give clear images of functional activity in the brain. Thus, we use a technique called phase and amplitude cancellation in which Dr. Knüttel here is a pioneer (2) (Fig. 1).

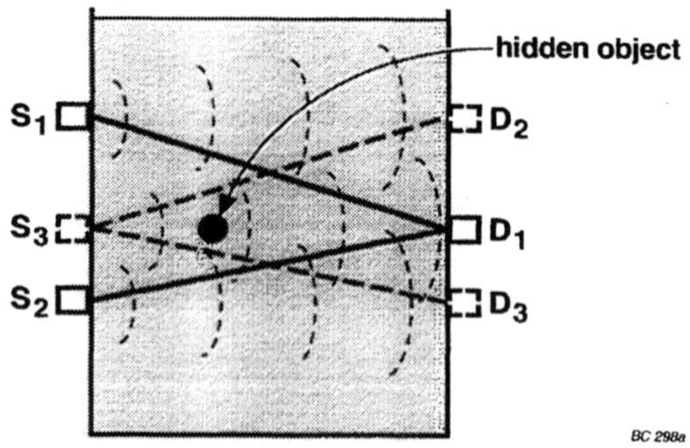

Fig. 1. Illustrating the geometry of amplitude and phase cancellation, i.e., two sources and a single detector, or two detectors and a single source. We use in this case both examples, two detectors and a single source for amplitude cancellation and two sources and one detector for phase cancellation.

We have either duplicate sources and a single detector or duplicate detectors and a single source so that the amplitude and phase characteristics can be nulled and the system becomes a differential. At the cross-over between zero phase and 180° phase, a 90° phase shift is obtained. This can be observed directly by remission on the human forehead by scanning the detector across the sources giving a very stable phase shift and a fairly symmetrical amplitude null. This is a way to enhance the imaging sensitivity .

Kyung Kang has shown the null plane can be bent to the left or the right depending upon whether an absorbing object is present (3) (Fig. 2). Thus, the imagers that we make are sensitized by using either phase or amplitude cancellation. Figure 3 shows the general geometry of 4 detectors and 9 sources the dual laser diode wavelength sources.

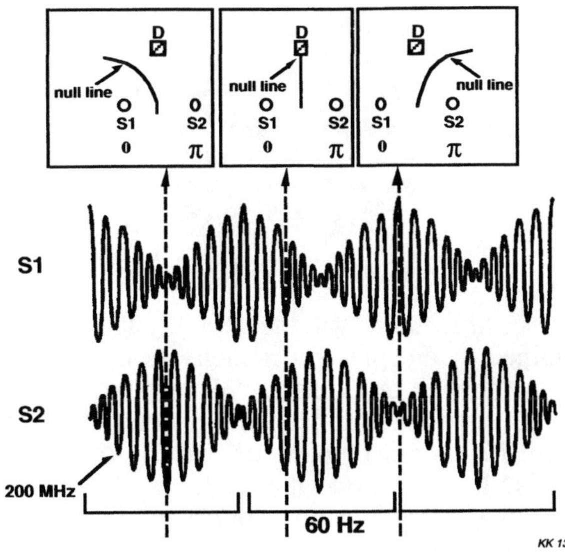

Fig. 2. Illustrating that the presence of an absorbing object bends the null plane towards the absorbing object because of the weaker signals from that sector.

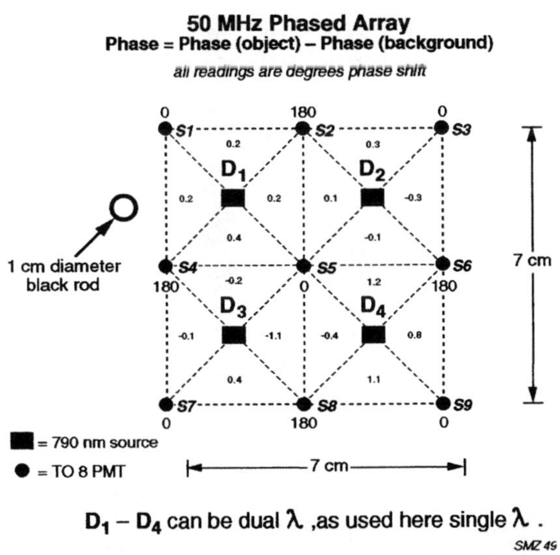

Fig. 3. Illustrating the geometry of sources and detectors used in a simple imaging system where each pair of sources is equidistant from the 4 detectors. This is used in the 50 MHz phased array system.

But I shall first talk about a single wavelength source that is very near the crossover between oxy and deoxy hemoglobin spectrum so that we measure essentially the BOLD phenomenon of NMR, namely increased blood volume (Fig. 3).

We time share the light sources and the detector output using "Radio Shack" switches which give 60 dB attenuation of the RF (Fig. 4).

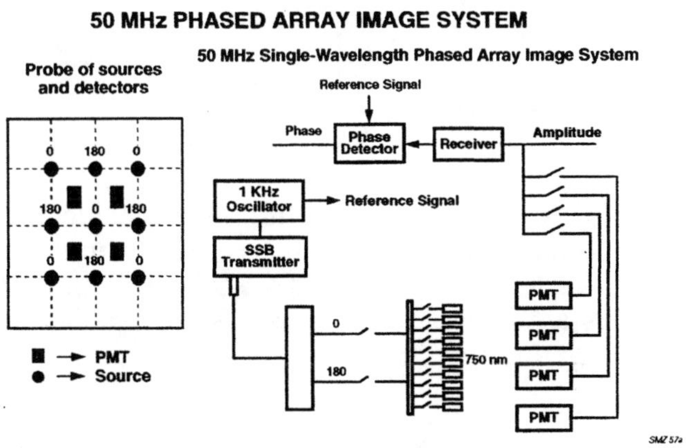

Fig. 4. Illustrating the electronic circuitry associated with a phased array system where a single sideband 50 MHz transmitter excites the appropriate laser diodes through a switch sequence and the signals detected by a single side band receiver are appropriately switched so that the phase signals correspond to appropriate source detector combinations.

The system is a single sideband phase modulation system of high sensitivity (~ 0.5 μvolts/kHz at the detector). With this system, we can pick up very small phase shifts. So how do we test in humans? We use parietal stimulation on the scalp of a fairly hairless subject, and when we stimulate, we evoke in the contralateral hemisphere a peak, which is about 4 mm wide and is co-registers with the image due to NMR at 4 Tesla.

The optical image is narrower with phased array than if we use phase only (Fig. 5). Then we get a rather broadened image. But you can see that the signals are quite different, 40° phase shift for phased array and 1 degree phase shift for the direct phase modulation system. So we have gained quite a lot by the 50 MHz modulation of the signal, and by using phase cancellation devices. We can increase the resolution by more sources and detectors, but eventually, we will come up to the limit calculated by Boas, O'Leary and Li, in our laboratory and that of many others. The times of acquisition are 8 sec.

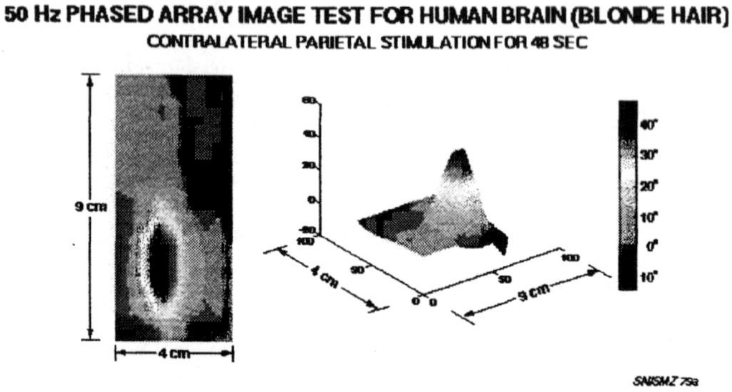

Fig. 5. Illustrating the resolution obtainable with the phased array system using parietal stimulation of an adult human shown as a 2D projection (left) and as a 3D amplitude analysis showing that the peak is around 4 mm.

Figure 6 shows some tests of high school students (many minority students) where we have done over 150 tests last summer. The test the students used was to spell backwards five letter words the best they could, more or less, by putting the word in short term memory using visual imagery for picturing the word and then reading backwards.

We used 30 second stimulation, 30 second rest. This caused a 30°, 40° phase shift in the prefrontal region. The pad went from eyebrow above the sinuses to the hairline and more or less from temple to temple. This gave phase shifts which you can see are about 40°. The subject became inattentive and the remnant image here is down by a factor of 10. So the method is extremely sensitive to what the subject is doing in the prefrontal region and therefore it has obvious applications to attention deficit and to other pathologies, even mild brain trauma.

The results is a histogram (Fig. 7) and this histogram is taken over a 4 week interval, gives roughly the same data, not exactly, but it does differ from the histogram of other ladies in this study. They had different histograms. We like to call this a "brain print".

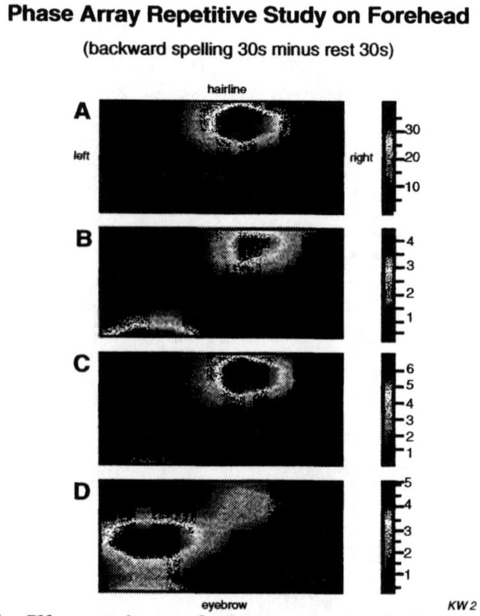

Fig. 6. Illustrative of the reproducibility usually obtained in the pre-frontal activation of a young female subject tested every 30 sec and rested 30 sec between teats undergoing 4 consecutive tests in this manner of backwards spelling of five letter words. It is seen that there is a highly consistent response in the first three intervals, on the fourth interval the response has decreased and shifted to another region, presumably by an attention shift of the subject.

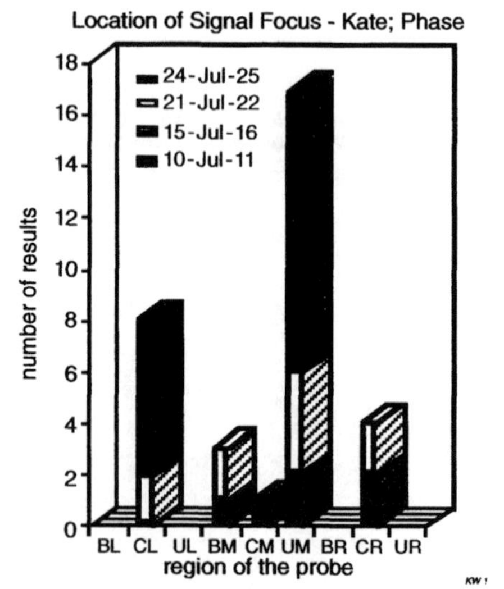

Fig. 7. Histograms of responses in four weeks of study of a particular individual. The response are assigned to 9, 2 cm squares extending from over the frontal sinus to the hairline and from temporal lobe to temporal lobe. the designations, UM, upper middle, etc., are self explanatory.

My colleague Tamura has shown that schizophrenics have lost symmetry in the response (4). Together with Dr. Scott Eleff who worked with me on NMR and is now at Hopkins and Godfrey D. Pearlson, M.D., Professor of Psychiatry and Chief of Psychiatric Neuro-imaging. there have used the Stroop test, for schizophrenics who have difficulty with decision-making. In Stroop, white is colored red and they are expected to name the word and not the color. Their difficulty in decision making evokes a large BOLD or blood volume change in the prefrontal region as compared to normal controls in which little change evokes a specificity which is tenfold for the schizophrenic (Fig. 8).

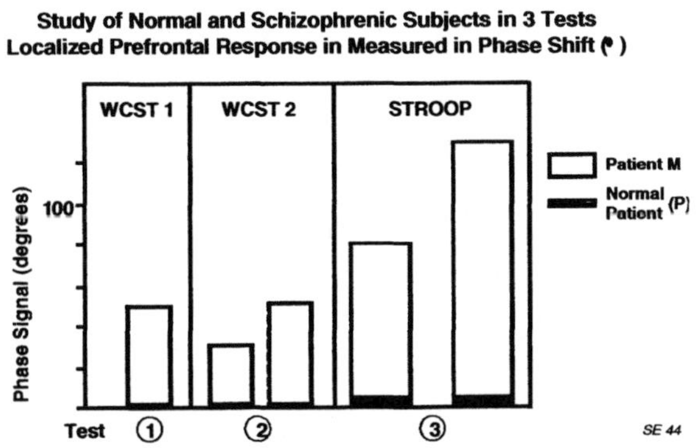

Fig. 8. Illustrative of the discrimination between a schizophrenia patient (yellow blocks) and a normal subject (blue blocks) in executing the Stroop test, a decision making test between prescribed directions in reporting the color of a word which would be painted in a different color than it spells. The normals require very little pre-frontal activation, while the schizophrenics require a much larger pre-frontal activation. Other tests are used, for example, the Wisconsin card sorting test.

Neonates have formerly been studied by NMR. We assembled had the first 7 inch NMR for studying neonates and it was a real job, first, to get them out of the incubator (often "bagged") into the NMR; second, to stabilize the baby and the NMR (often chloral hydrate was used); and thirdly, to wait out the time required to get the NMR spectrum which had been 15 or 20 minutes to get ^{31}P NMR data. But what we were not able to do was any brain functional activity. Now we were just able to do biochemistry of the phosphorus compounds. Since Dave Benaron has shown elegant images of brain damage in neonate brains, we wished to try functional imaging. Now we can put this imager pad on the parietal region (Fig. 9). By touching the contralateral foot, we obtain on the right side a response with a 90° phase shift due to the fact that the neonate has a thin skull, a very active response to functional activity. So this might be a quick and convenient method of testing whether or not the neonate is functionally active. And here, we again compare the phase modulation pictures with the phased array image and we achieve a better localization and a better signal-to-noise ratio.

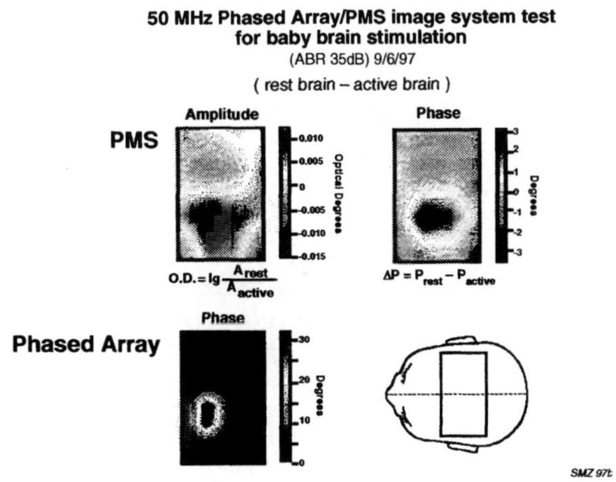

Fig. 9. Illustrative of the response of a full-term neonate to parietal stimulation shown below as a phase cancellation image; the two upper images are on the left, the amplitude signal and on the right, the phase signal taken in the usual manner without cancellation.

Sometimes the infant is not cooperative and not only will they show a parietal response to stimulation of the right finger, but then they will kick simultaneously as shown. When we stimulated the left finger, so there is an involuntary response and a voluntary response (Fig. 10).

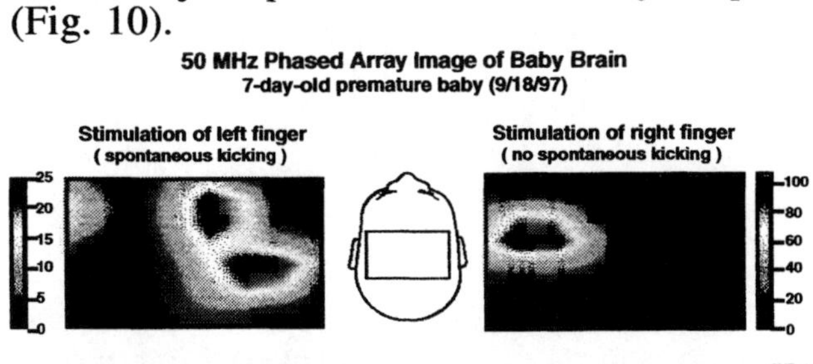

Fig. 10. Illustative of the response of a neonate to stimulation of the right foot in the left diagram and the right foot in the left diagram, except that in the left image, motion of the foot was observed as well and a double image is recorded.

Thus parietal stimulation has been useful but visual stimulation is also possible. In many cases where damage of the brain may have occurred, one would like to stimulation of particular parts.

At Hitachi Central Research, Drs. Koizumi and Maki have recently identified the speech area in neonates (Broca's Area) (unpublished data). Thus we can image different parts of the baby's head in normal and pathological infants. Also we get good images from the pre-term underweight babies in touching stimulation there is a characteristic increased blood volume signal. These signals are measured with respect to the contralateral side as reference and often with respect to a brain model, so they are incremental signals. The $\mu_a\mu_s'$ of the brain model is configured to be close to that of the subject that we are studying.

We are testing transabdominal signals to detect the baby's head therein, because it is thought that many of the problems of the baby develop in the 9 month stay in the uterus and not in the several hours stay in the birth canal for which the baby is uniquely prepared. Nimmi Ramanujam and Hong Long have shown that we can get signals through the uterine wall and the overlying tissue (Fig. 11). We are also testing imager so that we could follow the baby pre-partum and post-partum. Intra-partum have been very difficult due to large mechanical stresses.

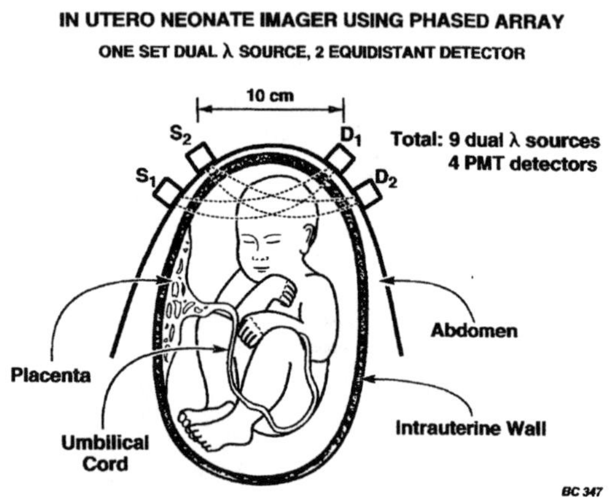

Fig. 11. Future studies using optical imaging will concentrate upon the body organs and one project underway is the detection of optical signals from the brain of the fetus in utero. At the present time imaging will be most valuable in the study of the fetus and the internal organs.

Breast Tumor Imaging

Another specificity, namely that of blood volume in the tumor is based upon Vaupel's data which show the angiogenesis of the tumor. Vaupel also shows with oxygen electrode probing of the tumor, it is largely hypoxic - the red part of the chart (Figs. 12, 13). Thus specificity for breast tumors might be achieved by an imaging system which used dual wavelengths and which scanned through this matrix of sources and detectors in a DC fashion.

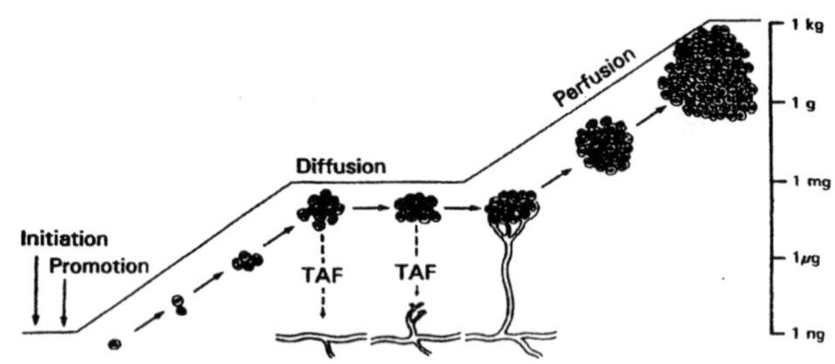

Fig. 12. Illustration of the acquisition of new blood vessels as a function of mass of a tumor, angiogenesis starts when the tumor is no larger than 1 mg and proceeds avidly thereafter as the tumor grows (5).

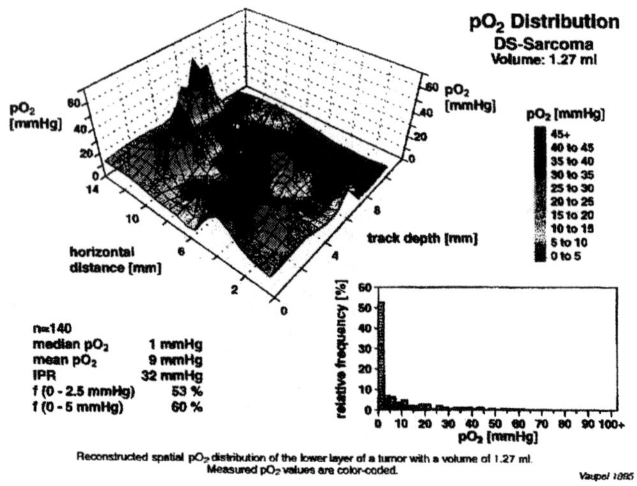

Fig. 13. Illustration of the large fraction of a tumor that is hypoxic as indicated by the red color measured by a platinum electrode inserted into the tumor volume and plotted as a three dimensional diagram (5).

Each source is illuminated for 50 μsec. and then we compare equidistant detectors and cancel the signals so that there is no signal on breast model. The sources and detectors are mounted on a pad that is flexible. There is no compression. The pad is held on the breast for 8 seconds and then on the contralateral breast for 8 seconds. We have had the co-operation of Emily Conant, M.D. who also provides the ultrasound, NMR and biopsy. The diagnosed breast tumors show increased blood volume due to the angiogenesis and deoxygenation hyperactive metabolism tumor area. The scale here is 9 x 4 cm so these images are about 2 cm in blood volume oxygenation, 4 cm in blood volume.

Figure 14 is exemplary of the high contrast that is obtainable between a tumor bearing right breast and a tumor free left breast. In this case, both are measured with respect to the breast model (containing no tumor). It is clear that there is an increase of blood volume signal and blood deoxygenation signal in approximately congruent places in the right breast, whilst on the same scale, the left breast shows scarcely any detectable signal. The fact that there is a possible ductile process coupled to the blood volume image for the right beast does not affect our score since we score only congruent images as shown in this case. In this case, both of these areas were congruent with the "suspicious mass" as identified by palpation, X-ray mammography, ultrasound, NMR and biopsy.

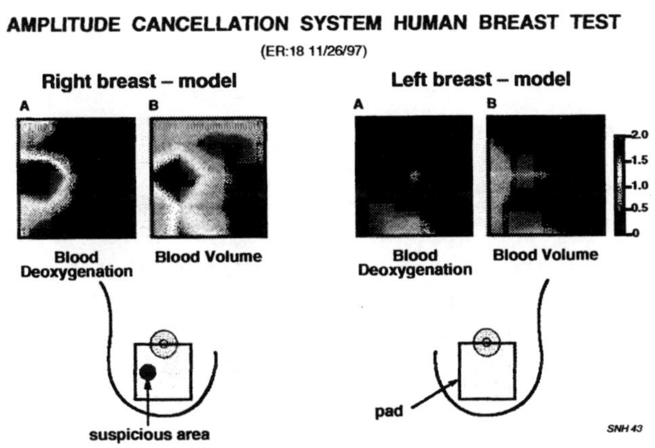

Fig. 14. Illustrative of the effectiveness of the simple algorithm, increased blood concentration and increased deoxygenation imaged optically. These two images (lef) are congruent and within the error, and also the pairs of maps are congruent with the "suspicious mass" as identified by palpation, X-ray mammography, ultrasound, NMR and biopsy.

These two images are measured with respect to a breast model and are in this sense "absolute". The two right hand images are from the contralateral tumor-free breast which shows no significant features.

We have completed a survey of 50 human breast containing suspicious masses and have summarized the data in a format which allows a display on the abscissa of the blood volume increase and on the ordinate display of the deoxygenation both localized to the image of the tumor (1.5 x 1.5 cm) and by the congruence of the images of deoxygenation and blood volume (Fig. 15). Thus, the data can be divided into 4 zones which contain the maximum number of cancers in Zone I, all the cancers in Zones I and II, and Zones II and IV generally contain cysts and fibro-adenomas, etc., i.e. all non-cancers. Zone I contains as well a dysplasic breast, a pre-cancer state in which there is already increased blood volume and increased deoxygenation. While this was not rated to be a full blown cancer, it was deemed to be a high risk state and worthy of watching. It would be noted in Zone II that there is one normal breast. This subject has to be recalled for further study, our data would have indicated a highly suspicious zone, congruent deoxygenation and blood volume.

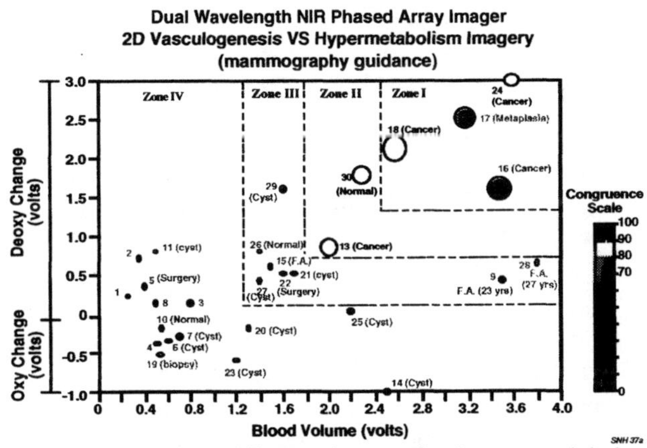

Fig. 15. An illustration of the sensitivity, specificity obtainable by plotting the blood concentration signal (abscissa) against the deoxygenation signal (ordinate). The data are divided into four zones for ROC calculation; zone I containing all but one of the tumors, and in addition, an extra signal due to a high risk hyperplastic pre-tumor region. The remaining zones containing increasing amounts of fibroadenoma, cysts, etc. non-cancer states.

The remainder of the studies of the 26 shown here and an additional 30 studied by Dr. Ornella Piazza were scored in Zones II and III, as cyst, fibranonma, etc. by virtue of the fact that there was a non-congruence of either deoxygenation or blood concentration signals. This discrimination has served us well and our criterion of congruent deoxygenation and blood volume increase, and congruence with the mammographical detected "suspicious mass" gives high sensitivity, specificity. If, however, the only *apriori* information is that there is a suspicious mass, the quality of the data are scarcely deteriorated, thus this method has possibility of affording screening information

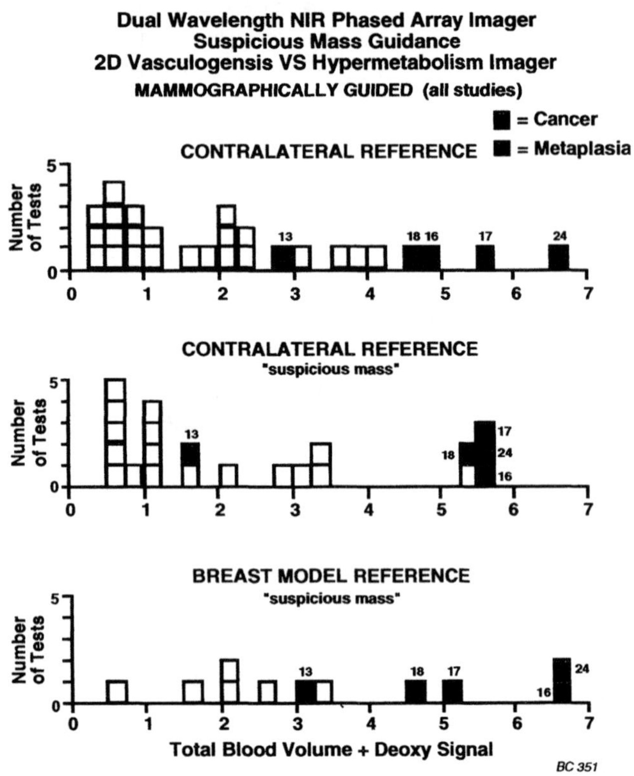

Fig. 16. Illustration of the sensitivity/specificity obtainable (top row) when the *apriori* information is based upon all available data (top), when the *apriori* information is limited to the fact that one breast contains a suspicious object (middle) and finally, sensitivity specificity obtainable when the reference material is not the contralateral breast but the model system (bottom) as indicated in Fig. 14 above)

In order to display the relationship of the specificity, sensitivity to congruence, we plotted Fig. 16, the same data in histogram form for the three conditions: 1) 20, all *apriori* information; 2) only a "suspicious mass" designated in the breast; and 3) instead of contralateral breasts as reference (which does remove signals from similarly located glandular/ductile structures) we use a model as reference and obtain very nearly the same result. In summary, the method appears to be robust, mainly relying upon the congruence of increased blood volume and increased deoxygenation in localized regions congruent with each other.

Dr. Joshua Fishkin put on one of Bruce Tromberg's data from breast cancer (6) which will show you it is not an isolated example to have an increased blood volume in a tumor and here is a breast tumor now scanned point-by-point along it with a source detector combination and here, quantitation of the $\mu_a\mu_s'$ and it shows it is about 50% at least greater in the tumor area as shown across here. Whilst, if we look closely there is a little bit of deoxygenation. Joshua also has an image of the tumor which by just varying the source and detector position, showed an image, similar to ours which identifies the tumor region. So I think there are many problems that were heretofore inaccessible in the study of photon migration patterns within the tissue. We have made some shortcuts to show that images can be obtained, we have hopes that the more sophisticated imagers which use many more source detector combinations and more complicated algorithms for image formation will be coming up soon.

Informed Consent

In all cases IRB approved informed consent was obtained from the subject or from the parent in the case of the minority students (University of Pennsylvania) and from the mother in the neonate studies (Allegheny University Hospital) and records are kept on File. Several hundred studies have been made under these protocols with no untoward effect.

The Future of Optical Imaging

The imaging of internal organs of not only tumors, but also of the organs themselves, would seem to be the next and appropriate step for optical tomography and one that closely follows the development of MRI. For example, current work on detecting and eventually imaging the fetus in urtero is suggested by Fig. 11 where

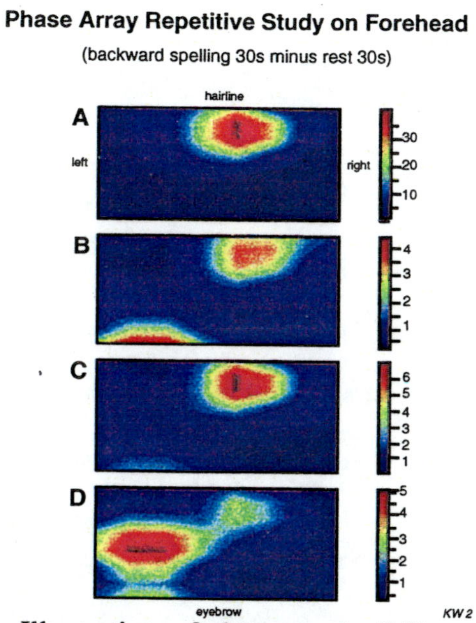

Fig. 6. Illustrative of the reproducibility usually obtained in the pre-frontal activation of a young female subject tested every 30 sec and rested 30 sec between teats undergoing 4 consecutive tests in this manner of backwards spelling of five letter words. It is seen that there is a highly consistent response in the first three intervals, on the fourth interval the response has decreased and shifted to another region, presumably by an attention shift of the subject.

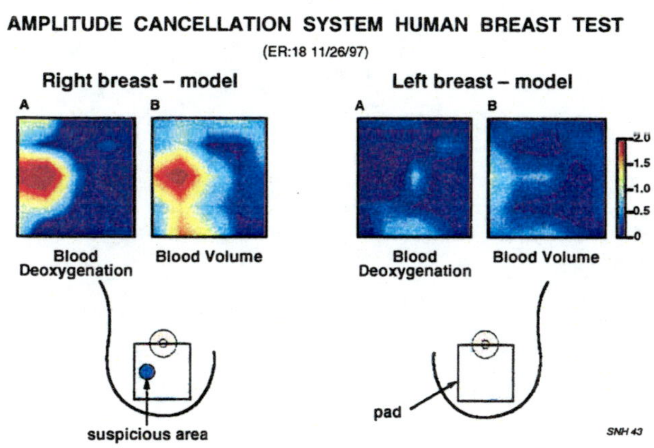

Fig. 14. Illustrative of the effectiveness of the simple algorithm, increased blood concentration and increased deoxygenation imaged optically. These two images (lef) are congruent and within the error, and also the pairs of maps are congruent with the "suspicious mass" as identified by palpation, X-ray mammography, ultrasound, NMR and biopsy.

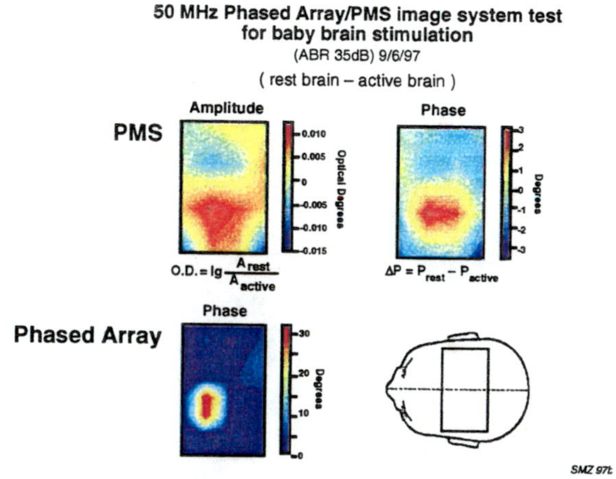

Fig. 9. Illustrative of the response of a full-term neonate to parietal stimulation shown below as a phase cancellation image; the two upper images are on the left, the amplitude signal and on the right, the phase signal taken in the usual manner without cancellation.

Sometimes the infant is not cooperative and not only will they show a parietal response to stimulation of the right finger, but then they will kick simultaneously as shown. When we stimulated the left finger, so there is an involuntary response and a voluntary response (Fig. 10).

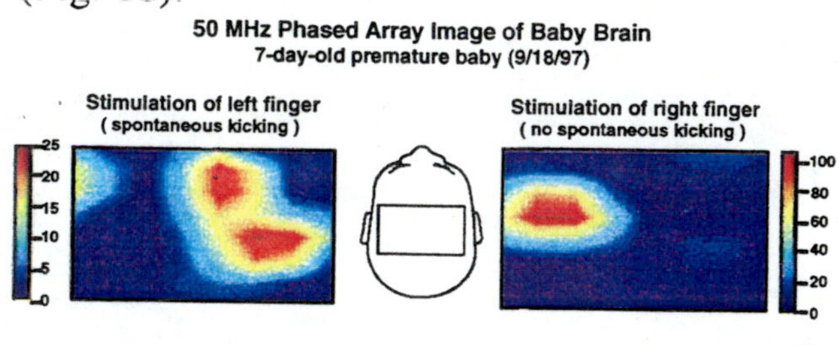

Fig. 10. Illustative of the response of a neonate to stimulation of the right foot in the left diagram and the right foot in the left diagram, except that in the left image, motion of the foot was observed as well and a double image is recorded.

signals related to the function of the neonate in utero can be obtained with CW systems currently, and will be quantified using a special version of the IQ system to determine oxygen saturation in the neonate brain and possible dangers of desaturation due to pathological states.

In summary, the track of optical tomography is now closely following that of MRI with less resolution but with greater affordability and patient convenience. This idea may be applied to the liver, lungs, kidney, etc. as the technology evolves.

ACKNOWLEDGMENTS

Supported in part by NIH 60182, CA72895.

REFERENCES

1. Yodh, this volume
2. Knüttel, A., Schmitt, J.M. and Knutson, J.R., "Spatial localization of absorbing bodies by interfering diffusive photon density waves," *Appl. Opt.* **32**, 381-389 (1992).
3. Kang, K.A., Chance, B., Zhao, S., Srinivasan, S., Patterson, E. and Trouping, R. (1993) Breast Tumor Characterization using Near-Infra-Red Spectroscopy. SPIE Proceedings Vol. 1888 Photon Migration and Imaging in Random Media and Tissues (Chance, B. and Alfano, R.R., eds.) *SPIE*, Bellingham, WA, pp. 487-499.
4. Hoshi, Y., Onoe, H., Watanabe, Y., Andersson, J., Bergstrom, M., Lilja, A., Langstom, B. and Tamura, M., "Non-synchronous behavior of neuronal activity, oxidative metabolism and blood supply during mental tasks in man," *Neurosci. Lett.* **172**, 129-133 (1994).
5. P. Vaupel, in *Biological Basis of Oncologic Thermotherapy*, Gautherie, M. ed., (Springer-Verlag, New York, 1990), pp. 73-134.
6. Fishkin, this volume

Regional Imaging of Brains using Near-infrared Diffusive Waves

Robert M. Danen [1, 2], Yong Wang [1], Xingde Li [2], C. Dean Kurth [3], William S. Thayer [1], and Arjun G. Yodh [2]

[1] NIM Incorporated, 3624 Market Street, Suite 508, Philadelphia, PA 19104
phone: (215) 387-6784, fax: (215) 387-6959, email: danen@lrsm.upenn.edu

[2] Dept. of Physics and Astronomy, University of Pennsylvania, Philadelphia, PA 19104

[3] Dept. of Anesthesiology, Children's Hospital of Philadelphia, Philadelphia, PA 19104

Abstract

We have developed a near-infrared instrument for functional imaging of localized regions of the brain. This paper presents results from model systems containing blood and clinical studies of pediatric subjects.

Key Words

Imaging; frequency-domain; brain; hydrocephalus.

NIM Regional Imager

We describe a near-infrared imaging instrument capable of providing spectroscopic information over a localized region of the brain [1]. Rather than develop a comprehensive optical imaging instrument for the whole brain, we instead developed an intermediate device which is faster and more directed than a "complete" brain imager [2,3], and improves upon purely spectroscopic near-infrared probes wherein the sample is assumed to be homogeneous. Our Regional Imager is a frequency-domain, multi-wavelength system with multiple source and detector positions. The Regional Imager offers the possibility for reconstruction of optical properties with linear resolutions of ~ 1 cm, beneath an area of ~ 40 cm^2 in size. Clinical applications include evaluation of critical care conditions such as hydrocephalus, hematoma, and hypoxic ischemic brain injury, as well as monitoring blood oxygenation during heart surgery and cognition studies [2-6]. Results from model systems containing blood and clinical studies of pediatric subjects, particularly those with hydrocephalus, are presented below.

Figure 1 shows a schematic of the Regional Imager [1]. Light is sent to and from the head via optical fibers, which are positioned against the head by the optical pad (Figure 1(B)). The pad is a semi-flexible, sturdy rubber sheet that holds 13 source fibers and 4 detector fiber bundles arranged in a 9.6 cm x 4.8 cm rectangular area. This gives separations (on a flat surface) between sources and detectors ranging from 2 to 7 cm. A two-stage optical-fiber multiplexer selects one of the two laser sources (i.e., 779 or 834 nm) and sends light from the selected source sequentially to the 13 source positions on the pad. The four detector bundles (6 mm diameter multi-mode fiber bundles) on the pad couple the diffusive waves, after they travel through the head, into four parallel electronic detection channels (see Y. Wang et al., these proceedings). A complete measurement at a single wavelength and single modulation frequency takes approximately 2.5 minutes.

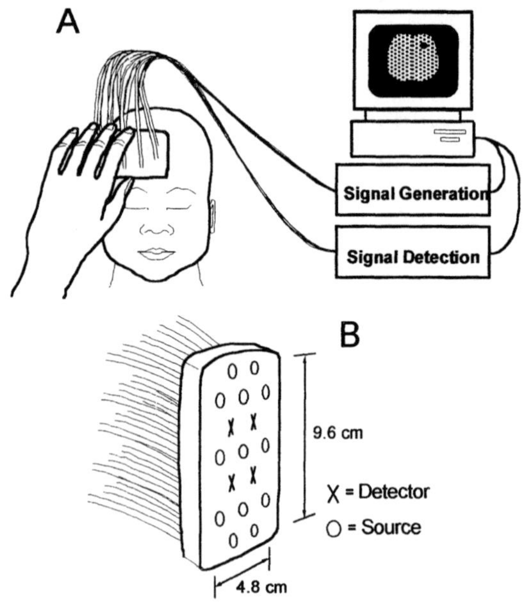

Figure 1: Schematic of the Regional Imager (A) and the source-detector geometry of the optical pad (B). The pad is semi-flexible and can conform to curves of the head. The optical pad can be placed at different locations on the head.

The Regional Imager employs intensity-modulated light to produce macroscopic waves of diffusing photons (at selectable radio frequencies ranging from 50 to over 400 MHz). Avalanche photodiodes (APD) detect the light after it travels through the head. Standard radio-frequency mixing techniques are used to determine the amplitude and phase of the transmitted diffusive wave. A personal computer records and displays data, including reconstructed images. The computer also controls the amplification of the signal leaving the APD by means of a programmable step attenuator to provide a needed dynamic range of $\sim 10^4$.

Image Reconstruction

For images, we employ two reconstruction techniques. One is a backprojection approach with a voxel weighting factor that is essentially the three-point Green's function of the medium underlying the pad using the average optical properties of the medium [7]. We use this approach because it is computationally fast and conceptually simple, and thus offers a quick two-dimensional image of the underlying tissue volume. The other technique is the Rytov algorithm. The Rytov algorithm provides greater image fidelity, including better depth resolution, than the backprojection method, but is computationally slower and more difficult to implement.

In both reconstruction approaches, we first combine measurements from all source-detector pairs to derive the average absorption and scattering coefficients of the underlying medium [1]. These average coefficients are found from the slopes of the lines describing the measured amplitude (more precisely, $\ln[\rho^2|U(\rho)|]$, where U represents the measured wave) and phase ($\phi = \tan^{-1}[\mathrm{Im}\{U\}/\mathrm{Re}\{U\}]$) as a function of the separation distance ρ between sources and detectors [8]. This follows from the solution of a homogenous semi-infinite medium with source and detector fibers on the surface of the medium [9,10].

For the backprojection reconstruction, we associate an absorption and scattering coefficient for each source-detector pair. This is accomplished by numerically solving the semi-infinite solution for the values of μ_a and μ_s' using the measured amplitude and phase from the source-detector pair. These values are then combined in a weighting scheme in order to determine μ_a and μ_s' in a planar slice of the turbid medium that is parallel to the optical pad. For each voxel in the slice, the optical properties determined for the different source-detector pairs are combined using weights that depend on the distance between the source and voxel, r_{sv}, and the distance between the voxel and detector, r_{vd}. The weight $Wt(r_{sv}, r_{vd})$ for each source-detector pair *sd* is essentially the three-point Green's function of the medium underlying the pad using the average optical properties of the medium. The optical properties in a voxel located at r_v are calculated using the following equations:

$$\mu_a(r_v) = \frac{\sum_{sd} Wt(r_{sv}, r_{vd}) \cdot \mu_a(\rho_{sd})}{\sum_{sd} Wt(r_{sv}, r_{vd})} \quad \text{and}$$

$$\mu_s'(r_v) = \frac{\sum_{sd} Wt(r_{sv}, r_{vd}) \cdot \mu_s'(\rho_{sd})}{\sum_{sd} Wt(r_{sv}, r_{vd})}, \qquad (1)$$

where $\mu_a(\rho_{sd})$ and $\mu_s'(\rho_{sd})$ are the optical properties determined from the source-detector pair *sd.* The sums are over all possible source-detector pairs on the pad. In practice, the voxels are 0.5 cm on a side and we compute these averages for all voxels in the planar slice. Equation 1 thus enables us to create an image of absorption and scattering for any planar slice in the medium beneath the detector plane. However, these backprojection reconstructions are insensitive to the choice of slice depth. Consequently, this method cannot be used to determine the depth of heterogeneities, although we are sensitive to relatively deep structures (~ 4 cm below surface under realistic tissue conditions).

We have begun implementing the Rytov algorithm for image reconstruction. The Rytov algorithm is a perturbation technique based on the presupposition that the gradients in the optical properties are relatively small (e.g., Kak & Slaney 1988). The measured wave is written as $U(\vec{r}) = U_0(\vec{r})\exp(\phi_s(\vec{r}))$, where $U_0(\vec{r})$ is the solution for a point source in a homogenous medium (incident wave) and $\phi_s(\vec{r})$ represents the scattered wave resulting from heterogeneity. To first order, the scattered wave may be written as

$$\phi_s(\vec{r}_s, \vec{r}_d) = \frac{-1}{D_0 \cdot U_0(\vec{r}_s, \vec{r}_d)} \int [v\delta\mu_a(\vec{r})G(\vec{r}, \vec{r}_d) \cdot U_0(\vec{r}_s, \vec{r})$$
$$- \delta D(\vec{r})\nabla G(\vec{r}, \vec{r}_d) \cdot \nabla U_0(\vec{r}_s, \vec{r})] d^3r, \qquad (2)$$

where $\vec{r}_s$ is the position of the source, $\vec{r}_d$ is the position of the detector, v is the speed of light in the diffusive medium, and D_0 is the diffusion coefficient for a homogenous medium. Here, $G(\vec{r}, \vec{r}_d)$ is the medium's Green's function, which is $\exp(ik_0|\vec{r} - \vec{r}_d|)/4\pi|\vec{r} - \vec{r}_d|$ for an infinite medium. The terms $\delta\mu_a(r)$ and $\delta D(r)$ represent variations in the absorption coefficient and diffusion coefficient at position r from the values of the homogenous medium. The first term within the integral describes variations in the absorption, and the second term represents variations in the scattering of the medium. The integral is over the tissue or sample volume.

We solve equation (2) for $\delta\mu_a$ and δD at each voxel, using our measurements to constrain ϕ_s [11]. The integral is digitized into a sum over voxels and then solved using an algebraic technique called Simultaneous Iterative Reconstruction Technique [7,11].

Model Studies of Blood

We have tested the Regional Imager on model systems containing blood, to model spatial variations of oxygenation in the brain. Here we present recent results using the Rytov perturbative algorithm. Our model consisted of a rectangular glass tank (aquarium with dimensions of 30 x 20 x 15 cm^3) filled with an Intralipid solution containing India ink. The optical properties of this solution were μ_a = 0.07 cm^{-1} and μ_s' = 5 cm^{-1}. The optical pad was placed from the top of the aquarium against the surface of the liquid so that the semi-infinite boundary condition was satisfied. We submersed two cylindrical test tubes filled with a solution of rabbit blood (0.5 mM Hb) into the Intralipid solution; they were oriented so that they pointed from the bottom of the aquarium towards the top surface, with their ends at a distance of 1.5 cm from the optical pad (figure 2). One test tube contained oxyHb blood; the other contained deoxyHb blood. Both test tubes contained an appropriate amount of Intralipid so that the μ_s' was the same as for the Intralipid solution in the aquarium. The test tubes had diameters of 1.3 cm and were separated by 3.5 cm.

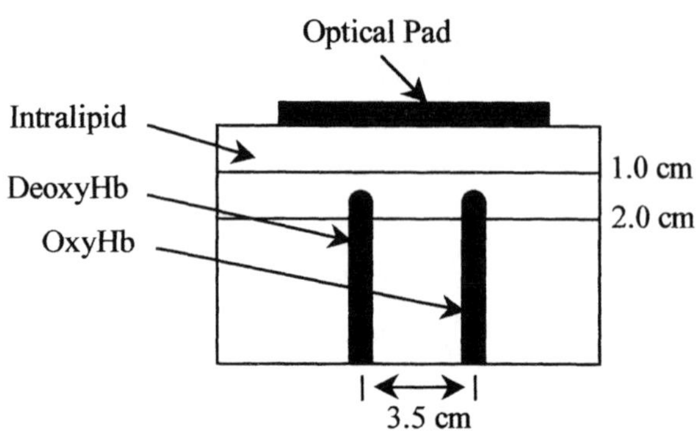

Figure 2. Side view of model system containing two test-tubes of blood. The test tubes diameters were 1.3 cm and their ends were 1.5 cm below the optical pad. The horizontal lines at depths of 1.0 and 2.0 cm below the optical pad define the planes of the reconstructed slices in figures 3 and 4.

Figures 3 and 4 show reconstructed images of μ_a for slices of the medium parallel to the optical pad at depths of 1.0 and 2.0 cm for wavelengths of 834 and 779 nm respectively. The data was obtained using a modulation frequency of 200 MHz. For the reconstruction, we used the Rytov algorithm with 3000 SIRT iterations to reconstruct these images [11]. Since the ends of the test tubes were located at a depth of 1.5 cm below the pad (figure 2), reconstructed slices of the medium at depths less than 1.5 cm should not reveal the blood samples whereas reconstructed slices for depths greater than 1.5 cm should reveal the samples. The slices at a depth of 1.0 cm (figs. 3a and 4a) accurately show no increased absorption above the background. The slices at a depth of 2.0 cm (figs. 3b and 4b) show increased absorption at the positions of the blood samples.

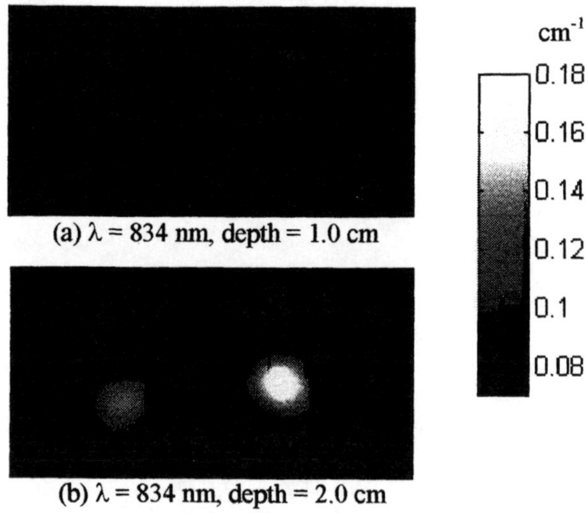

Figure 3. Reconstructed images of μ_a at wavelength of 834 nm for slice depths of (a) 1.0 cm and (b) 2.0 cm of an Intralipid solution containing two test tubes of blood, one with deoxyHb (left) and the other with oxyHb (right). Scale at right gives μ_a in cm^{-1}. The size of each image is 9.6 cm x 4.8 cm. The Intralipid solution in the aquarium had μ_a = 0.07 cm^{-1} and μ_s' = 5 cm^{-1}.

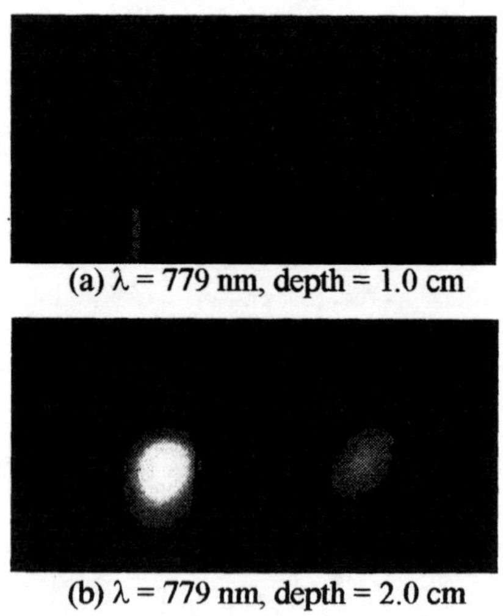

Figure 4. Same as figure 3, but for a wavelength of 779 nm. These images have the same scale as in figure 3.

For the slices at a depth of 2.0 cm, the Rytov reconstruction accurately determined the positions, widths, and separation between the blood samples. The full-width at half maximum (FWHM) measured from the reconstructed images is 1.3 cm for both blood samples; the actual width is 1.3 cm. The distance between the blood samples was 3.5 cm; the reconstructed images have separations of 3.75 cm.

The images accurately show increased absorption at 834 nm and decreased absorption at 779 nm for the oxyHb sample compared to the deoxyHb sample. (In images, the deoxyHb sample is to the left, oxyHb sample to right.) Table 1 lists the absorption coefficients of the two samples determined from the Rytov reconstructions, as well as the expected coefficients of the blood samples. For the concentration of blood used, we estimated the expected absorption coefficients using the extinction coefficients from [12]. For both samples and both wavelengths, the measured values given by the reconstruction are approximately 60% of the values calculated from the concentration of blood. The measured ratio of the absorption coefficients of oxyHb and deoxyHb agrees with the expected ratio at 834 nm and is within about 20% the expected ratio at 779 nm, suggesting that the Rytov algorithm can be used to establish the relative oxygenation for tissue.

The saturation calculated from the measured absorption coefficients, using the extinction coefficients from [12], is – 0.7% for the deoxyHb sample and 77% for the oxyHb sample. For these derivations, we subtracted the absorption coefficients of H_20 ($\mu_a(834) = 0.033\,\text{cm}^{-1}$ and $\mu_a(779) = 0.023\,\text{cm}^{-1}$ [14]) from the measured absorption coefficients [13]. The expected saturation is 0% for the deoxyHb sample and 100% for the oxyHb sample. The deoxyHb sample was prepared by mixing dithionite with the blood solution; the oxyHb sample was prepared by bubbling pure O_2 into the sample.

Table 1. Reconstructed and expected absorption coefficients of blood samples in figure 3.

	$\mu_a(834)$ cm^{-1}	$\mu_a(779)$ cm^{-1}
Reconstructed		
HbO_2	0.167	0.139
Hb	0.136	0.176
HbO_2/Hb	1.2	0.8
Expected		
HbO_2	0.28	0.20
Hb	0.23	0.32
HbO_2/Hb	1.2	0.6

The Rytov algorithm gives more accurate results than the backprojection method. For example, the reconstructed absorption coefficients given by the backprojection are at most 10% greater than the background, even for an infinitely-absorbing heterogeneity. For the measurements of the model reported here, the backprojection did not clearly resolve the two test tubes, nor could it determine the depth of the test tubes. Furthermore, the reconstructed FWHM given by the backprojection is generally twice the actual value. However, for certain situations we still use the backprojection because it gives us an immediate estimate of the variations in the sample beneath the pad in real time.

Clinical Studies

We have tested the Regional Imager on pediatric patients at the Children's Hospital of Philadelphia. Many of our initial studies consisted of patients with hydrocephalus. Hydrocephalus is an abnormal increase in the amount of cerebrospinal fluid (clear fluid – water) within the cranial cavity. The optical properties of water ($\mu_a = 0.025\,\text{cm}^{-1}$ and $\mu_s' = 0\,\text{cm}^{-1}$) are significantly less than for brain tissue ($\mu_a = 0.1 - 0.2\,\text{cm}^{-1}$ and $\mu_s' = 5\text{-}10\,\text{cm}^{-1}$). We found that the average optical properties of regions of the head containing large accumulations of this fluid are lower than for regions containing only brain tissue. Figure 5 shows bar graphs of our results, which include measurements of patients with hydrocephalus and patients without hydrocephalus. For our sample of patients, $\mu_a \leq 0.1\,\text{cm}^{-1}$ and $\mu_s' < 1.5\,\text{cm}^{-1}$ for regions of the head

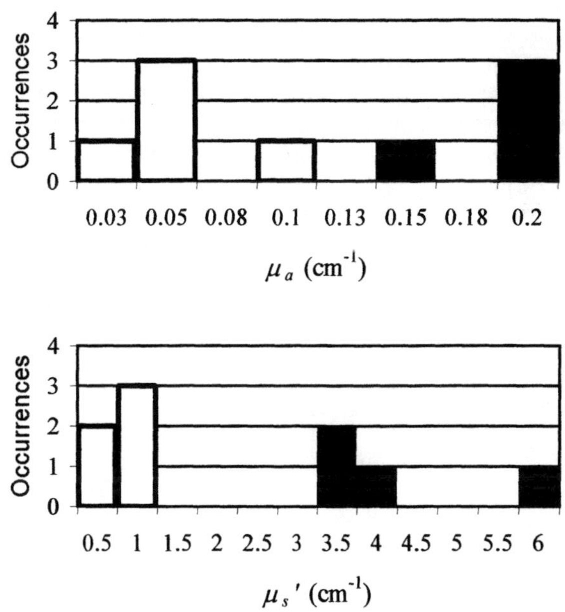

Figure 5: Occurrence of optical properties from a sample of 8 patients. Each occurrence corresponds to a different patient or a measurement from a different location on a patient's head. The unfilled bars are from measurements over regions of the head with excess fluid; the shaded bars are from measurements over only normal tissue, as determined from CT scans.

containing excess fluid, whereas $\mu_a \geq 0.14\,\text{cm}^{-1}$ and $\mu_s' > 3.0\,\text{cm}^{-1}$ for regions containing only brain tissue. These optical properties describe the entire medium below the pad and were calculated using measurements from all source-detector pairs on the pad as discussed above.

We report here an image of a patient with hydrocephalus who was undergoing surgery to have a shunt replaced. We measured the patient immediately before surgery while the patient was anesthetized. The optical pad was placed on a shaved region of the patient's head above the valve of the shunt (see Figure 6). The shunt is a plastic tube with an aluminum valve at one end (top); the valve had a teflon cover. The valve diameter was ~ 1cm, and was located ~1.2 cm inside the surface of the patient's head. The plastic tube went through the brain (away from the skull and optical pad) to a ventricle. Another plastic tube, the drainage tube, went from the valve towards the spinal column, just above the skull.

Figure 6 shows a backprojection image of μ_a from this study, reconstructed at a slice depth of 1.2 cm. The data was obtained using a modulation frequency of 100 MHz and wavelength of 834 nm. The image clearly shows the valve of the shunt: the peak in the upper-mid section of the image accurately matches the position of the valve. The feature at half the image peak extending ~3 cm from the peak (valve) towards the lower-left of the image traces the drainage tube. A reference measurement of an Intralipid-ink solution was used to determine the initial amplitude and phase of the sources.

These first *in-vivo* measurements using the Regional Imager are preliminary, but indicate the potential of the instrument. This work is supported by contract N44-NS-5-2317 from NINDS.

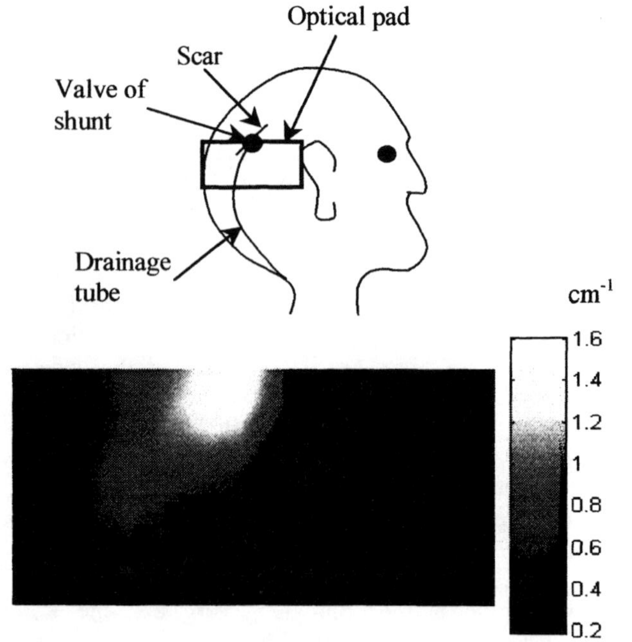

Figure 6: Image of shunt. The image shows the reconstructed μ_a of a region 9 cm x 4 cm in size, made using the backprojection technique for a slice depth of 1.2 cm. The diagram at top shows the orientation of the optical pad, which defines the region of the image, with respect to the head and the position of the shunt. The peak in the image accurately matches the position of the shunt valve; the region of intermediate absorption extending from the peak towards the lower-left traces the drainage tube.

References

1. R.M. Danen, Y. Wang, X.D. Li, W.S. Thayer, and A.G. Yodh (1998), "Regional imager for low resolution functional imaging of the brain with diffusing near-infrared light," Photochemistry and Photobiology **67(1)**, 33.
2. J. van Houten, D. Benaron, S. Spilman, and D. Stevenson (1996), "Imaging brain injury using time-resolved near-infrared light scanning," Pediatric Research **39**, 470.
3. M. Ishii, J. Leigh, and J. Schotland (1995), "Photon diffusion imaging in model and biological systems," in *Optical Tomography, Photon Migration and Spectroscopy of Tissue and Model Media: Theory, Human Studies and Instrumentation* eds. B. Chance and R.R. Alfano, SPIE 2389, p. 312.
4. M.Cope and D.T. Delpy (1996), "A system for the long-term measurement of cerebral blood and tissue oxygenation in newborn infants by near-infrared transillumination," Med. Biol. Eng. Comput. **26**, 289.

5. S.P. Gopinath, C.S. Robertson, R. Grossman, and B. Chance (1993), "Near-infrared spectroscopic localization of intracranial hematomas," J. Neurosurg. **79**, 43.
6. J. Brazy (1991), "Cerebral oxygen monitoring with near-infrared spectroscopy: clinical application to neonates," J. Clin. Monit. **7**, 325.
7. A.C. Kak and M. Slaney (1988), *Principles of Computerized Tomographic Imaging* (IEEE Press: New York).
8. A.G. Yodh and B. Chance (1995), "Spectroscopy and imaging with diffusing light," Physics Today **48**, 34.
9. R.C. Haskell, L.O. Svaasand, T. Tsay, T. Feng, M.S. McAdams, and B.J. Tromberg (1994), "Boundary conditions for the diffusion equation in radiative transfer," J. Opt. Soc. Am. **A 11**, 2727.
10. T.J. Farrell, M.S. Patterson, and B.Wilson (1992), "A diffusion theory model of spatially resolved, steady-state diffuse reflectance for the noninvasive determination of tissue optical properties *in vivo*," Med. Phys. **19**, 879.
11. M.A. O'Leary, D.A. Boas, B. Chance, and A.G. Yodh (1995), "Experimental images of heterogeneous turbid media by frequency-domain diffusing-photon tomography," Optics Letters **20 (5)**, 426. M.A. O'Leary (1996), *Imaging with Diffuse Photon Density Waves,* Ph.D. dissertation, Univ. of Penn. M.A. O'Leary, D.A. Boas, B. Chance, and A.G. Yodh (1995), "Simultaneous Scattering and Absorption Images of Heterogeneous Media Using Diffusive Waves within the Rytov Approximation," in *Optical Tomography, Photon Migration and Spectroscopy of Tissue and Model Media: Theory, Human Studies and Instrumentation,* eds. B. Chance and R.R. Alfano, SPIE 2389, p. 320.
12. W.G. Zijlstra, A. Buursma, and W.P. Meeuwsen-van der Roest (1991), "Absorption spectra of human fetal and adult oxyhemoglobin, de-oxyhemoglobin, carboxyhemoglobin, and methemoglobin," Clin. Chem. 37/9, 1633-1638.
13. M.A. Franceschini, S. Fantini, A. Cerussi, B. Barbieri, B. Chance, and E. Gratton (1997), "Quantitative Spectroscopic Determination of Hemoglobin concentration and Saturation in a Turbid Medium: Analysis of the Effect of Water Absorption," J. Biomedical Optics **2(2)**, 147.
14. G.M. Hale and M.R. Querry (1973), "Optical constants of water in the 200 nm to 200 micron wavelength region," Applied Optics **12(3)**, 555.

Frequency-Domain Optical Measurements *in Vitro* on Two- and Three-Layered Tissue-Like Phantoms, and *in Vivo* on Infant Heads

Maria Angela Franceschini[1], L. Adelina Paunescu[1], Sergio Fantini[1], Simone Pratesi[2], John S. Maier[1], Gian Paolo Donzelli[2], and Enrico Gratton[1]

[1]*Laboratory for Fluorescence Dynamics, Department of Physics, University of Illinois at Urbana-Champaign 1110 West Green Street, Urbana, Illinois 61801-3080 Phone (217) 244-5620, Fax (217) 244-7187*

[2]*Department of Pediatrics, University of Florence, 50132, Italy Phone: (39) 55-586835; Fax: (39) 55-570380;*

Abstract

We have performed frequency-domain optical measurements on the surface of two- and three-layered turbid media (*in vitro*) and on infant heads (*in vivo*). For the *in vitro* experiment we measured a number of combinations of one layer (thickness in the range 0.1-1.6 cm) on top of one semi-infinite block, with or without a clear layer (thickness of 0.1 cm) sandwiched in between. For the *in vivo* experiment, we did measurements on 20 preterm infants at the Meyer Children's Hospital in Florence, Italy. In the two-layered samples, we measured the effective absorption and reduced scattering coefficients with a multi-distance method based on data taken at source-detector distances in the range 1.5-4.5 cm. We found that the effective optical coefficients reproduce those of the underlying block if the superficial layer is less than about 0.4 cm thick, whereas they are representative of the superficial layer if it is more than about 1.3 cm thick. Larger source detector separations reduce the effect of the superficial layer on the recovery of the effective optical coefficients. Measurements in the presence of the clear layer gave results that did not reproduce, even qualitatively, the *in vivo* measurements on infant heads.

Key Words: (300.6500) Time resolved spectroscopy; (170.7050) Turbid media; (300.1030) Absorption; (170.0170) Medical optics and biotechnology.

1. Introduction

Near-infrared spectroscopy is a promising technique for the non-invasive study of biological tissues *in vivo*. To obtain quantitative information about the optical coefficients of tissues, one often makes the assumption of tissue homogeneity. This assumption can only be approximately valid in tissues, and it appears to be violated in the non-invasive study of the brain, where the inhomogeneity due to the cerebrospinal fluid (CSF), skull, and scalp can significantly affect the optical measurements. For this reason, several studies have been devoted to the investigation of light propagation in layered strongly scattering media.[1-3] In this contribution, we consider frequency-domain experimental results obtained *in vitro* on two- and three-layered media, and *in vivo* on infant heads.

2. Materials and Methods

For the *in vitro* experiment, we have used gelatin to prepare 5 blocks (about 5 cm thick) and 21 layers (thickness ranging from 0.1 to 1.6 cm). Different amounts of white paint and black India ink were added to the gelatin samples to obtain media with different optical properties. One of the 21 layers does not contain any paint or ink and is optically transparent in the visible/near-infrared spectral region (clear layer). Fig. 1 reports the optical properties of the 5 blocks (measured with the frequency-domain multi-distance method) at 750 nm. The blocks are labeled with progressive numbers from 1 to 5. The layers have the same optical properties of the corresponding blocks. The letters a, b, c, d, are used to indicate different thicknesses (the thickness increases from a to d). For the two-layered media study, we measured all possible combinations (100) of one superficial layer on top of one underlying block. For the three-layered media study, we measured 30 combinations where the clear layer is sandwiched between one superficial layer and one underlying block (we used the 5 blocks and layers 1a, 1b, 1c, 4a, 4b, and 4c).

The *in vivo* measurements were performed on 20 preterm infants at the Meyer Children's Hospital in Florence, Italy. We acquired data in four different positions on the head (temple, forehead, anterior fontanelle, and posterior fontanelle) with the infants resting in their bassinets.

The data were acquired at 750 and 840 nm with a frequency-domain (110 MHz), tissue spectrometer[4] (ISS, Inc. Model 96208). In the *in vitro* measurements we acquired data at source-detector distances (r) ranging from 1.5 to 4.5 cm. In the *in vivo* experiment we employed a flexible measuring probe (capable of fitting the shape of the infant heads) with source-detector distances ranging from 2.1 to 3.6 cm. The data were analyzed by using the solution to the diffusion equation for a semi-infinite, homogeneous medium. The purpose of this study is to investigate the effect of superficial layers on the performance of the frequency-domain multi-distance measurement protocol based on the equations for homogeneous media.

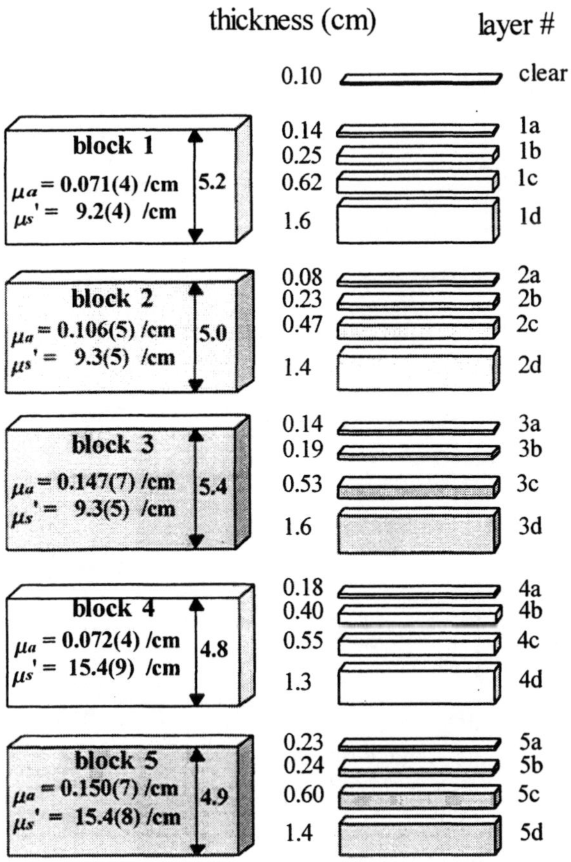

Fig. 1. The effectively semi-infinite blocks and the layers employed in our experimental study. The optical properties of each batch at 750 nm are indicated inside the blocks. The numbers in parenthesis give the error in the last digit of the corresponding parameter. The thickness of each block and layer is also reported in the figure. The clear layer is made of the same gelatin material, without adding any absorbing or scattering compound.

3. Results

Two-layered media experiment. When the thickness of the superficial layer is $\lesssim 0.4$ cm, the multi-distance method correctly recovers the optical properties of the underlying block. In the case of the thickest layers (1.3-1.6 cm), the recovered values of μ_a and μ_s' are close to those of the upper layer. Figure 2 shows the results for a superficial layer of intermediate thickness (0.55 cm, layer 4c) on top of the five different blocks. In this case, the measurement at larger source-detector distances accurately reproduces the block properties, even in the presence of a 0.55 cm thick layer. In all cases, we have found that by considering larger source-detector distances (circles in the figure), we get a more accurate recovery of the optical coefficients of the underlying block. However, this improvement is not substantial in the range of distances considered by us.

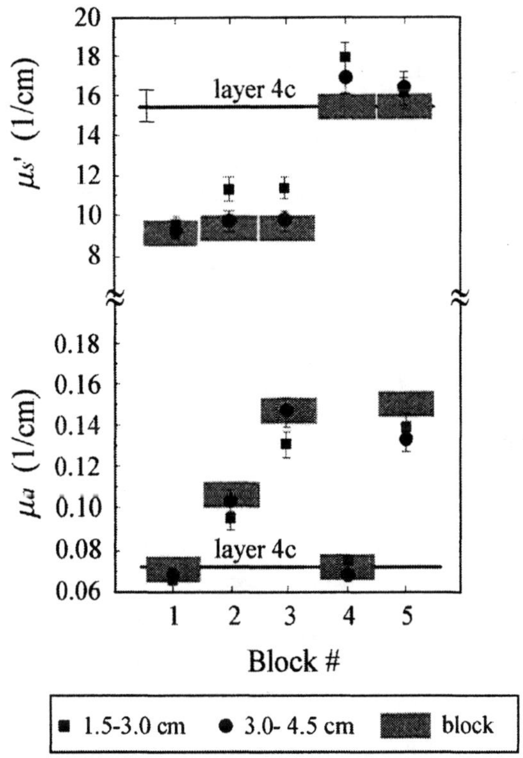

Fig. 2. Effective optical properties measured at 750 nm for the case of layer 4c (0.55 cm thick) on top at each of the five blocks. The squares are the measurements obtained with source-detector distances between 1.5 and 3.0 cm, while the circles are obtained from the range 3.0 to 4.5 cm. The thick horizontal bars represent the optical properties of the blocks. The thin horizontal lines indicate the optical properties of the layer.

To look at our measurements under a different perspective, we consider the r-dependence of the average value (dc) and phase of the modulated intensity. The symbols in Fig. 3 represent the experimental values of $\ln(r^2\mathrm{dc})$ and phase, for the case of layer 1c on top of block 5. In this case, the difference in the optical properties of the layer and the block is the largest among our samples. Furthermore, the layer thickness (0.60 cm) has an intermediate value between the cases where the effective optical properties reproduce those of the block (thickness $\leq$0.4 cm) or those of the layer (thickness $\geq$1.3 cm). For comparison, we also show the corresponding straight lines obtained in the semi-infinite geometry on block 1 (thin lines) and block 5 (thick lines). Since, to a first approximation, μ_a and μ_s' can be calculated from the slopes of $\ln(r^2\mathrm{dc})$ and phase vs. r, one can see that the presence of the superficial layer affects the recovered optical parameters (because it changes the slopes). Nevertheless, the experimental points are approximately aligned in the whole range of distances measured. Consequently, the multi-distance method can be applied even in the presence of a superficial layer. The effective absorption (reduced scattering) coefficient recovered in this layer/block configuration is $\mu_a = 0.18\ \mathrm{cm}^{-1}$ ($\mu_s' = 10.1\ \mathrm{cm}^{-1}$). These values deviate by 20% (-34%) from the coefficients of block 5, and by 154% (10%) from the coefficients of layer 1c.

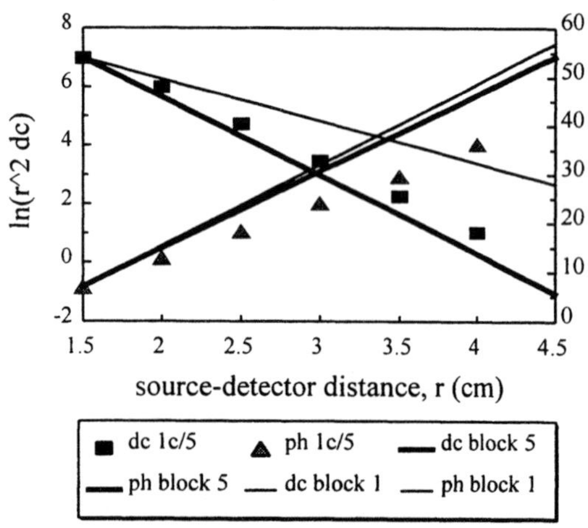

Fig. 3. Normalized experimental values of $\ln(r^2\mathrm{dc})$ and phase versus source-detector distance for the various samples indicated in the legend.

Three-layered media experiment. This experiment was conducted on phantoms where a 0.1 cm-thick clear layer was sandwiched between one superficial layer and one underlying block. Since the two layers and the block were made out of the same material there was no index of refraction mismatch at the interfaces.

In Fig. 4 we show the comparison between the $\ln(r^2\mathrm{dc})$ (panel (a)) and phase (panel (b)) measured in a three-layered medium (layer 4c on top, clear layer in the middle, block 3 at the bottom), in a two-layered medium (layer 4c on top of block 3), and in blocks 3 and 4 alone. The results in all of the layer/block configurations show the large effect of the clear layer on the optical data. From the r dependencies of the dc and phase, we can see that there is a light channeling effect through the clear layer as observed by Okada *et al.*[2] The dc intensity decreases slowly with distance, and the phase tends to flatten. Figure 4(b) shows that the phase increases by only 2.7 deg/cm in the presence of the clear layer, while it increases by 13.5 deg/cm in the absence of the clear layer.

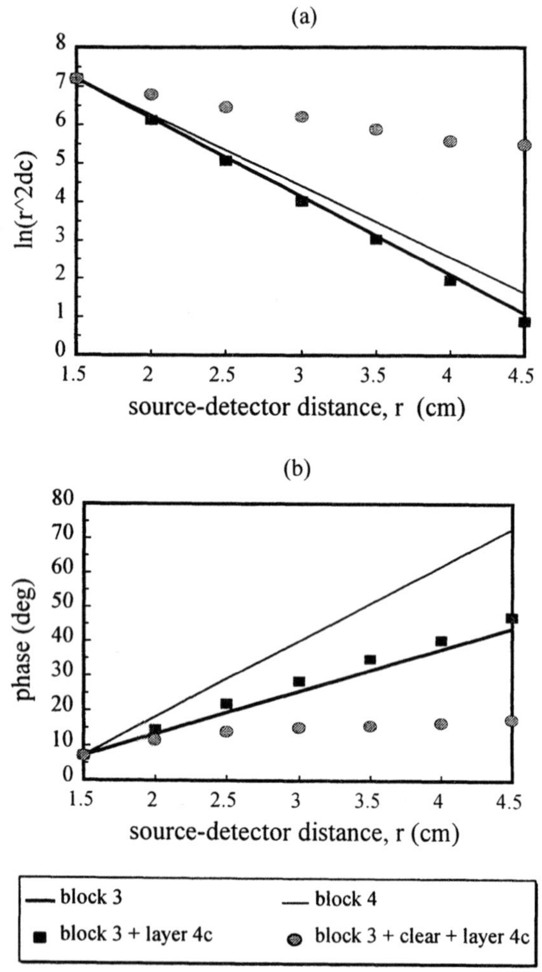

Fig. 4. Normalized experimental values of $\ln(r^2\mathrm{dc})$ (panel (a)) and phase (panel (b)) for the different cases indicated in the legend.

In vivo experiment. Figure 5 shows the comparison between the $\ln(r^2\mathrm{dc})$ and the phase measured in the previously considered three-layered medium (layer 4c on top, clear layer in the middle, block 3 at the bottom), and *in vivo* on the temple area of the head of a female infant (born on the 34th week of gestation; age: 12 days; weight: 1.9 kg;

head circumference: 30 cm). The propagation of light in infant heads does not seem to be affected by the channeling effect that we see in the phantom measurements in the presence of the clear layer. The average phase slope we measured in 20 infants in the temple area is 12±2 deg/cm. The average phase slope measured in the phantoms containing the clear layer (30 cases) is 4±2 deg/cm. The intensity (panel (a)) is also much more attenuated by increasing the source-detector distance in the infant head than in the *in vitro* case with the clear layer.

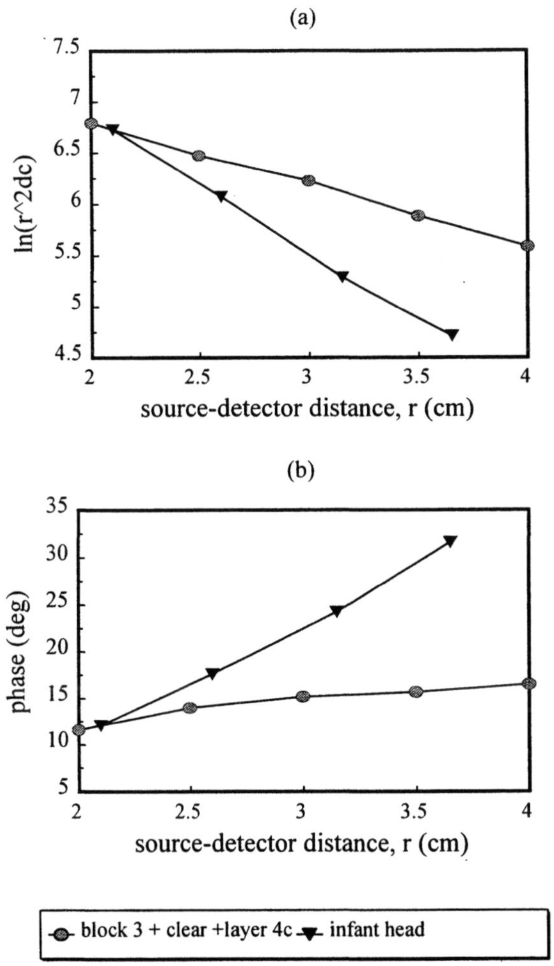

Fig. 5. Normalized experimental values of ln(r^2dc) (panel (a)) and phase (panel (b)) in a phantom containing a clear layer and *in vivo*. The optical properties of the infant head recovered from these data are $\mu_a = 0.087$ cm^{-1}, and $\mu_s' = 6.0$ cm^{-1} at 750 nm.

4. Discussion

For non-invasive optical studies of skeletal muscles, two-layered media constitute a model of the superficial skin/fat layer and the underlying muscle. Our study was based on the frequency-domain multi-distance method. The intensity modulation frequency was 110 MHz, the range of source-detector separations was 1.5-4.5 cm, and the range of optical properties was 0.071-0.15 cm^{-1} for μ_a, and 9.2-15.4 cm^{-1} for μ_s'. These experimental conditions must be kept in mind when considering our results reported in Section 3. In our experimental conditions, we found that a superficial layer up to 0.4 cm thick does not affect the accuracy of quantitative spectroscopy of the underlying material.

It has been suggested that three-layered media in a planar geometry may be taken as a rough model of the human head to predict the optical diffuse reflectance. In fact, the superficial layer, the clear layer, and the underlying block would mimic the scalp/skull, the cerebrospinal fluid, and the brain cortex, respectively.[2] However, our *in vivo* data on infant heads has not shown effects on the optical data similar to those observed in the three-layered media examined. Preliminary studies in spherical phantoms with the same layered structure (superficial layer on top, clear layer in the middle and spherical block in the bottom) show the same channeling effect observed in the planar geometry, so that a curved geometry does not significantly modify our results. An explanation for the failure of these simple models to describe the *in vivo* data on the infant head can be the fact that the CSF in the brain is not as geometrically homogeneous as the clear layer we made. A more irregular geometry, which includes regions where the clear layer is thinner, may reduce the light channeling effect. Furthermore, the CSF in the brain flows in a region which contains blood vessels, whose presence significantly contributes to light absorption.

5. Conclusion

<u>Two-layered media</u>:
Our results show that a superficial layer (i.e. skin, subcutaneous fat) less than about 0.4 cm thick does not affect the quantitative optical study of the underlying medium by the frequency-domain (110 MHz) multi-distance method.
<u>Three-layered media / infant head</u>:
We stress that our results do not demonstrate that the clear cerebrospinal fluid does not affect the diffuse reflectance measured on the head. However, our results indicate that to describe the propagation of light in the human head, one should consider more accurate models than a uniform clear layer sandwiched between two turbid media.

6. Acknowledgments

This work is supported by National Institutes of Health (NIH) grants RR03155, CA57032, and by Whitaker-NIH grant RR10966.

7. References

1. A. H. Hielscher, H. Liu, L. Wang, F. K. Tittel, B. Chance, and S. L. Jacques, "Determination of Blood Oxygenation in the Brain by Time Resolved Reflectance Spectroscopy (I): Influence of the Skin, Skull, and Meninges," Proc. SPIE **2136**, 15-25 (1994).
2. E. Okada, M. Firbank, M. Schweiger, S. R. Arridge, M. Cope, and D. T. Delpy, "Theoretical and Experimental Investigation of Near-Infrared Light Propagation in a Model of the Adult Head," Appl. Opt. **36**, 21-31 (1997).
3. A. Kienle, M. S. Patterson, N. Dögnitz-Utke, R. Bays, G. Wagnieres, H. van den Bergh "Investigation of Two-Layered Turbid Media with Steady-State, Frequency and Time Domain Reflectance," Proc. SPIE **3194**, 269-278 (1997).
4. M. A. Franceschini, D. Wallace, B. Barbieri, S. Fantini, W. W. Mantulin, S. Pratesi, G. P. Donzelli, and E. Gratton, "Optical Study of the Skeletal Muscle During Exercise with a Second Generation Frequency-Domain Tissue Oximeter," Proc. SPIE **2979**, 807-814 (1997).

Monitoring Brain Oxygenation Using Time-of-Flight Spectroscopy

Susan R. Hintz,[1] David A. Benaron,[1] Robert Robbins,[2] Joshua L. Duckworth,[3] Aileen L. Murphy,[3] John W. Price,[3] Frank W. H. Liu,[3] David K. Stevenson,[1] Wai-Fung Cheong[1,3]

[1]Biomedical Optics Group
Division of Neonatal and Developmental Medicine, Stanford School of Medicine
and the Hansen Experimental Physics Laboratory, Department of Physics
[2]Cardiovascular Surgery, Department of Surgery, Stanford School of Medicine
750 Welch Rd #315, Stanford University, Palo Alto, CA USA 94305
Phone 001.650.723.5711 Fax 001.650.725.8012, benaron@forsythe.stanford.edu

[3]Spectros Corporation, an NIH-supported engineering concern
4370 Alpine Road #203
Portola Valley, CA USA 94028
Phone 001.650.851.4040 Fax 001.650.851.4099

1. ABSTRACT

We measured oxygenation of the brain in animals during changes in oxygen delivery (hypotension, shock). We found that such changes were measurable, and detected one change before the problem was clinically noted, thus suggesting that such methods may allow for earlier intervention in the intensive care setting.

2. INTRODUCTION

We are developing real-time methods for obtaining data from biological tissues illuminated with light.[1,2,3] When the brain undergoes changes in oxygen delivery (during hypotension, shock, or activation such as thinking), the blood flow and the oxygenation of the blood in that part of the brain changes. This creates a signal that can be imaged and quantitated using time-resolved optical methods, which treat the transport of photons through tissue in a manner analogous to diffusion of particles though a medium. In this report, we discuss preliminary measurements of oxygenation in animal brain using our time-of-flight system.

3. METHODS

3.1 Device

For studying the diffusion of photons through tissue, we used a time-of-flight spectrophotometer, reported earlier.[4] Briefly, the system emits light from up to three laser diodes (785, 850, and 904 nm) into the subject tissue via 100 μm fiber optic cables, and then detects light entering a second fiber, placed at another location, as the scattered photons scatter and emerge from the sample. Total intensity is low, averaging just over 1 mW. The system is sensitive to light at losses of 90 dB, and thus can measure through about 5-6 cm of tissue, despite the low power used. Resolution of average photon path length is better than 500 μm. Time of flight curves are processed at several wavelengths and stored on disk. The advantage of this system over conventional spectroscopy is that the travel of photons trough tissue is analogous to a diffusion of heat. Diffusion-based equations[5,6] allow the effects of scattering and absorbance to be quantitatively separated. This time-of-flight system can use surface probes to monitor points on a subject's head, or multiple sensors may be collected into a headband to form images.[7] We have applied this clinically.[7,8,9,10]

3.2 Animal oxygenation monitoring

A piglet under study for the testing of bypass procedures was optically monitored. Optical sensors were placed bilaterally on the piglet's head. During the study, medical interventions occurred that resulted in

Reprinted from Advances in Optical Imaging & Photon Migration, 1998 Technical Digest, © 1998 Optical Society of America

hemodynamic changes in the animal, such as insertion of bypass cannulae into major vessels, initiation of heart/lung bypass, and alterations in the bypass flow rates. The mean time of flight and the rate of decay of the sample curve were processed by the time of flight software reported earlier. Oxygenation was calculated using diffusion based equations. The placement of the probes was as follows:

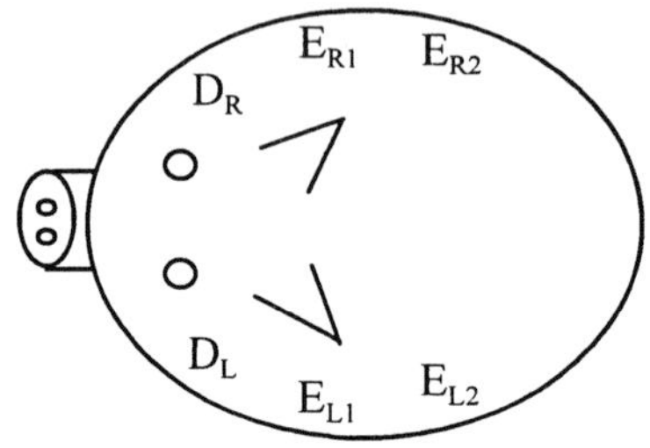

Figure 1. Arrangement of sensors on the head of the pig during cardiac surgery. Detectors (D_R and D_L, for right and left, respectively) were placed anteriorly and laterally, over the frontal/parietal region of the brain. Next, two emitters were placed on each side of the head, one pair at 1 cm (E_{R1} and E_{L1}) and one pair at 2 cm from each detector (E_{R2} and E_{L2}).

4. RESULTS

A plot of the oxygenation of the animal brain is shown in Figure 2, below. Briefly, at ***A***, the animal is intubated and stabilized. Oxygenation remains stable at 64.0 +/- 1.1% until ***B***, when the surgery begins. During the surgery, oxygenation falls slowly, averaging 62.7 +/- 2.3% until falling rapidly at clamping for initiation of bypass at ***C***, at which time saturation reaches a minimum of 45.6%. During initial heart-lung bypass, at high pump flows, oxygenation rises and peaks at 66.4% +/- 0.3%. By ***D***, when stable but low-flow (ischemic) bypass conditions are present, oxygenation stabilizes for nearly one hour at a low value of 49.3 +/- 1.5%. At ***E***, the pump flow is increased, resulting in rising oxygenation. However, despite increased flow, the TOFA measured values peak at ***F*** at 74.5 +/- 3.8%, and then begin to fall. Based upon the bypass settings, it was expected that the oxygenation at this time should be rising, or stable, but not falling. Therefore, this result was surprising. Although the source of this unexpected drop was not initially clear, it was soon discovered that one of the bypass cannulae had eroded through the wall of the aorta, resulting in massive internal hemorrhage and systemic hypotension. Resuscitation (including CPR at the time of a cardiac arrest) was initiated at approximately 10 minutes before ***G***. The heart rate was noted to return at time ***G***. This returning heart rate was also associated with a rising saturation from 46.1% to 55.2%, but then the fall in saturation continued, likely due a persistent hemorrhage, hypotension, and anemia, and bypass was terminated at ***H***, with a measured saturation at that time of 23.4%.

5. DISCUSSION

We have shown that the application of time-resolved methods to brain function is possible. During cardiac bypass, TOFA was able to yield saturation values near expected values for venous blood, and was able to non-invasively estimate changes in saturation that occurred, with the estimated changes proceeding in the direction and magnitude expected given the events occurring in the operating room. More significantly, TOFA detected a catastrophic event not otherwise initially clinically apparent. In our view, the most significant result of this experiment is the detection of a catastrophic event not immediately apparent to the medical team. The early detection of such an event during a human heart procedure would most likely have resulted in early initiation of treatment, and possibly avoided the death that ensued in this animal. Detection of a similar event during a human surgery would be compelling evidence of the utility of optical methods of brain monitoring in patients at high risk for brain injury due to ischemic or hypoxic conditions.

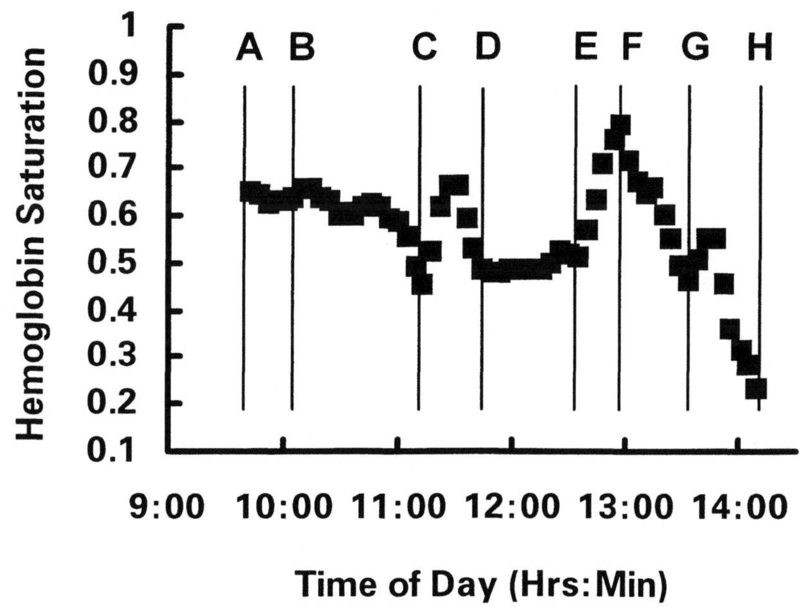

Figure 2. Oxygenation of a pig brain over time. Changes in oxygenation can be correlated to specific events in the animal's medical or surgical care. Of note, a drop in oxygenation, signaling the rupture of a major vessel, was detected by TOFA at ***F***, before it was detected by the surgeons.

Bedside availability may be critical for identifying existing or impending injuries during therapeutic windows that last only a few minutes or hours, and for making prognostic decisions that may affect care. Illness may preclude transport to an CT or MRI facility, as those patients requiring such imaging are most likely to be those too ill for safe transport. In addition, critical illnesses are by nature unstable and evolving, which further advocates continuous bedside monitoring, rather than intermittent imaging at a distant imaging center. This approach may therefore lead to development of clinically relevant optically-based devices that produce rapid, continuous, bedside measurements and/or images of oxygen sufficiency. Other groups have used optics to follow the changes in cerebral oxygenation in animals and in humans, and multiple clinical instruments are under development.[11]

6. REFERENCES

[1] D. A. Benaron, C. Contag, D. K. Stevenson, W.-F. Cheong. Tissue Optics. Science 276:2002-2003 (1997).

[2] D. A. Benaron, G. Muller, and B. Chance. A medical perspective at the threshold of clinical optical tomography. In: Medical Optical Tomography: functional imaging and monitoring, G. Muller, B. Chance, R. Alfano, eds. SPIE press, Bellingham WA, pp. 3-9 (1993).

[3] D. A. Benaron, C. Contag, P. Contag. Imaging Brain Structure and Function, Infection, and Gene Expression in the Body Using Light. Phil. Trans. Royal Society: Biological Sciences B 352: 755 - 761 (1997).

[4] D. A. Benaron, D. K. Stevenson. Optical time-of-flight and absorbance imaging of biologic media. Science 259:1463-1466 (1993).

[5] M. S. Patterson, B. Chance, B. C. Wilson. Time resolved reflectance and transmittance for the non-invasive measurement of tissue optical properties. Appl. Opt. 28:2331-2336 (1989).

[6] E. M. Sevick, B. Chance, J. Leigh, S. Nioka, M. Maris. Quantitation of time- and frequency resolved optical spectra for the determination of tissue oxygenation. Analyt. Biochem. 195:330-351 (1991).

[7] S. R. Hintz, D. A. Benaron, J. P. van Houten, J. L. Duckworth, F. W. H. Liu, S. Spilman, D. K. Stevenson, W.-F. Cheong. Stationary headband for clinical time-of-flight optical imaging at the bedside. Photochem. Photobiol., accepted pending revision (1997).

[8] J. P. van Houten, D. A. Benaron, S. Spilman, and D. K. Stevenson. Imaging brain injury using time-resolved near infrared light scanning. Pediatr. Res. 39, 470-476 (1996).

[9] D. A. Benaron, J. P. van Houten, W.-F. Cheong, E. L. Kermit, R. A. King. Early clinical results of time-of-flight optical tomography in a neonatal intensive care unit. SPIE 2389,582-596 (1995).

[10] W.-F. Cheong, J. P. van Houten, E. L. Kermit, T. R. Machold, D. K. Stevenson, D. A. Benaron DA. Pilot comparison of light-based optical tomography versus ultrasound for real-time imaging of neonatal intraventricular hemorrhage. Pediatric Res. 39:201A (1996).

[11] Devices are under development in our laboratory, as well as in the laboratories of E. Gratton (U. Illinois, Urbana, Illinois), B. Chance (U. Pennsylvania, Philadelphia, Pennsylvania), D. Delpy (U. College, London, England), M. Tamura and Y. Yamashita (Japan), Phillips Corporation (Eindhoven, Netherlands), and Siemens (Munich, Germany), among others.

Investigation of Cortical Spreading Depression in Rats by Near Infrared Spectroscopy: Scattering and oxygenation changes

Matthias Kohl, Ute Lindauer, Ulrich Dirnagl, Arno Villringer
Department of Neurology, Charité, Humboldt University,
Schumannstrasse 20/21, 10117 Berlin, Germany

Abstract

We used near-infrared spectroscopy to separate tissue scattering changes from changes in cerebral oxy- and deoxyhemoglobin and the redox state of Cytochrome-c-oxidase during cortical spreading depression (CSD) in barbiturate anesthetized rats. A separate term for the transport scattering coefficient (μ_s') was added in a modified Lambert-Beer equation where the differential pathlength factor (D_a) and its scattering equivalent (D_s) are included. During CSDs recorded through the skull of rats strong biphasic changes in μ_s' of about ± 0.1 mm^{-1} were observed that spread with a velocity of about 5 mm/min. It is shown that the DC potential recorded on the skull has a strong correlation with the scattering properties. This was found during CSDs as well as during the period of the anoxic cell depolarisation after sacrificing the animal.

Introduction

Near infrared (NIR) spectroscopy is widely used for the non-invasive monitoring of changes in oxy-and deoxyhemoglobin concentration (oxy-Hb, deoxy-Hb) in brain based on attenuation data [1, 2]. In addition, changes in light scattering have been observed in single neurons [3] and on the exposed brain [4] and might be linked to fast signals detected in humans [5]. However it has not been conclusively answered whether these scattering changes can be measured noninvasively through the skull as the separation from concurrent absorption changes (i.e. in chromophore concentration) proves difficult. Time resolved [6] and frequency domain methods [7-9] based on a diffusion equation analysis do provide this separation, though they are limited to homogeneous media and can formally not be applied to layered measurement geometries like the head.

Here we expand an established modified Lambert-Beer algorithm for the conversion of attenuation data into concentration changes by a scattering term. The differential pathlength factor was obtained by injecting a dye of known absorption spectrum. A simple diffusion equation approximation is included to derive the scattering equivalent of the differential pathlength factor.

CSD is a wave of synchronous neuronal excitation followed by sustained depolarisation of neuronal and glial cells. It is characterised by flattening of the EEG and by a deflection wave of the cortical direct current potential (DC), and is accompanied by an increase of regional cerebral blood flow and blood oxygenation [10]. The phenomenon of CSD has been associated with the pathophysiology of migraine and stroke [11, 12]. To test the method it is applied to measurements of the cortical spreading depression (CSD) in rat where it has been proposed that both pronounced changes in scattering and chromophore concentration occur.

Recently we have shown that strong scattering changes do occur during CSD and that they can be separated from changes in oxy-Hb, deoxy-Hb and the redox state of the intracellular enzyme Cytochrome-c-oxidase (Cyt-ox) [Kohl 98]. Here we discuss these results in more detail.

Modfied Lambert-Beer Equation

The wavelength dependence of the transport scattering coefficient $\mu_s'(\lambda)$ has been investigated in the NIR range. While the magnitude of μ_s' shows strong variations for different tissues, its spectral shape is rather similar. In *in vivo* studies Matcher et al. determined μ_s' by fitting the experimental time spread function to diffusion equation solutions and found μ_s' to be monotonically decreasing with wavelength with a slope of the order of 0.1 % per nm for various tissues [13]. Similar values were obtained in *in vitro* studies [14]. The assumption of this letter is that any change in μ_s' (λ) of the tissue does not alter its spectral form, i.e. $\Delta\mu_s'(\lambda) = \delta S \cdot \mu_{s0}'(\lambda)$ where δS is wavelength independent. This can be justified for NIR by Mie and Rayleigh scattering theory for the size and relative refractive index of the scattering components in tissue [15]. The attenuation changes $\Delta A(\lambda)$ measured at the light-source detector distance r can consequently be expressed in terms

OSA TOPS Vol. 21 Advances in Optical Imaging and Photon Migration, James G. Fujimoto and Michael S. Patterson (eds.)

of changes in the chromophore concentrations Δc_i (oxy-Hb, deoxy-Hb and the redox state of Cyt-ox.) and the fractional scattering change δS:

$$\Delta A(\lambda) = r \cdot \left(\sum_i [\varepsilon_i(\lambda) \cdot \Delta c_i] \cdot D_a(\lambda) + \delta S \cdot \mu_{s0}'(\lambda) \cdot D_s(\lambda) \right) \quad (1)$$

Here ε_i signifies the molar extinction coefficient of the ith chromophore [16]. $D_a = r^{-1} \cdot \partial A/\partial \mu_a$ and is called the differential pathlength factor as it corresponds to the ratio of the mean optical photon pathlength in the tissue and r [2]. Similarly, $D_s = r^{-1} \cdot \partial A/\partial \mu_s'$. Both D_a and D_s are functions of r and the tissue optical properties and therefore wavelength dependent and only approximately known. Knowledge of the wavelength dependence of $D_a(\lambda)$ and $D_s(\lambda)$ is crucial for a low crosstalk in the calculation of the chromophore and scattering changes. While there are experimental data for $D_a(\lambda)$ from time-of-flight measurements for various tissues [17] at large source-detector distances, there are no data for $D_s(\lambda)$ available.

Differential pathlength factor and its scattering equivalent

A diffusion theory analysis, however, allows a simple relationship between $D_s(\lambda)$ and $D_a(\lambda)$ to be derived. For a 'pencil beam' light source on a semi-infinite and homogenous halfspace characterised by μ_a and μ_s', the reflectance R can be written in the simplest approximation as

$$R(\mu_s', \mu_a, r) = \frac{1}{\mu_s'} \cdot \left(\frac{1}{\rho} + \mu_{eff} \right) \cdot \frac{\exp(-\mu_{eff} \cdot \rho)}{2\pi \cdot \rho^2} \quad , \qquad (2)$$

with $\rho = (r^2 + \mu_s'^{-2})^{1/2}$ and $\mu_{eff} = (3 \cdot \mu_a (\mu_a + \mu_s'))^{1/2}$ [6]. Analysing the partial derivatives of the attenuation A, defined as $A = \log(1/R \cdot F)$ where F is the surface area of the detector, reveals that for a given r D_s^{-1} is proportional to D_a:

$$D_s(\mu_s', \mu_a)^{-1} = \left(\frac{1}{r} \cdot \frac{\partial A}{\partial \mu_s'} \right)^{-1} \propto D_a(\mu_s', \mu_a) = \frac{1}{r} \cdot \frac{\partial A}{\partial \mu_a} \qquad (3)$$

This holds for the range of μ_a and μ_s' encompassing that of biological tissue ($1\ mm^{-1} < \mu_s' < 3\ mm^{-1}$, $0.005\ mm^{-1} < \mu_a < 0.03\ mm^{-1}$).

The product $D_a \cdot D_s$ has values of about 1/6 with variations of < 7 % (r = 5 mm) and < 1% (r = 20 mm) for the range $0.01\ mm^{-1} < \mu_a < 0.03\ mm^{-1}$ when μ_s' is held constant.

In the equation for the reflectance given above the Fresnel reflection at the boundary between medium and surrounding has been ignored. Different improved equations have been published that include so called mismatched boundary conditions [18]. The analysis of these solutions, however, confirms the simple relationship between D_a and D_s given in Eq. 3. The proportionality between D_a and D_s^{-1} breaks down when r is much smaller than 5 mm as the condition for the diffusion equation $r \cdot \mu_s' \ll 1$ is not fulfilled. In consequence, once the wavelength dependence of the differential pathlength factor $D_a(\lambda)$ is known, changes in the concentrations of oxy-Hb, deoxy-Hb, the redox state of Cyt-ox as well as in μ_s' can be calculated from Eq. 1 with a standard least square regression procedure.

Cortical spreading depression model and experimental setup

Male Wistar rats (250-300g) were anesthetized with sodium thiopental (Trapanal, Byk Gulden, Germany, initial dose 100mg/kg i.p., supplemental doses of 5mg i.p. as necessary), artifically ventilated and body temperature controlled. Arterial blood samples (A. / V. femoralis cannulated) were measured periodically. The CSD was elicited by topical application of a 150 mM KCl solution for 30 s at a small cranial window (dura mater removed) at the frontal cortex, and was monitored by an epidural calomel electrode (see Fig. 1) at the parietal cortex (DC potential). Details of the preparation are described elsewhere [10]. A standard CCD imaging spectrograph (Princeton Instruments) was used to measure attenuation changes through the intact skull for

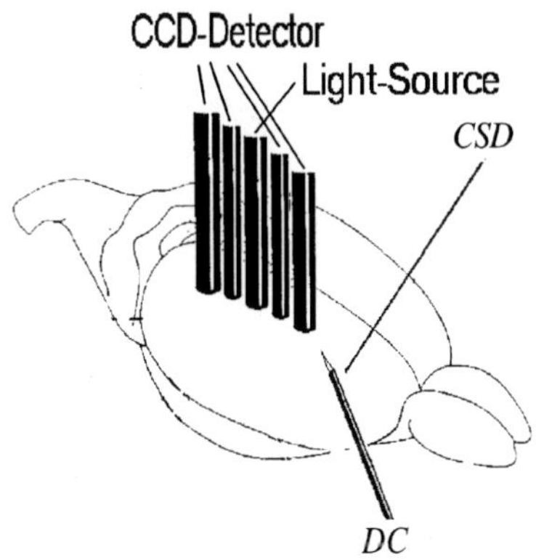

Fig. 1: Schematic drawing of the experimental setup on the rat's cortex. CSDs were elicited at a cranial window. The DC potential was monitored with a calomel electrods was monitored. 4 light collecting fibers were positioned in frontal and caudal direction with respect to the light delivering fiber (r = ± 2.5 and ± 5 mm).

wavelengths between 700 and 1000 nm with a sampling rate of 1 Hz. Four detector fibers (0.6 and 1 mm diameter) were positioned 2.5 and 5 mm in frontal and caudal direction with respect to the light delivering fiber (1 mm diameter). The fibers were stabilised by metal tubing that provided additional light shilding.

Results

Determination of the wavelength dependence of the differential pathlength factor

The wavelength dependence of the differential pathlength factor $D_a(\lambda)$ was determined by injecting a NIR dye intravenously. The measured attenuation change $\Delta A_{dye}(\lambda)$ was then devided by the absorption spectrum of the dye to give: $D_a(\lambda) \propto \Delta A_{dye}(\lambda)/\mu_{a,dye}(\lambda)$. The mean (n = 6) of the measured D_a versus λ relationship (see Fig. 2) is approximately flat for 800 - 900 nm and varies by about 15 % over the whole wavelength range with minima coinciding with the absorption bands of deoxy-Hb (760 nm) and water (970 nm).

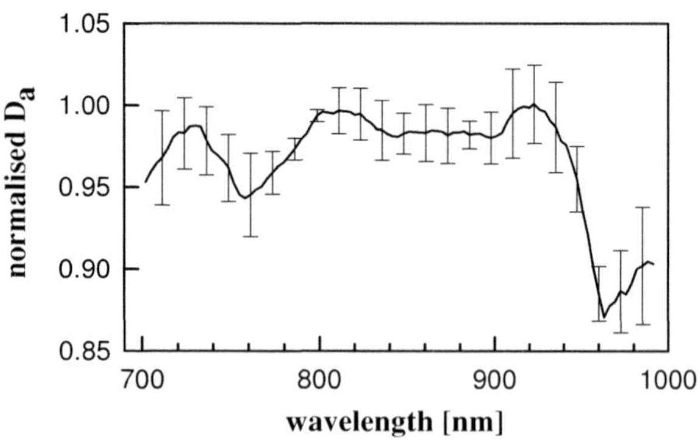

Fig. 2 Wavelength dependence of the differential pathlength factor $D_a(\lambda)$ measured on the skull of rats (n = 6) for r = 5 mm and normalised at 800 nm.The error bars give the standard error.

The chromophore and scattering changes were calculated from the attenuation data (700 - 920 nm) according to Eq. 1 incorporating the measured wavelength dependence of $D_a(\lambda)$ from Fig. 2 and $D_s(\lambda) = (6 \cdot D_a(\lambda))^{-1}$. As μ_a and μ_s' are only approximately known for tissue, exact values for D_a cannot be given. Based on Eq. 2 for r = 5 mm, an absolute value of D_a(800nm) = 3 (± 20%) can be estimated for $\mu_s' = 2$ mm^{-1} and 0.005 mm$^{-1} < \mu_a <$ 0.02 mm^{-1}. Therefore, all the calculated concentration and scattering changes discussed below have the same uncertainty of ± 20%. A decrease in $\mu_s'(\lambda)$ of 0.1% per nm was included.

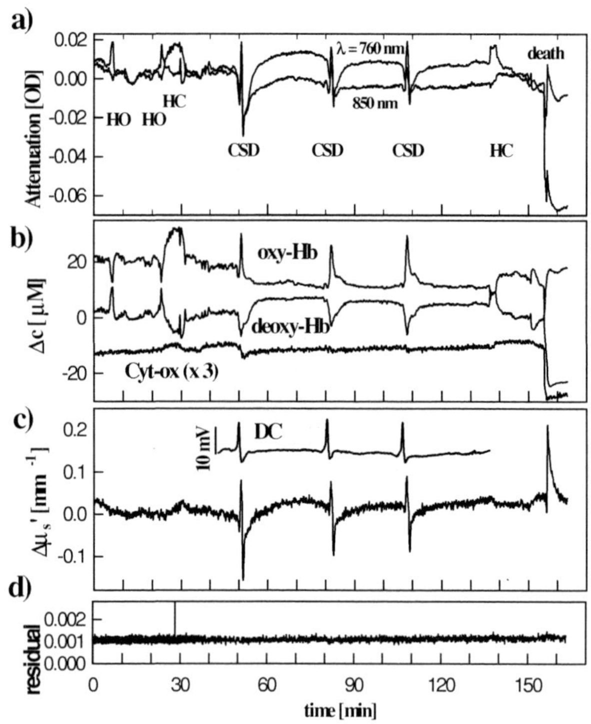

Fig. 3 Three cortical spreading depressions (CSD) recorded on a rat skull for a source-detector separation of 5 mm. a) Attenuation changes for λ = 760 nm and 850 nm. For the calculation of changes b) in oxy-Hb, deoxy-Hb and Cyt-ox and (c) in transport scattering coefficient at 800 nm ($\Delta\mu_s$') attenuation data from 700 to 920 nm were used. The DC potential was recorded with a calomel electrode. Data recorded during hypoxia (HO) and hypercapnia (HC) and the death are shown.

Scattering Changes during CSD

A typical example of the recorded attenuation data is shown in Fig. 3a. To test for possible crosstalk in the calculation of the chromophore and scattering changes, mild hypoxia (p_aO_2 minimum 70 mm Hg), hypercapnia (p_aCO_2 50-60 mmHg) and the period after killing the animal (i.v. injection of 1 ml of a 3 M KCl solution) were monitored. A part of the data is shown in Fig. 4a-c on an expanded scale. There are changes in oxy-Hb and deoxy-Hb (Fig. 3b, 4b) induced by the CSDs of about +15 μM and −7 μM, respectively. Only small changes in the Cyt-ox redox status were observed that are within the (crosstalk) errors left by the remaining uncertainties in $D_a(\lambda)$ and $D_s(\lambda)$. Strong biphasic changes in μ_s' with an initial increase of about 0.1 mm^{-1} followed by a steep decrease by approximately 0.2 mm^{-1} and a subsequent recovery to the baseline were calculated (Fig. 3c, 4c). The

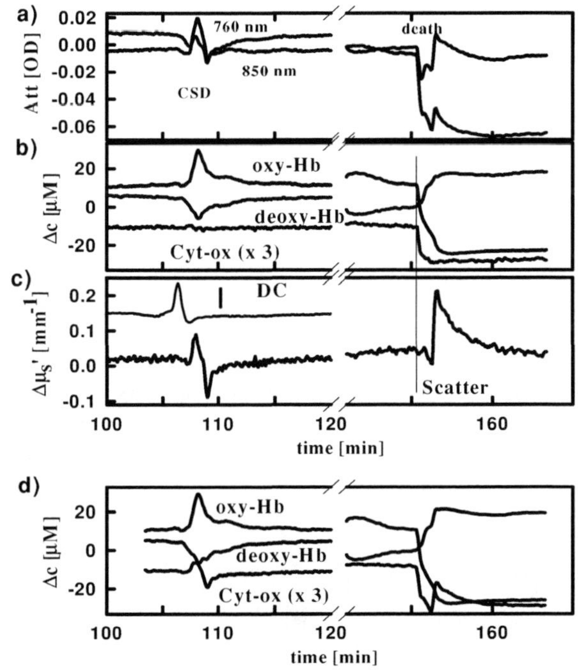

Fig. 4 Data of Figure 3 on an expanded scale for a CSD and the period after sacrificing the animal. Attenuation data in a), chromophore and scattering changes in b) and c). In d) the chromophore changes are shown when no scattering changes were included in the calculation.

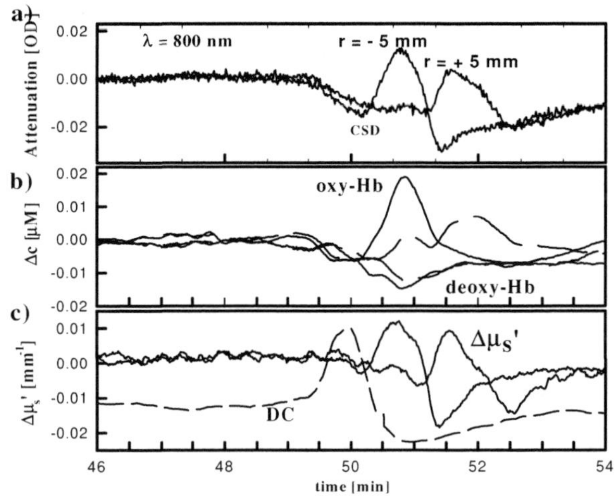

Fig.5 Spreading of the CSD recorded with two detectors at distance ± 5 mm with respect to the light delivering fiber. The different magnitude of the oxy-Hb and deoxy-Hb (b) signal might reflect the different vascularisation at the two detector positions. In c) the $\Delta\mu_s$' traces and the DC potential (dashed line) is shown. The data for the position - 5 mm were shown in Fig. 3 and 4.

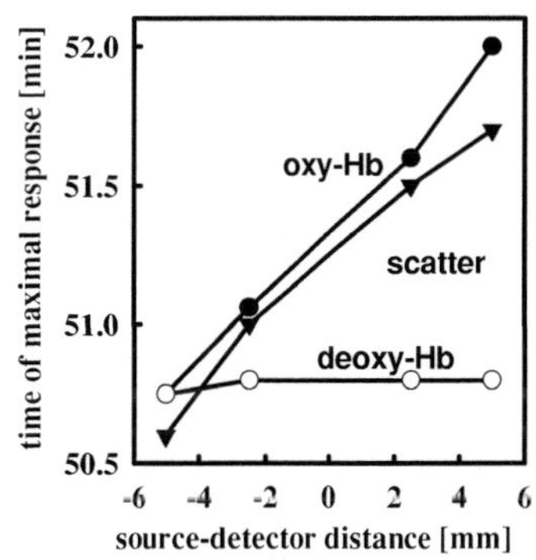

Fig. 6 Time of maximal oxy-Hb, deoxy-Hb and scattering signal during a CSD for the 4 source-detector positions.

biphasic time course of μ_s' is similar to that of the DC potential. During mild hypoxia and hypercapnia only tiny changes in scattering or Cyt-ox were observed. After killing the animal changes of approximately Δoxy-Hb = – 40 μM, Δdeoxy-Hb = + 18 μM and ΔCyt-ox = – 6 μM were calculated. μ_s' showed a rapid increase within 5 s by 0.20 mm^{-1}. This occurred about 1 min after the onset of the chromophore changes and was followed by a decrease to the baseline value with a decay constant of about 1.4 min when described by a simple exponential function.

When the scattering term was not included in the fitting procedure biphasic changes in the Cyt-ox concentration of about 5 mM amplitude were calculated during CSDs and changes of more than 7 mM after death (see Fig. 4d). The magnitude and the time course of this signal after cell death indicate that the attenuation data cannot be described by physiological chromophore changes only when the scattering term is omitted.

Spreading velocity of the CSD on the cortex

The spreading of the CSD was recorded with the four detector fibers that were positioned perpendicular to the CSD wave front. An example for 2 detectors is shown in Fig. 5. As only one light source was used the sampling volumes of the detectors overlap. Their centers are separated by 5 mm for the detectors positioned at ± 5 mm from the light source. Taking the time of the maxima in the scattering or the oxy-Hb traces (Fig. 6) a spreading velocity of the CSD of about 5 mm / min was obtained. While the scattering signal has about the same magnitude for the different detector positions, the oxy-Hb and deoxy-Hb signals are very different. This might be linked to variations in the vascularisation for the different sampling volumes.

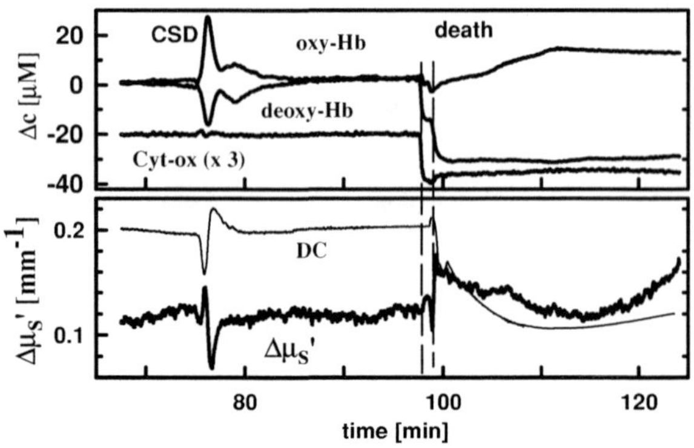

Fig. 7 Calculated changes in oxy-Hb, deoxy-Hb and Cyt-ox (a) and $\Delta\mu_s$' (b) during CSD and the period after sacrificing the animal. There is a clear correlation between $\Delta\mu_s$' and the DC potential. While this correlation is negative for the CSD, it is positve during the anoxid cell polarisation after death.

Correlation with the DC potential

Fig. 7 shows the oxy-Hb, deoxy-Hb, Cyt-ox and $\Delta\mu_s$' changes during a CSD and the period after sacrificing the rat. The DC potential was recorded throughout the whole experiment. There is a clear correlation between the DC potential and the scattering changes. While the DC and $\Delta\mu_s$' traces have opposite direction during CSD they have a positive correlation during anoxic cell depolarisation after death.

Discussion and Conclusion

Similar changes in chromophore concentrations and scattering were found in all 45 CSDs that were recorded on 10 rats. All CSDs showed biphasic variations in the transport scattering coefficient μ_s' between $\pm$ 0.05 and $\pm$ 0.15 mm^{-1}, an increase in Δoxy-Hb of 5 - 25 μM and a decrease in Δdeoxy-Hb of 3 to 20 μM. Within errors due to uncertainties in the differential pathlength factor (see Fig. 2) no changes in Cyt-ox were observed during CSDs.
The correlation with the DC potential rises more questions about the origin of the scattering signal. The DC potential is linked to the extracellular electrolyte concentration and has different sensitivities for different electrolyts like Na^+ and Ka^+. The DC versus μ_s' correlation is negative and positive for CSDs and cell death, respectively, and this might reflect the different nature of cell depolarisations.

Cortical spreading depression is a phenomenon that appears during pathophysiological conditions like stroke and migraine. Future will show whether scattering changes similar to those discussed here can be recorded in humans.

References

1. F. F. Jöbsis *Science* **198**, 1264 (1977).
2. D.T. Delpy, M. Cope, P. van der Zee, S. R. Arridge, S. Wray, J.S. Wyatt *Phys. Med. Biol.* **33,** 1433 (1988).
3. R. A. Stepnoski, A. LaPorta, F. Raccuia-Behling, G. E. Blonder, R. E. Slusher, D. Kleinfeld *Proc. Natl. Acad. Sci.* **88,** 9382 (1991).
4. D. Malonek, A. Grinvald *Science* **272**, 551 (1996).
5. G. Gratton, P. M. Corballis, E. Cho, M. Fabiani, D. C. Hood *Psychophysiol.* **32**, 505 (1995).
6. M.S. Patterson, B. Chance, B.C. Wilson *Appl. Opt.* **28**, 2331 (1989).
7. J. B. Fishkin, P. T. C. So, A. E. Cerussi, S. Fantini, M. A. Franceschini, E. Gratton *Appl. Opt.* **34**, 1143 (1995).
8. S. J. Masden, E. R. Anderson, R. C. Haskell, B. J. Tromberg *Opt. Lett.* **19**, 1934 (1994).
9. M. Kohl, R. Watson, M. Cope *Opt. Lett.* **21**, 1519 (1996).
10. T. Wolf, U. Lindauer, H. Obrig, J. Dreier, T. Back, A. Villringer, U. Dirnagl *J. Cereb. Blood Flow Metab.* **16**, 1100 (1996).
11. M. Nedergaard, A. J. Hansen, *J. Cereb. Blood Flow Metab.* **13**, 568 (1993).
12. M. Lauritzen *Brain* **117,** 199 (1994).
13. M. Kohl, U. Lindauer, U. Dirnagl, A. Villringer, Opt. Lett. *23*, April 1998.
14. S. J. Matcher, M. Cope, D. T. Delpy *Appl. Opt.* **36**, 386 (1997).
15. R. Simpson, J. Laufer, M. Kohl, M.E. Essenpreis, M. Cope, Proc SPIE 2979, 307-313 (1997).
16. R. Graff, J. G. Aarnoudse, J. R. Ziip, P.M. A. Sloot, F. F. M. de Mul, J. Greve, M.H. Koelink *Appl. Opt.* **31**, 1370 (1992).
17. S.J. Matcher, C.E. Elwell, C.E. Cooper, M. Cope, D.T. Delpy *Analyt. Biochem.* **227**, 54 (1995).
18. M. Essenpreis, C.E. Elwell, M. Cope, P. van der Zee, S.R. Arridge, D. T. Delpy *Appl. Opt.* **32**, 418 (1993).
19. A. Kienle and M.S. Patterson *J. Opt. Soc. Am.* A **14**, 246 (1997).

Development of Quantitative Imaging in Frequency-Domain Diffuse Optical Tomography for Breast Cancer Detection

Brian W. Pogue, Markus Testorf, Ulf L. Osterberg, Keith D. Paulsen
Thayer School of Engineering, Dartmouth College, Hanover, New Hampshire 03755
(E-mail: Pogue@Dartmouth.edu, tel: (603) 646-3861, fax (603) 646-3856)

Abstract

An automated system for near-infrared tomographic reconstruction imaging has been developed and is under investigation on phantoms and tissues. Calibration of measured data with the model-based image reconstruction approach is found to be critical but readily accomplished through automated parameter estimation using data collected in a homogenous reference medium. Results from phantom experiments also illustrate the importance of employing iterative (nonlinear) image reconstruction methods if quantitative values of the absorption coefficient are to be recovered. Initial studies on imaging human tissues will be discussed in this report.

Introduction

Near-infrared imaging of the breast to detect cancer tumors has been examined for almost 70 years (1, 2), with relatively poor clinical success (3, 4). However in recent years, the expansion of research in this area has grown significantly (5-7), and some of the problems associated with diaphanography or transillumination can partly be solved so as to improve the imaging resolution and detectability of tumors. This work centers on the development of a near infrared (NIR) breast cancer screening device using iterative reconstruction of the images for quantitative absorption imaging.

Some early studies on breast tissue imaging (8-12) have investigated the available contrast and concluded that the major difference between normal and carcinoma tissue is the increased vascularity and hence increased hemoglobin content. One study suggested a maximal contrast of 3 to 4 can be expected, based upon the increased absorption due to hemoglobin content (10). However these studies were based on estimates since the only precise measurements were obtained on excised tissues. A recent report by Tromberg et al. (13), has shown that NIR frequency-domain measurements performed in vivo on palpable breast cancers demonstrate an increased total hemoglobin and oxy-hemoglobin content over that of normal tissue by a factor of 2 -3, with no other consistent changes in scattering coefficient. Based upon these measurements it will be important to have a imaging device which can quantitatively image blood volume and perhaps hemoglobin oxygenation status in order to provide clinically useful results.

There have been a number of attempts to provide improved inverse imaging algorithms using either analytical projection (14), backprojection (15), analytic perturbation theory (16-18) or numerically based iterative methods (19-23). The latter approach is thought to provide the best potential for diffusive optical imaging in heterogeneous tissue, since it is the only method which can correct for the non-linear relationship between measured optical quantities and tissue optical properties. Iterative reconstruction methods

OSA TOPS Vol. 21 Advances in Optical Imaging and Photon Migration, James G. Fujimoto and Michael S. Patterson (eds.)

should be able to provide quantitatively accurate results, however they require longer computation times, and a relatively good initial guess of the image (24).

This work examines a prototype imaging system for optical mammography and demonstrates its quantitative calibration in tissue phantoms. The difference in quantitation between non-iterative and iterative imaging using tomographically obtained data will be examined. In addition, the potential benefits of increased sensitivity in slab-geometry tomography imaging, will be illustrated and contrasted with circular-geometry tomography and slab-geometry projection imaging. These basic results are then discussed within the framework of developing an optimized breast cancer imaging system.

Apparatus

A tomographic imaging system based upon frequency domain photon migration principles has been constructed and is described in detail in reference 25. The imaging array, shown in Figure 1, consists of 16 source fiber bundles and 16 detector fiber bundles arranged in a circle. A single diode laser source (800 nm wavelength with 100 mW power) is multiplexed into the source fibers with a translation stage, and the detector fiber bundles are multiplexed onto the photomultiplier tube in a similar manner. Image reconstruction is based upon a finite element diffusion theory forward calculation, with an iterative Newton-Raphson solver using Levenberg-Marquardt regularization (21,22) along with total variation minimization.

Figure 1. Photograph of the imaging array with 6 mm diameter fiber bundles arranged in a circle, with linear translation stages used to multiplex the source and detectors fiber bundles.

Results

The quantitative imaging ability of our system has been investigated using an early version of the instrument and is based on experiments in homogeneous and heterogeneous phantoms using Intralipid and India ink solutions mixed in an 86 mm diameter cylinder. A fitting algorithm has been developed to automatically fit the measurements of AC intensity and phase to a calculation with no heterogeneity in the optical property coefficients. This pre-fitting is essential in order to image tissue volumes where the average absorption and diffusion coefficients are not known apriori.

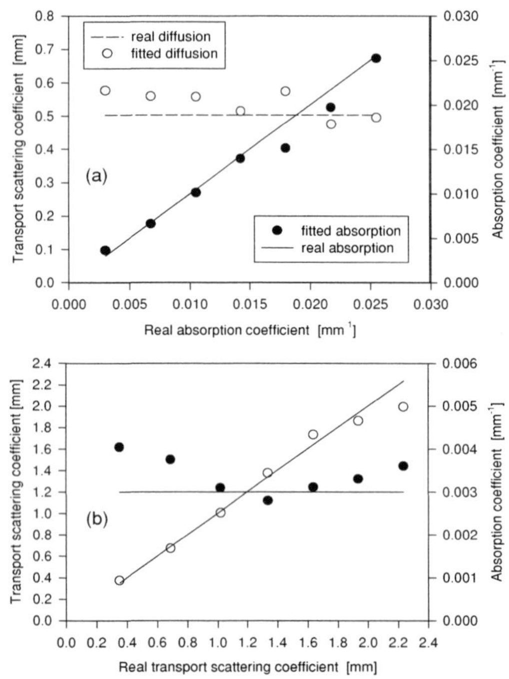

Figure 2. Calculations of absorption coefficient **(top)** and transport scattering coefficient **(bottom)** obtained from fitting a homogeneous calculation to data measured from homogeneous phantoms using an 86 mm diameter cylinder and Intralipid solutions.

Measurements in homogeneous phantoms with 7 different absorption coefficients were obtained and the fitted optical properties are shown in Figure 2 (top). A similar set of phantoms with increasing scattering coefficients were measured and fitted with the results shown in Figure 2 (bottom). On the same figures, the theoretical values of absorption and scattering are plotted, based upon the known scattering coefficient of Intralipid and our previously measured values for the absorption coefficient of our ink diluted mixtures.

Circular Phantom Studies

Heterogeneous phantoms with an absorbing object in the center were imaged, and a typical reconstruction is shown in Figure 3. The central absorbing object was a 25 mm mylar cylinder which could be filled with arbitrary concentrations of Intralipid and ink solution to vary the optical properties. The object's scattering coefficient was the same as the background, and the reconstructed image exhibits a nearly uniform (+/- 10%) diffusion property recovery. The absorption coefficient of the inhomogeneity was varied and images reconstructed for each sample, with the central peak value plotted in Figure 4, along with the real value. There is reasonable quantitative agreement to within 10-20% between the reconstructed and the actual quantities over a wide range of values, which are believed to be relevant to breast tissue imaging.

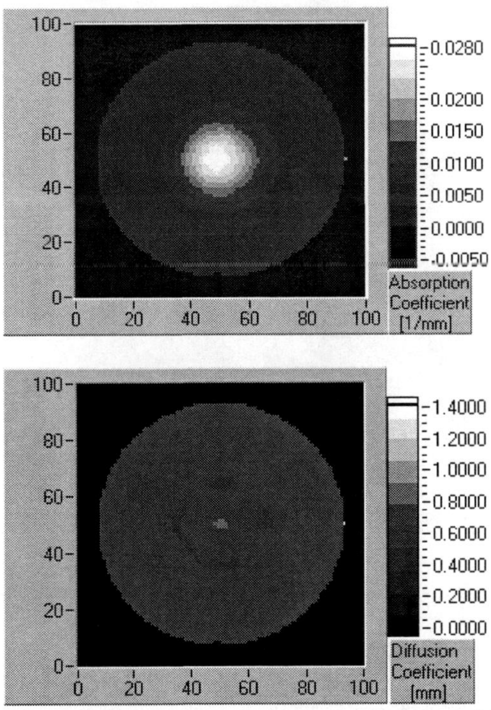

Figure 3. Reconstrution of a single absorbing object in the center of a tissue-simulating Intralipid phantom **(top)**. The diffusion image **(bottom)** is simultaneously reconstructed and for this test phantom, does not show any heterogeneity. The real object was 25 mm in diam and $\mu_a = 0.028\ mm^{-1}$ with μ_s' equal to the background.

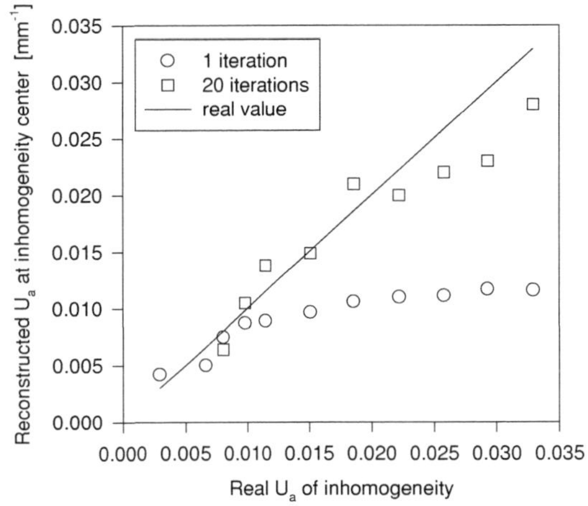

Figure 4. Reconstructed absorption coefficient at the center of the heterogeneity in the image, (as shown in Figure 2(top)), plotted against the actual absorption coefficient value. The values from 1 iteration and 20 iterations are included to illustrate the non-linearity in the inverse problem. Quantitatively accurate values can be approached after 5-20 iterations. The background optical properties in the rest of the phantom were $\mu_a = 0.0062\ mm^{-1}$ and $\mu_s' = 0.50\ mm^{-1}$.

Slab Phantom Studies

Currently, a slab imaging geometry is being investigated for mammographic imaging as well as the circular array approach as shown in Figure 4. The benefit of the slab (or compressed breast tissue) configuration is that a stronger light signal can be measured since less tissue is traversed allowing the photon path projections through the tissue to be narrower which potentially improves the resolution of the imaging modality. An image reconstruction can still be performed with slab geometry data, although it is not clear to what extent the lack of radial-symmetric projections through the tissue will affect the resulting image quality.

A simple test of the compressed slab geometry has been completed, using the same single object in the center of the phantom, as was imaged in Figure 3. The resulting reconstruction is shown in Figure 6. The image array also consisted of the same 16 sources and 16 detectors, but positioned in an alternating pattern as indicated in Figure 5. Some artifacts in the image appear in the upper surface of the absorption image, likely due to fiber positioning errors.

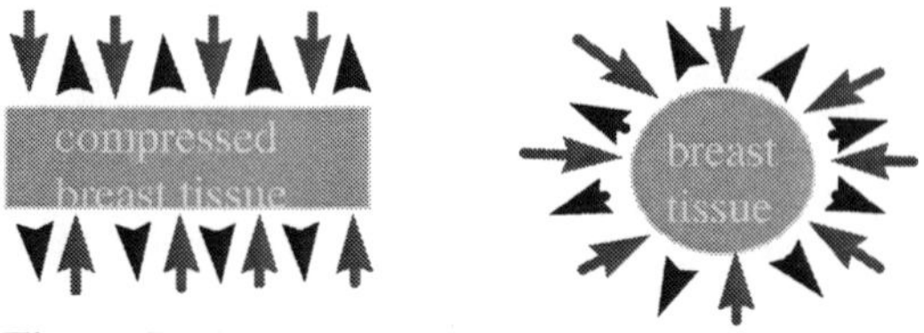

Figure 5. Comparison of two source-detector fiber array geometries which are being investigated for breast tissue imaging. The **left** geometry is referred to as a slab, while the **right** geometry is the circular. Source fiber locations are indicated by an inward facing arrow, and detector fiber locations with an outward facing arrow. In these figures 8 sources and 8 detectors are shown for simplicity, whereas 16 sources and 16 detector locations were used in the experimental work.

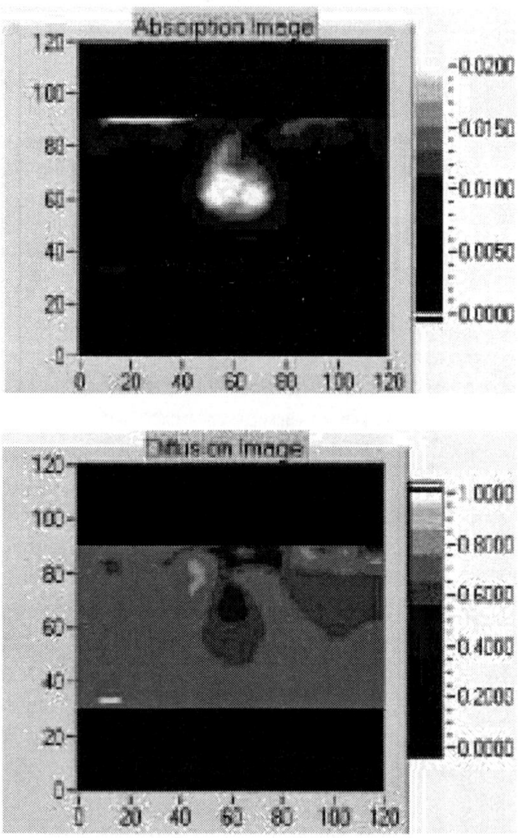

Figure 6. The reconstructed absorption **(top)** and diffusion **(bottom)** coefficient images, for a single absorbing heterogeneity within a 6 cm tissue-simulating slab., with background optical properties of $\mu_a = 0.0062$ mm^{-1} and D = 0.66 mm.

<u>Initial In Vivo Measurements</u>

The current imaging system is configured for mammographic imaging and is in the final stages of development prior to initiation of a clinical trial. As a first in vivo test the system was used to image through an adult human forearm to demonstrate its application to living tissue. The fiber array was located half way up the forearm of a volunteer, with the individual fibers in direct contact with the tissue. An imaging array consisting of 8 sources and 8 detectors was used here, (reduced from 16 x 16 due to size constraints) yielding 64 optical measurements of intensity and phase. The data acquisition time was 4 minutes, and the image reconstruction time was 4.5 minutes (for 3 itertations).

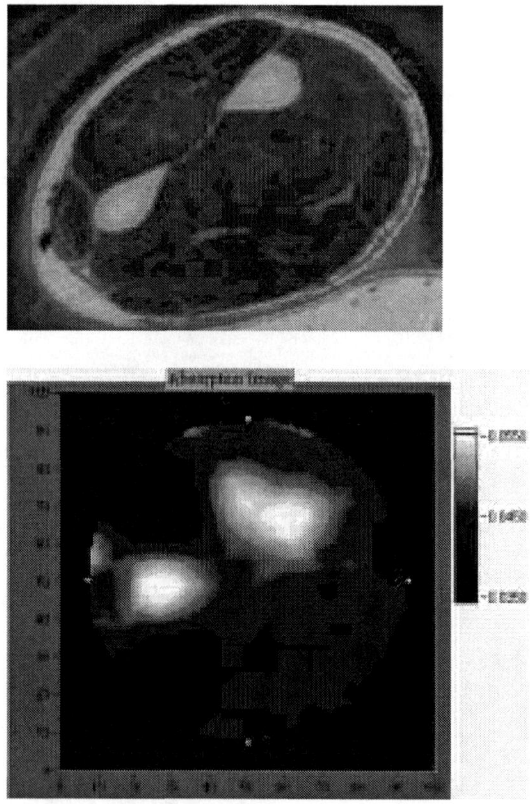

Figure 7. Initial in vivo image of an adult human arm is shown in this figure. The **top** image is a reference figure to show the geometry and orientation of the arm, while the **bottom** image is a reconstruction of the absorption coefficient image from the optical measurements through the arm. The locations of the two bright regions (high absorption) correlate well with the position of the two bones.

The top image in Figure 7 shows the relative geometry of the arm, while the reconstructed optical property image is shown in the bottom portion of the same figure. The presence of the bones is the dominant feature in this image. At the present time it is not possible to evaluate whether the image is quantitatively correct since the concentrations of hemoglobin, myoglobin, and the absorption coefficients of bone are not well established. It is also worth noting that the blood content is significantly higher in muscle than in breast tissue causing more attenuation of the light, and the contrast changes seen here between bone and muscle should not be important for breast imaging. Nevertheless, this initial in vivo experiment supports our hypothesis that non-invasive images can be obtained using near-infrared light with relatively good spatial resolution.

Discussion

The present work is focused on developing an automated tomography scanner for breast cancer detection and imaging. This application has been the focus of numerous studies, yet there are few reports of an automated data acquisition system and almost none which are coupled with iterative reconstruction imaging methods. The preliminary measurements shown here indicate that iteration of the images is essential to achieve quantitatively accurate heterogeneous images.

The potential for slab-geometry iterative reconstruction appears promising based on the pilot results presented here. However, a few artifacts in the reconstructed image need to be eliminated, and the system needs to be tested further with multiple objects at different locations and contrast levels.

The main advantage of this type of imaging is the realization of quantitatively accurate absorption images which will allow functional imaging of hemoglobin concentration. The current system illuminates with 800 nm light which is the isobestic point for oxy- and deoxy-hemoglobin. Future work will focus on the development of a multi-wavelength system for quantitative imaging of hemoglobin oxygen saturation imaging of breast tissue. Early measurements on human breast tissues are expected in the near future.

Acknowledgements

The authors would like to acknowledge the developmental work of Dr. Huabei Jiang and David Rinehart, as well as the advise of Dr. Michael Patterson in the course of this work. This project is supported by the NIH grant RO1CA69544 awarded by the National Cancer Institute.

References

1 Cutler, M., Surg. Gyn. Obst. *48:* 721-729, 1929.

2 Cartwright, C. H., J. Opt. Soc. Am. *20:* 81-84, 1930.

3 Bartrum, R. J., & Crow, H. C., Am. J. Roentg. *142:* 409-414, 1984.

4. Alveryd, A., Andersson, I., Aspegren, K., Balldin, G., Bjurstam, N., Edstrom, G., Fagerberg, G., Glas, U., Jarlman, O., Larsson, S.A., et al, Cancer; Diag. Treat. Res. *65:* 1671-1677, 1990.

5. Advances in Optical Imaging and Photon Migration. *In:* Opt. Soc. Am., Orlando, FL, 1994.

6. Medical Optical Tomography: functional imaging and monitoring. Washington: SPIE Publ., 1993.

7. Advances in Optical Imaging and Photon Migration. *In:* Opt. Soc. Am. Trends in Optics and Photonics, Washington, DC, 1996.

8. Watmough, D. J., Radiol. *147:* 89-92, 1982.

9. Navarro, G. A., Profio, A. E., Med. Phys. *15:* 181-187, 1988.

10 Profio, A. E., Navarro, G. A., Med. Phys. *16:* 60-65, 1989.

11. Peters, V. G., Wyman, D. R., Patterson, M. S., Frank G. L., , Phys. Med. Biol. *35:* 1317-1334, 1990.

12. Troy, T. L., Page, D. L., Sevick-Muraca, E. M., J. Biomed. Opt. *1:* 342-355, 1996.

13. Tromberg, B. J., Coquoz, O., Fishkin, J. B., Pham, T., Anderson, E. R., Butler, J., Cahn, M., Gross, J. D., Venugopalan, V., Pham, D., Phil. Trans. R. Soc. Lond. B. *352:* 661-668, 1997.

14. Fantini, S., Francecshini, M. A., Gaida, G., Gratton, E., Jess, H., Mantulin, W. W., Moesta, K. T., Schlag, P. M., Kaschke, M, Med. Phys. *23:* 149-157, 1996.

15. Walker, S., Fantini, S., Gratton, E., Appl. Opt. *36:* 170-179, 1997.

16. Kaltenbach, J. M., Kaschke, M.. *In:* G. Muller (ed.) Medical Optical Tomography: Functional Imaging and Monitoring. Bellingham, WA: SPIE Publ., 1993.
17. Barbour, R. L., Graber, H. L., Wang, Y., Chang, J. H., Aronson, R. *In:* G. Muller (ed.) Medical Optical Tomography: Functional Imaging and Monitoring, pp. 87-120. Bellingham, WA: SPIE Publ., 1993.
18. O'Leary, M. A., Boas, D. A., Chance, B., Yodh, A. G., Opt. Lett. *20:* 426-428, 1995.
19. Arridge, S. R., van der Zee, P., Cope, M., Delpy, D. T, Proc. SPIE. *1431:* 204-215, 1991.
20. Pogue, B. W., Patterson, M. S., Jiang, H., Paulsen, K. D., Phys. Med. Biol. *40:* 1709-1729, 1995.
21. Jiang, H., Paulsen K. D., Osterberg, U. L., Pogue, B. W., Patterson, M. S., J. Opt. Soc. Am. A. *13:* 253-266, 1996.
22. Paulsen, K. D., Jiang, H., Appl. Opt. *35:* 3447-3458, 1996.
23. Arridge, S. R., Schweiger, M, Phil. Trans. R. Soc. Lond. B. *352:* 717-726, 1997.
24. Pogue, B. W., Patterson, M. S., Farrell, T. J, Proc. SPIE. *2389:* 328-339, 1995.
25. Pogue, B. W., Testorf, M., McBride, T., Osterberg, U., Paulsen, K. D., Optics Express 1(13) 391-403.

Development and evaluation of the IRIS–OPTIscanner, a general–purpose optical tomographic imaging system

Randall L. Barbour[1,2], Randy Andronica[3], Quan Sha[1], Harry L. Graber[1] and Ira Soller[3]

[1]Department of Pathology, SUNY Health Science Center at Brooklyn

[2]Department of Electrical Engineering, Polytechnic University

[3]Scientific and Medical Instrumentation Center, SUNY Health Science Center at Brooklyn

Abstract

We have constructed a general purpose optical tomographic imaging device capable of evaluating a variety of body appendages. A novel functional feature of the instrument is its use of a geometrically *adaptive* scanning head, in the form of a mechanical iris, about which are attached optical fibers positioned in a circular array. By adjusting the pass–through diameter of the iris, optical fibers can be brought into gentle contact with a tissue target. This serves to improve optical coupling while simultaneously stabilizing the tissue against motion artifacts. In addition, in many instances it also permits gentle deformation of the tissue to a simple circular geometry which is favorable for data analysis. The instrument also employs a rotating filter array that serves to extend the dynamic range of measurement to $>10^{12}$ while retaining excellent precision (CV = 0.2–0.3%). Reconstruction results of laboratory phantoms and physiological imaging studies of the human forearm are presented.

Introduction

In recent years, the possibility of acquiring tomographic images of thick tissue structures from measurements performed in the NIR region has caught the attention of many [1,2]. Fueling this interest is the recognition of the significant cost, portability and sensitivity advantages that NIR measurements offer in comparison to established imaging methods (*e.g.*, MRI, CT, SPECT, PET). Until now, much of the development effort has been focused on algorithm refinement. Much progress appears to have been made, at least with analysis of computed data and measurements with simply structured laboratory phantoms [3–6]. Lagging has been serious attempts to develop practical clinical systems. Extension of measurement strategies suitable for evaluation of solid phantoms to clinical situations is not straightforward.

It is evident that tissue structures come in many geometries and are prone to motion artifacts. Accurate knowledge of external tissue geometry at the time of measurement may prove important, as many of the reconstruction methods currently proposed (*i.e.*, perturbation methods) require some form of comparative analysis to a defined reference state. It is well known that these methods can produce unstable solutions should the initial guess be severely inaccurate, which certainly could be the case in the absence of accurate knowledge of tissue geometry. This dilemma may prove problematic, as the external geometry of tissue is basically arbitrary and is not conveniently measured. It follows that it would be useful to adopt measurement strategies that obviate the criticality of prior knowledge of tissue geometry.

One approach would be to immerse the target tissue in an index–matching scattering fluid, thereby effectively extending the external boundary of the tissue. While potentially practical, this strategy effects a non–contact measurement, rendering it susceptible to motion artifacts. Attention to the latter is almost certainly necessary given that tissue attenuates light exponentially. Beyond this, the presence of an intervening scattering medium will certainly add to the measured attenuation and further reduce available resolution.

An alternative methodology would be to perform a contact measurement. While this likely would enable collection of improved signal levels it would not by itself necessarily resolve the geometry problem nor even the problem of motion artifacts. In principal, both could be simultaneously addressed by confining tissue to a fixed apparatus. This also it not without its limitations, given the range of tissue sizes and desirability of preserving a large view angle. The approach we have pursued has been to consider a design that would serve to: i. mechanically stabilize tissue, ii. bring optical fibers into direct tissue contact, iii. be geometrically *adaptable* to accommodate a range of tissue sizes. To this end we have incorporated a mechanical iris into the scanner head, about which optical fibers are attached in a circular array. Adjustment of the pass–through diameter of the iris causes the translation of the fibers in a radial direction. The device, shown in Figure 1, presently incorporates 18 fiber bundles, each serving to transmit and receive light. The current design allows for continuous adjustment of the iris opening over a diameter range of 3–15 cm. Other components of the system include a servo–controlled input multiplexer, an adjustable filter array that enables measurements over a large dynamic range, and a output coupler that couples the detector fibers to a CCD camera.

Figure 1 Photograph of iris and 18 attached sending/receiving optical fiber bundles.

Apart from the above considerations, the iris design would appear well–suited for optical measurements of tissue, as the unit also is capable of gently compressing tissue, causing it to assume a circular geometry. While skeletal features can restrict this capability, in certain cases we have found that the iris design readily permits examination of many features of the arms, the legs, and, the breast. In the breast case, the distendability of the tissue should allow for even greater compression should the need arise. Compared to planar compression techniques, radial compression has the added advantage of preserving view angle. The axial measurement geometry of the iris readily permits extension in the direction normal to the measurement plane. This is easily accomplished by arranging several iris units in parallel to form a stack, each of which whose openings is independently adjustable. This allows for specific contouring to different tissue sizes, facilitating interrogation of 3–D volumes.

Results

I. Calibration and Stability Studies

Figure 2 shows results of measurements determining the relative coupling efficiency of each optical fiber bundle. This measurement determines the efficiency with which light propagates through the system, normalized to the mean value of all the fibers.

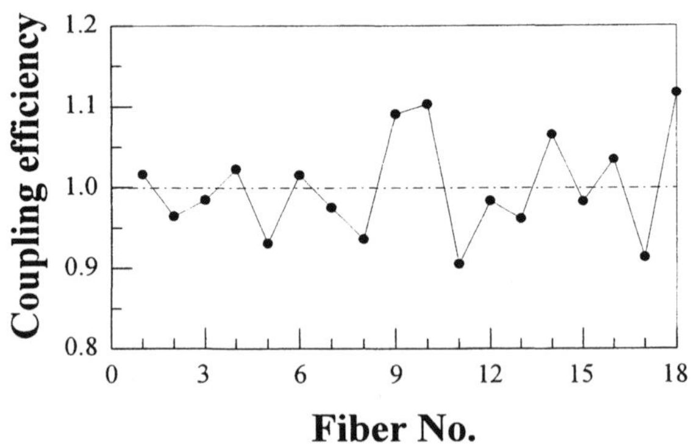

Figure 2 Optical fiber coupling efficiency.

Basically, light from a laser diode (780 nm) is directed to the input end of an 1–mm–diameter fiber that terminates at the opening of the iris. Light exiting from a target is collected by a total of 18 fiber bundles positioned at 20° intervals around the target. The collected light is then delivered to an adjustable filter assembly which contains a multi–segmented filter wheel designed to enable complete 360° simultaneous measurement from a target. Light

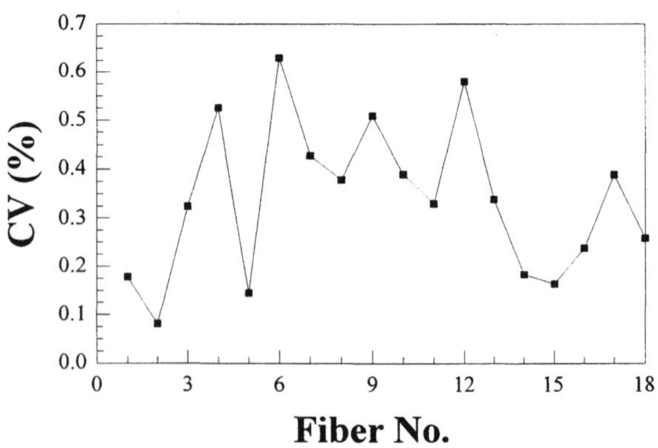

Figure 3 Measurement precision.

passing through the filters is collected by receiving bundles and directed to a CCD camera. Presently, the latter is accomplished by optical coupling. Use of a fiber taper is currently being incorporated.

Determination of the precision of measurement for each fiber bundle is shown in Figure 3. It can be seen that the coefficient of variation (CV) of measurement varies between <0.1–0.6%, with an average value of approximately 0.3% (n=5). Data were collected from a 9–cm–diameter delrin rod.

II. Tomographic imaging studies with laboratory phantoms.

Tomographic measurements were performed on three different laboratory phantoms: a homogeneous solid delrin rod measuring 9 cm in diameter, and two different arrangements of 2–mm–diameter black polyethylene tubing surrounding a brass rods. In one case two rods were present, four in the other. In the latter two cases, the rod assemblies were placed inside a latex laboratory glove filled with 1% Intralipid®. This was used to mimic the deformability of tissue. The assembly was then placed inside the iris, and the iris diameter adjusted until all fibers were in gentle contact (final dia. approx. 7 cm). Reconstructed images of these phantoms are shown in Figures 4–6. Data analysis was based on evaluation of an 18×18 data set (324 source–detector pairs) using a previously described iterative backprojection algorithm for perturbations in the absorption coefficient [7]. Figure 4 shows the image quality achieved from the delrin rod. The result is

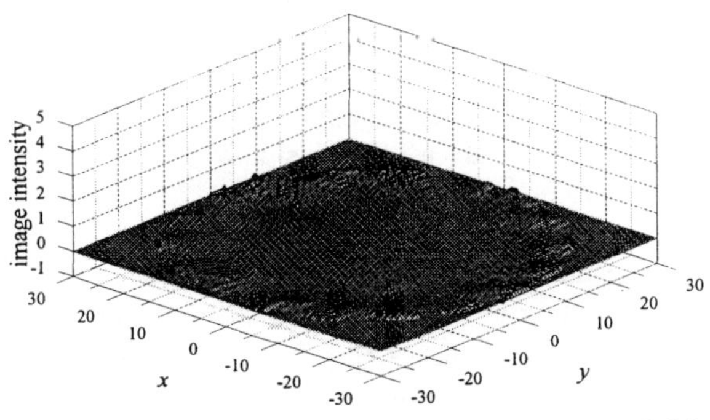

Figure 4 Reconstructed image of homogeneous delrin rod, relative to itself (*i.e.*, noise image).

essentially flat throughout the plane, with the exception of a boundary artifact associated with the source location. Data in Figures 5–6 are images of the laboratory glove containing 2 and 4 inclusions. With the exception of failing to resolve the two rods

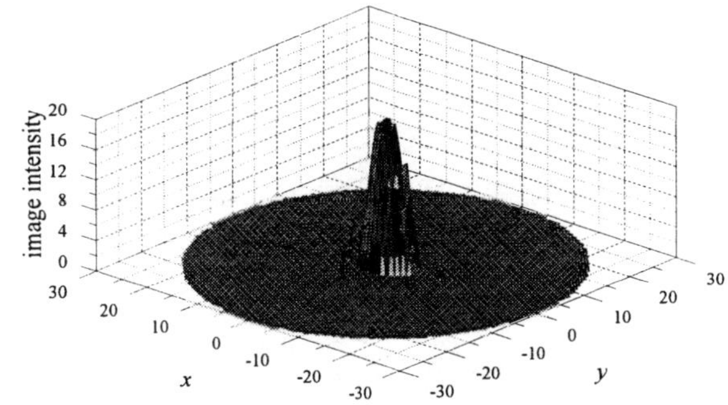

Figure 5 Reconstructed image of 2–rod phantom, relative to homogeneous reference medium.

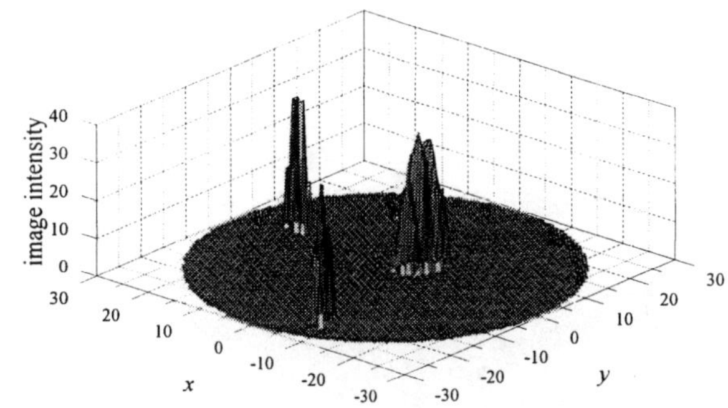

Figure 6 Reconstructed image of 4–rod phantom, relative to homogeneous reference medium.

positioned within 1.5 cm of each other (central region), the quality of the reconstructions are excellent. In both instances the solutions obtained were essentially free of any artifacts (positivity constraint used), and peak location had a maximum $x + y$ error of < 2 mm. In the case of the two rods located near the center, the position considered was the midpoint between them.

III. Physiological Imaging Studies of the Human Forearm.

Our principal objectives for these early studies was to ascertain the sensitivity and repeatability of measurement given a simple physiological perturbation. Two different protocols were followed, both involving tomographic measurements on the mid–forearm of a volunteer. In the first case we measured the difference in signal strengths obtained during a finger–flexor study. For each source position, measurements were made with the hand in the open and closed positions. The difference signals were stored and subsequently analyzed. Two separate sets of measurements were made and the reconstructed images are shown in Figures 7 and 8.

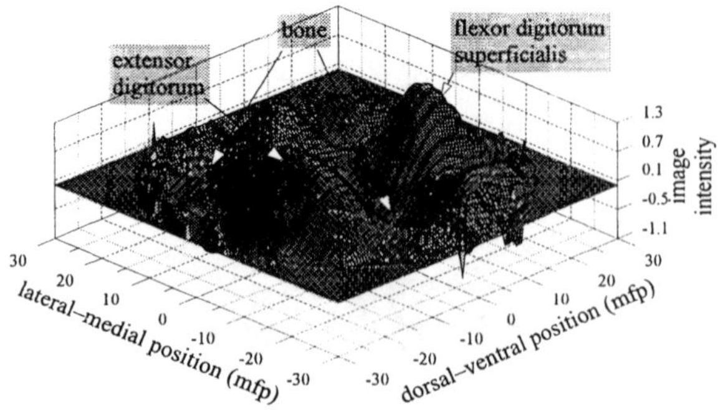

Figure 7 Image reconstructed from first set of finger–flexor measurements (target = forearm when fingers are flexed, reference = forearm when fingers are extended), with tentative identification of image peaks with anatomical structures.

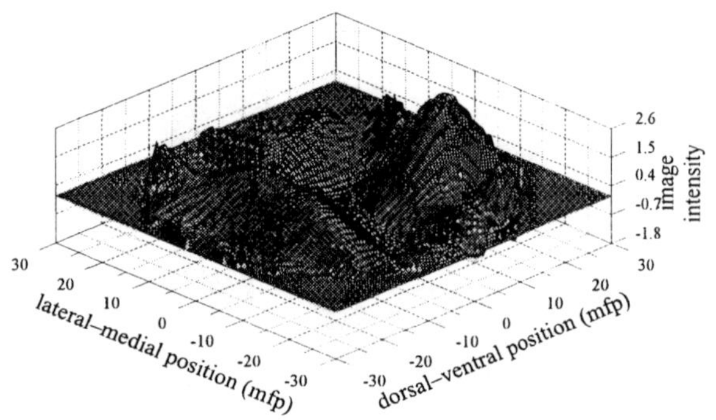

Figure 8 Image reconstructed from second set of finger–flexor measurements.

In the second experiment, a pressure cuff was place on the upper arm and inflated to 20 and 50 mm Hg. For these studies an initial set of measurements were performed without the cuff inflated, immediately followed by measurements with the indicated pressure. As with the finger–flexor study, the measured difference signal was evaluated. The reconstructed images are shown in Figures 9 and 10.

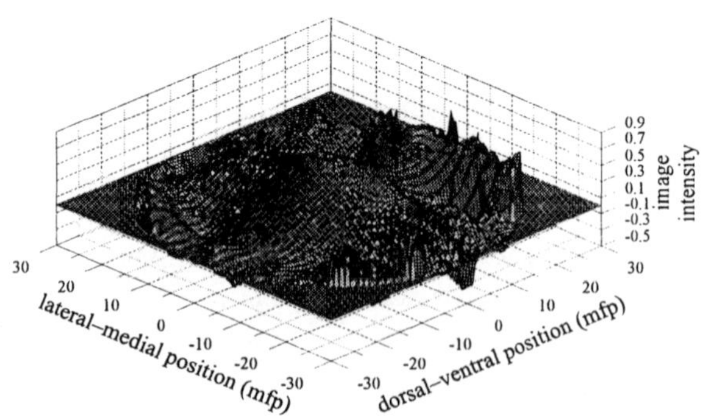

Figure 9 Image reconstructed from first set of pressure cuff measurements (target = forearm when cuff is inflated to 20 mm Hg, reference = forearm when cuff is deflated).

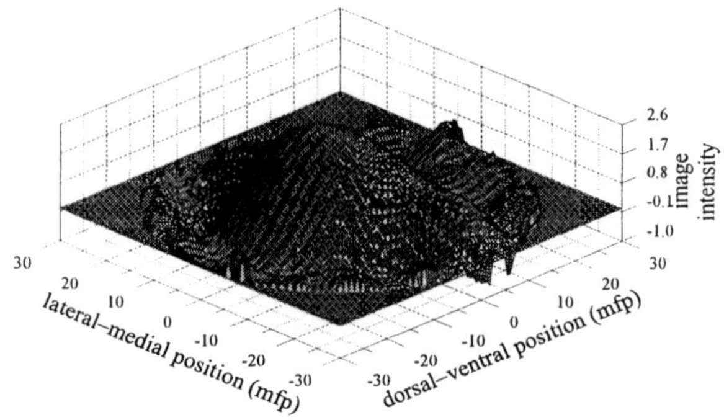

Figure 10 Image reconstructed from first set of pressure cuff measurements (target = forearm when cuff is inflated to 50 mm Hg, reference = forearm when cuff is deflated).

Discussion

In this report we have described the design and initial performance results of a novel geometrically adaptive tomographic imaging device well suited for investigation of a variety of body appendages. A key feature of the unit is a mechanical iris that serves to bring optical fibers into gentle contact with the tissue. While adjustable, the unit is mechanically quite stable at any pass–through diameter and is very effective in stabilizing the tissue against motion artifacts. Precision studies showed that the overall error of measurement was approximately 0.3% for light propagating through the entire system. While this is good, it is believed that much better performance can be achieved by using a fiber taper to couple light form the fibers to the CCD camera. In addition, further improvements can be made by adding a reference measurement and by using index matching fluid at the various interfaces whenever possible.

Results of phantom imaging studies are encouraging. The three cases examined, a homogeneous rod, two inclusions and four inclusions, all produced good quality results. Further improvements are expected by better matching of boundary conditions, and by computing perturbations in absorption and scattering simultaneously.

Results of the forearm studies are all also quite encouraging, but it is necessarily more difficult to render definitive interpretation. The anatomic assignments listed in Figures 6 and 7 were made based on the expected movement of muscle groups in the flexor study and known orientation of the arm in the apparatus. Recall that the computed image is a difference image based on measured values. In this regard it should not be expected that the structures seen would coincide as what would be expected by, say, MRI. No doubt definitive validation of these assignments will require much more careful study.

Putting aside these issues, there are specific features in the images that are noteworthy. First is the remarkable repeatability of the computed image. It deserves emphasizing that these images were derived from measurements made several hours apart, each requiring new subject setups with the apparatus. Second, we note that all incidences of high spatial frequencies are limited to the boundary. The features in the interior are all smooth. Similar results were obtained with the pressure cuff study. Definitive interpretation of these images is also difficult at this time. Nevertheless there are two significant features in the images. The first deals with the appearance of a broad peak in the middle of the image. We interpret this as pooling of blood in the core of the arm. This result is expected given the nature of the perturbation. The second is a quantitative observation. The maximum amplitude of the image at 20 mm Hg is approximately 25% that of the image at 50 mm Hg. This result coincides well with the observation that the maximal relative intensity change determined by measurement was approximately 25% for the former and >90% for the latter case. This finding also emphasizes the significant sensitivity the measurement has to variations in tissue hemodynamics.

Acknowledgments: This grant was support in part by NIH grant CA66184–A02.

References

[1] *OSA Tends in Optics and Photonics on Advances in Optical Imaging and Photon Migration*, R. R. Alfano, J. G. Fujimoto, eds., (Optical Society of America, Washington, DC 1996), Vol. 2.

[2] *Optical Tomography and Spectroscopy of Tissue: Theory, Instrumentation, Model, and Human Studies II*, B. Chance, R. R. Alfano, eds., Proceedings of SPIE Vol. 2979 (1997).

[3] M. Schweiger, S. R. Arridge, D. T. Delpy, "Application of the finite–element method for the forward and inverse models in optical tomography," *J. Mathematical Imaging and Vision*, Vol. 3, pp. 263–283 (1993); J. C. Hebden, S. R. Arridge, "Imaging through scattering media by the use of an analytical model of perturbation amplitudes in the time domain," *Applied Optics*, Vol. 35, pp. 6788–6796 (1996).

[4] M. A. O'Leary, D. A. Boas, B. Chance, A. G. Yodh, "Experimental images of heterogeneous turbid media by frequency–domain diffusing–photon tomography," *Optics Letters*, Vol. 20, pp. 426–428 (1995); X. D. Li, T. Durduran, A. G. Yodh, B. Chance, D. N. Pattanayak, "Diffraction tomography for biomedical imaging with diffuse–photon density waves," *Optics Letters*, Vol. 22, pp. 573–575 (1997).

[5] H. Jiang, K. D. Paulsen, U. L. Österberg, "Optical image reconstruction using DC data: simulations and experiments," *Physics in Medicine and Biology*, Vol. 42, pp. 1483–1498 (1996); H. Jiang, K. D. Paulsen, U. L. Österberg, B. W. Pogue, M. S. Patterson, "Optical image reconstruction using frequency–domain data: simulations and experiments," *J. Optical Society of America A*, Vol. 13, pp. 253–266 (1996); H. Jiang, K. D. Paulsen, U. L. Österberg, M. S. Patterson, "Frequency–domain optical image reconstruction in turbid media: an experimental study of single–target detectability," *Applied Optics*, Vol. 36, pp. 52–63 (1997).

[6] J. Chang, H. L. Graber, R. L. Barbour, R. Aronson, "Recovery of optical cross–section perturbations in dense–scattering media by transport–theory–based imaging operators and steady–state simulated data," *Applied Optics*, vol. 35, pp. 3963–3978 (1996); Y. Yao, Y. Wang, Y. Pei, W. Zhu, R. L. Barbour, "Frequency–domain optical imaging of absorption and scattering distributions by a Born iterative method," *J. Optical Society of America A*, Vol. 14, pp. 35–342 (1997); W. Zhu, Y. Wang, Y. Deng, Y. Yao, R. L. Barbour, "A wavelet–based multiresolution regularized least squares reconstruction approach for optical tomography," *IEEE Transactions on Medical Imaging*, Vol. 16, pp. 210–217 (1997).

[7] H. L. Graber, J. Chang, J. Lubowsky, R. Aronson, R. L. Barbour, "Near infrared absorption imaging of dense scattering media by steady–state diffusion tomography," in *Photon Migration and Imaging in Random Media and Tissues*, B. Chance, R. R. Alfano, eds., Proceedings of SPIE Vol. 1888, pp. 372–386 (1993).

CW Imaging Of Human Muscle Using Near Infrared Spectroscopy

H. Long, G. Lech, S. Nioka , S. Zhou and B. Chance
Department of Biochemistry and Biophysics
University of Pennsylvania, Philadelphia PA 19104

J. Park
Department of Radiology & Radiological Science
Vanderbilt University, Nashville TN 37232

ABSTRACT

Muscle hypoxia and fatigue are related to the generation of lactic acid in hypoxic regions of the muscle which may limit exercise capability. While generally thought that such regions are homogeneously extended over the muscle mass and activated by exercise, our recent results with a very simple muscle function imager show, to the contrary, that exercise limitation may involve specific rather than generalized portions of the muscle. Thus a simple, noninvasive optical device which is portable and efficacious, could be used to increase understanding of the contribution of particular muscles during different exercises. This could open new possibilities for the evaluation of muscle performance and fatigue of normal and pathological states.

KEY WORDS: (110,3080) infrared imaging; (300,6340) spectroscopy; (999,9999) muscle; (999,9999) mitochondria

INTRODUCTION

Six healthy sedentary males, age range 21 to 34 years, performed a leg-kick exercise and cycling ergometry on two separate occasions within a 3 week period. Eight athletes and three normal males, age range 20 to 26 years, performed a rowing ergometry test. A Continuous Wave Imager (CWI) probe was centered over the quadriceps muscle and secured by an elastic bandage. The levels of HbO_2 and Hb (deoxygenated) during exercise were calculated relative to the resting state. Absolute deoxygenation baseline was determined by cuff ischemia following cycling and rowing exercises.

PROTOCOL

Three protocols are described here: 1) The subjects performed a ramped maximal exercise protocol on a specially-designed leg-kick ergometer, which activated the quadriceps. The subject kicked against the ergometer with a frequency of 1 per second (60 rpm). 2) For the cycling exercise on a standard bicycle ergometer, the subject maintained a cycling speed of 60 rpm. 3) The subject was seated on a rowing ergometer and performed in accordance with his maximal effort.

Leg-kick Exercise Test

- Rest data collection (1 min)
- Exercise no added weight (2 min)
- Exercise at 40% of pretest maximum (2 min)
- Exercise at 75% of pretest maximum (2 min)
- Exercise at 95% of pretest maximum (2 min)
- Recovery from exercise (2 min)

Cycling Ergometry Exercise Test

- Rest data collection (2 min)
- Exercise no added weight (2 min)
- Exercise at 50 Watts (2 min)
- Exercise at 100 Watts (2 min)
- Exercise at 150 Watts (2 min)
- Exercise recovery - unloaded (3 min)
- 10 min rest
- Cuff ischemia (5 min)

Rowing Ergometry Exercise Test

- Rest data collection (2 min)
- Exercise at 20% Maximal rate (4 min)
- Exercise at 40% Maximal rate (4 min)
- Exercise at 60% Maximal rate (4 min)
- Exercise at 80% Maximal rate (4 min)
- Exercise at Maximal rate(4 min)
- 10 min rest
- Cuff ischemia (3 min)

OSA TOPS Vol. 21 Advances in Optical Imaging and Photon Migration, James G. Fujimoto and Michael S. Patterson (eds.)

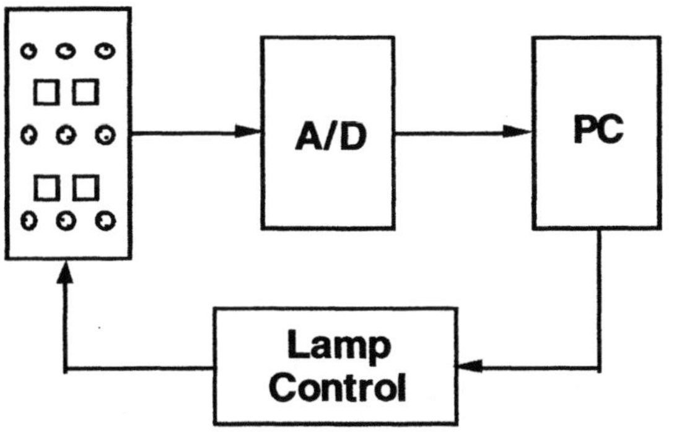

Figure 1 . Diagram of the imager. On the left are the 9 light sources for dual wavelength detector pad connected to A/D then PC. The computer controls the flashing of the lamp.

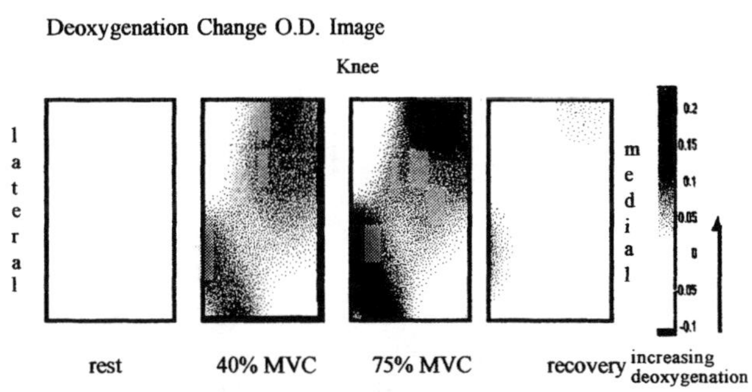

Figure 2. Demonstration of oxygenation changes at 4 workloads(separated by 2 min.) in a ramp "test" with no rest between workloads at 40 and 75% MVC. The more intense darker colors are due to deoxygenation, the lighter colors are due to oxygenation. The two positions where prominent oxygenation occurs are thought to be the lateralis and the medialis muscles.

The imager (fig. 1.) consists of nine light sources and four dual wavelength detectors It can be attached directly on the leg. It is coupled to an A/D board. A PC controls sequential flashing of the nine lights and data processing of outputs from the imager in terms of light absorption. Signals from each one of the lights to each one of the detectors are sequenced, affording 16 patterns which are coupled together in a simple back projection algorithm to create the images in Figs. 2,3,4,5&6.

RESULTS

CW Images of 3 quadriceps muscles during various exercise

One can observe the transition from rest to exercise showing peak absorbancy differences at two locations in the leg, in the region of the vastus lateralis and vastus medialis, which are shown as an intensity profile (fig. 2.). The workload initially 40%, is then increased to 75% MVC illustrating slight fatigue of the subject and increasing involvement in work by the rectus femoris. If one compares the images of maximal workload and recovery, one sees that the reoxygenation follows a nearly identical pattern of activation. The black & white figures are usually presented as color images which have better contrast.

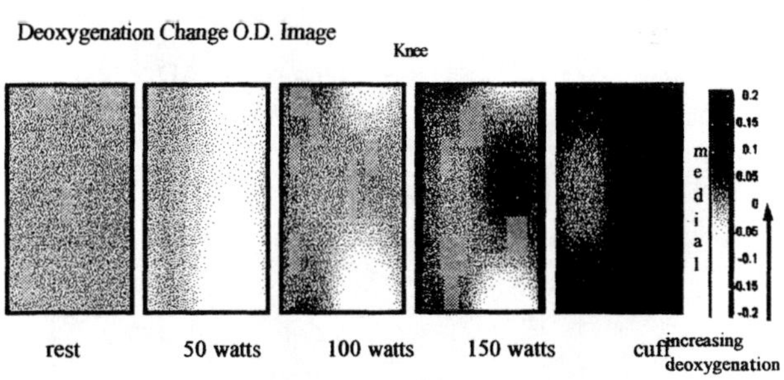

Figure 3. An example of exercise changes in bicycle ergometry, where the same color coding is used as in Figure 2 and the workload increased in the ramp test from 50 to 150 watts. In cuff ischemia, it would be expected that the leg would be uniformly deoxygenated. Thus the differences are due to different blood contents of the lateralis and the medialis.

Transition from rest to exercise (fig. 3.) shows that the peak absorbancy in the vastus medialis becomes increasingly deoxygenated with exercise intensity. After bicycle ergometry exercise. The cuff ischemia shows that the vastus medialis is the most vulnerable to cuff ischemia and shows the most rapid deoxygenation.

We can see peak absorbancy changes for deoxygenation on the left vastus laterialis and rectus femoris (fig. 4.). The large absorbancy changes during the highest power level (500 watts) shows the greatest deoxygenation. Deoxygenation during cuff ischemia occurs in the same areas as during exercise, but is an additional 20% higher than maximal exercise.

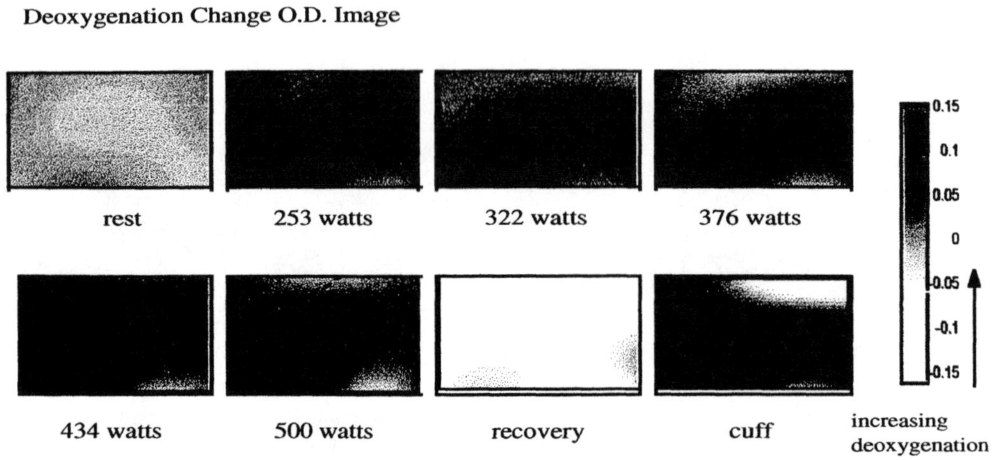

Figure 4. Rowing exercise by a nationally ranked college rower. The images start from the upper left rest and the exercise increases in a ramp test from 253 to 500 watts with increasing deoxygenation of the vastus lateralis and rectus femoris

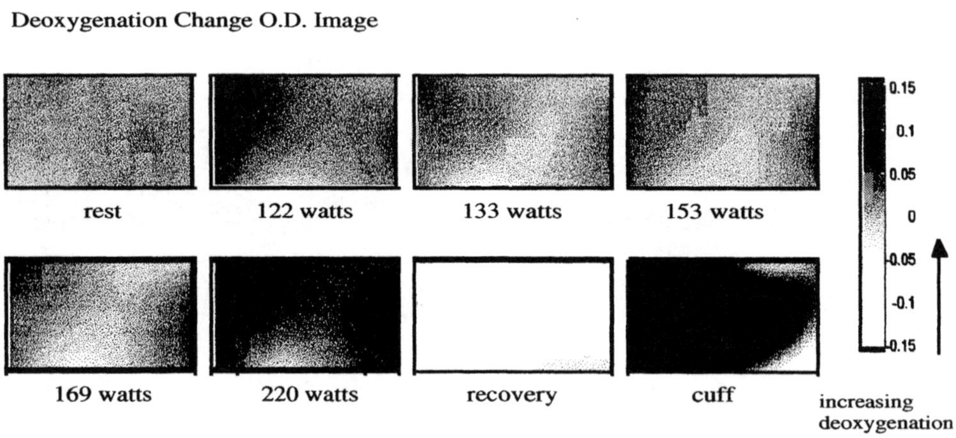

Figure 5. Rowing exercise by a normal subject. In this case, a ramp protocol is used up to the O_2 max., Appropriate for the normal subject, progressive deoxygenation is shown.

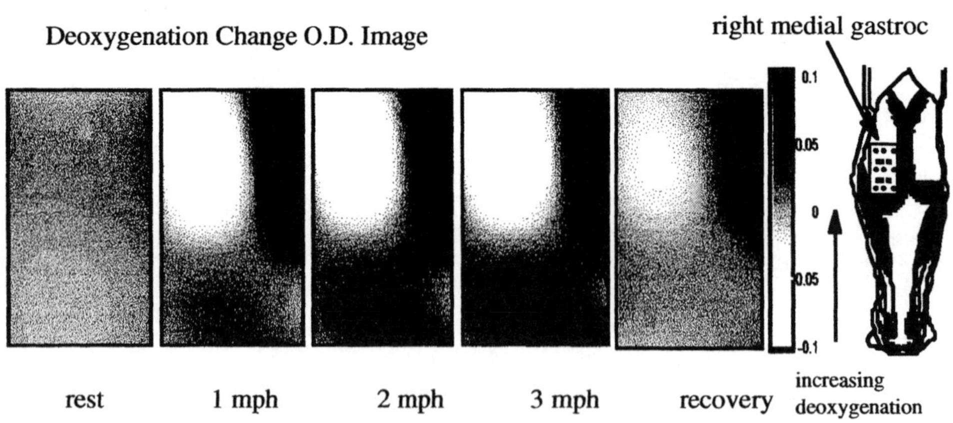

Figure 6. Treadmill exercise by a patient.

The image (fig. 5.) shows there are only small absorbancy changes (deoxygenation) in the left vastus laterialis and vastus medialis until maximal exercise. Then both of these muscles show more deoxygenation changes, but the rectus femoris still has minimal changes.

Cuff ischemia produces extensive deoxygenation in all three quadriceps muscles and is 80% greater than that of maximal exercise.

APPLICATIONS

With the imager (fig. 6.) one can determine whether specific muscle groups in patients with mitochondria myopathy exhibit abnormalities in their use of oxygen. This study could increase our understanding of the causes of fatigue and exercise limitations seen in these diseases. A patient reporting 19 years of exercise intolerance, performed treadmill walking at 1, 2 & 3mph. with the CWI on the right medial gastrocnemius. The image showed the paradoxical "oxygenation" which occurs during strenuous exercise in mitochondrial disease. However the gastrocnemius showed a split pattern of oxygen metabolism, with normal utilization (seen as deoxygenation) in part of the gastrocnemius while the adjacent part of the gastrocnemius showed the paradoxical oxygenation or limited oxygen utilization. This suggests heterogenicity in the expression of the disease.

CONCLUSIONS

1. The three muscle groups of the quadriceps (vastus medialis, vastus lateralis, rectus femoris) can always be distinguished during intense leg-kick and cycling exercise.
2. Each muscle group has a varying involvement depending on the exercise performed and the level of exercise intensity as is reflected by different levels of desaturation of oxygen hemoglobin.
3. Higher exercise intensity leads to greater desaturation and more muscles groups are mobilized.
4. Well-trained muscles of athletes utilize oxygen more efficiently during exercise than normal controls.
5. We believe that this simple imager offers a new and unique opportunity to study how oxygen delivery and oxygen utilization are activated in a localized fashion, how muscle groups participate during exercise, and how to better visualize the orchestration of muscle function
6. The NIR imager is suitable for clinic studies of muscle disease particularly in vascular and metabolic disease.

REFERENCES

1. Millikan, *el.al.* The oximeter an instrument for measuring continuously the oxygenation saturation of arterial blood in man. Rev. Sci. Instrum. 13, 434 (1942).
2. Smith *et.al.* Reperfusion hyperoxia in brain after circulatory arrest in humans. Anesthesiology. 73, 12-79 (1990).
3. Chance *et.al.* Recovery from exercise-induced desaturation in quadriceps muscles of elite competitive rowers. Am. J. Physical. 262, C1766-775 (1992).
4. Gratton, *et. al.* Feasibility of intracranial near-infrared optical scanning. Psychophysiology. 31, 211-215 (1994).
5. Ferrari, *et.al.* Cerebral and muscle oxygen saturation measurement by a frequency-domain near-infrared spectroscopic technique.SPIE 2388, 868-874 (1995).
6. Luo, *et.al.* Functional near-infrared image. SPIE in press (1997).

Compensated Transillumination of a Human Hand

Xingkun Wu and Gregory Faris
Molecular Physics Laboratory, SRI International, 333 Ravenswood Avenue, Menlo Park, CA 94025

Abstract

We demonstrate imaging through human tissue *in vivo* using compensated transillumination. Immersion in a matched scattering medium is used to dramatically improve image contrast.

Key Words

(170.3880) Medical and biological imaging; (170.5280) Photon migration; (110.7050) Turbid media

Transillumination is a simple technique for imaging through human tissue. The use of transillumination was reported as early as 1929.[1] Transillumination is performed by directing light through tissue and viewing the transmitted image either directly, or by using a camera. With the advent of cooler light sources and improved cameras, there was renewed interest in transillumination for mammography beginning in the 1970's.[2] However, image quality was not good, and transillumination could not compete with x-ray mammography.[3]

One limitation for transillumination is nonuniformity in thickness and geometry involved with any *in vivo* imaging. Because of the very large attenuation of light in tissue, small variations in the thickness of tissue across the image lead to large variations in intensity across the image. Variations in the location and orientation of the tissue surface cause variations in the optical coupling into and out of the tissue. Additionally, photons may be lost at tissue boundaries (edges) transverse to the imaging. Intensity variations due to these various effects obscure contrast changes within the tissue and limit the dynamic range in the image.

One approach to reducing intensity variation effects is to measure the optical path length and adjust the image contrast appropriately. This technique has been applied to mammography by using frequency domain techniques.[4] In this case, the phase of an amplitude modulated light source was used to determine the optical path length.

We apply an alternate approach to improving image uniformity for transillumination–to surround the imaged tissue with a medium of matching scattering and absorption properties. In this way, a uniform transillumination image is obtained and the maximum dynamic range is available to examine variations within the tissue. For a good match with the immersion medium, the human tissue will almost disappear, and the image will only show contrast due to small variations within the tissue. This is what we desire in imaging as internal variations are typically what is of interest, not the gross external variation associated with the tissue shape.

As a matching immersion medium we use a mixture of Ropaque and methylene blue dye. Ropaque is a suspension of hollow polymer spheres in water. We have used a frequency domain technique to measure the absorption and scattering properties of a human hand *in vivo*, and to match the properties of the medium to the hand. With the frequency domain system, we can independently determine the amount mismatch in absorption and scatter coefficients between the medium and tissue. At a good match there is very little difference optical transmission between tissue and the matching medium. Using the frequency domain system, we found an rf amplitude decrease of 4% and a phase variation of -0.4 degrees when a fingertip was inserted between the optical fiber bundles in the matching medium. Details of this matching are reported elsewhere.[5]

The apparatus for compensated transillumination is shown in Figure 1. A 100 W light bulb illuminates a Plexiglas box containing the matching medium. The light source is shielded such that light can only reach the camera by passing through the matching box. The face of the box opposite the lamp is imaged onto a camera. We have used both an intensified ccd camera and a simple film camera to acquire images.

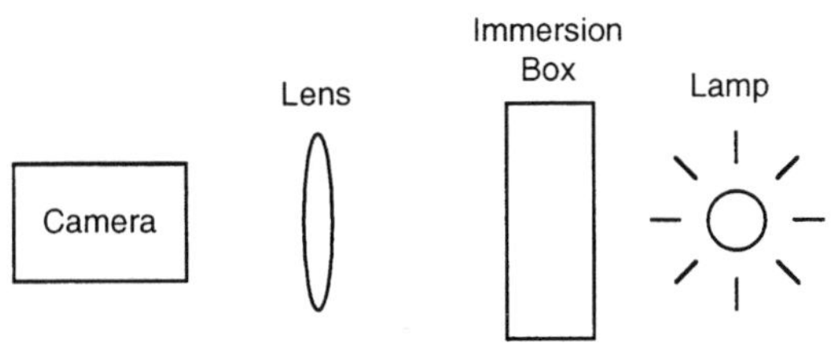

Figure 1. Apparatus for compensated transillumination.

An image of a human hand immersed in the matching medium is shown in Figure 2. This image was taken using the intensified ccd camera. Considerable detail is visible in the hand that cannot be seen in ordinary transillumination. The spaces between joints are visible in the fingers and within the main fleshy part of the hand.

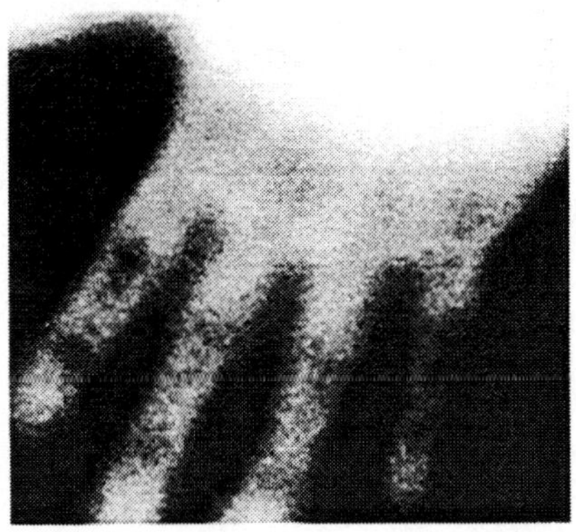

Figure 2. Image of a human hand taken using compensated transillumination.

When performed using film, this technique is extremely simple and inexpensive–the apparatus costs less than $100. This approach primarily improves contrast, not spatial resolution. Improvements in spatial resolution are possible using a scanned source detector or time resolved techniques. However, these changes add to the complexity and cost of the imaging system. Because of its simplicity and good image quality, compensated transillumination may hold promise for mammography or other applications.

References

1. M. Cutler, "Transillumination as an Aid in the Diagnosis of Breast Lesions," Surg. Synecol. Obstet. **48**, 721-729 (1929).
2. C. Gros, Y. Quenneville, and Y. Hummel, "Diaphanologie Mammaire," J. Radiol. **53**, 297-306 (1972); B. Ohlsson, J. Gendersen, and D. M. Nilsson, "Diaphanography: A Method for Evaluation of the Female Breast," World J. Surg. **4**, 701-707 (1980); C. H. Jones and S. P. Newberry, "Visualization of Superficial Vasculature Using a Vidicon Camera Tube with Silicon Target," Br. J. Radiol. **50**, 209-210 (1977).
3. R. F. Girolamo and J. V. Gaythorpe, "Clinical Diaphanography - Its Present Perspective," CRC Critical Reviews in Oncology/Hematology **2**, 1-31 (1989).
4. S. Fantini, et al. "Frequency Domain Optical Mammography: Edge Effects Corrections," Med. Phys. **23**, 149-157 (1996).
5. X. Wu, L. Stinger, and G. W. Faris, "Determination of Tissue Properties by Immersion in a Matched Scattering Fluid," Proc. SPIE **2979**, 300-306 (1997).

Phased array functional imaging of neonate's neurological disorders

S. Nioka. S. Zhou, EK. Anday*, W. Thayer**, D. Kurth+, M. Papadopoulos* Y. Chen and B. Chance

Dept. of Biochem/Biophysics, University of Pennsylvania, *Dept. of Pediatrics, Allegheny University, **NIM Inc., +Children's' Hospital of Philadelphia, Dept of Anesthesiology, Philadelphia, PA

ABSTRACT

We have developed an optical Imaging system, Dual Wavelength Phased Array system (DWLPA) that can be used for testing neonate's brain function. A phased array system consists of 16 sets of phase cancellation optodes, 0 and 180 degree laser sources with a detector in middle. For this purpose, 9 sets of lasers each with 780 and 830 nms (20 mwatt) and 4 detectors, (PMT-TO8 Hamamatsu) are placed on a probe, 4 x 9 cm. These optodes make 16 combinations of triangle; each has a function of phase cancellation. Thus all the probe areas were filled with 16 triangles and can be seen as sensitive to heterogeneity. The optodes separation is 2.5 cm for each array, and data information is taken by a detector in the middle when each two sources at 1 and 180 degree ignite at once. The actual images are made by subtracting from resting state to active state of cortex when physicians test for neurological examinations, hearing, touching, and visual display. The subtraction can cancel outcome of artifacts caused by inconsistent hair, probe pressure to the head, and slightly dislocated probe angles. The results show that like adults, we saw functions of cortices. The significant is that these blood volume changes by the stimulation can be detected much earlier stage of life than the development of EEG (electrical encephalogram) and Evoked potential. Thus the usefulness of this optical technology can be very beneficial. It implies that neonates already develop circulatory response from the metabolic needs of neurons in the area. We conclude that this technology, namely PA imaging system has a superior S/N ratio by the fact that it has 180 degree shift for 2.5 cm resolution. These methods can be used clinically to detect infant's neurological dysfunction before EEG and other test can be used.

Key Words
Optical Image, pediatrics, neurology, DWLPA

INTRODUCTION

A Dual Wavelength Phased Array (DWLPA) imaging system is one of the optical imaging systems that can be used for testing neonates brain function. It consists of arrays of frequency domain spectrometers, and gives appropriate spatial and temporal information when it is placed on the head. The principle of the device is described elsewhere (1). Here we present the structure of the device, mechanism, function, and performance.

Because oxygen concentration can be presented and the blood volume can be more accurately analyzed with two wavelength light sources, this DWLPA imaging system is more important than the previous single wavelength system. The importance of functional imaging has been recognized in studies of PET (2) and fMRI (3) as well as optical imaging system (4,5). These studies try to link an area of cortex with brain functions, and its relationship with some neurological disorders by means of metabolic activity indicated by oxygen concentration or glucose intake to the brain cells (2). Some complicated brain functions such as cognition are known to have hetero-modal association cortex, a system of parallel-distributed networks of circuits within the neocortex (6). This very important and complicated brain functionality can only be investigated with a device that detects metabolic alterations, such as oxygenation (3,4,5). In addition, we expect that lesions from strokes, underperfused areas can be detected with this device as well.

METHODS

Structure of DWLPA system

Contents. The DWL PA imager has 18 laser diodes (9 for each 750 and 830 nms) with a 20 mwatt source power (Class IIIb). We have shifted wavelength of 780 nms to 750 nms because at 750 nms the signal gain is more than twice due to the respective extinction coefficients. The wavelength has been tested and proved to shed enough light with 20 mwatt power on human heads. On the detection side, four PMT (TO8 Hamamatsu) are used. Two independent phase meters and 2 single side band (SSB) radio transmitters/receivers are used with 50 and 52 MHZ frequencies.

Probes. The design of this probe is similar to that in single PA imager. Because two laser sources are located in a point (previously one source), the size of the probe is large (5 cm x 10 cm) The source-detector separation is maintained at 2.5 cm. The 18 sources and 4 detectors are mounted on a thinner plastic rubber that is more bendable in order to fit the curvature of the subject's head. A sponge-like rubber material cushions the subject's skin and protects him/her from direct contact to the optodes. The structure of the optodes is shown in Figure 1.

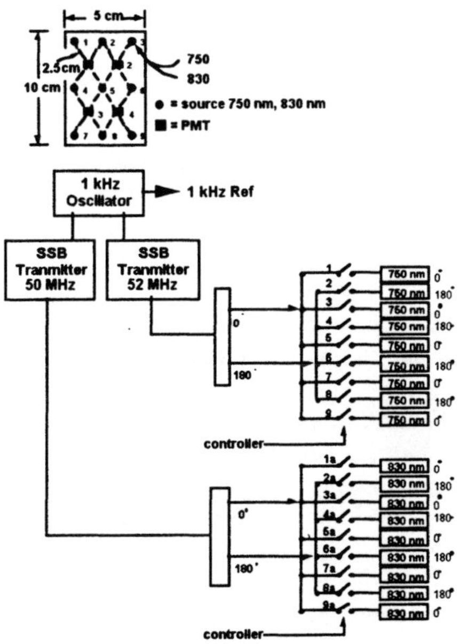

Figure .1 Schematic diagram of the probe and the source part of the DWLPA system.

Mechanism. As shown in a block diagram in Figure 1, frequencies of 50 and 52 are selected to separate signals from the 2 wavelengths. For better temporal resolution, it allows simultaneous injection of photons without cross talk. Single Side Band (SSB) technology is used for modulating 50 and 52 MHz frequencies for 750 and 830 nms. The phase shifts (0 and 180) are controlled by software through switches. The time sequence of run time for the 9 sources is controlled through software, one pair at a time for phase cancellation (Figure 2). For the detection, four PMT (R928) are coupled with SSB demodulator, SSB receiver with 1 kHz frequency (Figure 2).

The detector of in-phase and anti-phase signal establishes a null plane in a midpoint of two laser source locations and one detector. Such a null point and surrounding field is highly sensitive to perturbations by extremely small objects, of the order of the size of 70 microns volume and containing as little as 20 picomole of an absorber (Indocyanine green, ICG). In order to provide real time spatial localization with high sensitively (1 mm), we placed multiple numbers of this triangle of sensitive fields by the probe.

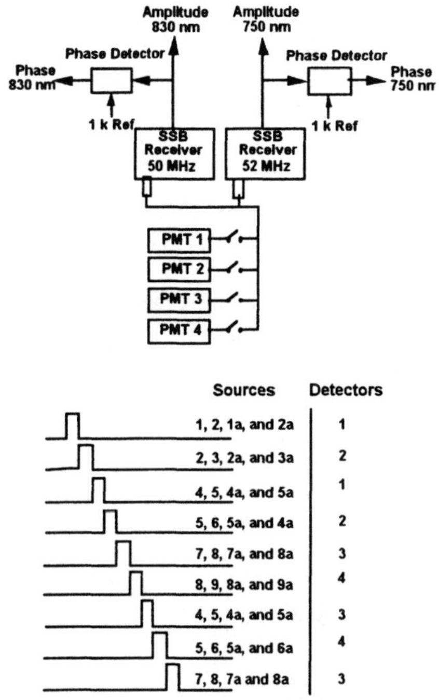

Figure 2. Schematic diagram of DWLPA imaging system, the time sequence of source engagement and the detector part.

RESULTS

Models

We have used a routine test model consisting of 1 % intralipid solution in an aquarium 1.5' x 8" x 1'. The probe was attached to the wider side of the container. A black dauring bar (diameter 8 mm) is placed at the depth of 1 to 2 cm. We tested locations of the black bar in the PA images. The results showed a satisfaction in locations and S/N ratios.

Human brain function studies

Adult human. We have performed finger-tapping test on human adults using the DWLPA imager. The finger-tapping test is one of the routine tests we often use on human adult brains in order to test performance of imaging devices. With the probe placed on the right parietal part of a human head, blood oxygenation and an increase in blood volume was observed when the subject tapped his left fingers were being tapped by him (images not shown). When the right fingers were tapped there was no signal depicted.

Neonate population. In human neonates, because the bone are much thinner and the head size smaller, the

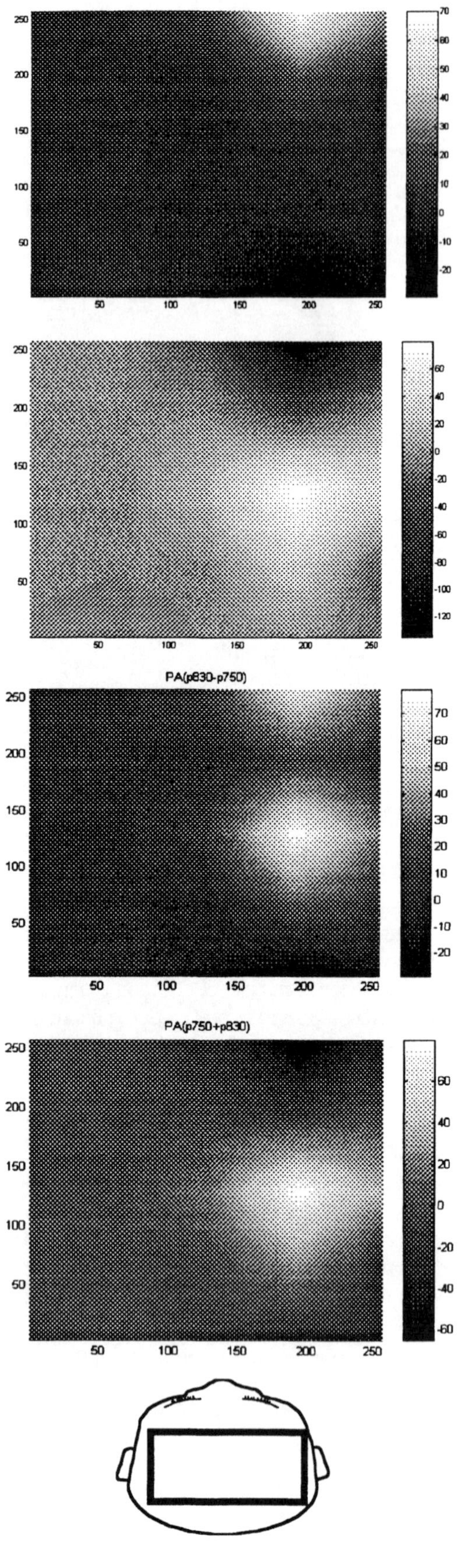

Figure 3. DWLPA images of full term neonate. From the top, Phase cancellation images (PA) at 830 and 750nms (two from the top), and oxygenation (830-750 nms) and blood volume (830 + 750 nms) are shown (two bottoms). This baby was touched in her left fingers by a physician. Note that there are oxygenation and blood volume increases in her right sensory cortex area.

source-detector separation is changed to an appropriate size, 2 cm. Since we can not demand testing such as finger tapping that needs cooperation with the patient, all the tests imposed upon them are passive. We have tried several stimulation, sensory, hearing and visual, and some reflex, such as sucking and startle. Many of their reactions are movement of their extremities, which also triggers the motor cortex. We show here: sensory reaction, some reflex such as startle, and passive movement of the extremities.

As shown in Figure 3, when neonates are examined for their sensory nerves by touching their extremities, oxygenation and blood volume increases are commonly observed in their ipsi-lateral sensory cortex. The oxygenation and blood volumes are presented by subtracting and adding the phase shift changes (absolute degrees).

This DWLPA imager can also detect raw data from a single source-single detector combination. These data can be used to calculate Ua and Us from AC amplitudes and phases. In the case of multiple functions subjects exhibit during testing, the raw data can be useful to locate the focuses of the activities. Frequently we have seen neonate's voluntary movements when we examined and touched them. Such raw data can be obtained by using a single source-detector mode. These are the phase and amplitude from each single source- single detector combination (PMS), shown in Figure 4. The images shown here are somatosensory cortices of both hemispheres (see the probe location) with a midline in the center. A physician touched her right fingers when she responded with startle reflex, left leg and right finger moving. There is oxygenation of both left and right hemispheres. It is interesting that there are not always matching oxygenation and blood volume increases in this time period. This may result from different time frames in each section of image pixel, during the 32 seconds of total time requirement. There is no doubt that more than a few parts of the somatosensory cortex are triggered in the startle reflex in neonate.

As we can see in the Figure 4, the amplitude (raw data; PMS) can itself show good S/N ratio. On the other hand, the raw phase shift shows a relatively small amount (~1 degree), as shown in the Figure 4. This may cause a poor S/N ratio. That is a good reason that we have a phase cancellation image, which shows much more detectable phase shifts as high as 40 degrees (Figure 3). In Figure 3, the sensory cortex responsible for right finger touching was detected with very high S/N ratio. We also looked at

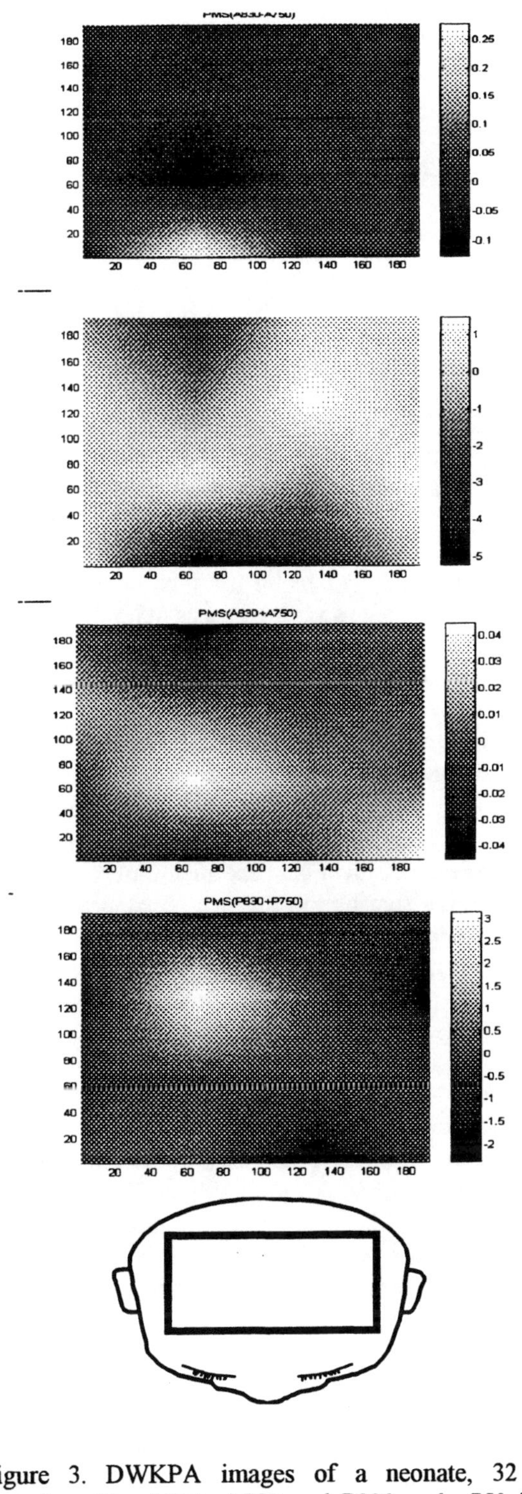

Figure 3. DWKPA images of a neonate, 32 weeks gestation. The A830, A750 and P830 and p750 indicate amplitude at 830, 750 nms and phase 830, 750 nms respectively. The subtraction and sum of these two wavelength data indicate oxygenation and blood volume respectively. This baby was touched on her right fingers and she startled in response. Note that there is oxygenation due to her responses.

some neonates with neurological disorders, such as brain edema after cardiac arrest during the delivery. We followed up his post-trauma course with our imaging system from day 1 to day 22 when he was discharged from the hospital. On the day the patient showed signs of brain edema, the response of finger touching was weak and with some movement of his extremity, he deoxygenated instead of oxygenation in the responsible cortex area. It showed that there was a dissociation of metabolism and the blood supply system due to brain edema, reduction of blood volume was evident and especially symptomatic.

CONCLUSION

As shown in the performance, the quality of images in the new device, DWLPA system has been equivalent to that of the old optical imagers using reflectance (3,4,5). We have only 1.5 cm spatial resolution, with the existing probe design with 2.5 cm source-detector separation. However improvement can be achieved by providing more sets of source-detectors. What we have gained here is a capability of imaging oxygenation of the tissues of interest, functions of brain activation, blood flow and thc balances between oxygen demands and oxygen supplies. With this function of the imager, we have extended many more applications for the use of the device in future.

ACKNOWLEGEMENT

We thank Ms. Xie for developing software of the device and Drs. Ravishankai, Piazza, and Cing Xue for help testing subjects. We also thank Ms. Shore for editing this manuscript. This research was supported partially by NIH contract N44-NS-5-2317.

REFERENCES

1. S. Zhou, C. Xie, S. Nioka, H. Liu, Y. Zhang, and B. Chance. "Phased array instrumentation appropriate to high-precision detection and localization of breast tumor" SPIE 2979 98-106. 1997.
2. L. Sokoloff et al. "Relation between physiological function and energy metabolism in the central nervous system" J. Neurochem. 27 13-26. 1997.
3. S. Ogawa, et al. Magnetic resonance in Med. 29 205-210. 1993.
4. A. Maki, Y. Yamashita, Y. Ito, E. Watanabe, Y. Mayanagi, and H. Koizumi. "Spatial and temporal analysis of human motor activity using noninvasive NIR topography" Med. Physics. 22 1997-2005. 1995.
5. Q. Luo, S. Nioka, and B. Chance. "Functional Near-Infrared Imager" SPIE. 2979 84-93. 1997.
6. J. Jonides, EE. Smith, RA. Koeppe, E. Awe, S. Minoshima, MA. Mintun. "Spatial working memory in humans as revealed by PET" Nature 363 623-625. 1993.

Breast tumor images of hemodynamic information Using a contrast agent with back projection and FFT enhancement

S. Nioka, SB. Colak*, X.Li, Y.Yang, and B.Chance

Dept. of Biochem/Biophysics, University of Pennsylvania, USA, and Phillips Laboratory*, Holland.

ABSTRACT

Recent TRS results of breast tumor images revealed to us that it may be difficult to diagnosis cancer from optical characteristics of tumors by solely absorption coefficient and scattering coefficient. Benign tumors develop blood vessels surrounding tumors as well as cancers do. Scattering coefficient can be lower or higher than in normal tissues, therefore it cannot be used for diagnosis. While there is very active work on characterizing cancers with optical methods, we sought a different approach to find a "cancer signature" in the area of hemodynamics. It is known that cancers will create hypoxic lesions as well as high oxygen lesion, and we hypothesized that these hypoxic and hyperoxic lesions attributed to heterogeneity of hemodynamics, i.e. some areas have a high velocity of blood circulation, and others have extravasation, causing high and low contrast agent clearance. We used an optically visible contrast agent; ICG injected intravenously into 6 tumor bearing women, and imaged their breasts with CWS system, which has 16 sets of source and detector in a circular holder. Time resolution was 8 seconds for 1 image thus enough to follow the time course of the ICG appearance-decay curves in each 16 x 16 source-detector combination. The data were saved in a PC and later analyzed using a back projected FFT enhanced image re-construction algorithm. The data suggest that indeed cancers have more heterogeneity as indicated by the time constant of ICG decay curves, and time of decay is generally longer. In contrast, in benign adenoma, the time constant was shorter. In conclusion, we have used the dynamics of ICG uptake and disappearance in cancer as compared to normal breast and benign tumor cases as a diagnostic approach.

KEY WORDS

Breast cancer, NIR, Optics, ICG, blood flow, clearance, back projection, Deblur

INTRODUCTION

Research Goal

In this investigation, we have focussed on cancer detection in the human breast using CW technology (Continuous Wave Spectroscopy and Imaging). The breast tumors have been characterized with the optical absorption and scattering coefficients (Ua and Us') in the Hospital of University of Pennsylvania (1) and by several others as well (2). According to these results, the breast cancers and other tumors both have a similar Ua, higher absorption than normal breasts. This allows detection of tumors, however because of the similarity in both cancer and benign tumors, we must look for some other optical measures to detect cancer as differentiated from benign tumors, such as hypoxia. While there is very active work on characterizing tumor tissues by optical methods in order to achieve sensitivity and specificity, we can gain an additional cancer signature in the area of extravasation. For this reason we focussed on the hemodynamic difference between cancers and other tumors and hypothesized that because of their leaky vessels (extravasation) cancers have a slower clearance of an optical contrast agent as compared with the normal blood vessels which do not permit extravasation.

Our proposed use of a light imaging system is to first use CWS for screening, detecting suspicious masses in a very cost and time efficient manner, with a very user friendly interface for both examiner and examinee. Once a mass is detected, to differentiate cancer by behavior of hemodynamics and leakage to cells extravasation in a lesion using a contrast agent is used.

Why use CWS system for cancer detection

There are new optical technologies we can adapt for imaging. After examining most of the optical methods for the imaging capability, we found CWS is sensitive enough to pursue for our objectives in this research. The temporal resolution that CWS system offers is the best among the current technologies, therefore most suitable for this application. In

addition, with a contrast agent the tumor delectability increases to compensate for it's low spatial resolution.
For this study with contrast agents, we used a CW Imager with circular probe location. We also improved the tumor delectability by using deconvolution (2D FFT) from the back projection algorithm described previously (3).

METHODS

CWS method with a contrast agent

Sixteen combinations of light sources (tungsten bulbs) and light detectors (silicon photodiodes) are located on a two dimensional circular line equi-distant from each other (11° angle apart, Figure 1). Within the circle, we placed a human breast or a breast model for calibration using 0.5% intralipid and ink, which have similar scattering and absorption coefficients to those in the breast tissue. This CW imager has been described elsewhere in detail (4). It takes only a few seconds to acquire data since one source is on, while all the detectors simultaneously detect the photons, 16 sources on a circle at a rate of 200 msec / source. The data acquisition for an image takes 4 - 8 seconds with 8x8 or 16x16 data points. A contrast agent, Indocyanine Green (ICG, (5)) 0.25 - 0.5 mg/kg, is injected intravenously into the human subject, from whom data are being collected continuously every few seconds over 10 - 15 min as the ICG absorption signal appears and disappears. Data are processed as OD changes due to ICG from the IV injection at the beginning. Some data points are averaged in several time windows and as a result, several images are reconstructed using the back-projection algorithm (3).

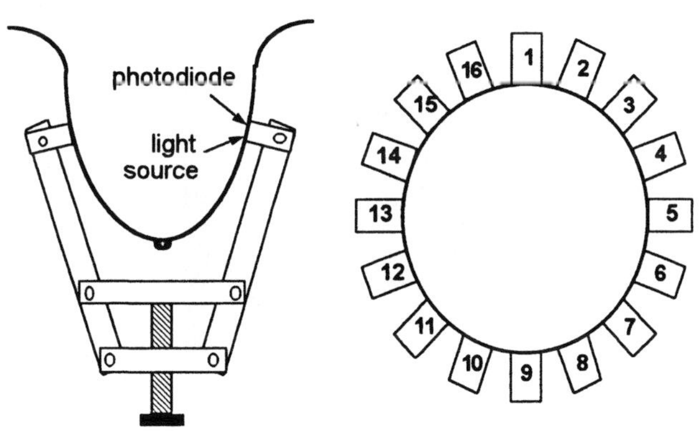

Figure 1. CWS imager. The scheme describes the vertical and horizontal cross sections of the 16 light source-detector device, holding a human breast. The circumference can be adjusted easily with setscrews. When the device fits a breast, the source detector relation remains unchanged.

Deblur technique

The most probable paths of the scattered light originating from a point source and reaching a point detector within tissue-like medium fall within a banana shaped region (6) with added distortions depending on the inhomogeneities (7) and the source-detector geometries (8). It is clear that the resolution and nonlinearities of imaging in diffusive heterogeneous medium depend on the width and shape of these banana path regions. The banana function varies greatly close to sources and detectors, but it has a relatively constant width in the middle, away from the source and detector positions. Therefore, it causes a spatially dependent broadening, blurring the projection images through diffuse media. Our tomography approach relies on the fact that we can apply standard X-Ray back projection techniques for tomography in such media and then apply semi-empirical image corrections (3) which are based on the shape of these banana regions found iteratively from previous reconstructions.

From their projections the tomographic images of objects in diffuse media can be improved by deblurring these images by a broadening function, and this function depends on the statistics of photon propagation in diffuse media. One of the easiest ways to handle the spatial dependence of the deblurring function is to use a radial coordinate transformation (9) to compensate for the spatial variation. Such approaches have already been applied in image restoration for medical images taken by CT and EIT. With such a coordinate transformation we basically would like to achieve an expansion of the space coordinates at large radial distances as compared to the small radial distances. To achieve such a transformation, the best is to work with the version of the image transformed from the (x,y) Cartesian coordinate space to the cylindrical space with expanded radial coordinates. After such a coordinate transformation, the image is Fourier transformed and divided by the Fourier transform of a single (space invariant) deblurring (PSF) function, which is also transformed into the cylindrical space.

In our experiments, we have applied a deconvolution technique to deblur the optical projection images of breast tissue taken with 16 sources and 16 detectors. Using deconvolution with a broadening function dependent on the physics of light propagation in such media as described above, we deblurred the images. Although, due to the existing noise in the data our success in deblurring was limited, we believe that this method provides a simple and fast tool to obtain real-time images of the interior of biological tissue for medical diagnostic imaging. We believe that this tomographic approach, together with the contrast agent enhanced imaging discussed here in this report, will provide a practical optical imaging tool for breast cancer screening and/or diagnosis in the near future.

RESULTS

Human Breast CW Imaging with a contrast agent, ICG

Cancer Case 1 (Figure 2). This patient has a cancer (6 cm x 3 cm) easily recognized from the surface, skin surface is rough like "orange skin" in lower lateral quadrant of the breast. Her breast images are shown in Figure 2 at 2, 5 and 8 min after the ICG injection. The images are expressed as ICG concentration since pre-injection absorption's are subtracted from post-injection. After injection of 0.25 mg/kg ICG intravenously, the peak absorption in the breast was observed after approximately 2 min, similar to normal breasts. However the lower lateral quadrant showed greater retention of the ICG compared to the rest of the breast (Arrow). This case shows the relative intake of ICG is increased after 8 min in the cancer lesion, while ICG has already cleared surrounding tissue.

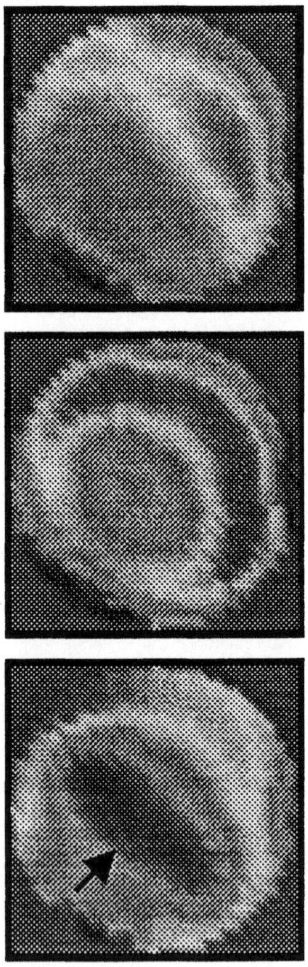

Figure 2. Optical images of a cancer bearing breast (case 1). The images indicate ICG absorption (dOD) 2 min, 5 min, and 8 min after the ICG IV injection from the top to bottom. The ICG location is indicated as higher absorption (darker). The images are relatively contrasted. The retention of ICG is seen at 8 min in the cancer located at lower lateral quadrant (6 x 3 cm) shown as a dark oval shape (indicated by an arrow).

Cancer Case 2, (Figure 3). This middle aged patient has a cancer (3 cm x 2 cm) easily palpable in the left lateral upper quadrant. In figure 3, it is shown at 8 o'clock. The images indicate absorption of ICG distribution at 2, 5 and 8 min after the IV injection ICG 0.25 mg/kg (from the top to bottom respectively). The ICG location is indicated as OD (dark) with an adjusted contrast. In this case, the cancer lesion at 8 o'clock has a higher intake of ICG at peak (2 min), suggesting high blood flow in this region. And as the blood ICG concentration decays, ICG is seen in the peripheral and in the cancer. At 5 min., most of the remaining ICG is in large vessels surrounding the breast tissue and in the lesion. This is a case that has a higher blood flow in the lesion. At 8 min, the ICG is trapped in the lesion and makes contrast clearer with time.

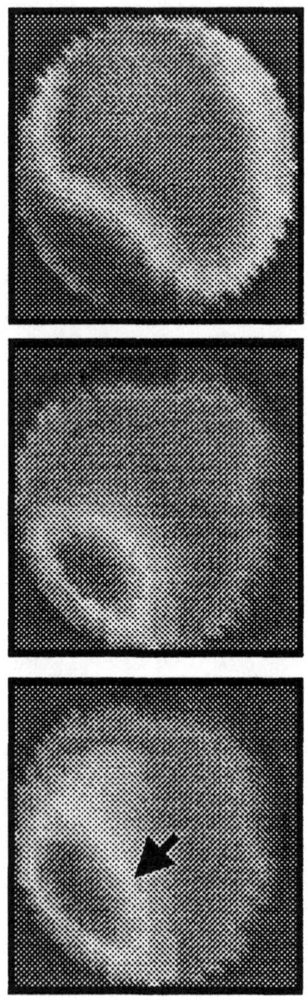

Figure 3. Optical images of a cancer bearing breast (case 2). A Cancer is located at 8 o'clock in this figure (2 x 3 cm,

indicated by an arrow). These images are coronal section at the depth of 1 inch from the chest wall. The images indicate ICG absorption (dOD) at 2 min, 5 min, and 8 min after the ICG injection (from the top to bottom). The images are auto-contrasted. In this case, the high concentration of ICG can be seen at an earlier time (2 min) and maintains high at 8 min.

Non-tumor bearing Case 3, (Figure 4). This patient has a normal breast with a small post biopsy lesion at 2 o'clock near the surface shown as a high ICG mark (Figure 4, top). These 3 pictures indicate ICG distribution at 2, 5 and 8 min from the top to bottom respectively. In this case, the ICG was already cleared from the normal breast tissue at 2 min and gone to venous phase. There is an accumulation of ICG in a biopsied lesion (arrow).

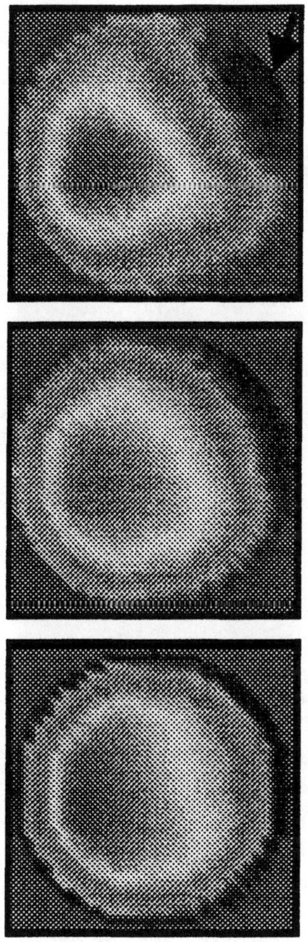

Figure 4. Non-tumor bearing breast ICG images. These 3 pictures indicate ICG distribution at 2, 5 and 8 min from the top to bottom respectively. An ICG accumulation at 2 min at 2 o'clock shows an inflammatory reaction due to biopsy done 2 weeks before (arrow).

The post-biopsied lesion suggests only inflammation because of high blood flow (quick clearance of ICG). The ICG at 2 o'clock cleared already at 5 min. The multiple rings made of different ICG concentrations in these images can be partially attributed to error caused by a boundary condition. At 8 min, the ICG shifted towards the outside of the breast tissue shown as a dark mark in the outline as observed by comparison to earlier times in this non-tumor bearing case.

CONCLUSION

There is a possibility that cancer can be distinguished from benign tumors by hemodynamics and tissue permeability (extravasation). This requires injection of a contrast agent and observation of a change. We are looking for a suitable contrast agent, which can permeate the cancer lesion and be retained longer than ICG. Those ideal agents have yet to be discovered (Licha personal communication), but it will be worthwhile to develop the technology when those agents are available for the purpose of testing and cancer detection in the human breast.

ACNOWLEDGEMNET

We thank to Ms. Xie for developing the software, to Ms.Li, and Xie for help examining patients, to Ms. Shore for editing the manuscript. This research is supported by NIH grants CA 60182 and CA 72895.

REFERENCES

1. KA. Kang, B.Chnace, S. Zhao, S. Srinivasan, E. Patterson, NS R. Troupin. "Breast tumor characterization using Near-Infra-red Spectroscopy," Proc.SPIE, 1888 487-499, 1994.
2. K. Suzuki, Y. Yamashita, K. Ohta, M. Kaneko, M. Yoshida, and B. Chance, "Quantitative measurement of optical parameters in normal breasts using time-resolved spectroscopy:in vivo results of 30 Japanese women," J. Biomed. Optics. 1 330-334. 1996.
3. SB. Colak, D.GPapaioarmou, QW. Hooft, and MB van der Mark, "Optical back projection tomography in heterogenous diffusive media," Proc. SPIE. 2626 306. 1996.
4. S. Nioka, Y. Yang, M. Shnall, S. Zhao, C. Xie, B. Chance, and L. Solin. "Optical imaging of breast tumor by means of continuous waves," Adv. Exp. Med. Biol. 411 227-232, 1997.
5. Cardio Green (CG), SterileIndocyanine Green USP. Becton Dickinson and Company 1990.
6. S.Feng, F.Zeng, B.Chance, Appl.Opt. 34. 3826-3837, 1995.

7. E.M.Sevick etal, in Medical Optical Tomography, eds.G.Muller, Proc. SPIE 1511 485, 1993.
8. A.Ya.Polishchuk, R.R.Alfano, in Optical Tomography, Photon Migration, and Spectroscopy of Tissue and Model Media, eds. B.Chance, R.R.Alfano, A.Katzir, Proc. SPIE 2389 6-9, 1995.
9. S.B.Colak, D.G.Papaioannou, G.W.'t Hooft, M.B.van der Mark, H.Schomberg, J.C.J.Paasschens, J.B.M.Melisses, N.A.J.J. van Asten, Appl.Opt. 36 180, 1997

In vivo Absorption and Scattering Spectra of Human Tissues in the Red and Near Infrared

Rinaldo Cubeddu, Antonio Pifferi, Paola Taroni, Alessandro Torricelli, and Gianluca Valentini

CEQSE-CNR and INFM-Dipartimento di Fisica, Politecnico di Milano, Piazza Leonardo da Vinci 32, 20133 Milano, Italy

Abstract

Absorption and scattering spectra of human breast, muscular and lipid-rich tissues were assessed *in vivo* between 610 and 1010 nm by means of time-resolved reflectance.

The absorption spectra show features typical of hemoglobin, water, and lipids, while the scattering coefficient decreases smoothly upon increasing wavelength.

Key Words

(170.1470) Blood/tissue constituent monitoring; (170.6510) Spectroscopy, tissue diagnostics; (170.6930) Tissue; (300.6500) Spectroscopy, time-resolved.

Introduction

Optical imaging techniques are gathering more and more interest as possible non-invasive means for the diagnosis of breast cancer [1-3]. However, the best illumination wavelength for imaging purposes has not been determined yet.

Based on recent studies [4-8], the overall image quality seems to improve upon increasing the scattering and absorption of the background medium. Consequently, it is of particular interest to collect data on the *in vivo* optical properties of breast. Such investigations should be extended not only to the near infrared, but also to the visible, or at least to the red region of the visible spectrum, since the high scattering and absorption values in the blue-green could lead to an excessive attenuation of the transmitted signal.

On the basis of these considerations, we started an *in vivo* study for the evaluation of absorption and transport scattering spectra, by means of time-resolved reflectance measurements. Data were collected not only from breast, but also from muscular and lipid-rich tissues (arm and abdomen, respectively), in order to try to identify the origin of the spectral features detected in the case of breast.

Materials and Methods

A synchronously pumped mode-locked dye laser (610-700 nm) and an actively mode-locked Titanium:Sapphire laser (700-1010 nm) were used as the illumination sources. A couple of plastic-glass fibers (1-mm diameter), placed parallel to each other at a relative distance of 2 cm, delivered the laser light into the medium and collected the reflected photons. A double microchannel plate photomultiplier and an electronic chain for time-correlated single-photon counting were used for the detection of time-resolved reflectance curves. Overall, the system transfer function was <150 ps FWHM. A small fraction of the incident beam was coupled to a 200 μm fiber and fed directly to the photomultiplier to account for any eventual time drift of the instrumentation. All the measurements were performed with an incident power <10 mW and the acquisition time was 4 s per wavelength.

The experimental time-resolved reflectance curves, collected every 5 nm, were fitted with the solution of the transport equation in the diffusion approximation for a semi-infinite homogeneous medium with the extrapolated boundary condition [9]. The theoretical curve was convoluted with the system transfer function and normalized to the area of the experimental curve. The fitting range included the points with a number of counts higher than 80% of the peak value on the rising edge of the curve and 1% on the tail. The best fit was reached with a Levenberg-Marquardt algorithm [10] by varying both μ'_s and μ_a to minimize the reduced χ^2.

OSA TOPS Vol. 21 Advances in Optical Imaging and Photon Migration, James G. Fujimoto and Michael S. Patterson (eds.)

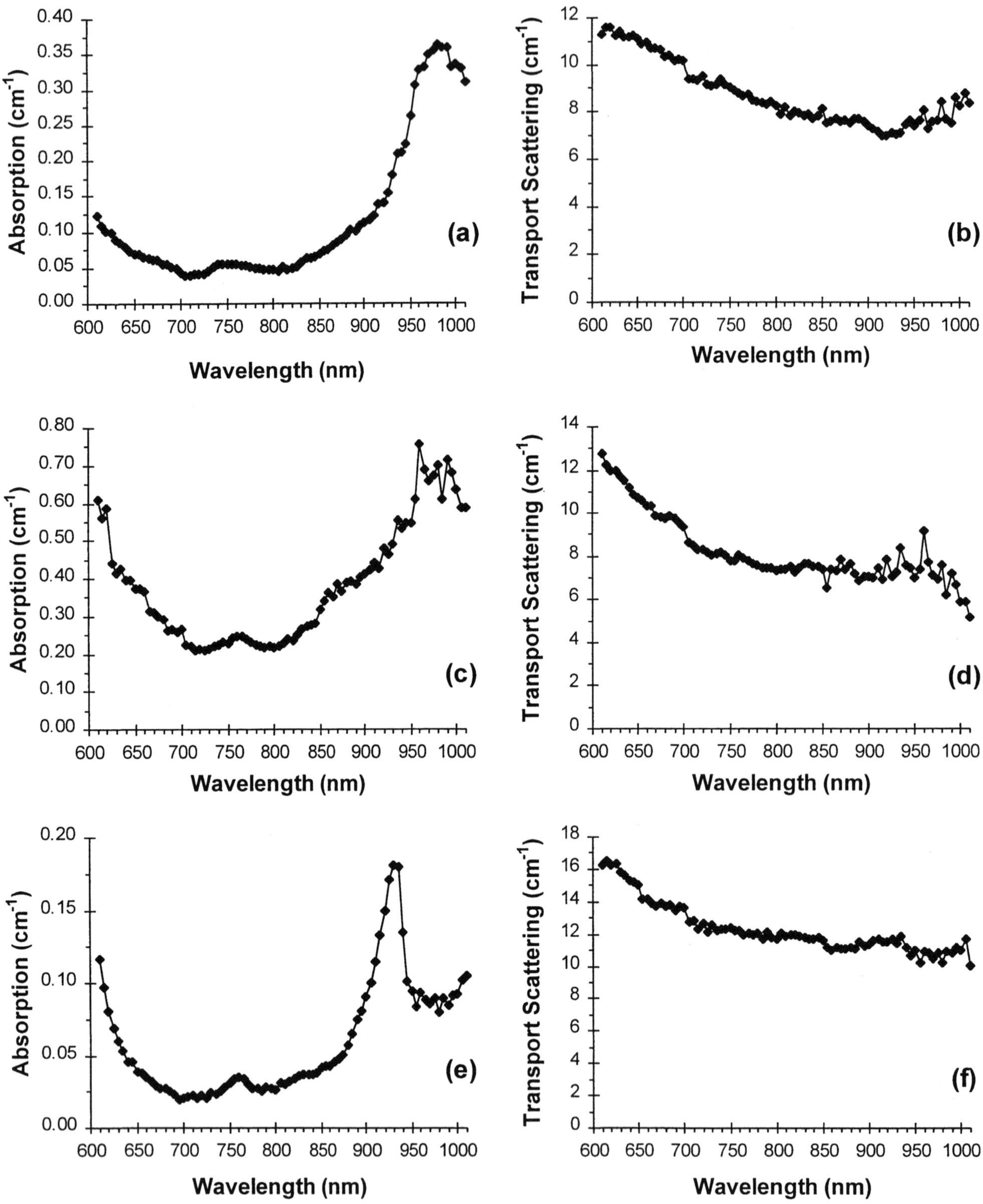

Figure 1. Typical absorption and transport scattering spectra of breast [(**a**) and (**b**), respectively], arm [(**c**) and (**d**)], and abdomen [(**e**) and (**f**)].

A more detailed description of the experimental set-up and of the fitting procedures can be found in Ref. 11.

Data were acquired from breast (at several sites, in the upper and lower middle region), abdomen (center region), and arm (bicipital region) of 5 healthy volunteers. At each site, two measurements were repeated on close positions.

Results

As shown in Fig. 1.a, in the red region of the visible spectrum, the absorption of breast progressively decreases from 0.1-0.15 cm^{-1} to 0.03-0.05 cm^{-1} upon increasing the wavelength. A similar spectral behavior is observed for the muscular tissue of the arm (Fig. 1.c), even though the absorption values are several times higher. For the adipose tissue of the abdomen (Fig. 1.e), the decrease is steeper at short wavelengths, while the absolute values of the absorption coefficients are comparable to the ones obtained for the breast.

In the near infrared, a small absorption peak is detected around 740-760 nm in the spectrum of the breast. A sharper peak, centered around 760 nm, is present even in the spectra of the abdomen and of the arm.

For longer wavelengths, in the case of the breast, a weak peak is observed at 830-840 nm. However, the spectrum is dominated by a major absorption peak around 970 nm. Depending on the volunteer, a further spectral feature can be detected around 930 nm as a shoulder or as a peak comparable to the one at longer wavelength.

In the long wavelength range, even for the arm the dominant absorption feature is the peak centered around 970 nm. On the other hand, the infrared absorption of the abdomen is characterized by a sharp peak at 930 nm, similar to what observed in some measurements on the breast.

For what concerns the behavior of the transport scattering as a function of wavelength, a progressive decrease is observed for all three tissue types considered, with no detectable spectral features either in the red or in the near infrared. Similar values of the transport scattering coefficient are detected in breast and arm (6-12 cm^{-1}), while the abdomen seems to be more strongly diffusive (12-18 cm^{-1}).

Discussion

Time-resolved reflectance spectroscopy with tunable laser sources is particularly appealing for the non-invasive optical characterization of biological tissues *in vivo*. If we take into account the absorption spectra of the main tissues components and we compare the results obtained from different tissue types, we can derive some information on the origin of the measured line shapes.

The rapidly decreasing absorption in the red region is likely influenced by the tail of the visible absorption of deoxyhemoglobin.

The absorption spectrum of deoxyhemoglobin is characterized also by a peak around 760 nm, which seems to be detected in all tissue types. A contribution at these wavelengths may also come from lipids, as suggested by the fact that significant absorption is detected in the case of lipid-rich abdomen.

Lipids are clearly detected also around 930 nm, as a peak or shoulder in the case of abdomen and breast, respectively.

The presence of oxyhemoglobin is not easily identified in the absorption spectra, since it is characterized by a broad absorption band centered around 900 nm, but by no sharp spectral features. However, its contribution is detected, especially at long wavelengths, as a high background in the strongly vascularized muscular tissue of the arm.

The dominant peak around 970 nm can be attributed to water, which is likely responsible also for the small peak or shoulder around 835 nm. Where the water content is higher (i.e. breast), even its absorption peak around 740 nm can be detected, as it is slightly blue-shifted with respect to deoxyhemoglobin and lipids (at 760 nm) and broadens the absorption feature observed in the range 740-760 nm.

If we consider the intra-subject variability, the optical properties of the breast seem to be more variable than those of the other tissue types examined. This behavior may be attributed to a more marked heterogeneity. For what concerns the inter-subject variability, significant differences were observed. In the case of the absorption spectra, this variance concerns not only the absolute values, but also the spectral shape, i.e. the relative weight of the different components.

Work is presently in progress to confirm the results obtained so far with a more accurate evaluation of the inter-subject variability and of the effects of the measurement position, and subsequently to extend the study to the assessment of the optical differences between pathologic areas and surrounding healthy tissues.

References

1. A. Yodh, B. Tromberg, E. Sevick-Muraca, and D. Pine eds., Diffusing photons in turbid media, Appl. Opt. 36, 9-231 (1997).
2. A. Yodh, B. Tromberg, E. Sevick-Muraca, and D. Pine eds., Diffusing photons in turbid media, J. Opt. Soc. Am. A 14, 136-342 (1997).
3. H. Moseley ed., Optical radiation technique in medicine and biology, Phys. Med. Biol. 24, 759-996 (1997).

4. A. H. Gandjbakhche, R. Nossal, and R. F. Bonner, "Resolution limits for optical transillumination of abnormalities deeply embedded in tissues," Med. Phys. 21, 185-191 (1994).
5. G. Mitic, J. Kölzer, J. Otto, E. Piles, G. Sölkner, and W. Zinth, "Time-gated transillumination of biological tissues and tissuelike phantoms," Appl. Opt. 33, 6699-6710 (1994).
6. K. M. Yoo, F. Liu, and R. R. Alfano, "Imaging through a scattering wall using absorption," Opt. Lett. 16, 1068-1070 (1991).
7. D. Contini, H. Liszka, A. Sassaroli, and G. Zaccanti, "Imaging of highly turbid media by the absorption method," Appl. Opt. 35, 2315-2324 (1996).
8. R. Cubeddu, A. Pifferi, P. Taroni, A. Torricelli, and G. Valentini, "Imaging with diffusing light: an experimental study on the effects of the background optical properties", Appl. Opt. (1998), in press.
9. R. C. Haskell, L. O. Svaasand, T.-T. Tsay, T.-C. Feng, M. S. McAdams, and B. J. Tromberg, "Boundary conditions for the diffusion equation in radiative transfer," J. Opt. Soc. Am. A 11, 2727-2741 (1994).
10. W. H. Press, S. A. Teukolsky, W. T. Vetterling, and B. P. Flannery, "Numerical recipes in C: The art of scientific computing" (New York, Cambridge University Press, 1992).
11. R. Cubeddu, A. Pifferi, P. Taroni, A. Torricelli, and G. Valentini, "Time-resolved imaging on a realistic tissue phantom: μ_s' and μ_a images versus time-integrated images," Appl. Opt. 35, 4533-4540 (1996).

Laser Doppler Imaging studies of tissue perfusion

Gert Nilsson, Department of Biomedical Engineering, Linköping University, S-581 85 Linköping, Sweden

Introduction

The cutaneous microcirculation possesses both temporal fluctuations and spatial variability. The temporal fluctuation can be of rhythmic origin such as vasomotion (Salerud et al 1983) or more stochastic in nature (Tenland et al 1983). The spatial variability has its origin in the heterogeneity of the cutaneous microvascular network, with richly perfused arteriovenous anastomosis in some areas and mainly consisting of capillaries in others. In order to be able to evaluate the cutaneous blood perfusion, a system that enables recording of both temporal fluctuations and spatial variability must be employed. Using Laser Doppler Perfusion Imaging (LDPI), several characteristics of the cutaneous circulation can be investigated, since this method allows for flow mapping of tissue perfusion over a pre-selected surface area to a depth of a few hundred micrometers as well as continuous measurement at discrete sites.

Operating principle

Laser Doppler Perfusion Imaging (Wårdell et al 1993) is a noninvasive technology that creates two-dimensional images of the tissue perfusion. A laser beam scans the tissue surface step-wise by means of a computer controlled stepper motor system positioned about 15 cm above the tissue surface (Fig 1). The images cover a size of about 12 x 12 cm. At each measurement site, the beam is arrested for about 50 msec. During this time period the backscattered light, partly Doppler shifted by the moving blood cells in the superficial microvascular network, is recorded by a photo-detector, positioned next to the stepper motors in the scanner head. On the surface of the photo-detector, frequency-shifted and non-frequency-shifted beams are mixed to form a photocurrent which is composed of a fluctuating portion superimposed on a stationary base level. The frequency content of the fluctuating portion is related to the average speed of the blood cells, while its magnitude is related to their concentration within the scattering volume. The average sampling depth in skin tissue is in the order of 0.2 mm. After signal processing, an output signal is generated that scales linearly with tissue perfusion defined as the product of the blood cell speed and concentration. This signal is stored in the computer memory and the beam is moved to the next measurement site. When all measurement sites (at maximum 4096 points) have been sampled, the stored values are displayed in the form of a color coded image on a computer monitor. This image can be further analyzed by use of the integrated analysis software or exported to other software packages for further analysis.

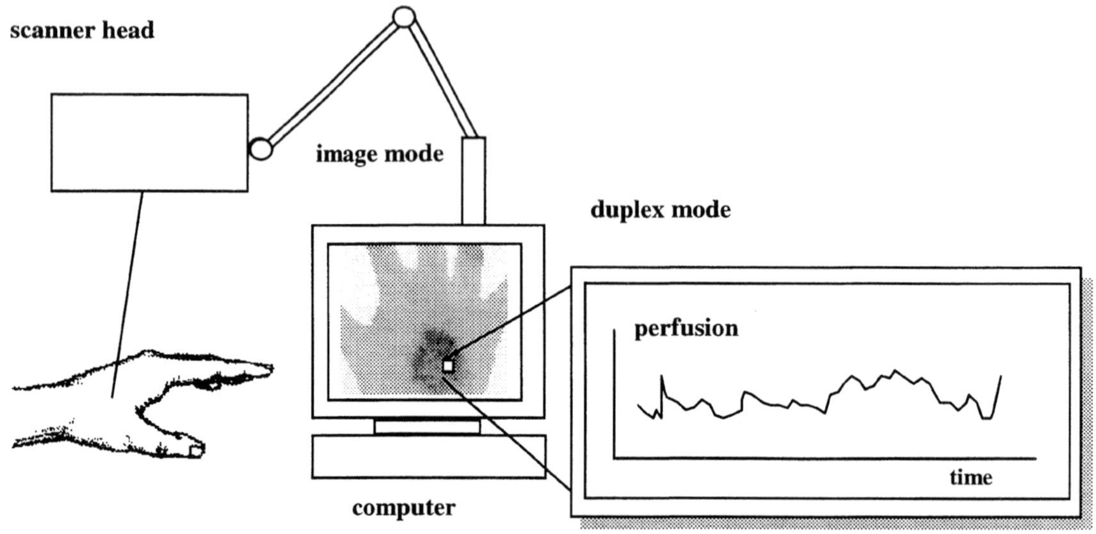

Fig. 1. Schematic diagram of the laser Doppler perfusion imager with duplex mode as an integral feature.

Recent development of the duplex mode (Wårdell et al 1996) combines the imaging features with continuous single- or multi-point measurement for tracking of the temporal variations in tissue perfusion. In the duplex mode, an overview image is generally first generated. Based on the information in this image, the laser beam is moved to an area of interest at which the temporal variations in tissue perfusion are continuously recorded and presented as a time trace. Spatial integration is attained by repeatedly scanning the laser beam over typically 4 or 16 adjacent but separate measurement points. The time trace represents the average perfusion recorded from these individual measurement point.

Image generation of the tissue perfusion with a Laser Doppler Perfusion Imager does not involve the use of dyes and tracer elements and offers the possibility of repeated investigations without physical contact with tissue.

More information about the laser Doppler perfusion imager PIM1.0 can be obtained from Lisca AB under address Lövsbergsvägen 13, S-589 37 Linköping, Sweden.

Assessment of cutaneous circulation

In the evaluation and treatment of a number of diseases and conditions that affect the skin, there is a need for cost-effective, objective and reliable medical technologies. The most important clinical applications for the emerging technology of laser Doppler perfusion imaging are likely to be found within the discipline of dermatology and may have a substantial impact on a variety of diverse fields including patch testing, screening for malignancies, multiple parameter assessment in association with cutaneous microdialysis and development of skin care products. A multi-national European project

(HIRELADO (High Resolution laser Doppler Perfusion Imaging in Dermatology, Project No IN10178I)) within the *European Commission* DG XIII/D *INNOVATION Programme* has therefore been initiated with focus on the following issues:

Standardisation of irritant and allergy patch testing. Today irritant and allergy patch testing is based on subjective procedures. The increasing number of potentially harmful chemicals we come in contact with in our daily life calls for more efficient and objective patch testing procedures to be developed.

Discrimination of malignant and benign skin neoplasmas. Ultraviolet radiation from the sun may produce malignant skin tumours such as melanoma, basal cell carcinomas and squamous cell carcinomas. A versatile and reliable method for discriminating malignant from benign neoplasmas is urgently needed in order to develop a standardised and cost-effective screening procedure for early detection of these conditions.

Visualisation of skin perfusion during evaluation by microdialysis of compounds which come in contact with the skin. Visualising the skin perfusion due to the tissue trauma caused by the introduction of the microdialysis fibre may help to interpret the measurement results as well as serving as an early indicator of a biological effect when a provoking substance is administrated via the microdialysis fibre.

Evaluation of skin perfusion in development of skin care products. Successful development of pharmaceutical and consumer products for skin care is dependent on the possibility to investigate both the beneficial and potentially harmful effects of the new products on the skin. Imaging of skin perfusion is a sensitive method for visualising and ascertain changes in the microvasculature.

The user needs within the four sub-projects listed above are not met by existing methods and procedures. Standardisation and harmonisation of the assessment and evaluation of various skin conditions and diseases are urgently required in order to meet these needs and to provide cost-effective and reliable alternatives to the procedures applied today.

References

Salerud, EG, Tenland, T, Nilsson, .E. and Öberg, P-Å. “Rhythmical variations in human skin blood flow”. Int J Microcirc: Clin & Exp, 2, 91-102, 1983.

Tenland, T, Salerud, GE, Nilsson, GE and Öberg, PÅ. “Spatial and temporal variations in human skin blood flow”. Int J Microcirc: Clin & Exp, 2, 81-90, 1983.

Wårdell, K, Jakobsson, AJ, Nilsson, GE “Laser Doppler Perfusion Imaging by Dynamic Light Scattering” IEEE Trans on Biomedical Engineering, Vol. 40, No 4, 309 - 316, 1993.

Wårdell, K and Nilsson, GE “ Duplex Laser Doppler Perfusion Imaging”, Microvascular Research 52, p171-182, 1996.

Time-domain optical mammography: results on phantoms, healthy volunteers, and patients

H. Rinneberg, D. Grosenick, H. Wabnitz, H. Danlewski
Physikalisch-Technische Bundesanstalt Berlin-Charlottenburg, Abbestraße 2-12, 10587 Berlin

K. Moesta, P. Schlag
Robert-Roessle-Klinik, Lindenberger Weg 80, 13125 Berlin-Buch, Germany

We have built a time-domain optical mammograph capable of recording optical mammograms within 3 - 5 min. A picosecond diode laser with an average output power of 5 mW provides laser pulses ($\lambda = 785$ nm) of 400 ps pulse width at a repetition rate of 80 MHz. The breast is gently compressed between two parallel glass plates and distributions of times of flight of photons are measured in transmission every 2.5 mm at about 2000 scan positions. Transmitted photons are collected by a 4 mm diameter fiber bundle, detected by a fast photomultiplier with a transit time spread of 110 ps followed by time-correlated single photon counting. Count rates up to 1 MHz allow to measure distributions of times of flight of photons N(x,y,t) of sufficient statistical accuracy at each scan position x,y within 100 ms only. For on-line visualization the total number of photons detected at each scan position is displayed.

Besides total photon counts different characteristic quantities such as photon counts in selected time windows, Fourier amplitudes, and phase shifts are derived from measured distributions of times of flight to generate optical mammograms. For illustration we show an optical mammogram of a compressed breast, 7.5 cm thick, with a tumor 2.5 cm in diameter, centered at x = -1.75 cm, y = 6.5 cm (s. Fig. 1). The tumor was also visible in X-ray and magnetic resonance mammograms. In addition, blood vessels can be discerned in Fig. 1. To generate the mammogram, the inverse of the fractional (f = 0.1) transmittance $T_n^f(x,y)$ defined as

$$T_n^f(x,y) = \frac{\int_0^{\tau_f} N(x,y,t)dt}{\int_0^{\tau_f} N(x_{ref}, y_{ref}, t)dt} = \frac{N_{0...f}(x,y)}{N_{0...f}(x_{ref}, y_{ref})}$$

has been plotted at each position x,y sampled. The upper integration limit τ_f is defined by

$$\int_0^{\tau_f} N(x_{ref}, y_{ref}, t)dt = f \int_0^{\infty} N(x_{ref}, y_{ref}, t)dt = N_{0...f}(x_{ref}, y_{ref}),$$

i.e. for f = 0.1 the time window of width τ_f corresponds to the first 10 % of all photons detected at the reference position ($x_{ref} = 0$, $y_{ref} = 0$) chosen. As can be seen from Fig. 1, at the location of the tumor fractional transmittance is reduced by about a factor of two compared to the reference position. In order to account for edge effects, mainly caused by a decrease in thickness of the breast, we have corrected measured distributions of times of flight N(x,y,t) according to the thickness d(x,y) of the breast.

Compared to frequency domain mammography at one modulation frequency, the additional information provided by time-resolved measurements may be exploited to deduce optical properties of the breast tissue and of the tumor in vivo. Modelling the compressed breast as a homogeneous infinite slab of thickness d = 7.5 cm, from measured distributions of times of flight transport scattering $\mu_s'(x,y)$ and absorption $\mu_a(x,y)$ coefficients of the breast were derived. Fig. 2 illustrates $\mu_s'(x)$ and $\mu_a(x)$ along the line $y = 6.5$ cm through the center of the tumor, derived from the same distributions of times of flight from which Fig. 1 was generated. At the center of the tumor the absorption coefficient is increased by about 10 %, whereas μ_s' is dominated by edge effects. It should be noted that only average optical properties can be deduced in this way. Diffraction of photon density waves by a spherical inhomogeneity modelling the tumor, might be a more realistic approach to deduce its optical properties.

In addition, we have measured optical properties of breast tissue in vivo at wavelengths ranging between 740 nm and 840 nm and at high lateral resolution. For this purpose distributions of times of flight through the breast of a healthy volunteer were measured every 2 mm along two orthogonal lines. The breast was compressed in craniocaudal direction to a thickness of 6 cm. As light source we used a mode-locked Ti:Sapphire laser, distributions of times of flight were recorded at high statistical accuracy, corresponding to total photon counts of $N_{0...1.0} = 4 \cdot 10^5$ at each location sampled. As can be seen from Fig. 3, the absorption coefficient μ_a exhibits a maximum at $\lambda = 760$ nm close to the absorption peak of deoxyhemoglobin. The transport scattering coefficient is seen to decrease monotonously with increasing wavelength.

In order to obtain information on the depth at which inhomogeneities are located we have recorded optical mammograms with the transmitting optical fiber and the receiving fiber bundle offset by different amounts Δx. As an example, Fig. 4 shows results obtained from five optical mammograms of a healthy volunteer recorded at displacements $\Delta x/\text{cm} = 0, \pm 1, \pm 2$, with the breast compressed in craniocaudal direction to a thickness of 5 cm. To generate mammograms (s. Fig. 4) corresponding to a depth of $z = 0$ cm, $z = -2.5$ cm and $z = -5$ cm, the five mammograms recorded were shifted by $\Delta x(z/d)$ and added subsequently, d being the thickness of the breast. As can be seen from Fig. 4, blood vessels close to the upper surface of the breast ($z = 0$) appear most clearly in the upper part of Fig. 4, whereas different blood vessels show up at $z = -5$ cm corresponding to the lower surface of the compressed breast.

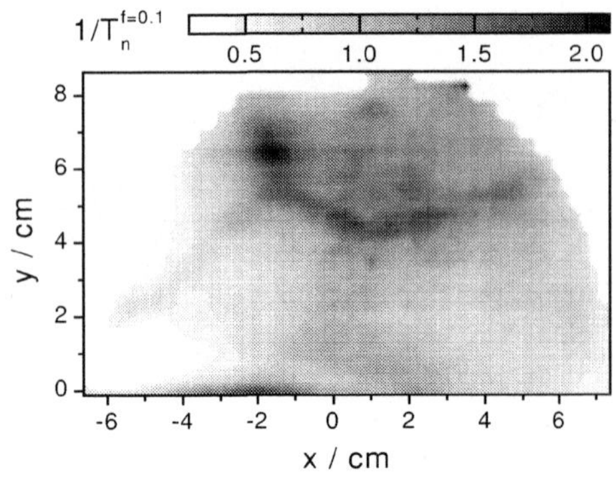

Fig. 1.
Optical mammogram (craniocaudal projection) of a breast with a tumor centered at x = -1.75 cm, y = 6.5 cm. Image based on reciprocal fractional transmittance $1/T_n^{f=0.1}$.

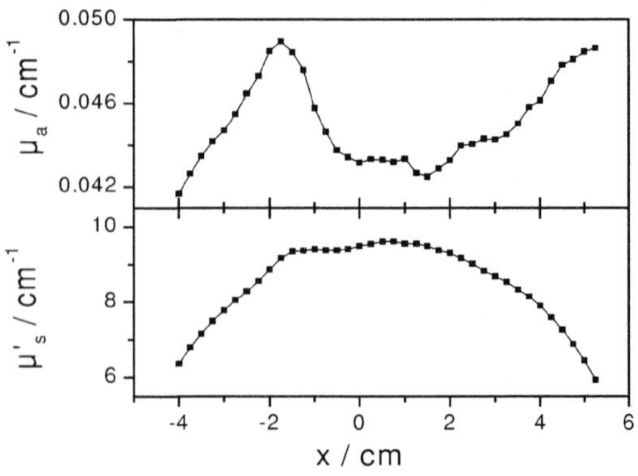

Fig. 2.
Transport scattering (μ_s') and absorption coefficients (μ_a) corresponding to line y = 6.5 cm of Fig. 1. Center of tumor at x = -1.75 cm, y = 6.5 cm.

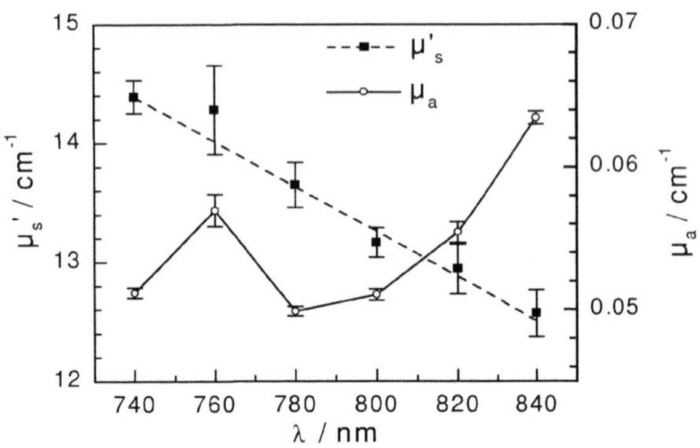

Fig. 3.
Transport scattering (μ_s') and absorption (μ_a) coefficients from line scans across the compressed breast of a healthy volunteer at various wavelengths. Data shown are mean values from 5 different locations, error bars (single standard deviations) represent distribution of transport scattering and absorption coefficients obtained from different locations.

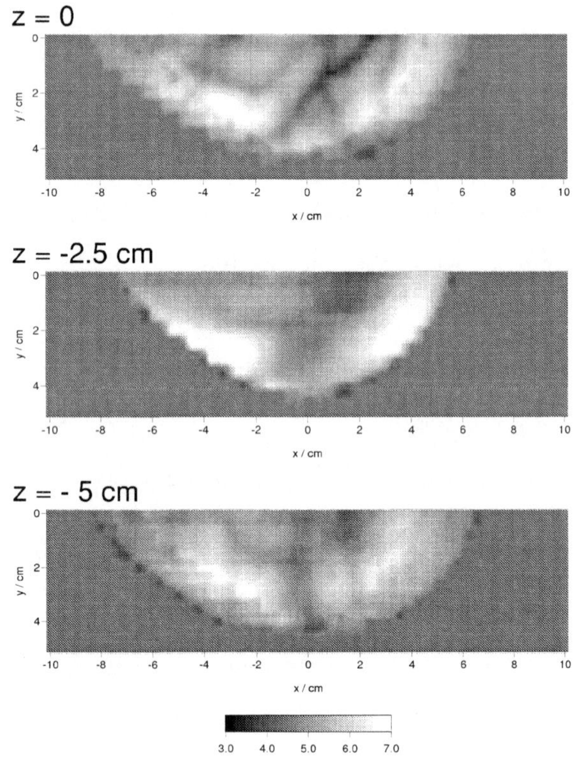

Fig. 4.
Optical mammograms of a breast of a healthy volunteer corresponding to different depths inside the breast (z = 0, z = -2.5 cm and z = -5 cm). Images have been derived from measurements at five different projection angles using total photon numbers $N_{0...1.0}$.

Diffraction Tomography with Diffuse Photon Density Waves: Clinical Studies and Background Subtraction

X. D. Li[†*], J. P. Culver[†], T. Durduran[†], B. Chance[*] and A. G. Yodh[†]
[†]*Department of Physics and Astronomy,* [*]*Department of Biochemistry and Biophysics*
University of Pennsylvania, Philadelphia, PA 19104

D. N. Pattanayak[†]
GE Corporate Research and Development, Schenectady, NY 12301

In recent years the application of near-infrared (NIR) diffusing light for biomedical diagnosis and imaging has gained favor due to its non-invasive nature, economy, and novel contrast relative to other diagnostics[1,2]. To this end a variety of techniques for imaging with diffuse light have been explored[3,4]. Most of these methods employ direct matrix inversion or iterative techniques for image reconstruction.

In this paper we use a recently developed imaging algorithm base on K-space spectral analysis[5] that is similar to conventional diffraction tomography[6]. It differs from least-square techniques such as ART and SIRT in that it is fast and non-iterative. In addition to providing information about the position and shape of hidden object(s), accurate optical properties of thin heterogeneities can be obtained from the projected images when information about their depth is available. We report here the first clinical images obtained by FFT analysis of DPDW measurements. In addition a new method for determining the background optical properties of an imaged sample and the influence of boundary conditions are discussed.

In uniform turbid media, the DPDW from a point source at $\mathbf{r_s}$ detected at position $\mathbf{r_d}$ has the form $U_0(\mathbf{r_d}, \mathbf{r_s}, \omega) = M_0 \exp(i\, k_0|\mathbf{r} - \mathbf{r_s}|)/(4\pi D_0|\mathbf{r} - \mathbf{r_s}|)$. Here ω is the angular source modulation frequency, M_0 is the ac amplitude of the source, $D_0 = v/(3\mu'_{so})$ is the diffusion coefficient (v is the speed of light in the medium), $k_0 = \sqrt{(-\mu_{ao}v + i\omega)/D_0}$ is the DPDW wave number, and μ_{ao} and μ'_{so} are respectively the homogeneous absorption and reduced scattering coefficients of the medium. In heterogeneous media, the total DPDW is a superposition of incident and scattered DPDW's. To the first order in the variation of optical absorption and reduced scattering coefficients, the scattered wave is

$$U_1(\mathbf{r_d}, \mathbf{r_s}, \omega) = \int_V T(\mathbf{r}, U_0(\mathbf{r}, \mathbf{r_s}, \omega))\, G(|\mathbf{r_d} - \mathbf{r}|)\, d^3r \ . \tag{1}$$

Here $G(|\mathbf{r_d} - \mathbf{r}|) = \exp(ik_0|\mathbf{r} - \mathbf{r_s}|)/(4\pi|\mathbf{r} - \mathbf{r_s}|)$ is the Green's function for a DPDW in a homogeneous medium, and $T(\mathbf{r}, U_0(\mathbf{r}, \mathbf{r_s}, \omega))$ is the inhomogeneity function. For absorbing objects $T_{abs}(\mathbf{r}, U_0(\mathbf{r}, \mathbf{r_s}, \omega)) = (-\delta\mu_a(\mathbf{r})\ v/D_0)\ U_0(\mathbf{r}, \mathbf{r_s}, \omega)$ and for scattering objects, $T_{scatt}(\mathbf{r}, U_0(\mathbf{r}, \mathbf{r_s}, \omega)) = (\delta\mu'_s(\mathbf{r})\ 3Do\ ko^2/v)U_0(\mathbf{r}, \mathbf{r_s}, \omega) - \nabla\ ln(\delta\mu'_s + \mu'_{so}) \cdot \nabla Uo(\mathbf{r}, \mathbf{r_s}, \omega)$ The integral is over the sample volume V.

In this paper, we consider a parallel-plane geometry as shown in Fig.1(a). For this case, utilizing the simple Weyl expansion of a Green's function[7], we can then take the 2-D spatial Fourier transform in the x-y plane of both sides of Eq.(1) and obtain the following relation (assuming a thin object at (x, y, z_obj))

$$\tilde{T}(p, q, z_{obj}, \mathbf{r_s}, \omega) \exp(-imz_{obj}) = \frac{2m}{i\,\delta z} \exp(-imz_d)\tilde{U_1}(p, q, z_d, \mathbf{r_s}, \omega) \ , \tag{2}$$

$\tilde{U_1}$ is the spatial Fourier spectrum of the scattered wave measured on the detection plane $z=z_d$, $m = (k_0^2 - p^2 - q^2)^{\frac{1}{2}}$ with $\mathrm{Im}(m) > 0$ and $\tilde{T}$ is the spatial Fourier spectrum the inhomogeneity function . The projection image can be obtained from the the 2-D inverse Fourier transform of $\tilde{T}$.

Experimentally the generation and detection of diffuse-photon density waves was carried out as described previously[5]. To demonstrate the feasibility of the FFT algorithm, we have performed amplitude and phase measurements on a tissue phantom with an absorbing target (Fig. 1). The amplitude and phase of the DPDW were recorded at each position for a total of 1024 points. We directly measured the amplitude and phase in the *homogeneous* medium to obtain $U_0(\mathbf{r_d}, \mathbf{r_s}, \omega)$.

An absorbing target made of resin, TiO_2 and absorbing dye with dimensions 1.5 x 1.5 x 0.4 cm^3 was submerged in the Intralipid mixture. The target had a μ_a = 0.10 cm^{-1} and a μ_s = 8.0 cm^{-1} similar to the background. The scattered wave $U_1(\mathbf{r_d}, \mathbf{r_s}, \omega)$ was obtained by subtracting the homogeneous background DPDW from the total measured signal. The 2-D Fourier transform of $U_1(\mathbf{r_d}, \mathbf{r_s}, \omega)$ leads to the inhomogeneity function $\tilde{T}(p, q, z_j, \mathbf{r_s}, \omega)$ in Eq.(2). The 2-D inverse Fourier transform of $T(p, q, z, \mathbf{r_s}, \omega)$ at $z = z_{obj}$ gives an accurate spatial map of the absorption variations $\delta\mu_a(x, y, z_{obj})$.

The reconstructed images from experimental data are shown in Fig. (1b-c). The object is well resolved . The reconstructed x-y position of the slice is about (-1.80, -0.25) cm, close to its true x-y positions (-1.6, -0.3) cm. The reconstructed absorption coefficient is well above the background noise level and close to the true value (e.g., $\mu_{a1}^{rec} \approx 0.14 \text{cm}^{-1}$). The FFT calculation is very fast and it takes less than 200 msec CPU time on Sun Sparc10 workstation.

This method has also been applied to clinical measurements of a breast tumor. The patient had a large fiberadenoma tumor in the left breast as determined by MRI and depicted schematically in Fig. 2a. Measurements were made in the compressed breast geometry for both the left and right sides. The detection fiber was scanned over a 9x9 cm grid with 0.6 cm steps for a total of 256 points. The homogeneous background U_o was obtained from measurements on a matching Intralipid phantom. Real space images of $\delta\mu_a(x, y)$ were obtained using the same algorithm as above. The difference of the reconstructed values of μ_a for the left and right breast (Fig. 2b) clearly depict the tumor. This demonstrates for the first time the utility of this image reconstruction technique in a clinical setting.

A continual problem for image reconstruction with DPDW's is accurate determination of the homogeneous background signal. Optimally one would obtain the background properties from the single data set of interest. A comparison was made of two methods for calculating the background optical properties of the heterogeneous sample depicted in Fig. 1. First the data was fit for μ_a and μ_s' assuming a homogeneous sample. The resultant μ_a and μ_s' values differed from the true values by -40% and -13% respectively. In Fig. 3 the deviations of the raw amplitude signal from a symmetric signal show which detectors are sampling the inhomogeniety most heavily. Improved values for the background properties can by obtained by fitting only the detectors that sample mostly background tissue. In this case the errors were only -25% in μ_a and +0.4% μ_s. We are currently working to optimize the selection of measurements from which to obtain background information.

The geometry used so far has been infinite. In practice, boundary effects present important problems. On the one hand, matching materials might be used to reduce the boundary effects. On the other hand, the boundary effects may be incorporated by adding a surface integral term of the form $\int_S U_{total}(\mathbf{r}, \mathbf{r_s}, \omega)(\frac{\partial}{\partial n} - \frac{1}{2D(r)})\, G(|\mathbf{r_d} - \mathbf{r}|)\, d^3r$ to the total detected photon density or using a better Greens function that vanishes on the extrapolated boundary. The FFT technique presented provides a basis for more complicated and realistic reconstruction methods to address these issues.

To conclude, we have successfully applied near field diffraction tomography, diffuse photon density waves, and FFT's to obtain projection images of hidden objects in highly scattering tissue phantom. The a approach has yielded clinical projection images with processing times much smaller than current collection times. We are currently working to incorporate in situ determination of background optical properties and to address more realistic account of the boundary conditions.

---

1. A. Yodh and B. Chance, Physics Today, March, 31(1995), and references therein.
2. See related studies in Advances in Optical Imaging and Photon Migration, R. Alfano, Ed., (OSA, Orlando, FL, 1996) and in Medical Optical Tomography: Functional Imaging and Monitoring, G. Müller, Ed., Vol.**IS11**, (SPIE Optical Engineering Press, 1993).
3. M. A. O'Leary, D. A. Boas, B. Chance and A. G. Yodh, Opt. Lett. **20**, 426(1995); C. P. Gonatas, M. Ishii, J. S. Leigh and J. Schotland, Phys. Rev. E **52**, 4361(1995); B. M. Pogue, M. S. Patterson, H. Jiang and K. D. Paulsen, Phys. Med. Biol. **40**, 1709(1995); H.B. Jiang, K. D. Paulsen, U. L. Osterberg, B. W. Pogue and M. S. Patterson, Opt. Lett. **20**, 2128(1995).
4. S. Fantini, M. A. Franceschini, G. Gaida, E. Gratton, H. Jess, W. W. Mantulin, K. T. Moesta, P. M. Schlag and M. Kaschke, Med. Phys. **23**, 149(1996).
5. X. D. Li, T. Durduran, A. G. Yodh, B. Chance and D. N. Pattanayak Opt. Lett. **22** 573 (1997).
6. E. Wolf, Opt. Commun. **1**, 153(1969); E. Wolf, in Trends in Optics, Anna Consortini, ed., (Academic Press, San Diego, 1996), p.83; J. Devaney, Inverse Problems **2**, 161(1986).
7. A. Baños, Dipole radiation in the presence of a conducting half-space, (Pergamon, New York, 1966) p.18.

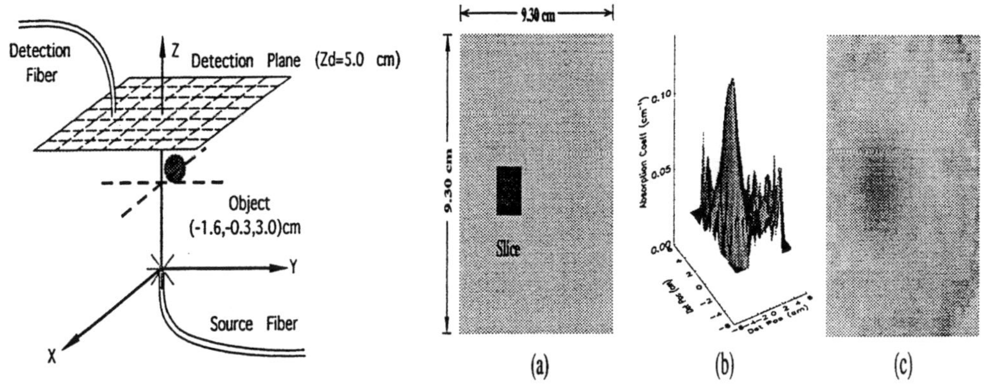

Fig. 1. (a)Experimental geometry: The source and detector fibers were placed in a large fish tank of 50L 0.75% Intralipid (μ_a=0.03 cm^{-1}, μ'_s=8.0 cm^{-1}). The source is fixed at the origin and the detector scans in a planar geometry at z=z_d=5.0 cm over a 9.3 x 9.3 cm^2 region. (b) shows the exact position of the slice. (c) and (d) illustrate the surface and planar images of an absorbing slice. The absorption coefficient and the position of the slice are recovered as shown in (c) and (d).

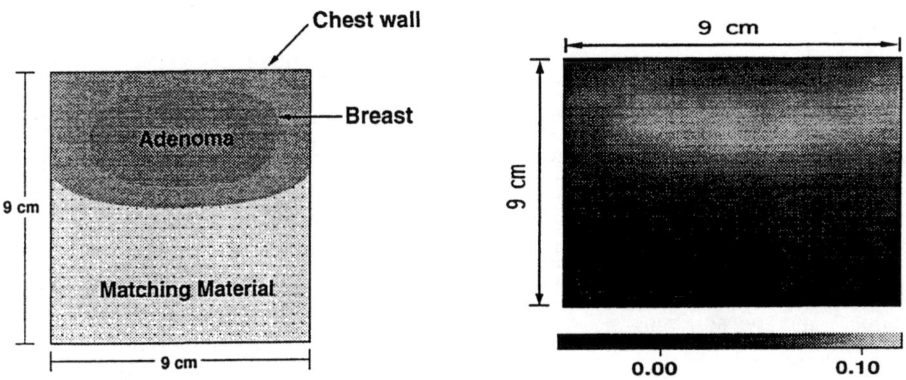

Fig. 2. Projected images of a patients breast using 2D K-space reconstruction algorithms: (a) The measurements were made in the compressed breast geometry with the tumor location independently determined by MRI. (b) Subtracting the reconstructed values of μ_s for the right breast from the left breast cancels out the modulations in the tissue optical properties near the rib cage thereby distinguishing the tumor more clearly.

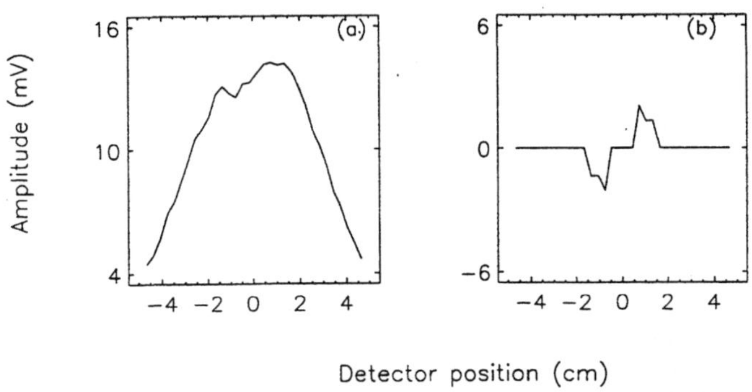

Fig. 3. (a) A one dimensional scan of the raw data for the absorbing target imaged in Fig. 1. (b) The deviations from a symmetric signal show which detectors are sampling the inhomogeniety most heavily. By fitting only the detectors that sample mostly background tissue, better background optical properties are obtained.

Concurrent multi-channel time-resolved NIR with MR mammography: Instrumentation and initial clinical results

Vasilis Ntziachristos[♠•], XuHui Ma[♠], Mitchell Schnall[♣], Arjun Yodh[♦], and Britton Chance[♠]

[♠]Department of Biochemistry Biophysics, [•]Department of Bioengineering,
[♣]Department of Radiology, [♦]Department of Physics
University of Pennsylvania
D501 Richards Bldg., Philadelphia PA19104-6089 USA
Tel. (215) 898 4387 Fax. (215) 898 1806

Abstract

We report the development of a multi-channel NIR imager-spectrometer based on the time-correlated single photon counting technique[1] for breast imaging. The instrument is capable of operating as a stand-alone modality or coupled to a MR scanner. Initial clinical results are also presented.

Keywords: NIR mammography, time resolved imager, NIR contrast agents.

Introduction

Optical mammography has gained an increased interest the last few years due to the expanding demand for alternative breast screening techniques[2] and the apparent suitability of the optical method for this application. An effective way to investigate the competence of NIR mammography as a clinically beneficial technique is to compare NIR images with images from other, well established modalities, such as X-ray Mammography or Magnetic Resonance Imaging (MRI)[3,4,5]. MRI is an excellent candidate for such an approach due to it's excellent sensitivity, the great anatomical information it renders and the fact that it employs non-ionizing radiation. The simultaneous acquisition of MR and NIR data can afford precise co-registration of images and examine the potential and the limitations of the optical technique. In return, by developing knowledge and experience on the NIR mammography capabilities, the MR specificity may be enhanced due to the additional information content of the simultaneous examination.

The time-resolved imager we have developed has the grat ability of acquiring images simultaneously with the MR examination. We have used specially designed soft compression plates that carry the optical fibers and the MR coils. We report on tests with the instrument on volunteers and patients. The NIRI examination protocol we designed runs in parallel with the standard MR breast imaging protocol at the Hospital of the University of Pennsylvania (HUP). In this report we have focused on the investigation of carcinoma contrast enhancement following intravenous administration of the Infracyanine[R] 25 (IC25), an NIR contrast agent.

Apparatus

Figure 1 depicts the block diagram of the instrument. The instrument uses two ~40μW, laser diodes pulsing at 5Mhz with a pulse width of ~50ps (a). The wavelengths used are 780nm and 830nm. A wavelength coupler (b) and a 19:1 signal splitter (c), are used to combine the two wavelengths and direct them to the 1x24 optical DiCon FiberOptics switch (e) and the reference branch (d). Graded index, 10m long optical fibers are used (f) to guide the light to the medial breast soft compression plate. Light collection is achieved by 8 step-index, 10m long fiber bundles (h). The fiber bundles are coupled to the lateral soft compression plate on one end and to the detector module on the other. The signal, after detection, is amplified and directed via the 8:1 router to the single photon counting, time-correlated system. The data are finally stored in a personal computer.

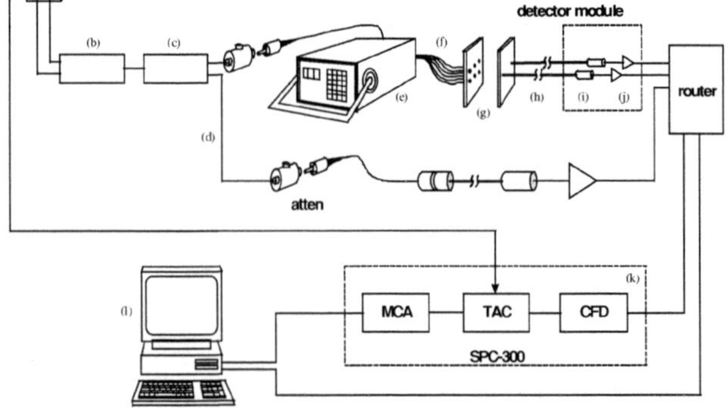

Figure 1: *Multi-channel, time correlated single photon counting NIR imager/spectrometer. See text for details*

We have tested the imager with three types of Hamamatsu Inc. detectors (i), the H5783-01multi-alkali PMT, the R4110U - 05MOD 8-channel MCP-PMT and the R5600U-50 GaAs PMT. We have chosen the R5600U-50 GaAs PMT for the breast examination due to the very attractive radiant sensitivity features. Figure 2 depicts the signal to noise (SNR) achieved by shining light through the breast of four volunteers (transmittance geometry) as a function of separation using the GaAs detector. All measurements are for one second of acquisition. Satisfactory signal to noise ratio is obtained even at 10 cm separations and it is typically further improved by acquiring data for longer times.

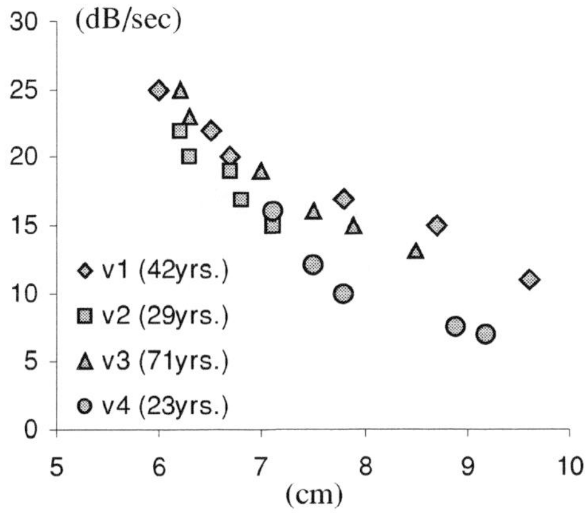

Figure 2: *Signal-to-noise ratio achieved using the GaAs PMT detector as a function of separation for one second of acquisition. The measurements are from four volunteers of different age and skin color.*

Clinical Set-up

Figure 3 shows the placement of the patient in the MR scanner. The plates are attached to the breast in such a way that it applies soft compression, minimizing movement and ensuring the full contact of the optical sources and detectors array with the breast surface.

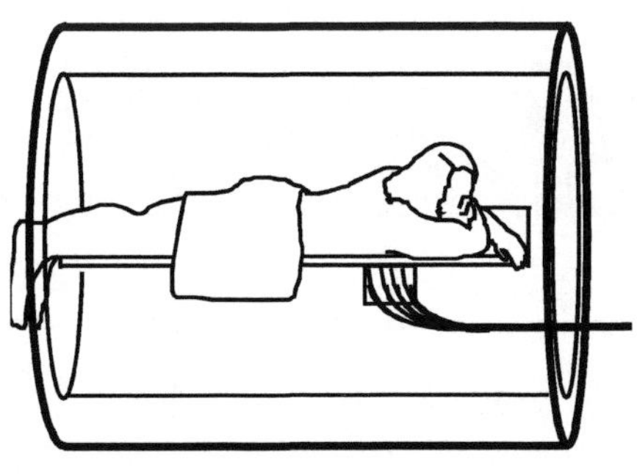

Figure 3: *Patient placement in the MR scanner and optical fibers attachment.*

Due to the long source fibers and fiber bundles the instrument is operated in a safe distance outside the Faraday cage of the scanner where minimal magnetic field exists. The breast soft compression plates used, carry the radio-frequency coils for the MRI, the optical probes for the NIRI and fiducials filled with water and copper sulfate ($CuSO_4$) which mark the position of the plates on the MR images for later retrieval of the exact geometry.

Examination Protocol

Patients already having a suspicious finding at the X-ray mammogram and scheduled for biopsy are recruited for an MR-NIR examination. The patients undergo the standard MR research protocol including pre- and post-gadolinium injection scanning. During the pre-gadolinium period an optical scan is acquired by switching light between all the different sources. Typical acquisition time is 10 – 20 sec per source. During the post gadolinium period we focus our attention to the volume that the suspicious region has appeared, using the information already produced by the pre-gadolinium MR images. We select the appropriate sources placed around that region and we acquire an image with an acquisition time of an average of 5-7 sec per source. At the end of the MR scanning, .2mg/kg of the NIR absorbing dye Infracyanine[R] 25 (IC25) are injected through the intravenous (i.v.) channel used for the gadolinium. We follow the initial rapid increase in absorption using one source repeatedly and we perform a scan for all the selected sources once a dye wash out plateau is reached (approximately three minutes after injection).

Theoretical approach

In order to investigate contrast enhancement due to the IC25 administration we use the pre IC25 scan as a baseline measurement and calculate the post-IC25 scan differential absorption coefficient for all source detector pair measurements with comparable separations. Absorption changes are calculated[6] using Eq.1

$$\Delta\mu_a = -\frac{2}{c\Delta t^2}\int_{t_1}^{t_2} \ln\frac{J_2(r,t)}{J_1(r,t)}\,dt \qquad (1)$$

where c is the speed of light in the medium, Δt is the time resolution of the pulse-height analysis[1] (PHA) of the multi-channel analyzer[1] (MCA), J_1 , J_2 is the photon current[7] measurement pre- and post-IC25 injection respectively. The interval $[t_1 , t_2]$ is the width of the J_1 time resolved curve. We have found that Eq.1 gives accurate $\Delta\mu_a$ calculation for small absorption changes[6] under the assumption of a constant background scattering coefficient, a scheme. The accuracy and sensitivity to absorption changes achieved with this method is less than 10^{-3} cm^{-1}.

Using such accurate determination of absorption changes we can construct simple projection images assuming a *projection plane* passing through the center of the suspicious lesion as shown in Fig. 4. We draw straight lines connecting those source detector pairs in space that have comparable separations. These lines cross the projection plane into points, shown in Fig. 4 as diamonds, forming a 2D spatial matrix. We assign the absorption differences, calculated with Eq.1 for each selected pair, to the associated matrix cell. This matrix is first interpolated to produce a rectangular grid and then interpolated again to improve the visual effect.

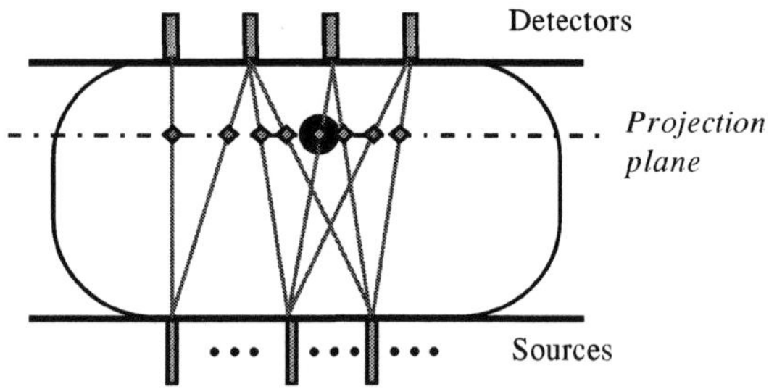

Figure 4. *Simple projection images demonstrating absorption or scattering absolute values or changes along source-detector straight lines helps us evaluate contrast enhancement due to IC2 injection.*

This procedure, although simplistic, is used as a first approach to examine the existence of detectable contrast at the detector plane and its correlation with the underlying pathology. Such discernible signal differences can further be employed into reconstruction algorithms reported earlier, for accurate 2D or 3D imaging and quantification.

In order to directly compare these NIR images with the corresponding MR images we simplify the MR multi-slice images by adding all the saggital slices (slices parallel to the projection plane) in order to create a single MR projection image as well. This image is geometrical similar to the one that an X-ray mammogram would hav presented and demonstrates Gd distribution along lines perpendicular to the chosen projection plane.

Clinical results

Figure 5 presents results from our first clinical trials. In this case the patient (70yrs old Caucasian) was diagnosed with an infiltrating ductal carcinoma approximately 8mm. Figure 5a depicts the MR saggital slice showing the cancerous lesion as a dark spot in the upper center of the image. During examination we focused on the area indicated with the dark rectangle around the cancerous lesion. Figure 5b shows the Gd-enhanced MR projection image of the outlined rectangle. Figures 5c and 5d show the NIR projection images at 780nm and 830nm. These images were produced using 6 sources and 8 detectors. The 780nm NIR image demonstrated more

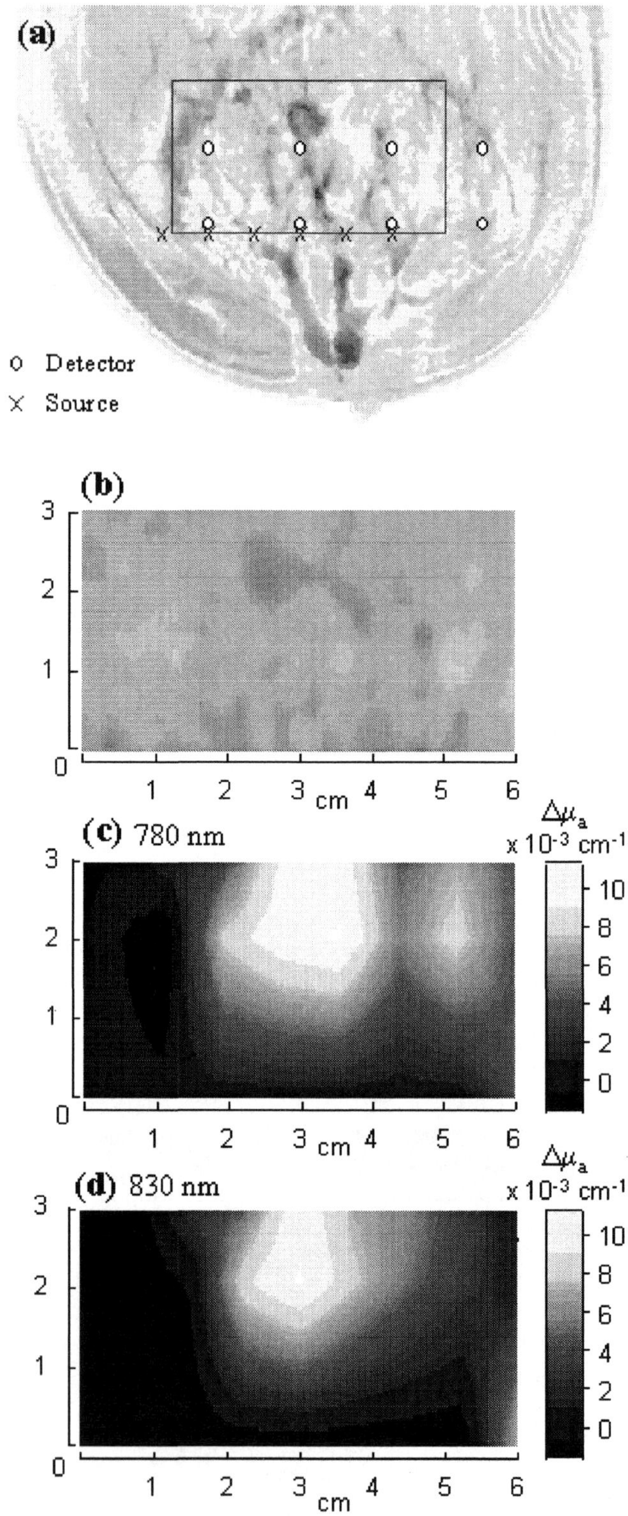

Figure 5. *a) MR saggital slice Gd enhanced. An infiltrating ductal carcinoma appears as a dark spot in the upper middle of the image. b) MR projection of the region of interest c), d780 and 830 NIR images, Infracyanine[R] 25 enhanced, in scale with 5b.*

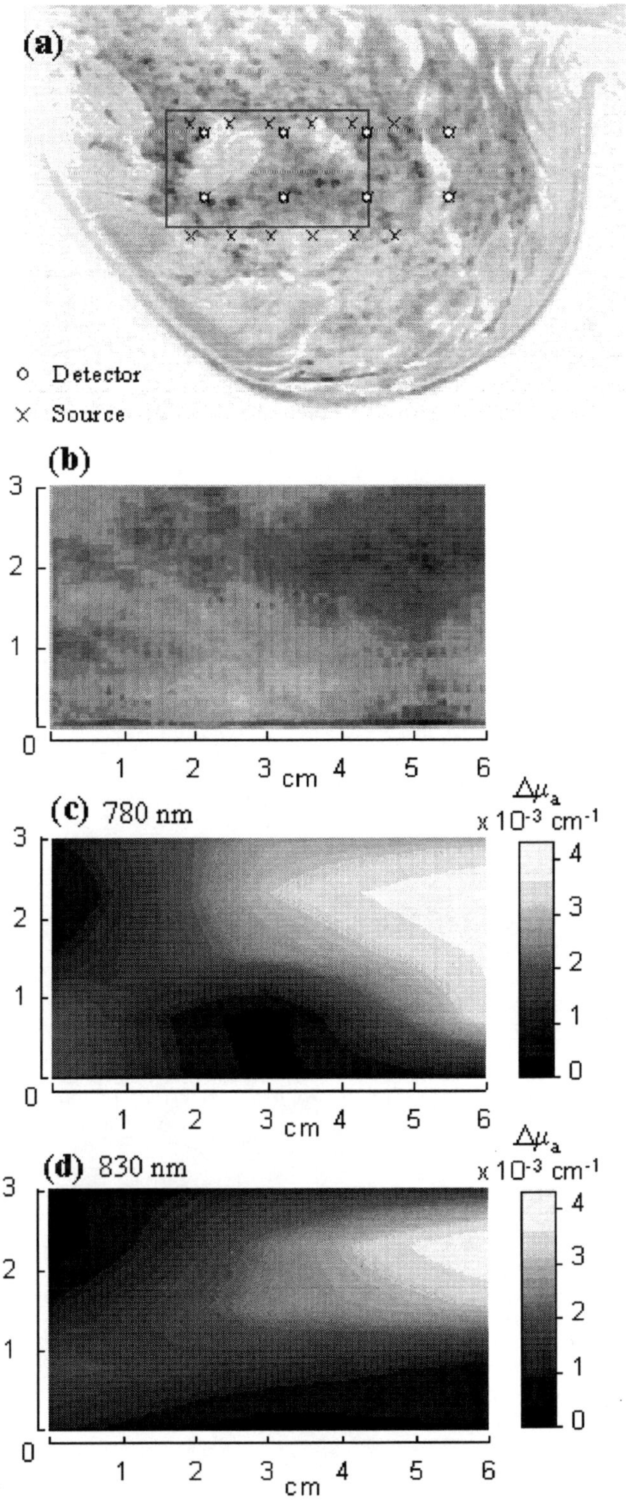

Figure 6. *a) MR saggital slice Gd enhanced. A group of cysts appear as a light gray area at the right side of the rectangle marking the NIR region of interest. b) MR projection of the region of interest c), d780 and 830 NIR images, Infracyanine^R 25 enhanced, in scale with 6b.*

contrast since Infracyanine^R 25 absorbs 12% more in the 780nm wavelength compared to the 830. The NIR images show very good correlation of the Gd distribution with the Infracyanine^R 25 distribution. The cancerous lesion shows a significant contrast enhancement of more than 10 times the background signal.

The second case we present here, belongs to a 45 yrs. old woman that had a palpable mass and was diagnosed having multiple cysts. The cysts are shown in Fig. 6a as a light gray area located at the right side of the region of interest (ROI), again indicated by a gray rectangle. The MR projection image of the ROI, depicted in Fig. 6b, shows no enhancement in the suspicious area as expected, however contrast appears at the upper right corner, indicating increased vascular concentration along the along the lines of projection. NIR Infracyanine^R 25 enhanced images show again a good correlation with the MR Image in terms of monitoring the distribution of the of enhancement.

Conclusion

The single-photon counting imager developed has shown to work harmoniously with the MR scanner. We have achieved satisfactory signal to noise ratio using the GaAs detector. Furthermore our first clinical attempts suggest accurate coregistration of Infracyanine^R 25 distribution with Gd distribution and have demonstrated a NIR contrast enhancement due to the contrast dye. Moreover, the small NIRI data set obtained so far shows coincidence with the sensitivity and specificity of the MRI, Gd examination.

Aknowledgements

We would like to thank Jean Mc Dermott and Norman Butler for their invaluable assistance in patient scheduling and examination and Thomas Connick for his help in building the coregistration plates. This work was supported by the National Institutes of Health grants No. CA 60182 and CA 72895.

References

1 Hamamatsu Editorial Committee, "*Photomultiplier tube*". 1st edition Hamamatsu Photonics K.K. March 1994

2 Mc Lelland, R. "*Screening for breast cancer. Opportunities, status and challenges*". In Recent Results in Cancer Research. Advances in Breast Cancer Detection, eds. S Brünner and B Langfeldt (Springer Verlag, Berlin), **119** pp.29-38 1990.

3 Franceschini MA. Moesta KT. Fantini S. Gaida G. Gratton E. Jess H. Mantulin WW. Seeber M. Schlag PM. Kaschke M. " *Frequency-domain techniques enhance optical mammography: initial clinical results*" Proc. Natl. Acad. Sci. USA **94**(12) pp.6468-73, 1997.

4 R.L. Barbour. H.L.Graber, J. Chang, SL S. Barbour, P.C. Koo, R. Aronson, "*MRI-Guided Optical tomography:*

Prospects and Computation for a new Imaging Method", IEEE Computational Science and Engineering , pp.63-77, 1995.

5 S. Nioka, M. Miwa, S. Orel, M Scnall, M. Haida, S.Zhao, B. Chance, "*Optical imaging of human breast cancer*", Oxygen transport to tissue XVI, New York 1994.

6 For details see Ntziachristos. V , Ma X., Yodh A., Chance B., "*Quantitation of functional motor cortex activity using time-resolved, spatially localized NIR spectroscopy*". TOPS on Biomedical Optical Spectroscopy and Diagnostics ('98).

7 M.S. Patterson, B. Chance, B.C. Wilson, "*Time-resolved reflectance and transmittance for the non-invasive measurement of optical properties*" Appl. Opt. **28**, pp.2331-2336, 1989.

Optical Characterization of Breast Tumors by Frequency-Domain Optical Mammography

Sergio Fantini[1], Scott A. Walker[1], Maria Angela Franceschini[1], Albert E. Cerussi[1], Julian Edler[1], K. Thomas Moesta[2], Peter M. Schlag[2], Michael Kaschke[3], and Enrico Gratton[1]

1 *Laboratory for Fluorescence Dynamics, Department of Physics, University of Illinois at Urbana-Champaign 1110 West Green Street, Urbana, Illinois 61801-3080*

2 *Robert Roessle Hospital and Tumor Institute, Humbodlt University, D 13122 Berlin, Germany*

3 *Carl Zeiss, Medical Optical Instruments, D 73446 Oberkochen, Germany*

Abstract

We present a method aimed at the non-invasive optical characterization of breast tumors. This method is based on using diffusion theory to model the propagation of intensity-modulated light in breast tissue. The tumor is assumed to be a spherical inhomogeneity. We report our first results obtained in the laboratory on a tissue-like phantom, and in the clinic on a patient affected by breast cancer. The *in vivo* result indicates that the malignant tumor considered has a higher absorption and a higher reduced scattering coefficient than healthy tissue.

Keywords

Medical optics instrumentation, Medical and biological imaging, Optical diagnostics for medicine, Imaging systems.

OCIS codes: 170.3830, 300.0300, 300.1030, 290.0290.

1. Introduction

Near-infrared light is a potentially effective probe to detect breast tumors non-invasively. By transilluminating the slightly compressed breast with intensity modulated light, we have found promising results in tumor detection [1]. Optical methods can proceed beyond the simple detection of optical inhomogeneities by providing information about the optical properties of such inhomogeneities. This article presents an approach aimed at the optical characterization of breast tumors by non-invasive optical methods in the frequency-domain.

2. Theory

Diffusion theory has been successfully employed to describe light propagation in turbid media such as tissue. The analytical solution of the diffusion equation for the photon density has been reported for homogeneous media [2], and for the case of media containing spherical [3] or cylindrical [4] inhomogeneities. We propose to use the analytical solution for one spherical object embedded in a uniform slab to extract information on the optical properties of breast tumors detected by frequency-domain optical mammography. This result is achieved by fitting the photon-density-wave amplitude and phase to the analytical solution. We performed the fit using the PMI (photon migration imaging) code [5].

3. Methods

Laboratory experiment

The basic features of the frequency-domain spectrometer are described in detail in Ref. [6]. Briefly, the light source is a laser diode emitting at 750 nm, whose intensity is modulated at a frequency of 120 MHz. The laser diode is coupled to a silica optical fiber 1 mm in core diameter. A second fiber (also 1 mm in core diameter) is used to collect the light signal. The two fibers, separated by 4.8 cm, are deeply immersed in a TiO_2 suspension having optical properties ($\mu_a = 0.020$ cm^{-1}; $\mu_s' = 8.7$ cm^{-1}) that match those of breast tissue in the near-infrared. The optical detector is a photomultiplier tube. Amplitude and phase of the modulated intensity at 120 MHz are obtained by heterodyning methods and digital filtering. An approximately spherical object (radius 0.6 cm, $\mu_a = 0.30$ cm^{-1}, $\mu_s' = 7.2$ cm^{-1}) made of hot melt glue was placed in between the source and detector fibers, at distances of about 1.9 and 2.9 cm, respectively. Frequency-domain data were taken while the source and detector fibers were scanned in tandem, always being collinear, along a line perpendicular to the source-detector direction (see Fig. 1(a)). The sample geometry is shown in Fig. 1(a). Figure 1(b) shows the amplitude and phase measured along the scanning line.

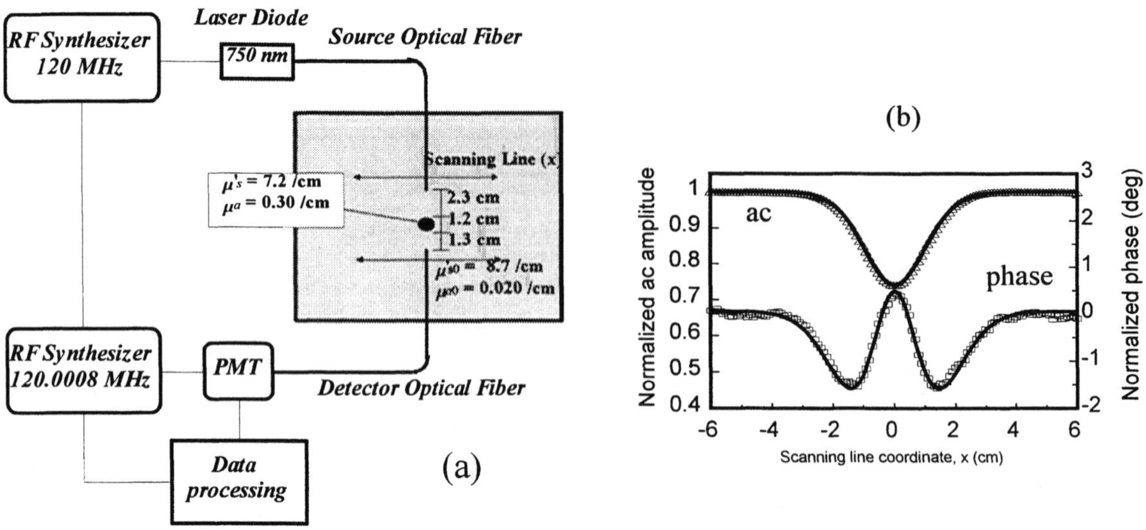

Figure 1. (a) Sample geometry for the phantom experiment. (b) The symbols are the normalized amplitude and phase measured along the scanning line. The lines are the best fit of diffusion theory to the data. The parameters obtained from the fit are reported in Table I.

Table I. Phantom experiment: comparison between the expected values of various parameters and the corresponding values recovered from the fit. The measured background optical properties are $\mu_{a0} = 0.020$ cm^{-1} and $\mu_{s0}' = 8.7$ cm^{-1}.

a (cm)		*x* (cm)		*z* (cm)		μ_a (cm^{-1})		μ_s' (cm^{-1})	
expected	fit	expected	fit	expected	fit	expected	fit	expected	fit
0.6	0.7	0	0	0.5	0.3	0.30	0.25	7.2	7.6

Clinical case

We acquired a frequency-domain optical mammogram on a 55-year-old woman affected by breast cancer. The light mammography apparatus (LIMA) developed at Carl Zeiss, Oberkochen, (Germany) and currently tested at the Robert Roessle Hospital and Tumor Institute, Berlin (Germany), is described in detail in Refs. [1,7,8]. It employs intensity modulated (110 MHz) laser diodes to acquire a two-dimensional projection image (corrected for edge effects) of the slightly compressed breast. The laser beam and the detector optical fiber, always kept collinear, are scanned in tandem. Fig. 2(a) shows the particular scanning line (the one showing the largest effect of the tumor on the optical data) considered in our analysis. By contrast with the laboratory experiment, where the data in the absence of the inhomogeneity are constant over the scanning line, the optical data on the breast show significant edge effects (larger amplitude, and smaller phase in locations closer to the breast edge, because of a reduced tissue thickness and other geometrical effects [8]). These edge effects must be accounted for, before the amplitude and phase data can be fitted with the analytical solution to the sphere-in-slab problem. We do that by fitting a smooth curve to the data out of the tumor region. This curve is shown in Fig. 2(b) for both the amplitude and the phase collected at 690 nm along the selected scanning line. These curves should be interpreted as being representative of the data we would have collected in the absence of the tumor (but in the presence of the edge effects). For this reason, we call them "background curves." We use the background curves to normalize the experimental data to account for the edge effects. Specifically, we divide the measured amplitude by the background amplitude, and we subtract the background phase from the measured phase. The resulting normalized data are shown in Fig. 2(c).

4. Results

The result of the fits of the diffusion equation solution to the experimental data are shown by the lines in Figs. 1(b) and 2(c). The fit in Fig. 1(b) has employed the solution for a sphere in an infinite medium, whereas the fit in Fig. 2(c) is based on the solution for a sphere in an infinite slab of thickness 4 cm. In both cases, the fitted parameters are the radius (a), the position (x along the scanning line, z along the source-detector line), the absorption (μ_a) and the reduced scattering (μ_s') coefficients of the sphere. In doing the fits, we have assumed no refractive index mismatch between the sphere and the surrounding medium, and we have used measured (for Fig. 1(c)) or literature [9] (for Fig. 2(c)) values for the host medium optical coefficients. The comparison between the values of the parameters recovered from the fit and the expected values is reported in Table I for the phantom experiment, and in Table II for the clinical case.

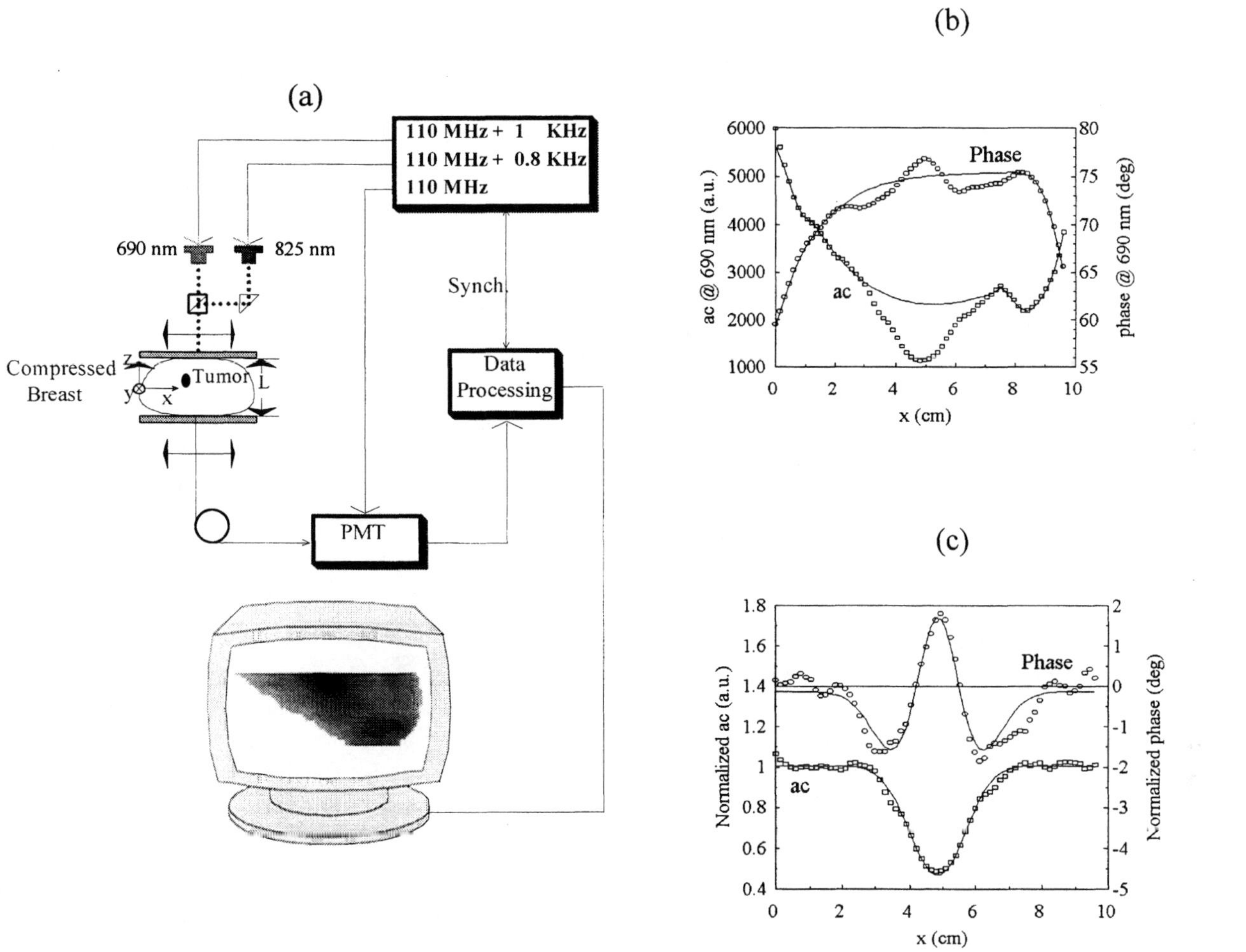

Figure 2. (a) Vertical cross-section of the slightly compressed breast and schematic diagram of the frequency-domain light mammography unit. The breast compression is $L = 4$ cm in the case reported here. (b) The symbols are the amplitude and the phase measured along the selected scanning line. The continuous lines are smooth curves through the points out of the tumor area. (c) The symbols are the normalized data. The lines are the best fit of diffusion theory to the data. The parameters obtained from the fit are reported in Table II.

Table II. Clinical case: comparison between the expected values of various parameters and the corresponding values recovered from the fit. The background optical properties are $\mu_{a0} = 0.03$ cm^{-1} and $\mu_{s0}' = 12$ cm^{-1} (from Ref. [9]).

a (cm)		x (cm)		z (cm)		μ_a (cm^{-1})		μ_s' (cm^{-1})	
expected	fit	expected	fit	expected	fit	expected	fit	expected	fit
0.8	1.1	4.9	4.9	0.5-1.0	0	–	0.070	–	15.2

5. Discussion

The results of the phantom experiments show that diffusion theory can be successfully employed to optically characterize inhomogeneities embedded in turbid media. The relatively small deviations between the predicted values and the fitted values of the parameters cannot be justified in terms of the instrumental uncertainties. Rather, they may be attributed to the non-perfect spherical shape of our inhomogeneity, and to a refractive index mismatch between the inhomogeneity ($n \sim 1.5$) and the background medium ($n_0 \sim 1.33$).

The results of the clinical case show that our non-invasive method is applicable on human breasts *in vivo.* The tumor size, the only fitted parameter on which we had knowledge (from the pathologist report), is in reasonable agreement with the size obtained from the fit. The tumor coordinate along the source-detector line (depth) deviates by several millimeters from the expected value estimated from the optical mammogram in the perpendicular projection. It may be more effective to obtain the depth coordinate by other methods (for instance by use of an off-axis detector fiber, or by perpendicular views of the breast), and take this value as a fixed parameter in our fit. The absorption and the reduced scattering coefficient of this malignant tumor (papillary cancer) are both greater than those of the surrounding healthy tissue.

It is worth stressing two points: (1) Our method does not provide a spatial reconstruction of the breast optical properties, but rather it takes a simplified approach based on dividing the breast into a spherical inhomogeneity (the tumor), and a uniform background (healthy tissue). (2) On the basis of recent studies [10,11], we expect that the simultaneous assessment of the optical properties and the size of tumors smaller than about 1 cm in diameter is not feasible. However, a parameter proportional to the tumor absorption coefficient can be obtained also for small tumors. Consequently, by using two or more wavelengths, it should always be possible to quantify the hemoglobin saturation of the tumor area, which is a physiologically relevant parameter.

6. Conclusions

We have presented a non-invasive method to obtain information on the optical properties of breast tumors *in vivo.* This approach has the potential to add specificity to frequency-domain optical mammography, which has already shown promising results in breast tumor detectability.

7. Acknowledgments

This research is supported by NIH grants RR03155 and CA57032.

8. References

[1] M. A. Franceschini, K. T. Moesta, S. Fantini, G. Gaida, E. Gratton, H. Jess, W. W. Mantulin, M. Seeber, P. M. Schlag, and M. Kaschke, "Frequency-Domain Techniques Enhance Optical Mammography: Initial Clinical results," Proc. Natl. Acad. Sci. USA **94**, 6468-6473 (1997).

[2] S. R. Arridge, M. Cope, and D. T. Delpy, "The Theoretical Basis for the Determination of Optical Pathlengths in Tissue: Temporal and Frequency Analysis," Phys. Med. Biol. **37**, 1531-1560 (1992).

[3] D. A. Boas, M. A. O'Leary, B. Chance, and A. G. Yodh, "Scattering of Diffuse Photon Density Waves by Spherical Inhomogeneities within Turbid Media: Analytic Solution and Applications," Proc. Natl. Acad. Sci. USA **91**, 4887-4891 (1994).

[4] S. A. Walker, D. A. Boas, and E. Gratton, "Photon Density Waves Scattered from Cylindrical Inhomogeneities: Theory and Experiments," Appl. Opt., *in press*.

[5] The PMI code is available on the world wide web at http://dpdw.eotc.tufts.edu/boas/PMI/pmi.html. The developers of the code are D. Boas, X. Li, M. O'Leary, B. Chance, A. Yodh, M. Ostermeyer, S. Jacques, G. Nishimura, and S. Walker.

[6] S. Fantini, M. A. Franceschini, and E. Gratton, "Effective Source Term in the Diffusion Equation for Photon Transport in Turbid Media," Appl. Opt. **36**, 156-163 (1997).

[7] M. Kaschke, H. Jess, G. Gaida, J. M. Kaltenbach, and W. Wrobel, "Transillumination Imaging of Tissue by Phase Modulation Techniques," in *Advances in Optical Imaging and Photon Migration*, Editor R. R. Alfano, Proc. OSA **21**, 88-92 (1994).

[8] S. Fantini, M. A. Franceschini, G. Gaida, E. Gratton, H. Jess, W. W. Mantulin, K. T. Moesta, P. M. Schlag, and M. Kaschke, "Frequency-Domain Optical Mammography: Edge Effect Corrections," Med. Phys. **23**, 149-157 (1996).

[9] H. Heusmann, J. Kölzer, and G. Mitic, "Characterization of Female Breasts *In Vivo* by

Time Resolved and Spectroscopic Measurements in Near Infrared Spectroscopy," J. Biomed. Opt. **1**, 425-434 (1996).

[10] D. A. Boas, M. A. O'Leary, B. Chance, and A. G. Yodh, "Detection and Characterization of Optical Inhomogeneities with Diffuse Photon Density Waves: a Signal-to-Noise Analysis," Appl.Opt. **36**, 75-92 (1997).

[11] G. Ledanois and J. Virmont, "Optical Medical Diagnostic and Imaging," in *Photon Propagation in Tissues III*, D. Benaron, B. Chance, and M. Ferrari, eds., Proc. SPIE **3194**, *in press*.

Optical imaging of breast tumor by using Dual wavelength amplitude cancellation system (phased array)

Shuoming Zhou, Shuning Huang, Chunhua Xie, Hong Long, Shoko Nioka, and Britton Chance

Department of Biophysics and Biochemistry
University of Pennsylvania, Philadelphia PA 19104
shuoming@mail.med.upenn.edu

INTRODUCTION

Early detection of the breast cancer can significantly decrease the mortality of this epidemic disease. Since NIH has recommended X-ray mammography for annual screening of woman over 50 years old for years, the mortality has indeed dramatically reduced. However there are some concerns about the use of X-ray mammography, we consider the X-ray to be an unsatisfactory technique for the screening purpose. X-ray mammography is insensitive to one fourth of cancer cases and itself is carcinogenic. It would not penetrate dense breasts, and we are unable to apply to younger subjects for woman under 40 years old. So if a person is in the niche of the under 40 group, the modality of X-ray mammography may a dangerous life-time accumulated dosage as well as a discomforting due to hard compression . In addition, the mammogram is turned white by the collagen fibers in the breast, and it is not possible to detect the cancer. The under 40 years old BRCA-1 population has a serious diagnostic problem.

We have chosen to developed an optical technique for breast cancer detection for the following reasons. We emphasize not only possible improvement of sensitivity of detection by amplitude cancellation technique , but also high specificity of cancer detection by functional imaging. It creates an oxygen image through hemoglobin oxygenation, a blood volume image and a scattering image, and therefore can be helpful in interpreting the cell function of lesions. Localization of absorbers/ fluorochromes deep within a highly scattering large body of tissue such as the human breast can be effected most precisely by photon diffusive waves(yodh, physical Review)that are in and out of phase amplitude modulated sources. The detector of in-phase and anti-phase signal establishes a null plane. Such a null is highly sensitive to perturbations by extremely small objects, of the order of the size of 70 microliters volume and containing as little as 20 picamole of an absorber (ICG).The dual wavelength amplitude cancellation imaged system(phased array) is applied to the detection the blood concentration and its oxygenation/deoxygenation state in human breast tumor .The sensitivity of the system is estimated by the detectable blood concentration and the corresponding oxygenation/deoxygenation of the blood in the 1 cm3 tumor phantom.We have done more than 100 experiments for human breast test .

Reprinted from Advances in Optical Imaging & Photon Migration, 1998 Technical Digest,

METHOD

We have shown that this system can be used in imaging using Dual wavelength amplitude cancellation image system (Figure 1).This device consists of 8 sources(750 nm, 830 nm) and 21 silicon photodioes for imaging capability. The sources are modulated by 50 us pulse and the pulse encode the dual wavelength sources(750 nm, 830 nm) . The image is acquired in 4 seconds.

Dual wavelength amplitude cancellation image system

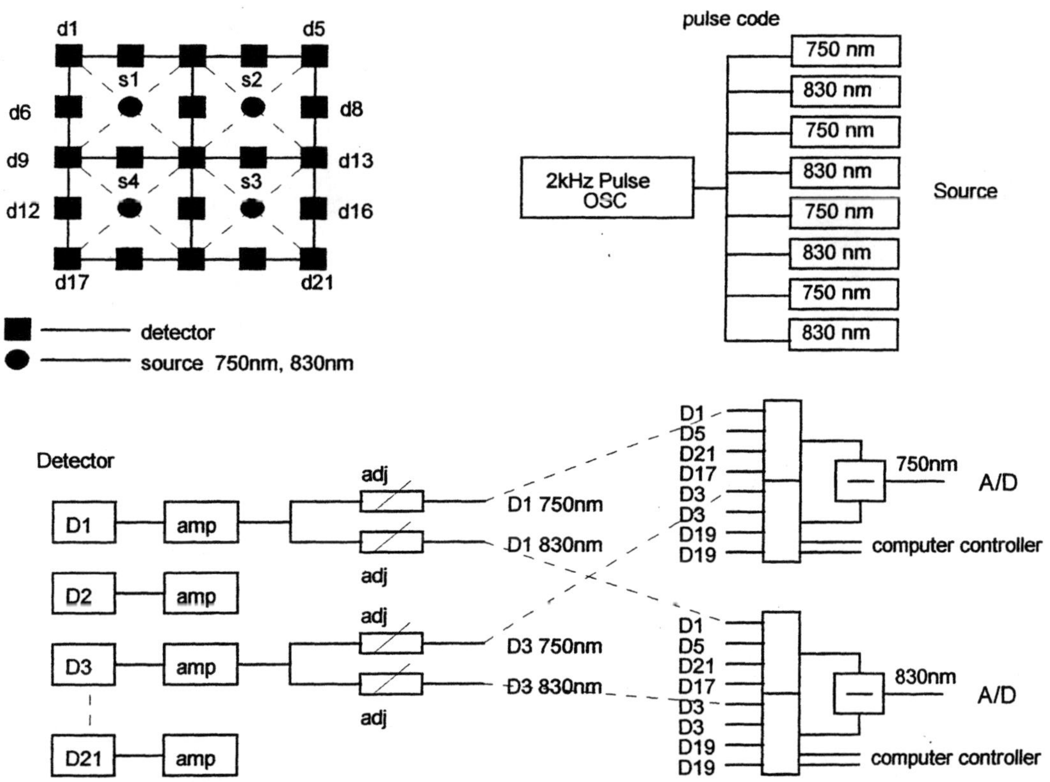

Figure 1. It shows schematic diagram of a dual wavelength amplitude cancellation image system .The 21 detectors and 8 sources are used to this image system.

EXPERIMENT

We have applied the dual wavelength amplitude cancellation image system to breast cancer study, and It can get the detection of blood volume and blood deoxygenation for breast tumor. This instrument give large signal from a small tumor for breast study. The experiment results are shown in figure.

Figure 2 shows that there are two tumors in the left side breast, and the right side breast is normal breast. The result has shown that there are two deoxygenating spots, and corresponding with blood volume increasing.

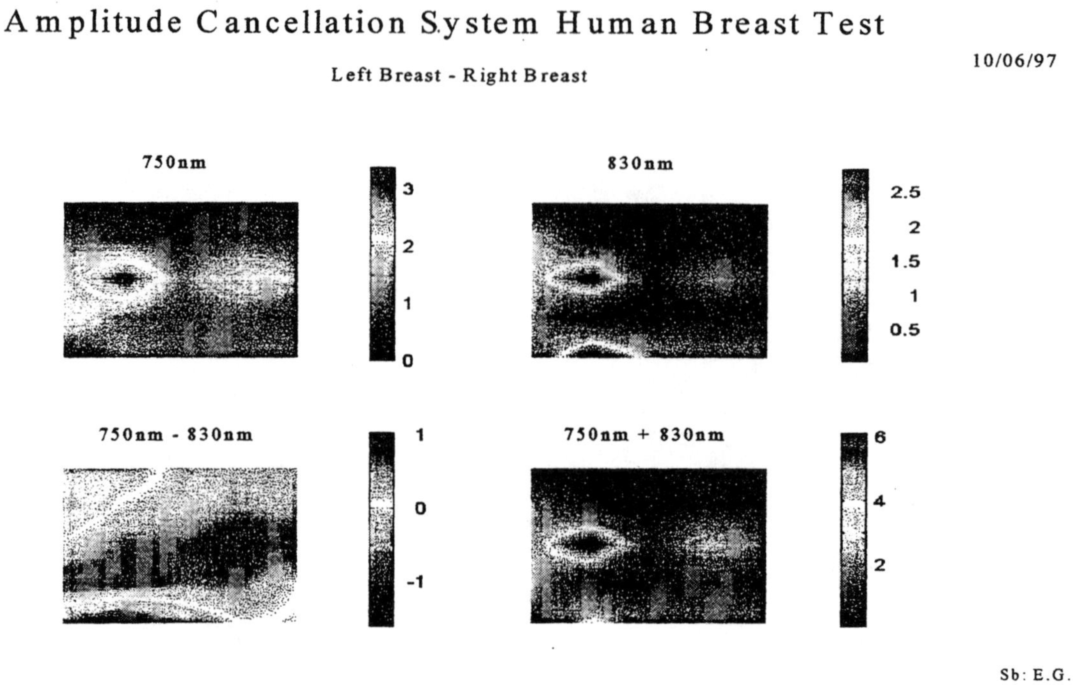

Figure 2.It is a patient for breast test.

REFERENCES

1.B.Chance,K.kang,L.He,J.Weng,and E.Sevrick Proc.Natl, Acad,Sci,USA,Vol.90.pp.3423-3427,APRIL 1993,MEDICAL SCIENCES.

2.S.Zhou,C.Xie,S.Nioka,H.Liu and B.Chance, phased array instrumentation appropriate to high-precision detection and localization of breast tumor.

3.B.Chance,K.kang L.He,H.Liu,and S.Zhou,Precision localization of hidden absorber in body tissues with phased-array optical system.

Quantitation of Brain Oximetry using a Model System and Piglet Brain.

Britton Chance, Hongyan Ma, Congwu Du[2], Carol Anderson[3], Gitesh Kumar[4], Joseph DeCesare[4]

Dept. of Biochemistry & Biophysics, University of Pennsylvania, Philadelphia, PA
[2]Center for Light Microscopy Imaging & Biotechnology, Carnegie Mellon University, Pittsburgh, PA
[3]Dept. of Neonatology, Allegheny University, Philadelphia, PA
[4]Perkin Elmer Corporation, Wilton, CT

The innovation of significant reliability in determining tissue optical properties offered by the 140 MHz IQ system has brought increased attention to optimum algorithms in relation to determined quantities of amplitude and phase measurements at multiple wavelengths and frequencies. Phase meters embodying these permutations are available. Most studies have focused on two wavelengths - amplitude and phase, and three wavelengths - phase only instruments. One key to the use of these instruments is the initial instrument calibration which is optimally carried out with intralipid ink models of known μ_a (absorption) and μ_s' (scattering). Data tables on the reproducibility in this determination for both dual wavelength, amplitude and phase determinations, as well as three wavelength, phase determination devices. Performance over the full range of hemoglobin oxygenation can only be obtained by the blood yeast model which can be configured to require infinite and semi-infinite approximations. Perturbation of scattering background, blood concentration and hemoglobin oxygenation state over the range representing animal and human brain models (μ_a 0.02-0.08 cm^{-1} and μ_s' from 6 - 12 cm)is essential and can readily be carried out with the above mentioned model for infinite and semi-infinite boundary conditions.

Previous studies have demonstrated that the absorption coefficient μ_a of tissue can be absolutely determined by measuring the amplitude and the phase delay of the photon density waves propagated through the tissue by frequency-domain technique. After computing μa^λ the saturation and concentration of hemoglobin can be obtained by the following equations.

$$[HbO_2] = \frac{\mu_a^{\lambda_1}\varepsilon_{Hb}^{\lambda_2} - \mu_a^{\lambda_2}\varepsilon_{Hb}^{\lambda 1}}{\varepsilon_{HbO_2}^{\lambda_1}\varepsilon_{Hb}^{\lambda_2} - \varepsilon_{HbO_2}^{\lambda_2}\varepsilon_{Hb}^{\lambda_1}} \quad \cdots\cdots (1)$$

$$[Hb] = \frac{\mu_a^{\lambda_2}\varepsilon_{HbO_2}^{\lambda_1} - \mu_a^{\lambda_1}\varepsilon_{HbO_2}^{\lambda_2}}{\varepsilon_{HbO_2}^{\lambda_1}\varepsilon_{Hb}^{\lambda_2} - \varepsilon_{HbO_2}^{\lambda_2}\varepsilon_{Hb}^{\lambda_1}} \quad \cdots\cdots (2)$$

$$Y = \frac{[HbO_2]}{[HbO_2 + Hb]} = \frac{\varepsilon_{Hb}^{\lambda_2} - \varepsilon_{Hb}^{\lambda_1} \cdot \frac{\mu_a^{\lambda_1}}{\mu_a^{\lambda_2}}}{(\varepsilon_{Hb}^{\lambda_2} - \varepsilon_{HbO_2}^{\lambda_2}) - (\varepsilon_{Hb}^{\lambda_1} - \varepsilon_{HbO_2}^{\lambda_1})\frac{\mu_a^{\lambda_1}}{\mu_a^{\lambda_2}}} \quad \cdots\cdots (3)$$

Another alternate method for quantitating saturation and concentration of hemoglobin is phase-only method, namely only using the measurement phase delay of the photon density waves. This method can minimize the experimental noise since the phase signal is affected less than amplitude signal by room light or other experiment interference in clinical environment. The principle of this method is based on the theoretical approximation that the phase shift can be simplified as a function of μa in some special case.

$$\frac{\mu_a^{\lambda 1}}{\mu_a^{\lambda 2}} \approx \frac{\theta^{\lambda 1} - \theta_o^{\lambda 1}}{\theta^{\lambda 2} - \theta_o^{\lambda 2}} \qquad for \qquad 2\pi f >> \mu_a^{\lambda 1, \lambda 2} \cdot c \qquad \cdots\cdots (4)$$

$$\frac{\mu_a^{\lambda 1}}{\mu_a^{\lambda 2}} \approx (\frac{\theta^{\lambda 1}}{\theta^{\lambda 2}})^2 \qquad for \qquad 2\pi f << \mu_a^{\lambda 1, \lambda 2} \cdot c \qquad \cdots\cdots (5)$$

Where $\theta_o^{\lambda} = -\sqrt{\frac{3\pi\mu_s' f}{C}}\rho$ is the phase delay caused by tissue scattering in the absence of absorption. Then the equation (3) can be rewritten as following:

$$Y \approx \frac{\varepsilon_{Hb}^{\lambda_1} - \varepsilon_{Hb}^{\lambda_2} \cdot \dfrac{\theta^{\lambda_1} - \theta_o^{\lambda_2}}{\theta^{\lambda_2} - \theta_o^{\lambda_2}}}{(\varepsilon_{Hb}^{\lambda_1} - \varepsilon_{HbO_2}^{\lambda_1}) - (\varepsilon_{Hb}^{\lambda_2} - \varepsilon_{HbO_2}^{\lambda_2})\dfrac{\theta^{\lambda_1} - \theta_o^{\lambda_2}}{\theta^{\lambda_2} - \theta_o^{\lambda_2}}} \qquad 2\pi f >> \mu_a^{\lambda_1, \lambda_2} \cdot c \quad \cdots (6)$$

$$Y \approx \frac{\varepsilon_{Hb}^{\lambda_1} - (\dfrac{\theta^{\lambda_1}}{\theta^{\lambda_2}})^2 \cdot \varepsilon_{Hb}^{\lambda_2}}{(\varepsilon_{Hb}^{\lambda_1} - \varepsilon_{HbO_2}^{\lambda_1}) - (\varepsilon_{Hb}^{\lambda_2} - \varepsilon_{HbO_2}^{\lambda_2})(\dfrac{\theta^{\lambda_1}}{\theta^{\lambda_2}})^2} \qquad 2\pi f >> \mu_a^{\lambda_1, \lambda_2} \cdot c \quad \cdots (7)$$

So the hemoglobin saturation (Y) can be directly determined from the two-wavelength phase measurement.

From our blood model and piglet experiment, we have found that the theoretical approximation ($2\pi f >> \mu_a^{\lambda_1, \lambda_s} \cdot c \;\; or \;\; 2\pi f << \mu_a^{\lambda_1, \lambda_s}$) are not practical for most biological applications and may cause calculation errors. Here we introduce an experiment ($\theta_{o,m}^{\lambda}$) to correct this error.

$$Y \approx \frac{\varepsilon_{Hb}^{\lambda_1} - \varepsilon_{Hb}^{\lambda_2} \cdot \dfrac{(\theta^{\lambda_1} - \theta_o^{\lambda_2}) - \theta_{o,m}^{\lambda_1}}{(\theta^{\lambda_2} - \theta_o^{\lambda_2}) - \theta_{o,m}^{\lambda_2}}}{(\varepsilon_{Hb}^{\lambda_1} - \varepsilon_{HbO_2}^{\lambda_1}) - (\varepsilon_{Hb}^{\lambda_2} - \varepsilon_{HbO_2}^{\lambda_2})\dfrac{(\theta^{\lambda_1} - \theta_o^{\lambda_2}) - \theta_{,,m}^{\lambda_1}}{(\theta^{\lambda_2} - \theta_o^{\lambda_2}) - \theta_{o,m}^{\lambda_2}}} \qquad 2\pi f >> \mu_a^{\lambda_1, \lambda_2} \cdot c \quad \cdots (8)$$

$$Y \approx \frac{\varepsilon_{Hb}^{\lambda_1} - (\dfrac{\theta^{\lambda_1} - \theta_{,m}^{\lambda_1}}{\theta^{\lambda_2} - \theta_{,m}^{\lambda_2}})^2 \cdot \varepsilon_{Hb}^{\lambda_2}}{(\varepsilon_{Hb}^{\lambda_1} - \varepsilon_{HbO_2}^{\lambda_1}) - (\varepsilon_{Hb}^{\lambda_2} - \varepsilon_{HbO_2}^{\lambda_2})(\dfrac{\theta^{\lambda_1}}{\theta^{\lambda_2}})^2} \qquad 2\pi f << \mu_a^{\lambda_1, \lambda_2} \cdot c \quad \cdots (9)$$

The measured phase shifts in saturation Y_1 and Y_2 can be acquired and then applied to equation (8) and (9) to calculate ($\theta_{o,m}^{\lambda_1}$) and ($\theta_{o,m}^{\lambda_2}$). The Y_1 and Y_2 can be measured by other invasive methods. The ($\theta_{o,m}^{\lambda_1}$) and ($\theta_{o,m}^{\lambda_2}$) can be transferred from small population to large population.

Since the phase-only two wavelength method needs to obtain the empirical ($\theta_{o,m}^{\lambda_1}$) and ($\theta_{o,m}^{\lambda_2}$), it is not very convenient for clinical application. We have developed 3 wavelength phase-only method. In other words , we don't the $\theta_{o,m}^{\lambda_1}$ and $\theta_{o,m}^{\lambda_2}$ if we add the third wavelength.

$$Y \approx \frac{(\varepsilon_{Hb}^{\lambda_1} - \varepsilon_{Hb}^{\lambda_2}) + (\varepsilon_{Hb}^{\lambda_3} - \varepsilon_{HbO_2}^{\lambda_2}) \cdot \dfrac{(\theta^{\lambda_1} - \theta^{\lambda_2})}{(\theta^{\lambda_3} - \theta^{\lambda_2})}}{(\varepsilon_{\Delta}^{\lambda_2} - \varepsilon_{\Delta}^{\lambda_1}) - (\varepsilon_{\Delta}^{\lambda_3} - \varepsilon_{\Delta}^{\lambda_2}) \dfrac{(\theta^{\lambda_1} - \theta_o^{\lambda_2})}{(\theta^{\lambda_3} - \theta_o^{\lambda_2})}} \qquad 2\pi f >> \mu_a^{\lambda_1,\lambda_2} \cdot c \qquad \cdots(10)$$

$$Y \approx \frac{(A-1) \cdot \varepsilon_{Hb}^{\lambda_3} - (A-C) \cdot \varepsilon_{Hb}^{\lambda_2} - (C-1) \cdot \varepsilon_{Hb}^{\lambda_2}}{-(A-1) \cdot \varepsilon_{\Delta}^{\lambda_3} + (A-C) \cdot \varepsilon_{\Delta}^{\lambda_2} + (C-1) \cdot \varepsilon_{\Delta}^{\lambda_2}} \qquad 2\pi f << \mu_a^{\lambda_1,\lambda_2} \cdot c \qquad \cdots(11)$$

Where $A \approx (\theta^{\lambda_1} / \theta^{\lambda_2})^2$, $C \approx (\theta^{\lambda_2} / \theta^{\lambda_3})^2$, $\varepsilon_{\Delta}^{\lambda_1,\lambda_2,\lambda_3} = \varepsilon_{HbO_2}^{\lambda_1,\lambda_2,\lambda_3} - \varepsilon_{Hb}^{\lambda_1,\lambda_2,\lambda_3}$

For three-wavelength phase-only PMS, the error is self-compensated and the hemoglobin saturation can be accurately determined by equation (10) or (11).

In the piglet brain, the tissue oxygen levels track between arterial and sagittal sinus levels in a way dependent upon the relative values of oxygen delivery, DO_2, and the tissue uptake, VO_2, tending towards VO_2 at high extraction levels and towards DO_2 in hyperemias. The quantitation of a tissue value have important impacts upon the management of brain hypoxia in neonates and adults, especially in cases where high oxygen extraction makes arterial values of poor diagnostic quality as is the case in many critical situations.

Optical coherence tomography and coherence imaging

High resolution imaging of *in vivo* cardiac dynamics using color Doppler optical coherence tomography

Siavash Yazdanfar[1], Manish D. Kulkarni[1], and Joseph A. Izatt[1,2]
Departments of [1]Biomedical Engineering and [2]Medicine
Case Western Reserve University, Cleveland, Ohio 44106
Tel: (216) 844-3298; Fax: (216) 844-8011

Optical coherence tomography (OCT) uses low-coherence interferometry to image internal tissue structures with micron-scale resolution [1]. OCT has become viable as a clinical diagnostic tool with the recent advent of high-power, low-coherence sources and near real-time image scanning technologies [2,3]. The high resolution (<20 μm in three dimensions) and dynamic range (>110 dB) of OCT allow for *in situ* tissue imaging approaching the resolution of excisional biopsy. A novel extension of OCT technology is color Doppler optical coherence tomography (CDOCT), which performs micron-scale resolution velocity flow mapping simultaneous with anatomical imaging [6,7]. In previous studies, quantification of flow within these vessels was achieved with velocity precision of less than 1 mm/s [6,7]. However, the minimum resolvable velocity in CDOCT is proportional to image acquisition rate [6]. Images must thus be acquired relatively slowly (>>1 s/image) in order to obtain high velocity resolution (<1 mm/s). This prevents imaging of dynamic structures due to motion artifact in living subjects. Thus the problem arises of overcoming motion artifact while retaining the capability of measuring flow in the microcirculation. We present a technique for motion-artifact-free gated reconstruction of dynamic flow in small vessels, and demonstrate its application in a beating *Xenopus laevis* heart.

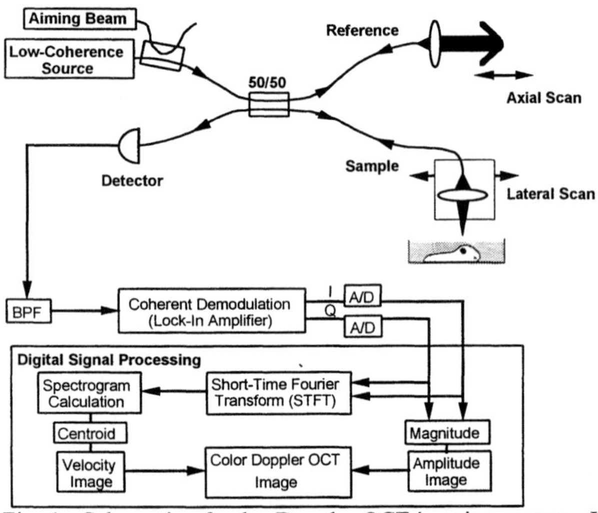

Fig. 1. Schematic of color Doppler OCT imaging system. I, in-phase; Q, quadrature; A/D, analog-to-digital converter. The sample probe beam is incident on the ventral surface of the specimen.

OCT uses low-coherence interferometry to obtain a map of highly localized reflection sites within tissue (Fig. 1). Axial ranging measurements (A-scans) are performed by scanning the reference arm length while synchronously recording the envelope of the detected interferometric signal. Two-dimensional cross-sectional images are formed by laterally scanning the probe beam across the sample. Doppler flow imaging using OCT is performed by using coherent detection to monitor moving scatterers within the sample [6]. The interferometric fringe frequency detected in OCT arises from the net sum of Doppler shifts generated by the moving reference mirror and moving scatterers in the sample. Coherent (phase-sensitive) demodulation of the detector current at the Doppler frequency induced by the constant motion of the reference arm results in the complex envelope of the interferogram. The local Doppler frequency generated by moving scatterers is estimated from the centroid of the localized power spectrum, obtained with the short-time Fourier transform (STFT), and related to the mean velocity of the scatterers.

The velocity resolution, $V_s^{\min}$, can be related to the image frame rate, R_f, by $V_s^{\min} = (\lambda_0/2n_t\cos\theta)\cdot(KLR_f/N\rho)$, where the physical parameters n_t, θ, and λ_0 represent the mean tissue index of refraction, the angle between the incident beam and direction of motion of scatterers within the sample, and the source center wavelength, respectively. The analysis parameters ρ, L, K, and N are the axial scanning duty cycle, the number of pixels per A-scan, the number of A-scans per image, and the STFT window size, respectively. Hence, for a given set of physical and analysis parameters, the desired frame rate compromises the minimum detectable velocity.

In vivo CDOCT measurements were performed in an experimental *Xenopus* model. With the sample beam incident on the ventral side of each specimen, OCT sections of the beating heart were acquired and analyzed for flow. The heartbeat of the specimen was measured using OCT "optical cardiograms" [4]. Reconstruction of the beating heart was performed by obtaining an OCT image which was oversampled in the lateral direction, ensuring that at least 5 A-scans were acquired per heartbeat while the sample was translated laterally by one focused sample probe beam spot size. From this image, separate time-gated cardiac image frames were extracted, each composed of A-scans occurring at the same segment of the cardiac cycle. Gating of the image data according to the heartbeat was performed by estimating the number of A-scans per heartbeat retrospectively, by selecting that value which eliminated motion artifact in the reconstructed frames. It must be noted that no *a priori* information was needed regarding the heart rate, although heart beating periodicity was required for the duration of image acquisition.

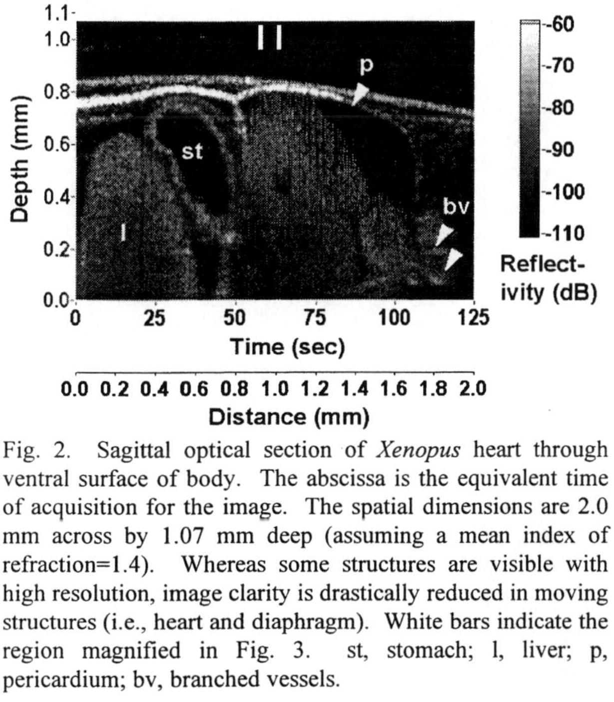

Fig. 2. Sagittal optical section of *Xenopus* heart through ventral surface of body. The abscissa is the equivalent time of acquisition for the image. The spatial dimensions are 2.0 mm across by 1.07 mm deep (assuming a mean index of refraction=1.4). Whereas some structures are visible with high resolution, image clarity is drastically reduced in moving structures (i.e., heart and diaphragm). White bars indicate the region magnified in Fig. 3. st, stomach; l, liver; p, pericardium; bv, branched vessels.

An oblique sagittal OCT section through the ventral surface of a specimen is shown in Fig. 2. Although stationary structures are clearly delineated in the OCT image, the motion artifact, indicated by alternating dark and light vertical bands in the center of the image, blurs the image data within the pericardium. The general shape of the heart is apparent, but large structures such as the distinct chambers are not resolvable.

Fig. 3 isolates a selected smaller time segment from the OCT image in Fig. 2, demonstrating the repetitive expansion and contraction of the heart that result in motion artifact. This figure illustrates that each heart beat comprised approximately five A-scans. Therefore, approximately every 5th A-scan occurred at the same segment of the cardiac cycle, during which the sample beam was translated less than one focused sample probe beam spot size.

Fig. 4 is a gated CDOCT reconstruction of the beating *Xenopus* heart performed using the five sequential time frames extracted from Fig. 2. Gating of the data was performed according to the procedure described earlier, in which it was determined that heart rate was 1.6 beats per second and T=5.00±0.01. Color Doppler flow processing was performed on the region of interest indicated by the rectangle surrounding the truncus arteriosus and branched vessels. In this black and white reproduction, both positive and negative Doppler shifts appear black. By assembling the sequential gated color Doppler images into a movie, it is clearly observed that the truncus arteriosus, the portion of the aorta immediately after exiting the ventricle, generated a positive pulsatile Doppler shift corresponding to the time of ventricular contraction. The Doppler processed vessel outside the pericardium, apparently a vein, generated a less pulsatile negative Doppler shift during the time of ventricular filling. Motion of the vessel walls also resulted in Doppler shifts in the backscattered light. In Fig. 4, the upper region of the arterial wall expands during systole, generating negative Doppler shifts that could be misinterpreted

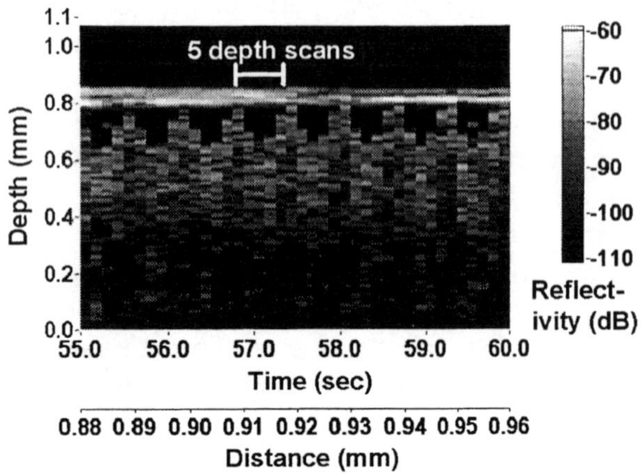

Fig. 3. A-scans extracted from Fig. 2 between the vertical white bars within the ventricle. One period of the cardiac cycle required exactly five (T=5.00±0.01) A-scans.

as flow. Hence, although motion artifact in the amplitude (OCT) image has been avoided, velocity artifacts are inevitable due to Doppler shifted backscatter from dynamic structures. This artifactual "clutter" is also prevalent in clinical Doppler ultrasonography.

In conclusion, we demonstrate reconstruction of the beating heart of the *Xenopus* by retrospective gating of the cardiac cycle. This method allows imaging of dynamic, periodic biological processes with high spatial acuity, without *a priori* knowledge of the rate of the processes.

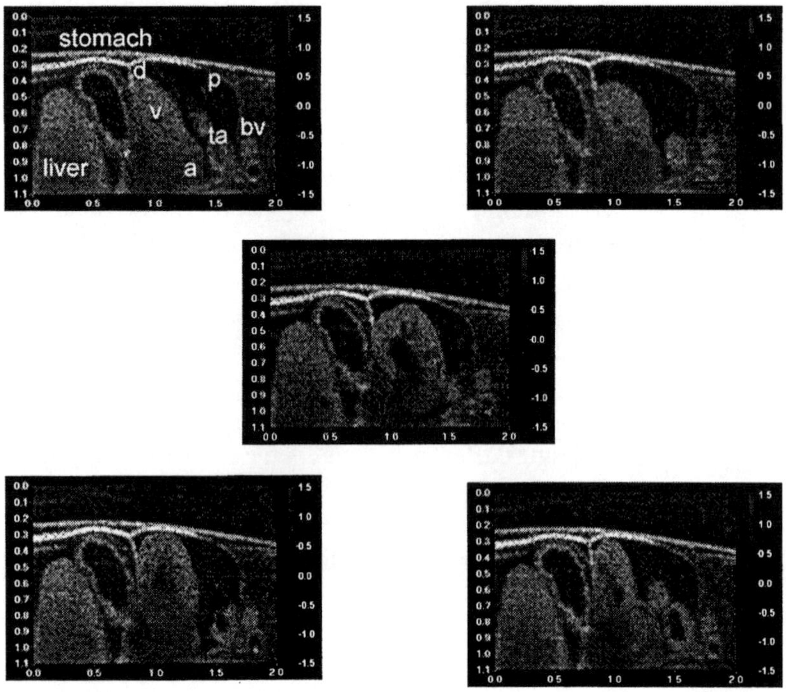

Fig. 4. Reconstruction of a beating *Xenopus* heart using the frame gating technique. Doppler processing is restricted to the region surrounding the truncus arteriosus, ta, and branched vessels, bv. v, ventricle; a, atrium; p, pericardium; d, diaphragm. In this black and white reproduction, both positive and negative Doppler shifts appear black.

References

1. D. Huang, et al., Science **254**, 1178-1181 (1991).
2. A. M., Sergeev, et al., *OSA Trends in Optics and Photonics on Advances in Optical Imaging and Photon Migration*, R. R. Alfano and J. G. Fujimoto, eds. (Optical Society of America, Washington, DC 1996) **2**, 196-199 (1996).
3. G. J. Tearney, et al., Science **276**, 2037-2039 (1997).
4. S. A. Boppart, et al., Proc. Natl. Acad. Sci. **94**, 4256-4261 (1997).
5. S. A. Boppart, et al., J. Neurosci. Methods **70**, 65-72 (1996).
6. J. A. Izatt, et al., Opt. Lett. **22**, 1439-1441 (1997).
7. Z. Chen, et al., Opt. Lett. **22**, 1119-1121 (1997).

In-Vivo Catheter-Based Imaging with Optical Coherence Tomography

Costas Pitris[1,2], Parth Patwari[1], Stephen A. Boppart[1,2], Brett E. Bouma[1], Gary J. Tearney[1], James G. Fujimoto[1], Mark E. Brezinski[3]

[1]Department of Electrical Engineering and Computer Science,
Research Laboratory of Electronics, Massachusetts Institute of Technology, Cambridge, MA 02139
[2]Harvard-MIT Division of Health Sciences and Technology,
Masachusetts Institute of Technology, Cambridge, MA 02139
[3]Cardiac Unit, Harvard Medical School,
Massachusetts General Hospital, Boston, MA 02114

Abstract

The high incidence and mortality rate associated with complications resulting from coronary artery atherosclerosis dictate an important and immediate need for a new, high resolution, diagnostic imaging technology that would allow characterization of artery microstructure with a micron scale resolution. OCT allows catheter based, cross-sectional imaging of blood vessels with a 10-20 μm resolution and is ideally suited for *in vivo* imaging of arterial structure. Recently, OCT imaging was extended to in vitro pathology, including multiple studies on atherosclerotic arteries. OCT has also been directly compared with a 30 MHz, IVUS transducer. In addition to qualitatively superior resolution with respect to imaging plaque, OCT was demonstrated to be quantitatively superior. *In vivo* imaging was performed on normal, New Zealand White rabbits using a novel rapid data acquisition system which allows imaging at 4 frames per second. This work, focusing on dynamic changes, demonstrates that high resolution intravascular imaging can be performed *in vivo* using optical coherence tomography. Images at 10 μm resolution were obtained, representing an improvement of over 10X that of 'state of the art' IVUS transducers.

Key Words: (170.4500) Optical coherence tomography; (170.4580) Optical diagnostics for medicine, (170.7180) Ultrasound diagnostics.

Introduction

The high incidence and mortality rate associated with complications resulting from coronary artery atherosclerosis dictate an important and immediate need for a new, high resolution, diagnostic imaging technology that would allow characterization of artery microstructure with a micron scale resolution. Despite the fact that the rupture of thin wall lipid filled plaques was established as the major cause of acute coronary syndromes [1], current imaging technologies can not identify these plaques prior to rupture and they are not adequate to guide interventional procedures, which usually only involve a few millimeters of tissue [2,3]. The current clinical technology with the highest resolution, 30 MHz high frequency ultrasound or IVUS, has an axial resolution of approximately 110 μm and can not identify critical plaque microstructure, which is generally below 100 μm. [3-5]

OCT, a new technology that allows catheter based, cross-sectional imaging of blood vessels with a 10-20 μm resolution, is ideally suited for *in vivo* imaging of arterial structure. [6] OCT is analogous to ultrasound, measuring the intensity of backreflected infrared light rather than acoustical waves. Instead of measuring the echo delay time, however, a low coherence Michelson interferometer is used to measure the pathlength and to determine the spatial origin of the backreflected light. In order to produce cross-sectional tomographic images, the light beam is scanned across the tissue to produce two and three dimensional data sets. This technique was successfully applied to intraocular imaging [7].

Recently, OCT imaging was extended to in vitro pathology in nontransparent tissue[8-10]. This work includes multiple studies on atherosclerotic arteries [8, 11,12]. Structures characteristic of unstable plaque, such as thin intimal caps, focal lipid collections, and fissures have been identified in vitro at 16 μm resolution[8]. Imaging completely through the width of a normal coronary artery has been achieved[11]. OCT has also been directly compared with a 30 MHz, IVUS transducer[12,13]. In addition to qualitatively superior resolution with respect to imaging plaque, OCT was demonstrated to be quantitatively superior

OSA TOPS Vol. 21 Advances in Optical Imaging and Photon Migration, James G. Fujimoto and Michael S. Patterson (eds.)

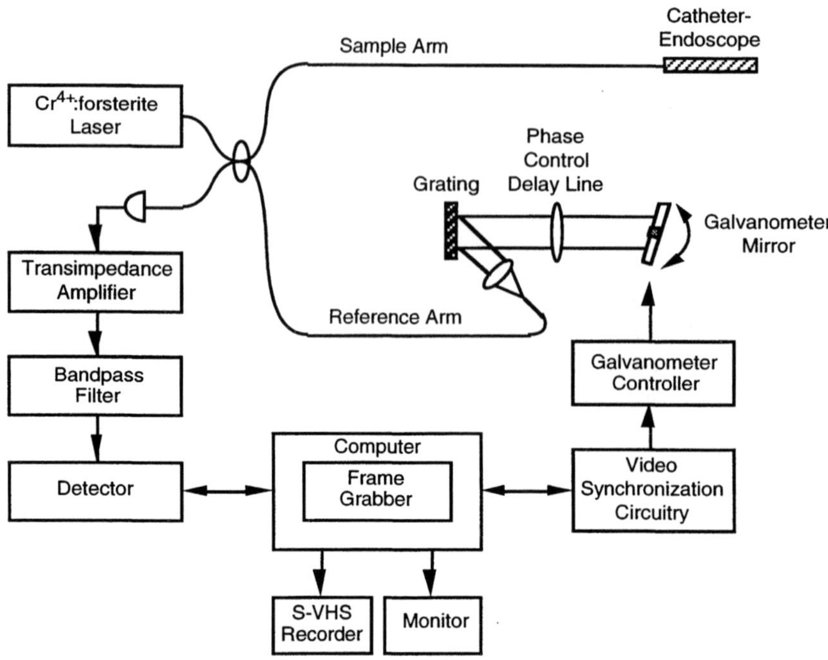

Figure 1. ***In vivo*** **OCT System**

with an axial resolution of 16±1 μm for OCT compared with 109±7 μm for IVUS.

Methods

In vivo imaging required several technical advances which have been recently described[14]. Extension of previous work to the dynamics of artery imaging was performed in this study [19]. A novel rapid data acquisition system which allows imaging at 4 frames per second was constructed. This system utilized a high speed scanning optical delay line, which is based on phase control techniques used in femtosecond pulse shaping[15]. A high power, low coherence, short pulse solid state Chromium:Forsterite laser was used in order to achieve the high resolution and sensitivity required at high acquisition rates[16,17]. A 2.9 French (1 mm), second generation OCT imaging catheter was developed for high speed transluminal imaging [18]. These components were integrated into an *in vivo* imaging system which synchronized the catheter motor, frame grabber, and reference arm to a master clock. A schematic of the catheter OCT imaging system is shown in Figure 1.

In vivo imaging was performed on normal, 12 week old, New Zealand White rabbits. After the animal was anesthetized, a mid-abdominal incision was performed and the abdominal aorta was isolated[19]. A 7 French Judkins guiding catheter, with a Y-connector on the proximal end, was introduced into the aorta using a micro-retractor. The guiding catheter was placed approximately 4 cm into the aorta. The OCT imaging catheter was introduced through the center of the guiding catheter. Imaging was performed over an approximately 8 cm length of aorta. Normal saline was manually injected through the side port of the Y-connector via a 50 cc syringe to remove the blood from the imaging plane of the catheter. After imaging, the animal was sacrificed and the imaged regions were excised and immersed in 10% formalin in preparation for routine histologic processing. Gross tissue registration was performed by suturing the guiding catheter in place after imaging. The specimens were blocked in paraffin, cut into 5 micron sections, and stained with hematoxylin and eosin [H&E] for microscopic examination.

Results

Figure 2, is a series of images from the aorta after a saline injection. Saline flushes were performed manually at a rate of approximately 2-3 cc/sec. An observation of particular importance in this study was the attenuation which resulted from the presence of blood. No significant structural information was obtained unless simultaneous injections of saline were performed. This likely results from the large numbers of scatters (i.e. red blood cells) present within blood[20]. Since the saline injections were performed

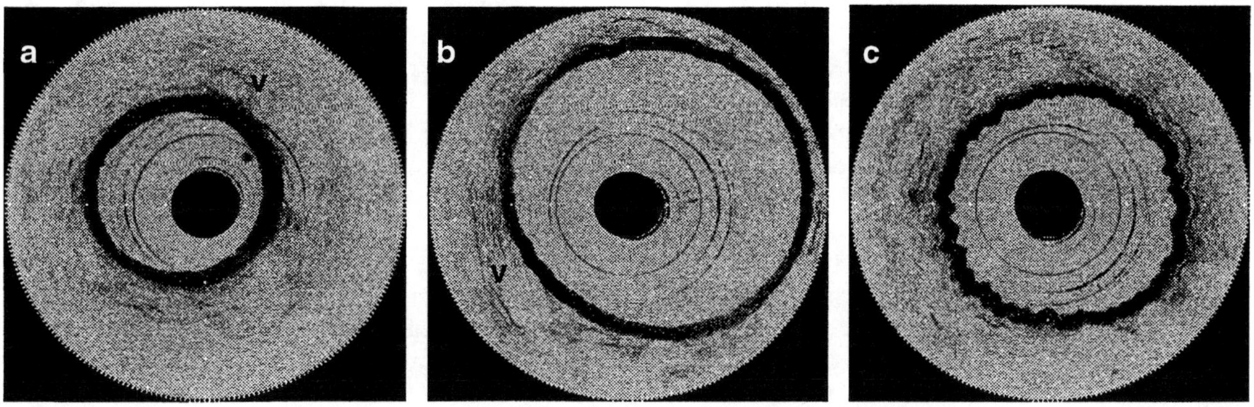

Figure 2. Rabbit aorta images after saline injection. Initiation of injection (a), maximal distention (b) and relaxation (c). Also visible is a vein adjacent to the aorta (v).

at a relatively low pressure and the volumes were small, this will likely be the preferred method for *in vivo* assessments over total vessel occlusion. If saline injections, a procedure routinely used in clinical setting, are required for *in vivo* patient imaging, clinicians will need to weigh the increased resolution and low cost against the likely increase in procedure time. In this image, the media and surrounding supportive structures are clearly identified. A structure consistent with adjacent vessels is imaged through the aorta (v). Figure 2c shows evidence of wall movement during fast relaxation after the injection of saline. Each image is 496 x 496 pixels and depicts a 4.6 mm diameter cross-section.

Conclusions

This work, focusing on dynamic changes, demonstrates that high resolution intravascular imaging can be performed *in vivo* using optical coherence tomography. Images at 10 μm resolution were obtained, representing an improvement of over 10X that of 'state of the art' IVUS transducers. High contrast was noted between the media and the surrounding supportive tissue. In addition, fine structural detail was noted within the surrounding supportive tissue. This work, taken together were previous studies examining in vitro atherosclerotic plaque, strongly suggest a role of OCT for high resolution intraarterial assessments of coronary atherosclerosis[10,14].Future technology development will focus on increasing data acquisition rates and integrating the system with low cost light sources. In this study, the acquisition rate was 4 frames per second. Some motion artifacts were noted at this rate. With future embodiments, video acquisition rates near video speed are likely. The light source used in this study was a mode-locked solid state laser. This source is too expensive and complex for routine clinical use. However, compact, inexpensive sources, such as doped fiber lasers or semiconductor diode sources, are currently under development and will likely be available within the next few years.

Acknowledgments

This research is supported in part by the National Institutes of Health, Contract NIH-9-RO1-EY11289-10 (JGF), the Medical Free Electron Laser Program, Office of Naval Research Contract N00014-94-1-0717 (JGF), the Whittaker Foundation Contract 96-0205 (MEB), 1RO1AR44812-01 (MEB), the National Institutes of Health, Contract NIH-1-R29-HL55686-01A1 (MEB).

References

1. Falk E. "Plaque rupture with severe pre-existing stenosis precipitating coronary thrombosis, characteristics of coronary atherosclerotic plaques underlying fatal occlusive thrombi." Br. Heart J. 50 127-134 (1983).
2. Fuster, V., L. Badimon, J.J. Badimon. and J.H. Chesebro. "The pathogenesis of coronary artery disease and the acute coronary syndromes." New Eng. J. Med. 326 242-250 (1992).
3. Peters, R.J., J. Ge, D.T. Linker, C.A. Visser, and P.G. Yock. "Observer agreement on qualitative analysis of intrcoronary ultrasound images." Circulation 90 I-551 (1994).
4. Yock PG, Fitzgerald PJ, Linker DT, Angelsen BAJ. "Intravascular Ultrasound Guidance for

Catheter-Based Coronary Interventions." J Am Col Card 17 39B-45B (1991).

5. Mudra H, Klauss V, Blasini R, Kroetz M, Rieber J, Regan E. "Ultrasound guidance of Palmaz-Schatz Intracoronary Stenting with a Combined Intravascular Ultrasound Balloon Catheter." Circulation 90 1252-1261 (1994).
6. Huang D, Swanson, EA, Lin CP, Schuman JS, Stinson WG, Chang, W, Hee, MR, Flotte, T, Gregory K, Puliafito CA, Fujimoto JG. "Optical Coherence Tomography. " Science 254 1178-1181 (1991).
7. Hee MR, Izatt, JA, Swanson EA, Huang D, Lin CP, Schuman JS, Puliafito CA, Fujimoto JG. "Optical coherence tomography of the human retina." Arch Opthalmol 113 325-32 (1995).
8. Brezinski ME, Tearney GJ, Bouma BE, Izatt JA, Hee MR, Swanson EA, Southern JF, Fujimoto JG. "Optical coherence tomography for optical biopsy: properties and demonstration of vascular pathology." Circulation 93 1206-1213 (1996).
9. Fujimoto JG, Brezinski ME, Tearney GJ, Boppart SA, Bouma B, Hee MR, Southern JF, Swanson EA. "Optical biopsy and imaging using optical coherence tomography." Nature Med 1 970-972 (1995).
10. Schmitt JM, Yadlowsky M, Bonner RF. "Subsurface imaging of living skin with optical coherence microscopy." Dermatology 191 93-98 (1995).
11. Brezinski, ME, Tearney GJ, Brett BE, Boppart SA, Hee MA, Swanson EA, Southern JF, Fujimoto JG. "Imaging of coronary artery microstructure with optical coherence tomography." Am J Card 77 92-93 (1996).
12. Brezinski ME, Tearney GJ, Weissman NJ, Boppart SA, Bouma BE, Hee MR, Weyman AE, Swanson EA, Southern JF, Fujimoto JG. "Assessing atherosclerotic plaque morphology: comparison of optical coherence tomography and high frequency ultrasound." Heart 77 397-403 (1997).
13. Tearney GJ, Brezinski ME, Boppart SA, Bouma BE, Weissman N, Southern JF, Swanson EA, and Fujimoto JG. "Catheter-based optical imaging of a human coronary artery." Circulation 94 3013 (1996).
14. Tearney GJ, Brezinski ME, Bouma BE, Boppart SA, Pitris C, Southern JF, Fujimoto JG. "*In vivo* endoscopic optical biopsy with optical coherence tomography." Science 276 2037-2039 (1997).
15. Heritage JP, Weiner AM, Thurston RN. "Picosecond pulse shaping by spectral phase and amplitude manipulation." Opt. Lett. 10 609-611 (1985).
16. Petricevic V, Gayen, SK, Alfano R.R.. "Laser action in chromium-activated forsterite for near-infrared excitation: Is Cr4+ the lasing ion?" Appl. Phys. Lett. 53 2590-2592 (1988).
17. Bouma BE, Tearney GJ, Bilinsky IP, Golubovic B, Fujimoto JG. "Self-phase modulated Kerr-lens modelocked Cr:forsterite laser source for optical coherence tomography." Opt Lett 21 1839-1841 (1996).
18. Tearney GJ, Boppart SA, Bouma BE, Brezinski ME, Weissman NJ, Southern JF, and Fujimoto JG. "Single Mode Scanning Catheter/Endoscope for Optical Coherence Tomography." Opt. Lett. 21 543-545 (1996).
19. Fujimoto JG, Boppart SA, Tearney GJ, Bouma BE, Pitris C, Brezinski ME. Ultrahigh Resolution "In Vivo Intraarterial Imaging with Optical Coherence Tomography," submited for publication to Circulation.
20. Steinke J.M., and. Shepherd A.P. 1988. "Comparison of Mie theory and the light scattering of red blood cells." Appl. Optics 27 4027-4033 (1988).

In Vivo Imaging of Human Internal Organs with an Integrated Endoscopic OCT System

G. V. Gelikonov, V. M. Gelikonov, A. M. Sergeev, F. I. Feldchtein,
R. V. Kuranov, N. D. Gladkova
Institute of Applied Physics of RAS, Nizhny Novgorod, Uljanov st, 46, Russia 603600
flx@ufp.appl.sci-nnov.ru
N. M. Shakhova, L.B. Snopova,
Nizhny Novgorod Medical Academy
I. A. Kuznetzova, V.V.Pochinko, Yu.P.Chumakov
Nizhny Novgorod Regional Hospital

Monitoring of epithelial and subepithelial layers of internal organs in the near infrared spectral range to the depth of 1.5-2 mm is one of the challenging problems targeted by modern methods of bioimaging since its solution promotes noninvasive early diagnosis for a wide range of inflammatory and tumor processes. Optical coherence tomography is a method potentially promising for observation human mucosa at these depths [1-4] though several optical and engineering problems have to be solved for successful endoscopic implementation of this technique. Among the problems, a reliable and convenient access of probing low-coherence radiation to the surface of internal organs and acquisition of OCT data in parallel with common endoscopic imaging are of primarily importance.

We have developed an endoscopic OCT system solving these problems by integration of the sampling arm of an all-optical-fiber Michelson interferometer into a standard endoscopic device using its biopsy channel to transmit the probing radiation to the surface of investigated tissue. By introducing the flexible sampling arm of a single OCT device intò different endoscopes we create a whole family of diagnostic tools suitable for studying of a variety of biotissues. This combination of the OCT system with the standard endoscopic equipment has proved to be quite convenient for clinical studies allowing doctors to perform usual observation of internal organs and in the case of interest to make simultaneously the noninvasive optical biopsy for any amount of tissue locations.

Implementation of an extended flexible arm of the OCT interferometer to provide the access to the studied tissue is based on the usage of polarization maintaining fibers for transportation of the low-coherence light. By this we eliminate polarization fadings connected with polarization distortions at bending of the endoscope arm. To probe the surface of an internal organ we have developed a miniaturized electro-mechanical unit controlling the lateral scanning process. This unit terminates the sampling arm of the fiber interferometer and has the size to fit the diameter and the curvature radius of the biopsy channel of endoscopes. The probe beam swinging on the tissue surface approaches 2 mm. The scanning unit and the extended part of the fiber interferometer inserted in the endoscope are sealed which allows common cleaning procedure of the device in clinical applications.

Our clinical studies have been focused on mucous membranes of hollow organs connected with the outer environment. We have chosen mucous membranes of upper and middle portions of gastrointestinal tract (esophagus and stomach), upper respiratory tract (larynx), urinary tract (urinarybladder and uretra) as well as mucosa of uterine cervix as typical localization for carcinomatous processes. Several tens of patients with different pathologies have been observed with OCT in parallel with routine endoscopical examinations. Fig. 1 shows typical *in vivo* images of healthy mucosa of larynx (a), esophagus (b), uterine cervix (c) and urinary bladder (d). Owing to the different backscattering intensity we can identify different mucosa layers, so that the stratified squamous (or transitional) epithelium is seen as a weakly scattering layer, while *lamina propria* , as a bright (mostly scattering) one. A peculiarity of the esophagus structure, associated with the food movement, consists in the presence of a muscular plate, so called muscularis mucosa, that is observed in the tomogram as a minimally backscattering layer, i.e. as a dark stripe. Besides, one can differentiate deeper layers that are submucosa (more backscattering) and muscular layers (less backscattering) in esophagus and urinarybladder. In *lamina propria* and submucosa one can find blood vessels and glands (cardiac glands in esophagus and cervical glands in uterine cervix), visible as dark elongated formations of different size.

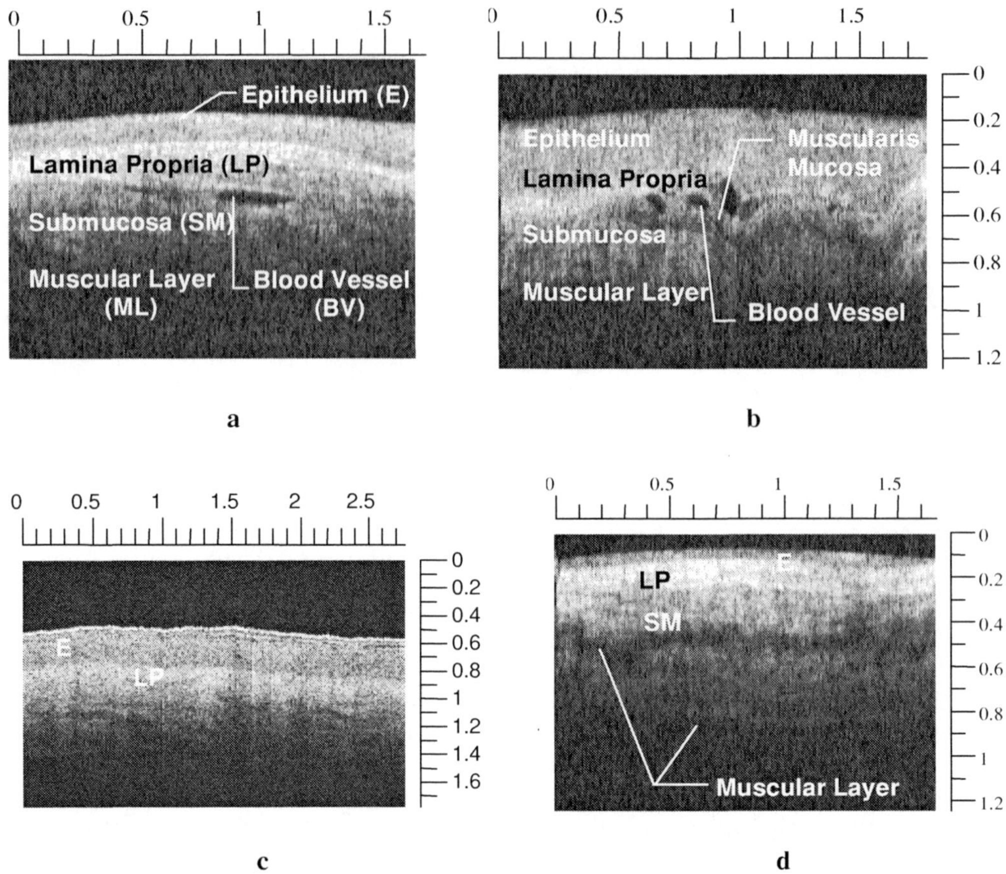

Figure 1. OCT image of healthy portion of larynx (a); esophagus (b); uterine cervix (c); urine bladder (d). Scales are given in mm.

Beside recording images and compiling albums of normal tissues, which is the first necessary step in OCT application in endoscopy, we anticipate the development of endoscopic OCT in the following directions:

- obtaining unique *in vivo* information on pathological alterations of biotissue structure in the depth of mucosa;
- search of correlation between OCT images and functional states of internal organs;
- OCT guidance of therapeutic and surgical treatment and assessment of adequacy of treatment in post-operative and healing processes.

Below we illustrate our approach to this development in gynecology and present results of first clinical studies in the above mentioned directions.

The modern gynecology defines background processes, cervical intraepithelial neoplasia (CIN), and microcarcinoma as pathological states of uterine cervix. The majority of medical mistakes are made in diagnosing CIN, or precancer, of uterine cervix. Difficulties in diagnosing of this state are connected with the absence of clinical manifestations and the minimal degree of alterations in the epithelial surface.

We have performed the OCT study in several tens of female patients with stated diagnosis of uterine cervix precancer. The obtained tomograms show correlation with the histologically known changes in architectonics of epithelium and underlying stroma in the case of uterine cervix neoplasia. It is known that the altered epithelial layers form so-called "pillars", "blocks", "crypts", while stroma approaches the epithelial surface as vertical bands. A series of tomograms in Fig. 2 reveals images with features similar to those mentioned above. While reading tomograms with CIN, most attention should be focused on the state of basal membrane that separates less scattering epithelial structures from more scattering stroma. The fact is that the next stage following CIN is a form of uterine cervix cancer. Microcarcinoma exhibits a destruction of basal membrane with invasion by tumor cells of underlying stroma. The disintegration of basal membrane seen in OCT images as the disappearance of a visual border between epithelium and stroma is a sign of invasion. An example of the basal membrane destruction is presented in Fig. 2c. Hence, OCT can noninvasively indicate areas with different states of neoplasia and invasion. This knowledge is particularly valuable for carrying out the excisional biopsy since it increases essentially the probability of taking biopsy material from exactly localized suspicious areas.

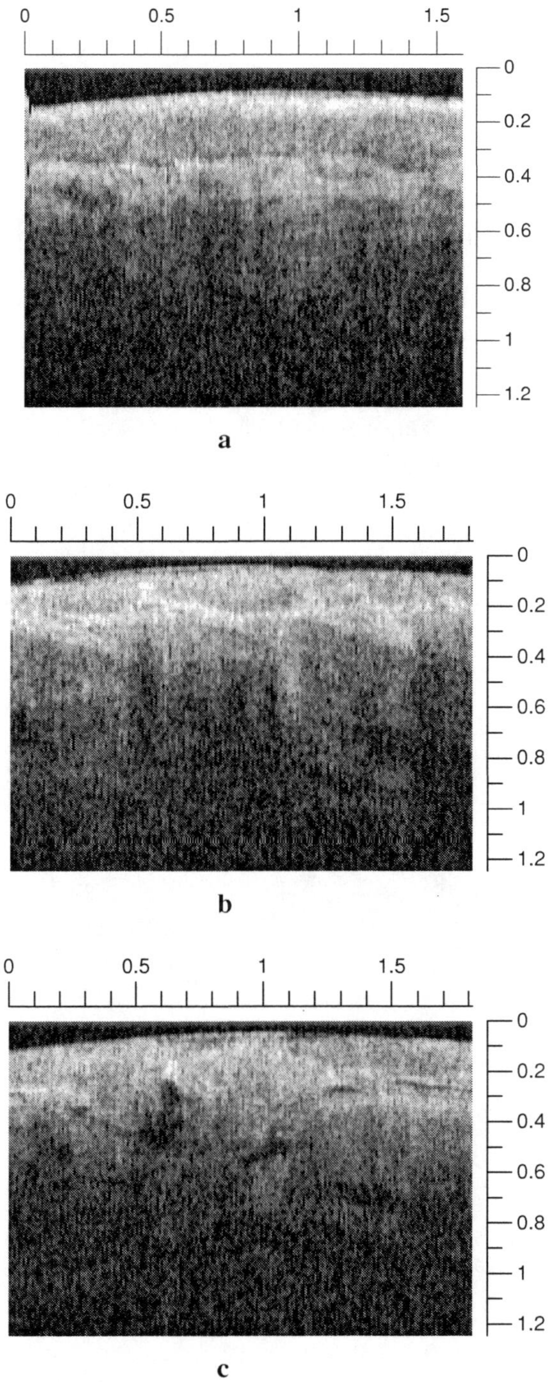

Figure 2. OCT images for patients with precancer of uterine cervix.

In the second direction, we report some examples with our observation on the correlation of OCT cervix images and functional states of the female genital system in general. These examples illustrate the fact that the tickness of the mucous membranes in all parts of female genital tract depends on the estrogen concentration. When the concentration of estrogens decreases, for example in hypoestrogenia of old-aged women, the stratified squamous epithelium becomes atrophied, the superficial cell layers are practically absent, the mucosa is represented by basal and parabasal cells, due to which the epithelial thickness is significantly reduced. Contrary, when the estrogen concentration increases the number and sizes of epithelial cells in uterine cervix and vagina are also increased, glycogen is accumulated in the cells, which leads to thickening of the mucous membrane. The tomograms in Fig. 3 easily show the differences in the epithelium of an old-aged woman and a young female patient with clinically pronounced hyperestrogenia.

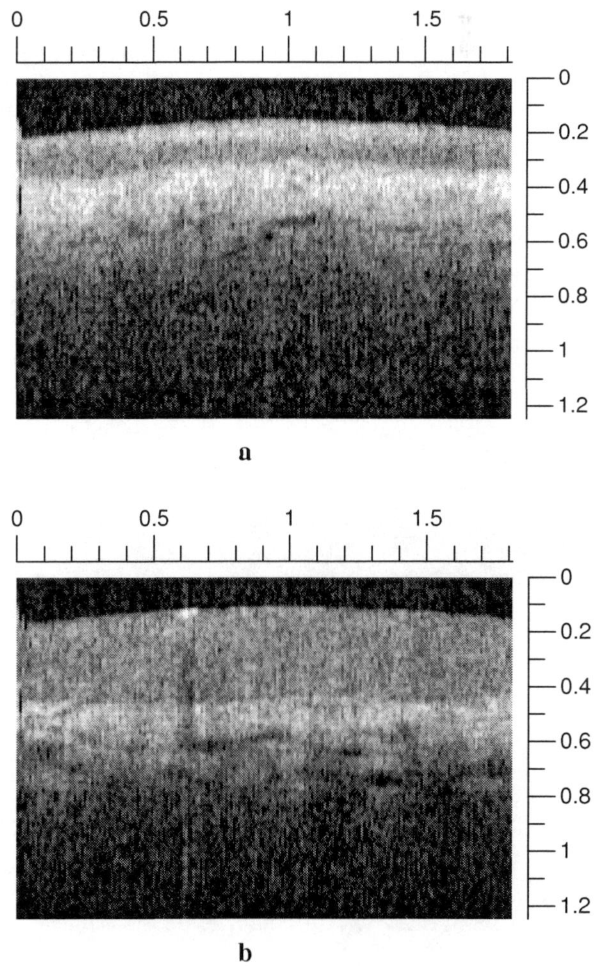

Figure 3. OCT image of uterine cervix of woman with hypoestrogenia (a) and hyperestrogenia (b).

More clearly the functional differences are seen in the tomograms of pregnant women. The histological study of uterine cervix in pregnant women reveals increased vascularization and stroma edema resulting from destruction of collagen fibers and accumulation of extracellular liquid that separates the connective tissue fibers. The OCT tomograms clearly exhibit the large number of dark spots and horizontal stripes (Fig. 4).

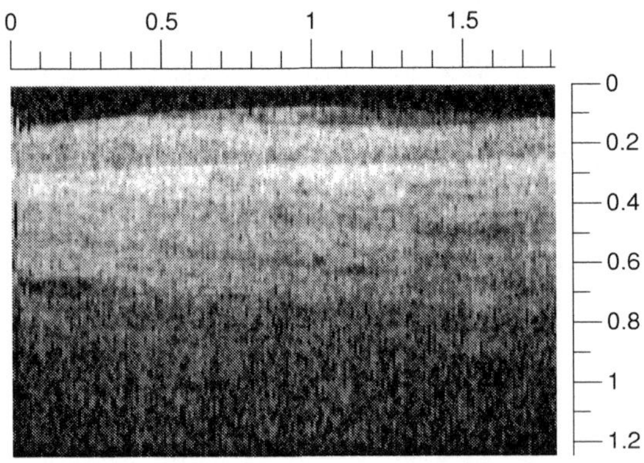

Figure 4. OCT image of uterine cervix of pregnant women.

Summarizing the above information, we can state that data obtained by OCT on the functional state of mucous membranes can be applied for evaluation of hormonal activity of ovaries.

The discovered potentialities of OCT for structural and functional imaging of human mucosa can be put as a basis for application of this method for guiding a treatment and assessing its adequacy during the healing process. We have preliminary demonstrated this capability in experiments on OCT monitoring of cryotherapy of uterine cervix. Cryotherapy (or cryodestruction) is used for organ-preserving treatment of different epithelial pathologies such as ectopy, intraepithelial neoplasia, etc. Under the action of low temperature there occurs a minute freezing of superficial tissue at the depth of few mm, resulting in the necrosis and subsequent slow regeneration of normal epithelium during the time of several weeks. The OCT diagnostics in cryotherapy can provide the exact pre-operative localization of the lesion borders, determination of the extension of cryodestruction zone beyond the visual limits, and monitoring of the regeneration process. In Fig. 5 we present the results of OCT guided and followed-up cryodestruction treatment in a female patient suffered of exocervix ectopy.

The tomograms have been recorded at the border of the visual freezing zone outside the cryoapplication area. Successive images of the same place are shown before the treatment (a), immediately (b) and 15 min. after (c) the applicator removing. In Fig. 5a the interface between the normal squamous epithelium and the thinned metaplastic epithelium followed by the ectopy zone (to the right beyond the image) is demonstrated.

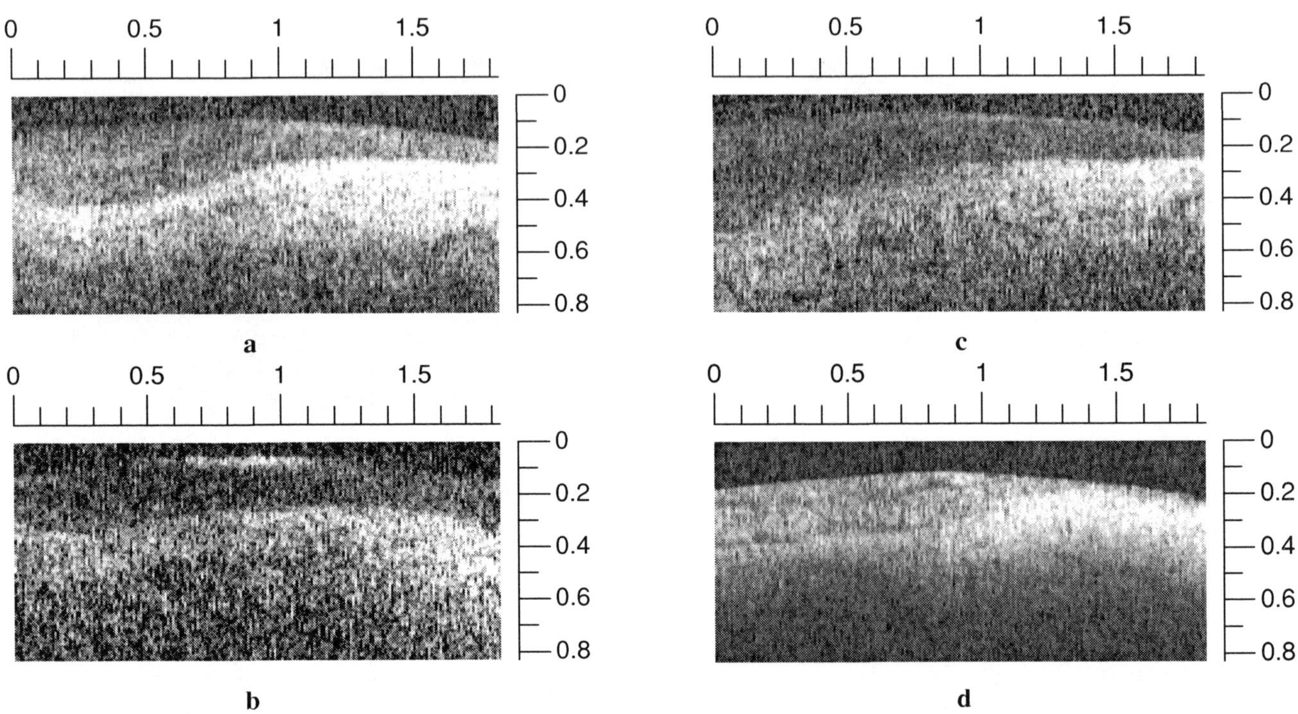

Figure 5. Successive images of uterine cervix cryodestruction: **a** -before cryodestructor application, **b**- immediately after, **c** - 15 min after, **d** - 2 days after

Fig. 5b shows swelling the epithelium layer and decreasing the light scattering which is presumably connected with accumulation of fluid and occurrence of the edema. (Note that in some cases we have observed that the swelling process is inhomogeneous with formation of islands weakly scattering light (see, e.g. Fig. 6)). Fig. 5c demonstrates a successive growth of edema and a gradual disappearance of the epithelium-stroma junction. The superficial tissue is getting more and more transparent, probably due to immersion of light scattering microobjects (cells and collagen fibrils) by the intercellular fluid. A tomogram of nearly the same place 2 days after the treatment is shown in Fig. 5d. One can obviously observe an edge of the necrosis zone that strongly scatters the light and hence is seen as a bright stripe. The *in vivo* observed process we are describing here is well in accordance with morphological data of tissue alterations at cryodestruction, that have been known from standard post-mortem histology.

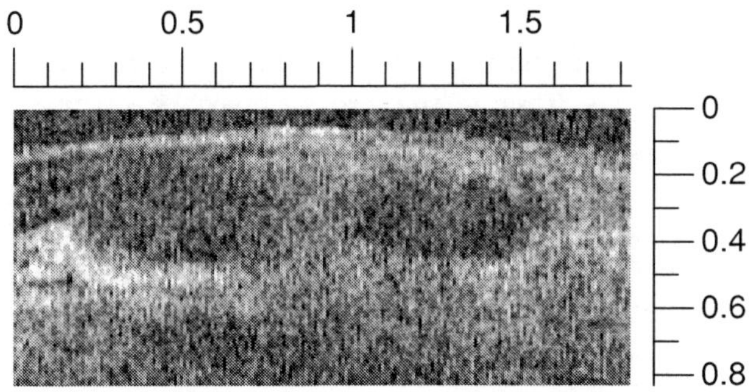

Figure 6. Island-type structure of uterine cervix edema following cryodestruction

Acknowledgments

The authors thank the Russian Basic Research Foundation for the financial support of this work. The authors are also grateful to the staff and management of Regional Hospital for the possibility to conduct this research study.

References

1. A. Sergeev, V.Gelikonov, G.Gelikonov , F.Feldchtein, K.Pravdenko, R.Kuranov, N.Gladkova, V.Pochinko, G.Petrova, N.Nikulin, "High-spatial resolution optical coherence tomography of human skin and mucous membranes", in: Technical Digest of Conference on Lasers and Electro-optics, (Optical Society of America) p.349 (1995)
2. J.G.Fujimoto, M.E.Brezinski, G.J.Tearney, S.A.Boppart, B.Bouma, J.A.Izatt, M.R.Hee, J.F.Southern, E.A.Swanson,, "Optical biopsy and imaging using optical coherence tomography," Nat. Med. 1 970 (1995).
3. J.A.Izatt, M.D.Kulkarni, H-W Wang, K.Kobayashi, M.V.Sivak, Optical coherence tomography and microscopy in gastrointestinal tissues IEEE Journal of Selected Topics in Quantum Electronics 2 1017-1028 1996
4. G.Tearney, B.Bouma, M.E.Brezinsky, N.J.Weismann, J.F.Southern, .G.Fujimoto Scanning sigle-mode fiber optic catheter-endoscope for optical coherence tomogarphy, Optics Letters 21 543-545 (1996)
5. F.I. Feldchtein, V.M. Gelikonov, G.V. Gelikonov, R.V. Kuranov, N.D. Gladkova, A.M. Sergeev, N.M. Shakhova, I.A. Kuznetzova, A.N. Denisenko, O.S. Streltzova, "Design and performance of an endoscopic OCT system for *in vivo* studies of human mucosa," in Technical Digest of Conference on Lasers and Electro-optics, (Optical Society of America, 1998).
6. A.M. Sergeev, V.M. Gelikonov, G.V. Gelikonov, F.I. Feldchtein, R.V. Kuranov, N.D. Gladkova, N.M. Shakhova, L.B. Snopova, A.V. Shakhov, I.A. Kuznetzova, A.N. Denisenko, V.V. Pochinko, Yu.P. Chumakov, O.S. Streltzova, "*In vivo* endoscopic OCT imaging of precancer and cancer states of human mucosa," Optics Express 1 No.13, 432-440 (1997).
7. A.M. Sergeev, F.I. Feldchtein, G.V. Gelikonov, R.V. Kuranov, V.M. Gelikonov, I.A. Kuznetzova, L.B. Snopova, N.M. Shakhova, A.N. Denisenko, V. Almazov, "Integrated endoscopic OCT system *and in-vivo* images of human internal organs," - Proc SPIE 3251 paper number 8 (1998).
8. N.M. Shakhova, I.A. Kuznetzova, N.D. Gladkova, L.B. Snopova, V.V. Pochinko, Yu.P. Chumakov, G.V. Gelikonov, V.M. Gelikonov, F.I. Feldchtein, R.V. Kuranov, A.M. Sergeev, "Endoscopic OCT for imaging of uterine body and cervix pathologies," Proc SPIE, 3251, paper number 43 (1998).
9. G.J. Tearney, M.E. Brezinski, B.E. Bouma, S.A. Boppart, C. Pitris, J.F. Southern J.G. Fujimoto, "*In vivo* endoscopic optical biopsy with optical coherent tomography," Science 276 2037-2039 (1997).

Velocity Estimation Accuracy in Color Doppler Optical Coherence Tomography

Manish D. Kulkarni, Ton G. Van Leeuwen, Siavash Yazdanfar, and Joseph A. Izatt
Department of Biomedical Engineering
Case Western Reserve University, Cleveland, Ohio 44106
Tel: (216) 844-3298; Fax: (216) 844 8011

Color Doppler optical coherence tomography (CDOCT) is a recent invention which permits spatially localized blood flow velocity mapping simultaneous with micro-structural imaging[1,2]. CDOCT is an extension of optical coherence tomography (OCT)[3] which performs high resolution sub-surface imaging of biological tissues using coherence ranging. In CDOCT, velocities of moving scatterers in the sample are estimated by measuring Doppler shifts in the backscattered light. Longitudinal velocity profiles are estimated from centroids of localized backscatter light spectra and are color coded for velocity magnitude and direction. Two-dimensional velocity maps are built up from adjacent longitudinal velocity profiles. In this paper we propose methods to minimize velocity estimation errors in CDOCT.

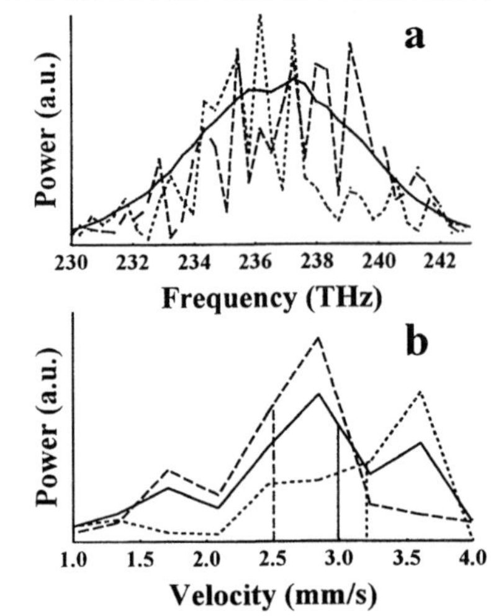

Fig. 1a. Two cross-power spectra (long and short dashed lines) measured from a turbid sample. Ensemble averaging the square of the cross-power spectrum using 500 A-scans recovers the source power spectrum (solid line). **b.** Doppler shifted spectra measured from a flowing intralipid solution. The individual detector current spectra (dashed lines) are spiky resulting in corrupted velocity estimates. Vertical lines indicate centroids of respective spectra. The solid curve is an average of 3 spectra. Its centroid provides a more accurate velocity estimate. The X axis (i.e., the frequency axis) is converted to the velocity axis using Doppler's law.

The Michelson interferometer, which serves as the basis for CDOCT, measures the cross-correlation function between the fields returning from the reference and sample arms. The Fourier transform of the cross-correlation is the cross-power spectrum[4] $\widetilde{S}_{is}(\nu)$ where ν is optical frequency. Placing a mirror in the sample arm provides the autocorrelation function, and its Fourier transform is the source power spectrum $\widetilde{S}_{ii}(\nu)$. In CDOCT, the detector current is coherently demodulated at the reference arm Doppler shift frequency f_r to provide its complex envelope. Let $R_{ii}(t)$ and $R_{is}(t)$ denote the complex envelopes of the currents measured with a mirror, and with a scatterer, respectively, moving with a velocity V_s in the sample arm. Then, $\widetilde{S}_{ii}(\nu)d\nu = S_{ii}(f - f_r + f_s)df$ and $\widetilde{S}_{is}(\nu)d\nu = S_{is}(f - f_r + f_s)df$ where the Fourier transforms of $R_{is}(t)$ and $R_{ii}(t)$ are $S_{is}(f + f_s)$ and $S_{ii}(f + f_s)$, respectively, f_s is the Doppler shift induced in light in the sample arm, $f = 2\nu(V_r - V_s)/c$, and V_r is the reference mirror velocity. Note that $S_{is}(f)$ and $S_{ii}(f)$ are the detector current spectra measured when the scatterer and the mirror, respectively, are not moving. The full-width half-maximum (FWHM) width of S_{ii} is given by $\Delta f = 2(V_r - V_s)\Delta\nu/c$ where $\Delta\nu$ is the FWHM width of $\widetilde{S}_{ii}(\nu)$. Thus the detector current electronic spectrum is directly proportional to the light source optical spectrum, scaled by the ratio of the speed of light and the net differential Doppler shift between the two arms.

Light scattered from a turbid specimen generates a speckle pattern consisting of the sum of partial reflections from many randomly distributed scatterers. As a result, the field scattered into a specific direction (such as backscatter) may be strongly frequency

Reprinted from Advances in Optical Imaging & Photon Migration, 1998 Technical Digest,

dependent. Therefore we may describe[5] the backscattered light spectrum $\widetilde{S}_{is}(\nu)$ detected by OCT as the source power spectrum modulated by a postulated specimen transfer function $\widetilde{H}(\nu)$. The transfer function encodes the spatial distribution of scatterers. Treating the light-specimen interaction as a linear-shift-invariant system, $\widetilde{S}_{is}(\nu) = \widetilde{S}_{ii}(\nu)\widetilde{H}(\nu)$. To illustrate this concept, in Fig. 1(a) we show two cross-power spectra computed by Fourier transforming complex axial reflectivity profiles (i.e., coherently demodulated A-scans) of an OCT image of a highly scattering specimen (0.48 µm diameter polystyrene microspheres embedded in a gel.) The individual cross-power spectra appear very spiky due to modulation of the source spectrum by the transfer function. Assuming that within a small volume of a tissue specimen, the locations of point scatterers can be considered as random and uncorrelated, then ensemble averaging of individual cross-power spectra recovers the source power spectrum. In Fig. 1(a) we show an estimate of the source power spectrum obtained by computing $\sqrt{E\{|\widetilde{S}_{is}(\nu)|^2\}}$ using 500 A-scans. Here $E\{x\}$ denotes the expected value of x.

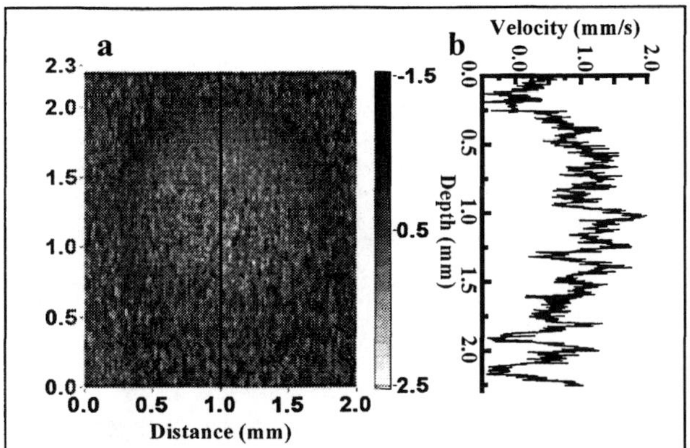

Fig. 2a. A velocity image of laminar flow of intralipid solution. Velocities were calculated using traditional centroid algorithm. The image is noisy, the estimates are inaccurate by a factor of >2 from actual velocities, and the lower surface of the tube is barely noticeable. **b.** The velocity profile along the black line in (a). The profile is non-parabolic in contrast to the theoretical prediction.

Due to the equivalence between the detector current spectrum and the optical spectrum, $S_{is}(f + f_S)$ is also strongly modulated. Therefore a Doppler shift estimated by computing the centroid of $S_{is}(f + f_S)$ as performed in current implementations of CDOCT[1,2] may be inaccurate. Preferably, f_S should be estimated by computing the centroid of $S_{ii}(f + f_S)$ which can be recovered by calculating $\sqrt{E\{|S_{is}(f + f_s)|^2\}}$. In Fig. 1(b) Doppler shifted spectra measured from a flowing 1% intralipid solution are shown, as well as an averaged spectrum. The random variations in the centroids of individual spectra around that of the averaged spectrum are clearly evident.

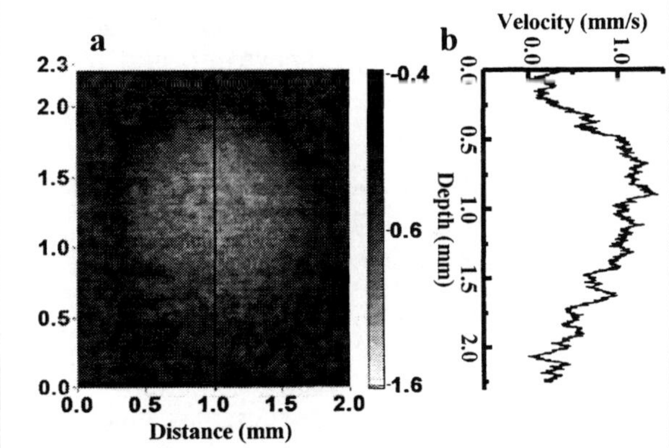

Fig. 3a. Velocity image obtained by computing centroids of the ensemble averages of detector current electronic spectra. Velocity noise is significantly reduced assisting in clearer visualization of the tube wall. The velocity estimates are still inaccurate by a factor of >2. **b.** The velocity profile along the black line in (a). The profile remains non-parabolic.

In order to test the improvement in velocity measurement accuracy provided by ensemble averaging of spectra, we acquired five velocity images of 1% intralipid solution flowing in a 1.59 mm diameter glass tube positioned at an angle θ (=58°) with respect to the probe beam. The block diagram of the schematic is provided in reference 1. In Fig. 2(a) one of the velocity images of flowing intralipid is shown. This image was computed by using the short time Fourier transform (STFT) algorithm implemented by performing a Fourier transform on the first 64 points of 2048 point A-scans, (providing 70 μm longitudinal velocity resolution), shifting the analysis window by a decimation factor of 1, and repeating. The velocity of scatterers was estimated from the centroid $\bar{f}_s$ of the spectrum magnitude according to $V_s = \bar{f}_s \lambda_0 / (2n_t \cos\theta)$ where n_t is the local tissue refractive index, and λ_0 is the source center wavelength. In Fig. 2, the velocity estimates for the deeper scatterers are very poor. The estimated peak velocity is 2.12 mm/s and is much less than 3.88 mm/s predicted theoretically for the measured flow rate of 3.84 μl/sec. In Fig. 3 we computed centroids of $\sqrt{E\{|S_{is}(f+f_s)|^2\}}$ which was estimated from a measurement of $S_{is}(f+f_s)$ from each of the 5 images at a given depth and adjacent measurements at the same depth (providing 60 μm lateral resolution). Thus we used 15 measurements of $S_{is}(f+f_s)$ to estimate $S_{ii}(f+f_S)$. The image is noticeably improved due to velocity noise reduction. However, the velocity profile (Fig. 3(b)) still does not resemble a parabola, and the estimate of peak velocity (1.44 mm/s) remains less than the theoretical prediction.

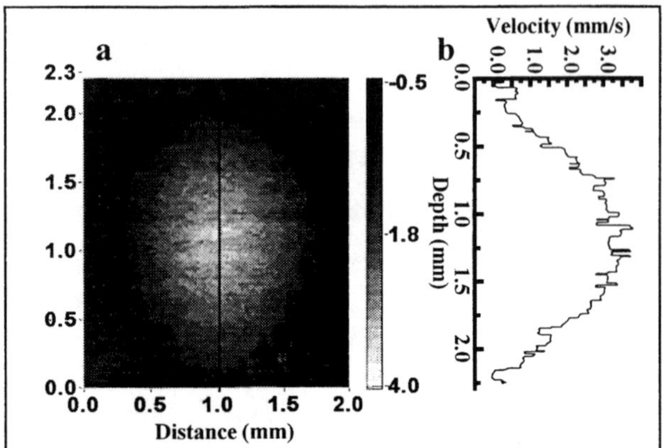

Fig. 4a. CDOCT image obtained using the proposed velocity estimation algorithm which achieves a dramatic increase in velocity estimation accuracy in deeper areas of the turbid solution. The wall of the conduit is clearly noticeable. **b.** The velocity profile along the black line in (a). The profile appears parabolic as predicted theoretically. The peak velocity estimate (3.78 mm/s) matches well with the theory (3.88 mm/s).

This discrepancy can be explained as follows. The detector current spectrum spans only a small part of the bandwidth provided by the detection electronics. Since the noise present in measured power spectra is always positive, the centroids of Doppler shifted power spectra will always under-estimate the true Doppler shift during STFT processing. Further, the performance of the centroid calculation degrades as a function of depth into the sample, where the available signal-to-noise ratio decreases (Fig. 2).

Therefore, we propose an improved velocity estimation algorithm. It scans the complete bandwidth provided by the detection electronics, locates the peak of the detector current spectrum, and uses the power at frequencies spanned symmetrically around the peak to compute the centroid. The span of the frequencies used in the centroid calculations is chosen to match the source spectral bandwidth. While this algorithm provides very accurate estimates in the presence of a reflection, it fails in the case of no reflection. Therefore, we used traditional centroid estimates for noisy areas. We developed a noise recognition algorithm which is capable of differentiating between the noise and signal. In Fig. 4 the result obtained using this new algorithm on averaged data used in Fig. 3 is displayed. Striking improvement is achieved in the velocity image. Note the increased visibility of velocities of particles in the deeper regions. This algorithm worked quite well on non averaged spectra as well.

In conclusion, we describe the theoretical basis for velocity estimation inaccuracies in CDOCT, and propose a new velocity estimation algorithm which greatly improves the velocity measurement accuracy of CDOCT.

References

1. J.A. Izatt et al., *Opt. Lett.* **22**, 1439-1441 (1997).
2. Z. Chen et al., *Opt. Lett.* **22**, 1119-1121 (1997).
3. D. Huang et al., *Science* **254**, 1178-1181 (1991).
4. A. Papoulis, *Systems and transforms with applications in optics*, (McGraw-Hill, NY, 1968).
5. M.D. Kulkarni et al., *Electron. Lett.* **16**, 1365-1367 (1997).

Optical Coherence Tomographic Imaging of *In Vivo* Cellular Dynamics

Stephen A. Boppart[1,2], Brett E. Bouma[1], Costas Pitris[1,2], James F. Southern[3], Mark E. Brezinski[4], James G. Fujimoto[1]

[1]Department of Electrical Engineering and Computer Science,
Research Laboratory of Electronics, Massachusetts Institute of Technology, Cambridge, MA 02139
[2]Harvard-MIT Division of Health Sciences and Technology,
Masachusetts Institute of Technology, Cambridge, MA 02139
[3]Pathology Department, The Children's Hospital, Boston, MA
[4]Cardiac Unit, Harvard Medical School,
Massachusetts General Hospital, Boston, MA 02114

Abstract

Optical coherence tomography (OCT) is an optical imaging technology capable of cellular resolutions. OCT is based on the detection of backscattered near infrared laser radiation and is analogous to ultrasound B-mode imaging except reflections of light are detected rather than sound. Using the *Xenopus laevis* (African frog) developmental biology animal model, *in vivo* mitotic activity of differentiating mesenchymal cells was imaged throughout the cell cycle. Individual cell nuclei and membranes were identified with OCT free-space resolutions of 5 μm. Cell migration was observed by tracking neural crest melanocytes within nontransparent *Xenopus* specimens which may be relevant in developmental and molecular biology. Cell morphology was confirmed with corresponding histology. Optical coherence tomography may have the potential for imaging *in vivo* cellular morphology in humans for the early detection of neoplastic changes.

Key Words: (170.4500) Optical coherence tomography; (170.1530) Cell analysis; (170.0180) Microscopy; (170.4580) Optical diagnostics for medicine.

Introduction

The ability to image tissue, *in vivo*, at the cellular and sub-cellular level would enable the identification of normal and diseased tissue without having to biopsy a specimen for histological processing. For example, the *in vivo* detection of early dysplastic and neoplastic changes would reduce the morbidity and mortality associated with cancer and other life-shortening diseases. Technologies including magnetic resonance, ultrasound, x-ray and laser-confocal microscopy have been applied to the problem, but have substantial limitations when applied clinically. Image acquisition times are prohibitively long for magnetic resonance and x-ray microscopy. Ultrasound requires contact or an index-matching gel and resolutions at the cellular level have not been demonstrated with clinical devices. Advances in high-speed, *in vivo,* laser-scanning confocal microscopy have provided researchers with a means of imaging cellular morphology in skin [1]. A limitation, however, is the inability to image deeply in scattering, nontransparent tissue. Imaging penetration is limited to a few hundred microns requiring specimens to be either relatively small or optically transparent.

Methods

Optical coherence tomography (OCT) is a technique that combines the high resolutions of optical techniques with an ability to reject multiply scattered photons [2]. Imaging at cellular resolutions up to millimeters deep is possible in non-transparent tissue [3]. OCT is somewhat analogous to ultrasound B-mode imaging; relying on the detection of backscattered infrared light from tissue rather than sound. OCT was initially applied to imaging the transparent structures of the human eye with unprecedented resolution [4]. The use of longer wavelengths has permitted imaging in more highly scattering tissue such as the skin [5] and the cardiovascular system [6]. A schematic of the instrument used for imaging at cellular resolutions in this study is shown in Figure 1.

A Kerr-lens modelocked Cr^{4+}:forsterite laser operating at 1280 nm center wavelength is used as a broad bandwidth source [7,8]. The broad self-phase modulated output of the laser is passed through dispersion-shifted fiber to shape and smooth the spectrum. The coherence length is inversely

OSA TOPS Vol. 21 Advances in Optical Imaging and Photon Migration, James G. Fujimoto and Michael S. Patterson (eds.)

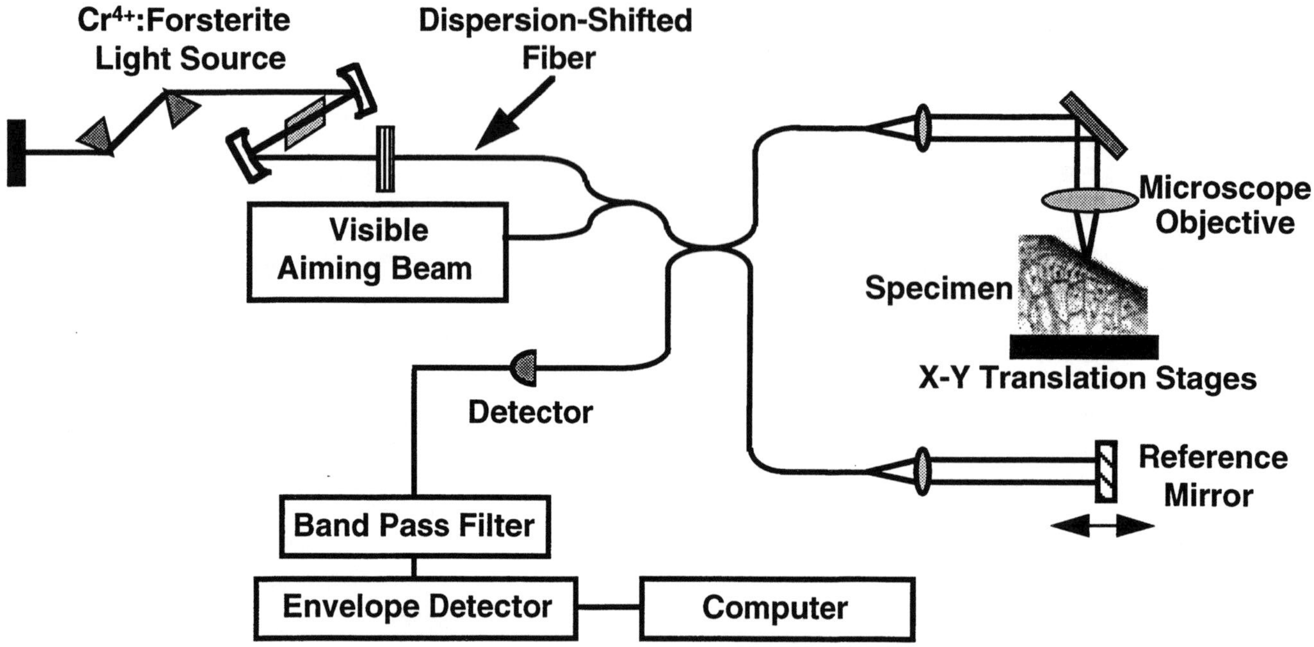

Figure 1. Schematic of OCT instrument for imaging at cellular resolutions.

proportional to the spectral bandwidth and was measured to be 5.1 μm in air. The coherence length represents the axial resolution while the transverse resolution is determined by the spot size within the tissue. A 632 nm visible aiming beam is combined with the infrared imaging beam to permit visualization of the scan location on the tissue. Anesthetized *Xenopus laevis* (African frog) tadpole specimens were placed on a computer-controlled X-Y translational stage for *in vivo* two- and three-dimensional OCT imaging. The sample-arm beam was focused with a 4.5 mm lens which produced a 9 μm diameter spot (100 μm confocal parameter). The incident power on the specimen was 1 mW. The reference arm contained a mechanical galvanometer with a retroreflector. An average index of 1.35 was measured for these specimens. The *in vivo* axial resolution was therefore 5.1 μm / 1.35 ≈ 3.8 μm. A two-dimensional cross-sectional image of the optical backscatter intensity was acquired in 10 s with a signal-to-noise ratio of 105 dB. Three-dimensional volumes were assembled from adjacent two-dimensional cross-sections. The logarithm of the image data was displayed in gray-scale.

Results

The *Xenopus laevis* is an established developmental biology animal model. Normal and abnormal development of organ systems have been previously characterized using OCT [9]. The *Xenopus* animal model was selected because of the high cell mitotic rate, the variety of cell sizes and types, the ability to image *in vivo* cellular dynamics, and the ease of care and handling. In this study, imaging at cellular resolutions was performed. Images were acquired perpendicular to the anteroposterior axis. Detailed cellular morphology could be readily imaged at these resolutions as is shown in Figure 2. Figure 2A is an image of mesenchymal cells immediately following cell division. Two daughter cells are shown (arrow). Cell membranes and individual cell nuclei are apparent. Figure 2A is one image from a sequence demonstrating *in vivo* imaging of cellular processes such as mitosis. By acquiring three-dimensional volumes at 10 minute intervals, individual cells could be imaged throughout the cell cycle. During development, melanin-laden neural crest cells migrate from the neural tube to remote locations where they differentiate and form various tissue types. Figure 2B is one figure from a sequence which tracks the migration of neural crest cells (arrows). Finally, other detailed morphology within the organism can be visualized at these resolutions including the intricate gill structure (fig. 2C) and a superficial artery (fig. 2D). Non-invasive imaging of viable cells was possible for up to 3 hrs. after which time the specimens were allowed to recover and proceed with normal activity.

Although some cells observed were as large as 100 μm, cell sizes ranged down to dimensions below the current resolution of OCT. This range of cell dimensions provided an evaluation of cell size limits for future human investigations. It has been suggested that the minimal resolution allowing differentiation of cellular detail in

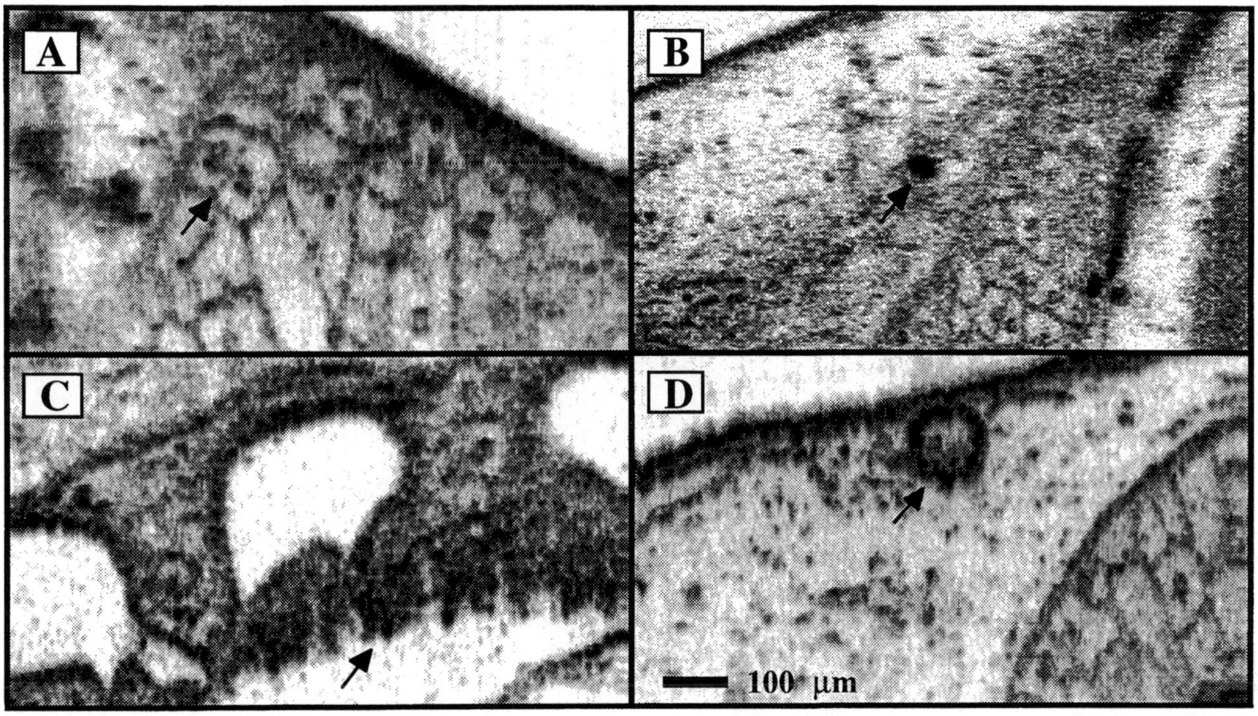

Figure 2. OCT images of cell morphology and dynamics in *Xenopus laevis*.

normal human myocardium is 5 μm [10]. In this study, cells as small as 20 μm could be imaged at 3.8 μm *in vivo* axial resolution. Thus, current OCT resolutions suggest a feasibility for *in vivo* imaging of normal cellular morphology in humans. In addition, both the size of cells and their organelles increase dramatically in most cancers. OCT imaging of neoplasms could resolve cell nuclei for determining clinically relevant criteria such as the number of cells undergoing division and the nuclear-to-cytoplasmic ratios.

Although *in vivo* resolutions in this study were as high as 3.8 μm, higher resolutions are possible by using broader spectral bandwidths or shorter wavelengths of light. A solid-state titanium:sapphire laser is capable of *in vivo* resolutions as high as 1.4 μm at a center wavelength of 800 nm [11,12]. This shorter wavelength, however, is scattered and absorbed more in biological tissue and the image penetration will be less. Recent advances in OCT technology have permitted imaging at high-speeds [13] and through various beam delivery instruments including a flexible catheter/endoscope, hand-held probe, laparoscope, and surgical microscope. Combining these advancements with high-resolution OCT imaging will enable a future diagnostic tool for early detection of dysplastic and neoplastic changes at the cellular level.

Conclusions

Our results demonstrate the use of OCT as a powerful tool for tracking cellular dynamics in living specimens for extended periods of time. This ability of OCT to generate cellular level imaging at high acquisition rates suggests its potential for the identification of early *in vivo* neoplastic changes. Utilizing OCT for cellular imaging may have the potential for improving patient outcome by the early detection of many pre-malignant states.

Acknowledgments

The authors would like to thank Dr. Hazel Sive and her laboratory personnel at MIT for providing us with specimens and advice on the care and handling of embryos. Dr. Brett Bouma is currently with the Wellman Laboratory of Photomedicine, Massachusetts General Hospital. This research is supported in part by grants from the Office of Naval Research, Medical Free Electron Laser Program, Grant N00014-97-1-1066, the National Institutes of Health, Contracts NIH-1-RO1-CA75289-01 (JGF, MEB), NIH-1-R29-HL55686-01A1 (MEB), NIH-RO1-AR44812-01 (MEB), the Whittaker Foundation Contract 96-0205 (MEB), the Air Force Office of Scientific Research, Grant F49620-95-1-0221, and the Air Force Palace Knight Program.

References

1. M. Rajadhyaksha, M. Grossman, D. Esterowitz, R.H. Webb, and R.R. Anderson, "*In vivo* confocal scanning laser microscopy of human skin: melanin provides strong contrast," J. Invest. Dermatol. 104 946-952 (1995).

2. D. Huang, E.A. Swanson, C.P. Lin, J.S. Schuman, W.G. Stinson, W. Chang, M.R. Hee, T. Flotte, K. Gregory, C.A. Puliafito, and J.G. Fujimoto, "Optical coherence tomography," Science 254 1178-1181 (1991).

3. S.A. Boppart, B.E. Bouma, C. Pitris, J.F. Southern, M.E. Brezinski, and J.G. Fujimoto, "Imaging cellular mitosis and migration in *Xenopus laevis* with optical coherence tomography," Submitted to Nature Medicine (1998).

4. M.R. Hee, J.A. Izatt, E.A. Swanson, D. Huang, J.S. Schuman, C.P. Lin, C.A. Puliafito, and J.G. Fujimoto, "Optical coherence tomography of the human retina," Arch. Ophthal. 113 326-332 (1995).

5. J.M. Schmitt, M. Yadlowsky, and R.F. Bonner, "Subsurface imaging of living skin with optical coherence microscopy," Dermatology 191 93-98 (1995).

6. M.E. Brezinski, G.J. Tearney, B.E. Bouma, J.A. Izatt, M.R. Hee, E.A. Swanson, J.F. Southern, and J.G. Fujimoto, "Optical coherence tomography for optical biopsy: properties and demonstration of vascular pathology," Circulation 93 1206-1213 (1996).

7. V. Petricevic, S.K. Gayen, and R.R. Alfano, "Laser action in chromium-activated forsterite for near-infrared excitation: Is Cr^{4+} the lasing ion?," Appl. Phys. Lett. 53 2590-2592 (1988).

8. B.E. Bouma, G.J. Tearney, I.P. Bilinsky, B. Golubovic, and J.G. Fujimoto, "Self-phase modulated Kerr-lens modelocked Cr:forsterite laser source for optical coherence tomography," Opt. Lett. 21 1839-1841 (1996).

9. S.A. Boppart, M.E. Brezinski, B.E. Bouma, G.J. Tearney, and J.G. Fujimoto, "Investigation of developing embryonic morphology using optical coherence tomography," Dev. Biol. 177 54-64 (1996).

10. P.A.N. Chandraratna, M.I. Awaad, P. Chandrasoma, and M. Khan, "High-frequency ultrasound: Determination of the lowest frequency required for cellular imaging and detection of myocardial disease," Am. Heart J. 129 15-19 (1995).

11. X. Clivaz, F. Marquis-Weible, and R.P. Salathe, "Optical low coherence reflectometry with 1.9 μm spatial resolution," Elec. Lett. 28 1553-1555 (1992).

12. B.E. Bouma, G.J. Tearney, S.A. Boppart, M.R. Hee, M.E. Brezinski, and J.G. Fujimoto, "High resolution optical coherence tomographic imaging using a modelocked Ti:Al_2O_3 laser source," Opt. Lett. 20 1486-1488 (1995).

13. G.J. Tearney, M.E. Brezinski, B.E. Bouma, S.A. Boppart, C. Pitris, J.F. Southern, and J.G. Fujimoto, "*In vivo* endoscopic optical biopsy with optical coherence tomography," Science 276 2037-2039 (1997).

Two-Dimensional birefringence imaging in biological tissue using phase and polarization sensitive optical coherence tomography.

Johannes F. de Boer

Beckman Laser Institute and Medical Clinic, University of California, Irvine, CA 92612

Thomas E. Milner

Biomedical Engineering Program, University of Texas, Austin, TX 78712

Martin J. C. van Gemert

Academisch Medisch Centrum, University of Amsterdam, Amsterdam, The Netherlands

J. Stuart Nelson

Beckman Laser Institute and Medical Clinic, University of California, Irvine, CA 92612

Polarization sensitive optical coherence tomography (PS-OCT) was used to obtain images of optical birefringence in biological tissues. Through simultaneous detection of two orthogonal polarization states of the signal formed by interference of light backscattered from the biological sample and a mirror in the reference arm of a Michelson interferometer, the optical phase delay between light propagating along the fast and slow axes of birefringence was measured. Simultaneous detection of both polarizations also permits reconstruction of the electro-magnetic wave backscattered from the sample. Inasmuch as any fibrous structure will influence the polarization of light, PS-OCT is a potentially powerful technique in the field of biomedical imaging. It allows rapid non-contact investigation of tissue structural properties through spatially resolved imaging of birefringence.

(170.4500) Optical Coherence Tomography, (260.1440) Birefringence

Summary

We present a combination of Optical Coherence Tomography[1] (OCT) and polarization sensitive detection[2] to record 2D images of the change in polarization of circularly polarized incoming light backscattered from a turbid birefringent sample. In contrast to conventional reflectance OCT, in which the amplitude of backscattered light in a single polarization channel is measured, we detected both polarization channels to characterize the sample birefringence as a function of depth. 2D maps of birefringence of biological materials can reveal important structural information that is difficult to resolve with other imaging techniques. Moreover, since many biological tissues are birefringent (e.g., tendon, skin, bone, cornea, cartilage, muscle and blood vessels), PS-OCT is important even for reflectance OCT. Conventional reflectance images can show pseudo structure when tissue birefringence rotates backscattered light to the polarization state orthogonal to the detected channel. The detected polarization channel is determined by the polarization state of the light from the reference arm, since orthogonal polarization states do not interfere. Only detection of both polarization channels gives a complete reflectance OCT image in birefringent biological tissue[3–5]. To demonstrate polarization-sensitive OCT, we present a 1 mm wide by 700 μm deep image of bovine tendon birefringence. For comparison, a reflectance OCT image, formed by the sum of both detected polarization channels is shown. Details of the experimental configuration can be found in Ref. 3. With an experimental configuration[2], described in Fig. 1, birefringence was measured in mouse muscle. This experimental configuration allows the reconstruction of the electro-magnetic wave of the light backscattered from the sample since the interference fringes of orthogonal polarization components are measured simultaneously. The measurement of both amplitude and relative phase of the two orthogonal polarization components will allow assessment of birefringence with higher accuracy than in experimental configurations, where only the amplitude of both polarizations is obtained.

Corresponding author: JFdB, Email: deboer@bli.uci.edu; Telephone: 949-824-3284; Fax: 949-824-8413
Articles can be downloaded from: http://www.bli.uci.edu/research/jd/deboer.html

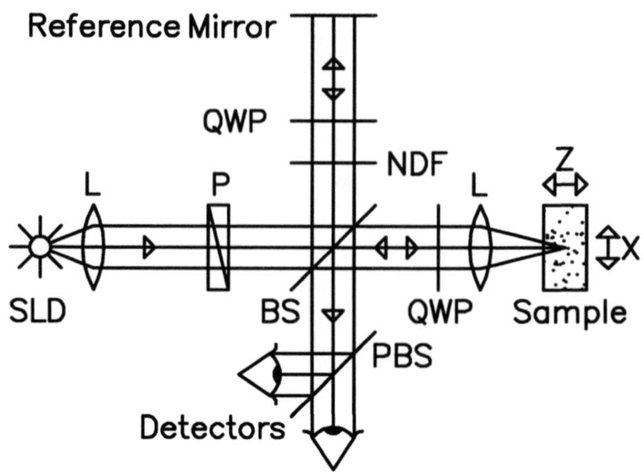

Fig. 1. Schematic of the polarization sensitive OCT system. *SLD*: superluminescent diode, 0.8 mW output power, central wavelength $\lambda_0 = 856\ nm$ and spectral FWHM $\Delta\lambda = 25\ nm$; *P*: polarizer; *BS*: polarization insensitive beam splitter; *PBS*: polarizing beam splitter; *QWP*: quarter wave plate, QWP in the reference arm is at 22.5^0, QWP in sample arm is at 45^0 with incident polarization; *NDF*: neutral density filter; *L*: lens. 2D images are formed by longitudinal movement of the sample with constant velocity $v = 1\ mm/s$ (z-direction), repeated after each 10 μm lateral displacement (x-direction).

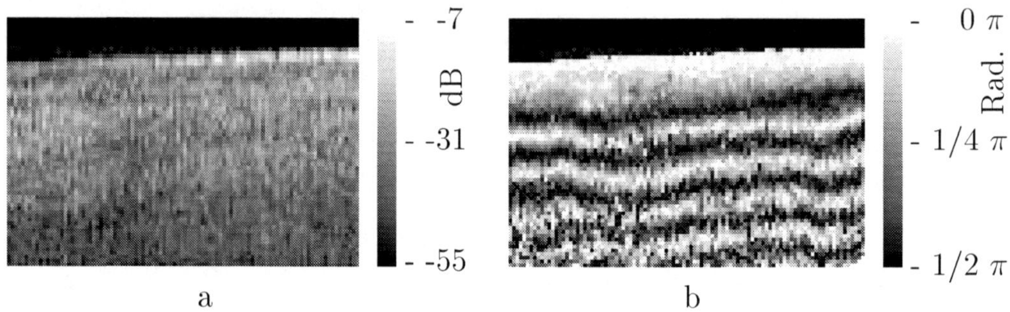

Fig. 2. Images of fresh bovine tendon 1 mm wide by 700 μm deep, constructed from the same measurement. a) Reflection image generated by computing $10 \log[I_H(z) + I_V(z)]$. The grey scale to the right specifies the signal magnitude. b) Birefringence image, generated by computing the phase retardation ϕ in Eq. (2). The grey scale at the right specifies the phase retardation. The banded structure, indicative of the birefringence, is clearly visible. Each pixel represents a 10 μm × 10 μm area.

Experimental results

Using the experimental configuration and signal demodulation described in Ref. 3 the images in Fig. 2 were obtained. The backscattered horizontal and vertical polarized intensities as a function of depth in the birefringent biological tissue are proportional to,

$$I_H(z) \propto \cos^2(k_0 z\delta); \qquad I_V(z) \propto \sin^2(k_0 z\delta), \tag{1}$$

where z is the sample depth; $k_0 = 2\pi/\lambda_0$ is the wave vector; δ is the difference in refractive indices along the fast and slow axes of the birefringent sample ($n_s - n_f$). The birefringence images were formed by calculating the phase retardation ϕ,

$$\phi = \arctan\sqrt{I_V(z)/I_H(z)} = k_0 z\delta, \tag{2}$$

and the reflection OCT images were formed by the sum of both polarization components,

$$I_b(z) = I_H(z) + I_V(z). \tag{3}$$

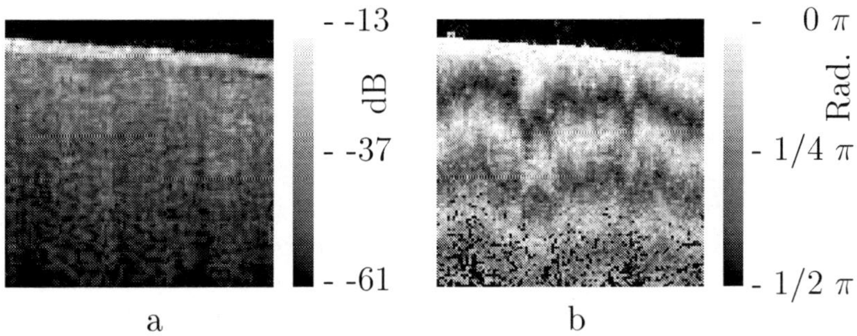

Fig. 3. Images of mouse muscle 1 *mm* wide by 1 *mm* deep, constructed from the same measurement. a) Reflection image generated by computing $10\log[I_H(z) + I_V(z)]$. The grey scale to the right specifies the signal magnitude. b) Birefringence image generated by computing the phase retardation ϕ in Eq. (2). The grey scale at the right specifies the phase retardation. The banded structure, indicative of the birefringence, is clearly visible. Each pixel represents a 10 μm × 10 μm area.

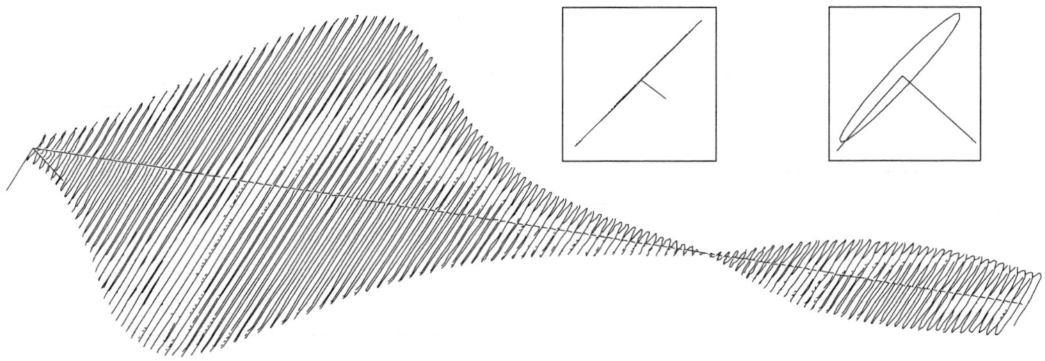

Fig. 4. Evolution of the wave of the reflected light propagating through birefringent mouse muscle. The electro-magnetic wave is reconstructed from the horizontal A_H and vertical A_V polarized component and relative phase of the interference fringes detected in the experimental scheme shown in Fig. 1. The displayed section is a small part of a longitudinal scan in the middle of the image shown in Fig. 3. Length of the section is 38 μm in a sample with refractive index $n = 1.4$. The beginning of the section shows the reflection from the sample surface modulated by the coherence envelope. The inserts show cross sections of the electro-magnetic wave over a full cycle perpendicular to the propagation direction taken at, respectively, 5.7 μm and 35.7 μm from the beginning of the section. As can be seen from the inserts the initial polarization is linear along one of the displayed axes, changing to an elliptical polarization state deeper in the tissue.

In Fig. 2 reflection and birefringence images of fresh bovine tendon are shown. The banded structure in Fig. 2b, indicative of birefringence, is clearly visible up to a physical depth of 700μm, while this structure is completely absent in the reflection image in Fig. 2a. This shows that the banded structure in the birefringence image is not an anatomical artifact in the tendon, but due solely to its birefringence. We determined the birefringence by the average distance between the first and second dark bands from the top of Fig. 2b over the full width (100 lateral scans). The average distance $\bar{z} = 116 \pm 13$ μm corresponded to a phase retardation $\phi = k_0\bar{z}\delta = \pi$ in Eq. (2), which gives an experimentally determined birefringence $\delta = 3.7 \pm 0.4 \times 10^{-3}$ of bovine tendon (predominantly type I collagen).

Figure 1 shows a schematic of the PS-OCT system that was used to acquire Fig. 3. The polarization state in each arm of the interferometer was computed using the Jones matrix formalism. The horizontal A_H and vertical A_V polarized components of the interference intensity between light in the sample and reference paths were detected separately. Since light from the reference arm was split equally into the horizontal and vertical polarization states, A_H and A_V were proportional to the light amplitude fields backscattered from the sample,

$$
\begin{aligned}
A_H &= \sqrt{R(z)}\cos(2k_0\Delta z)\, e^{-(\Omega\Delta z/c)^2}\cos(k_0 z\delta),\\
A_V &= \sqrt{R(z)}\cos(2k_0\Delta z+2\alpha)\, e^{-(\Omega\Delta z/c)^2}\sin(k_0 z\delta),\\
I_H &= |A_H|^2; \qquad I_V = |A_V|^2,
\end{aligned} \tag{4}
$$

where Δz was the optical path length difference between the sample and reference arms of the interferometer, z the depth of backscattered light from the sample, $R(z)$ the reflectivity at depth z, $k_0 = 2\pi/\lambda_0$, $\Omega = (\Delta\lambda\pi c)/(\lambda_0^2\sqrt{\ln 2})$, c the speed of light in a vacuum, δ the birefringence given by the difference in refractive indices along the fast and slow axes of the sample ($\delta = n_s - n_f$) and α the angle of the fast optical axis measured with respect to the horizontal. Fig. 3 shows a reflection and a birefringence OCT image of mouse muscle. The birefringence in muscle is caused by its highly fibrous structure.

In Fig. 4 the electric field of the electro-magnetic wave is shown for a small part of a longitudinal scan in the mouse muscle. The electric field is reconstructed from the horizontal A_H and vertical A_V polarized component and phase of the interference fringes as detected by the detectors (Eq. (4)). It does not reflect the actual polarization state in the sample, since after being backscattered from the sample the light has made a second passage through the quarter wave plate. Fig. (4) shows the change of the polarization from a linear to an elliptic state as a consequence of tissue birefringence.

In conclusion, we have demonstrated polarization sensitive optical coherence tomography that allows for spatially resolved imaging of birefringence in biological tissue. With an experimental configuration in which the interference fringes of both polarization components are measured simultaneously, the electric field of the electromagnetic wave was reconstructed. Algorithms in development that process both polarization components and their relative phase will allow better analysis of the polarization state changes in backscattered light, and consequently better characterization of birefringent structures.

Research grants from The Biomedical Research Technology Program and the Institute of Arthritis and Musculoskeletal and Skin Diseases of The National Institutes of Health, the Whitaker Foundation (WF 21025) and the Dermatology Foundation are gratefully acknowledged, as is Institute support from the U.S. Department of Energy, the National Institutes of Health, and the Beckman Laser Institute Endowment.

---

1. D. Huang *et al.*, Science **254**, 1178 (1991).
2. M. R. Hee, D. Huang, E. A. Swanson, and J. G. Fujimoto, J. Opt. Soc. Am. B **9**, 903 (1992).
3. J. F. de Boer, T. E. Milner, M. J. C. van Gemert, and J. S. Nelson, Opt. Lett. **22**, 934 (1997).
4. J. F. de Boer, T. E. Milner, M. J. C. van Gemert, and J. S. Nelson, Proc. Soc. Photo-Opt. Instrum. Eng. **3196**, 32 (1998).
5. M. J. Everett, K. Schoenenberger, B. W. Colston Jr., and L. B. Da Silva, Opt. Lett. **23**, 228 (1998).

En-face OCT imaging of the retina using path modulation introduced by the transversal scanning mirror

Adrian Gh. Podoleanu, George M. Dobre, David J. Webb and David A. Jackson
Applied Optics Group, School of Physical Sciences, University of Kent ,Canterbury, CT2 7NR, UK
Fax: +44 (0)1227 827558; e-mail@:A.G.H.Podoleanu@ukc.ac.uk

Abstract

The performance of an optical coherence tomography system using the phase modulation introduced by the transversal galvo-scanners is analyzed in terms of bandwidth, transversal resolution and image distortions. The method is illustrated by producing transversal and longitudinal images of the retina of the living eye.

Key Words

Optical coherence tomography, retinal imaging, Newton rings.

Introduction

Considerable interest has been recently generated by the application of low coherence interferometric techniques in the high resolution imaging of *in vivo* biological targets. The imaging of moving objects, demanding high acquisition rates, can only be achieved in an optical coherence tomography system (OCT) by fast modulation of the path imbalance.

In order to obtain en-face images using low coherence reflectometry, an X,Y-scanning device and a path imbalance modulator are generally employed [1,2]. The frequency of the signal modulating the path imbalance is subsequently used as a carrier for the processing technique. An interference signal arises only when the optical path difference between the object and the reference arm is less than the coherence length, l_c, of the optical source.

Fast modulation can be performed at present in fiberized configurations by stretching the fiber with a piezo-electric cylinder subject to rapid cycles of expansion and contraction, with the associated drawbacks of increased transmission loss and the introduction of birefringence. With electro-optic modulation techniques, dispersion compensation [3] is required in order to maximize the improvement in depth resolution offered by recently developed sources exhibiting very low coherence [4].

We present here a technique based on the modulation introduced by the scanner itself that not only offers a simpler alternative to the existing phase modulation techniques but is also dispersion free. It may be an ideal solution in special circumstances where a phase modulator is not available or its insertion into the system should be avoided.

We show that in an X-Y scanning OCT system using a galvanometric scanning mirror pair the optical path difference (OPD) between the sampled target voxel and the reference plane changes in relation to the scan angle and we investigate how the modulation generated by scanning procedure itself can be used in signal processing.

OCT set-up

A fiberized set-up based on a modified Michelson interferometer was constructed for the purpose of performing image acquisition and its schematic diagram is shown in Figure 1.

A retinal scan pattern resembling a raster is generated by applying either triangular or sawtooth signals (with appropriate ramp time ratios) to the X-galvo-scanner and to the Y-galvo-scanner.

Path modulation introduced by the transversal scanning mirror

In the schematic diagram of the OCT object arm, shown in detail in Figure 2, the galvanometric scanning mirror producing the deflection is assumed to be the X mirror, i.e. the deflection is in the x direction. For reasons of simplicity the target is first assumed to be a smooth plane, AP, perpendicular to the optical axis. When the eye model

EM is the object, lens L2 replaces the eye lens EL and AP replaces the retina R.

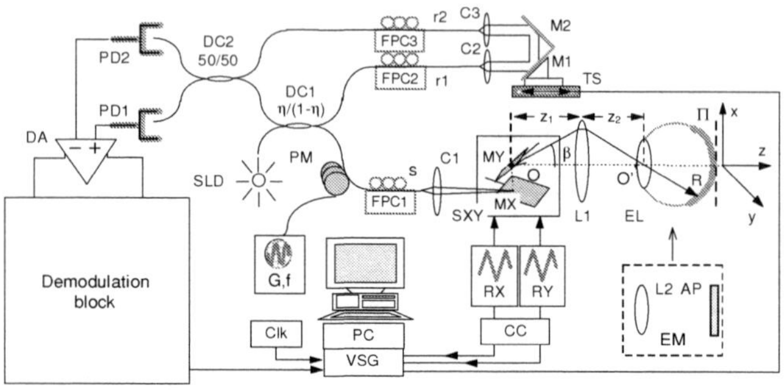

Figure 1. OCT experimental set-up used for retinal imaging.

SLD: Superluminescent diode (Superlum SLD-361); C1, C2, C3: microscope objectives; DC1, DC2: directional couplers; PM: 5 cm diameter piezo-cylinder with 2 turns of fiber acting as a phase modulator; G: sinusoidal generator, f = 30 kHz; M1, M2: mirrors; L1: convergent lens; FPC1,2,3: polarization controllers; SXY: orthogonal scanning mirror pair; MX(MY): mirror of the X(Y) scanner; DM: demodulation block; RX, RY: ramp generators; PD1, PD2: photodetectors; DH: monostable circuit for the VSG line trigger; CC: controlling circuitry; Clk: VSG clock generator; EL: eye lens; R: human retina; EM: eye model; L2: convergent lens; AP: aluminum plate; TS: computer controlled translation stage; PC: personal computer.

Distances z_1 between the axis of rotation of MX and L1 and z_2 between L1 and L2 are measured along the optic axis of the eye collimation arrangement. f_1 is the focal length of L1 and f_2 is the focal length of lens L2.

The object beam is deflected by mirror MX, and the reflection point may or may not be on the axis of rotation of MX. A parameter δ can be defined as the distance, measured along the optic axis of the object arm, between the axis of rotation of MX and the location of the point where the beam impinges on MX.

The relation between the OPD and the instantaneous deflection angle β of the X galvo-scanning mirror is given by [5]:

$$\mathrm{OPD} = -2\delta\beta + f_2 \frac{z_1^2}{z_2^2}\beta^2 \tag{1}$$

The description of the sampling function generated when the interferometer is coherence matched requires the identification and analysis of two separate cases, defined by whether or not the beam is centered with respect to the axis of rotation of the deflecting galvo-scanning mirror.

Centered beam case, $\delta = 0$

The centered beam case corresponds to the situation where the reflection point of the object beam is situated on the axis of rotation of the X galvo-mirror. In this case the beam displacement relative to the axis is $\delta = 0$.

Figure 2 illustrates the object beam leaving the exit mirror in A and crossing the optic axis in B. The optical paths of all ray trajectories between A and B are equal by virtue of A and B being conjugate points.

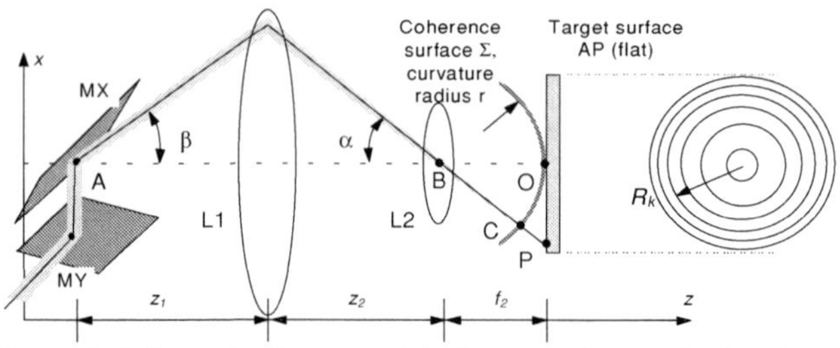

Figure 2. Schematic diagram of the beam trajectory in the object arm of the OCT.

The surface Σ is the three dimensional locus of those points situated at an optical distance from A equal to the optical length of the reference arm. In the diagram, the coherence surface Σ and the optic axis intersect in O. Surface Σ has the shape of a sphere of radius f_2, centered in B. For an angle β for which the target object is intersected in P, the beam is partly reflected (specular reflection) and partly backscattered towards B and directed on towards A in a reciprocal manner such that the combined onward and return journey is along the route A-B-P-B-A.

The position of the mirror in the reference arm, meanwhile, is kept constant, such that at any instant the reference beam travels the equivalent of an optical distance equal to A-B-O-B-A (in a straight line along the optic axis).

Based on (1), the OPD between the object and the reference arms can be simply expressed in this particular case in terms of the scan angle β and the distances z_1 and z_2:

$$\mathrm{OPD}(\beta) = f_2 \frac{z_1^2}{z_2^2}\beta^2 \tag{2}$$

The detected signal registers maxima and minima as the angle β (and implicitly α) is varied. Maxima are obtained when the OPD is an integer multiple of the wavelength, λ.

It can be shown [6] that the geometrical locus of the points on the surface S of maximum interference is described by rings of radius R_k given by:

$$R_k = \frac{2f_2 + k\lambda}{2}\sqrt{1 - \left(\frac{2f_2}{2f_2 + k\lambda}\right)^2} \tag{3}$$

where k is an integer. For small angles and low order k, (3) becomes:

$$R_k \approx \sqrt{k f_2 \lambda} \tag{4}$$

which shows that the locus of interference maxima is given by a similar relation to that describing Newton's rings. The interference image from the plane mirror is shown in Figure 3.

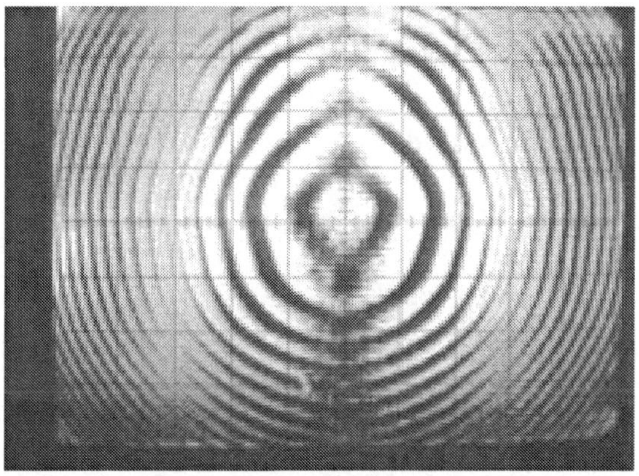

Figure 3. Image of a mirror in the centered beam case. Horizontal and vertical scale: 185 μm / div, Tx = 25 ms, U = 0.25 V

Given the limited coherence length l_c of the source, the Newton rings visibility is negligible beyond an area on the target given by:

$$A \approx \frac{\pi}{2} l_c f_2 \tag{5}$$

and the angle β for which |OPD| = l_c is β_C:

$$\beta_C = \sqrt{\frac{l_c}{f_2}\frac{z_2}{z_1}} \tag{6}$$

The circular Newton rings appearance of the sampling function in the centered case is due to the fact that the OPD introduced by the rotation of the galvo-mirror is proportional to the square of the deflection angle β (see equation (2)).

Off-centered beam case, δ ≠ 0

Displacing the beam by a distance δ results in an optical path change of δβ (Figure 4), expressed by the first term in (1) after the double pass of light (to and from the object) is accounted for [5].

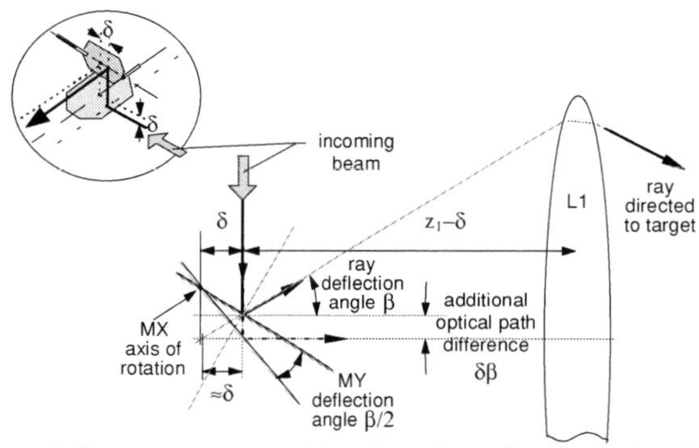

Figure 4. Ray geometry resulting from laterally displacing the beam with respect to the rotation axis of mirror MX (off-centered case). The inset shows the manner of introducing a displacement δ between the incoming beam and the X-mirror in an XY galvanometer scanning mirror pair.

It should be noted that equation (1) represents an approximation valid only for small values of β, sufficient for producing lateral scans a few millimeters wide across the retina.

The immediate consequence of the presence of the linear term in (1) is that the curvature of the fringes changes, with the sampling function approximately resembling a set of parallel lines, as shown in Figure 5.

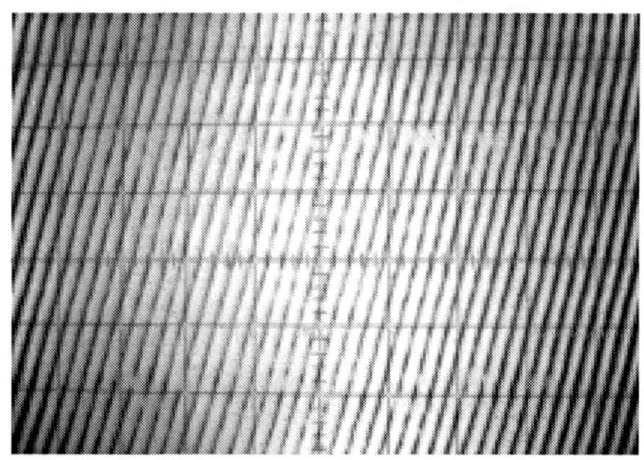

Figure 5. Horizontal image of a mirror. Horizontal and vertical scale: 29 μm / div, Tx = 25 ms, δ = 3 mm, U = 0.17 V

Bandwidth

Taking into account the size of a pixel defined on the target at the back of lens L2 of focal length f_2, for a beam diameter D and the number of pixels within the transversal scan width on the target, ΔX, the image bandwidth is given by [7]:

$$B_i = 0.41 \frac{F_x D}{\lambda} \frac{\Delta X}{f_2} \tag{7}$$

Spectrum of the modulation signal

Beam centered case

When the beam is scanned across the sampling function, shown in Figure 3, the photodetected signal is modulated at a frequency ν. The maximum value of this frequency, ν_g, is expressed as:

$$\nu_g = \frac{4\,G\,U}{T_x\lambda} f_2 \frac{z_1^2}{z_2^2} |\beta_m| \qquad (8)$$

where G is the scanner sensitivity, U is the scanner voltage amplitude, T_x is the X-galvanometer scanner ramp duration, λ is the central source wavelength and β_m is the maximum deflection angle β for which coherence is still obtained.

For $|\beta| > \beta_C$, $|\beta_m| = \beta_C$ and for $|\beta| < \beta_C$, $|\beta_m| = GU$. According to (8), this means that for $U < U_C = \beta_c/G$, ν_g has a quadratic dependence on U and for $U > U_C$ the dependence becomes linear.

This can be understood in the light of the fact that for |β| values above β_C, the explored areas correspond to |OPD| values outside coherence and consequently they do not contribute to the interference signal. Up to β_C, interference is obtained during the entire duration of a transversal scan, T_x. Above β_C, the time interval during which interference takes place, and also the period 1/ν of the modulation signal, are proportional to β_C/β and consequently both decrease with increasing U. As a result, ν_g becomes linear with U (Figure 6).

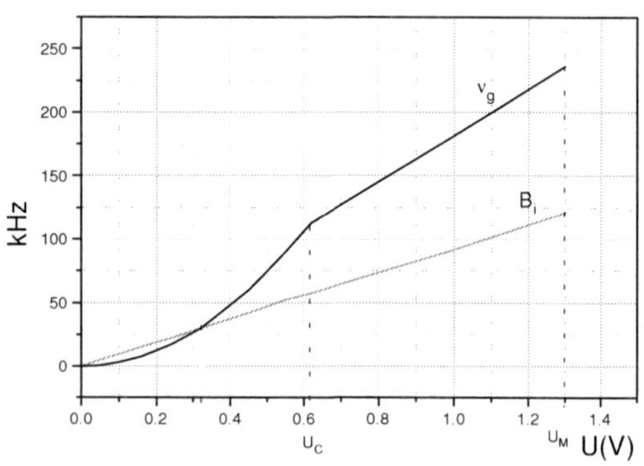

Figure 6. Variation of the signal frequency and information bandwidth with scanning mirror drive signal amplitude

For $U > U_C$, components up to ν_g are generated (from (6)and (8)):

$$\nu_g = 4G\frac{U}{T_x}\frac{z_1}{z_2}\frac{\sqrt{f_2 l_c}}{\lambda} \qquad (9)$$

The frequency spectrum of the photodetected signal contains components from 0 Hz up to ν_g, when the beam samples the maximum concentration of Newton rings (towards the extremities of the Newton rings shown in Figure 3).

As shown in Figure 6, when the horizontal field of view, Δα, is larger than 1.7° (value worked out with f_2 = 2cm), the frequency ν_g of the modulation signal due to the galvo-scanner exceeds the bandwidth of the image, B_i, and ν_g can be used as a carrier for the image signal. However, for an angle Δα less than 1.7°, a conventional modulation technique must be used.

Off-centered beam case

The interference signal due to the mirror scan is modulated at a frequency ν given by:

$$\nu = \frac{4GU}{T_x\lambda}\left(\delta - 4GUf_2\frac{z_1^2}{z_2^2}t\right) \qquad (10)$$

For δ = 0, the expression for the modulation frequency in the centered case is obtained. During a scan, the frequency ν varies between:

$$\nu_{min,max} = \nu_c \pm \nu_g = \frac{4GU}{T_x\lambda}\left(\delta \pm f_2\frac{z_1^2}{z_2^2}\beta_m\right) \qquad (11)$$

where ν_c is the central frequency and ν_g the frequency spread owing to the nonlinear OPD dependence on β in (1) (this is also the maximum component in the spectrum in the centered beam case).

For large enough values of δ the linear term in (10) becomes dominant. Consequently, there is a minimum value of δ that results in a linear increase of ν with the scan amplitude U. A second condition results from the requirement that the information bandwidth B_i in (7) should be less than the minimum frequency generated during a scan, i.e. $\nu_c - \nu_g > B_i$. It can be shown that the second condition prevails which leads to:

$$\delta > \frac{z_1}{z_2}\sqrt{f_2 l_c + 0.04D^2} \qquad (12)$$

The variability of the carrier frequency ν during scanning implies that the bandwidth of the band pass filter should be large enough to accommodate both the minimum ν_{min}-B_i and the maximum ν_{max}+B_i, i.e. its span should be $2(B_i+\nu_g)$ (this is only an approximate evaluation). In terms of noise, therefore, the method presented here is inferior to the conventional technique for larger scan angles.

The enlargement $\Delta\nu_o$ due to the optical linewidth is much smaller than that due to the second term in (11) and can be neglected in this first approximation treatment. However, for most applications, the information bandwidth B_i is larger than ν_g.

Images from the living eye

We have investigated the utility of both the centered and off-centered cases for retinal imaging. We have produced transversal (x,y) as well as longitudinal OCT (x,z) images using the phase modulation generated by the galvo-scanner only or using the modulation introduced by the galvo-scanner in addition to that of a piezo-cylinder in the reference arm.

We have not noticed any difference in quality between the images produced in this way and those produced with the conventional technique (reference mirror vibrated at 35 kHz and demodulation at the same frequency).

Little difference was noticed between the images obtained from the living retina when the beam was centered or displaced from the axis of the galvo-mirror for the conditions imposed by the performance of our devices: X-scanning rate up to 1000 Hz, image size up to 10^0, frame rates from 1 to 5 Hz and processing bandwidth about 150 kHz.

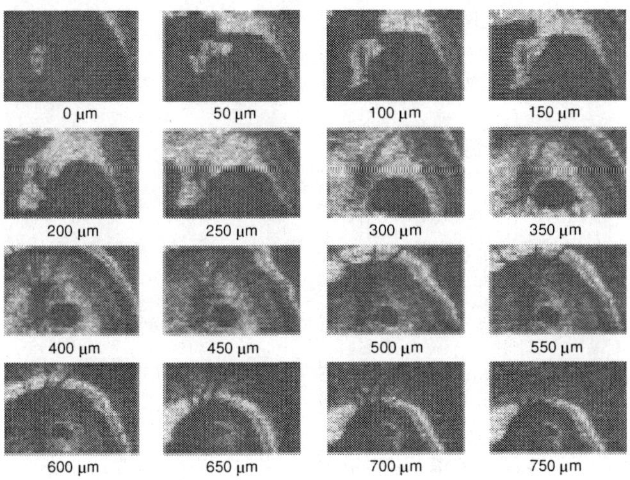

Figure 7. In-vivo en-face image of the human retina. Each frame obtained in 1 s. $\Delta x = \Delta y = 3$ mm.

A set of OCT transversal images of the optic nerve area, obtained at different depths by successively incrementing the reference arm length 50 μm at a time, is shown in Figure 7. The top of the retinal nerve layer is seen in the 0 μm image. The central parts of the 450 μm - 600 μm images correspond to the photodetector fiber layer. The high intensity areas in the 600-700 μm images show the retinal pigment layer.

Four other images from the foveal region of a different volunteer's eye are shown in Figure 8.

By driving only the X galvo-scanner and translating the reference mirror at a speed of 1 mm/s the longitudinal OCT image in Figure 9 was obtained. All images in Figure 7, Figure 8 and Figure 9 were obtained with a power of 90 μW incident on the eye; no averaging was applied.

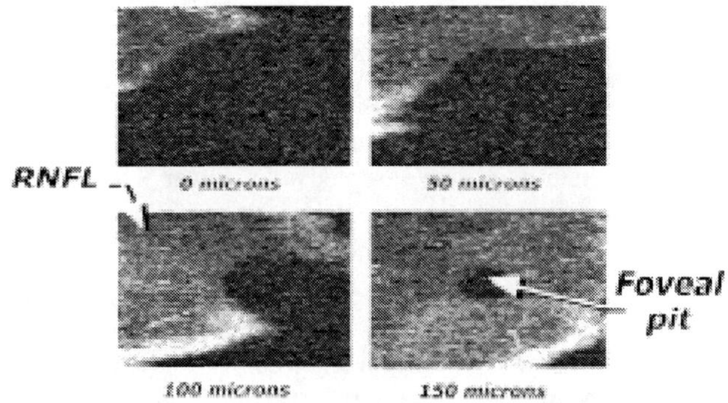

Figure 8. OCT images of the foveal region at different depths. Each frame obtained in 1 s. $\Delta x = \Delta y = 3$ mm.

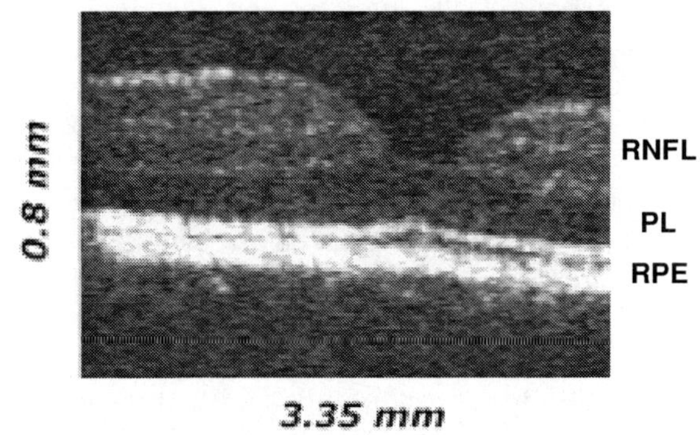

Figure 9. In vivo OCT longitudinal image of the human retina showing the foveal pit. Angular extent of 6.8° corresponds to a span $\Delta x = 3.35$ mm on the retina.

Conclusions

Transversal as well as longitudinal OCT images are obtained from the living retina using the phase modulation introduced by the galvo-scanner.

We have found that for a sufficiently large scan Δx images of similar quality are obtained with those using the conventional technique, where the optical beam is centered on the galvo-mirror, i.e. $\delta = 0$, and a separate device is used in the reference path for modulation.

The modulation spectrum is different in each of the two cases. When $\delta = 0$ (centered case), the spectrum is almost flat up to a maximum frequency, ν_g. When $\delta \neq 0$ (off-centered case), the spectrum is centered about a carrier ν_c, with side bands up to ν_g. The larger the horizontal image size, the higher the frequencies ν_g, ν_c and the image bandwidth, B_i. In the off-centered case, in order to avoid superpositions of spectra, a minimum value ν_c is needed to act as a carrier for the signal of bandwidth B_i. Obviously in this case the sampling function is better for imaging reasons because of a higher achievable transversal resolution. In the off-centered case, for a sufficiently large δ, the sampling is homogeneous across the entire target. In Figure 5 the period on the target is about 6 μm which means that the minimum feature which can be sampled and displayed using this sampling function is 12 μm. In the centered case details in the center of the image are sampled at a lower spatial rate. Consequently, the image pixel size depends on the path modulation method used and for the centered case is variable across the target.

All these comments are valid for a plane mirror which is perpendicular to the optical axis (Z) of the system. If the plane AP is inclined in relation to the Z axis, then the geometry of the sampling function changes: for the centered case the center of the Newton rings moves away from the center of the image and for the off-centered case the orientation of the lines of the sampling function changes. This situation is depicted in Figure 2, where the orientation of lines shows that either the mirror used as target is not perpendicular to the Z axis or the optical set-up is not perfectly centered.

Another issue concerns the shape of the surface sampled at coherence from the object volume. In the first approximation, the effect of the shift δ is that it changes the inclination of the plane sampled from the object by an angle $\varphi = \tan^{-1}(\delta/f_2)$.

The disadvantages of the method are: (i) no modulation arises when the object and the coherence surfaces coincide and the target becomes "invisible"; (ii) the pixel size varies across the target area; (iii) the modulation frequency and consequently the central frequency of the demodulation block are dependent on the image size; (iv) the sampling function modulating the image intensity can be visible when fine details of size comparable with its periodicity are investigated. Moreover, the sampling function is visible when imaging surfaces exhibiting specular reflection, as shown in [5]. However, most surfaces will not be specular so the sampling function will not appear in the image.

Acknowledgments

The authors would like to acknowledge the support of EPSRC and that of Dr Fred Fitzke from The Institute of Ophthalmology, London. A. Podoleanu acknowledges the support of the Wellcome Trust and that of the Whittaker Foundation.

References

1. J.G. Fujimoto, M.E. Brezinski, G.J. Tearney, S.A. Boppart, B. Bouma, M.R. Hee, J.F. Southern, E.A. Swanson, "Optical biopsy and imaging using optical coherence tomography", Nature Medicine 1 970-972 (1995).
2. A.F. Fercher, "Optical coherence tomography", J. Biomedical Optics 1 157-173 (1996).
3. Ch. K. Hitzenberger, W. Drexler, A. Baumgartner and A. F. Fercher, "Dispersion effects in partial coherence interferometry", in "Proceedings SPIE", SPIE 2981 29-36 (1997).
4. B. Bouma, G.J. Tearney, S.A. Boppart, M.R. Hee, M.E. Brezinski, J.G. Fujimoto, "High resolution optical coherence tomographic imaging using a mode-locked TiAl2O3 laser source", Opt. Lett. 20 1486-1488 (1995).
5. A. Gh. Podoleanu, G. M. Dobre and D. A. Jackson, "En-face coherence imaging using galvanometer scanner modulation" Opt. Lett. 23 147-149 (1998).
6. A. Gh. Podoleanu, G. M. Dobre, D. J. Webb, D. A. Jackson, "Coherence Imaging by Use of a Newton Rings Sampling Function", Opt. Lett. 21 1789-1791 (1996).
7. A. Gh. Podoleanu, M. Seeger, G. M. Dobre, D. J. Webb and D. A. Jackson, "Transversal and Longitudinal Images from the Retina of the Living Eye Using Low Coherence Reflectometry ", J. Biom. Opt. 3 12-20 (1998).

Contrast optimization in full-field optical coherence microscopy

E. Beaurepaire and A.C. Boccara
Laboratoire d'Optique Physique, Ecole Supérieure de Physique et Chimie Industrielles, Centre National de la Recherche Scientifique UPR A0005, 10 rue Vauquelin, F-75005 Paris, France.
phone : 33 1 40 79 46 03 fax : 33 1 43 36 23 95 e-mail : boccara@optique.espci.fr

F. Amblard
Laboratoire de Neurobiologie et Physiologie, Ecole Supérieure de Physique et Chimie Industrielles, Centre National de la Recherche Scientifique UMR 7637, 10 rue Vauquelin, F-75005 Paris, France.

M. Lebec, L. Blanchot and H. Saint-Jalmes
Laboratoire d'Etudes et de Recherche en Instrumentation Signaux et Systèmes, Université Paris 12-Val de Marne, F-94010 Créteil, France.

Abstract

We demonstrated recently a system enabling one to acquire a head-on (X-Y) OCT image in parallel on a CCD camera without scanning. We summarize the principle of the detection and we address some experimental issues influencing the contrast of the images in this arrangement.

Key words

(110-4500) Optical coherence tomography; (110-0180) Microscopy; (040-1520) CCD, charge-coupled device

Optical coherence tomography (OCT) makes use of a short coherence light source and a Michelson interferometer to isolate photons undergoing a single backscattering event in the tissue, at a given depth [1-3]. Most of the current OCT systems generate X-Z cross sections of the specimen (Z being the optical axis and X one of the transverse dimensions). Usually, in order to eliminate the need for Z-scanning in the sample arm, the excitation beam is not highly focused in the specimen, so that the depth of focus is kept greater than the Z-dimension of the image[4-7]. This leads in practice to images with lateral resolutions on the order of 10-30 µm.

We have introduced recently an alternative system, where a parallel detection scheme is used to acquire a head-on (X-Y) image at a given depth without scanning [8-9]. The lateral resolution (2 µm) is limited only by the optics and the camera pixel size, and the axial sectioning ability (8 µm) is half the coherence length of the source.

We briefly describe of the microscopy setup and summarize the general principle of the detection used and its application to the OCT context. We then discuss the effects of balancing the relative amount of light sent to the interferometer arms on the image contrast in our system, and the effects of signal averaging.

Microscopy setup

Our system is implemented on a commercial microscope body (Olympus BX60) equipped with a 10x, 0.25 NA objective and a polarizing beam splitter (PBS) which separates the beam into 2 orthogonally polarized components, thereby constituting a Michelson interferometer as long as suitable polarizers are placed before and after the PBS . The light source is an infrared LED (20 mW, 840 nm, 15 µm coherence length, delivering 300 µW to the sample) which is linearly polarized. Fig. 1 shows the optical path for a single point of the field stop. The PBS separates the beam into 2 components with orthogonal polarization which are focused by the objective.

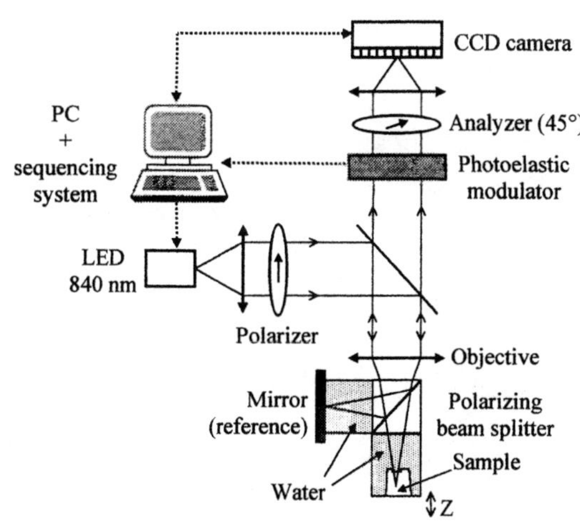

Figure 1. Overview of the setup and image of one point through the optical part.

One point at the field stop thus corresponds to two points imaged by the objective, one being focused in the sample volume and the other on the reference mirror surface. The reference mirror is positioned so that the zero-path difference plane matches the plane of focus in the sample. In our experiment, this condition is approximately met when imaging down to depths of several hundred micrometers in the sample, due to the presence of an index medium (water) in the interferometer arms. The presence of water also reduces stray reflections at the interfaces.

After reflection / backscattering and recombination by the PBS, the coherent part of the beam emerging from the objective lens is elliptically polarized. A measurement of its polarization state provides the value of the relative amplitudes of the orthogonal components. This measurement is achieved by using a photoelastic birefringence modulator to introduce a periodic (50 kHz) phase shift between the two orthogonal components and an analyzer. Finally the beam is focused onto a CCD array (Dalsa CA-D1 256x256 pixels, maximum speed 200 images/s).

Detection

To achieve lock-in detection at a frequency much higher than the CCD readout rate, we use an original detection principle which involves a modulation of the light source (secondary modulation) [10-12].

We summarize the principle of this detection and how it is applied in our OCT scheme.

General principle of the detection

We consider a modulated experiment, where the signal incident on a single CCD pixel is described as :

$$E(t) = e_0 + 2\sum_{q=1}^{+\infty} e_q \times \cos(\omega_0 q t + \zeta_q) \qquad (1)$$

We introduce a secondary modulation M(t), on the light source (LED), such as :

$$M(t) = m_0 + 2\sum_{q=1}^{+\infty} m_q \times \cos(\omega_0 q t + \xi_q) \qquad (2)$$

An even number of sequences 2P are integrated on the CCD, corresponding to different phase shifts applied to M(t). These sequences are integrated over an integer number of modulation periods K. The signal integrated on a pixel during the pth sequence is expressed as :

$$\begin{aligned} S_p &= \int_0^{KT_0} E(t) M\left(t + \frac{p}{2P} T_0\right) dt \\ &= KT_0 e_0 m_0 + 2KT_0 \sum_{q=1}^{+\infty} e_q m_q \times \cos\left(2\pi \frac{pq}{2P} + \xi_q - \zeta_q\right) \end{aligned} \qquad (3)$$

When no harmonics superior to P are present in E or M, the knowledge of (S_p) for $0 \le p < 2P$ gives access to (e_p, ζ_p) for $0 \le p < P$.

Application to the OCT experiment

If one is imaging a reflecting surface within a scattering sample, at any given time, the signal incident on a pixel of the CCD detector is of the form:

$$E(t) = \eta\varepsilon\left(A_R^2 + A_I^2 + A_C^2 + 2A_R A_C \times \cos\left(\varphi + \psi_M \sin(\omega_0 t)\right)\right) \qquad (4)$$

where η is the quantum efficiency of the detector, ε is the overall transmission of the detection path, A_R^2 is the signal reflected by the reference mirror, $A_C^2\exp(i\varphi)$ is the signal coherently backscattered by the reflecting surface, A_I^2 is the intensity of the incoherent signal detected on the pixel, and $\psi_M\sin(\omega_0 t)$ is the phase difference introduced by the birefringence modulator.

Introducing Bessel functions J_i and using the series expansion :

$$\exp(iz\cos\theta) = J_0(z) + 2\sum_{k=-\infty}^{+\infty} i^k J_k(z)\cos(k\theta) \qquad (5)$$

E(t) can be expressed in the general form of equation 1 with :

$$\begin{cases} e_0 = \eta\varepsilon\left(A_R^2 + A_I^2 + A_C^2 + 2A_R A_C J_0(\psi_M) \times \cos\varphi\right) \\ e_{2k+1} = -\eta\varepsilon A_R A_C J_{2k+1}(\psi_M) \times \sin\varphi \\ \zeta_{2k+1} = -\pi/2 \\ e_{2k} = \eta\varepsilon A_R A_C J_{2k}(\psi_M) \times \cos\varphi \\ \zeta_{2k} = 0 \end{cases} \qquad (6)$$

and a square secondary modulation M(t) (Figure 2) can be expressed in the form of equation 2 with :

$$\begin{cases} m_0 = \alpha \\ m_q = 2\sin(\pi\alpha q)/\pi q \\ \xi_q = -2\pi q\beta \end{cases} \qquad (7)$$

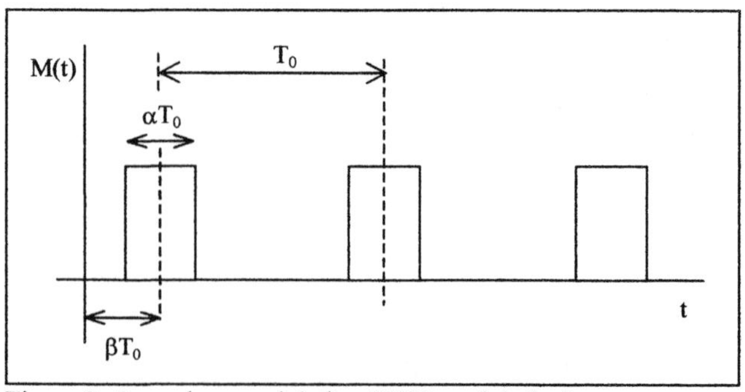

Figure 2. Secondary modulation (LED modulation).

The signal integrated during K modulation periods of the pth sequence can then be expressed as :

$$S_p = KT_0 \times \Big[e_0 m_0 + \frac{2}{\pi}\cos\varphi \times \eta\varepsilon A_R A_C \sum_{k=1}^{\infty} \Big\{ J_{2k}(\psi_M) \times \sin(\pi\alpha 2k) \times \cos\Big(\frac{2kp\pi}{P} - 4k\pi\beta\Big)\Big\} - \frac{2}{\pi}\sin\varphi \times \eta\varepsilon A_R A_C \sum_{k=0}^{\infty} \Big\{ J_{2k+1}(\psi_M) \times \sin(\pi\alpha(2k+1)) \times \cos\Big(\frac{(2k+1)p\pi}{P} - \frac{\pi}{2} - 4k\pi\beta\Big)\Big\}\Big] \quad (8)$$

We commonly use two measurement schemes, with 2P=2 or 4 sequences :

Two-sequence scheme :
Setting 2P=2, α=½ and β=¼ leads to a measurement where:

$$\begin{aligned} S_0 - S_1 &= A_C \sin\varphi \times \frac{4}{\pi} KT_0 \eta\varepsilon A_R \sum_{k=0}^{\infty} \frac{J_{2k+1}(\psi_M)}{2k+1} \\ &\propto A_C \sin\varphi \end{aligned} \quad (9)$$

Computing $|S_0 - S_1|$ gives an estimate of the coherent backscattered signal.

Four-sequence scheme :
Setting 2P=4, α=¼, β=0 and introducing :

$$\begin{cases} \Sigma = \frac{1}{4}(S_0 + S_1 + S_2 + S_3) \\ C = \frac{1}{4}(S_0 - S_1 + S_2 - S_3) \\ Z = \frac{1}{2}(S_3 - S_1) \\ S = \frac{1}{2}(S_2 - S_0) \end{cases} \quad (10)$$

leads to a measurement where :

$$\begin{cases} \Sigma = e_0 m_0 + \frac{2A_R A_C \eta\varepsilon}{\pi}\cos\varphi \sum_{l=1}^{\infty} \frac{J_{4l}(\psi_M)}{4l} \\ C = A_C \cos\varphi \times \frac{2A_R \eta\varepsilon}{\pi} \sum_{l=0}^{\infty} \frac{J_{4l+2}(\psi_M)}{4l+2} \\ Z = 0 \\ S = A_C \sin\varphi \times \frac{2A_R \eta\varepsilon}{\pi} \sum_{l=0}^{\infty} \frac{J_{2l+1}(\psi_M)}{2l+1} \end{cases} \quad (11)$$

The amplitude of the coherent signal can then be computed from the evaluation of C and S.

Contrast

50% beam splitters are commonly used in the OCT systems reported to date. Such an arrangement would severely limit the amount of contrast that can be obtained on CCD cameras, due to their limited dynamic range. Since the amount of light backscattered by the sample is generally much smaller than the amount of light reflected from the reference mirror, we optimize the ratio of the two fluxes by rotating the angle of the input polarization.

Let us consider the model situation where images are obtained from a tilted reflecting boundary inside a diffusing sample. If a_0 is the amplitude of the excitation signal incident on the polarizing beam splitter and α is the angle between the direction of the incident polarization and the PBS axis, then a signal of amplitude $a_0\cos(\alpha)$ illuminates the sample and a proportion of amplitude $a_0\sin(\alpha)$ is sent in the reference arm. If we characterize the reference mirror by a reflection coefficient of 1 and the sample by a “coherent reflection coefficient” r_c and an “incoherent reflection coefficient” r_i, the minimum and maximum signal detected by the CCD along the boundary (in the absence of modulation) can be expressed as :

$$I_\pm = a_0^2\left(\sin^2\alpha + \left(r_c^2 + r_i^2\right)\cos^2\alpha \pm 2r_c\cos\alpha \times \sin\alpha\right) \quad (12)$$

An estimate of the contrast is then given by :

$$\gamma = \frac{I_+ - I_-}{I_+ + I_-} = \frac{\rho 2r_c \sin\alpha\cos\alpha}{\sin^2\alpha + \left(r_c^2 + r_i^2\right)\cos^2\alpha} \quad (13)$$

where a factor ρ is introduced to take into account contrast losses due to optical defects such as the aberrations introduced by the PBS plane faces. Contrast is maximized for a value of α which depends on r_c and r_i. As an example, we imaged a dark glass/water interface and measured the fringe contrast. The coefficient of reflection at the interface was 0.07 (n_{glass}=1.53). Figure 3 shows the contrast as a function of the polarization angle along with curves calculated from equation (13).

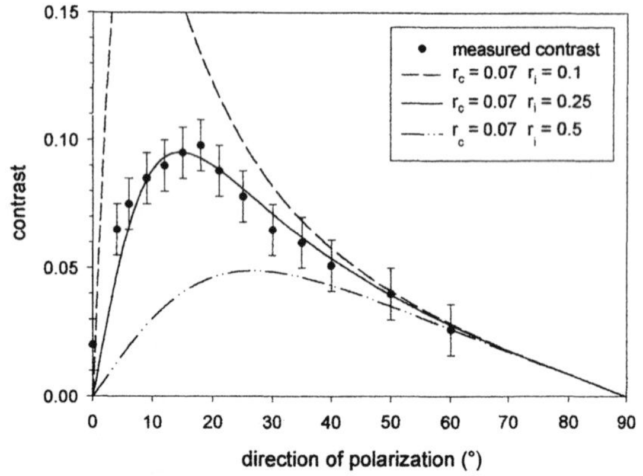

Figure 3. Contrast as a function of the direction of polarization.

Signal averaging

We commonly optimize the signal-to-noise ratio in our images by averaging the data over several measurements. Our software allows us to process the images in either of two manners. The sequences (S_p) can be accumulated in 2 (4) buffers and then processed as described above, or image pairs (quadruplets) can be processed on the fly and the resulting data be accumulated. The first type of treatment is less computationally intensive and enables one to accumulate data from the camera at full frame rate, but it is valid only when the sample can be considered as "static", e.g. in some vegetal tissues[8]. On the other hand, if scatterers' motion inside the sample cause the optical phase (φ) in a given pixel to vary from one image pair (quadruplet) to the other, the signal has to be computed in real time and summed.

Conclusion

We introduced a geometry and detection technique that uses a polarization interferometer and enables one to acquire a full head-on OCT image in parallel on a CCD detector. We find that the contrast can be optimized by rotating the direction of polarization of the excitation light, and that a suited strategy for signal averaging has to be chosen depending on the sample being imaged. We point out that contrast optimization is an important issue in our system since CCDs have a limited dynamic range as compared to point detectors.

Acknowledgments

This work was supported by the Direction Générale de l'Armement (stipend to E.B.) and by the Centre National de la Recherche Scientifique ("Ultimatech" program).

References

1. J. A. Izatt, M. R. Hee, G. M. Owen, E.A. Swanson and J. G. Fujimoto, Opt. Lett. 19, 590 (1994).
2. D. Huang, E. A. Swanson, C. P. Lin, J. S. Schuman, W. G. Stinson, W. Chang, M. R. Hee, T. Flotte, K. Gregory, C. A. Puliafito and J. G. Fujimoto, Science 254, 1178 (1991).
3. S. A. Boppart, M. E. Brezinski, B. E. Bouma, G. J. Tearney and J. G. Fujimoto, Develop. Biol. 177, 54 (1996).
4. G. J. Tearney, B. E. Bouma, S. A. Boppart, B. Golubovic, E. A. Swanson and J. G. Fujimoto, Opt. Lett. 21, 1408 (1996).
5. C. B. Su, Opt. Lett. 22, 665 (1997).
6. J. Ballif, R. Gianotti, P. Chavanne, R. Wälti and R. P. Salathé, Opt. Lett. 22, 757 (1997).
7. G. J. Tearney, M. E. Brezinski, B. E. Bouma, S. A. Boppart, C. Pitris, J. F. Southern and J. G. Fujimoto, Science 276, 2037 (1997).
8. E. Beaurepaire, A. C. Boccara, M. Lebec, L. Blanchot and H. Saint-Jalmes, Opt. Lett. 23, 244 (1998).
9. E. Beaurepaire, P. Gleyzes, M. Lebec, L. Blanchot, H. Saint-Jalmes, and A. C. Boccara, presented at SPIE BiOS '98, January 24-30, 1998, San Jose, CA.
10. A. C. Boccara F. Charbonnier, D. Fournier and P. Gleyzes, French patent FR90.092255 and international extensions (June 1990).
11. P. Gleyzes, F. Guernet and A. C. Boccara, J. Optics (Paris) 26(6), 251 (1995).
12. P. Gleyzes, A. C. Boccara and H. Saint-Jalmes, Opt. Lett. 22, 1529 (1997).

Optical Birefringence Imaging of Dentin and Enamel

Xiao-jun Wang[1], J.Y. Zhang[1], T.E. Milner[2], J.F. de Boer[3], Y. Zhang[4], D.H. Pashley[4], and J.S. Nelson[3]

[1]Department of Physics, Georgia Southern University, Statesboro, GA 30460
Tel: (912)681-5503; Fax: (912)681-0471; Email: xwang@gasou.edu

[2]Biomedical Engineering Program, The University of Texas at Austin, Austin, TX 78712

[3]Beckman Laser Institute and Medical Clinic, University of California, Irvine, CA 92612

[4]Department of Oral Biology-Physiology, Medical College of Georgia, Augusta, GA 30912

Abstract

Tomographic images of human dentin and enamel are obtained using OCT and polarization sensitive OCT. Strong birefringence effect is detected in enamel. Light propagation along dentinal tubules is also observed. Index of refraction for both dentin and enamel has been measured.

Keywords: (999.9999) OCT, (170.1850) dentistry, (290.3030) index measurements, (260.1440) birefringence.

Introduction

Tomographic imaging of dental materials is an effective method for investigation of optical and structural properties of tooth substance. Understanding the origin of these properties is necessary for the development of new optical methods for caries diagnosis and the measurement of tooth color. For example, the detailed studies of light propagation in tooth tissues are needed for the application of laser techniques on teeth.

Optical coherence tomography (OCT), an interferometric imaging technique to characterize the optical properties of various materials, is based on coherent cross-correlation detection of the interference fringe intensity of light backscattered from a test sample. This technique was initially introduced for characterizing closely spaced reflections in optical components used in the telecommunications industry [1,2]. In the past a few years, OCT has been applied to investigate highly scattering biological tissues [3-6]. The polarization-sensitive OCT has also been reported [7,8]. In contrast with conventional OCT, in which the magnitude of backscattered light as a function of depth is imaged, the polarization-sensitive OCT uses the backscattered light to image the magnitude of the sample birefringence as a function of depth. The birefringent images of biological materials can reveal important structural information that is difficult to resolve using conventional OCT or other imaging techniques. In this paper, we present optical low coherence images from dentin and enamel of human teeth obtained using polarization-sensitive OCT, which records images of the change in polarization of circularly polarized incoming light backscattered from dentin and enamel of human teeth. Strong birefringent effect is observed in enamel. Light propagation along dentinal tubules is clearly seen. Group index of refraction for both dentin and enamel is determined from the OCT images.

Experimental

A schematic of the polarization-sensitive OCT instrumentation is shown in Figure 1. Continuous near infrared light (I mW at λ_0=862 nm) emitted by a superluminescent diode (SLD) passes through a Glan-Thompson polarizer to have a pure vertical direction before division into reference and probe beams by a beam splitter BS. Light in the probe arm passes through a quarter-wave plate QWP1 oriented at 45° to the vertical to provide circularly polarized incident light. Reflected light from the sample, after passing again QWP1, is in arbitrary elliptical polarization, determined by the birefringence of the sample. Light in the reference arm, after double passing QWP2 oriented at 22.5° to the horizontal, linearly polarizes at 45° to provide equal reference power in the vertical and the horizontal detection. Polarizing beam splitter PBS splits the recombined probe and reference beams into the corresponding photo-receivers PR1 (horizontal) and PR2 (vertical). Optical power in the reference arm of the interferometer is attenuated by a neutral-density filter NDF to reduce intensity fluctuations and thus achieve a higher signal to noise ratio. Optical phase of SLD light in reference arm of the interferometer is modulated (1000 Hz) by a piezoelectric transducer PZT attached to the reflection mirror.

The sample is mounted on a high-precision motorized X-Y stage and scanned parallel (lateral direction) to the sample surface. A scanning scheme, in which a small axial increment (~10 microns) is made perpendicular to the film after each lateral scan, eliminates noise due to Doppler shift fluctuations caused by variation in the axial translation velocity and allows construction of a two-dimensional image. 1-D stage is designed to track beam focus for each new axial position and to maintain a zero optical pathlength difference between the focus in the sample and reference mirror so that the magnitude of interference is maximized[9].

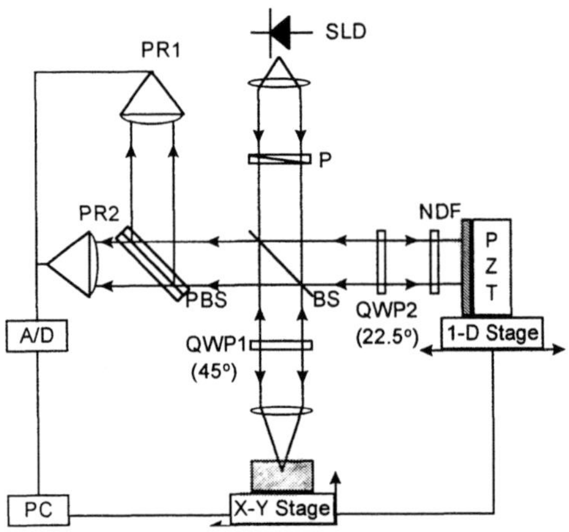

Figure 1. Polarization-sensitive OCT instrumentation. SLD, superluminescent diode; P, Polarizer; BS, beam slitter; PBS, polarizing beam slitter; QWP's, quarter-wave plates; NDF, Neutral density filter; PR's, photo-receivers; PZT, piezoelectric transducer. A/D, analog-to-digital converter.

It is well known that human tooth consists of three main tissues: enamel, dentin, and pulp. As shown in Figure 2, the bulk of the tooth is formed by the semitransparent dentin, which is thickest in the crown and gradually tapers as it reaches the apex of the root. Its outer surface is covered, in the region of the crown, by a thin and transparent layer of enamel. Enamel consists almost entirely of calcium salts in the form of large apatite crystals. The microcrystals form enamel prisms or rods with 4-6 μm transverse dimension oriented normally to the tooth surface [10]. Interprismatic space substance is enamel protein, which is up to 2% enamel volume. As in bone, the substance of dentin consists of an organic (30 vol%), mostly the collagen fibrils, an inorganic (50 vol %) part, where the components are incorporated into hydroxyapatite crystals, and water (20 vol %). The main dentin structural component is micron-sized dentinal tubules, which radiate with an S-shaped curve from the pulp cavity toward the periphery. In the innermost part of dentin, near the pulp, their diameter is 3 to 4 μm with density of 7.5×10^4 mm^{-2}; in the outer portion, they become narrower with a lower density of 3×10^4 mm^{-2}. Sample disks are prepared from the middle one third of the crowns of extracted human third molars and sliced by means of an Isomet saw. The thickness of the disks varies from 300 μm to 1.0 mm.

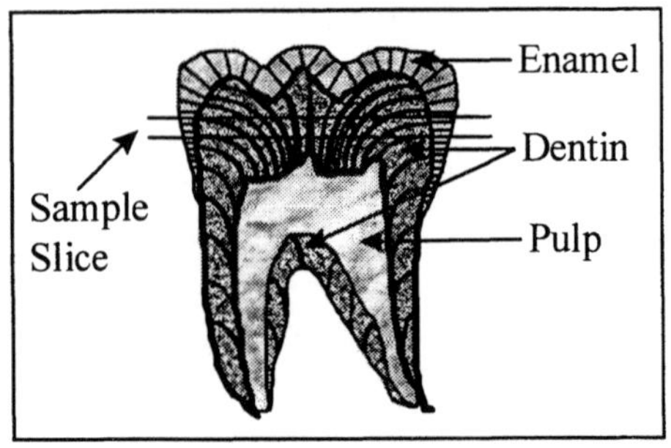

Figure 2. A schematic of sagittal section of human molar.

Results and discussions

Figure 3 shows a polarization-sensitive OCT image from enamel over a line from the surface of enamel to the boarder of enamel and dentin. A, B, and C correspond the images acquired by PR1, detecting the horizontal component light as a function of depth z, $I_H(z)$, which correspond to the same polarization of the light source, by PR2, detecting the vertical component, $I_V(z)$, and by the combination of PR1 and PR2, $I_H(z)+I_V(z)$, which is same to the conventional OCT. Part D is a false color birefringent image constructed by computing $\mathrm{sgn}[I_H(z)-I_V(z)]10\log|I_H(z)-I_V(z)|$[8]. The bottom lines on each image are the reflections from the glass substrate. The broken banded structure due to the birefringence can be clearly seen on part D and can be attributed to the enamel prisms in the form of cut-cylindrical fibers oriented normally to the tooth surface. There is no birefringent effect observed on dentin from our disk samples.

Figure 4 shows an OCT image ($I_H(z)+I_V(z)$) recorded over a grid of points lying on a diameter of the thin disk (~300 μm). It has been shown that the scattering processes in enamel and dentin can be described by a linear combination of oriented/forward and diffused/isotropic components [10,11]. Since the forward scattering strongly dominates the light scattering process in dentin[11], the isotropic scattering component, yielding detected backscattering light is very weak. To obtain more information of the light propagation in the substance, a reflected mirror is placed underneath the sample. Lines at the right and left ends of the image are reflections from air-mirror interface without sample. The areas close to the lines are recorded from the enamel ring of the disk, where the diffused component is well observed. The reflections from the front (R1) and end (R2) surfaces can be seen. The line next to the end surface (R3) is the displaced images of the mirror. The dark areas between R2 and R3 are the air gap between them. The bottommost line (R4) is the multiple reflections between R2 and R3. The whole central part is the image from dentin. The top surface curves a little due to the cutting. The strongly curved image in the bottom is the reflection of the mirror. Since the distribution of collagen fibrils is uniform over dentin, the strongly curved image can be attributed to the light flow along the dentinal tubules. As shown in Figure 2, the tubules in the disk center have the minimum length and they become longer and longer as the radius increases. Around the boarder of dentin and enamel, there are no tubules connecting the front and back surfaces, resulting in the weakest signal observed. This shows the wave-guided propagation along the tubules dominates the light flow over the scattering from the tubules at different orientation.

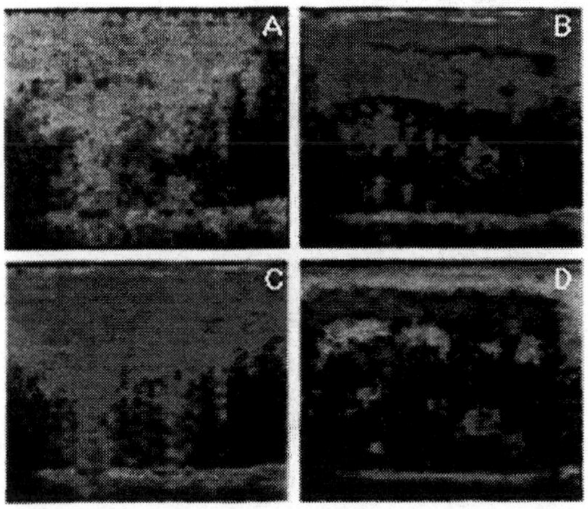

Figure 3. OCT images with different polarization. A, horizontal $I_H(z)$; B, vertical $I_V(z)$, C, both $I_H(z)+I_V(z)$, and D, birefringent image $\mathrm{sgn}[I_H(z)-I_V(z)]10\log|I_H(z)-I_V(z)|$. (1.2 mm wide by 350 μm depth each).

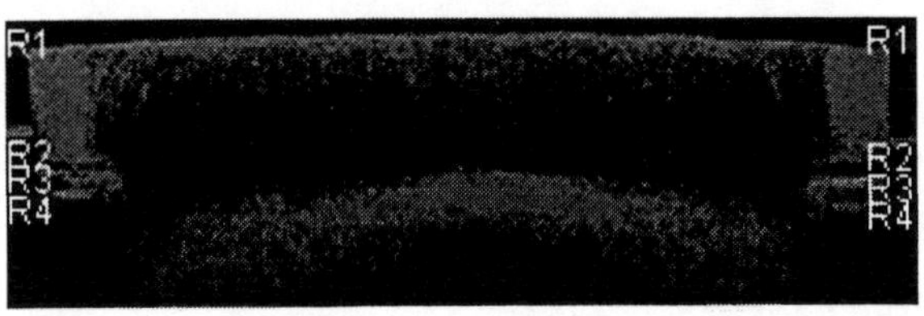

Figure 4. OCT images recorded along a linear grid over a diameter of the disk sample (10.8 mm wide by 600 μm depth).

The group refractive index and physical thickness can be simultaneously determined from OCT images[12,13], The average value for the group refractive index of enamel and dentin at λ_0=862 nm is, respectively, 1.62 ± 0.02 and 1.50 ± 0.02.

Acknowledgements

This work was supported by the National Institutes of Health training grant (XJW).

References

1. K. Takada, I. Yokohama, K. Chida, and J. Noda, *Appl. Opt.* **26**, 1603-1606 (1987).
2 B.L. Danielson and C.D. Whitenberg, *Appl. Opt.* **26**, 2836-2842 (1987).
3. D. Huang, E.A. Swanson, C.P. Lin, J.S. Schuman, W.G. Stinson, W. Chang, M.R. Hee, T. Flotte, K. Gregory, C. Puliafato, and J. G. Fujimoto, *Science* **254**, 1178-1181 (1991).
4. J.M. Schmitt, A. Knüttel, and R.F. Bonner, *App. Opt.* **32**, 6032-6042 (1993).
5. J. A. Izatt, M.R. Hee, G.M. Owen, E.A. Swanson, and J.G. Fujimoto, *Opt. lett.* **19**, 590-592 (1994).
6. X.J. Wang, T.E. Milner, J.S. Nelson, R.P. Dhond, W.V. Sorin, and S. A. Newton, *Opt. Lett.* **20**, 524-526 (1995).
7. M.R. Hee, D. Huang, E.A. Swanson, and J. G. Fujimoto, *J. Opt. Soc. Am.* **B9**, 903-908 (1991).
8. J.F. de Boer, T.E. Milner, M.J.C. van Gemert, and J.S. Nelson, *Opt. Lett.* **22**, 934-936 (1997)
9. Z. Chen, T.E. Milner, D. Dave, and J.S. Nelson, *Opt. Lett.* **22**, 64-66 (1997).
10. G.B. Altshuler and V.N. Grisimov, *Lasers and medicine, SPIE* **1353**, 97-102 (1989).
11. D. Fried, R.E. Glena, J.D.B. Featherstone, and W. Seka, Appl. Opt. **34**, 1278-1285 (1995).
12. W.V. Sorin and D.M. Baney, *IEEE Phot. Tech. Lett.* **4**(12), 1404-1406 (1992).
13. X.J. Wang, T.E. Milner, M.C. Chang, and J.S. Nelson, *J. Biomcd. Opt.* **1**, 1 5 (1996)

A Novel Application of Coherent Detection Imaging Method for Diagnosis of Rheumatoid Arthritis in Near Infrared Region

B. Devaraj, Y. Watanabe, M. Takeda, M. Usa, T. Yuasa[1], T. Akatsuka[1] and H. Inaba[2]

Biophotonics Information Laboratories, 2-2-1 Matsuei, Yamagata 990, Japan. Tel. +81-236-47-3100, Fax. +81-236-47-3108.

[1]Faculty of Engineering, Yamagata University, 4-1-16 Jonan, Yonezawa, Yamagata 992, Japan.
[2] Tohoku Institute of Technology, 35-1 Yagiyama-Kasumicho, Taihaku ku,
Sendai 982, Japan, and Biophotonics Information Laboratories.

Introduction

Rheumatoid arthritis (RA) is a systemic disease that affects soft tissues in the joint regions. It is prevalent over a fairly large cross-section of the human population with an increasing incidence in recent times. The earliest and most frequently affected joints include the proximal interphalangeal (PIP) and metacarpal (MC) articulations of the human finger. Subtle changes in the soft tissues of these joint regions characterize the onset of RA. Early diagnosis of RA is therefore important as the soft tissue changes usually occur during the first two years where therapeutic intervention could alleviate or stall the progression of RA.

The most commonly used practice for diagnosis and characterization (progression of the disease, outcome of treatment, etc.) of RA is dependent onradiological analysis. Inspite of the vast improvements in both technique and analysis over half a century, radiographic methods are not reliable for detection of slight changes in soft tissues of the joint region. Recently, the magnetic resonance imaging (MRI) method is proposed as an alternative, but again the present status for the early detection of RA leaves much scope for improvement.

We had earlier reported high resolution laser computed tomographic images of the PIP joints of healthy volunteer's forefingers [1,2]. The images were obtained with the coherent detection imaging (CDI) method that is based on the optical heterodyne detection technique [3]. In this paper, we compare laser computed tomographic images of the PIP joint of normal and patients with RA.

Coherent Detection Imaging (CDI)

CDI is an interferometric technique that has been used to obtain transillumination images of centimeter order thick biological tissues with submillimeter resolution [1-4]. The experimental arrangement based on a Mach-Zehnder interferometer is displayed in Fig. 1. Unlike other reported coherent detection methods such as OCT/OCM [5,6] and holography [7], CDI uses single frequency laser light sources (coherence length ~ meters). The shot noise limited detection capabilities in combination with the antenna properties of heterodyne receivers [3], allows the CDI method to detect the weak and high directional emerging components from a highly diffusive forward scattering media such as biological tissues. Coherence gating ability of the CDI method discriminates and selects the coherence retaining photons that also preserve polarization states with respect to the incident photons [2,3]. As each scattering event contributes to loss in coherence and changes in phase, polarization and direction, the coherence gating method as in CDI detects the minimally scattered directional photons that retain their original characteristics after passing through the tissue. Therefore, it is possible to use the conventional X-ray CT algorithms for image reconstruction and obtain high resolution laser tomographic images.

Experimental methods

Continuous wave, single frequency, diode pumped Nd:YAG laser operating at 1.064 μm was used in our measurements. The beam was well collimated to 0.7 mm in diameter and the incident power on the finger was ~ 10 mW. The signal beam illuminated the middle finger PIP joint (right hand) of the patients/healthy volunteers after being mounted on a translational-rotational stepper motor stage. The emergent signal beam was then mixed with the local oscillator beam and impinged on a InGaAs photodiode to generate an intermediate frequency (i.f.) at 50 kHz, that was fed to fast Fourier transform analyzer interfaced to a computer. The i.f. was recorded for every 500 μm translational step over 10° rotational step. Image reconstruction was carried out by the filtered back projection using the 18 projection data. The dynamic range of our measurement system is 140 dB for an incident power of 0.1 mW. The displayed images in Fig. 2, are reconstructed with optical signals ranging in the power of 10^{-15} W to 10^{-17} W.

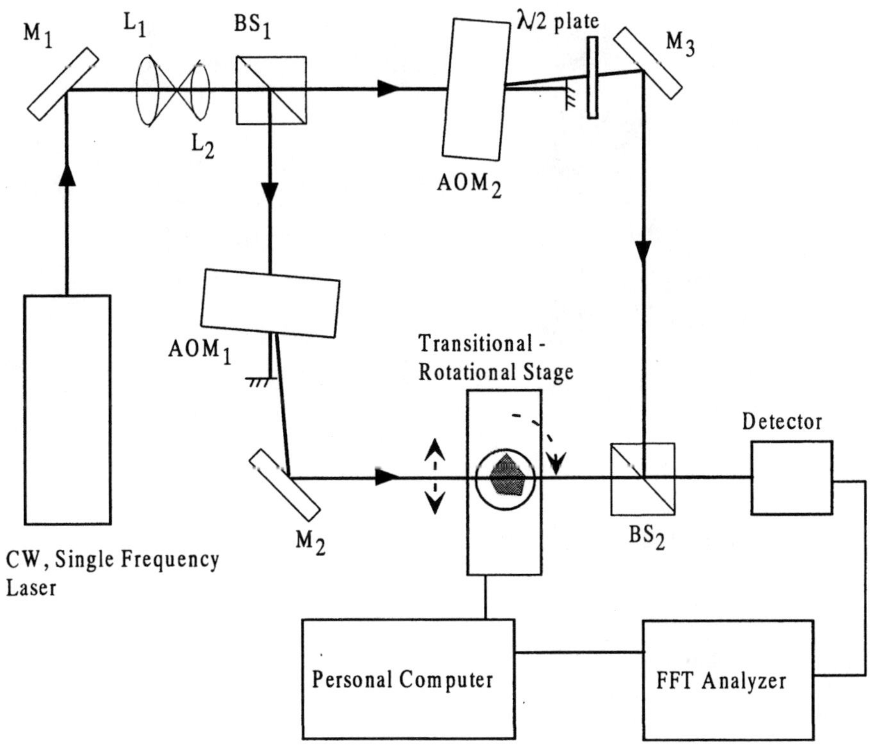

Fig. 1 Schematic of the CDI system. L_1 and L_2 are collimating lenses, M_1, M_2 and M_3 are mirrors, BS_1 and BS_2 are beam splitters and AOM_1 and AOM_2 are acuostooptic modulators.

Fig.2(a) is the reconstructed laser CT image obtained at the PIP joint of a patient's finger. The central region in the image corresponds to the bone tendon sheath complex. In the region above the bone tendon sheath complex, the inflamated extensor expansion is clearly evident. This structure is not differentiated in the image of the PIP joint region of a healthy volunteer, Fig. 2(b). For comparison, we also analyzed the radiograph and X-ray CT images of the patient's PIP joint and observed that although bone erosion and swelling of the joint were evident, there was no indication of the inflamated extensor expansion.

In conclusion, we demonstrate for the first time, the possibility of early diagnosis of rheumatoid arthritis using the CDI method. Further case studies and comparison with laser spectroscopic CT images obtained at other wavelengths could lead to a novel and relatively safe imaging modality for rheumatoid arthritis.

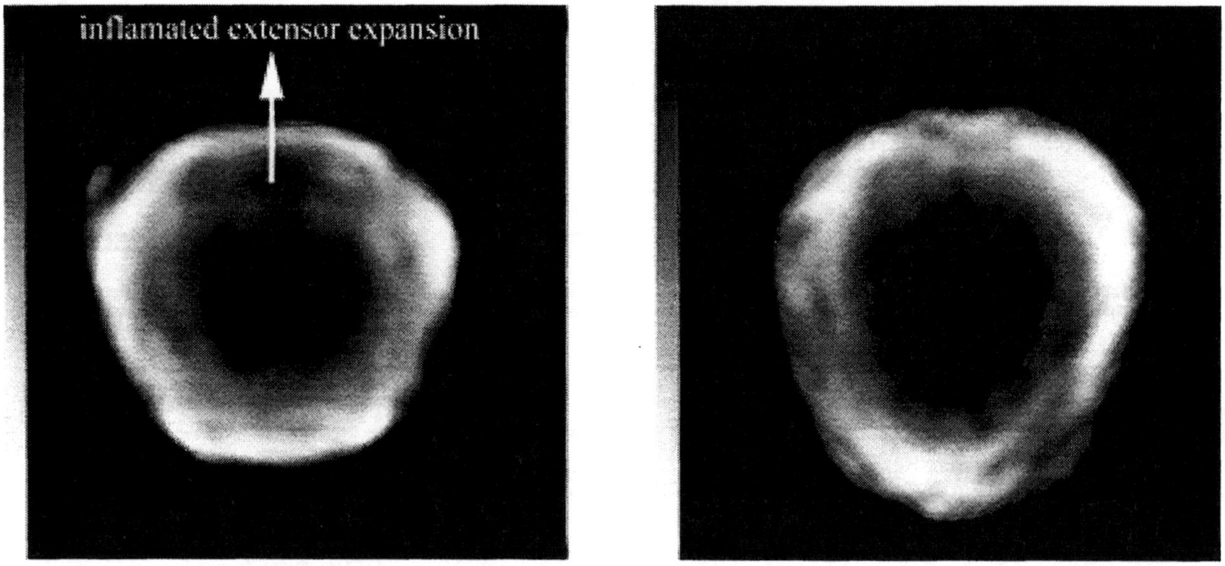

(a) (b)

Fig. 2 Laser CT images of the PIP joint of (a) patient with RA, and (b) normal subject

References
[1] B. Devaraj et al., Appl. Phys. Lett., vol. 69, 3671 (1996).
[2] B. Devaraj et al., IEEE J. of Special Topics in Quantum Electron., vol. 2, 1008 (1996).
[3] H. Inaba in G. Muller et al. (eds.), Medical Optical Tomography: Functional Imaging and Monitoring, SPIE Institutes for Advanced Optical Technologies Series, vol. IS 11, 317, SPIE Opt. Eng. Press, Bellingham, WA (1993).
[4] B. Devaraj et al., Electron. Lett., vol. 31, 874 (1995).
[5] D. Huang et al., Science, vol. 254, 1178 (1991).
[6] J.A. Izzat et al., Optics and Photonics News, vol. 8, 41 (1997).
[7] S.C.W. Hyde et al., Optics Lett., vol. 20, 2320 (1996).

Phase Contrast Microscopy with Digital Holography

Etienne Cuche, Patrick Ruffieux, *Patrick Poscio and Christian Depeursinge

Institut d'Optique Appliquée, EPFL, CH-1015 Lausanne, Switzerland.
*Institut de Recherches en Ophtalmologie (IRO), CH-1950 Sion 4, Switzerland.
E-mail: etienne.cuche@epfl.ch

Abstract

We present a new application of digital holography for phase contrast microscopy. This new holographic imaging technique uses a CCD camera as recording device and a numerical method for the hologram reconstruction. Preliminary results with a biological sample are presented.

Key Words

(180.0180) Microscopy; (040.1520) CCD, charge-coupled device; (090.0090) Holography; (170.1530) Cell analysis.

Introduction

We present a new holographic imaging technique which uses a CCD camera as recording device and a numerical method for the reconstruction. In comparison with classical holography which uses photographic plates as recording media, this digital application of holography presents the advantages of a higher acquisition rate (video frequency) and a higher reconstruction rate (up to 3 Hz actually). An other possibility to perform holography with a high acquisition-reconstruction rate consists in using photorefrative crystals as recording media [1,3]. This considerable decrease in the processing time opens interesting perspectives for new applications of holography. In particular in biomedicine where it should be applied for tomographic imaging [2], for vision trough turbid media [1,3], for deformations analysis [4] and for optical biopsies [5]. An interesting development for further clinical applications has been presented by O. Coquoz et al. [6] who have developed a digital holographic endoscope.

In this paper we present a new feature of digital holography for cellular imaging. A similar digital holographic microscope has been proposed by K. Boyer et al. [7] in the Fourier geometry for amplitude contrast imaging at several wavelengths. Here, we propose a new method in the Fresnel geometry and demonstrate its ability for phase contrast microscopy.

Methods

The experimental setup for the recording of digital holograms is presented in Fig. 1. A 10 mW, linearly polarized HeNe laser is used as light source and the combination of a variable neutral density filter (NF), a half wave-plate ($\lambda/2$) and a polarizing beam splitter (PBS) allows the adjustment of the intensities entering in the reference and object arms of a Mach-Zender interferometer.

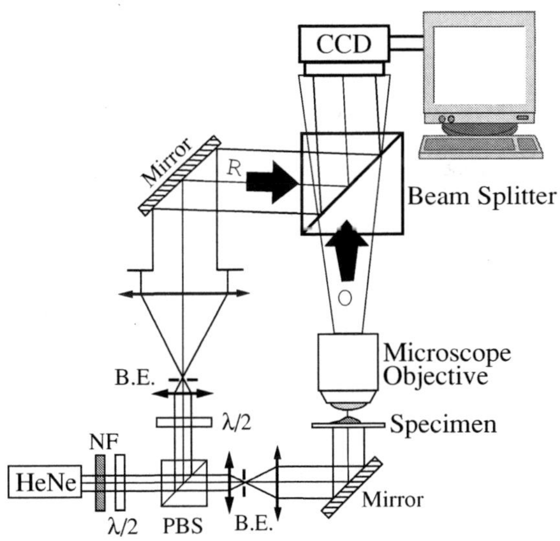

Figure 1. Experimental setup.

At the exit of the interferometer, a standard 8 bits CCD camera (Hitachi Denshi KP-M2) records the interference between a reference wave $\boldsymbol{R}$ and an object wave $\boldsymbol{O}$ which produces the hologram intensity:

$$I_H(x, y) = |\boldsymbol{R}|^2 + |\boldsymbol{O}|^2 + \boldsymbol{R}^*\boldsymbol{O} + \boldsymbol{R}\boldsymbol{O}^* \qquad (1)$$

In order to obtain a plane wave as reference, a beam expander (B.E.) including a pinhole is introduced in the reference arm and enlarges the beam diameter to about 2 cm. A similar system which produces a smaller beam diameter (~5 mm) is inserted in the object arm. In order to perform holography in the off-axis geometry, the mirror in the reference arm is oriented in such a way that the reference wave reaches the CCD camera with a small incidence angle.

The configuration in the object arm is presented in Fig. 2. For holographic microscopy, the distance between the object and the microscope objective (d_o) must be set in such a way that a diffracted field reaches the recording media [8]. For this reason, the magnified image of the object is not focused on the recording device as usually in classical microscopy, but behind the CCD at a distance d. This can be observed in Fig. 3 which presents an example of hologram recorded in this geometry and where we can see that the object appears out of focus on the inteferogram. The situation presented in Fig. 2 can be considered as equivalent to a holographic configuration without magnification optics with an object wave emerging from the magnified image and not from the object itself. The reason why the image is not focused on the CCD is evident because this would signify a zero hologram-object distance and this situation is not suitable for off-axis holography [9].

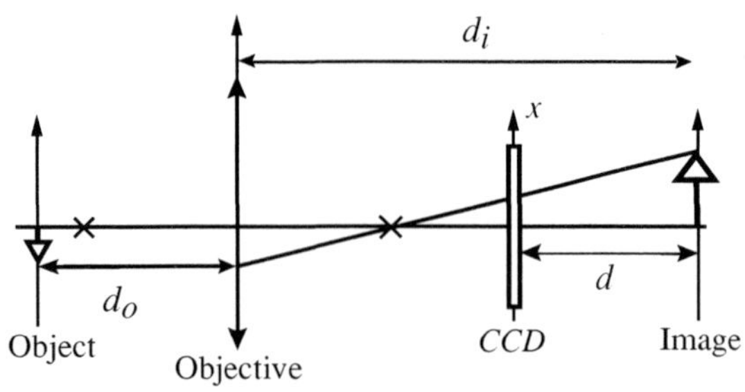

Figure 2. Experimental configuration for holographic microscopy.

For our purpose of phase contrast imaging, it is important to realize that the insertion of a microscope objective into the object arm of the interferometer modify the phase distribution in the image plane. For a single lens imaging system in the Fresnel approximation, the introduced phase aberration is quadratic and can be expressed as follow [9]:

$$\exp\left\{\frac{i\pi}{\lambda d_i}\left(1-\frac{d_o}{d_i}\right)r_i^2\right\}, \tag{2}$$

where d_o is the object-objective distance, d_i the objective-image distance and r_i represents the distance to the optical axis in the image plane.

A typical example of digital hologram is shown in Fig. 3. The image contains $N\times N$=512×512 pixels and 256 gray-levels, the CCD area is 4.85×4.85 mm^2. The curvature of the interference fringes is due to the quadratic phase aberration of the microscope objective and the fringe pattern is not concentric but translated at the top left corner of the hologram because of the off-axis geometry.

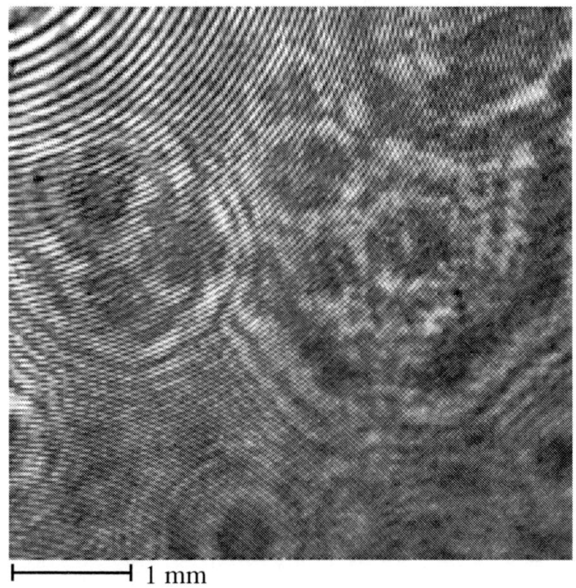

Figure 3. Example of digital hologram.

The numerical method that we are using for the hologram reconstruction has been first proposed by Yaroslavski et al. [10] and has been applied recently by Schnars et al. [11] for the reconstruction of digital holograms recorded with a CCD camera. Here we present a modified version of the same algorithm which has been adapted for phase contrast imaging. The computations consists in the calculation of the wavefront that would be diffracted by the same hologram during a standard optical reconstruction when the developed photographic plate is illuminated by a replica of the reference wave. The algorithm simulates the propagation of the reconstructed wave front Y, from the hologram plane $0xy$ to an observation plane $0\xi\eta$. Basically, we are using a discrete expression of the Fresnel integral:

$$\mathrm{Y}(m\Delta\xi, n\Delta\eta) = A\exp\left(\frac{i\pi}{\lambda D}(m^2\Delta\xi^2 + n^2\Delta\eta^2)\right)$$
$$DFT\left[\mathbf{R}(k,l)I_H(k,l)\exp\left(\frac{i\pi}{\lambda d}(k^2\Delta x^2 + l^2\Delta y^2)\right)\right]_{m,n}, \tag{3}$$

where, λ is the wavelength, $A=exp(i2\pi d/\lambda)/(i\lambda d)$ is a complex constant, k,l,m,n are integers and DFT is the discrete Fourier transform operator. $\Delta x\times\Delta y$ represents the area of a pixel in the hologram plane (CCD). The distance between the reconstruction plane $0\xi\eta$ and the hologram is defined by the reconstruction distance d. In order to obtain a focused reconstructed image, d must be equal to the length of the optical path between the CCD and the magnified image of the object (see Fig. 2) during the hologram recording. The pixel size in the reconstruction plane is defined by $\Delta\xi$ (or $\Delta\eta$) and is related to the pixel size of the

CCD (Δx) and to the number of pixels in one dimension (N) by the following relation [10,11]:

$$\Delta\xi = \frac{\lambda d}{N\Delta x} = \frac{\lambda d}{L}, \quad (4)$$

where L is the size of the CCD camera. This pixel size represents the resolution of the reconstructed image and defines the resolution of the entire imaging system if no microscope objective is used. When a microscope objective is introduced in the object arm as presented in Fig. 2, the resolution of the reconstructed image is equal to the diffraction limit of the microscope objective.

Five successive stages are necessary for a complete hologram reconstruction:

1) The reconstructed wavefront in the hologram plane $\Upsilon(x,y)$ is calculated by multiplication of the digitized hologram intensity $I_H(k,l)$ by an array of complex numbers $\boldsymbol{R}(k,l)$ which acts as a digital reference wave. In order to obtain a correct reconstruction of the phase distribution, the wavefronts of this digital reference wave must precisely reproduce those of the experimental reference wave $\boldsymbol{R}(x,y)$. As we are working in the off-axis geometry, with a plane wave as reference, we define:

$$\boldsymbol{R}(k,l) = \exp\left(\frac{2\pi i}{\lambda}(k\Delta x\sin\theta\cos\varphi + l\Delta y\sin\theta\sin\varphi)\right), \quad (5)$$

where θ is the incidence angle (with respect to the optical axis) and φ the azimuthal angle of the wave vector.

2) The product $I_H(k,l)\boldsymbol{R}(k,l)$ is multiplied by a quadratic phase factor in the hologram plane coordinates: $\exp\{i\pi(x^2+y^2)/(\lambda d)\}$.

3) Computation of the discrete Fourier transform (DFT) by means of a Fast Fourier Transform (FFT) algorithm.

4) The result of the Fourier transform is multiplied by a quadratic phase factor in the observation plane $\exp\{i\pi(\xi^2+\eta^2)/(\lambda D)\}$ which contains a new parameter D. In the case without microscope objective, this parameter is simply equal to the reconstruction distance ($D=d$) but in the general case, we have to multiply the reconstructed wavefront by the complex conjugate of the phase aberration of the objective. Thus, if we consider Eq. 2, we have:

$$\frac{1}{D} = \frac{1}{d} - \frac{1}{d_i}\left(1 - \frac{d_o}{d_i}\right), \quad (6)$$

5) The obtained result $\Upsilon(m\Delta\xi,n\Delta\eta)$ is an array of complex numbers. Amplitude and/or phase contrast images can be obtained by calculating its modulus $((\mathrm{Re}(\Upsilon)^2+\mathrm{Im}(\Upsilon)^2))$ and/or its argument $(\mathrm{atan}(\mathrm{Re}(\Upsilon)/\mathrm{Im}(\Upsilon)))$ respectively.

Using a personal computer with a 200 MHz Risk processor, the computation time is ~ 1 sec. for the reconstruction of an entire hologram (512^2 pixels) and ~ 0.33 sec. for a half-size hologram (256^2 pixels). If phase contrast imaging is not desired, steps 1 and 4 of the reconstruction procedure are not necessary because these operations do not modify the amplitude of the reconstructed wave front.

Altogether, four parameters (d, θ, φ and D) must be introduced in algorithm (3). Their values could be measured on the experimental setup but phase contrast microscopy requires such a high precision in their evaluation ($\leq$ 1 cm for d, ~0.01 mrad for θ, ~0.001 rad for φ and $\leq$ 0.1 mm for D), that their values are iteratively estimated during the reconstruction process.

Finally, it must be noticed that the reconstructed images contain three different components; a zero order of diffraction, a twin image and a real image [2]. As a consequence of the off-axis geometry, these components are reconstructed at different locations in the observation plane. Here (Figs. 4, 5 and 6), we present only the real images.

Results and discussion

In this section we present the results obtained by numerical reconstruction of the hologram presented in Fig. 3. The specimen for this experiment was an eptithelial cell from the human oral mucosa which has been simply placed on a glass plate. A microscope objective of magnification ×20 and numerical aperture NA=0.35 has been used. The values of the reconstruction parameters are d=53.2 cm, θ=8.51 mrad, φ=0.265 and D=19.82 cm. As biological cells are almost transparent, the reconstructed image in amplitude contrast (Fig. 4) reveals less details than the phase image (Fig. 5), the contrast of which is due to differences in the refractive index and/or the thickness of the specimen.

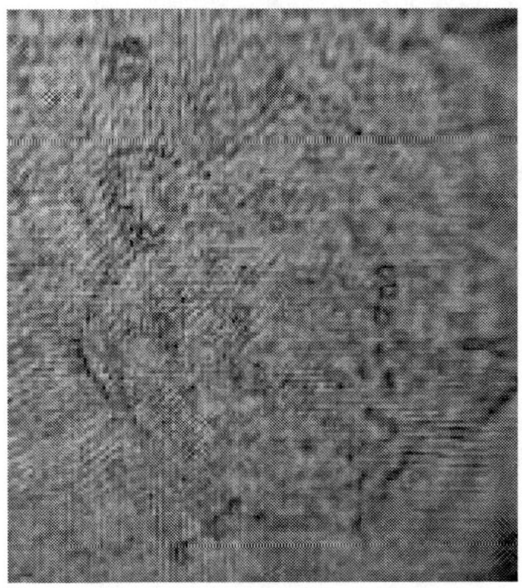

Figure 4. Amplitude contrast.

In the phase contrast image of Fig. 5, we can observe intracellular structures such as the nucleus or small inclusions. This image is similar to those which can be obtained with other standard techniques for phase contrast microscopy such as the Zernicke [12] or the Nomarski [13] methods. The advantage of our technique lies in the fact that the reconstructed phase distribution can be directly used for quantitative measurements.

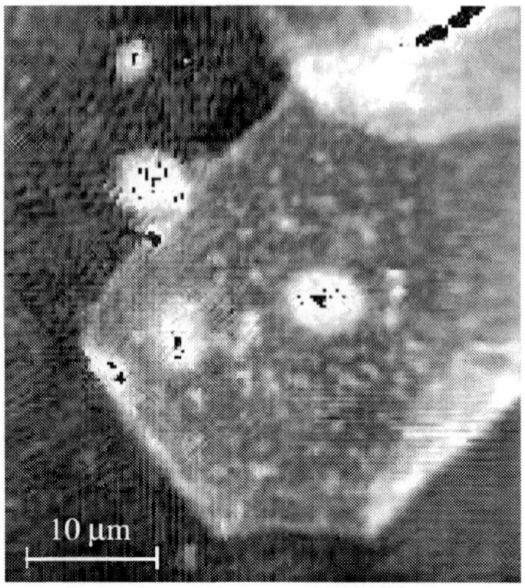

Figure 5. Phase contrast.

The image presented in Fig. 6 illustrates the importance of the quadratic phase aberration correction. This result has been reconstructed with the same values for D and d as if no microscope objective was inserted in the object arm. We can see that concentric fringes appear in the reconstructed image. In standard interference microscopy, this problem is solved optically using a Linnik or a Mirrau interferometer.

Figure 6. Phase contrast without correction of the quadratic phase aberration of the microscope objective.

Conclusion

We have demonstrated that a digital hologram recorded with a standard CCD camera can be numerically reconstructed with amplitude and phase contrast by means of a fast Fresnel transform algorithm. In order to obtain a correct phase reconstruction, the hologram must be multiplied by a digital reference wave, the wavefronts of which must precisely fit those of the wave which has been used as reference for the hologram recording. In addition, if a microscope objective is used, the reconstruction method must include a correction of the quadratic phase aberration of the microscope objective.

References

1. S.C.W. Hyde, et al., "Depth-resolved holographic imaging through scattering media by photorefraction", Opt. Lett. 20, p. 1331-1333 (1995).
2. E. Cuche, P. Poscio, C. Depeursinge. "Optical Tomography at the Microscopic scale by means of a Numerical Low Coherence Holographic Technique", SPIE Proceedings vol. 2927, (Vienna, 1996) p. 61-66.
3. S.C.W. Hyde, et al., "Sub-100-μm depth-resolved holographic imaging through scattering media in the near infrared", Opt. Lett. 20 p. 2330-2332 (1995).
4. U. Schnars, "Direct phase determination in hologram interferometry with use of digitally recorded holograms", J.Opt.Soc.Am.A 11, p. 2011-2015 (1994).
5. E.B. de Haller, G. von Bally, C. Depeursinge, "High-resolution holography and biopsy: preliminary results", Bioimaging 3, p. 76-87 (1995).
6. O. Coquoz, et al., "Performances of endoscopic holography with a multicore optical fiber", Appl. Opt. 34, p. 7186-7193 (1995).
7. K. Boyer, et al., "Biomedical three-dimensional holographic microimaging at visible, ultraviolet and X-ray wavelenght", Nature Medicine 2, p. 939-941 (1996).
8. M. Pluta, "Holographic Microscopy", in Advances in optical and electron microscopy Vol. 10. (Academic Press, London, 1987).
9. J.W. Goodman, "Introduction to Fourier Optics", (Mc Graw-Hill, San Francisco, 1968).
10. L.P. Yaroslavskii, N.S. Merzlyakov, "Methods of Digital Holography". (Consultants Bureau, New York, 1980).
11. U. Schnars, W. Jüptner, "Direct recording of holograms by a CCD target and numerical reconstruction", Appl. Opt. 33, p. 179-181 (1994).
12. M. Pluta, "Non-standard Methods for Phase Contrast Microscopy", in Advances in optical and electron microscopy Vol. 6. (Academic Press, London 1975).
13. M. Pluta. "Nomarski's DIC microscopy: a review", SPIE Proceedings vol. 1846 (Warsaw, 1992) p. 10-25.

Dynamic Low Coherence Interferometry as a tool for imaging and quantification of Brownian motion within highly scattering media

K. Bizheva, A. Siegel and D.Boas

Electro-Optics Technology Center, Tufts University, 4 Colby St., Medford, MA 02155
tel. (617) 627 3136, fax. (617) 627 3151, kbizheva@tufts.edu , dboas@tufts.edu, andy.siegel@ir.lnco.com

Dynamic light scattering (DLS) has been used extensively during the last few decades for characterization of the structural and dynamical properties of optically dilute media [1,2]. Information about the sample properties is extracted either from the power spectrum or the temporal correlation function of the single-scattered light. In this case both the scattering angle and the polarization of the scattered wave are well defined, in contrast to the case of multiply scattered light when those parameters are partially or completely randomized and information about the properties of the examined object is difficult or impossible to obtain. As DLS is based on detecting single scattered light, probing highly scattering media is problematic.

The demand for a method that would allow *non-invasive* study of turbid materials has led to the development of diffusing wave spectroscopy (DWS). This method is based on light diffusion in optically thick media [3,4] and has been used to determine the dynamical properties of highly scattering materials like colloidal suspensions [5], foams [6], and human tissue [7]. Although both DLS and DWS can be used to obtain information about the structural and dynamical properties of turbid media, their application is valid only in the two extremes of single scattered and diffusive light. A detailed quantitative analysis of the few scattering events regime could provide significant insight about the resulting degradation of image contrast and resolution.

In this paper we demonstrate that dynamic low coherence interferometry (DLCI) can be used for non-invasive imaging of Brownian motion within turbid media. We show that a single axial scan contains sufficient information for determination of both the optical and the dynamical properties of highly scattering, but weakly absorbing media. We also find that with one axial scan we can measure the transition from the single to the multiple scattering regime and that the transition is a function of the scattering anisotropy of the material. Moreover our experiments show that DLCI is very sensitive to multiply scattered light, thus it may prove to be an excellent tool for studying the few scattering events regime.

A schematic of a typical single-mode fiber optic LCI system is shown in fig.1. The interferometer is illuminated with light from a broadband source (usually a super-luminescent diode). The optical properties of the sample generate a distribution of optical path-lengths in the sample arm, while the path-length in the reference arm is determined solely by the position of the retroreflector. Interference is observed only when the optical path-length difference between the reference and the sample arms is within the coherence length of the source. The axial (path-length) resolution of the measurement is determined by the coherence length of the light source, while the lateral resolution is depends on the optics used to focus the probe beam in the sample. For a fixed position of the reference mirror the photo-current at the detector is proportional to [8,9]:

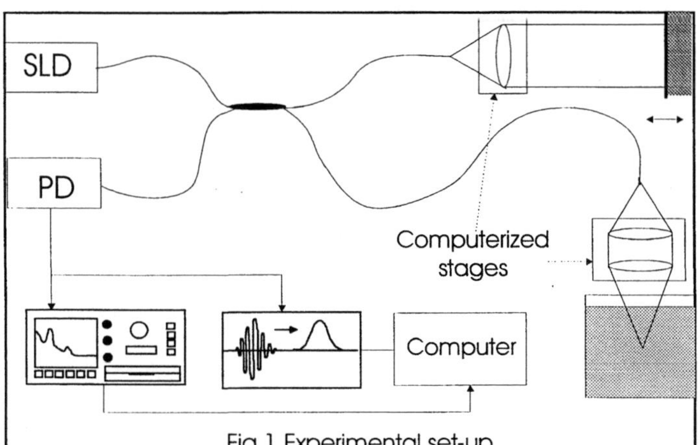

Fig.1 Experimental set-up
SLD - super-luminescent diode, PD - photo-detector

$$i_d \propto I_r + I_s + I^{fluct}(z,t) \qquad (1)$$

where I_r (I_s) is the intensity from the reference (sample) arm and $I^{fluct}(z,t)$ is the cross term (or optical heterodyne term) resulting from the coherent mixing of E_r with the path length distribution of $E_s(z)$. Temporal fluctuations in the cross term may result from changes in the amplitude and/or phase of E_r or E_s. When the reference mirror is held fixed, the amplitude and the phase of E_r do not change over time. Any dynamics in the sample such as flow or Brownian motion, result in a distribution of Doppler frequency shifts in the light propagating through the medium, which causes phase fluctuations in $E_s(z)$. The power spectrum derived for a system undergoing Brownian motion is a Lorentzian and is proportional to the heterodyne term [1,2]:

$$P(f) \propto \frac{\sqrt{I_r I_s(z)}}{1+\dfrac{(2\pi f)^2}{D_B q^2(\theta)}} \qquad (2)$$

where $I_s(z)$ is the intensity of light that has traveled a path length of $2z$ through the sample, D_B is the self-diffusion coefficient of the scattering particles and is given by $D_B = k_B T/(3\pi\eta a)$ where k_B is the Boltzmann constant, T is the temperature, η is the viscosity of the suspending liquid, and a is the hydrodynamic diameter of the scattering particle. The momentum transfer of the light scattering event is $q = 2\,k\,\sin(\theta/2)$, where θ is the light scattering angle defined by the geometry of the optical instrument.

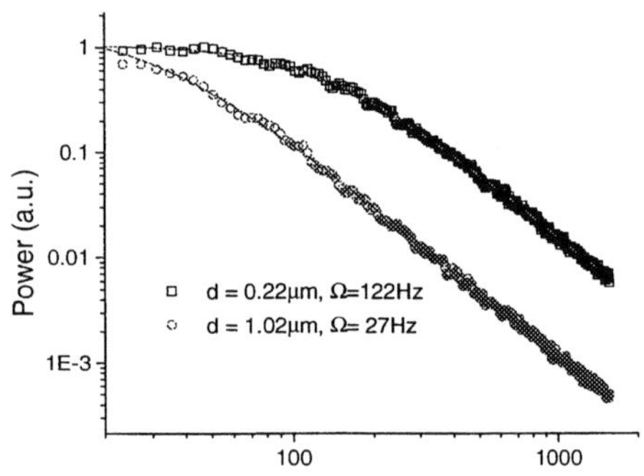

Fig.2 Power spectrum for two monodisperse suspensions of PLS microspheres measured at 200um inside the sample.

For our experiments we used an 850nm SLD at with a coherence length of 30μm. Microscope objectives of 10X (NA=0.25) were used to collimate the light beams in both arms of the interferometer, while a 16X (NA=0.32) was used for focusing the beam in the sample. The collimating and focusing optics were mounted on computer operated stages, which allowed accurate control of the positions of both the focal point in the medium and the reference mirror. The power spectrum was measured with an FFT spectrum analyzer as a function of depth in the sample in steps of 33μm. The position of the reference mirror was correspondingly adjusted to select a path-length in the medium equal to twice the depth of the focal point. By aligning the focus of the beam with the coherence gate we were able to maximize the rejection of multiply scattered light

The samples used in our experiments were prepared by suspending polystyrene microspheres (PS) of mean diameters d=0.22 μm and d=1.02 μm (Bangs Labs, Inc.) in distilled water. The volume fractions of the particles in each monodisperse suspension (4% and 0.5% respectively) were chosen to provide a photon scattering length in the sample of about 100 μm. The sample holder was made out of low thermal conductivity PMMA and the temperature of the suspensions was measured prior to and after each complete depth scan.

To demonstrate that LCI can be used to image and quantify Brownian motion within highly scattering media, we measured the power density spectrum for two separate monodisperse PS suspensions at a depth of 200 μm inside the sample (fig.2). The acquired data was fit with a Lorentzian (eq.2) as indicated by the solid lines through the data points. The linewidths obtained from these fits were used to determine the experimental Brownian diffusion coefficient for the particles, and the corresponding particle sizes. The results from the calculations for the microsphere diameters (0.222μm and 0.997μm) were within 5% of the values cited by the manufacturer. Multiple experiments with different size PS spheres showed the same accuracy.

Fig.3a shows the linewidth determined from the Lorentzian fits for a suspension of 0.22μm diameter microspheres as a function of the depth. It was observed that the linewidth does not change significantly (within 5% of the mean value) as the focal point was translated to a depth of 500 μm into the suspension. In the depth range from 500 to 900 μm, the linewidth broadened nonlinearly

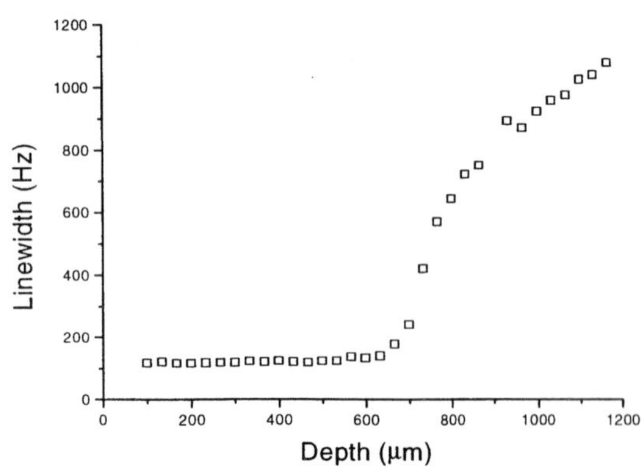

Fig3a The linewidth determined from the Lorentzian fits as a function of the depth in the sample.

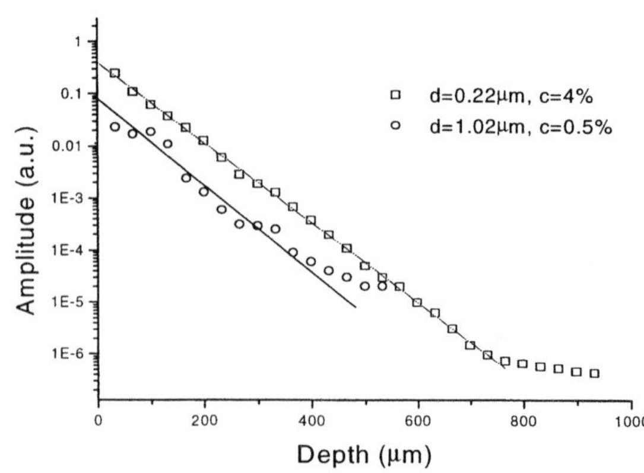

Fig.3b The amplitude determined from the Lorentzian fits as a function of the depth in the sample.

due to detection of multiply scattered light. Beyond 900 μm the linewidth was found to increase linearly with the depth, which is in a good agreement with the DWS theory. From the slope of graph in the diffusion regime one can determine the reduced scattering coefficient of the sample μ_s', since the slope is equal to the product $\Omega_0\mu s$', where Ω_0 is the linewidth measured in the single scattering regime.

Fig.3b shows the amplitude determined from the Lorentzian fits as a function of depth in the suspension. As expected, it decays exponentially with the path-length in the single scattering regime and with an extinction coefficient equal to the scattering coefficient of the medium. Knowing μ_s' and μ_s one can easily determine the scattering anisotropy of the sample. In our case the experimentally determined optical properties of the suspensions were within 10% of the values calculated from Mie theory.

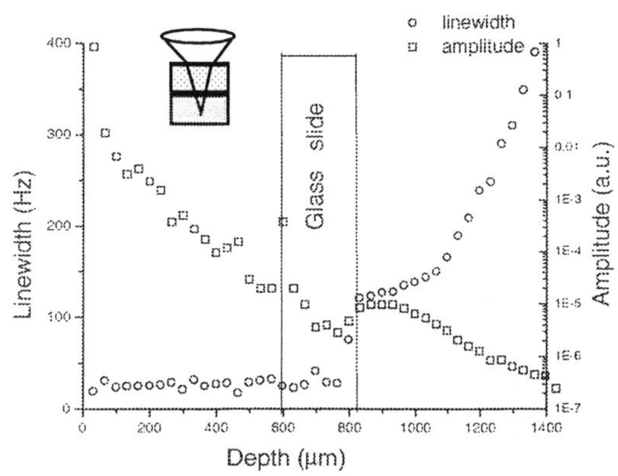

Fig.4 The amplitude and the linewidth determined from the fits for the two-layer system showed on the upper left corner of the graph.

To demonstrate that LCI is able to map spatial variations in the particle Brownian diffusion coefficient, we used a two-layer cell, the cross-section of which is shown in the inset of fig.4. A 580 μm thick layer of 1.02 μm PS microspheres (0.5% volume fraction) was sandwiched between two 150 μm cover slips, and placed on top of 4% solution of 0.22 μm PS spheres. Both the linewidth and the amplitude, determined from the fits to the experimental data, are graphed as functions of the depth in the sample on fig.4. The two layers can be clearly distinguished either from the Fresnel reflections at the glass/solution interfaces appearing in the amplitude data, or from the abrupt changes in the Lorentzian linewidths. Note that the thickness of the glass slide appears to be greater than its actual thickness since we have not corrected for the refractive index of the glass slide. No signal was expected within the cover slip since no moving scatterers ware present. The existence of the signal within the glass slide was probably due to detection of multiply scattered light outside of the confocal volume.

In summary we have shown that LCI can be used for imaging and quantification of Brownian motion within highly scattering media. The technique has potential applications both as a tool for non-invasive characterization of the dynamical and structural properties of optically thick materials as well as a method for studying the effect of multiple scattering on image contrast and resolution in confocal and optical coherence microscopy.

1. B. Berne and R. Pecora, *Dynamic light scattering : with applications to chemistry, biology and physics* (Willey, New York, 1976).
2. R. Pecora, editor, *Dynamic light scattering: applications of photon correlation spectroscopy* (Premium press, New York, 1985).
3. D. Pine, D.Weitz, P. Chaikin, E. Herbolzheimer, "Diffusing wave spectroscopy", Phys. Rev. Lett., 60, 1134, (1988).
4. G. Maret, P. Wolf, "Multiple light scattering from disordered media, the effect of brownian motion of scatterers", Z. Phys. B, 65, 409 (1987)
5. J. Zhu, J. Durian,J. Miller, D. Weitz and D. Pine, "Scaling of transient hydrodynamic interactions in concentrated suspensions", Phys. Rev. Lett. 68, 2559 (1992)
6. J. Durian, D. Weitz and D. Pine, "Scaling behavior in shaving cream", Phys. Rev. A 44, R7902 (1991)
7 . D. Boas "Diffuse photon probes of structural and dynamical properties of turbid media:theory and medical applications" PhD dissertation, 1996
8. M. J. Yadlowsky, J. M. Schmitt and R. F. Bonner, "Multiple scattering in optical coherence microscopy," Appl. Opt. **34:**5699-5707 (1995).
11. J. A. Izatt, M. R. Hee, G. M. Owen, E. A. Swanson and J. G. Fujimoto, "Optical coherence microscopy in scattering media," Opt. Lett. **19:**590-592 (1994).

Measurement of Smoothed Wigner Phase Space Distributions in Multiple Scattering Media

A. Wax and J. E. Thomas

Physics Department, Duke University, Durham, NC 27708-0305

Abstract

We use a simple heterodyne method to measure smoothed Wigner phase space distributions for light transmitted through a turbid medium. A three component distribution is seen to emerge from the sample, consisting of a ballistic peak, a narrow diffractive pedestal and a broad background. The narrow pedestal is found to decay with an extinction coefficient that is 0.65 as large as that of the ballistic component. These results are in excellent agreement with a simple analytical model that treats diffractive scattering explicitly, but includes large angle scattering only as a loss.

Key Words: multiple scattering, turbid media, coherence, Wigner distributions, heterodyne detection

Introduction

Propagation of optical coherence in multiple scattering media is currently receiving renewed attention because of its importance to biological and medical imaging based on optical fields [1]. It has been suggested by John [2] and independently by Raymer [3] that new venues for medical imaging may be based on measuring the coherence properties of scattered light using Wigner phase space distributions.

In 1932 Wigner [4] introduced a wave-mechanical phase space distribution function that is applicable to coherence tomography. This distribution plays a role closely analogous to a classical phase space distribution in position and momentum. For a wave field varying in one spatial dimension, $\mathcal{E}(x)$, the Wigner phase space distribution is given by [5]

$$W(x,p) = \int \frac{d\epsilon}{2\pi} e^{i\epsilon p} \langle \mathcal{E}^*(x+\epsilon/2)\mathcal{E}(x-\epsilon/2)\rangle, \quad (1)$$

where x is the position, p is a wavevector (momentum), and $\langle\ldots\rangle$ denotes a statistical average. It is easy to show that $\int dp\, W(x,p) = \langle|\mathcal{E}(x)|^2\rangle$, the position distribution of the intensity, and $\int dx\, W(x,p) = \langle|\mathcal{E}(p)|^2\rangle$, the corresponding momentum distribution. These results suggest that $W(x,p)$ is analogous to a classical phase space distribution in x and p. However, Eq. 1 shows that the Wigner distribution is Fourier transform related to the two-point mutual coherence function, and therefore is sensitive to the spatially varying phase and amplitude of the field.

Wigner phase space distributions obey rigorous transport equations, which are derivable from the underlying optical wave equations. Thus, measurement methods based on Wigner distributions can be placed on a firm theoretical footing and permit detailed study of the propagation of optical *coherence* in multiple scattering media for comparison with theory. Hence, studies of Wigner distributions will impact many current imaging methods, such as optical coherence tomography (OCT) [6,7] and optical coherence microscopy (OCM) [8].

In this paper, we briefly describe a heterodyne method for measuring smoothed Wigner phase space distributions. We then present data for multiple small angle scattering in a medium consisting of 5.7 micrometer radius polystyrene spheres suspended in a water-glycerol mixture. A narrow pedestal is observed in the Wigner phase space distributions that is seen to decay more slowly than the ballistic light. We present the results of an analysis that shows how this pedestal arises from multiple diffractive scattering [9].

OSA TOPS Vol. 21 Advances in Optical Imaging and Photon Migration, James G. Fujimoto and Michael S. Patterson (eds.)

Scheme

The scheme of the heterodyne experiments has been detailed previously [10]. In the experiments, a helium neon laser beam is split into a 1 mW input beam to the sample and a 1 mW local oscillator (LO) beam that is offset in frequency by 10 MHz. Both the LO beam and the transmitted beam from the sample are each passed through an acromatic input lens and mixed at a 50-50 beam splitter. Technical noise is suppressed by employing a standard balanced detection system with the detectors in the Fourier plane of the input lenses. The beat signal at 10 MHz is measured with an analog spectrum analyzer. An important feature of the experiments is that the analog output of the spectrum analyzer is squared using a low noise multiplier. The multiplier output is fed to a lock-in amplifier which subtracts the mean square signal and noise voltages with the input beam on and off [11]. In this way, the mean square electronic noise and LO shot noise are subtracted in real time. In this case, the lock-in output is directly proportional to the *mean square* beat amplitude, $|V_B|^2$. We have shown previously [10] that $|V_B|^2$ can be elegantly written in terms of the Wigner coherence functions for the LO, W_{LO}, and for the transmitted field, W_S. In one dimension, $|V_B|^2(x_M, p_M)$ is

$$\propto \int dx\, dp\, W_{LO}(x - x_M, p - p_M)\, W_S(x, p) \qquad (2)$$

Here, x is a position and p is a wavevector in the plane of the detector. Information about the transmitted phase space distribution, $W_S(x,p)$, is obtained by measuring the beat intensity as a function of the shift in the LO center position, x_M, and transverse momentum (angle), p_M. From Eq. 2, we see that the mean square beat signal is the convolution integral of the LO and signal field Wigner phase space distributions. Hence, the mean square beat signal that is obtained in real time in the experiments is a *smoothed* Wigner distribution [12]. W_S is measured over a range in position of ± 1 cm, and over a range in momentum of ± 300 mrad. The phase space $(x - p)$ resolution is determined by the spatial width and diffraction angle of the LO.

Results

We have measured the smoothed Wigner distribution for the field transmitted through a turbid medium consisting of latex microspheres (n=1.59) in a mixture of 25% glycerol and 75% water [9]. This mixture is chosen to produce neutral buoyancy for the microspheres. The index of refraction for the medium (n=1.36) results in a relative refractive index of 1.17 for the latex microspheres compared to the medium. Large spheres are specifically selected to give highly forward peaked scattering. The spheres used in this experiment have a radius of 5.7 micrometers. The Mie solution [13] for the above parameters gives a total scattering cross section σ_s that is 1.98 times the geometrical cross section, i.e., $\sigma_S = 202\,\mu\text{m}^2$. It is assumed that there is no absorption. The differential scattering cross section shows a central diffractive peak that is nominally gaussian. Data were taken for varying concentrations of spheres in water/glycerol in a precision cell of path length $L = 10$ mm.

In order to prevent the input source size from limiting the angular scattering characteristics, we have measured the smoothed Wigner distributions for transmission of a large diameter input beam through the sample ($a = 3.8$ mm half width at $1/e$ intensity). The LO beam is small in diameter ($a_o = 0.38$ mm) compared to the input beam, so that the output Wigner distribution is nearly independent of position x over the LO diameter. Further, the momentum width of the local oscillator is small compared to the width of the momentum dsitribution of the scattered light. In this case, the measured mean square heterodyne beat signal $|V_B|^2(x_M = 0, p_M)$ determines approximately the true transmitted Wigner distribution of the scattered light, $W_S(x = 0, p_M)$, excluding the ballistic contribution that is sharply peaked in the forward scattering direction.

For fixed LO position $x_M = 0$, the transmitted Wigner distributions were measured for varying concentrations ranging from $\rho = 0.4 \times 10^6/\text{cm}^3$ to $\rho = 6 \times 10^6/\text{cm}^3$ of 5.7 μm radius spheres. A typical plot is shown in Fig. 1 for a concentration of $\rho = 2 \times 10^6/\text{cm}^3$.

The scattering distribution has three components: a ballistic peak (not shown), a central narrow pedestal and a broad background. The transverse momentum p is given in units of the optical wavevector in air k_o. The position x is in mm. This figure shows data points which are taken for angles $\theta = p/k$ in the range from -225 to +75 mrad. However, the fitted curves (solid line) are done for the data outside ± 30 mrad, as interference with the gaussian tail of the ballistic light distribution causes these points to be unreliable. The ballistic peak is attenuated at this concentration by $\exp(-4)$ and is not shown in this figure as it is orders of magnitude larger than the scattered light. The half width at $1/e$ intensity of the ballistic peak is 0.4 mrad, which is much narrower

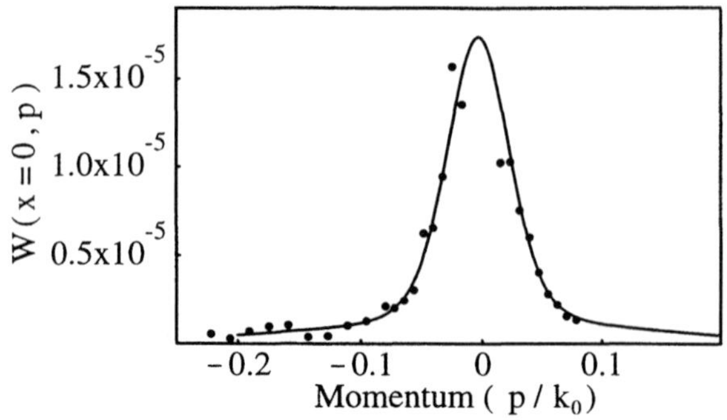

Figure 1: Smoothed Wigner distribution $\simeq W(x = 0, p)$ for large input beam (linear scale, ballistic contribution not shown) for $\rho = 2 \times 10^6/\text{cm}^3$. Note the narrow central pedestal arising from diffractive scattering. The solid line shows the theoretical prediction.

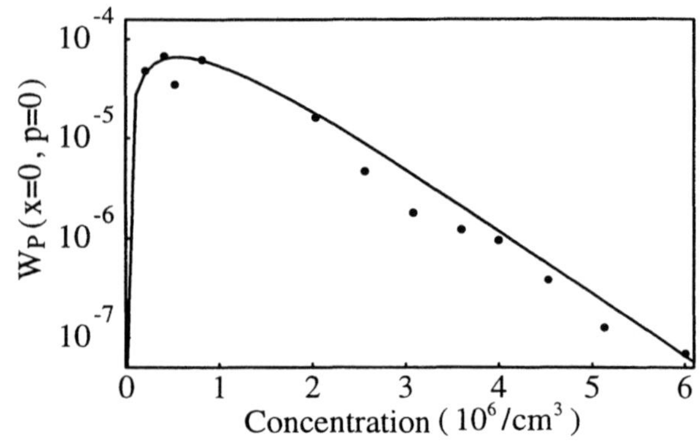

Figure 2: Amplitude of the narrow pedestal arising from multiple diffractive scattering as a function of scatterer concentration, ρ. The solid line shows the prediction using no free parameters.

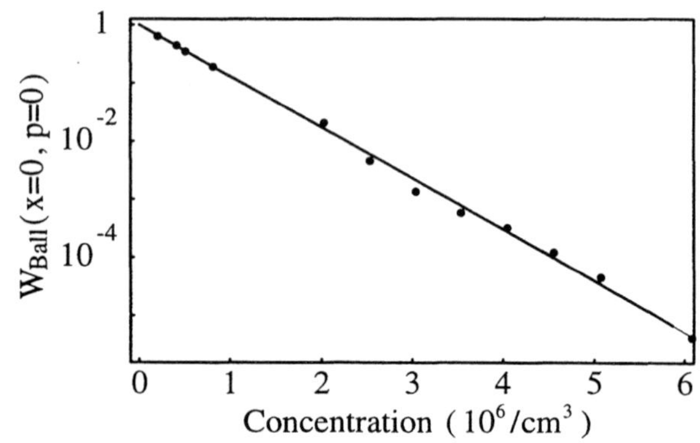

Figure 3: Amplitude of the ballistic component as a function of scatterer concentration, ρ. The solid line shows the prediction for exponential decay with the extinction coefficient determined from the total scattering cross section.

than the widths of the scattered light distributions.

The experimental angular distributions, $W_S(x = 0, p)$, are fit to two component curves (solid lines) consisting of a narrow central pedestal and a broad background. As discussed above, the ballistic contribution is ignored, as it is localized near the origin. The central pedestal in the two component curve arises from multiple diffractive scattering. The shape of this component (normalized to 1 at $p = 0$) is determined using the theory described in the next section. The amplitude of this component is left as a free parameter and is later compared to the predicted value. The broad background, which is not of interest here, is modeled as a Gaussian distribution in p with an undetermined amplitude.

To explore how the amplitude of the central pedestal varies with concentration, the maximum pedestal amplitudes for each of the fitted curves are shown in Fig. 2 as a function of scatterer concentration. These are compared to the predicted amplitude (solid line) with no free parameters. For comparison, the amplitude of the ballistic peak as a function of scatterer concentration is shown in Fig. 3.

Over most of the range of scatterer concentrations, the amplitude of the central pedestal, Fig. 2, is found to decay exponentially with an attenuation coefficient that is 65% of that found for the ballistic.

Discussion

The data obtained in the experiments exhibit a number of interesting features that can be understood in terms of a simple model for multiple diffractive scattering. In the simplest approximation, the transport equation for the Wigner distribution takes a form that is analogous to that for the specific intensity of radiative transport theory. In this transport equation, we include multiple diffractive scattering via a collision integral, but we treat large angle scattering as a loss [9]. In addition, the input beam is assumed large in diameter compared to the the LO beam, so that the system exhibits approximate translation invariance across the output face of the sample [14]. These conditions are approximately satisfied in the experiments.

With these simplifying assumptions, the calculated Wigner distribution for the diffractive pedestal can be cast in the following form

$$W_P(L, 0, p) \propto e^{-\mu_B L} \sum_{n=1}^{\infty} P_n \frac{\exp[-p^2/(n\theta_o^2 k^2)]}{\pi n \theta_o^2 k^2}. \quad (3)$$

Note that propagation is in the $\mathbf{z}$ direction, and

the solution is given for $x, y = 0$; $p_y = 0$. For each term, there is a normalized gaussian momentum distribution of width $\theta_o k\sqrt{n}$. This describes a random walk distribution for n diffractive momentum changes of magnitude $\theta_o k$, where θ_o is the diffraction angle and k is the optical wavevector. $P_n = \exp(-\mu_D L)(\mu_D L)^n/n!$ is a Poisson distribution describing the probability of n scatterings, with $\bar{n} = \mu_D L$ the mean number of scattering events, and μ_D the extinction coefficient due to diffractive scattering. With $\exp(-\mu_D L)$ incorporated into the probability of n scatterings, the pedestal decays as $\exp[-\mu_B L]$. The factor $\mu_B = \mu_T - \mu_D$ is just the extinction coefficient arising from large angle scattering (and absorption). From the Mie solution, we find that $\mu_B = 0.65\mu_T$, where μ_T is the ballistic extinction coefficient.

It is not difficult to understand why the diffractive pedestal decays more slowly than the ballistic signal. Diffractive momentum changes tend to scatter photons within the momentum distribution of the diffractive pedestal. Only large momentum changes are effective in scattering photons outside the momentum distribution of the pedestal. Hence, the pedestal decays with the attentuation coefficient μ_B that arises from the large angle scattering cross section. By contrast, both diffractive and large angle momentum changes are effective in scattering photons out of the momentum distribution of the ballistic light, which is very narrow. Hence, the ballistic distribution must decay with the extinction coefficient μ_T corresponding to the total scattering cross section.

Conclusions

In conclusion, we have demonstrated heterodyne measurement of smoothed Wigner phase space distributions for light that has undergone multiple diffractive scattering in a turbid medium. This method provides high angular resolution and is well suited for direct measurement of momentum distributions in the near forward scattering direction. The data are well fit by a theoretical model that assumes the Wigner phase space distributions obey an approximate transport equation that is identical in structure to the usual transport equation for the specific intensity. By incorporating transverse momentum changes arising from multiple diffractive scattering and treating large angle classical scattering as a loss, we find very good agreement with the data.

These experiments show that multiple diffractive scattering can produce a narrow pedestal that decays more slowly than the ballistically transmitted component. At high concentrations of scatterers, if the diffractive pedestal is comparable to the ballistic component, it can modify the apparent intensity of a probe beam compared to that expected for exponential attentuation of ballistic light. This is consistent with the anomalous probe intensity observed by Yadlowsky et al., in OCM [15].

The heterodyne method for measurement of Wigner distributions has a high dynamic range and can be used to study transmitted and reflected light in dense turbid media. Although the present work concentrated on the momentum, the joint evolution of the position and momentum distribution can be studied by this method. In the present experiments, which employ cw lasers, the small angle scattering distributions are ovewhelmed by the diffuse background as the concentration of scatterers is increased. Hence, the decay of the pedestal arising from multiple diffractive scattering could not be studied at very high concentrations with correspondingly high dynamic range. However by performing similar experiments with a low coherence light source, diffuse scatter can be substantially suppressed and the evolution of the diffractive pedestal can be followed to much higher scatterer concentrations. Experiments employing this method to measure Wigner distributions for low order scattered light currently are underway in our laboratory.

Acknowledgements

It is a pleasure to thank Prof. M. G. Raymer for many stimulating conversations on the physics of Wigner phase space distributions and coherence measurement in multiple light scattering. We also thank Prof. R. Glauber for suggesting the regime used for the measurements. This research has been supported in part by the National Science Foundation and the National Institute of Health.

References

[1] A. Yodh and B. Chance, "Spectroscopy and Imaging with Diffusive Light," Physics Today, **48**, 34 (1995).

[2] S. John, G. Pang, and Y. Yang, "Optical coherence propagation and imaging in a multiple scattering medium," in *Proceedings: Conference on Optical Tomography, Photon Migration, and Spectroscopy of Tissue and Model Media*, San Jose, CA, 5-7 February, 1995, eds. B. Chance and R. R. Alfano, (SPIE 1995).

[3] M. G. Raymer, C. Cheng, D. M. Toloudis, M. Anderson, and M. Beck, "Propagation of Wigner coherence functions in multiple scattering media," preprint, for *Proceedings: Advances in Optical Imaging and Photon Migration*, Orlando, FLA, March 18-20 (1996).

[4] E. P. Wigner, "On the quantum correction for thermodynamic equilibrium," Phys. Rev. **40**, 749-759 (1932).

[5] M. Hillery, R. F. O'Connel, M. O. Scully, and E. P. Wigner, "Distribution functions in physics: fundamentals," Phys. Rep. **106**, 121-167 (1984).

[6] D. Huang, E. A. Swanson, C. P. Lin, J. S. Schuman, C. A. Puliafito, and J. G. Fujimoto, "Optical coherence tomography," Science **254**, 1178-1181 (1991).

[7] J. A. Izatt, M. D. Kulkerni, K. Kobayashi, M. S. Sivak, J. K. Barton, and A. J. Welch, "Optical Coherence Tomography for Biodiagnostics," Optics and Photonics News **8** (5), 41-47, 65 (1997).

[8] J. A. Izatt, M. R. Hee, G. M. Owen, E. A. Swanson, and J. G. Fujimoto, "Optical coherence microscopy in scattering media," Opt. Lett. **19**, 590-592 (1994).

[9] A. Wax and J. E. Thomas, "Measurement of Smoothed Wigner Phase Space Distributions for Small Angle Scattering in a Turbid Medium," to appear in JOSA A, July, 1998.

[10] A. Wax and J.E. Thomas, "Optical Heterodyne imaging and Wigner phase space distributions," Opt. Lett. **21**, 1427 (1996).

[11] Related methods have been used previously. See for example, G. L. Abbas, V. W. S. Chan and T. K. Yee, "A dual-detector optical heterodyne receiver for local oscillator noise suppression," IEEE J. Lightwave Tech. **3**, 1110 (1985).

[12] The mean square beat is positive definite and takes the form of a smoothed Wigner distribution. See N. D. Cartwright, "A non-negative Wigner-type distribution," Physica, **83A**, 210-212 (1976).

[13] C. F. Bohren and D. R. Huffman, *Absorption and scattering of light by small particles* (Wiley, Interscience, New York, 1983).

[14] The idea of making measurements of Wigner distributions in this regime was suggested to us by Prof. R. J. Glauber, Harvard University, Dept. of Physics, Cambridge, MA 02138, private communication.

[15] M. J. Yadlowsky, J. M Schmidt, and R. F. Bonner, "Multiple scattering in optical coherence microscopy," Appl. Opt. **34**, 5699-5707 (1995).

Video Rate Optical Coherence Tomography

Andrew M. Rollins, Manish Kulkarni, Siavash Yazdanfar and Joseph A. Izatt

Department of Biomedical Engineering
Case Western Reserve University
Cleveland, OH 44106 USA
(216)844-3298; Fax:(216)844-8011
amr9@po.cwru.edu

Abstract

We have developed an optical coherence tomography system which can image up to video rate. The system utilizes a high power broadband source and real time image acquisition hardware and features a high speed scanning delay line in the reference arm based on Fourier-transform pulse shaping technology. We detail the operation of the delay line and demonstrate real time imaging of *in vivo* human skin and *Xenopus laevis* embryo and *in vitro* samples of porcine gastrointestinal tissue

Key Words

(170.4500) Optical coherence tomography; (170.3880) Medical and biological Imaging; (120.3890) Medical optics instrumentation.

Introduction

Optical Coherence Tomography (OCT) is a new noninvasive imaging technique which allows microscopic tomographic sectioning of biological samples [1, 2]. OCT fills a valuable niche in imaging of tissue ultrastructure, providing optical biopsies with high spatial resolution (~10 μm) and sensitivity (>100 dB) *in vivo* with no contact needed between the probe and the tissue. An OCT image (B scan) is built up as a series of adjacent axial depth scans (A scans). Depth is gated by low coherence interferometry where the sample is in one arm of a Michelson interferometer, and a scanning delay line is in the other arm. With fiber optic implementation, OCT can be easily adapted to minimally invasive diagnostic imaging technologies such as endoscopy and laparoscopy. In order to be appealing to the endoscopy community, however, OCT must provide the clinician with near real-time imaging.

Several technical challenges must be overcome in order to image at high speed with OCT. First, a high power optical source is required to provide adequate illumination in a short amount of time. Second, a high speed scanning delay line is required in the reference arm of the interferometer in order to produce image data quickly. Finally, image data must be acquired, processed, and displayed in real time.

Signal to noise ratio (SNR) in the shot noise limit is proportional to the optical power illuminating the sample and inversely proportional to the signal bandwidth. The signal bandwidth is proportional to the image acquisition rate, therefore, in order to maintain SNR, any increase in image acquisition rate must be accompanied by a proportional increase in source optical power.

A high speed scanning delay line in the reference arm of the OCT system is needed in order to acquire images at a high rate. The scan must also be long enough to be useful. A 3mm (free space) scan, typically used in OCT, corresponds to a delay of approximately 100 nsec. The delay should scan continuously rather than in discrete steps if Doppler shift and spectral information is to be preserved. The delay line should also scan at a constant rate. Any nonlinearity in the scan results in a nonconsistant signal center frequency and bandwidth and also warping of the image.

Theory

OCT signal

OCT measures the interferometric cross correlation of the light returning from the reference and sample arms. A mirror in both arms gives the interferometric autocorrelation of the source. This autocorrelation signal characterizes the response of the OCT system. The bandwidth of the detector current is a function of both the optical source spectral bandwidth $\Delta\nu$ and the group delay scan speed V_g:

$$\Delta f = \frac{\Delta \nu V_g}{c}, \qquad (1)$$

where c is the speed of light. The center frequency of the signal is given by the Doppler shift formula:

$$f_D = \frac{2V_\phi}{\lambda_0}, \qquad (2)$$

where V_ϕ is the scan speed of the phase delay and λ_0 is the center wavelength of the source. The full information bandwidth of the signal must be detected in order to maximize spatial resolution, while it is desirable to reject frequencies outside of the information bandwidth in order to maximize system SNR. Signal to noise ratio of OCT in the shot noise limit is [3]:

$$SNR = \frac{\eta \lambda_0 P_r R_s}{hcBW}, \qquad (3)$$

where η is the detector quantum efficiency, P_r is the reference arm power, R_s is the reflectivity of the sample, *h* is the Planck constant, and BW is the detector signal bandwidth. Table 1 illustrates the tradeoff between SNR in the shot noise limit and frame rate for a given optical power.

Table 1. The tradeoff between frame rate and dynamic range for systems imaging 250 lines per frame by varying the scan rate of the delay line.

Frame Rate (frames per second)	Dynamic Range (shot noise limit)
4	108.8 dB
8	105.7 dB
16	102.7 dB
32	99.7 dB

High Speed Scanning Delay Line

The delay line we chose to implement was developed for femtosecond pulse measurement [4,5], and has recently been applied to OCT [6]. The delay line consists of a grating-lens pair in a double-passed Fourier-domain pulse shaping configuration with a flat mirror serving as a spatial phase filter which imposes a linear phase ramp in the frequency domain (Fig 1). It is a well known property of the Fourier transform that a phase ramp in the frequency domain corresponds to a group delay in the time domain:

$$x(t-t_0) \leftrightarrow X(\omega)e^{-i\omega t_0}. \qquad (4)$$

In our delay line, the incident angle was such that the center wavelength λ_0 of the diffracted beam was normal to the grating and the entire grating was in the focal plane of

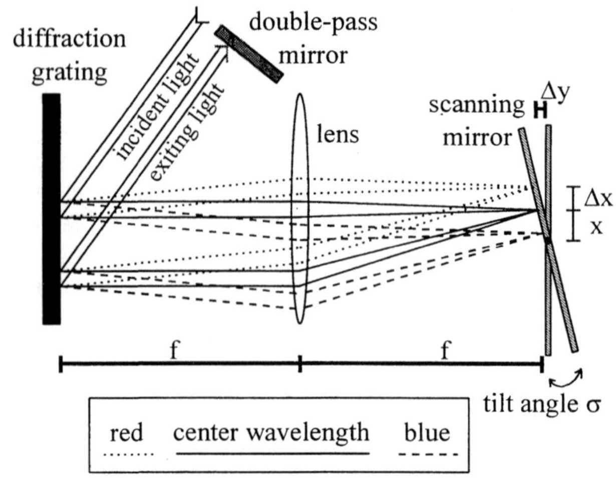

Figure 1. Schematic of the Fourier domain optical delay line. The incident, collimated broadband light is diffracted from the grating and spectrally dispersed. The lens spectrally collimates and spatially focuses the dispersed beam. The scanning mirror imposes a linear phase ramp on the spectrum and redirects the light back through the lens which spectrally focuses and spatially collimates the beam. The beam then diffracts in a reverse manner from the grating and returns toward the source collimated and undispersed. The double-pass mirror returns the light back through the identical path.

the lens. The mirror pivot can be offset from the center wavelength by an arbitrary distance *x*. The single-pass phase delay as a function of wavelength λ for a given mirror tilt angle σ is given as:

$$\phi(\lambda) = \frac{4\pi\sigma x}{\lambda} + \frac{4\pi\sigma f \Delta\lambda}{p\lambda}, \qquad (5)$$

where *f* is the focal length of the lens, *p* is the pitch of the grating, and $\Delta\lambda$ is defined as λ-λ_0. From Eq. (5) we can calculate the phase delay:

$$t_\phi = \frac{2\sigma x}{c}, \qquad (6)$$

where *c* is the speed of light. This corresponds to a free space distance d_ϕ which we will call the phase lag:

$$d_\phi = 2\sigma x. \qquad (7)$$

We can also calculate the group delay:

$$t_\phi = \frac{2\sigma x}{c} + \frac{2\sigma f \lambda_0}{cp}, \qquad (8)$$

which corresponds to a free space distance d_g which we will call the group lag:

$$d_\phi = 2\sigma x + \frac{2\sigma f \lambda_0}{p}. \qquad (9)$$

This expression for group lag can alternatively be derived by simply evaluating the change in path length experienced by the center wavelength as a result of an arbitrary mirror tilt angle, but with less insight into the operation of the device as a leg in a Michelson interferometer, as it is used in OCT. It can be seen that the group lag is equal to the phase lag plus an additional term that is a function of the properties of the lens, grating, and light source. Both the phase lag and the group lag are proportional to the tilt angle of the scanning mirror, so an angular scan of the mirror effects a scan of both phase and group lag. The second term in the expression for group lag is dominant in our configuration, so the group lag is much larger than the phase lag for a given angle σ. This feature is key to the power of this delay line. The mirror need only be moved physically a small amount to generate a large group lag. It is possible to adjust the offset x for a desired center frequency without significantly affecting the group lag.

Methods

A block diagram of our system is shown in Figure 2. Our system utilizes a recently commercialized high-power broadband source (AFC Technologies, BBS 1310 B-TS) centered around 1310 nm with 47 nm FWHM spectral width which provides 10 mW of optical power. We implemented our delay line with low-cost resonant scanners with resonant frequencies of 2kHz and 4kHz. We used a 1 inch square grating with 600 lines per mm, and a 50 mm focal length lens.

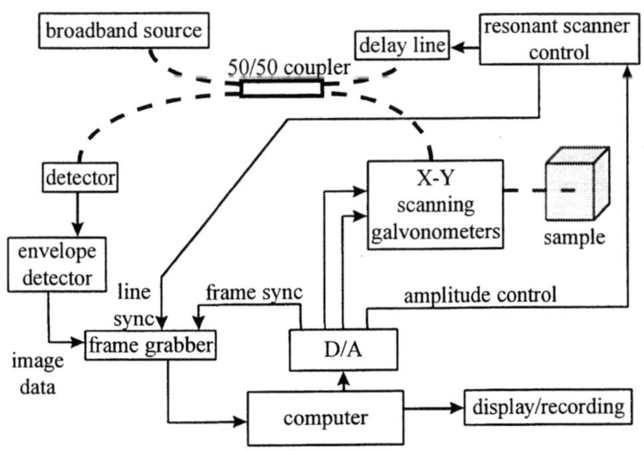

Figure 2. Schematic of the high speed OCT system. The dashed lines represent optical paths and the solid lines represent electronic paths

We double-passed the system in order to double the effective delay and to recouple the returning light into the fiber In order to minimize warping of the image and nonlinearity of the center frequency with time, we imaged during the middle 2/3 of the one sided scan. We detected the signal with a high speed photodiode detector module and demodulated with an envelope detecting logarithmic amplifier. Digitization and display was accomplished in real time with a variable scan monochrome frame grabber and computer. Lateral scanning was accomplished with an X-Y galvanometric scanning unit, providing for arbitrary sample sectioning and volume scanning. The images were recorded as digital movie files or directly to video tape.

Results

Figure 3 shows single movie frames of the beating heart of a *Xenopus laevis* embryo recorded at 4, 8, 16, 32 frames per second with the 2kHz scanner, demonstrating the trade off between image size and image acquisition rate for a fixed A-scan rate. Notice that although the sample was moving rapidly, no motion artifacts are evident in the images. Figure 4 shows a series of frames of *in vivo* human thick skin (fingertip) recorded at 8 frames per second. A sweat duct is indicated in the images at different depths as the image slice followed it toward the surface. Such structures may be overlooked in single, still images, but can be easily visualized in real-time dynamic imaging.

Conclusions

In conclusion, we have implemented an OCT system capable of imaging at up to video rate. This system features a high-speed Fourier-domain variable-phase optical delay line and utilizes a commercially available optical power source and image acquisition hardware. We have demonstrated imaging at various frame rates up to video rate. Real time imaging has allowed visualization of dynamic events without motion artifacts as well as visualization of features and structures in physiological samples which may not be striking in still images.

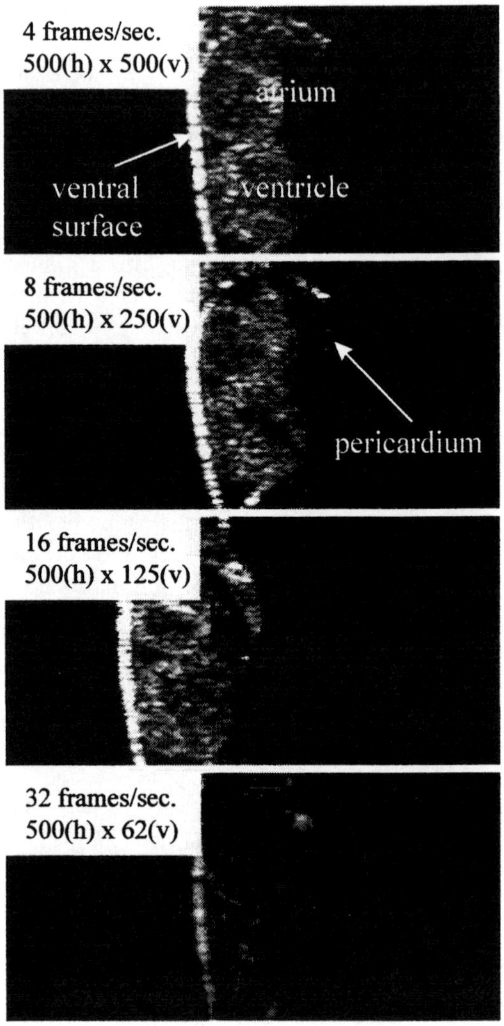

Figure 3. Single frames from recordings of beating *Xenopus* embryo heart at 4, 8, 16, and 32 frames per second. Images are 2.1mm (h) by 1mm (v). Probe light was incident from the left. The heart was beating at about 2 beats per second. Notice the decreasing lateral resolution as the frame rate increases.

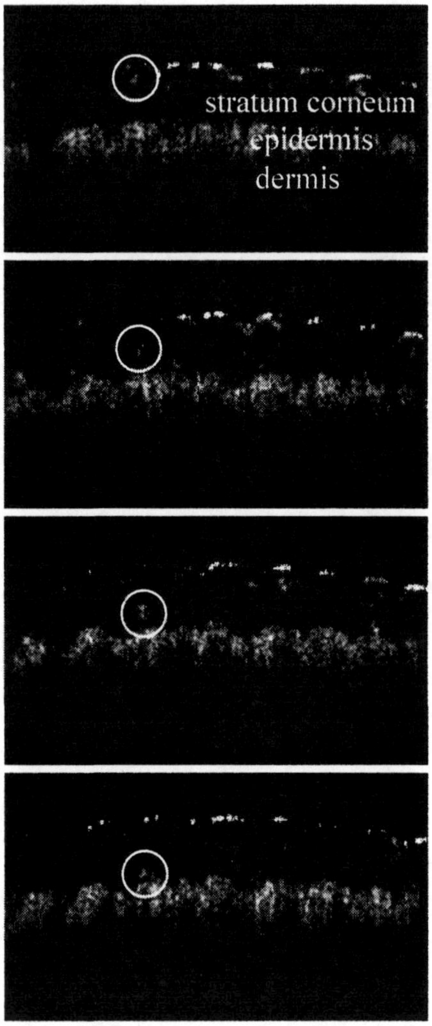

Figure 4. A series of frames from a recording of human thick skin (fingertip) *in vivo* with glycerine index matching. The circles indicate a sweat duct which the image slice follows from the epidermis through the stratum corneum to the surface. The images were recorded at 8 frames per second with 250 lines per frame and are 4mm (h) by 2.1mm (v).

Acknowledgments

We would like to acknowledge the contributions of Ton van Leeuwen, and Jop Perree, and the financial support of Olympus America, Inc.

References

1. Huang, E. A. Swanson, C. P. Lin, et al. "Optical Coherence Tomography," Science 254, 1178-1181 (1991).
2. A. Izatt, M. D. Kulkarni, H. Wang, et al. "Optical Coherence Tomography and Microscopy in Gastrointestinal Tissues," IEEE J. Sel. Top. Quant. Elect. 2, 1017-1027 (1996).
3. A. Swanson, D. Huang, M. R. Hee, et al. "High-speed optical coherence domain reflectometry," Opt. Lett. 17, 151-153, 1992.
4. Weiner, D. E. Leaird, J. S. Patel, J. R. Wulbert. "Programmable femtosecond pulse shaping by use of a multielement liquid-crystal phase modulator," Opt. Lett. 15, 326-328 (1990).
5. F. Kwong, D. Yankelevich, K. C. Chu, et al. "400-Hz mechanical scanning optical delay line," Opt. Lett. 18, 558-560 (1993).
6. Tearney, M. E. Brezinski, B. E. Bouma, et al. " In Vivo Endoscopic Optical Biopsy with Optical Coherence Tomography," Science 276, 2037-2039 (1997).

Peculiarities of Optical Formation and Filtering the Images of 3D Objects in Coherent Light

Yu. V. Chugui

Laboratory for Technical Vision

Technological Design Institute of Scientific Instrument Engineering
Siberian Branch of the Russian Academy of Sciences
41 Russkaya str., 630058 Novosibirsk, Russian Federation

Abstract

Application of classical optical dimensional methods based on shadow (projecting) image model for 3D bodies, in practice, can bring out to the significant errors. Since the existing strict and approximate solutions for diffraction and image problems of 3D bodies are extremely difficult for engineering applications, author previously suggested constructive method for their calculation which is simple physically obvious and sufficiently accurate. It is based on the model of equivalent diaphragms according to which the problem of light diffraction on 3D bodies is reduced to the calculation of diffraction phenomena on the plane objects (of zero thickness) located in space in Kirchhoff-Fresnel approximation. This method was developed analytically for study the formation and high-frequency filtering (contouring) of typical element of 3D objects like extended symmetric edge with perfectly absorbing and reflecting inner surfaces. The obtained theoretical results are provided by the optical experiments.

Key Words

(050.1960) Diffraction theory; (110.2990) Image formation theory; (100.6890) Three-dimensional image processing; (070.6110) Spatial filtering.

1. Introduction

Existing methods for optical dimensional inspection are based on, as a rule, the forming the shadow object images of the objects and on finding in the images the distances between the coordinates corresponding to boundaries (edges) of these ones. It is known [1] that for plane objects (of zero thickness) in coherent light the geometrical position of the one-dimensional edge (half-plane) corresponds to the image coordinate where the light intensity is 25% of its value in the illuminated areas far from the edge. This fact makes the measurement more complicated. Therefore there is the tendency to use in measurements the high-frequency (contoured) images formed by coherent-optical methods [2]. In this case the profile of the edge image being symmetric has the shape like two main maxima with a dip in the middle. It is essential that for one-dimensional edge the coordinate of the dip corresponds to geometrical position of the object edge.

For the 3D bodies (especially for microobjects) which represent a wide class of objects, the shadow model appears to be approximate and may lead to considerable errors. Since the existing strict [3, 4] and approximate solutions [5] for diffraction problem are extremely difficult for engineering applications, an urgent necessity appears in creating constructive (simple, physically obvious and at the same time sufficiently strict) theory the forming the Fraunhofer diffraction patterns (spectra) and images (including filtered out) of 3D-objects. Such constructive method for evaluating Fraunhofer diffraction patterns of constant thickness with flat inner surfaces based on the model of equivalent diaphragms was first suggested [6] and studied in detail [6, 7].

The essence and experimental verification of the constructive method for calculation of diffraction Fraunhofer patterns of the extended objects like 3D symmetrical edge and slit of constant thickness with perfectly absorbing and perfectly reflecting inner surfaces are given below. Using the obtained equations for Fraunhofer diffraction patterns peculiarities of the formation and filtering the image of 3D objects in coherent light were investigated theoretically and experimentally.

2. The Constructive Method for Calculation of Diffraction Patterns of Extended Objects

First of all, consider in detail the supposed constructive calculation method. It is based on the model of equivalent diaphragms (distributions) for 3D objects of constant thickness with flat inner surfaces [Fig. 1]. According to this model the front and rear faces of the object give the main contribution into the field. The influence of their inner surfaces on the diffraction component is supposed to be negligible small and, therefore, results in the absorption or reflection of the waves diffracted on the front face of the object. As a result, the problem of light diffraction on volumetric bodies is reduced to the analysis of diffraction phenomena on the plane objects (described by binary or bipolar boundary functions) which are located in space. It permits to apply the standard Fourier-optical methods for the calculation in the Kirchhoff-Fresnel approximation.

Fraunhofer diffraction patterns of such typical elements of extended bodies as volumetric edge and 3D slit with the absorbing and reflecting inner faces were under investigation [7]. In case of the 3D *perfectly absorbing asymmetric edge* [Fig. 1(a)], according to the model of the equivalent diaphragms, the boundary functions are the following: $f(x) = Y(x)$, $g(x_1)=Y(x_1 - C)$.

The spectrum is described as follows:

$$F_1(\theta) = \pi\delta(\kappa\theta) + \frac{\tilde{Y}(d\theta - C)}{j\kappa\theta} + \tilde{Y}(C)\frac{e^{-jk\theta C}}{jk\theta}\, exp\left(\frac{jkd\theta^2}{2}\right), \quad (1)$$

where $\delta(z)$ is the Dirack delta-function, $\tilde{Y}(z) = e^{j\pi/4}(\lambda d)^{-\frac{1}{2}} \int_{-\infty}^{+\infty} Y(x)\, exp\left[\frac{jk(x-z)^2}{2d}\right] dx$ is the Fresnel transform of the Heaviside step function $Y(z)$ [8] ($k = 2\pi/\lambda$ is the wave number, λ is the wavelength).

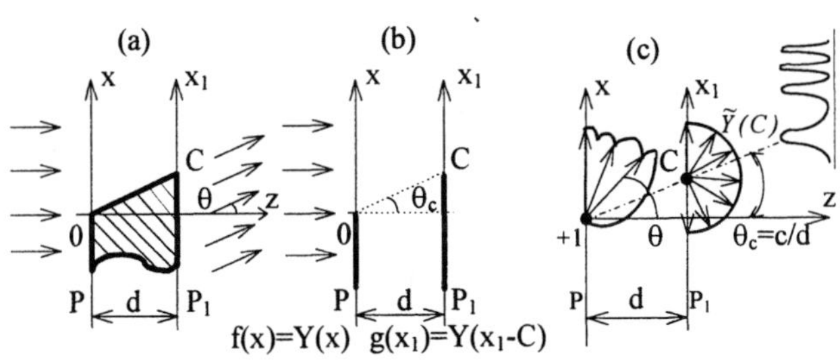

Figure 1. The 3D *perfectly absorbing asymmetric edge* (a), its equivalent diaphragm model (b), and the generalized two-point source model of diffraction (c).

The first term of (1) describes the contribution to the diffraction pattern from a direct (transmitted) light beam and the second one represents light diffraction by the front face of the object with regard to vignetting of diffracted waves by the back face. The third term corresponds to the secondary light diffraction by the back face, and it is defined by the value of the amplitude of the field $\tilde{Y}(C)$ (the result of light diffraction by the front face at the location of the edge of the back object face).

Model of field formation (accurate to the factor $k\theta$) can be presented in the form of the generalized delta sources with the corresponding amplitude and radiation patterns at the boundary points x=0 and x=C [Fig. 1(c)]. It is essential that the radiation pattern of the first source unlike the second one is anisotropic and is determined by the function $\tilde{Y}(z)$. In this case the behaviour of the Fraunhofer diffraction pattern of 3D asymmetric edge is completely determined by the peculiarities of wave interference of the both generalized delta sources.

From the analysis of the spectrum (1), when $C=0$ (case of the 3D *symmetric edge*) it follows that the 3D effects become apparent for the angles of diffraction $\theta_0>\theta_{cr}$, where $\theta_{cr} = \sqrt{\lambda / d}$ is the critical angle, under which the 3D effects are most significant. If $\theta_0<<\theta_{cr}$, the field $F_1(\theta)$ approaches to the plane object spectrum. Thus, the influence of the extension of the volumetric edge on its spectrum is determined by the parameter $N=\theta_{cr}/\theta_0$, being equal to the Fresnel zone number $D/\sqrt{\lambda d}$ in the domain $D=\lambda/\theta_0$ under investigation. 3D effects are weakly expressed in the spectrum, if $N >> 1$ and vise versa, at $N << 1$ their influence is essential.

Now let us consider the light diffraction by the 3D *perfectly absorbing asymmetric slit* front and back holes of which have different widths D and D_1 (let $D > D_1$), so that centres of these holes are shifted by b. In this case the boundary functions are the following:

$$f(x) = \mathrm{Rect}\left[\frac{x}{D}\right]; \qquad g(x_1) = \mathrm{Rect}\left[\frac{(x_1-b)}{D_1}\right].$$

Making use of (1) one can obtain the following formula for the spectrum of the 3D slit [Fig. 2(a)]:

$$\begin{aligned} F_2(\theta) = \frac{1}{jk\theta}\Bigg\{ & e^{jk\theta D/2}\,\tilde{\mathrm{Rect}}\left(\frac{d\theta - 0.5D - b}{D_1}\right) \\ & - e^{-jk\theta D/2}\,\tilde{\mathrm{Rect}}\left(\frac{d\theta + 0.5D - b}{D_1}\right) \\ & + e^{jkd\theta^2/2} e^{jk\theta(0.5D_1 - b)}\,\tilde{\mathrm{Rect}}\left(\frac{b - 0.5D_1}{D}\right) \\ & - e^{jkd\theta^2/2} e^{-jk\theta(0.5D_1 + b)}\,\tilde{\mathrm{Rect}}\left(\frac{b + 0.5D_1}{D}\right)\Bigg\}, \end{aligned} \quad (2)$$

where $\tilde{\mathrm{Rect}}(z)$ is the Fresnel image of the rectangular function $\mathrm{Rect}(z)$ [1]. The generalized point model in the case of diffraction on a 3D slit has four sources with the corresponding radiation patterns [Fig.2(b)]. Their interference determines the distinctive peculiarities being introduced into spectrum by the extension of the object. For example, in case of light diffraction on a symmetric volumetric slit ($D=D_1$, $b=0$) volumetric effects in the low-frequency domain of its spectrum reduce to the addition of spectrum modulation by low-frequency ("Fresnel") oscillations with the minima when $\varphi_m = \sqrt{(2m+1)\lambda/d}$ $(m = 0,1,2..)$ and to the decrease of frequency of high-frequency ("Fraunhofer") oscillations

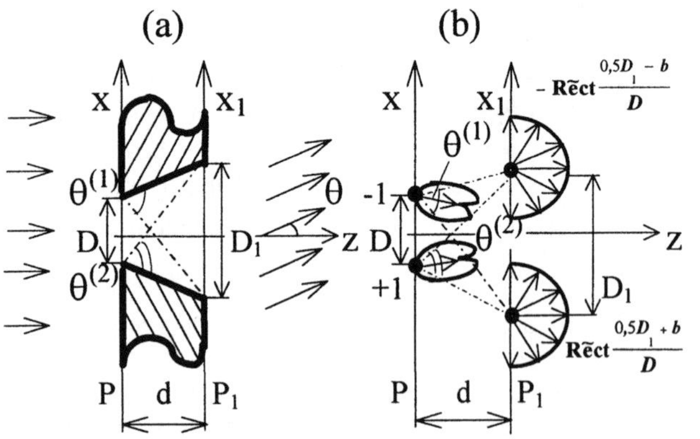

Figure 2. The 3D *perfectly absorbing asymmetric slit* (a) and its generalized four-point source model of diffraction (b).

with characteristic size $\theta_l^{(0)} = \lambda/D$ [Fig. 3]. In so doing, the speed of their frequency variation is determined by the value of the parameter $N=D/\sqrt{\lambda d}$.

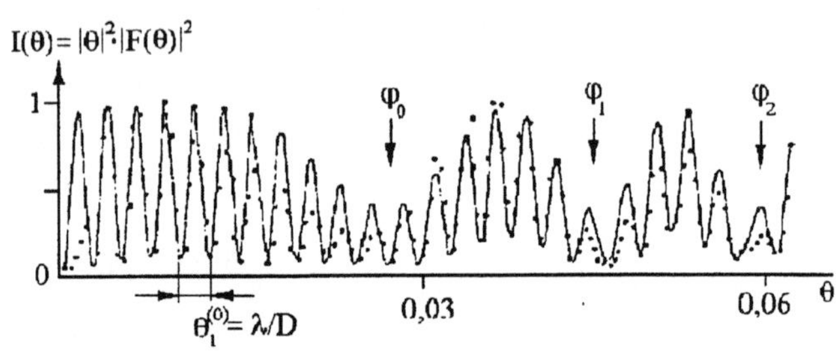

Figure 3. Experimental results: spectrum of the 3D symmetric perfectly absorbing slit with d=1 mm and D=0.277 mm. Points are experimental data, solid line is calculated spectrum.

Under the spectrum calculating of 3D *perfectly reflecting* objects, one should take into account the reflecting effects of diffracted waves by the inner surfaces of the object. We have investigated in detail the light diffraction on the extended symmetric objects made of metal [Fig. 4(a)] [9]. Due to phase change of the reflected wave by π, according to the model of the equivalent diaphragms [Fig. 4(b)], boundary functions for this 3D object are the following: $f(x)=\mathrm{sign}(x)$, $g(x_1)=Y(x_1)$. The spectrum of 3D *perfectly reflecting symmetric edge* is described by the following function:

$$F_2(\theta) = \pi\delta(k\theta) + \frac{2Y(d\theta)}{jk\theta} \tag{3}$$

It is essential that in consequence of the phase change of the diffracted reflected wave onto reverse one, the component corresponding to the secondary light diffraction on the back face is absent [Fig. 4(c)].

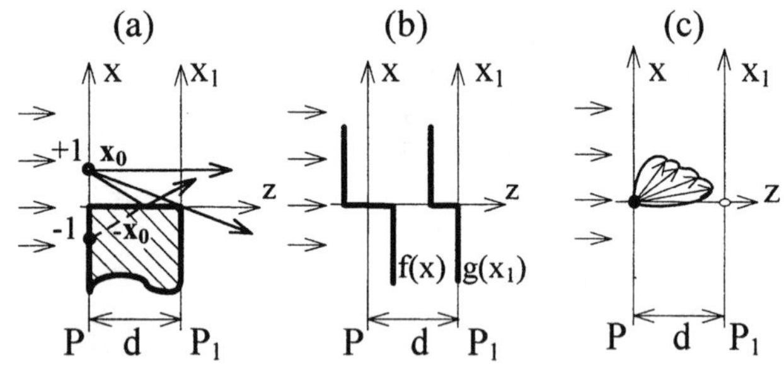

Figure 4. The 3D *symmetric perfectly reflecting edge* from metal (a), its equivalent field distribution model (b) and the generalized one-point source model (c).

By using (3), it is not difficult to obtain the spectrum of 3D *perfectly reflecting symmetric metallic slit* when taking into account the single wave reflection from their inner surfaces:

$$F_3(\theta) = \frac{2}{jk\theta}\left\{ \begin{array}{l} e^{jk\theta D/2}\, R\tilde{e}ct\left(\dfrac{d\theta - 0.5D}{D}\right) \\ -\, e^{-jk\theta D/2}\, R\tilde{e}ct\left(\dfrac{d\theta + 0.5D}{D}\right) \end{array} \right\} \tag{4}$$

It is essential that for $N >> 1$, the diffraction field is formed by the "remainders" of the Fresnel images $R\tilde{e}ct\,(z)$ and therefore the contrast of the Fraunhofer oscillations will fall when moving from the center of the spectrum to its periphery.

3. Imaging the 3D Perfectly Absorbing Edge

Let us consider the standard coherent-optical system for forming and filtering the object images [Fig. 5] [1]. As the extended object one choose a typical element of volumetric objects – 3D symmetric edge of thickness *d* with perfectly absorbing inner surface. One can assume that the front object face is placed in the plane *P* and the inner one touches the optical axis of the system.

The images of the front and back faces, described by the output fields accordingly $\hat{f}(x)$ and $\hat{g}(x_1)$, are observed in P_3 and P_4 planes. The image $\hat{f}(x)$ can be obtained as a result of the inverse Fourier transformation of the spectrum $F_1(\theta)$ (see Eq. (1) under parameter C=0). As for the image $\hat{g}(x_1)$ one may calculate this field using standard procedure [1] by inverse Fourier transformation of distribution $F_1(\theta)exp(\frac{-j\kappa d\theta^2}{2})$.

Two cases of imaging the 3D edge were studied: ideal (aberration free and diffraction unlimited) and diffraction-limited. In first case for the output fields $\hat{f}(x)$ and $\hat{g}(x_1)$ in the planes P_3 and P_4 we have the following:

$$\hat{f}(x) = 0.5Y(x) + \frac{j}{2\pi x}Y(x)\otimes exp(\frac{-jkx^2}{2d}) + 0.5\tilde{Y}^*(x)$$

$$\hat{g}(x_1) = \tilde{Y}(x_1)\cdot Y(x_1), \qquad (5)$$

where $\tilde{Y}^*(z)$ is the function which is complex conjugated to $\tilde{Y}(z)$. It can be seen that the distribution $\hat{f}(x)$ consists of three components. The first and second ones correspond to the focused and blurred images of the front face, while the third component describes the defocused image of the back face.

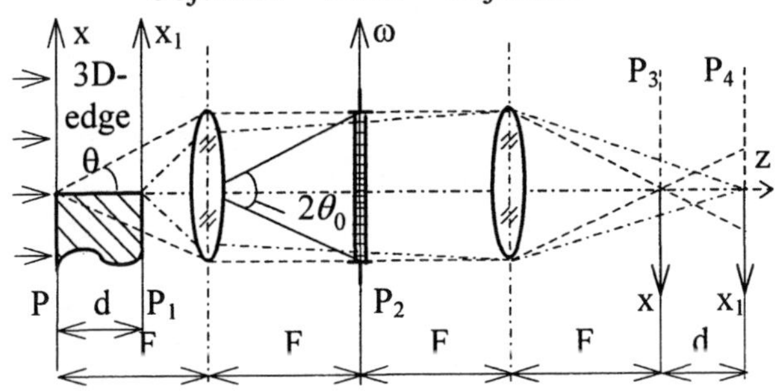

Figure 5. Coherent optical scheme of projecting /spatial filtering system.

As for diffraction-limited imaging we have investigated two cases when 3D effects are weakly expressed and very strong. These effects correspond to the following conditions: $N = \theta_{cr}/\theta_0 >>1$ and $N << 1$, where $2\theta_0$ is angular size of aperture diaphragm (filter) [Fig. 5].

It is shown that for weak 3D-effects $(N >>1)$ the output fields are approximately equal:

$$\hat{f}_{\omega_0}(x) \approx \hat{g}_{\omega_0}(x) = \hat{f}(x)\otimes h(x) = U_{\omega_0}(x)$$

$$- e^{j\pi/4}\sqrt{\lambda d}\, sin(\omega_0 x)/(2\pi^2 x), \qquad (6)$$

where $U_{\omega_0}(x) = Y(x)\otimes sin(\omega_0 x)/(\pi x)$ is the classical expression for diffraction-limited image of the one dimensional edge, $\omega_0 = k\theta_0$. If we fix the threshold at the level 0.25 (the case for plane objects), then for $N >>1$ the error of finding the edge position depends on the Fresnel zone size only: $\varepsilon = \sqrt{\lambda d}/(2\sqrt{2\pi}) \approx 0.13\sqrt{\lambda d}$. For example, assuming $d=100\ \mu m$ and $\lambda = 0.63\cdot 10^{-3}$ mm error of position ε is $0.9\ \mu m$.

4. Imaging the 3D perfectly reflecting edge

Now set into the system a typical element of volumetric metal objects – 3D *perfectly reflecting symmetrical edge* of thickness *d* with an inner plane surface parallel to optics axis [Fig. 4(a)]. The fields in the planes P_3 and P_4 for ideal imaging are the following:

$$\hat{f}(x) = Y(x) + j\left[Y(x)\otimes \exp(-jkx^2/2d)/\pi x\right],$$

$$\hat{g}(x_1) = \tilde{sign}(x_1)\cdot Y(x_1), \qquad (7)$$

where $\tilde{sign}(x) = 2\tilde{Y}(x) - 1$.

It is essential that consequence of the phase change of the diffracted reflected wave onto reverse one, the component corresponding to the secondary light diffraction on the back face is absent. Using (7) the expressions for the output fields were derived for *diffraction-limited* case. It is shown that the error of finding the edge position is twice as large as in the 3D perfectly absorbing edge, i.e. $\varepsilon = \sqrt{\lambda d}/(\sqrt{2\pi})$.

5. High-Frequency Filtering the 3D Edge Image

Specific features of optical processing of these objects were studied on the example of high-frequency filtering (contouring) of volumetric symmetric edge by square filter $H(\omega) = (\omega/\omega_0)^2 Rect(\omega/2\omega_0)$, which corresponds to the Laplace operator [1, 8].

As is known, the filter action reduces to calculation of the second derivative of the input distribution. Being one of the most important in Fourier-optical measurement systems, contouring operation leads to selection of the informative regions which correspond to object boundaries. In so doing, in case of contouring of a plane edge, a central part of its contouring image has a shape of two nearly located maxima separated by the principle minimum [2]. Its coordinate strictly corresponds to geometrical position of the object edge.

In case of contouring of volumetric bodies, another situation is observed. Influence of extension brings about to the shift of the principle minimum of the contoured image of extended edge and energy pumping from one maximum into another till one of them disappears. In case of the 3D *perfectly absorbing object* with slight volumetric effects (the depth of focusing of the system is significantly larger than d), the fields in the planes of face images of the diffraction-limited optical system differ weakly from each other and have the form:

$$\hat{f}(\bar{x}) \approx \hat{g}(\bar{x}) = \frac{\bar{x}\cos\bar{x} - \sin\bar{x}}{\bar{x}^2} - \frac{e^{j\pi/4}}{\pi^2 N}\frac{2\sin\bar{x} - 2\bar{x}\cos\bar{x} - \bar{x}^2\sin\bar{x}^2}{\bar{x}^3}, \quad (8)$$

where $\bar{x} = \omega_0 x$. Position of the main minimum of the function $I(\bar{x}) = |g(\bar{x})|^2$ is displaced with respect to that one of the geometrical edge of the object by the value $\varepsilon = 0{,}25\sqrt{\lambda d}$. In this case, the light intensity in this point differs from zero and equals to $I(\varepsilon)/I_{max}\ (d=0) = 0{,}55\ N^{-2}$. Diagrams of the function $I(\bar{x})$ for different values $N >> 1$: $N = \infty(a)$, $N = 16(b)$, $N = 8(c)$, are presented as an example in Fig. 6. Energy transfer from one maximum into another is observed along with the shift of the position of main minimum and the contrast fall.

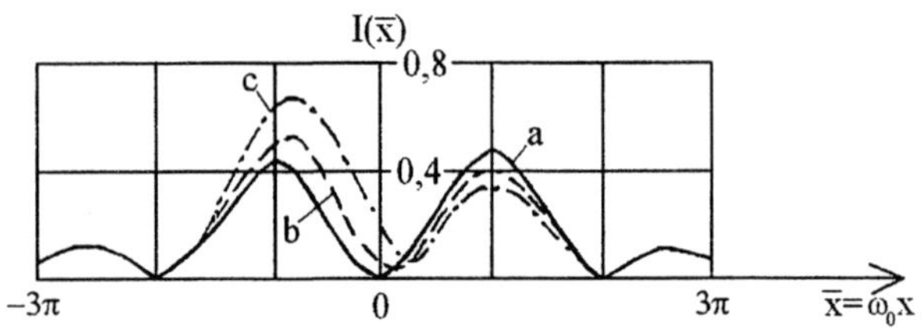

Figure 6. The intensity distribution in the contoured image of 3D *perfectly absorbing edge* for different values $N >> 1$: $N = \infty(a)$, $N = 16(b)$, $N = 8(c)$.

In the case of significant volumetric effects ($N << 1$) for the fields in the 3D edge images of the front and back faces, one can obtain:

$$\hat{f}(\bar{x}) = \frac{e^{j\pi/2}}{\pi}\left[\frac{2\sin(\bar{x}/2)}{\bar{x}^2} - \frac{e^{j\bar{x}/2}}{2\bar{x}}\right] + \frac{N^2}{8\pi^2} \quad (9a)$$

$$\hat{g}(\bar{x}) = \frac{1}{2\pi}\left[\frac{\bar{x}\cos\bar{x} - \sin\bar{x}}{\bar{x}^2} + e^{-j\pi/4}\frac{N}{2\pi^2}\frac{\sin\bar{x}}{\bar{x}}\right] \quad (9b)$$

It is evident that the profiles of the contours of the front and back faces of the volumetric edge ($N << 1$) differ significantly from each other. While the distribution $|\hat{f}(\bar{x})|^2$, (due to the filtering properties of the front face) has the form of the single maximum [Fig. 7], the profile of the contoured image of the back face is similar to profile of contoured image of the plane edge consisting of two main maxima with the dip in the middle and with numerous lateral lobes. In so doing, the observed shift of the central minimum is $\varepsilon = 3N/(\pi\sqrt{2}\omega_0)$ and its intensity is $I(\varepsilon) = 0.5N^2$ that, for example at $d=1$ mm, $N=0.3$, is equal to 0.3 μm and 5.5 % respectively. As for the light intensities of the main maxima of distributions, they are 4 times smaller than those of the contoured plane edge.

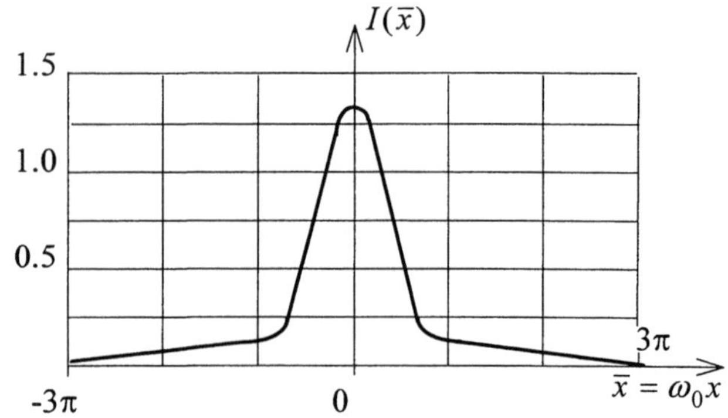

Figure 7. Light intensity distribution in the contoured image of the front face of the 3D *perfectly absorbing edge* for $N=0$.

Similar calculations are carried out for 3D edge with *perfectly reflecting* inner surface. It is shown, that for insignificant volumetric effects ($N >> 1$) the structure of the contoured images of the front and back faces is mostly identical to the case of the *perfectly absorbing edge* considered earlier. The difference is in the fact that the shift of the basic central minima of the contoured image of the edge with respect to the geometrical edge of the object is twice larger. It is explained by the absence of secondary light diffraction on the back face of *perfectly reflecting extended* object. On this reason, for significant volumetric effects ($N << 1$), the contoured image of the back face has a form of a single maximum (instead of double one for the perfectly absorbing case). As for the field in the contoured image of the front face, it coincides (within insignificant coefficient) with the field (9a).

6. Results of the Optical Experiments

In our experiments a steel gauge with extent d=1 mm served as tested 3D edge. This test was illuminated by He-Ne laser. The focal length of the Fourier objective was 180 mm. In order to vary parameter N in a wide range, a set of binary spatial filters was used like opaque circle screen of radius R_0 (removing low frequencies in the spectrum) and of the iris diaphragm of radius R_1 (cutting high frequencies), so ratio R_1 / R_0 was constant. Values of R_0 were changed from 1 to 10 mm ($N = 0.27 \div 2.7$). The observed images were recorded by the photo-diode line sensor.

Fig. 8 demonstrates experimental results of contoured image of the absorbing 3D edge (front face) under weak volumetricity (*N=1.6*). For this purpose the inner surface of object was covered with soot from burned camphor.

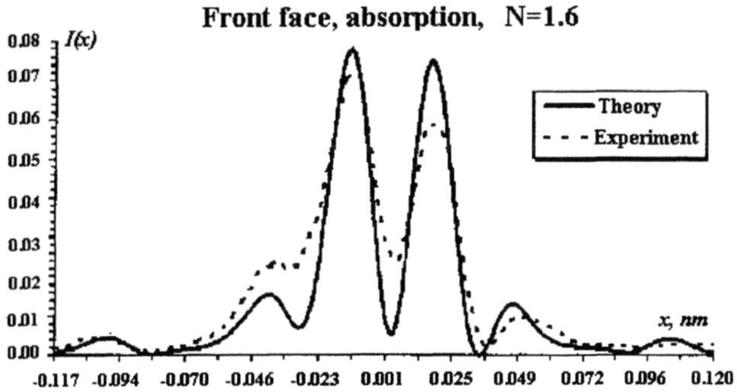

Figure 8. Experimental results of the high frequency filtering the *absorbing extended edge* image under weak 3D effects: front face, d=1 mm, *N=1.6*.

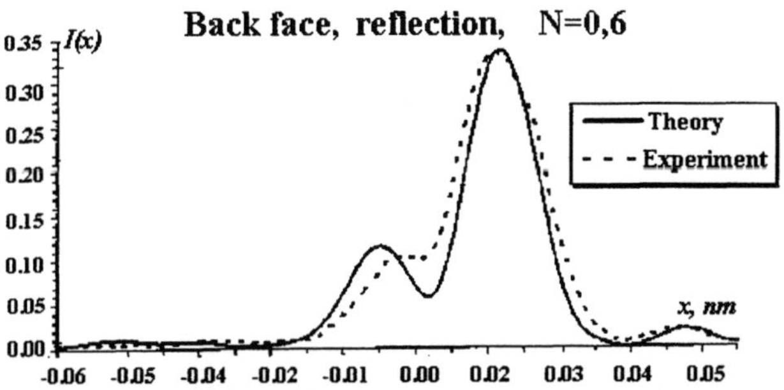

Fig. 9. Comparison of the theory and experimental data for the *contoured reflecting 3D edge* image under strong volumetric effects: back face, d=1 mm, *N=0.6*.

The satisfactory agreement with experiment was observed. Difference of the intensity of the basic maxima for the absorbing edge is caused by incomplete absorption of the object surface.

Fig. 9 shows filtered image of the reflecting 3D edge (back face) under strong 3D effects (*N=0.6*). For the reflecting edge the coincidence is found to a greater degree.

As a whole there is an accordance between theoretical calculations and experimental data.

7. Conclusion

Results on formation of the shadow and contoured images of the typical extended object as 3D edge perfectly absorbing and reflecting in ideal and diffraction-limited optical systems are presented in this report. These are based on the model of equivalent diaphragms according to which the problem of light diffraction on 3D bodies is reduced to more simple case: to the calculation of diffraction phenomena on the plane objects (zero thickness) located in space in Kirchhoff-Fresnel approximation.

It is shown, that under dimensional measurements, the use of the shadow model for the extended objects (even in case of weak volume) can bring about to the measurement errors which for example at small 3D-effects is proportional to the size of the Fresnel zone. In case of significant volumetric effects, the structure of the fields in the shadow and contoured images of the 3D edge sharply differs from the plane edge.

Peculiarities of coherent optical processing of 3D objects are investigated in detail by the example of high-frequency filtering (contouring) of volumetric edges with perfectly absorbing and reflecting inner surfaces. It is shown when volumetric effects are slight the structure of the contoured images of the front and back faces of 3D edge mostly reminds the profile of the plane edge. In this case the influence of extension brings about to the shift of the position of the basic minimum of contoured image proportional to the size of the Fresnel zone. Under appreciable 3D effects the image profile of the front face has a single maximum (instead of double one). Here the structure of the contoured image of the back face depends on reflectance of the inner surface changing from the form with the double maximum (perfectly absorbing) to the form with single one (perfectly reflecting case).

The obtained results can be generalized in case of objects of more complicated form and can be applied first of all for 3D microscopy, Fourier optics of 3D objects, optical metrology and industrial inspection.

References

1. J.W. Goodman. Introduction to Fourier Optics (McGraw-Hill, New York, 1968).
2. Birch, "A spatial frequency filter to remove zero frequency", Optica Acta, 15 (2) 113-119 (1968).
3. M. Born, E. Wolf. Principles of Optics (Pergamon Press, New York, 1964).
4. H. Honl, M. Maul, K. Westpfal. Theory der Beuding (Springer-Verlag, Berlin, 1961).
5. J.B. Keller, "Geometrical theory of diffraction", J. Opt. Soc. Am. 52 116-130 (1961).
6. Yu.V. Chugui, V.P. Koronkevich, B.E. Krivenkov, S.V. Mikhlyaev, "Quasi-geometrical method for Fraunhofer diffraction calculation for three-dimensional bodies", J. Opt. Soc. Am. 71 (4) 483-489 (1981).
7. Yu.V. Chugui, B.E. Krivenkov, "Fraunhofer diffraction by volumetric bodies of constant thickness", J. Opt. Soc. Am. **A** 6 (5) 617-626 (1989).
8. Papoulis, Systems and Transforms with Application in Optics (McGraw-Hill, New York, 1968).
9. Yu.V. Chugui, "Constructive theory of formation and filtering the optical images and Fraunhofer diffraction patterns of 3D opaque object of constant thickness in coherent light", in Proceedings of the SPIE. Three Dimensional Microscopy: Image Acquisition and Processing 3, 2655 (1996).

High Flow Velocity Imaging using Color Doppler Optical Coherence Tomography

Ton G. van Leeuwen, Manish D. Kulkarni, Siavash Yazdanfar, Andrew M. Rollins, Joseph A. Izatt
Departments of Biomedical Engineering and Medicine, Case Western Reserve University.
Address: University Hospitals of Cleveland, Division of Gastro-Enterology, 11100 Euclid Ave, Cleveland, Ohio 44106-5066. Tel: 216 844 3298 Fax: 216 844 8011

We and others have recently described a novel optical Doppler flow imaging technique, termed color Doppler optical coherence tomography (CDOCT), which is capable of micron-scale spatial resolution, two-dimensional tomographic velocity mapping in turbid media[1,2]. CDOCT was recently applied for blood flow mapping in sub-surface vessels in living tissues. The technique is an extension of optical coherence tomography (OCT). In OCT, longitudinal ranging measurements (A-scans) are performed using a fiber-optically integrated scanning white-light Michelson interferometer with resolution determined by the source coherence length. Cross-sectional images are captured by performing sequential ranging measurements while scanning the probe beam laterally across the sample (B-scan).

Doppler flow monitoring using OCT is based on the principle that Doppler shifts f_s in light backscattered from moving objects in a biological sample (e.g., flowing erythrocytes in blood vessels) either add to or subtract from the fixed Doppler frequency f_r associated with the moving reference mirror of the interferometer, depending upon the flow direction. For a scanning interferometer, $f_r=2V_r \bullet \nu_0/c$, where V_r is the reference mirror speed, ν_0 is the source center frequency, and c is the light velocity. Correspondingly, a scatterer in the sample moving with a velocity V_s induces a Doppler shift in the sample arm light by the frequency $f_s=2V_s\cos\theta \bullet n_t\nu_0/c$, where θ is the angle between the sample probe beam and the direction of motion of the scatterer, and n_t is the local tissue refractive index.

Each scatterer with velocity Vs in the sample contributes a time-dependent interferometric detector current of the form:

$$\tilde{\imath}_d(t) = A(t)\cos[2\pi(f_r - f_s)t + \varphi(t)] \qquad (1)$$

Here A(t) is the time-dependent amplitude of the detected signal, and φ(t) is the detector current phase term. Coherent (i.e., phase sensitive) demodulation of the detector current at a given f_{demod} is performed by multiplying $\tilde{\imath}_d(t)$ with $\cos(2\pi f_{demod})$ and subsequently low-pass filtering the product. Coherent demodulation of the detector current at frequency f_r provides the complex envelope of the interferometric signal given by

$$i_d(t) = A(t)\exp\{-j[2\pi f_s t + \varphi(t)]\} \qquad (2)$$

The optimal demodulation filter bandwidth B for high SNR is approximately given by $2(\Delta\lambda/\lambda)f_{demod}$, with $\Delta\lambda$ the bandwidth of the lightsource, and λ the center wavelength.[3] This enables tissue and flow imaging with high resolution (<15 μm in three dimensions) and high dynamic range (>100 dB). However, if the Doppler induced frequency shift f_s by the moving object is larger than the demodulation filter bandwidth B, the amplitude of the demodulated signal severely decreases, and flow imaging is not feasible. Typically the bandwidth of the demodulator is around 3-5 kHz, which reduces the measurable flow range to a few tens of mm/s (depending on the angle θ). In arteries and veins, peak flow velocities can easily reach hundreds of mm/s. In those vessels, blood will not be visible, and its flow cannot be estimated.

In this paper we describe two new methods to obtain high flow velocity images without compromising the high dynamic range of OCT. These methods included both analog and digital (software) demodulation of the detector current $\tilde{\imath}_d(t)$ at multiple frequencies corresponding to velocities of interest in the sample. Each demodulation at a frequency $f_{demod} = f_r + f_x$ and subsequent low-pass filtering will result in a signal given by

$$\tilde{i}_d(t) = A(t)exp\{-j[2\pi (f_s - f_x)t + \varphi(t)]\} \quad (2)$$

which reflects the signal from particles with velocities

$$V_x = (f_x \pm B)c/(2\cos\theta \bullet n_t v_0). \quad (3)$$

To test the methods in vitro, we used a conventional OCT system with a moving reference arm (V_r= 40 mm/s), and a lightsource with a bandwidth of 50 nm around its center wavelength of 1300 nm. The object was a glass capillary with an inner diameter of 0.97 ± 0.01 mm, which was positioned (in water) at an angle of 72° to the incident object beam. The flowing liquid was 1% intralipid solution, which had a constant flow of 4.0 ± 0.2 ml/min during the experiments. Theoretically, this (laminar) flow had a maximal peak velocity V_s of 180 ± 10 mm/s in the center of the lumen. Thus, for these experimental conditions, f_r was 60 kHz, and the expected frequency shift f_s due to the peak velocity was 118 ± 6 kHz.

For the hardware method, the demodulation was performed by a lock-in amplifier with an avalaible maximum demodulation frequency of 250 kHz. We obtained OCT images (B-scan, 100 • 1024 points = 1.0 • 2.0 mm) of the capillary at 60 kHz demodulation frequency under two conditions: no flow (figure 1A) and 4.0 ml/min flow (figure 1B). In the latter, it is obvious that the information on flow is lost. By additional imaging with demodulation frequencies of 80, 120, 160 and 180 kHz, five bands were captured given by equation 3. Summing the five images results in concentric equi-velocity bands (figure 1C).

In the software method, the photodiode signal was digitized with a frequency of 1.23 MHz. Due to the large amount of memory necessary to capture one A-scan (65536 points), only 50 columns were used to obtain a B-scan of 1.0 • 2.13 mm. Every A-scan of the image was off line demodulated by software at 5 demodulation frequencies (60, 90, 120, 150 and 180 kHz). When the five images are superimposed, a similar result is obtained (figure 1D).

For both methods, as shown in figures 1C and D, the image consists of 5 bands. The outer band is the signal obtained with the demodulation frequency being equal to the Doppler frequency induced the reference arm (60 kHz). This reflection comes from non moving objects, in this case from the capillary wall, and from objects which induce frequency shifts less then the bandwidth of the low-pass filter (2 kHz). The next inner band is obtained with the next higher demodulation frequency, thus imaging particles that induce Doppler shifts corresponding with the difference in demodulation frequency. The middle disc was obtained with a demodulation frequency of 180 kHz and reflects that part of the flow that reaches the peak velocity. This demodulation frequency is 120 kHz shifted compared to f_r, the Doppler frequency of the reference arm, which corresponds to the expected frequency shift of 118 ± 6 kHz due to the theoretically calculated peak velocity of 180 ± 10 mm/s. By implementing a color for each equivelocity signal, a color Doppler OCT image is obtained. Thus, both methods are useful for mapping high flow velocities. However, each method has its advantages and disadvantages.

In conclusion, we reported two new methods for extending CDOCT to high flow rate, which facilitated micron-scale resolution OCT mapping of fast flowing intralipid solutions. This technique may be useful in monitoring blood flow in vivo.

References

1. J. A. Izatt, M. D. Kulkarni, Siavash Yazdanfar, et al. *Opt. Lett.* **22**, 1439-1441 (1997).
2. Z. Chen, T. E. Milner, D. Dave, and J. S. Nelson, *Opt. Lett.* **22**, 64-66 (1997).
3. J. A. Izatt, M. R. Ree, D. Huang et al, *SPIE* **1877**, 136-144 (1993).

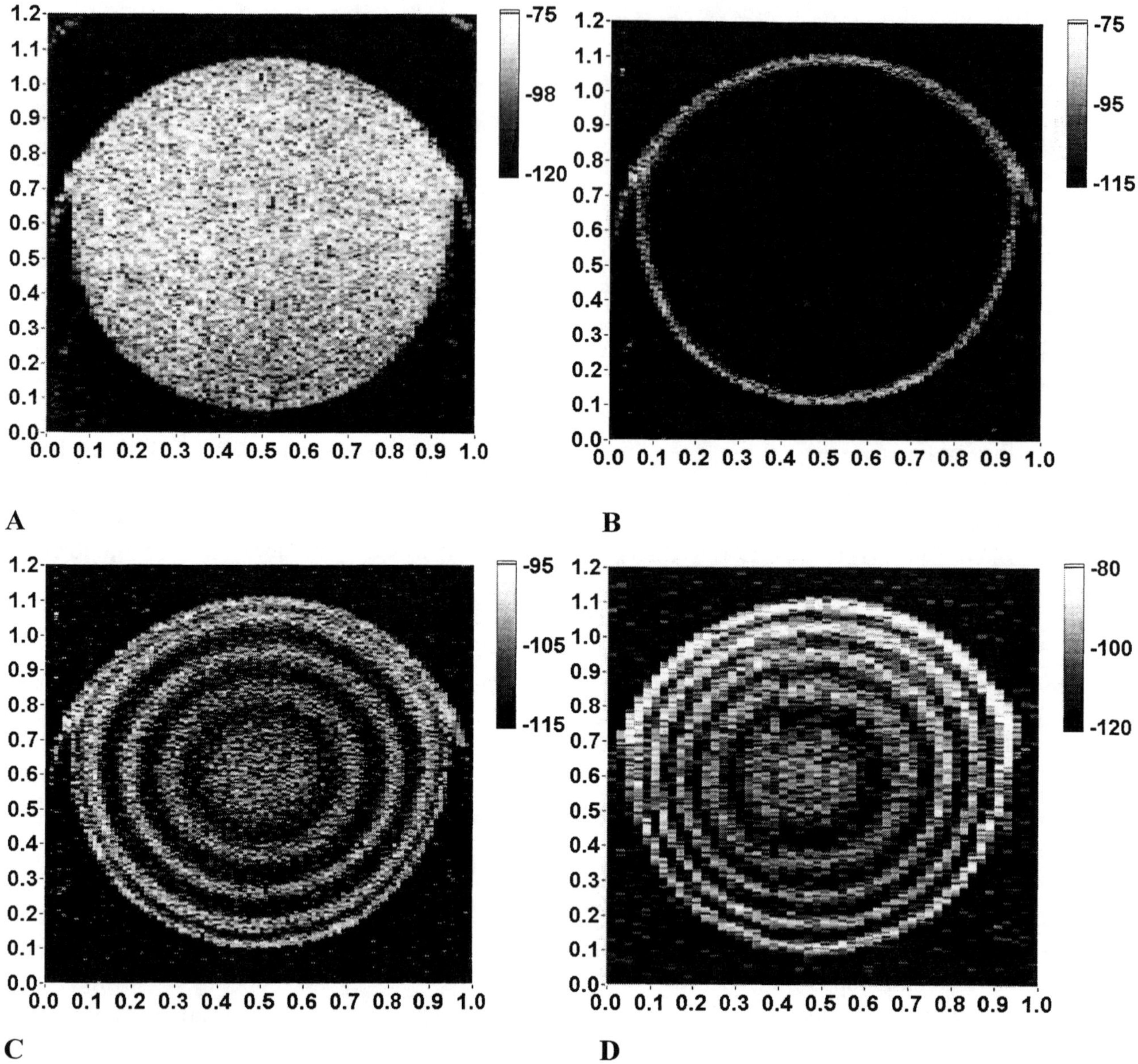

Figure 1. Cross-sectional magnitude-only OCT images of capillary tubes (0.97 mm diameter) filled with 1% intralipid solution.

A: no flow, using the signal demodulated at 60 kHz, B: 4 ml/min. flow, demodulated at 60 kHz only, C: 4 ml/min flow, sum of five separate images obtained at demodulation frequencies of 60, 80, 120, 160 and 180 kHz, D: 4 ml/min flow, image obtained from one undemodulated B-scan, and off-line demodulated (software) at frequencies of 60, 90, 120, 150 and 180 kHz.

Note that the incident beam was at an angle of 72° relative to the flow direction, resulting in an elliptical shaped image. Furthermore, due to internal reflections at the boundary between inner wall and intralipid solution, parts of the lumen are not visualized, which causes the tube to appear smaller than the 0.97 mm inner diameter.

Noninvasive spectroscopic computed tomography of plants *in situ* using near infrared lasers

Y. Sasaki, M. Usa, B. Devaraj, M. Yamada[1], and H. Inaba[1]

Biophotonics Information Laboratories, 2-2-1 Matsuei,
Yamagata 990, Japan.
Tel. +81-236-47-3100, Fax. +81-236-47-3108.

[1]Tohoku Institute of Technology, 35-1 Yagiyama-Kasumicho, Taihaku-ku, Sendai 982, Japan, and Biophotonics Information Laboratories.

Introduction

In recent years, optical imaging methods have shown promise as a noninvasive, relatively safe, portable and inexpensive modality for a wide range of applications in biomedicine. The merits of using light sources for biomedical imaging are well recogonized and the efforts of researchers in this area have been well documented [1,2]. Despite extensive activity in biomedical imaging, the biomedical optics community has not explored possibile of applications of optical imaging methods in plant sciences. Imaging of plant tissues such as leaves, stem and root has demonstrated that valuable information on the physiological aspects of plants can be obtained that have far reaching consequences in agriculture, enviornmental studies and forestry. For example, magnetic resonance imaging (MRI) reported for the quantitative analysis of sucrose content in intact seeds [3], for studying the dynamics of seed germination and growth [4], and for analyzing changes in the physical state of water during cold acclimation and freezing in wheat plants [5].

The tissue structure of plants is however different from those of the animal tissues. Generally speaking, plant cells have a hard and thick cell wall that are not present in animal cells, fiberous/woody tissues provide the skeleton and strength, have a different set of pigments such as chlorophyll, carotenoids and phytochromes, and have specialized vessels to transport water and nutrients. These differences apart, plant tissues are also a heterogenous and highly scattering media with a large water content similar to that of animal tissues. Optical imaging of plants would therefore pose similar problems that are faced in biomedical imaging at the present.

Our group had earlier proposed and demonstrated the feasibility of discriminating and selecting the highly directional coherence retaining photons emerging from tissues by using the optical heterodyne detection based coherent detection imaging (CDI) method [6,7]. In the present paper, we apply the CDI method to obtain the first laser CT images of potted cactus and coconut plants *in situ* , in the near infrared wavelength regions.

Reprinted from Advances in Optical Imaging & Photon Migration,
1998 Technical Digest,

Experimental Methods

A schematic of the experimetal arrangement is shown in Fig. 1. Continuous wave, single frequency, $Ti{:}Al_2O_3$ laser operating at 830 nm and diode-pumped Nd:YAG lasers operating at 1.064 μm and 1.319 μm, respectively, were used in our measurements. The beam was well collimated to 0.7 mm in diameter before being launched into the interferometer. Samples were mounted on a translational-rotational stepper motor stage and placed in the signal arm of the interferometer. The emergent signal beam was then mixed with the local oscillator beam and impinged on a InGaAs photodiode (silicon photodiode for the $Ti{:}Al_2O_3$ laser) to generate an intermediate frequency (i.f.) at 50 kHz, that was fed to fast Fourier transform analyzer interfaced to a computer. Scanning was performed in the T-R mode (similar to that of X-ray CT) during which the i.f. was recorded for each step. Image reconstruction was carried out by the filtered back projection. Our measurement system has a dynamic range of ~140 dB and very low signal power to the order of 10^{-17} W could be detected.

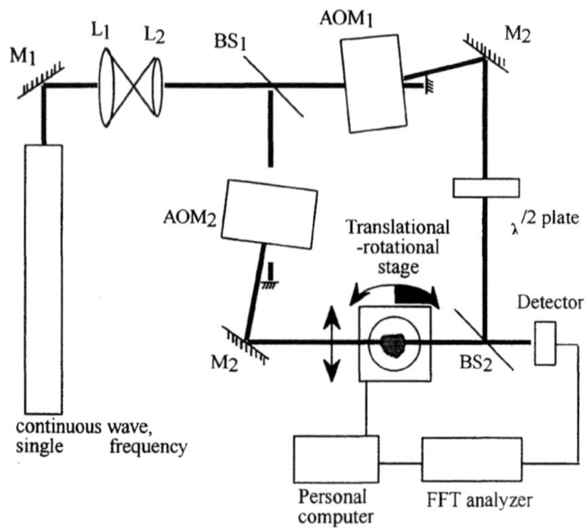

Fig. 1 Schematic of the coherent detection imaging (CDI) system. L1 and L2 are collimating lenses; BS1 and BS2 are beam splitters and AOM1 and AOM2 are acuosto-optic modulators.

Results

In Fig. 2, laser CT images of a cactus plant obtained at 830 nm and 1.319 μm, respectively, are shown for comparison. For reference, photographs of the sample (a) and a section cut along the measurement plane (b) are also displayed. Two kinds of tissues can be clearly differentiated in the laser CT image obtained at 830 nm in Fig. 2(c). The inner most region of soft tissues having lower light attenuation corresponds to the fibrous tissues that provide mechanical strength to the plant, and the outer region corresponds to thick, higher attenuating tissues that prevent water loss and withstand enviornmental stress. In the laser CT image obtained at 1.319 μm shown in Fig. 2(d), the water transporting vessels are clearly differentiated from the surrounding tissues due to higher absorption of water at 1.319 μm when compared to 830 nm.

In Figs. 3(c) and (d), laser CT images of a potted coconut plant obtained at 1.064 μm and 1.319 μm, respectively, are displayed for comparison. The layered internal structure of the stem is clearly differentiated at both these wavelengths. Incident power requirements however, were 1 mW at 1.064 μm and 55 mW at 1.319 μm. Here again the water transporting vessels exhibit higher attenuation and is clearly differentiated, especially at 1.319 μm.

Conclusions

We demonstrate the feasibility of observing the physiological and anatomical aspects of plants *in situ*, by laser CT images, for the first time, using the CDI method. Our results also indicate that functional monitoring of plant structure and physiology by optical imaging methods could find widespread applications in agriculture, horticulture, forestry and enviornmental studies in the near future.

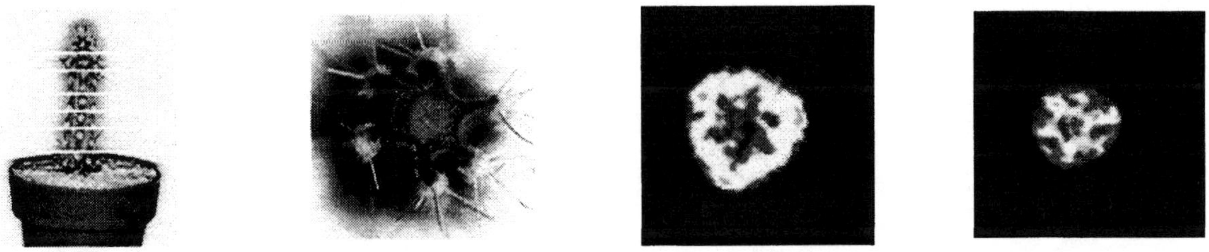

(a) (b) (c) (d)

Fig. 2(a) Photograph of the cactus plant sample, (b) cut section of the sample exposing the measurement palne, (c) laser CT image obtained at 830 nm, and (d) laser CT image obtained at 1.319 μm, respectively.

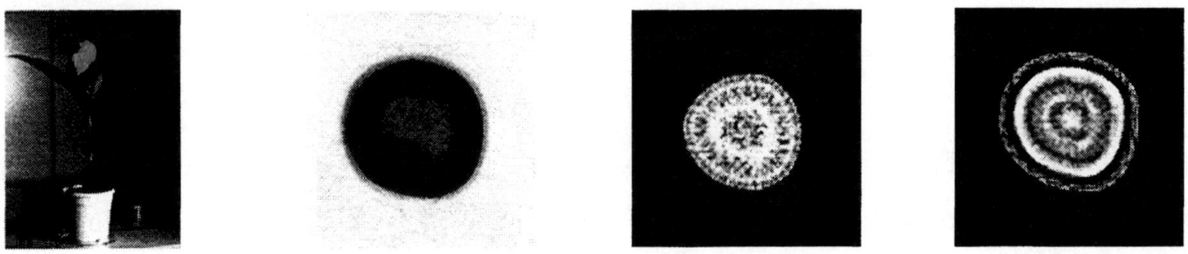

(a) (b) (c) (d)

Fig. 3(a) Photograph of the potted coconut plant sample, (b) cut section of the sample exposing the measurement palne, (c) laser CT image obtained at 1.064 μm, and (d) laser CT image obtained at 1.319 μm, respectively.

References

[1] G. Müller, et al., (eds.), Medical Optical Tomography: Functional Imaging and Monitoring, SPIE Institutes for Advanced Optical Technologies Series, vol. IS 11, SPIE Opt. Eng. Press, Bellingham, WA (1993).
[2] R.R. Alfano and J.G. Fujimoto, (eds). Trends in optics and photonics on advantages in optical imaging and photon migration, vol. 2, Optical Society of America, Washington, DC (1996).
[3] Y.T. Tse et al., Protoplasma, vol. 194, 54 (1996).
[4] P.A. Bottomley et al., Proc. Natl. Acad. Sci., vol. 83, 87 (1986).
[5] M.M. Millard et al., Plant Cell and Enviornment, vol. 18, 535 (1995).
[6] H. Inaba in reference [1], p.317 and reference cited therein.
[7] B. Devaraj et al., IEEE J. of Special Topics in Quantum Electron., vol. 2, 1008 (1996).

Optical tomography by means of a numerical low-coherence holographic technique

Etienne Cuche

Institut d'Optique Appliquée, EPFL, CH-1015 Lausanne, Switzerland
Tel: ++41 21 693 37 42, Fax: ++41 21 693 37 01, E-mail: etienne.cuche@epfl.ch

Patrick Poscio

Institut de Recherches en Ophtalmologie (IRO), CH-1950 Sion 4, Switzerland
Tel: ++41 27 203 59 71, Fax: ++41 27 203 59 70, E-mail: patrick.poscio@iro.vsnet.ch

Christian Depeursinge

Institut d'Optique Appliquée, EPFL, CH-1015 Lausanne, Switzerland
Tel: ++41 21 693 61 77, Fax: ++41 21 693 37 01, E-mail: christian.depeursinge@epfl.ch

Introduction

Standard Optical Coherence Tomography (OCT) setups involves scanning mechanisms with focused beams and point detectors. Such scanning procedures are time consuming and these techniques are thereby limited in the point of view of their imaging capabilities in clinical conditions. A natural way to suppress the lateral scanning consists in recording, on a two-dimensional detector, a hologram resulting from the interference between a low coherence reference wave and the coherent part of the light backscattered by the sample. Such applications of holography in biomedicine have already been proposed about two decades ago but haven't been much developed, essentially because of non-practical aspects of classical holography with photographic plates.

Moreover, one of the greatest technical limitation lies in the choice of a two dimensional detector which must combines the qualities of a high acquisition rate because of intrinsic movements occurring in living tissues and a good sensitivity because of the low intensity of the backscattered light. From this point of view, CCD camera appear as an ideal 2D detectors for biomedical applications and we present here a new low-coherence technique which involves a CCD camera as recording device and a fast numerical method for the reconstruction.

Reprinted from Advances in Optical Imaging & Photon Migration,
1998 Technical Digest, © 1998 Optical Society of America

Material and method

The experimental setup (see Fig. 1) basically consists of a Michelson interferometer with extended beams and a low-coherence light source (a Ti:Al_2O_3 mode-locked laser with 80 fs pulse duration). The light backscattered by the object forms the object wave ***O*** and a reference wave ***R*** is reflected by a mirror M the orientation of which is not perpendicular to incident light in order to perform holography in the *off-axis* geometry. In front of the object, an optional microscope objective MO can be introduced for microscopic investigations.

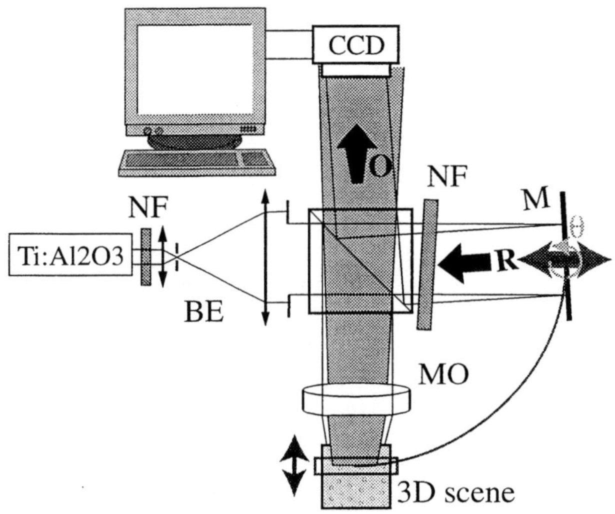

FIG. 1 Experimental setup

The position of the mirror along the axis of the reference arm defines the position of a zone in the depth of the 3D scene from which the emerging backscattered photons are coherent with the reference's photons and participate so to the hologram formation. Thus the tomography of the 3D scene can be performed by a single scan of the mirror along the reference arm axis of the interferometer.

The holograms resulting from the interference between ***R*** and the coherent part of the object wave ***O*** are recorded by a standard black and white CCD camera and digitally transferred to a computer by means of a frame grabber card. The holographic reconstruction is numerically performed by means of a fast discrete Fresnel transform algorithm (computation time: ~ 0.33 sec. for a 256 × 256 hologram).

Typical Result

Figure 2 illustrates an example of optical tomography with a 3D scene consisting in two crossed reflective test target with 200 lines/mm and separated in depth by about 75 μm. As magnification optics, a single lens of 30 mm focal length has been used. The images present only a part of the entire reconstructed plane which contains also a zero order of diffraction and a twin image as usually in *off-axis* holography. The 10 presented images have been numerically reconstructed from a sequence of 10 holograms successively recorded after enhancement with 15 μm steps of the reference's optical path.

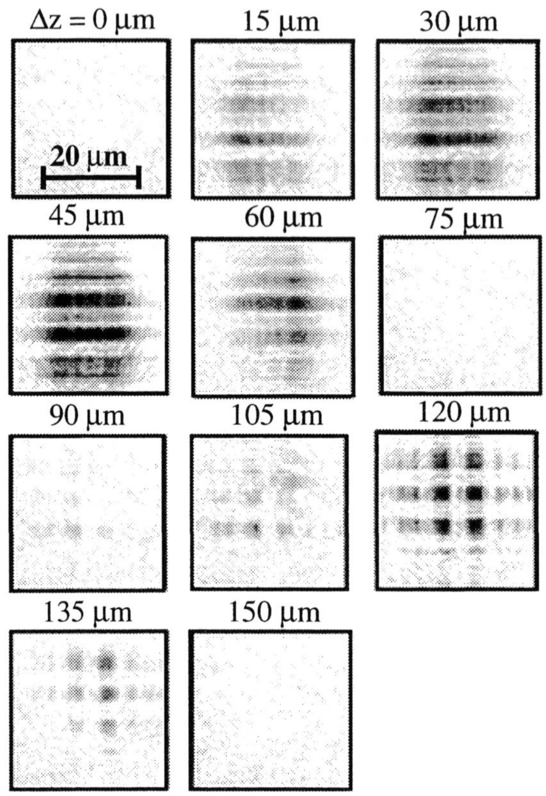

FIG. 2: Example of optical tomography

Conclusion

The numerical low coherence holographic technique allows the tomographic imaging of a 3D scene without lateral scanning. The depth resolution is defined by the coherence length of the source (here ~ 30 μm) and the transverse resolution by the diffraction limit of the magnification optics.This numerical approach of holography presents the advantages of a high acquisition rate (at least 25 images/sec) and a high reconstruction rate (~ 3 images/second actually).

Two and Three Dimensional Imaging of Normal and Osteoarthritic Cartilage Microstructure with Optical Coherence Tomography

Juergen M. Herrmann[1], Costas Pitris[1], Brett E. Bouma[1], Stephen A. Boppart[1], James G. Fujimoto[1] and Mark E. Brezinski[2]

[1]Department of Electrical Engineering and Computer Science , Research Laboratory of Electronics, MIT, Cambridge, MA 02139
[2]Department of Medicine, Harvard Medical School, MGH, Boston, MA 02114

The hallmark of osteoarthritis (OA) is degeneration of cartilage. Recently, multiple investigators have shown that the progression of articular cartilage damage can be modified by medical or surgical intervention. This has resulted in heightened interest in high resolution imaging of the microstructure within joints since the greatest impact on OA occurs at early stages. These microstructural changes offen occur on scales less than 100 μm. To assess these abnormalities, conventional radiography, ultrasound (US), computed tomography (CT), magnetic resonance imaging (MRI) and arthroscopy have been applied; with limited results. The main problem of these imaging methods is that the resolution is too low to detect microstructural changes.

Optical coherence tomography (OCT) has the potential to overcome this limitation and was shown to resolve abnormalities in the microstructure of cartilage(1). The typical axial resolution of OCT is 5-15μm and therefore about 10x greater than that of clinical CT, US or MRI. Optical coherence tomography is similar to ultrasound, but the intensity of backreflected light is measured rather than acoustical waves. Figure 1 shows a schematic of the OCT system. The principles behind OCT have been previously described(2,3). A light beam is focused into the tissue and the echo time delay of the light backscattered from different depths is measured by low coherence interferometry. Interference is observed only when the optical paths of the reflected light from reference and sample arms are matched to within the coherence length of the light source. To produce cross-sectional tomographic images, the light beam is scanned across the tissue. The resulting two or three dimensional data set of the optical reflectance properties of the tissue is represented as a grayscale or false color image. Light sources including a superluminescent diode (SLD) light source and a femtosecond Cr:forsterite laser(4) were used. The median wavelength λ and bandwidth $\Delta\lambda$ of the SLD were 1300nm and 50nm. The laser was operating at 1280nm with bandwidths of 65nm or 130nm. The axial resolution is defined by the coherence length:

$$\Delta Z = \frac{2\ln 2}{\pi}\frac{\lambda^2}{\Delta\lambda}$$

The acquisition rate was about 40 seconds for one cross-sectional image. Recently, a fast OCT system has been demonstrated which acquires images at 4-8 frames/s (5). Three dimensional reconstructions were performed with IPLab Spectrum 3.0 and NIH-image 1.60 on a Power Macintosh 9500/200.

Reprinted from Advances in Optical Imaging & Photon Migration, 1998 Technical Digest, © 1998 Optical Society of America

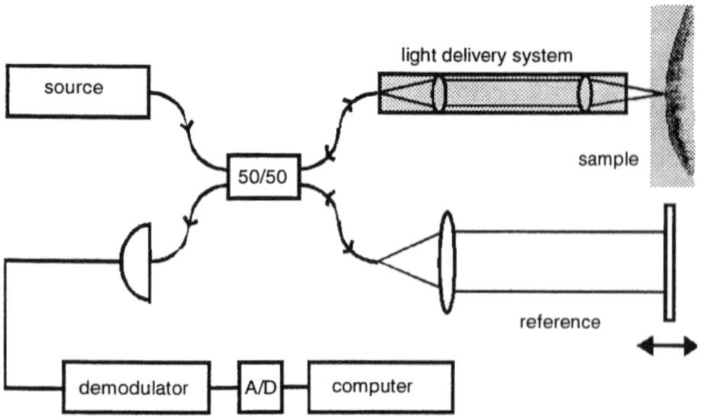

Figure 1 Schematic of the OCT system.

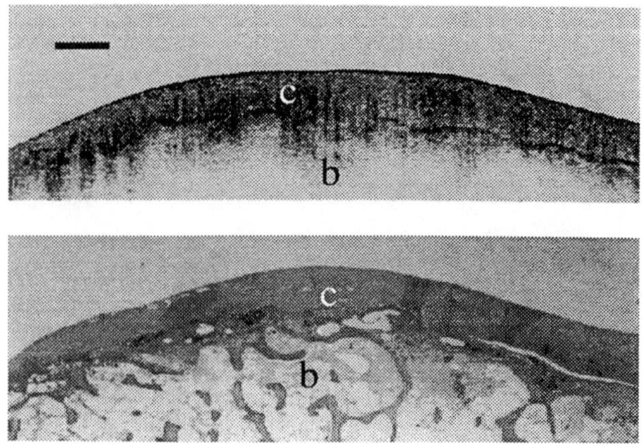

Figure 2 Comparison of OCT image (bar=500μm) of proximal surface of interphalangeal joint of little toe with histology (below).

Sites from knee, ankle, hip, elbow, interphalangeal, metatarsal, intervertebral and clavicular joints were imaged. The articular cartilage was obtained from postmortem examinations or postsurgical limb resections. For imaging, segments smaller that 10cm by 10cm were placed on a x-y-translation stage. Figure 2 shows an OCT image of the proximal surface of an interphalangeal joint of the little toe. The cartilage(c) bone(b) interface is sharply defined. This is confirmed by the corresponding histology. Figure 3 shows a 3D data set of a interphalangeal joint of a second toe under different views. 50 cross-sections were acquired with a lateral spacing of 100μm. The whole scanned area was 5mm x 5mm x 2.5mm (length x width x depth). Figure 4 shows a surface plot of the data.

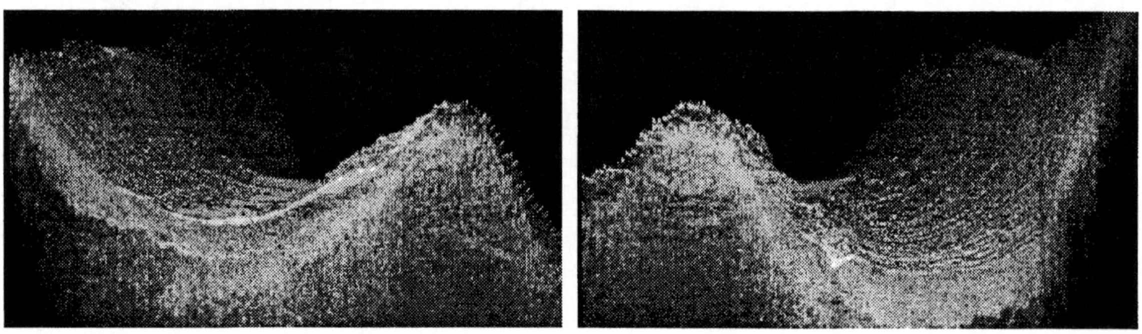

Figure 3 3D reconstruction of toe joint.

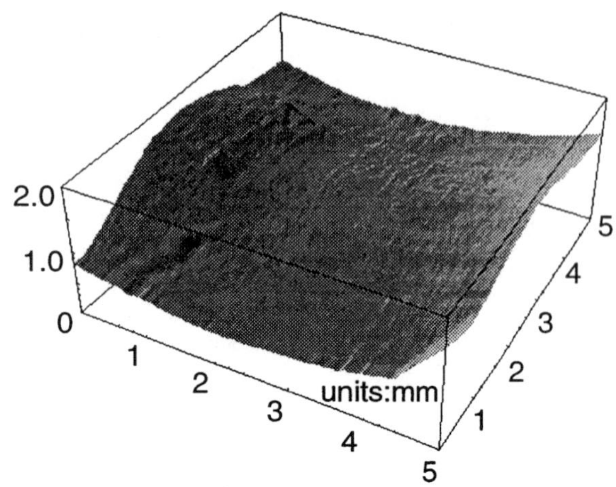

Figure 4 Surface plot of toe joint.

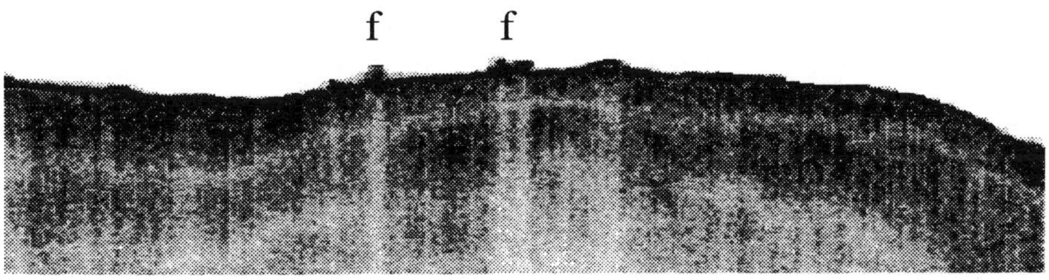

Figure 5 OCT image of diseased part of hip joint.

Figure 5 shows an OCT image of cartilage from a diseased hip joint. At the surface, fibrillations(f) can be identified, where cartilage thinning is noted. The low-signal bands (white) in the image are caused by birefringence of the collagen within the cartilage, changing with the polarization state of the incident light. The OCT images of cartilage are sensitive to the polarization of the incident light. Since the change in the organization of the birefringent collagen fibers is one of the early markers of OA, an important diagnostic role of polarization spectroscopy is suggested.

The results in this work show that OCT can image microstructural details of articular cartilage with high resolution. Abnormalities like cartilage thinning and fibrillation were identified. Several reasons, in addition to the high resolution, make OCT attractive for intraarticular imaging. First, OCT can be performed at high speeds; 4-8 frames per second. Second, the OCT technology is fiber based, allowing easy integration into small arthroscopes. Third, no transducing medium is needed as in US. And fourth, the system is small, compact and low cost. In conclusion a need for high resolution intraarticular imaging exists and OCT, both through microstructural imaging and polarization spectroscopy, represents a promising new technology to perform this imaging task.

Dr. J. Herrmann is visiting from the Physics Institute, University of Erlangen-Nürnberg,. Germany. Dr. Brett Bouma is currently with the Wellman Laboratory, Massachusetts General Hospital. The authors would like to thank Dr. James Southern and Dr. Guillermo Tearney for helpful discussion and comments. We also thank Ms. Jeanine Dudac and Mr. Jason Corbin for their help in the acquisition and handling of tissue.

1 J. M. Herrmann, C. Pitris, B. E. Bouma, S. A. Boppart, J. G. Fujimoto and M. E. Brezinski, High Resolution Imaging of Normal and Osteoarthritic Cartilage with Optical Coherence Tomography, submitted to the Journal of Rheumatology (1997).

2 D. Huang , E. A. Swanson, C. P. Lin, J. S. Schuman, W. G. Stinson, W. Chang, M. R. Hee, T. Flotte, K. Gregory, C.A. Puliafito, J.G. Fujimoto, Optical coherence tomography, Science 254:1178-1181 (1991).

3 M. E. Brezinski, G. J. Tearney, B. E. Bouma, J. A. Izatt, M. R. Hee, E. A. Swanson, J.F. Southern, J. G. Fujimoto, Optical coherence tomography for optical biopsy: Properties and demonstration of vascular pathology, Circ 93:1206-1213 (1996).

4 B. E. Bouma, G. J. Tearney, I. P. Bilinsky, B. Golubovic, and J. G. Fujimoto, Self-phase-modulated Kerr lens modelocked Cr:Forsterite laser source for optical coherence tomography, Opt. Lett. 21, 1839-1841, (1996).

5 G. J. Tearney, M. E. Brezinski, B. E. Bouma, S. A. Boppart, C. Pitris, J. F. Southern, J. G. Fujimoto, In vivo endoscopic optical biopsy with optical coherence tomography, Science 276:2037-2039 (1997).

Optical Low-Coherence Tomography Using an Injection-Seeded Optical Amplifier and a High-Speed A-Scan Rotating Plate Technique

Xiaonan Ma and Chin B. Su
Dept. of Electrical Engineering
Texas A&M University
College Station, Tx 77843
Tel: (409) 845-7584

Lihong Wang
Bioengineering Program
Texas A&M University
College Station, Tx 77843-3120

Optical coherence tomography (OCT) using superluminescent diodes and femtosecond lasers have been used for imaging of highly scattering media[1-5]. However, superluminescent diodes, though compact, provide relatively low optical power output. Additionally, the optical path length of the reference arm must be scanned at high -speed for *in-vivo* measurements. A standard technique for high-speed scanning the optical path length of the reference arm is by stretching an optical fiber coiled around a cylindrical piezoelectric transducer[5].

In this paper, we use a 1.3 μm wavelength semiconductor optical amplifier as the light source for OCT. The optical amplifier amplifies the injected light from a broadband light source operating at low power level to achieve a cw output power of about 15 mW. Since the optical amplifier is seeded by a light source operating at low power, the noise level of the output is significantly reduced. In addition, the displaced spectral peaks of about 30 nm between the two devices provided a broader spectrum for reducing the spatial resolution to about 20 μm despite the high optical power output. For fast longitudinal A-scan, we use a rotating square plate for repetitive scan of the optical path length[6]. This technique provides simplicity, high precision and high repeatability, which are important considerations for producing undistorted images. The formula for the optical path length change Δl as a function of rotation angle θ is given by,

$$\Delta l = d[\sqrt{n^2 - \sin^2(\theta)} + 2\sin^2(\frac{\theta}{2}) - n] \qquad (1)$$

Where d is the dimension of the square plate and n is the index of the optical plate.

Calculation shows that the maximum Δl achievable is 1.5 mm for a 0.5 inch plate. The path length change Δl, though nonlinear in θ, can be converted to a linear scale by the computer software. A schematic drawing of our OCT system is shown in figure 1.

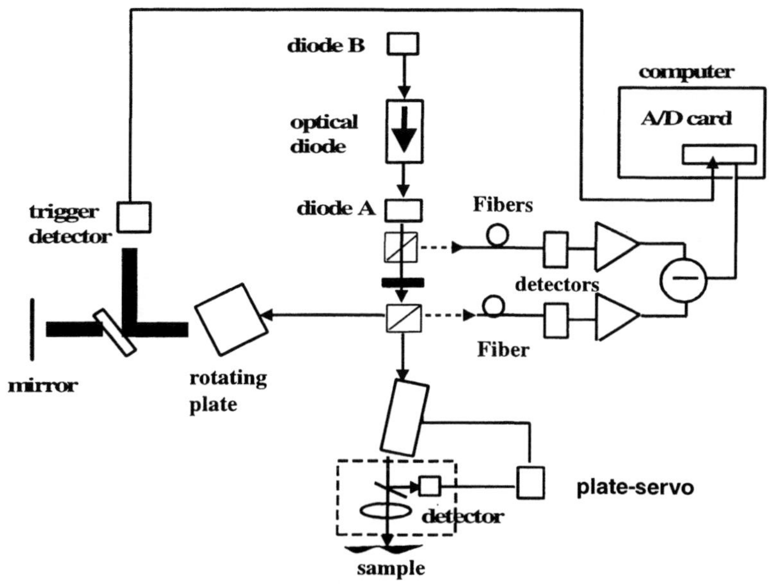

Figure 1. Schematic diagram of the experimental system. Diode A is the optical amplifier pumped by diode B. Whenever the facet of the plate is perpendicular to the reference beam, the trigger detector commands the A/D card to initiate data acquisition.

An important feature is the periodic optical path length scan as the plate rotates. Each rotation period provides a maximum of 8 longitudinal A- scans. In this work, we set the rotation period at 8 millisecond , allowing a maximum of 1 millisecond/scan. Thus, a frame consisting of 500 columns of pixels would take only 0.5 second. The lateral scan is based on a servo technique that locks the displaced signal beam to the center of a lateral moving microscope objective. The parallel displaced signal beam is initiated by a small angle rotation of an elongated plate. This prevents signal fading for smooth surfaces. The optical output is collimated by a miniature grin-lens producing a beam diameter of 0.5 mm. The small beam diameter prevents excess beam spreading near the focal plane while maintaining a spot size of about 8 μm. This feature is important for revealing more features as a function of scan depth. At the detector, the light collecting fibers are single-mode fibers for spatial filtering.

In vivo measurement was performed on the hand region. Figure 2a shows a typical 1 mm x 1 mm cross-sectional image of the fore-thumb.

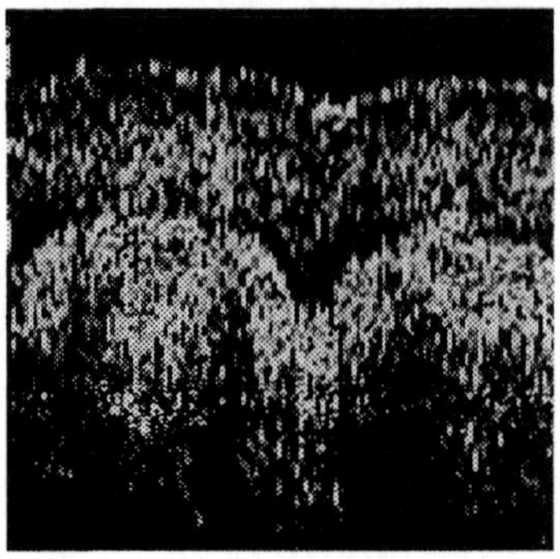

Fig.2a. 1 mm x 1mm image of the fore-thumb. Signals are shown as white.

Fig.2b. 120um x 70 um image of a dermis region. The feature may have cellular origin.

The surface of the skin is shown at the top. The stratum corneum is just below, followed by the interface between the stratum corneum and the malpighii (shown in white). The strongest signal occurs at the malpighii - dermis interface. The dent shown in the figure is likely due to the finger print mark of the thumb region. Figure 2b is a 120 μm x 70 μm expanded image of a dermal region. We speculate that the structural details observed may be of cellular origin as the sizes of the structures are about 10 ~ 20 μm.

In conclusion, an injection -seeded semiconductor optical amplifier with significantly improved power, noise and bandwidth performances was used as a light source for OCT. A new A-scan technique based on the reference beam traversing a rotating optical plate was used for achieving high-speed optical path length scan. *In vivo* measurements of the hand region yielded clear images of the stratum corneum, the malpighii, and the dermis. Structures, possibly cellular in origin, were also observed.

References

1. E. A. Swanson, D. Huang, M.R. Hee, J.G. Fujimoto, G.P. Lin, and C.A. Puliafito, " High-speed optical coherence domain reflectometry" , Opt. Lett., 17, 151 (1992).
2. J.A. Izatt, M.R. Hee, E.A. Swanson, C.P. Lin, D. Huang, J.S. Schuman, C.A., Puliafito, and J.G. Fujimoto, Arch. Ophthalmol., 112, 1584 (1994).
3. B. Bouma, G. Tearney, S.A. Boppart, M.R. Hee, M.E. Brezinski, J. G. Fujimoto, " High-resolution optical coherence tomography imaging using a mode-locked Ti:Al_2O_3 laser source", Opt. Lett., 20, 1486 (1995).
4. Z. Chen, T. Milner, S. Srinivas, X. Wang, A. Malekafzali, M. van Gemert, and J. Nelson, " Noninvasive imaging of *in-vivo* blood flow velocity using optical Doppler tomography", Opt. Lett., 22, 1119 (1997).
5. G.J. Tearney, B.E. Bouma, S.A. Boppart, B. Golubovic, E.A. Swanson, and J.G. Fujimoto, " Rapid acquisition of in vivo biological images by use of optical coherence tomography", Opt. Lett., 21, 1408 (1996).
6. Chin B. Su, " Achieving variation of the optical path length by a few millimeters at millisecond rates for imaging of turbid media and optical interferometry" , Optics Lett., 22, 1119 (1997).

Imaging and microscopy including nonlinear optics and light/sound interaction

High resolution Real-Time 3-D Imaging through Turbid Media using Photorefractive Holography

R. Jones, M. Tziraki, N. P. Barry, S. C. W. Hyde, D. Parsons-Karavassilis and P. M. W. French,
Femtosecond Optics Group, Physics Department,
Imperial College of Science, Technology and Medicine, Prince Consort Road, London SW7 2BZ, U.K.
Tel : 0171-594-7788, Fax : 0171-594-7782, E-mail : paul.french@ic.ac.uk

K. M. Kwolek and D. D. Nolte
Department of Physics, Purdue University, West Lafayette, Indiana 47907-1396

M. R. Melloch
School of Engineering, Purdue University, West Lafayette, Indiana 47907-1396

Abstract

A real time, high-resolution coherence gating system for imaging through turbid media has been demonstrated, which exploits photorefractive holography. The fast response time (<0.4msec) and the high optical quality of photorefractive GaAs/AlGaAs MQW devices allows images to be acquired, in real-time, direct to a video cassette recorder. The spectral versatility of this system has been demonstrated by recording depth-resolved images at wavelengths from 800-850 nm.

KEYWORDS: photorefractive holography; scattering; Multiple Quantum Well, coherence imaging

Introduction

Near infrared radiation offers the potential for medical diagnostic and functional imaging in biological tissue but unfortunately such imaging usually suffers severe degradation due to multiple scattering. Much research has focused on methods to form images through thin tissue depths using ballistic (unscattered) light, which is normally obscured by the scattered diffuse background. The work reported here has been motivated by a desire to develop imaging systems for "in vivo" applications such as screening for skin cancer, for which rapid image acquisition time is a priority. To this end we have concentrated on developing "whole-field" imaging techniques, which acquire all the pixel information in parallel, while accepting compromises in achievable resolution and sensitivity compared to confocal scanning techniques. The imaging technique discussed here uses holography in photorefractive media [1] as a means of discriminating in favour of the unscattered light that retains coherence with a reference beam. This is related to coherence gating techniques previously reported by various groups including optical coherence tomography [2] as well as to conventional, electronic and light-in-flight holography [3]. It fundamentally differs from other techniques, however, in that this coherent detection system is, to first order, insensitive to the incoherent diffuse light background. Images may be acquired using back-scattered light in a time-of-flight reflection geometry, exploiting low coherence interferometry to reconstruct 3-D images. The use of reusable photorefractive holographic media, such as bulk Rhodium-doped barium titanate (Rh:$BaTiO_3$) crystals or photorefractive multiple quantum well (MQW) devices allows for fast holographic image recording and the potential for real-time read-out.

Photorefractive holography with Rh:$BaTiO_3$

We first demonstrated depth-resolved photorefractive holography using the apparatus shown in Figure 1, with a mode-locked Ti:sapphire laser producing 3 ps, or later 100 fs [4] pulses used as the source. The holographic recording medium was a crystal of Rhodium-doped Barium Titanate (Rh:$BaTiO_3$) which permits the hologram to be read out by a Bragg-matched beam and recorded onto a CCD camera. As with any holographic imaging system, holograms could only be written when the interferometer arm lengths were matched to within the coherence length of the source (~ 30 µm). Using a short coherence length source therefore provided depth resolution, which we demonstrated using a 3-D test object consisting of a series of concentric aluminium cylinders, ranging from 1 to 5 mm in diameter and separated in depth by 100 μm. This object was unpolished and provided no significant specular reflection, but back-scattered ~ 8 % of the incident light. By varying the delay of the reference arm, a set of depth-resolved images sufficient to reconstruct a 3-D image of the object could be obtained. The depth resolution of the images was 100 μm, while the transverse spatial resolution was approximately 30 μm as predicted from the pulse duration and beam geometry [5].

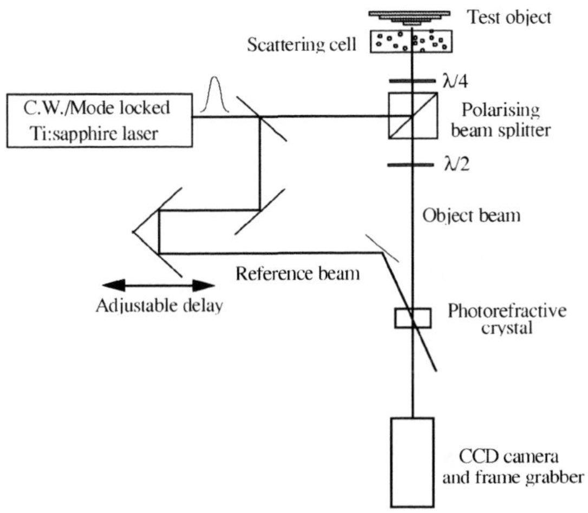

Figure 1: Experimental configuration for holographic imaging through a scattering medium using a bulk Rh:BaTiO$_3$ crystal

Figure 2 shows reconstructed images of the 3-D test object recorded through an aqueous suspension of polystyrene spheres of (a) 0, (b) 12 and (c) 14 MFP scattering depth. It can be seen that while the signal to noise ratio of the image decreases with scattering depth, the spatial resolution does not. Using a USAF test chart as the object, we obtained an image of 100 μm bars through 16 MFP of scattering solution. The minimum intensity required in the object beam to record a usable hologram was 10 nW/cm^2 (for 300 seconds integration).

Figure 2: Computer reconstructions of 3-D objects using depth resolved holographic images recorded in Rh:BaTiO$_3$. Images recorded through scattering medium of (a) 0 (b) 12 and, (c) 14 MFP scattering depths.

The discrimination against a diffuse background noise signal was investigated quantitatively by simulating an incoherent scattered light background using a third beam for which 100 fs pulses were delayed by more than their coherence length, relative to the signal and reference beam pulses. It was possible to record a useful hologram in the presence of an incoherent background which was 10^6 times more powerful than the signal beam [6]. This dynamic range exceeds that of many detectors and was ultimately limited by scattering off inhomogeneities in the crystal and by beam fanning. Unfortunately, the long response time of Rh:BaTiO$_3$ means that image acquisition times as high as 100's of seconds must be used when imaging through strong scattering media and this is clearly unacceptable for most biomedical applications. There is therefore a requirement to use photorefractive media with a much faster response. Two such media are bulk semiconductors, such as Cadmium Telluride (CdTe), and photorefractive GaAs/AlGaAs multiple quantum well (MQW) devices. The latter have been the subject of our recent research, which is described in the next section.

Photorefractive holography with GaAs/AlGaAs MQW Devices

The strong, narrow excitonic absorption features of MQW's lead to a large optical non-linearity [7] when an electric field is applied, parallel or perpendicular to the wells, this is due to the Franz-Keldysh or the quantum confined Stark effects respectively. Our photorefractive MQW devices were grown at Purdue University as in reference [8] and are used in the Franz-Keldysh geometry. The high carrier mobility's result in very fast response times and we have measured the response time of these devices while imaging through scattering media to be faster than 0.4 ms [8]. This permits images to be captured much faster than standard video rate (30 frames/sec) and so images can be directly recorded on a conventional video cassette recorder which we have demonstrated.

The photorefractive MQW devices were grown in a Varian GEN-II molecular beam epitaxial chamber on a GaAs substrate. The STG 10% Al barrier structure, grown at 600 °C, consists of an $Al_{0.5}Ga_{0.5}As$ stop-etch layer beneath a 1.5 μm 100 Å $Al_{0.1}Ga_{0.9}As$/75 Å GaAs MQW region sandwiched between $Al_{0.1}Ga_{0.9}As$ buffer layers. Deep defects were introduced by proton implantation at a double dose of 10^{12}/cm^2 at 160 keV and 5 x 10^{11} /cm^2 at 80 keV to make the device semi-insulating throughout the MQW region. The defects provide traps for photorefractive space-charge gratings. The MQW surface was protected with wax and lift-off of the film was performed using a highly selective (>10^8) 12 % HF etch to dissolve the AlGaAs layer. The sample was then van der Waals bonded to a glass substrate. Two gold contacts were evaporated on the top of the sample to apply a transverse electric field parallel to the quantum-well layers. None of these processing steps are considered exotic and these devices can potentially be manufactured at low cost.

In this work the MQW were used in the transverse Franz-Keldysh geometry, with a large applied field of the orders of 10's of kV/cm applied in the plane of the quantum wells, parallel to the optical vector being written. During photorefractive holographic recording, carriers are produced at the bright regions of the intensity gratings and drift to shield the applied field in these regions. The Franz-Keldysh effect is the broadening of the exciton peak via field-ionisation, which is due to the large applied field. In the bright regions where the carriers have shielded the applied field, the absorption is not effected. However in the dark regions, where the field is still strong, the absorption decreases at the exciton peak. Therefore the optical interference pattern is transferred to a spatial modulation of absorption and, therefore, refractive index

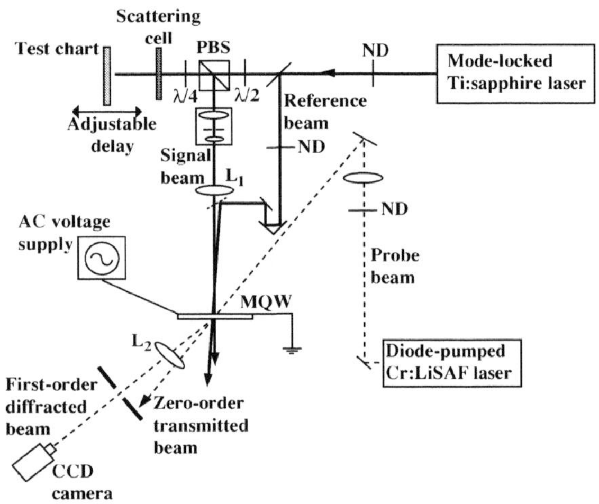

Figure 3 : The experimental configuration of the MQW holographic imaging system

Figure 3 shows the experimental set-up used to record and reconstruct 3-D holograms using the MQW devices. The image beam and the reference beam, which could be at any wavelength shorter than the absorption band edge at 850 nm, interfered at the MQW device with typical intensities of $4mW/cm^2$ and $2mW/cm^2$ respectively. The minimum intensity used, to date, to record a hologram is $200\ nW/cm^2$. It is anticipated that this may be reduced by a further factor of 10. These writing beams were separated at an angle of 1.4° which correspond to grating spacing of 31 μm. A sinusoidal AC field of peak amplitude of 8 kV and frequency of 2KHz was applied transversely across the MQW device. This resulted in a more homogeneous electric field, compared to an applied DC field, and so gave a clearer hologram. The hologram was reconstructed using a tunable Cr:LiSAF diode pumped laser tuned to the MQW device's exciton peak (~ 850 nm) to maximise the hologram diffraction efficiency. The +1 or -1 diffraction order of the holographic read-out beam was recorded on a low-cost CCD camera (bit depth of 8). For video recording, we acquired the image directly from the CCD camera to a domestic video cassette recorder (VCR). We used a frame grabber (Video Logic Captivator) to 'freeze' some of these video images or to record them as .avi files.

With no scattering medium, the reconstructed image was measured to have a transverse resolution of 13μm and the depth resolution was determined to be 60μm. The transverse spatial resolution was measured to be 38 line pairs per mm. We note that this spatial resolution is superior to that of $Rh:BaTiO_3$, which requires a relatively large angle between the beams writing the holograms and so compromises spatial resolution due to walk-off effects. This issue is discussed further in reference [5]. The test chart was readily imaged through a scattering cell of 13 MFP scattering depth. When the 3-D test object was imaged through a scattering cell corresponding to 10 MFP, the images shown in Figure 4 were acquired. This figure also includes a computer-generated reconstruction.

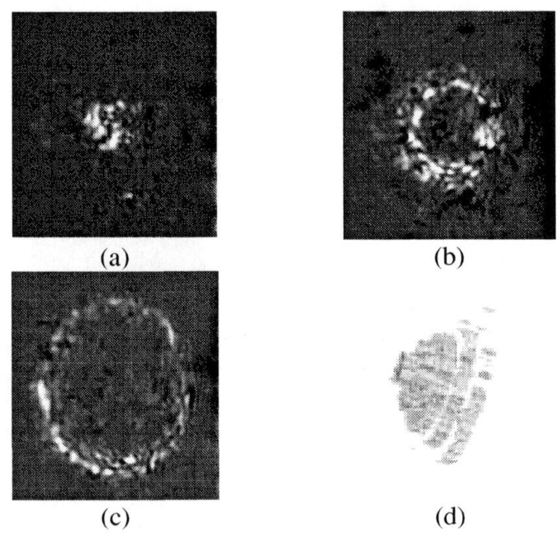

Figure 4 : Showing (a) - (c) depth-resolved layers of 3-D object through 10 MFP of turbid media (d) computer reconstruction.

Real time depth-resolved imaging

Figure 5(a) shows the real time reconstructed image of a USAF test chart, as it was read-out from the MQW device. This image has not been corrected for static background "noise" (which arises from light scattered by fixed inhomogeneities in the MQW device surface) by subtracting the light field read out in the absence of the hologram. A typical background image is shown in figure 5(b). The corrected 'clean' holographic reconstruction is shown in figure 5(c). The fast response times of the MQW devices allowed us to record useful, high resolution holograms to a conventional video cassette recorder with no signal processing. Figure 5(d) shows an image grabbed from a video signal recorded without background subtraction. Background subtraction could be subsequently applied to 'frozen' video images by subtracting a background image acquired in the same way, as is shown in figures 5(e), (f).

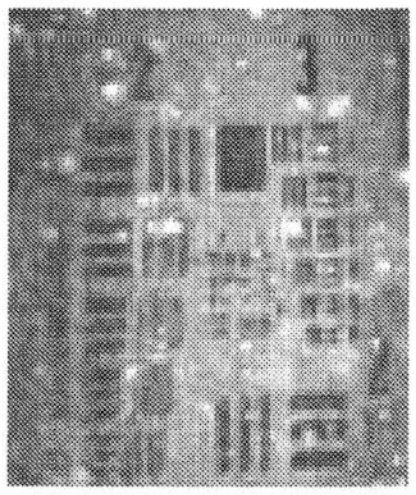

(a) Frame-grabbed image without any correction.

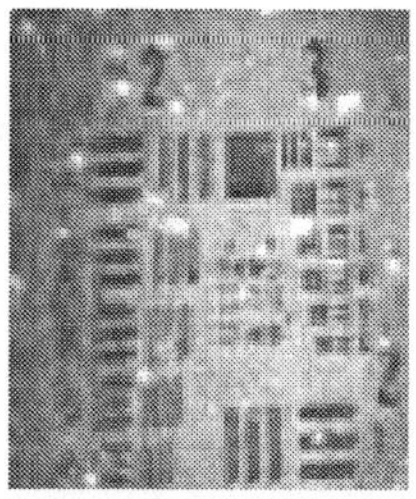

(d) Direct video image without any correction.

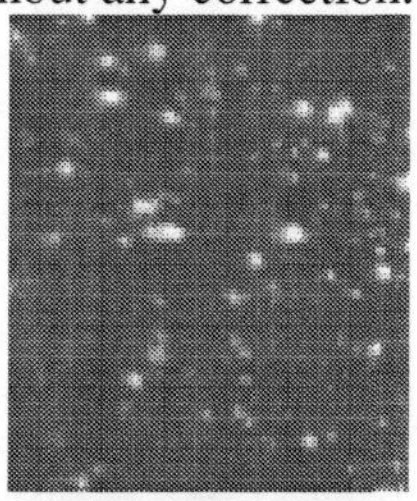

(b) Background

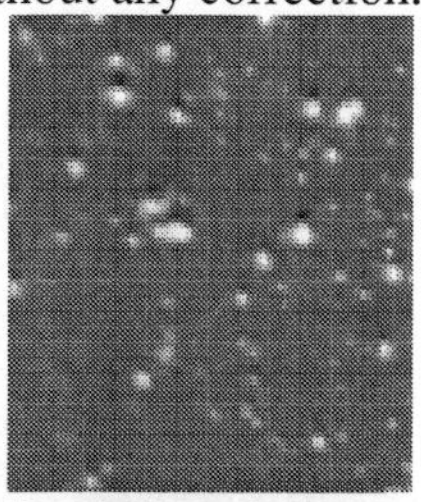

(e) Background

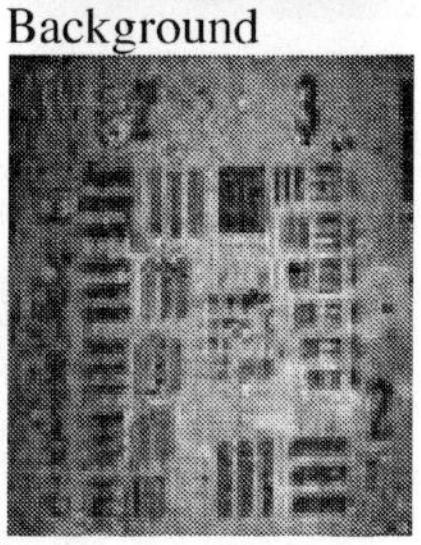

(c) Frame-grabbed image with background correction

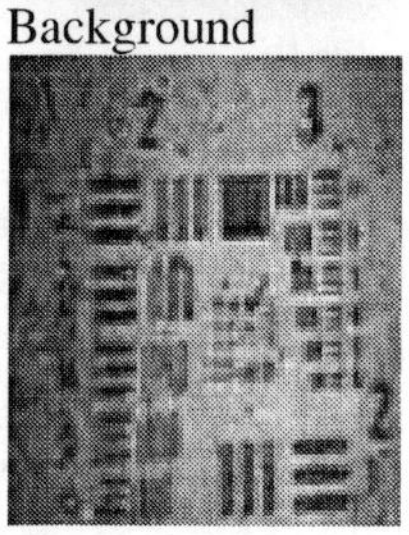

(f) Video image with background correction

Figure 5: Holographic reconstruction of an USAF test chart showing 50μm bars recorded: (a), (b), (c) to a CCD camera. (d), (e), (f) direct to VCR.

It will be seen that no significant difference can be observed comparing the background-subtracted image of figure 5(c), recorded with the frame-grabber, and the corresponding video image of figure 5(f). The real-time depth-resolved imaging represented by Fig.5(b) is possible because of the extremely fast response of MQW devices and because no computational signal processing is required, as in the case of electronic holography [8].

(a)

(b)

Figure 6: Computer reconstruction of 3-D object from holograms recorded with (a) frame grabber, (b) VCR

Figure 6 shows reconstructions of the 3-D test object from (a) frame-grabbed images and (b) images acquired directly to the VCR. Video clips of real time holograms have been submitted for publication to Optics Express [9]. This real-time depth-resolved imaging facility can be important for applications including microscopy and endoscopy of living/moving subjects. The response time of the MQW devices can be engineered to be less than 1 μs and so the maximum frame-rate of this technique is limited only by the rate of the camera acquiring the reconstructed images.

Depth-resolved imaging through scattering media

A comparison of the images recorded through liquid and solid scattering media was undertaken. A solid phantom was constructed using a suspension of silica spheres in araldite resin, in a similar manner as Firbank et al. [10]. Figure 7 shows reconstructions of the USAF test chart imaged (a) directly, (b) using holography but with no scattering medium, (c) through 8.6 MFP (double pass) of solid scattering medium and (d) through 8.6 MFP liquid scattering phantom.

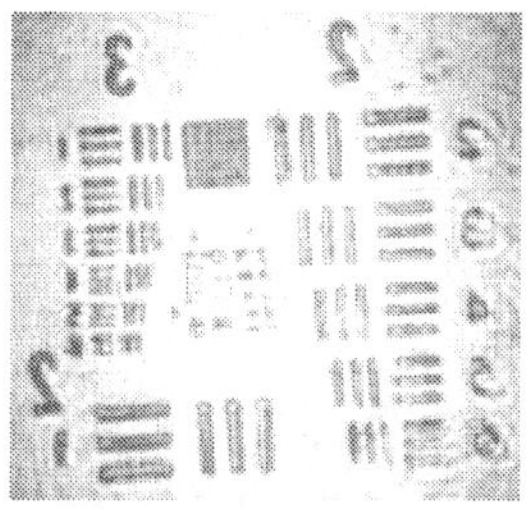

(a)

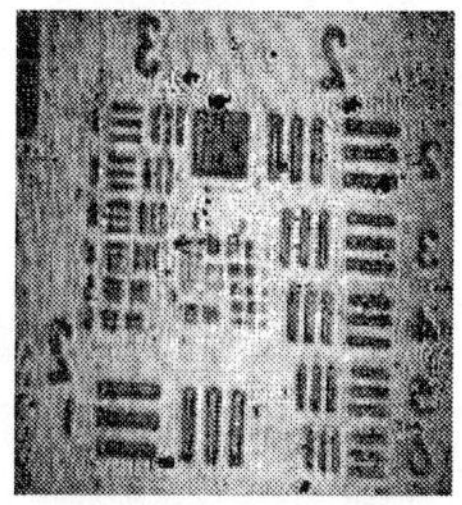

(b)

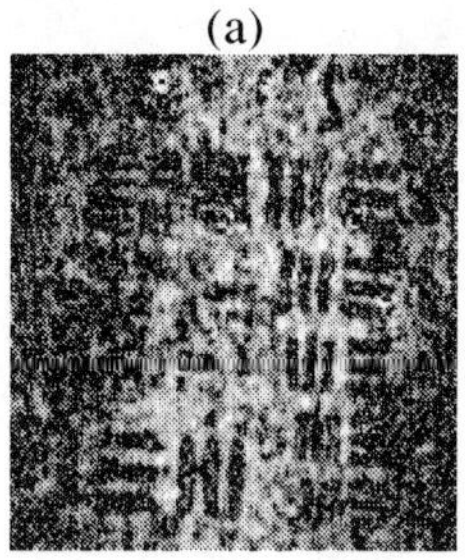

(c)

(d)

Figure 7: (a) The direct image of an USAF test chart and (b) its holographic reconstruction. The holographic reconstruction of the USAF test chart through 8.6 MFP (double pass) of (c) solid and (d) liquid scattering phantom

The factors determining the differences in image quality when imaging through the solid and liquid phantoms are not fully understood. For the liquid scattering phantom, the Brownian motion of the scattering centres results in a *time-varying* noise fringe patterns which average to a uniform background level over the image acquisition time. As it has been mentioned before, the photorefractive media respond to a spatial derivative of intensity, rather than its magnitude, so this uniform background does not effect the reconstructed image. For the static solid scattering phantom, the 'instantaneous' noise fringe patterns do not average to a uniform background and therefore can degrade the reconstructed image. This can be consider as "inter-pixel cross talk" and is a problem for *any* whole-field imaging technique. Possible approaches to counter this cross-talk include using spatial low coherence sources [11], this will limit the extent scattered photons can move transversely away from their original position before they become incoherent with the reference beam. In effect this limits the speckle size and will hopefully improve the image quality.

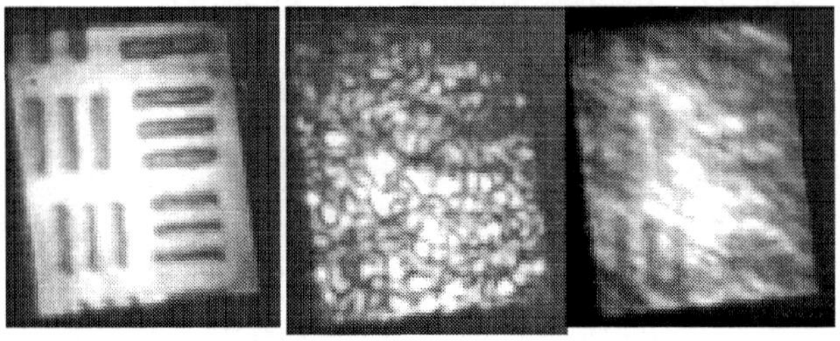

(a) (b) (c)

Figure 8: (a) The image of group 1 # 2,3 bars viewed directly and the transmitted image of the USAF test chart through 4.9 attenuation lengths of (b) static and (c) moving chicken breast.

Tissue is made up of cells and particles of different sizes, with varying absorption as well as scattering properties, and is much more complex than the phantoms used in most experiments. Therefore the next scattering medium investigated was a thin slice of chicken tissue breast (9.8 attenuation lengths double pass) ,placed between two microscope slides. An image of group 1 # 3 bars, as imaged through directly through the optical system, is shown in Figure 8. These bars correspond to a resolution of 70 μm at the MQW device. When the chicken breast was placed before the test chart, figure 8(b) was obtained through the system: no image can be made out and all that is detectable is speckle. The chicken was then shaken by hand to simulate a moving scattering media and the image in Figure 8(c) was obtained. As can be seen the USAF test can be once more made out due to the averaging of the speckle on the CCD camera. This illustrates the dramatic difference between imaging through moving and static scattering phantoms.

Spectrally-resolved imaging

One of the most important advantages of an optical imaging system is that it can provide spectroscopic information, which can lead to chemically-specific functional imaging. The photorefractive MQW devices offer the ability to record images at any wavelength shorter than the exciton peak at 850 nm. This is demonstrated in figure 9 which shows a series of reconstructed holographic images of the USAF test chart, half of which had been covered by a high pass cut-off filter which only transmits radiation above 815nm and half by a filter transparent to all the wavelengths used.

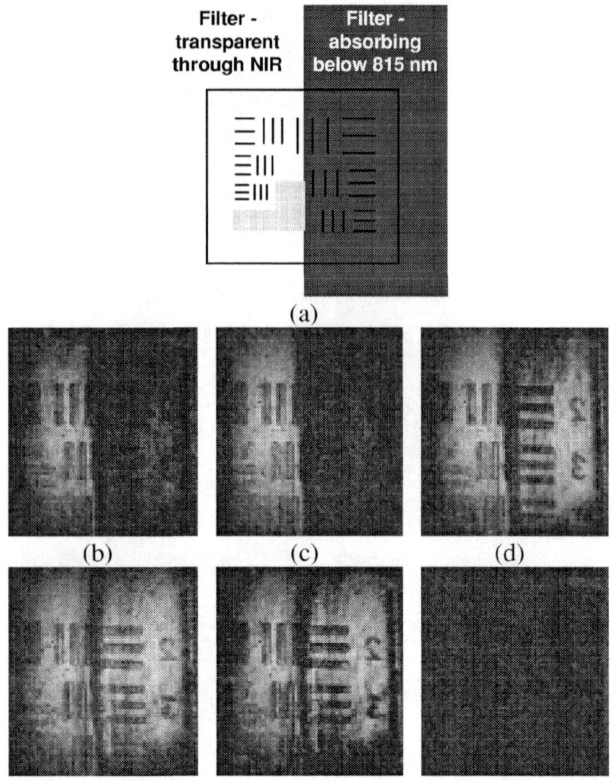

(a)

(b) (c) (d)

(e) (f) (g)

Figure 9. (a): schematic of USAF test chart partially covered by cut-off filter and (b)-(g): reconstructed images as the recording radiation was tuned from 800 nm to 860 nm.

The images were recorded as the operating wavelength of the Ti:Sapphire laser was tuned from 800 to 860 nm. Below 815 nm, only the portion of the test chart beneath the transparent filter was seen, while above the exciton peak wavelength of 850 nm, no hologram was recorded. It should be noted that any visible radiation could be used to record the holograms and that, given red, green and blue recording radiation, RGB images could be recorded and read out (using the same read-out beam) into the CCD camera. This would permit full colour reconstructed depth-resolved images of 3-D objects, including moving (living) subjects. The fast response of the MQW devices means that it would be quite feasible to record the depth-resolved RGB images sequentially.

To extend the spectral coverage above 850 nm, it would be necessary to use photorefractive MQW devices with an absorption band edge at longer wavelengths. MQW technology has already been developed for telecommunications applications and so this is quite feasible. Alternatively it is possible to use bulk semiconductor media such as CdTe as the photorefractive recording medium. In a preliminary experiment we have demonstrated depth-resolved holographic imaging using CdTe:Mn,V (supplied by Brimrose Corporation), which is a sensitive photorefractive medium up to 1.5μm with a response time similar to the MQW devices.

Conclusions

In conclusion, we have demonstrated the use of a photorefractive holographic imaging system as a coherence gate to discriminate against scattered light and provide whole field depth-resolved 2-D images through turbid media without the need for pixel-by-pixel transverse scanning. Using $Rh:BaTiO_3$, we have recorded depth resolved images through up to 16 MFP of turbid medium, with sub-100μm depth and transverse spatial resolution. Unfortunately the long integration times required for $Rh:BaTiO_3$ are a disadvantage for real world applications.

We have also demonstrated the use of MQW devices as a holographic recording medium and presented 3-D holographic images with transverse spatial and depth resolutions better than 50μm. The response time of the devices (determined to be less than 0.4 ms) makes them suitable for real-time imaging and direct depth-resolved image acquisition to a video recorder has been demonstrated. We anticipate many applications for rapid high-resolution 3-D imaging including 3-D microscopy e.g. of living subjects and motion analysis. To these ends, further improvements in the field of view and optical quality are anticipated as the technology matures.

We have applied this MQW based holographic imaging system to image through up to 13 MFP of scattering medium, with 50 μm transverse and 60 μm depth resolution. We note that 13 scattering MFP "penetration depth" corresponds approximately to a depth in tissue of ≥ 0.7 mm. This could be increased in practice by using a higher incident intensity at the tissue or by optimising the wavelength of the radiation. Initial results have been demonstrated comparing solid, liquid and tissue scattering media and the problem of speckle due to inter-pixel cross-talk has been highlighted. This may be overcome using spatially incoherent sources. Because holograms may be written in the MQW devices at any wavelength shorter than that of the exciton peak, it is possible to acquire spectrally-resolved or colour 3-D images using this technology. This has been demonstrated and is the subject of ongoing work. To further extend the spectral coverage of this technique, we have also demonstrated depth-resolved imaging through turbid media using CdTe:Mn,V, which has the advantages of NIR sensitivity and a fast response time.

Acknowledgements

Funding for this research was provided by the UK Engineering and Physical Sciences Research Council (EPSRC) and EU Training and Mobility of Researchers (TMR) program. M. Tziraki acknowledges an EU TMR. Studentship, and R. Jones and Duncan Parsons-Karavassilis acknowledge EPSRC studentships. This work has also been supported by the UK Defence Evaluation and Research Agency (DERA).

References

[1] A. V. Mamaev, L. I. Ivleva, N. M. Polozkov, V. V. Shkunov, "Photorefractive visualisation through opaque scattering media", *Conf. Lasers and Electro-Optics* 1993, paper CFK6, pp 632-634

[2] D. Huang, E. A. Swanson, C. P. Lin, J. S. Schuman, W. G. Stinson, M. R. Hee, T. Flotte, K. Gregory, C. A. Puliafito and J. G. Fujimoto, "Optical Coherence Tomography", *Science,* **254**, pp 1178-1181 (1991)

[3] N. Abramson, "Light-in-flight recording by holography", *Opt. Lett.*, **3**, pp 121-123 (1978)

[4] S. C. W. Hyde, R. Jones, N. P. Barry, J. C. Dainty and P. M. W. French, "Sub - 100 μm depth-resolved holographic imaging through scattering media in the near infra-red", Opt. Lett., **20** pp 2330-2332 (1995)

[5] S. C. W. Hyde, N. P. Barry, R. Jones, J. C. Dainty, P. M. W. French, K. M. Kwolek, D. D. Nolte and M. R. Melloch, IEEE JSTQE Special Issue on Lasers in Medicine and Biology, **2** pp 965-975 (1996)

[6] N. P. Barry, S. C. W. Hyde, R. Jones, J. C. Dainty and P. M. W. French, "High background holographic imaging using photorefractive barium titanate"*Electron. Lett.*, vol 33, pp1732-3 (1997)

[7] D. D. Nolte and M. R. Melloch, "Photorefractive effects and materials", chapter 7, D. D. Nolte ed. (Kluwer Academic Publishers, Dordrecht, 1995)

[8] R. Jones, S.W.C Hyde, M.J.Lynn, N.P. Barry, J.C. Dainty and P.M.W. French , "Holographic storage and high background imaging using photorefractive multiple quantum wells", Appl. Phys. Lett. **69** pp 1837-1839 (1996)

[9] R. Jones, M. Tziraki, and P. M. W. French, K. M. Kwolek and D. D. Nolte, M. R. Melloch, "Direct-to-video holographic 3-d imaging using photorefractive multiple quantum well devices", Submitted for publication in the special issue on Spatial Light Modulators in Optics Express.

[10] M. Firbank, "The Design, Calibration and Usage of a Solid Scattering and Absorbing Phantom for Near Infra Red Spectroscopy", Ph.D. thesis, University of London,

[11] E. N. Leith, C. Chen, H. Chen, Y. Chen, J. Lopez, P.-C. Sun and D. Dilworth, " Imaging through scattering media using spatial incoherence techniques", Opt. Lett., **16** pp 1820-1822 (1991)

3D- Photoacoustic Imaging of Blood Vessels.

C.G.A. Hoelen, R. Pongers, A. Dekker, F.F.M. de Mul (*)

University of Twente, Dept. Appl. Physics, POBox 217, 7500 AE Enschede, the Netherlands
tel/fax xx.31.53.489.3160/1105; e-mail: mul@tn.utwente.nl (*) corresponding author

Keywords: (110.7170) Ultrasound, (110.6960) Tomography, (170.0170) Medical Optics, (170.5120) Photoacoustic Imaging, (290.7050) Turbid media.

Abstract

We used Photoacoustics to image optically absorbing structures (blood vessels) in tissue samples. With 532 nm light depths down to 6-9 mm were reached. Samples were a 10 % solution of Intralipid-10% or real tissue (chicken breast), containing capillaries with blood or a variable dilution of Evans blue. Small PVdF piezoelectric hydrophones were used for detection in scanning mode. The sensitivity of the technique was demonstrated by the detection of two single red blood cells situated on a glass plate. The lateral resolution is limited by the diameter of the detector (200 μm in our case). The depth resolution is about 10 μm.
This study was undertaken to develop Photoacoustic Tomography of tissue for the localisation of blood vessels, *e.g.* for the detection of blood concentrations (*angiogenesis*) around tumors.

1. Introduction

In highly scattering media, like dermal and hypodermal tissue, the wavelength-dependent average reduced scattering coefficient not only determines the penetration of light into the medium but also the possibility to localize optical absorbers. Much research has been performed in developing photon transillumination techniques for tissue, however with varying results. For photoacoustic (PA) signal generation however (Figure 1), the signal amplitude is determined by the local fluence rate only. Pulsed mode operation generates the highest PA signal amplitudes. In this way in principle blood vessels or tumours may be imaged in vivo by choosing the appropriate wavelength. By measuring the pressure in a detection plane as function of position and time, a PA image can be constructed. The acoustic flight time is related to the position in depth and the angle of observation.

In medical ultrasonic imaging systems piezoelectric transducers act as ultrasound source and receiver. Here images of interfaces between media with different acoustic impedances are constructed, with a resolution of about 100 μm (at 25 MHz). The PA technique, however, makes it possible to reach a depth resolution of 10 μm, provided different light absorption coefficients are present.

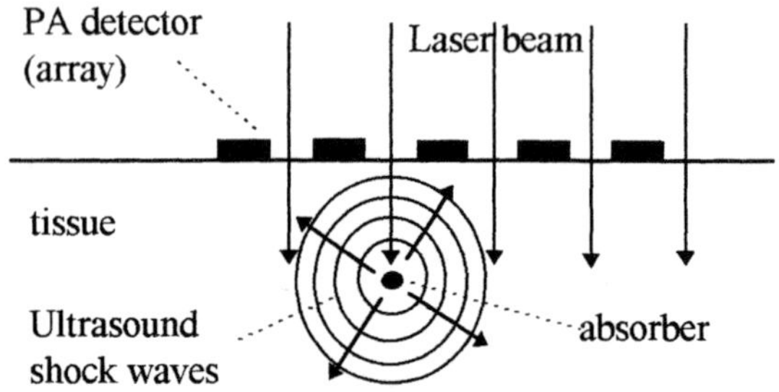

Figure 1. Photoacoustics: a laser light pulse is absorbed, by which a temperature rise and adiabatic expansion are induced, followed by an acoustic shock wave propagating to the surface.

The PA-technique has been used before to determine optical properties of tissue layers and to detect optical inhomogeneities in tissue phantoms [1,5]. In those studies a relatively large detector was used. For some

simple cases of a single absorber with mm dimensions, photoacoustic signals have been used to reconstruct a twodimensional image of this absorber. In that study [5] detection took place in a circle around the object.
In this study we will show a new application with images of real vessels with submillimeter sizes. We show that PA can detect hidden absorbing structures in tissue, like blood vessels. Due to scattering and absorption in the surrounding media the maximum reachable depth is limited to about 10 mm at about 550 nm light and several cm at about 850 nm light. The sensitivity will differ due to varying absorption contrast between blood and surrounding medium: about a factor of 100 at 550 nm to about 10 at 850 nm. In Table 1 some characteristic optical properties of blood and tissue are summarized.

Table 1. Typical scattering and absorption properties of blood and tissue at relevant wavelengths.

Scattering (μ_s ; μ_s') and Absorption (μ_a) coefficients [1/mm]	μ_a	μ_s	μ_s'
Blood at λ=577 nm	32	50	1
Dermis at λ=577 nm	0.03	21	3
Blood at λ=850 nm	1.2	30	0.5
Dermis at λ=850 nm	0.01	5	1

μ_s': reduced scattering coefficient;
$\mu_s' = \mu_s(1-g)$; $g = <\cos\theta>$; θ = polar scattering angle.

We recorded PA-signals at 532 nm, reaching to about 6-9 mm deep in tissue.
Normally in photon-transillumination methods (either in time or frequency domain), where light scattering is used to reveal the position of hidden structures, the scattering of light by the medium has an extremely disturbing influence on the imaging. In PA however, this scattering is very advantageous, since it will diffuse the light fluence conveniently and make the photoacoustic process more homogeneous.

In Figure 2. we present a comparison between actual methods for imaging hidden (absorbing and/or scattering) structures in turbid media like tissue.

Considerable effort is being made in the field of Photon-Transillumination (either direct transmission or reflection), in which in realistic clinical situations and for simple geometries a resolution of about 5 mm is obtained at a depth of a few cm. On the other hand, with Optical Coherence Tomography a resolution of about 10 μm can be obtained, but the method is capable of reaching about 1 mm deep only. It is seen that Photoacoustics will fit into the gap between those methods, either with green light to depths of about 10 mm, or with infrared light (about 850 nm) at larger depths.

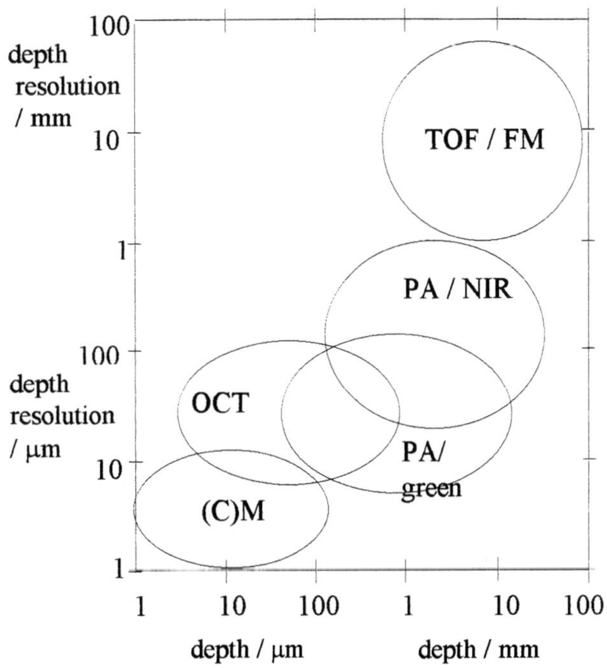

Figure 2. Comparison of imaging methods for hidden structures in turbid media (tissue):

(C)M:	(Confocal) Microscopy
OCT:	Optical Coherence Tomopgraphy
PA / green:	Photoacoustics (green light)
PA / NIR:	Photoacoustics (near infrared)
TOF / FM:	Photon-Transillumination:
	- time-of-flight / freq. Modulation

2. Measurements and Image Retrieval

We used 8 ns light pulses from a frequency-doubled Nd-YAG laser (wavelength 532 nm) operating at 10 Hz repetition frequency. The sample material was either 10% dilutions of Intralipid-10% or real tissue, in the form of a block chicken breast with blood capillaries. Both sample materials have reduced scattering coefficients $\mu_s' \approx 1$/mm at 532 nm, typical for tissue. The absorption coefficients was about 10^{-3} and <0.1/mm respectively. At 532 nm the absorption of blood is 30 mm^{-1} . Ultrasound attenuation in tissue is about 0.12 /(cm.MHz), resulting in a penetration depth of 3.3 mm at 25 MHz. Because blood vessels have more or less cylindrical geometry, the amplitude attenuation due to cylindrical propagation, which is proportional to $\sqrt{(1/r)}$ (r = propagation distance), is expected to dominate absorption and scattering losses. A glass fibre with 600 μm core diameter (Figure 3) is used to illuminate the sample with pulse energies of 70-80 μJ. The effective diameter of the PVdF-piezo film detector used here is 200 μm and its thickness is 9 μm (PVdF = polyvinylydene-difluoride). A Tektronix TDS-220 oscilloscope samples the detector signal at 1 Gsamples/s. The photodiode provides the trigger signal for the oscilloscope. Neutral density filters (Schott) are used to reduce the intensity. The sample is placed in a container filled with intralipid suspension, by which the acoustic coupling medium is water. The relative position of the detector is array-scanned with respect to the fiber and the sample

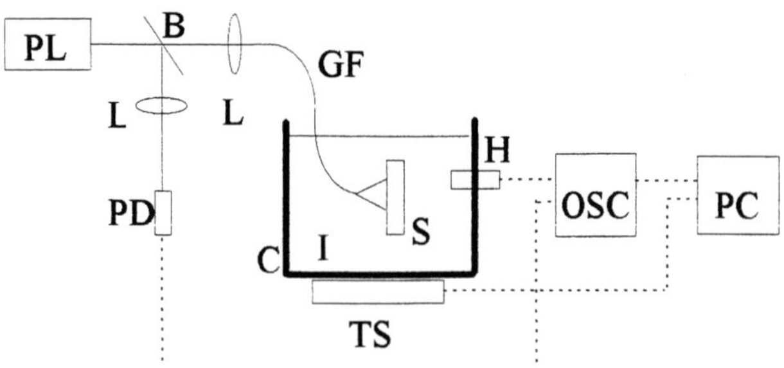

Figure 3. Set-up [2-4] used in this study for photoacoustic imaging of absorbing structures in a scattering medium. Illumination from behind or from aside. PL: pulsed laser; BS: beam splitter; L: lens; PD; photo diode; GF: glass fiber; C: cuvette; TS: translation stages; I: intralipid dilution; S: sample with absorbers; H: piezo hydrophone; OSC: oscilloscope; PC: computer. The fiber can be rotated around the sample, to enable measurements from behind, aside and front.

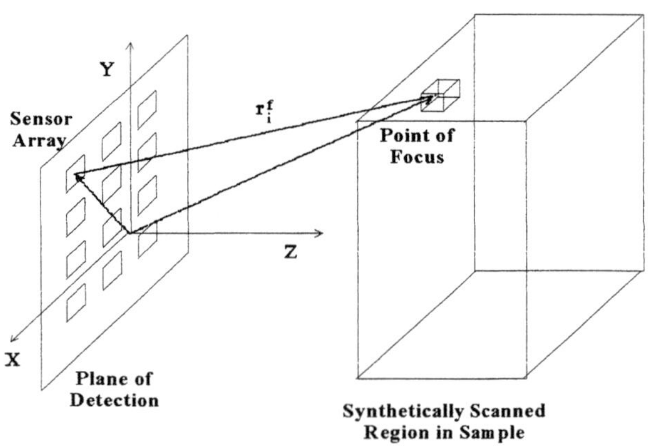

Figure 4. The general idea about image retrieval. For each acoustic point source (voxel) in the sample, the detector array is focussed onto that voxel by time-shifting of the signals according to the acoustic time-of-flight.

For detection of PA-signals of individual erythrocytes the cells were positioned on a microscope slide in a PBS-medium and illuminated from below using a 50 μm fiber.

Image retrieval algorithm.

For the image retrieval a weighted delay-and-sum focussed beam forming algorithm was used, modified to accomodate spherical waves. See Figure 4.

Each detector array element will record one single signal as a function of acoustical time-of-flight. In principle this signal will contain information from all acoustical source voxels within the numerical aperture of the element, but in general at different times. In the image retrieval algorithm, for each voxel, each detector signal is shifted over a time period determined by the flight time of the acoustical shock wave from that voxel to that element. In this way the detector array is focussed onto that particular voxel. In doing so, we are able to "window" the time signal around the expected arrival time of the signal from that voxel. The width of the window is given by the typical width of an acoustical signal from a spherical source with dimensions equal to those of the voxel. Taking only the part of the signal within the window into account reduces the computing time considerably. Besides this correction using the flight times, the algorithm also performs a correction of the amplitude of the signal to optimize the signal-to noise ratio. The latter correction is determined by: the attenuating effect of the voxel-detector distance on the amplitude, and by the numerical aperture function of that detector. Finally a correction for the spherical shape of the acoustical waves has to be inferred. This is necessary here, because unlike the seismic case, detection will take place in the near field.

3. Results And Discussion

Here we present some examples of imaging experiments. In Figure 5 an example is given of a photoacoustic signal originating from a point-like source. The source consist of two adjacent erythrocytes positioned on a glass plate in an aqueous environment. It is seen that the signal shape is bipolar Gaussian, indicating spherical sources.

To indicate the resolution that can be obtained, Figure 6 contains the PA-image of two words, made out of thin carbon threads. The depth resolution (10 μm) turns out to be much better than the lateral resolution, which is about 100-200 μm. The latter is determined by the dimensions of the individual detector elements we use presently, and by the choice of the weight factors in the reconstruction algorithm.

In Figure 7 the photoacoustic image of two blood vessels (inner diameter 200 μm) in chicken breast tissue is shown. Preferentially the back side of the vessel is imaged (pressure and decompression pulse). This is due to destructive acoustic interference of the signals from different points in the inside of the vessel within the numerical aperture of the detector, but at slightly different detection distances.

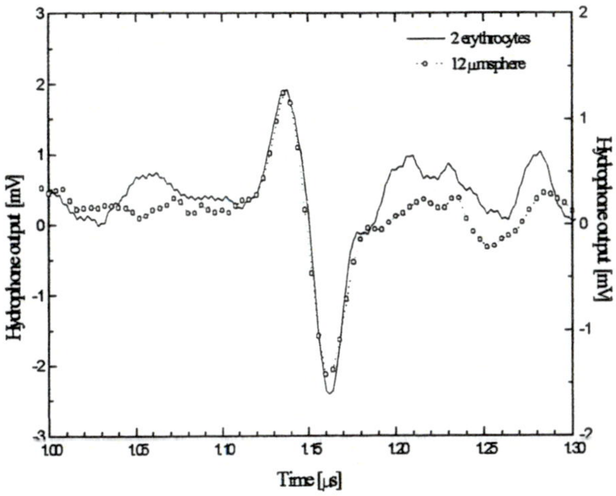

Figure 5. Photoacoustic response generated by a 12 μm blue polystyrene sphere in PBS (dotted line) and by two adjacent red blood cells (diameter ≈ 10 μm) in water (solid line). Illumination through 50 μm core glass fiber. Acoustic detection distance: 1.7 mm (= 1.15 μs × 1500 m/s).

Figure 6. Depth resolution: photoacoustic image of two transparant layers containing words "TOP" and "UT" made out of carbon threads, each with thickness 12 μm, height approx. 3 mm, in 10% Intralipid-10%. Depth resolution 10 μm. Depth of TOP and UT-plane: 1.52 and 1.76 mm respectively.

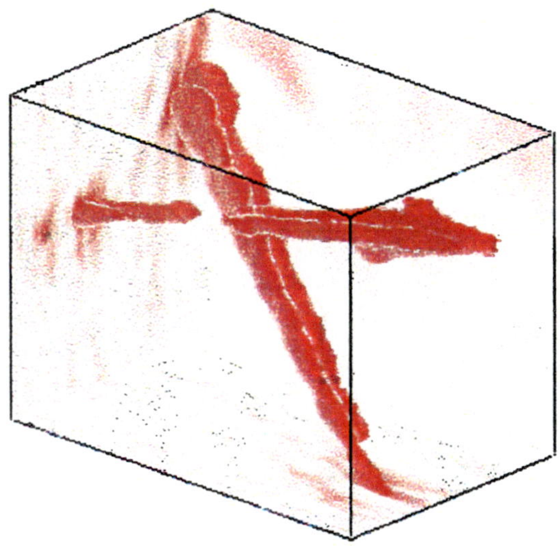

Figure 7. Photoacoustic image of two blood capillaries (diameter about 200 μm) in 5 mm thick chicken breast sample positioned in 10%-Intralipid-10%. Lateral dimensions: 3.0 x 3.0 mm. Shown: depth z = 1.1-2.2 mm (= distance to PA-detector). Illumination from right side, fiber facet at z = 3.5 mm. Acoustic detection at left side (z=0). Averaged: 16 pulses from Nd-YAG-laser, pulse energy 80 μJ. Seen: response from illumination side of capillaries. Preferentially the backside of the capillaries is imaged, together with a reflection at the outer vessel boundary.

4. Conclusions

We showed that Photoacoustics can be a powerful tool to image hidden structures in tissue, especially in cases there is a relative contrast in absorption between layers or structures in tissue. The resolution obtained was about 10 μm in depth. The lateral resolution is mainly determined by the dimensions of the detectors, which were here about 200 μm.

The beam-focussing algorithm, being our modification of the beam forming algorithm to accomodate spherical waves, appears to work properly in retrieving the sources from the acoustic signals. We recorded PA-signals from a depth of 9 mm. We imaged deep lying blood vessels in real tissue samples, at depths up to 6 mm at 532 nm laser light. We expect soon to reach much deeper (a factor of about 5) when using near-infrared light.

As a method for imaging hidden structure in turbid media like tissue, photoacoustics will fill the gap between

Optical Coherence Tomography and the various Photon-Transillumination methods, either in time-domain or in frequency domain. This concerns both the depth that can be reached and the depth resolution that can be obtained.

References

1. A.A. Oraevsky, R. Esenaliev, S.L. Jacques, S. Thomson, and F.K. Tittel, "Lateral and z-axial resolution in laser optoacoustic imaging with ultrasonic transducers", Proceedings of Optical Tomography, Photon Migration, and Spectroscopy of Tissue and Model Media: Theory, Human Studies, and Instrumentation, SPIE Vol. 2389, pp. 198-208, 1995.
2. C.G.A. Hoelen, G. Hamhuis, N. Pagoulatos, I. van den Ham, F.F.M. de Mul, and J. Greve, "Photoacoustic location of optical absorbers in phantom tissue", Proceedings of Optical and Imaging Techniques for Biomonitoring II, SPIE BIOS Vol.2927, pp. 142-153, 1996.
3. C.G.A. Hoelen, R. Pongers, G. Hamhuis, F.F.M. de Mul, J. Greve, "Photoacoustic blood cell detection and imaging of blood vessels in phantom tissue"; Proceedings of Optical and Imaging Techniques for Biomonitoring III, SPIE-BIOS Vol.3196, 1997.
4. C.G.A. Hoelen, F.F.M. de Mul, R. Pongers, A.Dekker, "Three-dimensional photoacoustic imaging of blood vessels in tissue", Optics Letters, 28,3, April 1998.
5. Kruger,P. Liu, Y. Fang, C.R. Appledorn, "Photoacoustic Ultrasound (PAUS) - Reconstruction Tomography", Med.Phys. 22 (1995) 1605-09,.,

----

Frequency-Swept Ultrasound-Modulated Optical Tomography

Lihong V. Wang and Geng Ku

Optical Imaging Laboratory, Biomedical Engineering Program
Texas A&M University, College Station, Texas 77843-3120, USA
http://biomed.tamu.edu/~lw

Abstract

A novel frequency-swept (chirped) ultrasound modulated optical tomography technique was developed to image scattering media. Laser light was modulated by a frequency-swept ultrasonic wave to add spatial information to the modulated optical signal. The modulated optical signal was received by a photomultiplier tube and was heterodyned with a reference modulation signal. Frequency analysis was employed to obtain spatial resolution along the ultrasonic axis. A spectrum can be converted to a 1D image. Multiple spectra were obtained and were used to form a 2D image of the scattering medium.

Keywords

Ultrasonic modulation, frequency sweep, optical tomography, turbid media.

Introduction

Nonionizing optical tomography of biological tissues including scattering medium has been an developing research area. It has the potential to provide new imaging tools for tumor detection in vivo functionally.[1] Several optical imaging techniques being investigated include time-gated or time-resolved optical imaging, frequency-domain optical imaging, and optical coherence tomography. The time-resolved and frequency-domain research have achieved several millimeter resolution for approximately 5 centimeter thick tissues. OCT has achieved high resolution but is limited to a penetration depth of a couple of millimeters into biological tissues.

Ultrasound-modulated optical tomography that combine the relatively transparent ultrasonic radiation with the strongly scattering optical radiation has been studied as well.[2-4] We developed a novel imaging technique: frequency-swept (chirped) ultrasound modulated optical tomography of turbid media.[5] A frequency-swept ultrasonic wave was adopted to add spatial information on transmitted laser light. The combination of ultrasound and light allowed us to image objects buried inside turbid media.

Experimental

Frequency sweep means that the frequency of the ultrasound varies with time as described by

$$f(t) = a + bt$$

where f(t) is the instantaneous frequency of the ultrasound, a is the starting frequency and b is the sweep rate.

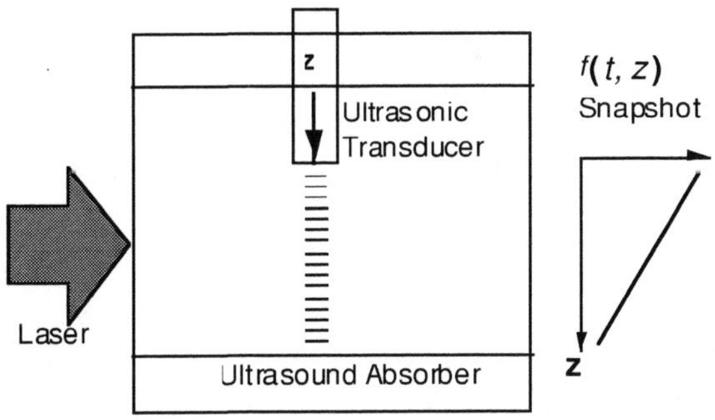

Fig. 1. Instantaneous frequency distribution along the ultrasonic axis (z).

The frequency-swept ultrasonic wave propagated vertically along the z axis. If we freeze the time and take a snapshot of the propagating ultrasonic wave, the instantaneous frequency varies linearly along the z axis as shown in Fig. 1. Because the instantaneous frequency of the ultrasonic source increases with time and it takes time for the ultrasonic wave to propagate downward, the instantaneous frequency in the snapshot decreases as z increases. In other words, the farther down from the ultrasonic transducer, the lower the instantaneous frequency.

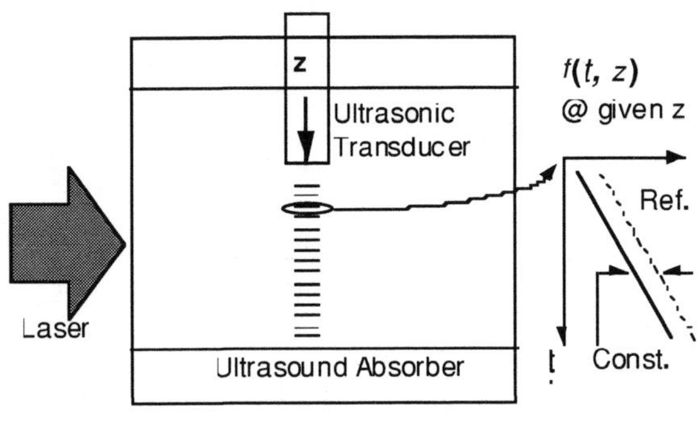

Fig. 2. Instantaneous frequency distribution vs. time.

Figure 2 shows a frequency distribution versus time at an arbitrary point on the ultrasonic axis. The ultrasonic wave propagates vertically along the z axis, passing through this point. The ultrasound frequency at this point varies with time linearly as shown by the solid straight line in Fig. 2. Light at this point is modulated by the frequency along this line. The instantaneous laser light frequency shift caused by the ultrasonic modulation is the frequency of ultrasound. The frequency shift can be detected as the modulated light and unmodulated light was heterodyned in the optical detector. We mixed this signal with a reference modulation signal. The modulation signal was also frequency-swept as shown by the dashed line in Fig. 2. The sweep rate of the reference modulation signal is the same as the signal for the ultrasonic wave but has a different starting frequency. The instantaneous frequency difference between the modulated optical signal at this point and the reference signal is a constant, independent of time. This constant is related to the distance between this point and the ultrasonic transducer. The larger the distance, the greater the frequency difference. The heterodyned frequency spectrum yielded image information along the ultrasonic axis.

A block diagram of the experimental setup is shown in Fig. 3. A frequency-swept signal was produced using a function generator and then was amplified in power and amplitude by a power amplifier and a transformer. The amplified signal was applied to an ultrasonic transducer. The ultrasonic transducer converted electric power into acoustic power. The ultrasonic wave propagated vertically into a scattering medium, which was contained in a glass tank. An ultrasonic absorber was placed at the bottom of the tank to prevent ultrasonic reflection from the water-glass interface.

A He-Ne laser beam, at the wavelength of 633nm, after being broadened to 15mm, illuminated the scattering medium perpendicularly to the ultrasound column. A photomultiplier tube (PMT) collected some transmitted light and converted the optical power into electric signal. The PMT was modulated in gain by a modulation signal produced from another function generator. The modulation signal was also frequency-swept but had a frequency difference with the signal for the ultrasonic transducer. The electronically heterodyned signal was band-pass filtered and then amplified using a filter and an amplifier, respectively. The amplified signal was recorded by a digital oscilloscope and then transferred to a computer for post-processing.

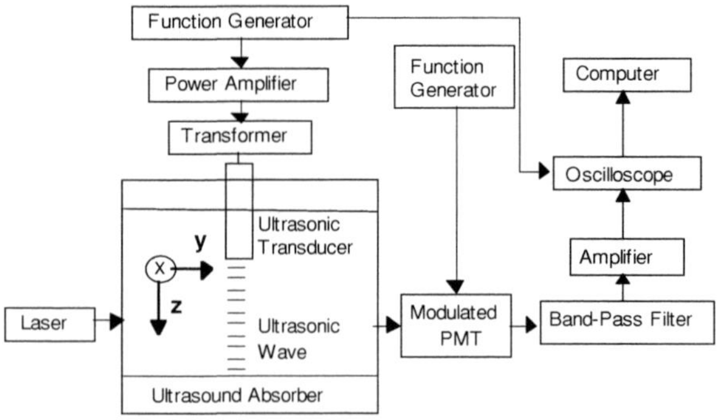

Fig. 3. A block diagram of the experimental setup.

An object was buried in the middle plane of the tank to simulate a tumor. The object was 50~60mm apart from ultrasonic transducer and was translated in the tank perpendicularly to both the laser beam and the ultrasonic column. A time-domain signal was recorded at each stop. Fast Fourier transform (FFT) was used to obtain the spectra of the recorded time-domain signals.

The frequency spectra yielded imaging information for the zone of interest selected by the band-pass filter. A frequency in the spectra corresponded to the frequency difference between the instantaneous frequency on ultrasonic axial position in the zone of interest and the instantaneous frequency of the modulation signal to the PMT, which was related to the frequency sweep rate and the ultrasonic propagation time. The frequency deference was independent of time and was related to the distance between the position on ultrasonic axis and ultrasonic transducer. A frequency in the spectrum was converted into a distance from the transducer surface to a point in the zone of interest. There was a one-to-one correspondence between the frequency in the spectrum and the position in the zone of interest. In other words, a frequency spectrum could be converted into a 1D image of the scattering medium along the ultrasonic axis. A 2D image of a scattering medium was obtained by combining the spectra for all the stops of the object.

Results and Discussion

A 2D image of an object buried in a turbid medium was shown in Fig. 4. The cross section of the object was ~ 6 mm × 2 mm in size. The scattering coefficient and anisotropy of the scattering medium at 633 nm wavelength were 0.21 cm^{-1} and 0.73, respectively. The length of the tank along the laser light was 17 cm.

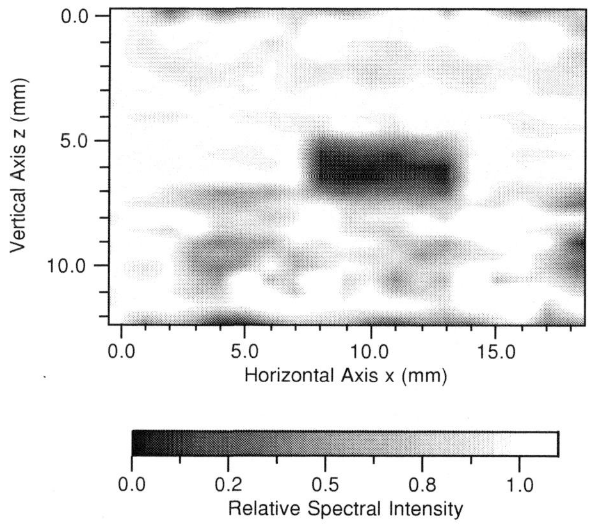

Fig. 4. A 2D tomographic image of an object buried in a scattering medium.

The object is clearly visible with a good contrast. The image of the object has a cross section comparable to that of buried object. The location is correct as well.

In order to examine the resolution of the image detail, Fig. 5(a) and Fig.5(b) show the 1D images of the object in the scattering medium cross the center of the object. The lateral resolution is determined by focal spot size of the ultrasound. In our experiment, the ultrasonic focal diameter is 1.2 mm. Axial resolution is determined by the bandwidth of the frequency sweep. We use a chirp swept from 7 MHz to 10 MHz. The corresponding axial resolution should be 1.5 mm.

Conclusion

We have experimentally proved the concept of frequency-swept ultrasound-modulated optical tomography. Images with ~1.5 mm resolution were obtained. The resolution is scaleable with the frequency sweep and the ultrasonic focal spot. A 1D image was obtained with a single time domain signal. 1D scanning cross a object yielded an image of the cross section of the object. The technique is currently limited to ballistic imaging and is being extended for diffuse light imaging. Because ultrasonic modulation of diffuse light with single frequency ultrasound has been observed [2-4, 6, 7], ultrasonic modulation of diffuse light by use of frequency-swept ultrasound should be detectable as well.

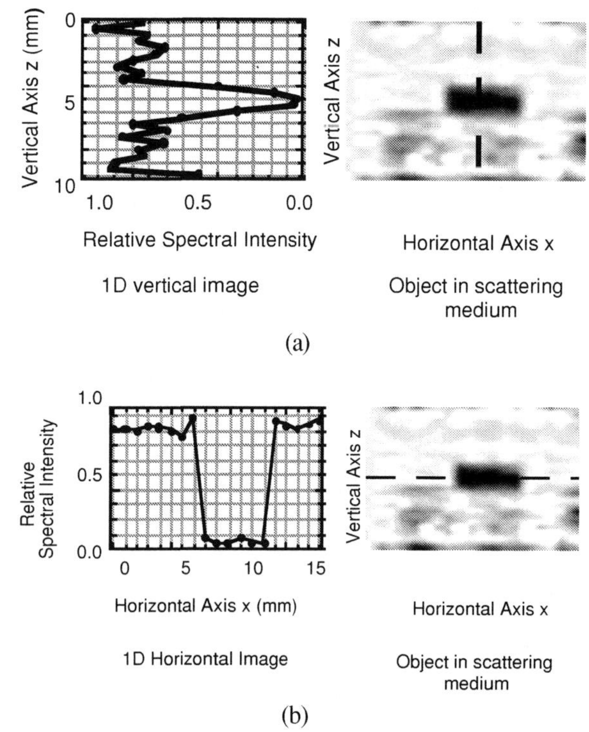

Fig. 5. (a). 1D image vertically across the center of the object. (b). 1D image horizontally across the center of the object.

Acknowledgment

Thanks to X. Zhao for experimental assistance in electronics. This project was sponsored by the National Institutes of Health grants R29 CA68562 and R01 CA71980.

References

1. R. R. Alfano (ed.), Proc. of Advances in Optical Imaging and Photon Migration, (1996), Optical Society of America.
2. L.-H. Wang, S. L. Jacques, and X.-M. Zhao, Opt. Lett. **20**, 629-631 (1995).
3. M. Kempe, M. Larionov, D. Zaslavsky, and A. Z. Genack, J. Opt. Soc. Am. **4**, 1151-1158 (1997).
4. L.-H. Wang and X.-M. Zhao, Appl. Opt. **36**, 7277-7282 (1997).
5 L. V. Wang and G. Ku, Opt. Lett. (in press)
6. F. A. Marks, H. W. Tomlinson, and G. W. Brooksby, in *Proc. Soc. Photo-Opt. Instrum. Eng.*, **1888**, 500-510 (1993).
7. W. Leutz and G. Maret, Physica B **204**, 14-19 (1995).

Acousto-Photonic Imaging

Charles A. DiMarzio
Center for Electromagnetics Research
Northeastern University
Boston, Massachusetts 02115
Phone: (617) 373-2034
Fax: (617) 373-8627
email: cdimarzio@lynx.dac.neu.edu

Thomas Gaudette
Center for Electromagnetics Research
Northeastern University
Boston, Massachusetts 02115
Phone: (617) 373-8570
Fax: (617) 373-8627
email: tgaudett@lynx.dac.neu.edu

1. INTRODUCTION

We are developing a new technique for imaging in optically diffuse media, by combining photon–migration imaging *[Boas, et. al.]*, with ultrasound modulation *[Wang et. al.]*. The ultrasound mixes with the photon density wave in the medium to produce new photon density waves at the sum and difference frequencies. The ultrasound beam is scanned in three dimensions to provide spatial resolution. We propose a model in which the interaction occurs through the change in optical properties of the diffusive medium resulting from the compression of the medium by the ultrasound.

2. ANALYSIS

Starting with the diffusion equation,

$$\nabla \cdot (D \nabla \phi) - \frac{\partial \phi}{\partial t} - \alpha \phi = -q c_0, \qquad D = \frac{c_0}{3n\left[\mu_s (1-g) + \mu_a\right]} \qquad \alpha = \frac{\mu_a c_0}{n}, \tag{1}$$

we postulate that the effect of the acoustic wave on the photon density wave results from changes in absorption and scattering coefficients caused by the change in density. We will model this by a sinusoidal fluctuation in the diffusivity, D and decay time, α;

$$D = D_0 + D_1 e^{i\left[\vec{k}_a \cdot \vec{r} - i\omega_a t\right]} \qquad \alpha = \alpha_0 + \alpha_1 e^{i\left[\vec{k}_a \cdot \vec{r} - i\omega_a t\right]}. \tag{2}$$

Note that D can not be assumed constant and thus must not be brought outside the spatial derivative in Equation 1.

We assume a harmonic solution in the flux density with terms of the carrier frequency of the photon density wave and the positive and negative mixing terms of the

Reprinted from Advances in Optical Imaging & Photon Migration, 1998 Technical Digest, © *1998 Optical Society of America*

carrier with the acoustic wave, along with DC and ω_a terms and all the complex conjugates (negative frequency terms), not shown;

$$\phi = \phi_c e^{i[\vec{k}_c\cdot\vec{r} - i\omega_c t]} + \phi_{c-} e^{i[\vec{k}_{c-}\cdot\vec{r} - i\omega_{c-} t]} + \phi_{c+} e^{i[\vec{k}_{c+}\cdot\vec{r} - i\omega_{c+} t]} + \ldots. \tag{3}$$

The zeroth–order equation, away from any sources, is

$$\nabla \cdot (D_0 \nabla \phi_c) - \frac{\partial \phi_c}{\partial t} - \alpha_0 \phi_c = 0, \tag{4}$$

with the usual solutions *[Svaasand and Tromberg]*. We collect the remaining terms by frequency, moving terms that look like the diffusion equation to the left, obtaining several equations such as

$$\nabla \cdot (D_0 \nabla \phi_{c+}) - \frac{\partial \phi_{c+}}{\partial t} - \alpha_0 \phi_{c+} = -\nabla \cdot (D_1 \nabla \phi_c) + \alpha_1 \phi_c. \tag{5}$$

The terms on the right are an effective source at the frequency $\omega_c + \omega_a$,

$$[qc_0]_{c+} = \vec{k}_a \cdot \vec{k} D_1 \phi_0 + \vec{k} \cdot \vec{k} D_1 \phi_0 - \alpha_1 \phi_0. \tag{6}$$

This new source is a true diffuse source inside the medium with no boundaries nearby to raise questions about the effective source location and boundary effects. Then with $k = 2\pi/\lambda$,

$$[qc_0]_{c+} = \left[\left(\frac{4\pi^2}{\lambda_{pdw}\lambda_a \cos\gamma} + \frac{4\pi^2}{\lambda_{pdw}^2} \right) D_1 - \alpha_1 \right] \phi_0 \tag{7}$$

For a volume compression factor C, $\delta\mu_s = C\mu_s$ and $\delta\mu_a = C\mu_a$ so

$$D_1 = \frac{\partial D}{\partial \mu_s}\delta\mu_s + \frac{\partial D}{\partial \mu_a}\delta\mu_a = \frac{c_0 C}{3n[\mu_s(1-g)+\mu_a]} \qquad \alpha_1 = \frac{\mu_a C c_0}{n}.$$

Then the source is

$$[qc_0]_{c+} = \left[\left(\frac{4\pi^2}{\lambda_{pdw}\lambda a \cos\gamma} + \frac{4\pi^2}{\lambda_{pdw}^2} \right) \frac{1}{3n[\mu_s(1-g)+\mu_a]} - 1 \right] \frac{\mu_a C c_0}{n} \phi_0. \tag{8}$$

For example, using $\mu_s(1-g) = 40/cm$, $\mu_a = 0.1/cm$, $\lambda_{pdw} = 20cm$, $\lambda_a = 0.1cm$, and $C = 10^{-6}$,

$$[qc_0]_{c+} = \phi_0 3000/\text{sec}, \tag{9}$$

which means that 3×10^{-5} of the carrier is converted to sideband signal in the 10ns cycle of the photon density wave.

3. EXPERIMENTS

The photon density wave source consists of a Melles Griot 690nm diode laser at 30mWatts, amplitude modulated by a HP 8647A generator at 70MHz. The phantom tank consists of a clear plastic box 15cm wide by 15 cm high and 6cm thick. Two, Hammamatsu C53331-03, avalanche photodiodes are set on the other side of the phantom tank with 1cm separation. One is connected to a HP–8553B spectrum analyzer to measure the sideband signal–to–noise ratio directly. The other goes to a HP–86016A oscilloscope's channel 1, with channel 2 used as the triggering and connected directly to the generator. With the oscilloscope we measure the absolute phase at the carrier frequency. The acoustic source is an NDT ultrasound transducer 1" in diameter with a 1.5" focal length and center frequency of 2.3MHz, driven continuously by a sine wave generated by a HP–3325A generator connected to a ENI Power Systems Model 325LA RF power Amplifier. The transducer can be scanned in a transverse direction to the photon density wave by a computer–controlled stepper motor and manually moved in a vertical direction.

Two types of experiments were conducted. The first are validation experiments with varying optical properties and acoustic power. For example, detected signal power is linear in acoustic power, as expected.

The second type of experiments imaged objects including a plastic frosted ball, a balloon filled with only 10% Intralipid, and a balloon filled with the same mix of intralipid and water as the phantom with added India ink.

4. CONCLUSIONS

Modulating a photon density wave with focussed acoustic energy causes sidebands on the carrier which provide the opportunity to use spatial information from both the carrier and the sidebands. The carrier provides the ability to make observations of optical absorption and scattering inside tissue, while the ultrasound offers the ability to focus the point of interaction, with high spatial resolution. Ultimately we could, for example, overlay hemoglobin and oxyhemoglobin maps over an ultrasound image.

The perturbation method shows that the sidebands are the result of a perfect diffusive–wave source, without recourse to complicated boundary conditions and assumptions. The source can be deep in the sample, near the region of interest. It can also be far enough from boundaries so that the validity of the diffusion equation is unquestioned there.

5. REFERENCES

Boas, D. A., L. E. Campbell, and A. G. Yodh, "Scattering and Imaging with Diffusing Temporal Field Correlations," *Physical Review Letters 75*, No. 9, 28 August 1995. Pp. 1855-1858.

Wang, Lihong, Steven L. Jacques, and Xuemei Zhao, "Continuous–Wave Ultrasonic Modulation of Scattered Laser Light to Image Objects in Turbid media," *Optics letters 20*, No. 6, March 15, 1995. Pp. 629-631.

Svaasand, O. P. and B. K. Tromberg. "On the Properties of Optical Waves in Turbid Media," In *Future Trends of Biomedical Applications of Lasers, Proc. SPIE 1525*, 1991. Pp. 41–50.

A Multidetector Approach to Ultrasonic Speckle Modulation Imaging

S. Lévêque and A.C. Boccara
Laboratoire d'Optique Physique, Ecole Supérieure de Physique et Chimie Industrielle, Centre National de la Recherche Scientifique UPR A0005, 10 rue Vauquelin, F-75005 Paris, France
phone : 33 1 40 79 46 03 fax : 33 1 43 36 23 95 e-mail : boccara@optique.espci.fr

M. Lebec, L. Blanchot and H. Saint-Jalmes
Laboratoire d'Etudes et de Recherche en Instrumentation Signaux et Systèmes, Université Paris 12 - Val de Marne, F-94010 Créteil, France
phone : 33 1 45 17 14 28 fax : 33 1 45 17 14 92 e-mail : lebec@univ-paris12.fr

Abstract

Detectivity of acousto-optic imaging has been improved by monitoring a full speckle pattern modulation using a CCD camera and a suitable signal processing.

Key Words : Ultrasonic Speckle Modulation, Optical Imaging, Scattering Media, Parallel Processing

Introduction

The aim of ultrasonic modulation of the light intensity propagating through scattering media is to tag the photon field inside the sample and thus to introduce a new kind of contrast related to the local optical contrast which can reveal structures buried in a turbid medium [1,2].

The axial resolution of the technique seems to be defined by the ultrasound focal spot and the depth of field by the axial expansion of the focal zone.

Working with diffuse light emerging from the turbid medium is a great advantage in term of energy reaching the detector compared with coherent methods which are limited to 20 to 30 photon mean free paths (≈ 1 mm).

In a typical experiment the source is a laser and the output of the sample exhibits a speckle pattern whose grain-sizes depends on the geometrical parameters of the experiments such as the sample-detector distance, the size of an intermediate diaphragm, etc... (usual sizes are in the 10 μm-100 μm range) and the detector add the signals generated by each of the N speckle grains reaching its surface ($N \approx 10^4$ to 10^8). The overall average power is giving rise to the DC signal is thus N times the power of one grain, whereas the modulated signal increases as $\sqrt{N}$.[3]

So if the signal is shot noise limited the signal to noise ratio will be independent of the detector size, whereas the signal to noise ratio increases as $\sqrt{N}$ if the noise is independent of the photon flux (e.g. electronic noise or thermal noise)
at last if the noise is related to laser intensity fluctuations the signal to noise ratio will decrease as $1/\sqrt{N}$.

In order to avoid such problems and to take advantage of the full information reaching the detector surface we propose here to use a multiple detector (CCD camera) and a parallel detection (lock-in) scheme. We will thus get an amplitude value for each pixel (and not a random complex number with an uniform phase distribution in the $[-\pi,+\pi]$ interval) and we add all these elementary signals to get our final signal.

Experimental Set-up

Our experimental scheme is close to one described in Wang et al.[1], the photomultiplier and the signal processing being replaced by a CCD camera and a parallel lock-in detection. More precisely a 10 mw monomode laser diode at 780 nm irradiates the scattering medium (we have used various more or less representative phantoms but we switched recently to animal tissues such as turkey breast).

The ultrasonic field is induced by a focusing transducer (Panametrics, focal length 75 mm and diameter 19 mm or focal length 69 mm and diameter 38 mm) working at 2,2 MHz.

We then use an original lock-in detection technique to measure the amplitude a_1 and phase φ_1 of the first harmonic of the modulated signals simultaneously from each pixel of the CCD array.

A lock-in amplifier is an efficient solution to improve the Signal-to-Noise ratio (S/N) but the high modulation frequency (2.2 MHz for our ultrasonic transducer) is not compatible with a 2D CCD working only at 50 Hz. To solve this problem, the proposed method [4] uses synchronous excitation instead of the conventional synchronous detection (lock-in). In our experiment the laser diode is modulated at 2.2 MHz, while the 2D CCD array is used as low pass filter. The signal is obtained from a four phase shifted light excitations, and we thus record four images, each corresponding to a given phase shift. By computing linear combinations of these images, we obtain images proportional to $a_1.\cos(\varphi_1)$ and $a_1.\sin(\varphi_1)$, and so for each pixel, we know a_1 the amplitude of the modulated light and φ_1 the phase. For example, if the modulated signal is supposed to be $A(t) = a_0 + a_1 \cos(2\pi f_0 t + \varphi_1)$, the four signals S_0,S_1,S_2 and S_3 obtained for a four phase experiment (0° ,90°, 180° and 270° with respect to the ultrasonic transducer) are given by :

$$S_0 = KT_0(\frac{a_0}{4} + \frac{\sqrt{2}}{\pi} a_1 \cos(\varphi_1))$$

$$S_1 = KT_0(\frac{a_0}{4} + \frac{\sqrt{2}}{\pi} a_1 \sin(\varphi_1))$$

$$S_2 = KT_0(\frac{a_0}{4} - \frac{\sqrt{2}}{\pi} a_1 \cos(\varphi_1))$$

$$S_3 = KT_0(\frac{a_0}{4} - \frac{\sqrt{2}}{\pi} a_1 \sin(\varphi_1))$$

where K is the number of period T_0 for the integration. By arranging these signals we can obtain for each pixel the amplitude and the phase :

$$\sqrt{(\frac{1}{4KT_0}(S_0 - S_2))^2 + (\frac{1}{4KT_0}(S_1 - S_3))^2} = \frac{\sqrt{2}}{2\pi} a_1$$

and $\varphi_1 = Arc\tan(\frac{S_1 - S_3}{S_0 - S_2})$

We started tests using turkey breast : the ultrasonic transducer was seated on a one dimension translation stage, which permitted to scan the sample perpendicularly to the laser beam while the rest of the system was fixed, including the sample and the laser beam. We supposed the diffusion of light strong enough to illuminate the whole sample which was typically 1 cm thick. The translation stage was scanned at 1 mm per step. Then we obtain a one-dimensional image for different objects by computing for each point the sum of the individual amplitude of each pixel.

Results and discussion

As first tests using solid objects we checked the lateral resolution which was found in agreement with the predicted ultrasonic focal spot size (typically 3 to 5 mm).

Then we have successfully checked the validity of our approach in term of signal statistics :

- Fig 1 shows the image of the speckle field recorded on our DALSA camera (256x256 pixels). The speckle grain size has been chosen in order to match our pixel size (16 μm).

- Fig. 2 indicates a clear linear grain of the signal as a function of the number of pixels indicating that we are indeed adding the individual amplitudes. Let us underline that this is only possible because our signal to noise ratio is larger than one, unless the incoherent noise will also be added in the same way.

To demonstrate that random noise is not strongly influencing the signal in our working conditions we have computed the difference between two successive recording of data. Fig.2 (lower part) indicates that the noise does not accumulate when summing the data of the individual pixels.

- We also computed the phase of the signal [fig.3] through the various pixels of the camera and we found an uniform distribution in the [-π,+π] interval. This can be explained if we consider that the ultrasonic field expands over many acoustic wavelengths.

Finally replacing the camera system by a single detector (0.7 mm^2) and a commercial lock-in detector we have found a reduction in the relative modulation amplitude decrease (AC over DC signal) in good agreement with the expected value (about 50).

We performed an experience where we first scan the turkey breast to get the ultrasound-modulated signal value. Then we injected a small volume of black oil paint to simulate a buried object with an acoustic impedance close to the turkey breast tissues value, and ran a new scan. We reconstructed the one-dimension image by calculating the ratio between the two sums of the individual amplitudes (fig 4). We observed a decreased of the ultrasound-modulated light, due to the black paint which trapped the light modulated and then reduced the signal integrated on the CCD.

In conclusion, these preliminary experiments have demonstrated the ability of our system to take advantage of the entire information contain in a full speckle field. We hope that this approach will soon make the ultrasonic speckle modulation imaging more efficient.

References

[1] L. Wang, S.L. Jacques and X. Zhao «Continuous-wave ultrasonic modulation of scattered laser light to image objects in turbid media » Optics Letters, **20**, 629 (1995)

[2] M. Kempe, M. Larionov, D.Zaslavsky and A.Z. Genack « Acousto-optic tomography with multiply scattered light » J. Opt. Soc. Am. A, **14**, 1151 (1997)

[3] L. Wang and X. Zhao « Ultrasound-modulated optical tomography of absorbing objects buried in dense tissue-simulating turbid media » Applied optics **36**,7277(1997)

[4] P.Gleyzes, F.Guernet et A.C. Boccara « Profilomètre picométrique. II. L'approche multi-détecteur et la détection .synchrone multiplexée . ». J. Optics (Paris) **26** 251,(1995)

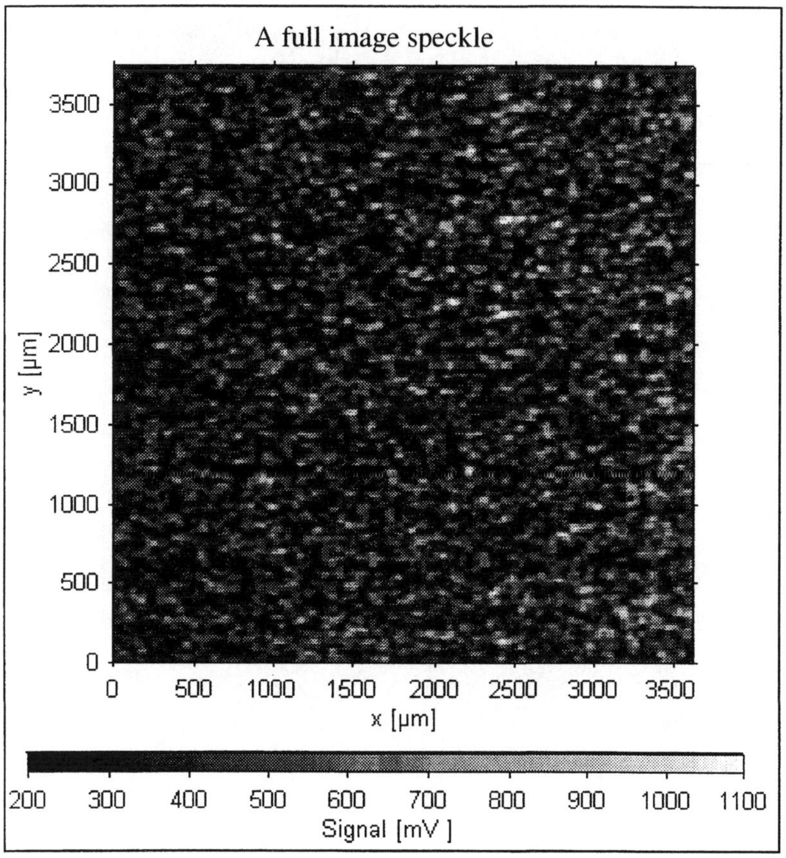

Figure 1

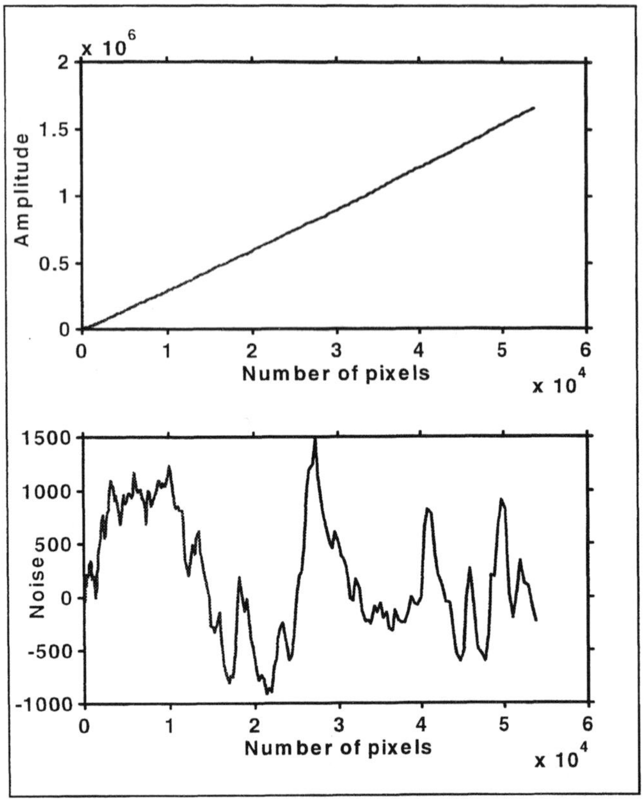

Figure 2

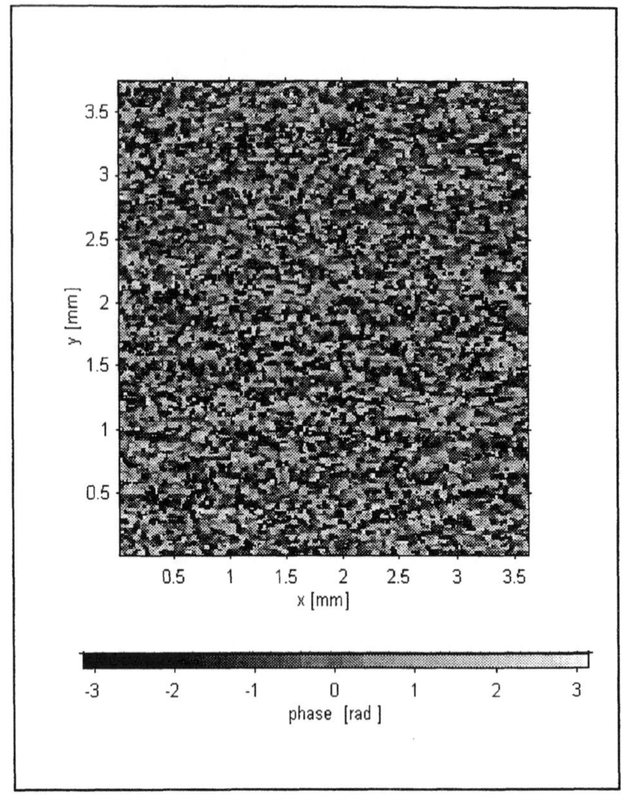

Figure 3

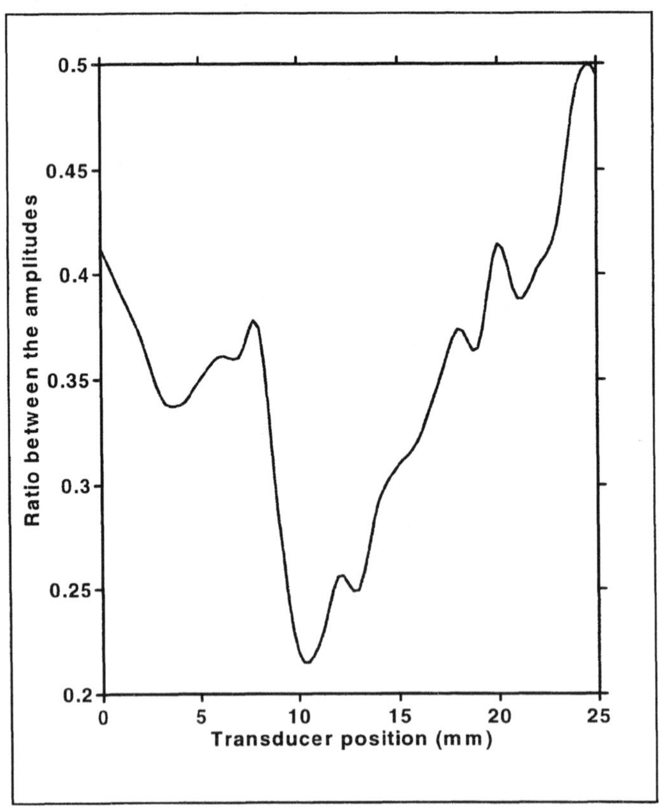

Figure 4

Imaging Turbid Media Using Sonoluminescence

Qimin Shen and Lihong V. Wang
Optical Imaging Laboratory, Biomedical Engineering Program
Texas A&M University, College Station, Texas 77843-3120

Abstract

Sonoluminescence, which was generated internally in the medium by cw acoustic wave, was used to image an object buried in turbid media in two dimensions. The image had a high contrast and spatial resolution of ~2-3 mm. The spatial resolution was limited by the focal-spot size of the acoustic wave, and one could improve the resolution by tightening the focus.

Key Words

Tomography, scattering, sonoluminescence.

Introduction

Optical tomography is an extraordinarily hot topic in recent years because of its advantages of noninvasion, nonionization, and potential contrast for biomedical diagnosis [1]. Several techniques being investigated include time-resolved optical imaging, frequency-domain optical imaging, optical coherence tomography, photo-acoustic tomography, and ultrasound-modulated optical tomography. All of these approaches use an external light source, sometimes a laser.

Ultrasonic generation of light called sonoluminescence (SL), which was first discovered in 1934 [2], has attracted a large amount of attention in this decade since single-bubble sonoluminescence (SBSL) was found in 1990 [3-11]. Although the deep mechanism of SL is still in development, it is well known that light is emitted when tiny air bubbles driven by ultrasound in liquid collapse. The light is visible by naked eye for SBSL in a lighted room and for multiple-bubble sonoluminescence (MBSL) in a darkened room [8]. In this paper, we demonstrate, what is to our knowledge the first time, a new optical tomography technique: sonoluminescent tomography (SLT) for imaging turbid media.

Experimental Considerations

Turbid media (phantoms) were prepared to simulate the optical properties of tissues by mixing dominantly scattering Intralipid in distilled water. Sample I was a liquid Intralipid phantom with a certain amount of Luminol. The Luminol was added to enhance the sonoluminescence so that the sonoluminescence column in clear water could be easily pictured. The corresponding reduced scattering coefficient $\mu_s' = 1.24\ cm^{-1}$ at the wavelength of 400 nm. Sample II was also a liquid Intralipid phantom but without Luminol. This was used to demonstrate that the Luminol was not necessary for imaging. Meanwhile, Trypan Blue dye was added as absorbers. The corresponding $\mu_s' = 8.5\ cm^{-1}$ and the absorption coefficient $\mu_a = 0.02\ cm^{-1}$. The phantoms were sequentially contained in a 400-ml beaker. The beaker was held on a two-dimensional translation stage. A ~1 cm^3 pink rubber cube was buried in the center of the beaker near the focal plane of the ultrasound as the object to be imaged. The distance between the bottom of the cube and the bottom of the beaker was 4.0 cm. The motorized translation stage, controlled by a PC, was able to scan along both the x and y axes, which formed a plane perpendicular to the ultrasonic axis. While the beaker was scanned, the rest of the system, including the optical and ultrasonic systems, was fixed. The ultrasonic transducer (Panametrics, V314-SU) with a focal length of 36.8 mm, a focal diameter of 3 mm and a focal zone of 34.4 mm was

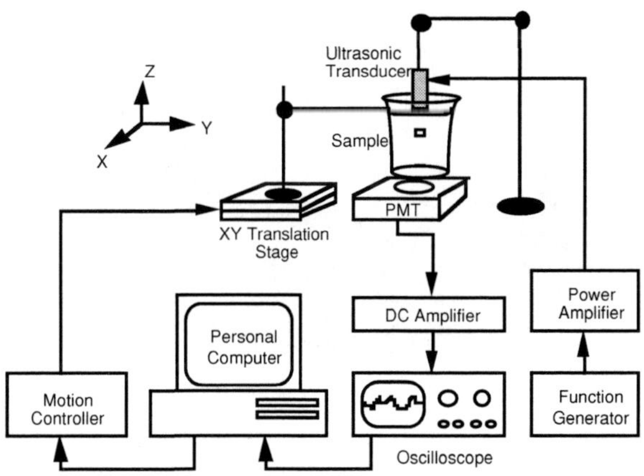

Figure 1. Schematic diagram of the experimental setup for imaging turbid media using sonoluminescence.

placed vertically on the turbid media, and was driven by a 1-MHz sinusoidal wave from a function generator (Stanford Research System, DS345) via a power amplifier (Mini-

Circuits, TIA-1000-1R8). A photomultiplier tube (PMT) (Hamamatsu, R928) beneath the beaker detected the DC signal. The signal was amplified by a low noise preamplifier (Stanford Research System, SR560) and was recorded using an oscilloscope (Tektronix, TDS 640A). Finally, a PC acquired the data from the oscilloscope through a GPIB interface (National Instruments, PCI-GPIB). A block diagram of the experimental setup is shown in Fig. 1.

Results and Discussions

Initially, we applied the ultrasound to a clear solution containing the Luminol. The SL could be seen by the naked eye, but it was so faint that it was only visible in a darkened room because of the MBSL. It could be pictured by long time exposure (Fig. 2) [12]. It can been seen that the length of the SL column was ~3.5 cm, which matched the focal zone of 3.44 cm of the ultrasonic field. The width of the SL column was ~3 mm, which matched the focal diameter of 3 mm of the ultrasonic field.

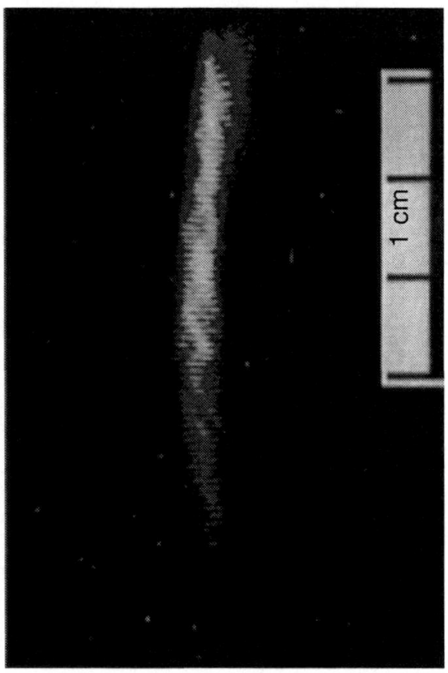

Figure 2. Photograph of sonoluminescence in clear water.

Then, the Intralipid was added into the solution and the cube was suspended in the middle of the beaker. By recording the sonoluminescence intensity as a function of the peak-to-peak voltage applied to the ultrasonic transducer in Intralipid phantom and in distilled water (Fig. 3), we could see clearly that there was a little difference in SL intensity of two different media. The peak-to-peak voltage represented the acoustic power emitted from the transducer. Both curves showed a threshold of acoustic power. When the acoustic power was below the threshold, there was almost no sonoluminescence. Otherwise, when the acoustic power was above the threshold, the sonoluminescence intensity increased rapidly.

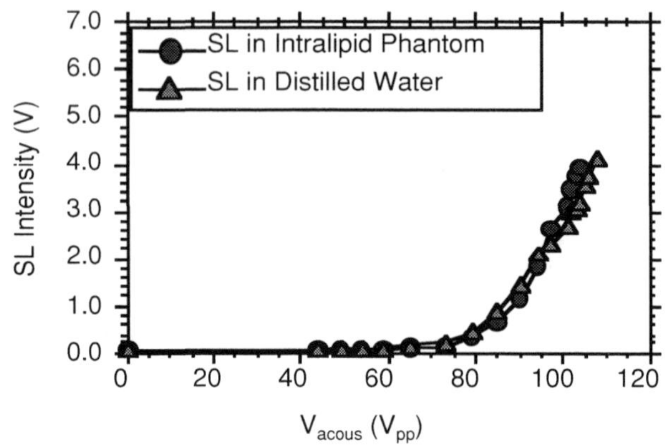

Figure 3. Sonoluminescence intensity versus the driving voltage on the ultrasonic transducer.

By raster scanning the beaker in the x-y plane line by line with a step size of 1 mm, a two-dimensional picture of a pink rubber cube in phantom I was imaged (Fig. 4).

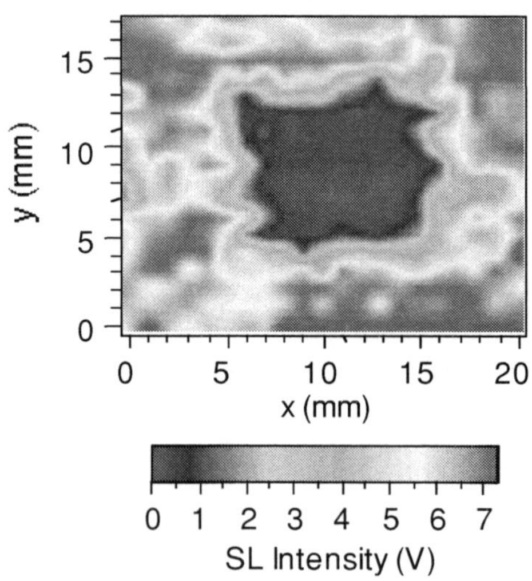

Figure 4. Two-dimensional image of a ~1 cm^3 pink rubber cube in sample I.

When the object was moved towards the ultrasound focus, the SL signal dropped quickly, and vice versa. The spatial resolution of the image was ~2-3 mm. It was expected that the difference between either optical properties or ultrasonic properties of the object and that of surrounding media changed the intensity of SL signal detected by the PMT

As a comparison, we also got a 2-D image of the phantom II in which there was no Luminol (Fig. 5). The contrast and the spatial resolution were as the same.

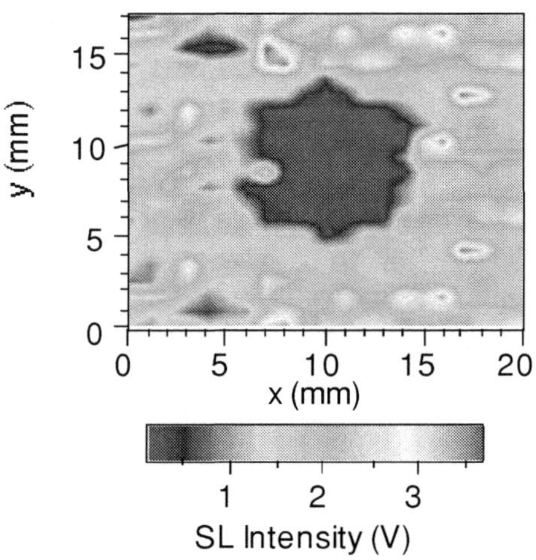

Figure 5. Two-dimensional image of a 8 mm × 8 mm × 7 mm pink rubber cube in sample II.

The contrast of the SL images was based on the difference between the optical and ultrasonic properties of the objects and those of the surrounding medium. The objects used in our experiments were optically opaque and ultrasonically absorbing. When the object was moved toward the ultrasonic focus, the SL intensity dropped quickly because of (1) reduction of the ultrasonic field at and below the focus; (2) no SL generation of the object; and (3) partial blocking of the SL signal by the object.

Although the objects used in this experiment had both optical and ultrasonic contrast relative to the background medium, images of objects may be obtained using sonoluminescence based on several contrast mechanisms in general. As depicted in Fig. 2, the SL generation is affected by the local acoustic power. When an object has ultrasonic contrast, the SL signal originating from the object will differ from that originating from the background medium. When an object has contrast in optical properties, such as absorption coefficient, scattering coefficient, or scattering anisotropy, the SL signal from object will be attenuated due to the light propagation through the object.

Although there are potentially harmful effects caused by cavitation, the threshold of ultrasound intensity leading to volume lesions is very high. The damage threshold in spatial-peak-temporal-peak (SPTP) power was reported to be 400 W/cm^2 and 900 W/cm^2 at 1 MHz for dog brain tissue and dog thigh muscle, respectively [13]. The peak pressure in our experiment was ~2.0 bars at the ultrasonic focus, corresponding to an SPTP power of 1.3 W/cm^2, which was far below the damage threshold. The peak pressure was also far less than the 23-bar safety limit set by the U.S. Food and Drug Administration, which is usually conservative [14]. SLT uses ultrasonic waves to drive pre-existing microbubbles between 5 μm to 50 μm in size and does not generate new cavitations. The acoustic power in SL is much less than that required to generate new cavitations. It was reported that irradiation using acoustic pressures less than a critical value does not cause tissue damage, even if generation of cavitations was observed.[15]

Conclusions

In conclusion, we demonstrated and developed a novel optical tomography technique using sonoluminescence. Both the spatial resolution and the contrast of the image were excellent and may be improved further. The SLT technique is expected to find applications in biomedicine and other fields involving turbid media.

Acknowledgment

Thanks to X. Zhao for her experimental assistance in electronics. This project was sponsored by the National Institutes of Health grants R29 CA68562 and R01 CA71980.

Detailed contact information: Qimin Shen, Ph.D. Phone: (409) 845-4487, Fax: (409) 845-4450, E-mail: QShen@pop.tamu.edu; Lihong V. Wang, Ph.D. (Corresponding author), Phone: (409) 847-9040, Fax: (409) 845-4450, E-mail: LWang@tamu.edu, URL: http://biomed.tamu.edu/~lw.

References

1. R. R. Alfano and J. G. Fujimoto, editors, *Advances in Optical Imaging and Photon Migration,* TOPS Vol. 2 (Optical Society of America, Washington, D.C., 1996).
2. H. Frenzel and H. Schultes, "Luminescenz im ultraschallbeschickten Wasser", *Z. Phys. Chem.*, B 27, 421-424 (1934).
3. B. P. Barber and S. J. Putterman, "Observation of synchronous picosecond sonoluminescence", Nature 352, 318-320 (1991).
4. E. B. Flint and K. S. Suslick, "Sonoluminescence from Alkali-Metal Salt Solutions", J. Phys. Chem. 95, 1484-1488 (1991).
5. L. A. Crum and S. Putterman, "Sonoluminescence", J. Acoust. Soc. Am. 91, 517 (1992).
6. C. C. Wu and P. H. Roberts, "Shock-Wave Propagation in a Sonoluminescencing Gas Bubble", Phys. Rev. Lett. 70, 3424-3427 (1993).
7. W. C. Moss, D. B. Clarke, J. W. White, and D. A.

Young, "Hydrodynamic simulations of bubble collapse and picosecond sonoluminescence", Phys. Fluids 6, 2979-2985 (1994).

8. L. A. Crum and R. A. Roy, "Sonoluminescence", Science, 266, 233-234 (1994).
9. C. Eberlein, "Sonoluminescence as Quantum Vacuum Radiation", Phys. Rev. Lett. 76, 3482-3845 (1996).
10. J. B. Young, T. Schmiedel and W. Kang, "Sonoluminescence in High Magnetic Fields", Phys. Rev. Lett. 77, 4816-4819 (1996).
11. B. P. Barber, R. A. Hiller, R. Löfstedt, S. J. Putterman, and K. R. Weninger, "Defining the unknowns of sonoluminescence", Phys. Reports 281, 65-143 (1997).
12. L. V. Wang and Q. Shen, "Sonoluminescent tomography of strongly scattering media", Opt. Lett. 23, 561-563 (1998).
13. F. J. Fry, N. T. Sanghvi, R. S. Foster, R. Bihrle, and C. Hennige, "Ultrasound and microbubbles: their generation, detection and potential utilization in tissue and organ therapy — experimental", Ultrasound Med. Biol. 21, 1227-1237 (1995).
14. T. A. Whittingham, "The safety of ultrasound", Imaging 6, 33-51 (1994).
15. S. Daniels, T. Kodama and D. J. Price, "Damage to red blood cells induced by acoustic cavitation", Ultrasound Med. Biol. 21, 105-111 (1995).

New Optical Difference Approaches for Sub-Surface Imaging of Tissues

S. G. Demos*, W. B. Wang, J. Ali and R. R. Alfano

Institute for Ultrafast Spectroscopy and Laser, New York State Center for Advanced Technology for Ultrafast Photonic Materials and Applications, Department of Physics, The City College and Graduate School of the City University of New York, 138 Str. & Convent Ave., New York, NY 10031.

*Presently at Lawrence Livermore National Laboratory, PO Box 808, L-411, Livermore, CA 94551.

Abstract

Subsurface imaging approaches are described based on the wavelength and polarization dependence of the backscattered light from different tissue components under polarized illumination and the optical characteristics of a luminous object relative to the absorption and emission characteristics of the host tissue.

Key Words: Light Propagation in Tissues, Polarization, Emission, Contrast Agents, Medical and Biological Imaging, Subsurface imaging.

Noninvasive biomedical optical imaging has attracted significant efforts of many research groups[1]. Optical coherence tomography[2], confocal microscopy, and two-photon excitation microscopy[3] imaging techniques provide high resolution subsurface images with the drawback that the imaging depth is small on the order of 1 mm or less. It has been demonstrated that light propagating inside a scattering medium retains some of its initial polarization information[4]. The relative intensity of the two polarization components of the backscattered light from various tissue types is different depending on the light scattering properties of each tissue type and the illuminating wavelength.[5]

In this work, new subsurface imaging approaches are employed and tested allowing for imaging and detection in the backscattering geometry of "interesting" tissue components having diameters on the order of 1 mm and located a few millimeters to a few centimeters underneath the surface of the tissue. Three imaging techniques are described. The spectral polarization difference technique allows for beneath the surface imaging on the order of few millimeters based on polarization gating. The image quality and/or detection ability of a luminous object containing contrast agent(s) that are embedded a few centimeters below the surface of the tissue can be improved using emission or excitation difference imaging.

The experimental setup used in this work is shown in fig. 1. The illumination laser beam was obtained from a dye laser, synchronously-pumped by a mode-locked, pulse-compressed Nd:YAG laser tunable in the 570-635 nm and 690-780 nm spectral regions using a R6-G and Pyridine 1 dyes, respectively. The dye laser beam was expanded to a diameter of 10-cm and used as the illuminating source with an average intensity of ~5 mW/cm^2. A cooled CCD camera equipped with a 50 mm focal length camera lens was used to record the retroreflected image components. While the camera was focused on the surface, the depth of field of the zoom lens allowed for imaging below the surface without any further adjustment of the lens.

For the spectral polarization difference imaging method, a first polarizer is located in the laser beam pathway to ensure illumination with linearly polarized light. A second polarizer is positioned in front of the CCD camera with its polarization orientation perpendicular to that of the illumination. Images are obtained using various illumination wavelengths so that the depth of propagation of the photons, before they are backscattered and used to form the image, is different due to differences in absorption and scattering.

For the imaging approaches that employ contrast agents, small breast tissue lesions were embedded in cardio green dye which exhibits strong absorption in the 720-820 nm spectral region and emits in the 800-860 nm spectral region. These tissue lesions were then sandwiched between breast chicken tissue layers at different depths underneath the surface of the tissue sample. A laser -line interference filter positioned in the laser beam pathway ensures monochromatic illumination while a laser-line notch filter positioned in front of the CCD lens excludes any backscattered light from reaching the detector. A second filter follows to select the desirable spectral range of the emitted light by the sample to be used for imaging. In the excitation difference technique, images recorder are formed by photons emitted in the 820-950 nm spectral region under different wavelength laser illumination. In the emission difference technique, images recorder are formed by photons emitted in different spectral regions under same wavelength laser illumination.

Fig. 2 shows polarized images of the palm of a human hand under 580 nm linearly polarized laser illumination. Fig. 2a shows the parallel polarization image (polarization orientation of analyzer in front of the CCD is parallel to polarization orientation of the illumination) where the details of the papillary ridges of the skin can be seen. Fig. 2b shows the perpendicular polarization image component where the structures of the skin are not observable. The reason for the differences in the two polarization image components arises from the fact that the photons that are backscattered and are perpendicularly polarized must first undergo sufficient scattering events to lose their polarization information. Photons backscattered from the surface of the tissue have less opportunity to depolarize due to scattering. As a result, the backscattered photons from the tissue surface and the initial layers beneath the surface are recorded predominantly in the parallel polarization image component. The perpendicular polarization image component is formed predominantly by photons that reached inside the tissue to certain depth before they emerged in the backscattering direction. Consequently, the parallel image component contains the image information of the surface of the tissue while the perpendicular image contains only the diffused image (depolarized image) where very little image information of the surface is present. The image arising from depolarized photons is equally distributed in both polarization image components. Subtracting the two image polarization components, the depolarized image components are canceling-out and a clearer image of the skin structures emerges[6].

When a scattering medium such as tissue is illuminated with linearly polarized light, the depolarized image component contains mostly subsurface image information and very little information about the surface. The axial volume of information depends on the depth of penetration of light before backscattering which is a function of the scattering properties of the medium and the absorption characteristics. A longer wavelength will penetrate deeper inside the tissue due to less absorption and scattering. The average depth of penetration depends on the wavelength and type of tissue. Based on these concepts, imaging of structures at different depths inside the tissue can be achieved by recording the perpendicular polarization images for different illuminating wavelengths (λ_1, λ_2) (in order to avoid the polarized surface image component and reach different mean penetration depths[7]) using times of exposure $t(\lambda_i)$ such that the image intensity from the outer layers of the tissue is normalized to approximately equal intensity for both illuminating wavelengths. Subtraction of these images obtained under different illuminating wavelengths cancels-out the image components from the outer layers of the tissue and the new image contains predominantly information from the inner layers of the tissue where the longer in wavelength illuminating light reaches more efficiently. Selecting the appropriate illuminating wavelengths, one can effectively reach and image different "depth zones" inside the tissue.

This polarization technique is illustrated by the images displayed in Fig. 3. The sample used for the demonstration of this subsurface imaging technique was prepared using fat and breast chicken tissue. A

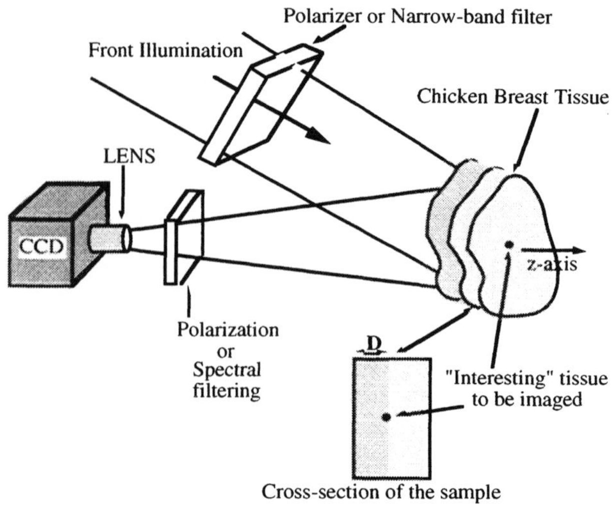

Figure 1. Schematic diagram of apparatus utilized for backscattering subsurface imaging. Details are provided in the text.

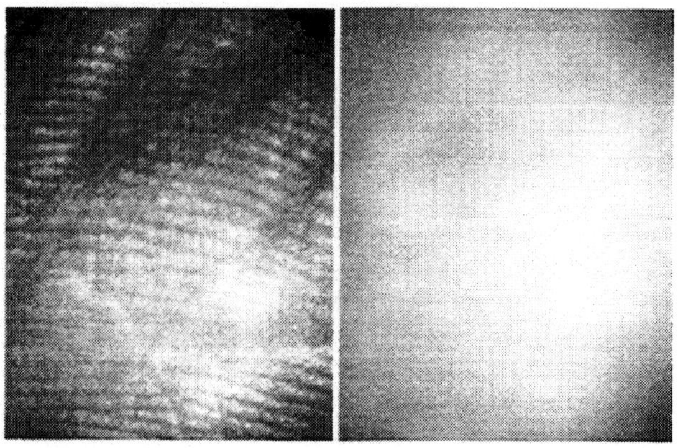

Figure 2. The parallel polarization image component (left) and the perpendicular polarization image component (right) of a human palm under 580 nm polarized laser illumination.

small piece of fat tissue (~5X5X2) mm^3 was positioned in the middle of a 3X3X1 cm^3 (1-cm thick) breast chicken tissue layer. A 3.5 mm thick layer of breast chicken tissue was then positioned on the top of the fat containing breast chicken layer. The sample was placed between two glass slides and it was slightly compressed to a uniform thickness. 580 and 630 nm linearly polarized illumination was used and the perpendicular polarization images were obtained in the backscattering geometry (see fig. 1). Fig. 3a shows the perpendicular polarization image under 580 nm illumination. Fig. 3b shows the perpendicular polarization image under 630 nm illumination at the same position. The exposure times were determined by measuring the temporal profile of the perpendicular component of a backscattered pulse at the same wavelengths. The peak intensity of the backscattered pulse is due to reflection of photons from the outer layers of the tissue. The times of exposure were adjusted so that the peak intensities of the temporal profiles under 580 and 630 nm pulsed illumination are equal. The 630 nm photons reach a larger penetration depth than the 580 nm photons because of lower absorption inside the tissue. Subtraction of the 580 nm image from the 630 nm image leads to a new image arising from photons that propagated to deeper layers of the tissue and reached the fat lesion. This principal is demonstrated in fig. 3c where the outline of the fat lesion emerges as a brighter in intensity structure. For comparison with the actual image of the fat lesion, the perpendicular image of the sample with the 3.5 mm thick outer layer removed is shown in fig. 3d.

The subsurface imaging technique described in this work based on subtraction of the perpendicular images obtained under different wavelengths can be used to highlight differences in absorption by blood or scattering due to the presence of different types of tissue at different depths zones inside a tissue.[7] A structure inside the tissue that exhibits higher absorption will emerge in the polarization difference image darker in intensity. A structure that scatters and depolarizes light more will appear in the polarization difference image brighter than the rest of the image as demonstrated in Fig. 3c. With gradual change of the illuminating wavelengths one may display structures at different depths. The maximum imaging depth depends on the scattering and absorption characteristics of the tissue sample. It has been shown that human breast tumor tissues scatter and depolarize light more than normal tissue[5]. Therefore, the subsurface imaging technique described here may be very useful in detecting malignant tumor lesions underneath the surface of a normal tissue.

Contrast agents have been introduced in optical imaging to enhance the ability to image an object hidden in a scattering medium and improve the image quality. An object located inside a scattering medium such as tissue is made luminous and viewed using its

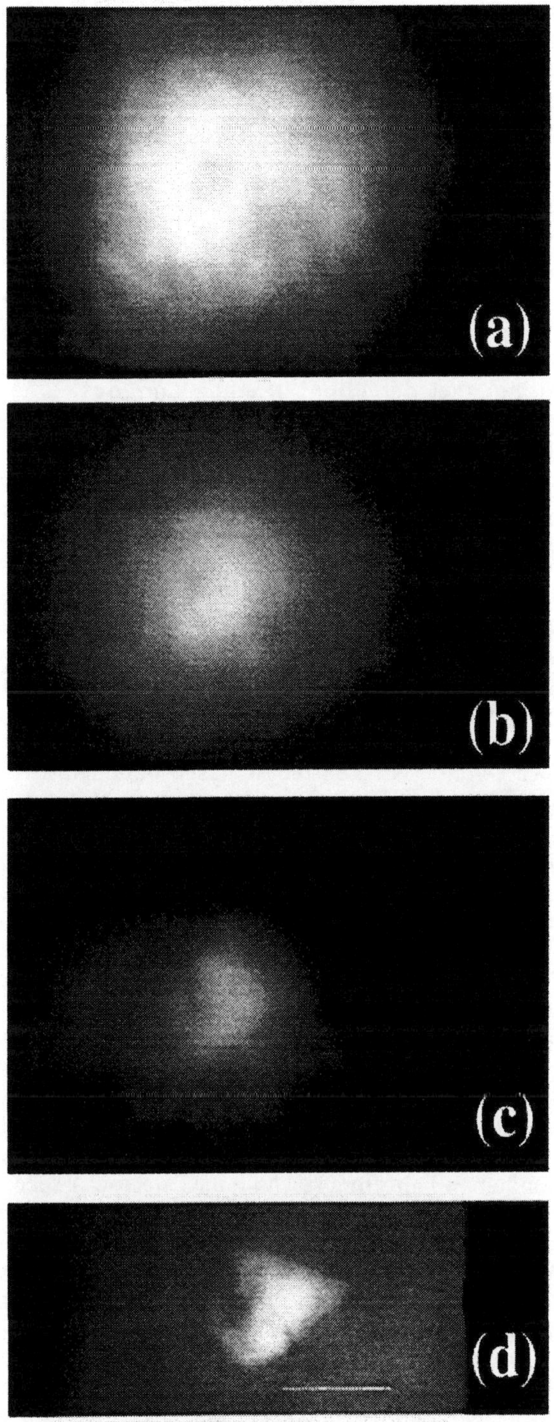

Figure 3. Backscattering images of the perpendicular polarization component of the breast chicken tissue containing a fat lesion 3.5 mm beneath the surface a) under 580 nm illumination, b) under 630 nm illumination, c) after normalization and subtraction of the 580 nm image from the 630 nm image, and d) after removing the 3.5 mm outer layer to expose the fat tissue lesion.

emission light by introducing a fluorescence dye. When tissue is illuminated by light, part of the light is absorbed by the tissue giving rise to emission by the tissue predominantly at longer wavelengths from the illuminating wavelength. When a luminous object embedded into the tissue is illuminated by an external to the tissue light source, the presence of the luminous object may be revealéd from the emitted light by the object superimposed to the emitted light by the tissue. When the object is very small and/or it is embedded deep into the tissue, emission by the tissue coinciding in the same spectral region with the emission by the luminous object may be orders of magnitude higher making the detection of the luminous object impossible. This work demonstrates approaches to address this problem.

The contrast agent utilized is the cardio green dye. This dye was chosen because its absorption and emission spectra are different than that of the tissue under investigation. The difference in the absorption at different wavelengths and the difference in the emission under the same excitation between tissue and contrast agent is used to enhance the visibility of the luminous object using emission difference detection and excitation difference detection, respectively. The animal tissue sample used for demonstration is a ~5X5X4 cm^3 chicken breast tissue obtained at the market. The sample is composed of ~5X5X1 cm^3 tissue layers to form the 4 cm thick sample. The illumination and imaging direction was approximately parallel to the 4 cm thick z-axis of the sample. A small breast tissue lesion ≈ 1-2 mm in diameter was embedded in cardio green dye and then sandwiched between breast chicken tissue layers. The sample was illuminated with a 10 cm diameter laser beam in a direction slightly off the z-axis. A CCD camera equipped with the 50 mm camera lens was used to capture the image in the backscattering geometry. A laser-line notch filter positioned in front of the camera lens was used to exclude any backscattered illumination light from reaching the detector. A second filter is used to select the desirable spectral range of the emitted light by the sample to record the image. This filter is a 790 nm and a 830 nm narrow band interference filter for the emission spectral difference imaging experiments or an 830 nm long pass filter for the excitation spectral difference imaging experiments.

In the emission difference imaging method, the sample is illuminated with monochromatic light of wavelength λ_E and two images are recorded using two narrow band filters in front of the CCD detector centered at wavelength λ_1 and λ_2, respectively. Wavelengths λ_1 and λ_2 are selected so that while the emission by the contrast agent (luminous object) is very different at λ_1 and λ_2, the difference in the emission by the tissue sample is approximately the same and in any case much smaller than that of the dye. Subtraction of the two images when normalized

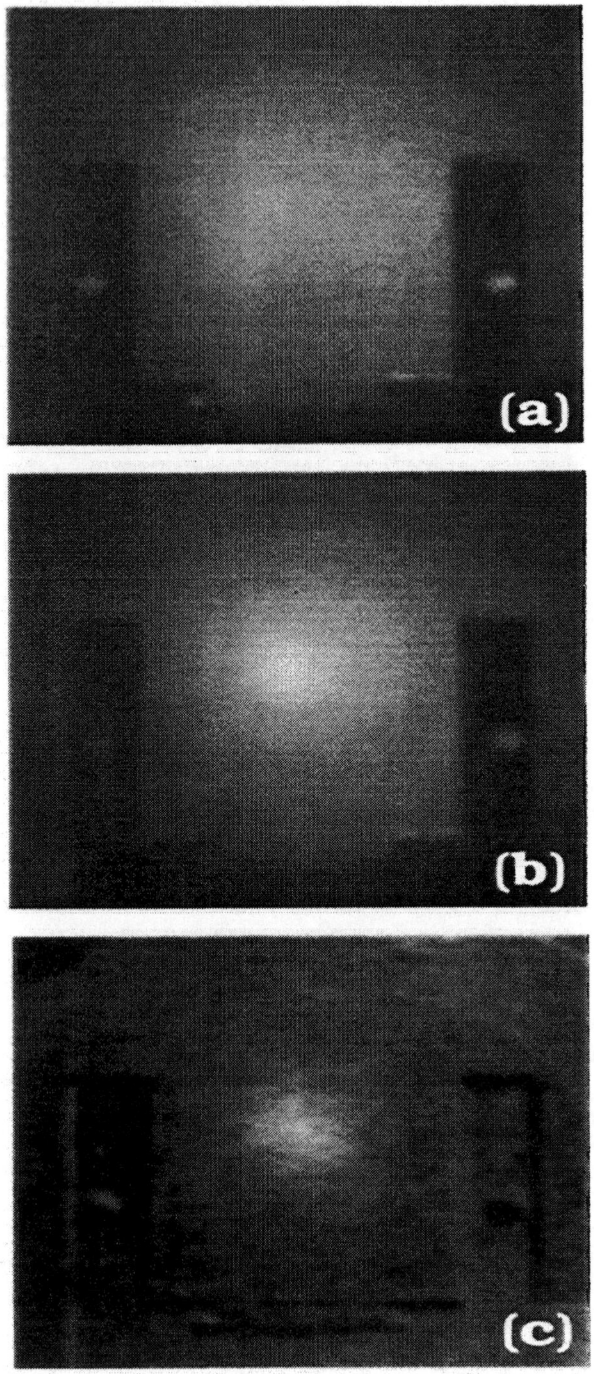

Figure 4. Images of a luminous object (~1-mm diameter chicken tissue that was embedded in cardio green dye) placed inside a breast chicken tissue sample 3.1 cm underneath its surface obtained under 630 nm illumination using (a) a 790 nm and (b) an 830 nm narrow band interference filters in front of CCD; (c) is the difference image following normalization of the intensity due to tissue emission.

to cancel out the image arising from the tissue leads to improvement of the visibility of the luminous object since the background signal arising from the tissue is reduced or eliminated. The above described principals were applied to a tissue sample where the luminous object (~1 mm diameter breast tissue lesion embedded in cardio green dye) is located 3.1 cm below the surface. The wavelengths used for illumination and image formation were λ_E=630 nm, λ_1=790 nm and, λ_2=830 nm. The emission by the tissue at 790 and 830 nm is not very different while the intensity of the emission by the contrast agent is different by a factor of ≈3.This allows for enhancement of the visibility of the luminous object after removal of the contribution in the image component arising from tissue emission using subtraction of the images. This is demonstrated in fig. 4 where the image of the tissue sample containing the luminous object is shown when the imaging wavelength was 790 nm (fig. 4a), 830 nm (fig. 4b) and, the image obtained after normalization and subtraction of the images at 790 and 830 nm (fig. 4c). The tissue sample was held in place by a metal holder which appears as a reversed Π object. In the images shown in fig. 4, the presence of the luminous object in figs. 4a and 4b is not clear and it is very difficult to determine while the presence of the luminous object in fig. 4c is conclusive and easy to see. This demonstrates that using the emission difference imaging technique one can clearly see the presence of the luminous object which otherwise is very difficult or impossible to determine.

In the absorption difference detection, the sample is illuminated with monochromatic light at two different wavelengths λ_1 and λ_2 and two images are recorded. A band pass filter is positioned in front of the CCD detector so that the image is formed by emitted light only from a specific spectral region $\Delta\lambda$. λ_1 and λ_2 are selected so that the absorption by the tissue is approximately the same for both wavelengths while the absorption by the contrast agent (luminous object) is very different at λ_1 and λ_2. As a result, the recorded emission strength of the tissue is approximately the same but the emission of the luminous object is very different. Subsequent subtraction of the acquired images leads to improvement of the visibility of the luminous object since the background signal arising from the tissue is reduced. This technique is demonstrated in fig. 5 where the luminous object (~2 mm diameter) was positioned 1.5 cm below the surface. The illumination wavelengths were λ_1=710 nm and λ_2=760 nm. An 830 nm long-pass filter was positioned in front of the CCD camera. The absorption by the tissue at 710 and 760 nm is not very different while the absorption by the dye and subsequent emission by the luminous object is different by a factor of ~2. The images of the tissue sample containing the luminous object for illuminating

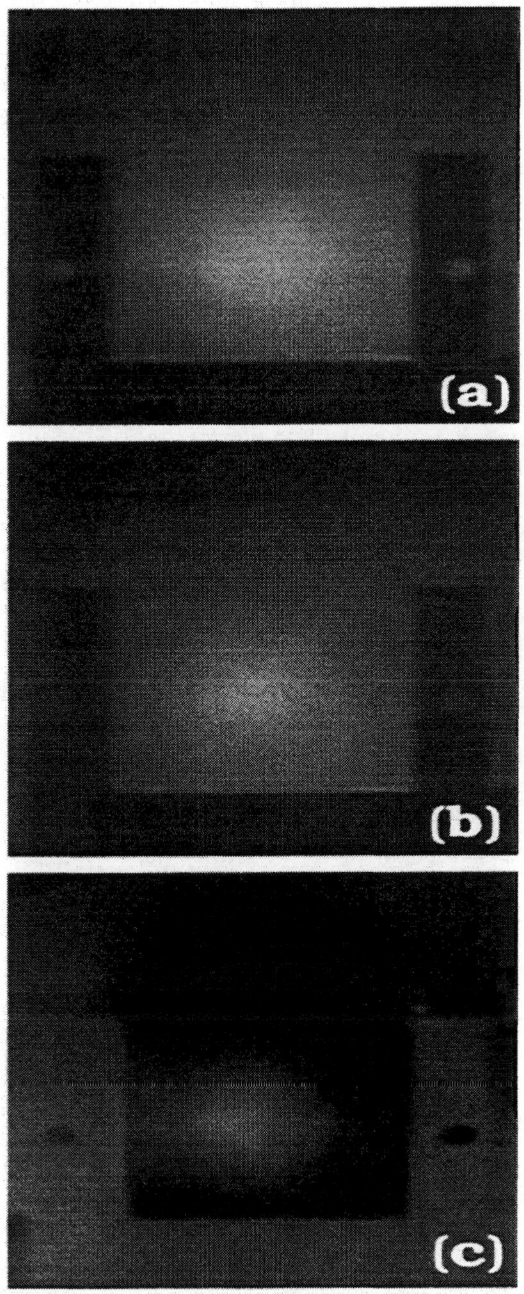

Figure 5. Images of a luminous object (~1-mm diameter chicken tissue that was embedded in cardio green dye) placed inside a breast chicken tissue sample 1.5 cm underneath its surface obtained under (a) 710 nm and, (b) 760 nm illumination with an 830 nm long pass filter in front of CCD; (c) is the difference image following normalization of the intensity due to tissue emission.

wavelengths 710 nm and 760 nm are shown in figs 5a and 5b, respectively. The image obtained after normalization and subtraction of the images obtained under 710 and 760 nm illumination is shown in fig. 5c. The presence of the luminous object in figs. 5a and 5b is not apparent while its presence is clearly denoted in fig. 5c.

The absorption difference and excitation difference imaging methods described in this work can help to increase the visibility of the luminous object located inside tissue. The overall result is the improvement of the visibility and image quality of the luminous object by canceling out the contribution of the tissue emission in the image.

In conclusion, subsurface imaging approaches were discussed in this work demonstrating the ability to detect and image in the backscattering geometry an "interesting" tissue component located below the surface of the tissue. The maximum imaging depth when using the spectral polarization difference technique and the appropriate illumination wavelengths may be up to 1 cm. The excitation or emission difference imaging technique have demonstrated ability to image a luminous object located up to a few cm below the surface. The drawback, however, for the enhancement of the imaging depth is the reduction of the image resolution.

This research is supported by New York State Science and Technology Foundation, Mallinckrodt Medical Inc., and Mediscience Technology Corp..

References:

1. S. K. Gayen and R. R. Alfano, "Emerging optical biomedical imaging techniques", Opt. and Phot. News, **7**, No 3, 17-22, (1996) and references therein.
2. D. Huang, E. A.Swanson, C. P. Lin, J. S. Schuman, W. G. Stinson, W. Chang, M. R. Hee, T. Flotte, K. Gregory, C. A. Puliafito, and J. Fujimoto,"Optical coherence tomography", Science, **254**, 1178-1181 (1991).
3. W. Denk, J. H. Strickler, and W. W. Webb, "2-photon laser scanning fluorescence microscopy", Science, **248**, 73-76, (1990).
4. S. G. Demos and R.R. Alfano, "Temporal gating in highly scattering media by the degree of optical polarization", Optics Lett., **21**, 161 (1996).
5. S. G. Demos, H. Savage, Alexandra S. Heerdt, S. Schantz and R.R. Alfano, "Polarization Filter for Biological Optical Imaging" Photoch. and Photob., **66,** 821 (1997).
6. S. G. Demos and R.R. Alfano, "Optical fingerprinting using polarisation contrast improvement", Electr. Lett., **33**, 582 (1997).
7. M S. Patterson, S. Andersson-Engels, B. C. Wilson, and E. K. Osei, "Absorption spectroscopy in tissue-simulating materials: a theoretical and experimental study of photon paths", Appl. Opt., **34**, 22-30 (1995)

Nonlinear Optical Histological Spectroscopy and Imaging of Biological Tissues

Yici Guo, Feng Liu, Q. Z. Wang, H. E. Savage,* N. Zhadin, P.P. Ho, S. Schantz,* and R. R. Alfano

Institute for Ultrafast Spectroscopy and Lasers and New York State
Center for Advanced Technology for
Ultrafast Photonic Materials and Applications
Departments of Electrical Engineering and Physics
The City College and Graduate School of The City University of New York
New York, New York 10031

**Memorial Sloan Kettering Cancer Center*
425 East 67 Street
New York, N.Y. 10021

Abstract

We report on nonlinear optical (NLO) spectroscopy and tomographic imaging using second harmonic generation (SHG) and two-photon excitation of fluorescence (TPF) in highly scattering native animal and human tissues.

Key words

(190.4180) Multiphoton process, (170.5810) Scanning microscopy, (170.6510) Spectroscopy, tissue diagnostics, (170.7050) Turbid media.

Second harmonic generation from native biological tissues has demonstrated the potential for optical histology based on symmetry dependence of tissue structures on the second order nonlinear optical $\chi^{(2)}$ tensor components[1-2]. Two photon and three photon excitation of fluorescence from native fluorophores in tissues provides assessment on physiological state related chemical content variations[3-6]. SHG and TPF signals can be used to image subsurface structures up to two scattering lengths below the sample surface due to the coherence constraint in the excitation process. The illumination wavelength used in these two NLO techniques in the NIR spectral region provides deeper penetration into the tissue, compared with conventional linear optical approaches using UV and visible light. SHG and TPF signals are localized at the focal region due to the quadratic dependence on the excitation intensity, leading to optical sectioning without restraint to a confocal pinhole in front of the photo-detector.[6]

The back-reflection geometry is used to demonstrate noninvasive optical histology and histochemical analysis of surface and subsurface structures in native animal and human tissues.

The experimental setup is shown in Fig. 1. Two femtosecond laser systems were used to excite NLO signals from the tissue samples. An amplified colliding-pulse mode-locked dye laser system generated 100 fs pulses at 625 nm, with a repetition rate

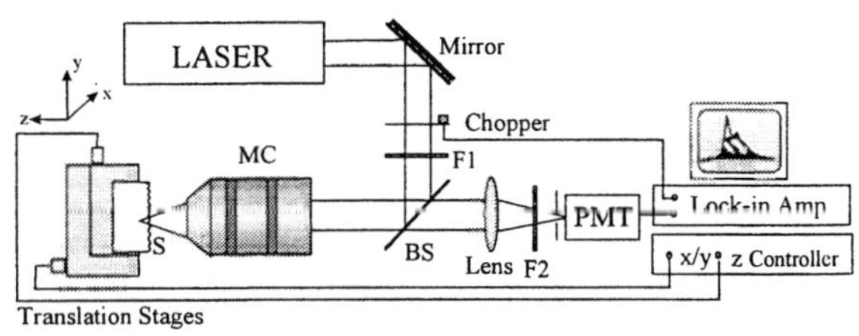

Fig. 1. Schematic of the experimental setup: MC, microscope objective; BS, beam splitter; F1, F2, long and narrow bandpass filters, respectively; PMT, Photomultiplier tube.

of 6.5 kHz. A Ti:Sapphire laser system generated 120 fs laser pulses at 780 nm. A 27X microscope objective was used to deliver the pump light and collect the signal. The axial SHG and TPF profiles versus depth were measured by a scan of the sample with a stepping-motor-controlled translation stage, and the lateral scan was performed with a second translation stage. A photomultiplier tube and a computer controlled lock-in amplifier were used to record the data. SHG and TPF tomographic images were obtained by scanning the laser focus in a two dimensional fashion along the lateral (x-y plane) and the depth (z) directions, while accumulating the NLO signal intensity point by point inside the tissue.

Fig. 2 shows the gray scale SH tomographic image of a sample comprised of a tendon-like membrane of fascia attached to a muscle chicken tissue. The salient feature of this figure shows a fast increase of the SH signal just beneath the tissue surface and a slow decrease deeper into the tissue.

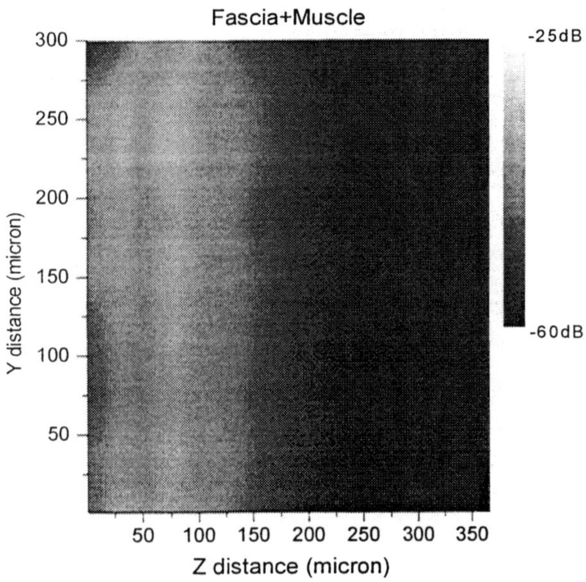

Fig. 2. 2D SHG depth-image of fascia membrane attached to the chicken muscle tissue. Image obtained under 625nm, 100fs laser excitation.

Multiple interfaces are resolved from the observed apexes along the z (depth) direction appeared before and after the dominant maximum.

Fig. 3 shows a raster (x,y) scanned TPF image of tryptophan distribution in a muscle chicken tissue. The muscle fibers were orientated along the z axis. The round shape of cross section of the muscle fibers is clearly displayed. Fig. 4 shows the SHG and TPF spectral responses obtained from a normal and a cancerous human breast tissues under 625nm femtosecond laser excitation. The spectral curves show a lower SHG intensity at 312.5 nm and a higher TPF intensity at 345 nm from cancerous breast tissue, compared with those from its normal counterpart.

A 2D lateral SHG image of a breast tissue sample comprised of partial tumor and partial normal tissue is shown in Fig. 5, in which the tumor

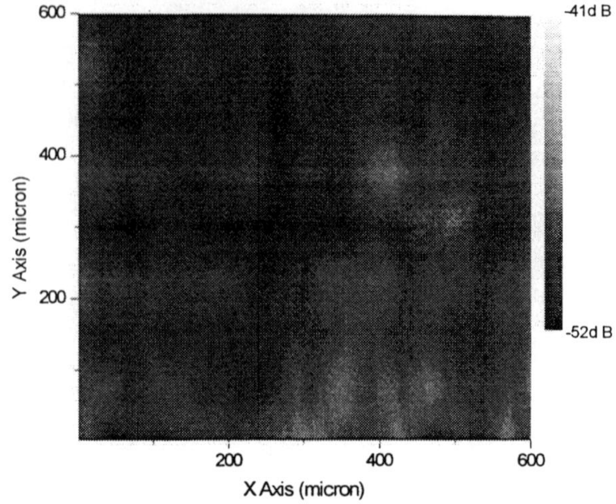

Fig. 3. Lateral TPF image of tryptophan distribution of chicken muscle tissue with fiber's long axis aligned along the z axis. Image obtained under 625nm, 100fs laser excitation.

area is marked by the lower SHG intensity response on the upper right quadrant. This signal reduction shows the randomness of the structure in the tumor region. Lower SHG intensity regions are also seen in the normal part of the tissue sample, especially on the lower left quadrant. This observation shows the spread of malignancy into the normal tissue.

SHG tomography reveals the morphological changes on a microscopic level. TPF imaging provides distribution of native molecular components beneath tissue surface. Compared with the reflectance based imaging techniques, the signal intensity in SHG relies inversely on the quadratic power of the index of refraction. Therefore, the reflectance change can be

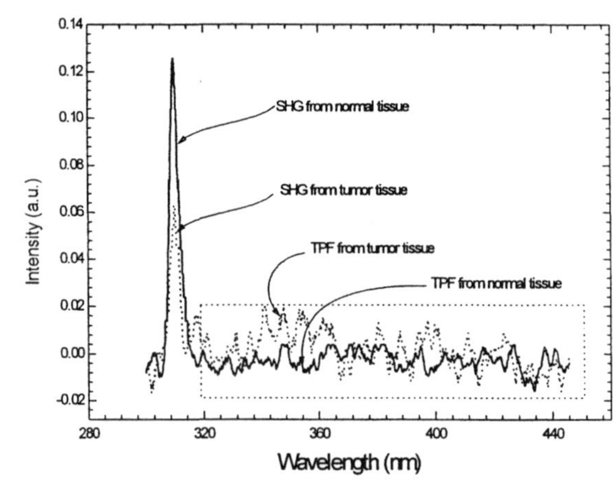

Fig. 4. SHG and TPF spectra obtained from human breast normal and cancerous tissues under 625nm, 100fs laser excitation.

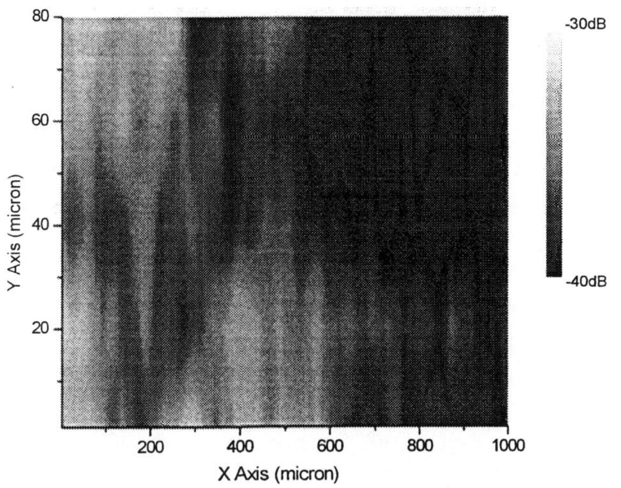

Fig. 5. Lateral SHG image of human breast tissue under 780nm 120fs laser excitation. The tumor site is indicated by the dark region in the upper right.

highlighted more than that in a linear optical method such as the optical coherence tomography.[7] Furthermore, SHG offers the additional advantage in its symmetry sensitivity due to dependence on the local $\chi^{(2)}$ tensor[2]. The SH response reflects on the symmetry of the microstructures and could provide additional information regarding to the state of the tissue. This property indicates the potential to detect epithelial abnormalities associated with morphological distortions, such as dysplasia, frank cancer, and precursors of malignancies.

Acknowledgment

This research is supported in part by grants from Mediscience Technology Corp., New York State Science and Technology Foundation, The U. S. Air Force Office of Scientific Research and DOE.

References

1. Yici Guo, P. P. Ho, H. Savage, D. Harris, P. Sacks, S. Schantz, Feng Liu, N. Zhadin and R. R. Alfano, "Second-harmonic tomography of tissues," Opt. Lett. **22**, 1323 (1997).
2. Y. R. Shen "*Principles of Nonlinear Optics*" Chapter 25, John Wiley and Sons, (1984).
3. W. Denk, J. H. Strickler and W. W. Webb, "Two-photon laser scanning fluorescence microscopy," Science **248**, 73 (1990).
4. D. W. Piston, B. R. Masters and W. W. Webb, "Three-dimensionally resolved NAD(P)H cellular metabolic redox imaging of the in situ cornea with two-photon excitation laser scanning microscopy," J. of Microsc. **178**, 20 (1995).
5. S. Maiti, J. B. Shear, R. M. Williams, W. R. Zipfel and W. W. Webb, "Measuring serotonin distribution in live cells with three photon excitation," Science **275**, 530 (1997).
6. B. R. Masters, P. T. C. So and E. Gratton, "Multiphoton excitation fluorescence microscopy of in vivo human skin," Biophys. J. **72**, 2405 (1997).
7. D. Huang, E. A. Swanson, C. P. Lin, J. S. Schuman, W. G. Stinson, W. Chang, M. R. Hee, T. Flotte, K. Gregory, C. A. Puliafito and J. G. Fujimoto, "Optical coherence tomography," Science **254**, 1178 (1991).

Laser Capture Microdissection (LCM) and the Future of Molecular Pathology

Robert F Bonner, Ph.D.

Laboratory of Integrative and Medical Biophysics, National Institute of Child Health and Human Development, National Institutes of Health , Bethesda, MD 20892-5766, ph. (301)-435-1946, fax (301)-496-6608, email: bonner@helix.nih.gov

With the revolution in molecular biology and its application to the understanding of human disease, we stand at the threshold of a new era of molecular medicine which offers more specific and effective therapies based on sophisticated molecular diagnosis of an individual patient's disease. Powerful treatments will be applied in protocols specific to the set of molecular changes that has occurred within the diseased cells. For example, if we can identify the set of specific molecular changes that have occurred in the cancer cells obtained by biopsy or surgery, we should be able to select a more effective treatment with minimal adverse side effects. Currently the human genome is being mapped and the list of all expressed human genes will before long be nearly complete. Currently, the major scientific and medical challenge is to understand the molecular events that drive normal tissue morphogenesis and the progression of pathologic lesions in actual tissue. Using polymerase chain reaction (PCR), and high-throughput, automated micro-hybridization arrays and mutation screening methods, DNA or RNA once extracted from tissue biopsies can be analyzed with a parallel panel of hundreds or even thousands of genetic markers. Tissues are complex mixtures of many different cell types all interacting with neighbors by complex biochemical and physical signaling which affects the molecular changes in each individual cell. In cancer, a critical spontaneous mutation in one cell may lead to the proliferation of a particular premalignant form, which in term may mutate to a localized malignancy, followed by another set of mutations leading to an invasive cancer cells resistance to a class of standard cancer drugs. If we fail to isolate and purify pure subgroups prior to our molecular analysis, the specific critical molecular changes in the most difficult to treat but critical cancer cells will be masked by the contaminating cells. Microscopic examination of the tissue sections by a trained pathologist can readily identify the critical cancer lesions, but new separation methods are needed if powerful multiplex molecular methods are to be applied to their analysis. Similarly, the task of determining critical gene expression patterns in studies of development, normal function, and disease progression depends on the extraction of specific cells from their complex tissue milieu. In multiplex molecular testing methods, (mixed populations)*garbage in* leads to *garbage out*. Laser capture microdissection (LCM) has been developed to provide a rapid, reliable method to procure pure populations of targeted cells from specific microscopic regions of tissue sections for subsequent molecular analysis. Initially LCM is being used by the research community in order to understand the complex molecular changes in normal development and in the progression of disease pathologies. In the longer term, we are rapidly developing means of integrating LCM with molecular analysis techniques in order to provide a future technology for routine molecular diagnosis of clinical specimens.

Cell sorting has long been used in isolating cell grown in tissue culture or from cellular suspensions such as a patients blood, providing purified sub-populations for molecular analysis. However, digestion of most complex tissues into individual cells is not practical. Further such a bulk process would still mix cells identified as the small class by flow cytometry which may have undergone different mutation sequences or had significantly different environments which can induce important changes in gene expression and post-transciptional modification of proteins. In LCM, a transfer surface is placed onto the tissue section and then focally bonded to the targeted tissue, allowing it to be selectively removed for molecular analysis. In the LCM adapted microscope, the pathologist or researcher views the tissue and selects microscopic clusters of cells for analysis, then activates a laser within the microscope optics. The pulsed laser beam is absorbed within a precise spot on the transfer film immediately above the targeted cells. At this precise location the film melts and fuses with the underlying cells of choice. When the film is removed, the

Reprinted from Advances in Optical Imaging & Photon Migration,
1998 Technical Digest, © 1998 Optical Society of America

chosen cell(s) remain bound to its undersurface, while the rest of the tissue is left behind.

LCM offers a number of advantages. The cells of interest are transferred from the tissue slide to the polymer film. The separate fragmentation step used in conventional microdissection or digestion step necessary for flow cytometry and the resulting contaminating debris are avoided. Secondly, only the targeted cells are affected with a precision of transfer which can approach 1 µm. Thus, the remaining tissue on the slide remains fully accessible for further capture, allowing comparative molecular analysis of adjacent cells. The exact morphology of the procured cells is retained and held on the transfer film, so transfer images constitute a diagnostic record of the cells undergoing molecular analysis. With LCM, the procurement of specific cells from a complex tissue section is reduced to a routine method amenable to widespread research and clinical diagnostic use.

The thermoplastic polymer film contains special infrared absorbing dyes to allow use with near-IR gallium arsenide laser diodes, which are easily and economically integrated into normal microscopes. The polymer and its substrate do not absorb the visible wavelengths used to examine and target cells within the stained tissue section in the microscope. The chosen thermoplastic polymer has a steep decrease in viscosity with rising temperature, so that mild focal heating is sufficient to cause it to flow into the interstices of the targeted tissue and rapidly form a bond which is stronger than the bond between the tissue section and the glass slide. In comparison to laser tweezers, LCM is capable of holding individual targets much more strongly which allows their physical extraction from a tissue with strong intercellular adhesive forces. Furthermore, the bonding process occurs within 20-50 milliseconds, allowing a statistically significant number of cells to be rapidly separately targeted and bonded to the transfer film, before the pooled collection is physical separated from the tissue or cytology specimen and placed in a chamber for molecular analysis. The mild, brief thermal transients experienced by the tissue in the LCM capture process do not damage DNA, mRNA, or proteins. Thermoplastic bonding avoids chemical reactions which can crosslink biological molecules in the tissue and alter subsequent molecular analysis particularly those which require quantitative transcription (PCR and RT) or enzymatic function. Focusing the laser diode beam with the microscope objectives used for imaging easily allows focal activation of the polymer in spots as small as 4 µm. The transferred tissue is firmly embedded in the film, but at the same time is fully accessible to aqueous solutions necessary for the extraction of DNA, RNA, and proteins from it.

NIH has developed computer software to control the tissue microdissection process and to store data records, including digital images of the microdissected cells before and after transfer. This record can be integrated with subsequent molecular analysis results, which is critical in research and clinical applications. We are currently developing computer-based automation of the entire LCM process including targeting specific cells, capture, and molecular analysis. Many of these molecular analysis techniques have included fluorescence detection for rapid analysis which is critical in a clinical environment. In general such optical methods will offer advantages in speed and cost. Rapid, automated multiplex molecular analysis may be critical for the rapid detection and identification of infectious organisms and particular mutations that are drug-resistant (e.g., drug resistance TB while the patient is still in the clinic) and for complicated molecular analysis in the operating room to guide surgical decisions.

LCM-based molecular analysis of histopathological lesions can be applied to any disease process which is accessible through tissue sampling. Examples include mapping the field of genetic changes associated with the progression of microscopic premalignant cancer lesions, analysis of gene expression patterns in multiple sclerosis, atherosclerosis, and diabetes, infectious microorganism diagnosis, typing of cells within disease foci (such as tuberculosis granuloma), and analysis of genetic abnormalities *in utero* from selected rare fetal cells in maternal fluids. Once we can reliably provide linear RT-PCR amplification of microdissected RNA, it will be practical to prepare cDNA libraries and microarray probes from LCM samples of normal ovarian surface epithelium, regenerating muscle or nerve, developing embryologic organs, healing wounds, pancreas islets, glomeruli, parathyroid cells, and subcellular elements. LCM allows routine analysis of DNA, RNA, or protein within selected pure cells from complex mixtures such as tissue sections or cytological smears. An important area for future development will be the adaptation of specialized immunoassay, enzymatic and 2-D gel panels to LCM samples.

Quantitative procurement of targeted tissue requires careful process control to ensure that the bond between the targeted tissue and the activated polymer exceeds the adherence of the cells to the slide. As the target size is reduced to that of single cells, the accuracy of targeting decisions becomes more difficult and may require immunohistochemical or molecularly specific fluorescent labels (such as FISH) and improved sample imaging to identify the cell population of interest. In order to apply LCM to routine clinical diagnosis, integration of LCM procurement with higher throughput, ultrasensitve molecular analysis will be required. This could be accomplished by extraction and molecular analysis in microvolumes using microfluidics-based PCR, RT-PCR, and capillary-gel electrophoresis. We are currently developing methods to integrate LCM with the subsequent molecular analysis steps in a seamless manner.

References:

1. M. Emmert-Buck *et al.*, *Science* **274**, 998-1001 (1996).
2. R. F. Bonner *et al., Science* **278**:1481-3 (1997)

Two-Photon Optical Biopsy Of Thick Tissues

Peter T. C. So[1], Barry R. Masters[2], Enrico Gratton[2], Irene E. Kochevar[3]

[1]Dept. of Mechanical Engineering, Massachusetts Institute of Technology, 77 Mass. Ave. Cambridge, MA 02139. Phone: (617)253-6552, Fax: (617)258-9346, Email: ptso@mit.edu
[2]Laboratory for Fluorescence Dynamics, Dept. of Physics, University of Illinois at Urbana-Champaign, IL 61201.
[3]Wellman Laboratories of Photomedicine, Dept. of Dermatology, Massachusetts General Hospital, Boston, MA

Keywords: Two-Photon Microscopy, (170.2520) Fluorescence microscopy, (170.1870) Dermatology, (170.6900) Three-dimensional microscopy, (170.3660) Light propagation in tissue

Abstract

Two-photon optical biopsy has been performed in human and mouse skin. Both cellular and extracellular tissue components have been visualized based on endogenous fluorescence.

Introduction

An assessment of tissue physiological states requires morphological information with sub-cellular details. This sub-cellular structural information is traditionally acquired from histological sections of excised tissue. Much of the cellular biochemical information is lost during the surgical and fixation processes. More importantly, the biopsy procedure is clinically undesirable due to its invasive nature. The development of a non-invasive optical technology that can acquire cellular structural and biochemical signals from within deep tissue will bring major advances in clinical diagnosis technology.

The opacity of typical tissues is a major obstacle in non-invasive optical tissue diagnosis. Today, one of the most promising approaches is two-photon excitation microscopy introduced by Denk et. al. [1]. Fluorophores can be excited by the simultaneous absorption of two photons each having half the energy needed for the excitation transition. Since the two-photon excitation probability is significantly less than the one-photon probability, appreciable two-photon excitation occurs only at the focal point, a region of high temporal and spatial concentration of photons. Two-photon excitation allows 3-D tissue structures to be imaged with resolution comparable to confocal microscopes but with a number of significant advantages when the sample's absorption and scattering coefficients are high: (1) The typical scattering and absorption coefficients in the infrared spectral range is over an order of magnitude less than the near UV or the blue-green region. The use of infrared excitation in the two-photon microscope minimizes the attenuation of the excitation signal. (2) Confocal microscopy uses the emission pinhole aperture to reject out of focus light. Inside deep tissue, scattering of the signal photons is inevitable. The consequent path deviation results in a significant loss of these photons at the confocal pinhole. The collection geometry for the fluorescence photons is less critical in the two-photon case where a large area detector can be used without a pinhole aperture. Most of the forward scattered photons can be retained. (3) Two-photon excitation minimizes tissue photo-damage. Conventional confocal techniques obtain 3-D resolution by limiting the observation volume, but fluorescence excitation occurs throughout the hourglass-shaped light path. In contrast, two-photon excitation limits the region of photo-interaction to a sub-femtoliter volume at the focal point.

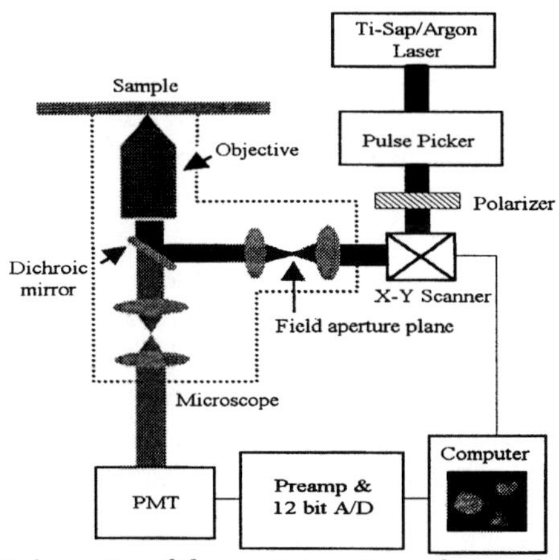

Fig. 1. Schematic of the prototype two-photon deep tissue microscope

Reprinted from OSA TOPS Vol. 22, Biomedical Optical Spectroscopy and Diagnostics / Therapeutic Laser Applications, Eva Sevick-Muraca, Joseph A. Izatt, and Marwood N. Ediger (eds.)

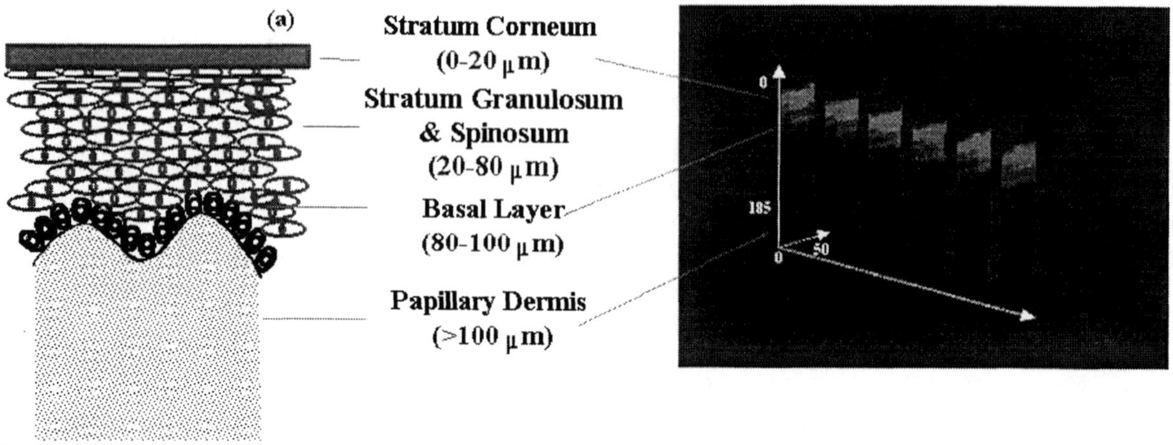

Figure 2. (a) A schematic of human skin anatomy. (b) Cross section images of human skin in vivo: 730 nm excitation. X, and Y orthogonal slices are shown. Units are shown in μm.

For non-invasive two-photon tissue biopsy, fluorescent labeling with extrinsic probe is undesirable and is often difficult. Fortunately, many tissue components are autofluorescent. Endogenous cellular proteins such as β-nicotinamide-adenine dinucleotide phosphate (NAD(P)H) and a number of flavoproteins are fluorescent [2,3]. Extracellular matrix components such as collagen [4] and elastin [5] can also be excited using UV excitation. Unfortunately, many of these biological molecules are poor fluorophores with low extinction coefficient and quantum efficiency. Therefore, successful non-invasive two-photon imaging of deep tissue structures requires careful optimization of detection optics.

Experimental Methods

A deep tissue microscope based on two-photon excitation has been constructed (Fig. 1). The laser used in this microscope is a mode-locked Titanium-Sapphire laser (Mira 900, Coherent Inc., Palo Alto, CA). The beam expanded laser light is directed into the microscope via a galvanometer-driven x-y scanner (Cambridge Technology, Watertown, MA). Images are generated by raster scanning the x-y mirrors. The excitation light enters the Zeiss Axiovert 110 microscope (Zeiss Inc., Thronwood, NY) via a modified epi-luminesence light path. The scan lens is positioned such that the x-y scanner is at its eye-point while the field aperture plane is at its focal point. Since the objectives are infinity-corrected, a tube lens is positioned to re-collimate the excitation light. The scan lens and the tube lens function together as a beam expander which over-fills the back aperture of the objective lens. The dichroic mirrors are custom made short pass filters (Chroma Technology Inc., Brattleboro, VT) which maximize reflection in the infrared and transmission in the blue-green region of the spectrum. The objective used in most of these studies is a Zeiss Fluar 40X with numerical aperture of 1.25. This objective is chosen for its high numerical aperture and its high throughput. The high numerical aperture is essential for the formation of a diffraction limited spot on the order of 0.1 femtoliter. The objective axial position is controlled by a piezo-stage (PI Inc., Auburn, MA). The axial resolution of this system is about 1 micron with a range of over 100 micron. The microscope field of view is about 50-200 μm on a side and the typical frame rate is about two seconds. The fluorescence emission is collected by the same objective and transmitted through the dichroic mirror along the emission path. An additional barrier filter is needed to further attenuate the scattered excitation light because of the high excitation intensity used. A de-scan lens is inserted between the tube lens and the photomultiplier tube (PMT). The number of available fluorescence photons is always a limiting factor in non-invasive deep tissue imaging. On one hand, the scattering and absorption properties of thick tissue quickly diminishes both the excitation and the emission light. On the other hand, we always seek to use the lowest excitation power possible to limit photodamage to the tissue. In this low light situation, we have implemented a single photon counting signal detection system. The fluorescence signal at each pixel is detected by a R5600-P PMT (Hamamatsu, Bridgewater, NJ) which is a compact single photon counting module with high quantum efficency of 12% at 500 nm and 25% at 400 nm. The dark noise of this PMT is less than twenty photon counts per second at room temperature. This high signal to noise ratio is crucial for high sensitivity detection. A 100 MHz single photon counting discriminator (F-100T, Advanced Research Instruments, Boulder, CO) converts single photon bursts into TTL pulses. The number of collected photons are counted using a home built interface circuit and the number is transferred to the data acquisition computer.

Fig. 3: Mouse ear dermal structure (from left to right): stratum corneum (0 μm), epidermal cell layer (10 μm), basal cell layer (15 μm), dermal layer (30 μm), dermal layer (50 μm).

Results

Cellular structural information has been obtained from in vivo human skin at the lower surface of the forearm [6]. Two-photon microscopy imaging was accomplished by exciting cellular auto-fluorescence from endogenous chromophores which primarily consist of NAD(P)H. The production of NAD(P)H is associated with the cellular metabolism and the intracellular redox state. Typically, a higher metabolic rate generates a higher concentration of NAD(P)H and produces brighter images. For these experiments, the Ti-Sapphire laser is tuned to 730 nm. The power incident upon the skin surface ranges from 10 to 20 mW. Z-section images were taken at depths of every five microns. Three dimensional images were re-constructed using Spyglass Slicer software. Fig. 2b displays the x-z planes of the reconstructed volume. The brightly fluorescent layer at the top corresponds to the stratum corneum. The second bright layer about 100 μm below the skin's surface has an "egg carton" like morphology which most likely corresponds to the basal cell layer at the dermal-epidermal junction. The higher fluorescence from this layer correlates well with the high activity level of the basal cells. Fig. 2a shows a schematic of the skin structure as confirmed by either confocal or electron microscopy; a good correlation between the two-photon results and the known skin physiology is observed.

Mouse skin structure at the tip of its ears has also been visualized using two-photon microscopy. The imaging was performed within four hours of the euthanasia procedure. In addition to visualizing cellular distribution in the epidermal layer, we have succeeded in imaging collagen/elastin like structures in the dermal layer. Fluorescence images of mouse epidermal/dermal structures are shown in Fig. 3.

Conclusion

We have demonstrated the feasibility of imaging deep tissue structures from a mouse model and an in vivo human subject. Most importantly, we have demonstrated that endogenous chromophores can provide sufficient contrast for the visualization of deep tissue structures. In particular, cellular morphology in deep tissue can be imaged based on NAD(P)H cytoplasmic fluorescence. The extracellular matrix structure can be mapped based on collagen or elastin autofluorescence. The ability to simultaneously visualize cellular and extracellular matrix structures in tissues is critical for the understanding of biomedical processes such as carcinogensis and tissue regeneration.

References

1. W. Denk, J. H. Strickler and W. W. Webb, "Two-photon laser scanning fluorescence microscopy," Science, 248, 73-76 (1990).
2. B. Chance and B. Thorell, "Localization and kinetics of reduced pyridine nucleotide in living cells by microfluorometry," J. Biol. Chem. 234, 3044-3050 (1959).
3. B. Chance and B. Schoener, "Fluorometric studies of flavin component of the respiratory chain" in Flavins and Flavoproteins, E.C. Slater, ed., (Elsevier Publishing Company, Amsterdam, 1966), 510-528.
4. M. K. Dabbous, "Inter- and intramolecula cross-linking in tyrosinase-treated tropocollagen," J. Bio. Chem., 241, 5307-5312 (1966).
5. F. S. LaBella, and W. G. Lindsay, "The structure of human aortic elastin as influence by age," J. Gerontol., 18, 111-118 (1963).
6. B. R. Masters, P. T. C. So, and E. Gratton, "Multi-Photon Excitation Fluorescence Microscopy and Spectroscopy of *In Vivo* Human Skin," Biophys. J., 72, 2405-2412 (1997).

Tissue Microscopy and Spectroscopy: A Two-Photon Approach

Barry R. Masters[1], Peter T. C. So[2], William Mantulin[1], Enrico Gratton[1]

[1]Laboratory for Fluorescence Dynamics, Dept. of Physics, University of Illinois at Urbana-Champaign, IL 61201. Phone: (703) 522-4671, Fax (703) 525-1426
[2]Dept. of Mechanical Engineering, MIT, 77 Mass. Ave. Cambridge, MA 02139.

Keywords: Two-Photon Microscopy, (170.2520) Fluorescence microscopy, (170.4470) Ophthalmology, (170.6900) Three-dimensional microscopy, (170.3660) Light propagation in tissue

Abstract

Using two-photon excitation, we have demonstrated that fluorescence spectroscopy of endogenous chromophores such as NAD(P)H can be performed in human skin and rabbit cornea.

Introduction

Two-photon technology has provided an unprecedented opportunity to study tissue biochemistry using fluorescence spectroscopy. Until the advent of two-photon and confocal methods, biochemical properties of tissues can only be determined in bulk. This is clearly insufficient, since tissue biochemistry is regulated on the sub-cellular level. The capability to monitor biochemistry with sub-micron resolution is needed to understand fully tissue physiology.

In order to perform spectroscopy in thick tissue, it is necessary to optically section thick, highly scattering tissues such as *in vivo* human skin. A method that can accomplish this task is confocal microscopy. Two apertures or pinholes are arranged in conjugate planes; one in front of the light source and one in front of the detector. This arrangement results in improved optical lateral and axial resolution. It is the improved axial optical resolution that results in the enhanced optically sectioning capability. This confocal principle has been exploited in a real-time, slit scanning, confocal microscope that is used clinically for optical biopsy and diagnostics in ophthalmology. The tandem-scanning confocal microscope, which operates in real-time, was developed by Petran and co-workers to image optical sections of thick, highly scattering biological materials [1]. Corcuff and co-workers have adapted the tandem scanning confocal microscope for in vivo skin imaging in the reflected light mode [2]. A point scanning laser confocal microscope has also been developed for imaging *in vivo* skin [3].

While these developments in confocal microscopy have succeeded in imaging deep tissue structures in highly scattering biological tissue up to a depth of about 200 to 500 μm, the success in using confocal microscopy to obtain biochemical information using fluorescence spectroscopy is, however, limited. Common fluorophores (especially endogenous ones) absorbs in the ultraviolet or blue/green spectral region. While typical tissue is relatively transparent in the infrared spectral region, tissue absorption is high in the blue/green and ultraviolet region. Confocal microscopes using shorter excitation wavelengths are limited to a penetration depth on the order of 50 μm. Moreover, when ultraviolet fluorescence microscopy is used to image living cells and tissues, it was observed that cell viability was compromised and that there was considerable photobleaching

This difficulty in tissue spectroscopy has been circumvented by the invention of two-photon microscopy [4]. The theoretical foundation for two-photon excitation was first developed by Maria Goeppert-Mayer as the subject of a Ph.D. thesis at Göttingen, Germany in 1931 [5]. The physical principle is the simultaneous absorption of two photons; two photons of wavelength in the near infrared induce an electronic transition that normally requires an ultraviolet photon. In microscopy applications, the diffraction-limited focusing of the femtosecond laser beam result in two-photon excitation process only in the region of focus. Outside of the focal region there is insufficient photon density to induced any significant absorption and hence little signal. This is the origin of the optical sectioning. **The optical sectioning is derived from the non-linear optical response of the sample material and an aperture in front of the light detector is not required**. For deep tissue spectroscopy, the two-photon method circumvents the need for ultra-violet

Reprinted from OSA TOPS Vol. 22, Biomedical Optical Spectroscopy and Diagnostics / Therapeutic Laser Applications, Eva Sevick-Muraca, Joseph A. Izatt, and Marwood N. Ediger (eds.)

excitation and uses infrared radiation for much deeper tissue penetration. Furthermore, the fluorescence photons generated deep within the sample must travel through the highly scattering medium in order to reach the detector. In the confocal case, the photon path deviation resulted from scattering will prevent the emitted photon from passing through the confocal pinhole. In the two-photon case, most of these scattered photons can be collected since no pinhole aperture is required and a large area detector can be used.

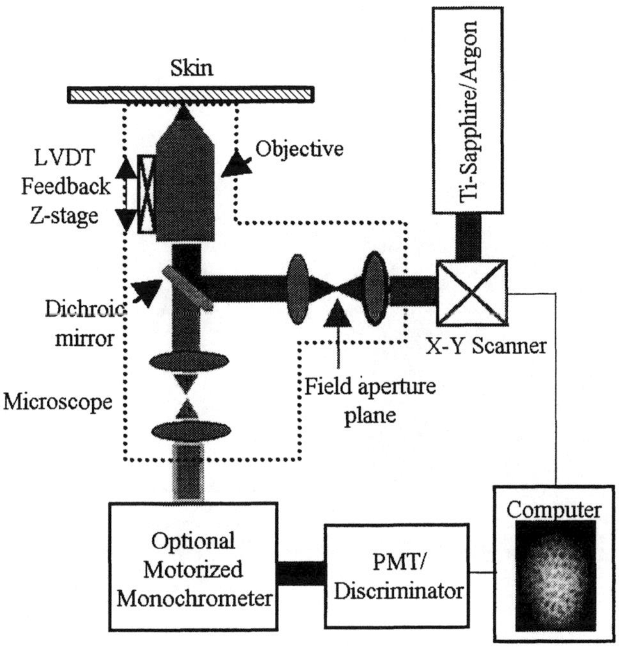

Figure 1: Schematics of the two-photon deep tissue microscope

Experimental Method

The schematic of the multi-photon scanning microscope is shown in Fig. 1. A femto-second Ti-Sapphire laser (Mira 900, Coherent Inc., Palo Alto, CA) tuned to 730 nm is used. The power incident upon the sample in on the order of 10-15 mW. The scanning is achieved via a galvanometer-driven x-y scanner (Cambridge Technology, Watertown, MA). The scan parameters are chosen such that 256x256 pixel images are generated with a pixel residence time of 40 μs corresponding to a frame rate of about 3 seconds. The excitation light enters the Zeiss Axiovert 35 microscope (Zeiss Inc., Thornwood, NY) via a modified epi-luminescence light path. The light is reflected by the dichroic mirror to the objective (Zeiss 40X Plan-Fluor, 1.3 N.A., oil). The fluorescence signal is collected by the same objective, transmitted through the dichroic mirror and barrier filter (3 mm BG-39, CVI Laser, CA) and re-focused on the detector. The fluorescence signal at each pixel is detected by a photomultiplier tube (PMT). For intensity measurements, a low noise R5600-P PMT (Hamamatsu, Bridgewater, NJ) is used for single photon counting. The output of the PMT is converted to single photoelectron pulses via a low noise amplifier/discriminator (Pacific AD6, Concord, CA). The number of photons collected at each pixel is counted and digitally recorded by the control computer (Intel 486, Gateway). A 3-D imaged stack can be obtained by controlling the microscope objective height with a stepper motor; the microscope objective position is monitored using a linear variable differential transformer (Schaevitz Engineering, Camden, NJ). Measurements of the tissue emission spectrum were performed at selected points. A motor driven monochromator is inserted between the microscope and the photomultiplier tube. The monochromator is driven at a constant speed of 1 nm per second. The wavelength resolution is set by the monochromator slit to 2 nm. The x-y scanner is set to remain at the origin during spectral measurement.

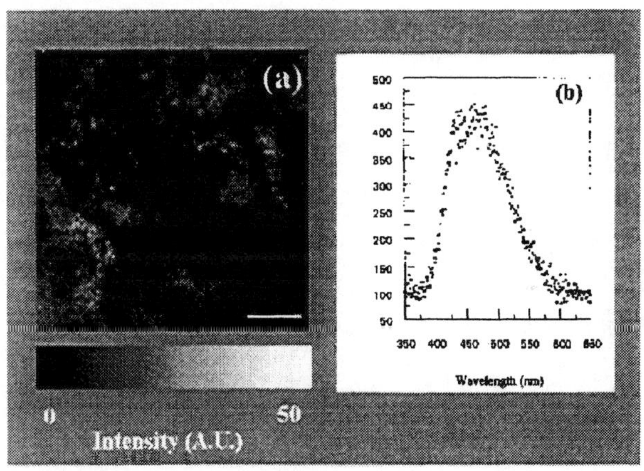

Figure 2: (a) An autofluorescence image of a cell layer in the epidermis. (b) The fluorescence spectrum associated with this cell layer.

Results

Cellular metabolism can be non-invasively monitored by the technique of redox fluorometry. The probes are the intrinsic fluorescent pyridine nucleotides, and the flavoproteins. The pyridine nucleotides, NAD(P)H are excited in the region 365 nm and fluoresce in the region 400-500 nm. The flavoproteins are excited in the region 450 nm and fluoresce in the region 500-600 nm. Typically the fluorescence is imaged following ultraviolet excitation, and the changes in fluorescence intensity follow changes in cell and tissue oxidative metabolism [6]. The advent of the two-photon approach allow redox fluorometry to be performed using infrared radiation from a sub-femtoliter volume at a selected location inside deep tissue.

We have used this methodology to investigate in vivo human skin spectroscopic properties [7]. Fluorescence microscopy was performed at selected locations at the under side of the human forearm. Two-photon microscopy has imaged individual cell layers in the epidermal layer using their endogenous autofluorescence (Fig. 2). Moreover, fluorescence emission spectra were obtained from these selected cell layers. The fluorescence emission maxima is consistent with the primary fluorophore being NAD(P)H. A slight bimodal distribution indicates either the presence of heterogeneity in NAD(P)H environment or other endogenous chromophores.

Fluorescence spectrum has also been obtained from the cornea of a freshly excised rabbit eye. We have been able to optically dissect the rabbit cornea and locate distinct cell layers. From these individual cell layers, autofluoescence spectra were measured. The fluorescence intensity from the basal cell layer is found to be the most intense which is consistent with the high activity level of the basal cells (Fig. 3a). The spectra is again consistent with NAD(P)H fluorescence but there is no bimodal distribution as observed in the skin (Fig. 3b).

Conclusion

We have demonstrated the feasibility of performing spectroscopy study of thick living tissues. This methodology will provide an opportunity to integrate our knowledge of tissue structural and function by simultaneously monitoring tissue cellular morphology and their underlying biochemical driving forces.

References

1. M. Petran, M. Hadravsky, M.D. Egger, and R. Galambos, "Tandem scanning reflected light microscope," J. Opt. Soc. Am, 58, 661-664 (1958).
2. P. Corcuff, C. Bertrand, and J.L. Leveque, "Morphometry of human epidermis *in vivo* by real-time confocal microscopy," 285, 475-481, (1993)
3. M. M. Rajadhyaksha, M. Grossman, D. Esterowitz, and R.H. Webb, "*In vivo* confocal scanning laser microscopy of human skin: melanin provides strong contrast," J. Invest. Dermatol., 104, 946-952 (1995).
4. W. Denk, J. H. Strickler and W. W. Webb, "Two-photon laser scanning fluorescence microscopy," Science, 248, 73-76 (1990).
5. M. Göppert-Mayer, M., "Über Elementarake mit zwei Quantensprungen," Ann. Phys. (Leipzig), 5, 273-294, (1931).
6. B. Chance and B. Thorell, "Localization and kinetics of reduced pyridine nucleotide in living cells by microfluorometry," J. Biol. Chem. 234, 3044-3050 (1959).
7. B. R. Masters, P. T. C. So, and E. Gratton, "Multi-Photon Excitation Fluorescence Microscopy and Spectroscopy of *In Vivo* Human Skin," Biophys. J., 72, 2405-2412 (1997).

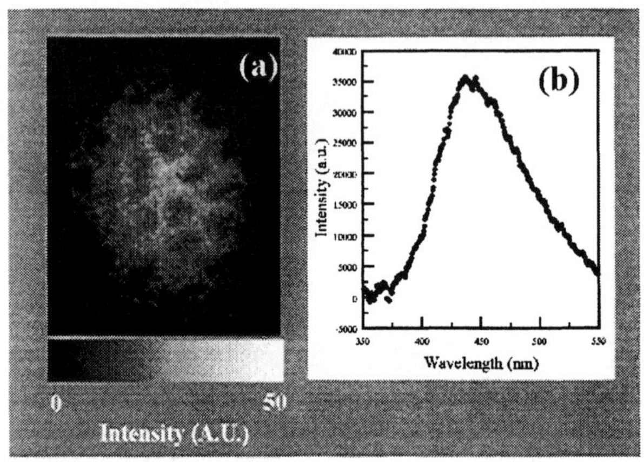

Figure. 3: (a) An autofluorescence image of the basal cell layer in the cornea. (b) The fluorescence spectrum associated with this cell layer

Bibliography

Bibliography

This is a partial list of OSA references on topics from the Advances in Optical Imaging and Photon Migration 1998 topical meeting. It is intended to provide a starting point for further reading. These references were compiled by searching OSA's Optics Infobase of publications from 1990-1997. Abstracts are included for references from 1995-1997. For your convenience the references are categorized by the search terms: **Acousto-optic imaging, Confocal microscopy, Imaging and Scattering media, Medical imaging, Optical coherence tomography, Optical imaging, Photon migration, Time-resolved imaging, and Tissue optics.**

If you would like single copies of the full articles, please contact OSA Customer Service.

Acousto-optic imaging

Gao, Guan-Hong; Lin, Zhong; "Acousto-optic supermultispectral imaging," *Appl. Opt.*, Vol. 32, pg. 3081 (1993).

Hull, Edward L.; Michael G. Nichols, and Thomas H. Foster, "Localization of Luminescent Inhomogeneities in Turbid Media with Spatially Resolved Measurements of cw Diffuse Luminescence Emittance," *Appl. Opt.* 37, No. 13, pp. 2755-2765 (1998).

We present a steady-state method for localizing a source of luminescence (i.e., fluorescence or phosphorescence) buried in a semi-infinite turbid medium with unknown optical properties. A diffusion theory expression describing the emittance of an isotropic point source is fit to spatially resolved surface measurements of the diffuse emittance from the luminescent source. The technique reports the location of the center of a 6.0-mm-diameter, fluorophore-containing spherical bulb embedded in a liquid phantom with an accuracy of 1.0 mm or better for source depths as great as 40.0 mm. Monte Carlo data are analyzed to investigate the range and the possible sources of error in the reconstructed source depth.

Kempe, M.; M. Larionov; D. Zaslavsky; A. Z. Genack; "Acousto-optic tomography with multiply scattered light," *J. Opt. Soc. Am. A*, Vol. 14, pg. 1151 (1997).

We have investigated the modulation of the optical field transmitted through a colloid of polystyrene spheres by a narrow quasi-cw ultrasound beam. Measurements of the scale dependence of the heterodyne modulation signal at the acoustic frequency are obtained for samples that are up to 140 scattering lengths thick. A calculation of the modulation signal predicts the possibility of tomographic imaging, which is confirmed experimentally.

Wachman, Elliot S.; Wen-hua Niu; Daniel L. Farkas; "Imaging acousto-optic tunable filter with 0.35-micrometer spatial resolution," *Appl. Opt.*, Vol. 35, pg. 5220 (1996).

Image blur in acousto-optic tunable filters (AOTF's) has been a persistent problem. Here we describe the connection between transducer structure and image blur and experimentally measure it by using a 5-cm 12 degrees-cut TeO_2 crystal of our design. With these quantitative results, we develop an image-processing method that minimizes AOTF-related image degradation. The combination of long crystal design and image processing results in substantially improved image contrast and spatial resolution relative to conventional AOTF imaging devices. We present high-magnification images of fluorescent actin fibers in cells in which we obtain a resolution of approximately 0.35 μm, representing the first successful use of an AOTF for ultra-high-resolution microscopy. Further improvements are also predicted.

Wang, Lihong and Xuemei Zhao, "Ultrasound-modulated optical tomography of absorbing objects buried in dense tissue-simulating turbid media," Appl. Opt. 36, No. 28, pp. 7277-7282 (1997).

Continuous-wave ultrasonic modulation of scattered laser light was used to image objects buried in tissue-simulating turbid media. The buried object had an absorption coefficient greater than the background turbid medium. The ultrasonic wave that was focused into the turbid media modulated the laser light that passed through the ultrasonic field. The modulated laser light that was collected by a photomultiplier tube reflected the local mechanical and optical properties in the zone of ultrasonic modulation. Objects buried in the middle plane of 5-cm-thick dense turbid media were imaged with millimeter resolution through the scanning and detecting alterations of the ultrasound-modulated optical signal. The optical properties of the dense turbid media included an absorption coefficient of 0.1 cm-1 and a reduced scattering coefficient of 10 cm-1 and were comparable with those of biological tissues in the visible and near-IR ranges. The dependence of the ultrasound-modulated optical signal on the off-axis distance of the detector from the optic axis and the area of the detector was studied as well.

Confocal microscopy

"Confocal imaging system for microscopy," *Appl. Opt.*, Vol. 35, pg. 2731 (1996).

Arons, E.; Leith, E., "Coherence confocal-imaging system for enhanced depth discrimination in transmitted light," Appl. Opt. 35, 2499 (1996).
A system for image-plane holography that uses a spectrally and spatially broad radiation source is analyzed and shown to provide depth discrimination in excess of that offered by conventional confocal imaging. The enhanced depth discrimination is a consequence of two factors: the simultaneous reduction of both the spatial and the temporal coherence of the source, and the special property of certain grating interferometers of presenting a spectrally dispersed light beam to the object while forming achromatic fringes at the hologram-recording plane.

Dickensheets, D.L.; Kino, G.S., "Micromachined scanning confocal optical microscope," Opt. Lett. 21, 764 (1996).
We have constructed a miniature confocal optical microscope for monochromatic imaging that uses single-mode fiber illumination and a two-phase off-axis zone plate objective lens. The scanning mechanism consists of two micromachined silicon torsional scanning mirrors with orthogonal axes of rotation. The objective lens is made of fused silica and has a N.A. of 0.24 at lambda =0.6328 µm , with a 1.0-mm working distance. The device is side looking, with die dimensions of 1.2 mm \times 2.5 mm \times 6.5 mm. We have measured 1.0- µm resolution over a 100- µm field of view.

Ding, Zhihua; Guiying Wang; Min Gu; Zhijiang Wang; Zhifeng Fan; "Superresolution with an apodization film in a confocal setup," *Appl. Opt.*, Vol. 36, pg. 360 (1997).
We put forward the idea of implementing a confocal setup to suppress the large sidelobes spreading all through the field outside the central core. They are produced when an apodization film is imposed on an ordinary lens. Furthermore, this avoids the imaging quality degradation caused by nonaxial points and non-Gaussian-plane points. Also a configuration for achieving three-dimensional superresolution is depicted and discussed.

Dunn, Andrew K.; Smithpeter, Colin; Welch, Ashley J.; Richards-Kortum, Rebecca, "Sources of contrast in confocal reflectance imaging," Appl. Opt. 35, 3441 (1996).
The relationship between optical properties and image contrast in confocal imaging is investigated. A Monte Carlo simulation has been developed to analyze the effects of changes in scattering, index of refraction, and absorption in a three-layer medium. Contrast was calculated from the computed signal-to-background ratios for changes in tissue optical properties. Results show that the largest source of contrast is changes in refractive index.

Fukano, Takashi; Yamaguchi, "Simultaneous measurement of thicknesses and refractive indices of multiple layers by a low-coherence confocal interference microscope," Opt. Lett. 21, 1942 (1996).
We propose a novel technique for simultaneous measurement of layer thicknesses and refractive indices of multiple layers. It is based on a combination of a confocal microscope and low-coherence interferometry. We derived an expression for the geometrical thickness and the refractive index of each layer from both tracing of a marginal ray accepted by a microscope objective and optical path matching conditions. Experimental verification of this method is illustrated by several samples that have a maximum of 13 layers. The geometrical thicknesses and refractive indices thus derived agreed well with those measured by a micrometer or cited from the literature.

Gmitro, Arthur F.; Aziz, David; "Confocal microscopy through a fiber-optic imaging bundle," *Opt. Lett.*, Vol. 18, pg. 565 (1993).

Gu, Min; Sheppard, Colin J. R.; "Analysis of confocal microscopy under ultrashort light-pulse illumination: comment," *J. Opt. Soc. Am. A*, Vol. 11, pg. 2742 (1994).

Gu, Min; Sheppard, Colin J. R.; "Effects of defocus and primary spherical aberration on images of a straight edge in confocal microscopy," *Appl. Opt.*, Vol. 33, pg. 625 (1994).

Gu, Min; Tannous, T.; Sheppard, J.R., "Effect of an annular pupil on confocal imaging through highly scattering media," Opt. Lett. 21, 312 (1996).
We report on confocal scanning imaging through highly scattering media. Various practical effects including those of the annular pupils and the size of the confocal pinhole as well as of the numerical aperture of objectives on the image quality are examined experimentally. The combination of an annular objective with a finite-sized detector may prove advantageous for improving image quality.

Hell, Stefan W.; Stelzer, Ernst H. K.; Lindek, Steffen; Cremer, Christoph; "Confocal microscopy with an increased detection aperture: type-B 4Pi confocal microscopy," *Opt. Lett.*, Vol. 19, pg. (1994).

Hell, Stefan W.; Utz, Axel R.; Schrader, Martin; Hanninen, Pekka E.; Soini, Erkki; "Pulsed laser fluorophore deposition: a method for measuring the axial resolution in two-photon fluorescence microscopy," *J. Opt. Soc. Am. A*, Vol. 12, pg. 2072 (1995).

We report the generation of thin (<50 nm) fluorescence layers on a glass substrate in the focal region of a high-numerical-aperture lens by employing subpicosecond pulses of 0.75- TW/cm^2 peak intensity. The conditions for generating the fluorescence layers are described. We find that the fluorescence molecules at the glass interface are less affected by bleaching than those in the surrounding area. The fluorescence layers are suited for measuring and monitoring the axial resolution of two-photon fluorescence microscopes. Key words: two-photon fluorescence microscopy, axial resolution, 4Pi confocal microscopy.

Hewlett, S. J.; D. T. Fewer; E. M. McCabe; "Influence of source coherence and aperture distribution on the imaging properties in direct-view microscopy," *J. Opt. Soc. Am. A*, Vol. 14, pg. 1066 (1997).

We discuss the imaging properties of the bright-field direct-view microscope employing source and detector arrays with an arbitrary distribution of pinholes. In general, we show that coherent laser illumination will result in a significant degradation in the imaging, compared with the more traditional use of an incoherent light source. We discuss the imaging in terms of the partially coherent transfer function and illustrate our general predictions by considering the specific responses to straight-edge, point, and planar objects.

Juskaitis, R.; Rea, N. P.; Wilson, Tony; "Semiconductor laser confocal microscopy," *Appl. Opt.*, Vol. 33, pg. 578 (1994).

Juskaitis, R.; Wilson, Tony; Reinholz, F.; "Spatial filtering by laser detection in confocal microscopy," *Opt. Lett.*, Vol. 18, pg. 1135 (1993).

Kawata, Y.; Fujita, K.; Nakamura, O.; Kawata, Satoshi, "4Pi confocal optical system with phase conjugation," Opt. Lett. 21, 1415 (1996).

We propose a 4Pi confocal system that incorporates a phase-conjugate mirror into the system to solve the alignment problem and automatically correct the spherical aberration associated with the conventional 4Pi confocal system when observing a thick sample. The proposed system is experimentally verified by use of cerium-doped barium titanate crystal as the phase-conjugate mirror. The full width of the central peak is \similar 0.15 μm . We also theoretically verify the advantage of automatic correction of the spherical aberration by calculating the intensity distribution at the focal spot.

Kempe, M.; A. Z. Genack; W. Rudolph; P. Dorn; "Ballistic and diffuse light detection in confocal and heterodyne imaging systems," *J. Opt. Soc. Am. A*, Vol. 14, pg. 216 (1997).

The detection of ballistic and diffuse light in confocal and heterodyne imaging systems in transillumination is studied experimentally and theoretically. We find an optimum pinhole size for ballistic light detection and diffuse light rejection for confocal imaging. The ratio of ballistic and diffuse light is found to be determined primarily by sample parameters and aberrations introduced by the sample. For sample and illumination characteristics that are typical for biomedical imaging, the limits of ballistic light detection in confocal imaging are close to the noise limits of standard detectors. Heterodyne detection with narrow-bandwidth light can extend these limits, depending on the spatial and the temporal coherence properties of the transmitted scattered light.

Kempe, M.; Rudolph, Wolfgang; "Analysis of confocal microscopy under ultrashort light-pulse illumination: reply to comment," *J. Opt. Soc. Am. A*, Vol. 11, pg. 2744 (1994).

Kempe, M.; Rudolph, Wolfgang; "Analysis of confocal microscopy under ultrashort light-pulse illumination," *J. Opt. Soc. Am. A*, Vol. 10, pg. 240 (1993).

Kulawiec, Andrew W.; Moore, Duncan T.; "Quantitative phase imaging in confocal microscopy by optical differentiation," *Appl. Opt.*, Vol. 33, pg. 6582 (1994).

Liang, Minhua; Robert L. Stehr; Andrew W. Krause; "Confocal pattern period in multiple-aperture confocal imaging systems with coherent illumination," *Opt. Lett.*, Vol. 22, pg. 751 (1997).

We have built a programmable multiple-aperture confocal imaging system that uses a spatial light modulator (a Texas Instruments digital micromirror device). Excellent axial resolution and confocal imaging quality have been experimentally demonstrated with this system, even when the distance between adjacent ON pixels is four times the size of the pixel aperture. By contrast, typical pinhole-based systems (e.g., a Nipkow disk) employ an adjacent aperture distance-to-size ratio of . We have achieved improvement over such systems by in both light-utilization efficiency and confocal image rate without adding more components and extra processing time.

Lindek, Steffen; Cremer, Christoph; Stelzer, Ernst H.K., "Confocal theta fluorescence microscopy with annular apertures," Appl. Opt. 35, 126 (1996).

In a confocal theta fluorescence microscope, two objective lenses with circular apertures are used, one to illuminate the sample and the other to detect the emitted light at an angle to the illumination axis. We show that annular illumination and detection apertures lead to a reduction in the extent of the point-spread function. A spatial resolution improved by more than 50% can be achieved with a central obstruction blocking the inner 80% of the diameter. For the limit of a very narrow annular aperture and a numerical aperture of 0.75, the volume at half-maximum of the point-spread function is reduced from 15 to 5 aL. A mixed setup with an annular illumination aperture and a circular detection aperture is also considered.

Lindek, Steffen; Ernst H. K. Stelzer; "Optical transfer functions for confocal theta fluorescence microscopy," *J. Opt. Soc. Am. A*, Vol. 13, pg. 479 (1996).

Optical transfer functions are presented for the confocal theta fluorescence microscope, which uses an orthogonally placed objective lens to detect the fluorescence emission. Therefore the transfer functions do not exhibit cylindrical symmetry about the illumination axis. We show that in an ideal noise-free system, confocal theta microscopy increases the cutoff spatial frequency along the illumination axis by a factor of 3.5 to 3.7 for a numerical aperture of 0.75. In a system with 10% noise the improvement by confocal theta microscopy is even greater, between a factor of 4.1 and 4.4. This explains the improved three-dimensional imaging properties of theta microscopy. Key words: confocal, microscopy, fluorescence, transfer function, theta.

Lindek, Steffen; Christoph Cremer; Ernst H. K. Stelzer; "Confocal theta fluorescence microscopy with annular apertures," *Appl. Opt.*, Vol. 35, pg. 0126 (1996).

In a confocal theta fluorescence microscope, two objective lenses with circular apertures are used, one to illuminate the sample and the other to detect the emitted light at an angle to the illumination axis. We show that annular illumination and detection apertures lead to a reduction in the extent of the point-spread function. A spatial resolution improved by more than 50% can be achieved with a central obstruction blocking the inner 80% of the diameter. For the limit of a very narrow annular aperture and a numerical aperture of 0.75, the volume at half-maximum of the point-spread function is reduced from 15 to 5 aL. A mixed setup with an annular illumination aperture and a circular detection aperture is also considered.

Masters, Barry R.; Thaer, Andreas A.; "Real-time scanning slit confocal microscopy of the *in vivo* human cornea," *Appl. Opt.*, Vol. 33, pg. 695 (1994).

Neil, M. A. A.; R. Juskaitis; T. Wilson; "Method of obtaining optical sectioning by using structured light in a conventional microscope," *Opt. Lett.*, Vol. 22, pg. 1905 (1997).

We describe a simple method of obtaining optical sectioning in a conventional wide-field microscope by projecting a single-spatial-frequency grid pattern onto the object. Images taken at three spatial positions of the grid are processed in real time to produce optically sectioned images that are substantially similar to those obtained with confocal microscopes.

Poon, T-C.; "Three-dimensional Fluorescence Microscopy by Optical Scanning Holography," *Opt. Phot. News*, Vol. 8, pg. 22 (1997).

Schlingloff, Gregor; Hans-Jürgen Kiel, and Andreas Schober, "Microlenses as Amplification for CCD-Based Detection Devices for Screening Applications in Biology, Biochemistry, and Chemistry," Appl. Opt. 37, No. 10, pp. 1930-1934 (1998).

We describe the use of microlenses as amplification for CCD-based detection devices. The possible amplification of a signal in a reaction chamber of a nanoplate is estimated with a first-order pproximation. This value was proved with a commercially available microlens and with a specially constructed microlens array made of glass hemispheres. Possible applications of this approach to amplification are given.

Schmitt, J. M.; Knuttel, A.; Yadlowsky, Michael; "Confocal microscopy in turbid media," *J. Opt. Soc. Am. A*, Vol. 11, pg. 2226 (1994).

Schmitt, J.M.; Ben-Letaief, K., "Efficient Monte Carlo simulation of confocal microscopy in biological tissue," J. Opt. Soc. Am. A 13, 952 (1996).

A variance-reduction technique is described that greatly improves the efficiency of Monte Carlo simulations of reflection-mode confocal microscopy in anisotropically scattering media. The efficiency gain is large enough that the performance of confocal microscopes probing as deep as 5 scattering lengths can be simulated with a desktop computer. We use the technique to simulate the response of a true confocal microscope probing biological tissue, a problem that has been impractical to undertake by using conventional Monte Carlo methods. Our most important finding is that operation of a confocal microscope in the true confocal mode enables much more effective rejection of undesired scattered light than operation in the partially coherent mode, but the maximum probing depths of microscopes operated in either mode are similar (2-3 scattering lengths) in practice because of sensitivity limitations.

Schrader, M.; M. Kozubek; S. W. Hell; T. Wilson; "Optical transfer functions of 4Pi confocal microscopes: theory and experiment," *Opt. Lett.*, Vol. 22, pg. 436 (1997).

We measure the point-spread function in the two main configurations of 4Pi confocal microscopy as well as in the traditional confocal arrangement and derive the optical transfer functions from the experimental data. The optical transfer functions are in good agreement with their theoretical counterparts. We find a 3.5- to 5-fold increased axial bandwidth of the 4Pi confocal microscope and hence confirm the enhanced spatial-frequency content of 4Pi images.

Sheppard, Colin J. R.; Gu, Min; "Aberration compensation in confocal microscopy," *Appl. Opt.*, Vol. 30, pg. 3563 (1991).

Sheppard, Colin J. R.; Gu, Min; "Edge-setting criterion in confocal microscopy," *Appl. Opt.*, Vol. 31, pg. 4575 (1992).

Sieracki, Christian K.; Levey, Christopher G.; Hansen, Eric W., "Simple binary optical elements for aberration correction in confocal microscopy," Opt. Lett. 20, 1213 (1995).
When a confocal fluorescence microscope with a high-numerical-aperture oil-immersion objective is focused deep into an aqueous medium, aberrations result that degrade image quality. We have designed and fabricated a simple two-level binary phase mask that partially corrects these aberrations, improving axial resolution. We present the design and some confirming results.

Smithpeter, Colin L.; Andrew K. Dunn, A. J. Welch, and Rebecca Richards-Kortum, "Penetration Depth Limits of *In vivo* Confocal Reflectance Imaging," Appl. Opt. 37, No. 13, pp. 2749-2754 (1998).
We present experiments to predict the maximum penetration depth at which typical biological structures in amelanotic tissue can be detected with confocal microscopy. The detected signal is examined as the signal source strength (index of refraction mismatch), the source depth, and the medium scattering coefficient are varied. The detected background produced by scattering outside the focal volume is examined as the medium scattering coefficient, the depth in the medium, the dimensionless pinhole radius, &ngr;p, and the shape of the scattering phase function are varied. When the system approaches ideal confocal performance (&ngr;p ≃ 3), the penetration depth is limited by the signal-to-noise ratio to approximately 3-4 optical depths (OD's) for a 0.05 index mismatch. As &ngr;p increases to 8, the penetration depth is limited by the signal-to-background ratio and is dependent on the scattering coefficient. At us= 100 cm-1 (ls = 100 um) and an index mismatch of 0.05, the maximum penetration depth is approximately 2 OD.

Soini, Juhani T.; Martin Schrader, Pekka E. Hänninen, and Stefan W. Hell, "Image formation and data acquisition in a stage scanning 4Pi confocal fluorescence microscope," Appl. Opt. 36, No. 34, pp. 8929-8934 (1997).
We describe the three-dimensional (3-D) image formation and data acquisition in a stage scanning 4Pi confocal fluorescence microscope with the use of two-photon excitation. The 3-D point-spread functions of the 4Pi confocal and regular confocal microscope are measured and compared. Particular emphasis is given to the data acquisition procedure. 4Pi confocal microscopy results in a point-spread function that is 4 times sharper than that of a regular confocal microscope, ultimately leading to superior 3-D imaging of translucent fluorescent specimens. For a two-photon excitation wavelength of approximately 800 nm, we obtain an axial resolution of 140 nm.

Tiziani, H.J.; Achi, R.; Kramer, R.N.; Weigers, L., "Theoretical analysis of confocal microscopy with microlenses," Appl. Opt. 35, 120 (1996).
Scanning confocal microscopy is well established and applied frequently in biomedical science and more recently in engineering disciplines. For technical applications a confocal principle based on microlens arrays was developed. The principle permits a high depth resolution on a large field. A theoretical analysis of this principle together with some experimental results are presented.

Tiziani, Hans J.; Uhde, H.-M.; "Three-dimensional image sensing by chromatic confocal microscopy," *Appl. Opt.*, Vol. 33, pg. 1838 (1994).

Torok, P.; Laczik, Z.; Sheppard, C.J.R., "Effect of half-stop lateral misalignment on imaging of dark-field and stereoscopic confocal microscopes," Appl. Opt. 35, 6732 (1996).
There are several ways to realize dark-field imaging in confocal microscopy. In a recent paper [J. Microsc. 181 260-268 (1996)] we suggested a simple modification of a commercial confocal microscope to incorporate dark-field imaging. This modification involved an aperture stop covering half of the entrance pupil of the objective lens. Now we investigate the lateral misalignment of the aperture stop for dark-field and stereoscopic confocal microscopes. We show the effect of lateral alignment of the half-stop on the point-spread and transfer functions and also examine the detected signal from a sloping plane reflector. Lateral and axial resolution values are given from theoretical data.

Torok, P.; Varga, P.; Konkol, A.; Booker, G.R., "Electromagnetic diffraction of light focused through a planar interface between materials of mismatched refractive indices: structure of the electromagnetic field. II," J. Opt. Soc. Am. A 13, 2232 (1996).
We consider the electromagnetic diffraction occurring when light is focused by a lens without

spherical aberration through a planar interface between materials of mismatched refractive indices, which focusing produces spherical aberration. By means of a rigorous vectorial electromagnetic treatment developed previously for this problem by Torok et al. [J. Opt. Soc. Am. A 12, 325 (1995)], the time-averaged electric energy density distributions in the region of the focused probe are numerically evaluated for air-glass and air-silicon interfaces as functions of lens numerical aperture and probe depth. Strehl intensity, lateral and axial sizes, and axial location of the probe are shown to be regular functions for low numerical apertures and probe depths but irregular functions for high numerical apertures and probe depths. An explanation to account for these occurrences is presented that also explains some previous experimental results of confocal microscopy.

Verveer, Peter J.; Thomas M. Jovin; "Efficient super-resolution restoration algorithms using maximum a posteriori estimations with application to fluorescence microscopy," *J. Opt. Soc. Am. A*, Vol. 14, pg. 1696 (1997).

We present efficient algorithms for image restoration by using the maximum a posteriori (MAP) method. Assuming Gaussian or Poisson statistics for the noise and either a Gaussian or an entropy prior distribution for the image, corresponding functionals are formulated and minimized to produce MAP estimations. Efficient algorithms are presented for finding the minimum of these functionals in the presence of nonnegativity and support constraints. Performance was tested by using simulated three-dimensional (3-D) imaging with a fluorescence confocal laser scanning microscope. Results are compared with those from two existing algorithms for superresolution in fluorescence imaging. An example is given of the restoration of a 3-D confocal image of a biological specimen.

Weise, Wieland; Pavel Zinin; Tony Wilson; Andrew Briggs; Siegfried Boseck; "Imaging of spheres with the confocal scanning optical microscope," *Opt. Lett.*, Vol. 21, pg. 1800 (1996).

The scalar theory of image formation for a strongly spherical object in reflection confocal optical microscopy is considered. The image contrast is derived from the far-field scattering amplitude. Experimentally obtained images of the sphere are seen to be in good agreement with the theoretical prediction.

Wilson, Tony; Masters, Barry R.; "Confocal microscopy: Introduction to the feature issue," *Appl. Opt.*, Vol. 33, pg. 565 (1994).

Wilson, Tony; R. Juskaitis; M. A. A. Neil; M. Kozubek; "Confocal microscopy by aperture correlation," *Opt. Lett.*, Vol. 21, pg. 1879 (1996).

Most confocal microscopes do not produce images in real time with nonlaser light sources. The tandem scanning confocal microscope does produce such images but, because the pinhole apertures of the Nipkov disk must be placed far apart to reduce cross talk between neighboring pinholes, only 1% or less of the light available for imaging is used. We show that, by using aperture correlation techniques and relaxing the requirement to obtain a pure confocal image directly, one can obtain real-time confocal images with a dramatically increased (25% or even 50%) light budget.

Imaging and Scattering media

Anderson, G. E.; Liu, Feng; Alfano, R. R.; "Microscope imaging through highly scattering media," *Opt. Lett.*, Vol. 19, pg. (1994).

Bashkansky, M.; Adler, C. L.; Reintjes, J.; "Coherently amplified Raman polarization gate for imaging through scattering media," *Opt. Lett.*, Vol. 19, pg. (1994).

Bhmer, Michael; Heiko Hinrichs; Klaus D. Hinsch; Jan Kickstein; Ren Netter; Christof Surmann; "Optical Coherence Tomography for the Analysis of Three-dimensional Flow Fields," *Opt. Phot. News*, Vol. 8, pg. 19 (1997).

Cerussi, Albert E.; John S. Maier; Sergio Fantini; Maria Angela Franceschini; William W. Mantulin; Enrico Gratton; "Experimental verification of a theory for the time-resolved fluorescence spectroscopy of thick tissues," *Appl. Opt.*, Vol. 36, pg. 0116 (1997).

Fluorescence spectroscopy provides potential contrast enhancement for near-infrared tissue imaging and physiologically correlated spectroscopy. We present a fluorescence photon migration model and test its quantitative predictive capabilities with a frequency-domain measurement that involves a homogeneous multiple-scattering tissue phantom (with optical properties similar to those of tissue in the near infrared) that contains a fluorophore (rhodamine B). After demonstrating the validity of the model, we explore its ability to recover the fluorophore's spectral properties from within the multiple-scattering medium. The absolute quantum yield and the lifetime of the fluorophore are measured to within a few percent of the values measured independently in the absence of scattering. Both measurements are accomplished without the use of reference fluorophores. In addition, the model accurately predicts the fluorescence emission spectrum in the scattering medium. Implications of these absolute measurements of lifetime, quantum yield, concentration, and emission spectrum from within multiple-scattering media are discussed.

Chan, Kin Pui; Makoto Yamada, and B. Devaraj; Humio Inaba; "Optical imaging through highly scattering media by use of heterodyne detection in the 1.3-μm wavelength region," *Opt. Lett.*, Vol. 20, pg. 492 (1995).

Optical imaging through highly scattering media was studied in the 1.3- μm wavelength region by use of a continuous-wave Nd:YAG laser and an optical heterodyne detection technique. We measured and compared the extinction coefficients of the fat emulsion Intralipid-10% at 0.80, 1.064, and 1.319 μm and demonstrated that the low scattering at 1.319 μm will permit optical imaging through highly scattering media, which otherwise may not be achieved. A possible use of water absorption at 1.319 μm to image the interior structure of biological tissues is also presented and discussed.

Chang, Jenghwa; Harry L. Graber; Randall L. Barbour; Raphael Aronson; "Recovery of optical cross-section perturbations in dense-scattering media by transport-theory-based imaging operators and steady-state simulated data," *Appl. Opt.*, Vol. 35, pg. 3963 (1996).

We present a useful strategy for imaging perturbations of the macroscopic absorption cross section of dense-scattering media using steady-state light sources. A perturbation model based on transport theory is derived, and the inverse problem is simplified to a system of linear equations, W Delta urbations, and Delta R is the vector of detector readings. Monte Carlo simulations compute the photon flux across the surfaces of phantoms containing simple or complex inhomogeneities. Calculation of the weight matrix is also based on the results of Monte Carlo simulations. Three reconstruction algorithms-conjugate gradient descent, projection onto convex sets, and the simultaneous algebraic reconstruction technique, with or without imposed positivity constraints-are used for image reconstruction. A rescaling technique that improves the conditioning of the weight matrix is also developed. Results show that the analysis of time-independent data by a perturbation model is capable of resolving the internal structure of a dense-scattering medium. Imposition of positivity constraints improves image quality at the cost of a reduced convergence rate. Use of the rescaling technique increases the initial rate of convergence, resulting in accurate images in a smaller number of iterations.

Chang, Jenghwa; Wenwu Zhu; Yao Wang; Harry L. Graber; Randall L. Barbour; "Regularized progressive expansion algorithm for recovery of scattering media from time-resolved data," *J. Opt. Soc. Am. A*, Vol. 14, pg. 306 (1997).

Reconstructions of the absorption cross sections of dense scattering media from time-resolved data are presented. A progressive expansion (PE) algorithm, similar to a layer-stripping approach, is developed to circumvent the underdeterminedness of the inverse problem. An overlapping scheme, which uses detector readings from several consecutive time intervals, is introduced to reduce the propagation of reconstruction errors that occur at shallower depths. To reduce the sensitivity of the PE algorithm to noise, a regularized progressive expansion (RPE) algorithm is proposed, which incorporates regularization techniques into the PE algorithm. The PE and the RPE algorithms are applied to the problem of image reconstruction from time-resolved data. The test media were isotropically scattering slabs containing one or two compact absorbers at different depths below the surface. The data were corrupted by additive white Gaussian noise with various strengths. The reconstruction results show that the PE and the RPE algorithms, when they are combined by proper overlapping, can effectively overcome the underdeterminedness of the inverse problem. The RPE algorithm yields reconstructions that are more accurate and more stable under the same noise level.

Chen, Zhongping; Thomas E. Milner; Digant Dave; J. Stuart Nelson; "Optical Doppler tomographic imaging of fluid flow velocity in highly scattering media," *Opt. Lett.*, Vol. 22, pg. 64 (1997).

An optical Doppler tomography (ODT) system that permits imaging of fluid flow velocity in highly scattering media is described. ODT combines Doppler velocimetry with the high spatial resolution of low-coherence optical interferometry to measure fluid flow velocity at discrete spatial locations. Tomographic imaging of particle flow velocity within a circular conduit submerged 1 mm below the surface in a highly scattering phantom of Intralipid is demonstrated.

Chiang, Hai-Pang; Chang, Wei-Sheng; Wang, Jyhpyng; "Imaging through random scattering media by using cw broadband interferometry," *Opt. Lett.*, Vol. 18, pg. 546 (1993).

den Outer, P. N.; Nieuwenhuizen, Th. M.; Lagendijk, Ad; "Location of objects in multiple-scattering media," *J. Opt. Soc. Am. A*, Vol. 10, pg. 1209 (1993).

Duncan, Michael D.; Mahon, Rita; Tankersley, E. Lawrence; Reintjes, John F.; "Time-gated imaging through scattering media using stimulated Raman amplification: erratum," *Opt. Lett.*, Vol. 17, pg. 625 (1992).

Duncan, Michael D.; Mahon, Rita; Tankersley, Lawrence L.; Reintjes, John F.; "Time-gated imaging through scattering media using stimulated Raman amplification," *Opt. Lett.*, Vol. 16, pg. 1868 (1991).

Faris, Gregory W.; Banks, Michael; "Upconverting time gate for imaging through highly scattering media," *Opt. Lett.*, Vol. 19, pg. (1994).

Gleyzes, P.; A. C. Boccara; H. Saint-Jalmes; "Multichannel Nomarski microscope with polarization modulation: performance and applications," *Opt. Lett.*, Vol. 22, pg. 1529 (1997).

Using a two-dimensional CCD array, a photoelastic modulator, and a commercial Nomarski microscope, we built a differential polarization interferometer. Multichannel lock-in detection of the signal allowed us to reach a shot-noise limit corresponding to 5-pm path differences for a few seconds of recording time. This interferometer was used as a surface profilometer as well as a system for imaging through scattering media.

Gu, Min; T. Tannous; J. R. Sheppard; "Effect of an annular pupil on confocal imaging through highly scattering media," *Opt. Lett.*, Vol. 21, pg. 312 (1996).

We report on confocal scanning imaging through highly scattering media. Various practical effects including those of the annular pupils and the size of the confocal pinhole as well as of the numerical aperture of objectives on the image quality are examined experimentally. The combination of an annular objective with a finite-sized detector may prove advantageous for improving image quality.

Hall, David J.; Jeremy C. Hebden; David T. Delpy; "Imaging very-low-contrast objects in breastlike scattering media with a time-resolved method," *Appl. Opt.*, Vol. 36, pg. 7270 (1997).

An investigation was performed of the effectiveness of a time-resolved method for imaging very-low-contrast features embedded in highly scattering media. Experiments employed slabs of breastlike material into which were inserted small cylindrical objects having either a scattering or an absorption coefficient of 4, 2, 1.5, and 1.1 times greater than the surrounding medium. An attempt was made to quantify the degree of contrast produced by each object. The results indicate that time-gating is far more effective at enhancing the contrast of the scattering inhomogeneities than of the absorbing inhomogeneities. This observation is shown to agree with a diffusion-based model, which also predicts that time-gating can decrease the contrast of absorbing inhomogeneities unless very short time-gates can be employed.

Hebden, Jeremy C.; Maria Tziraki; David T. Delpy; "Evaluation of the temporally extrapolated absorbance method for dual-wavelength imaging through tissuelike scattering media," *Appl. Opt.*, Vol. 36, pg. 3802 (1997).

An independent assessment is described of a dual-wavelength imaging technique, known as the temporally extrapolated absorbance method (TEAM), proposed by Yamada et al. [Opt. Eng. 32, 634-641 (1993)]. The technique involves recording the temporal distribution of light transmitted across a scattering medium at two carefully chosen wavelengths at which the scattering properties of the medium are assumed to be identical. The objective is to image internal structure that absorbs more strongly at one wavelength than it does at the other. A simple theoretical treatment of TEAM is presented that employs a perturbation model of photon transport. This indicates that despite the lack of a secure theoretical basis, the technique may provide a potentially effective ad hoc method of generating images of highly scattering media. The method was also evaluated experimentally by using, for the first time to our knowledge, a single object and two wavelengths. A single-projection, two-dimensional image was obtained of a solid phantom with optical properties representative of breast tissue. The results exhibited good agreement with the theoretical model, and a small embedded feature that absorbs 3.5 times as strongly as the surrounding medium at one wavelength was revealed successfully.

Hebden, Jeremy C.; Simon R. Arridge; "Imaging through scattering media by the use of an analytical model of perturbation amplitudes in the time domain," *Appl. Opt.*, Vol. 35, pg. 6788 (1996).

A method of generating images through highly scattering media is presented that involves comparing measurements of the time-dependent intensity of transmitted light with an analytical model describing the sensitivity of that intensity on localized changes in optical properties. A least-squares fitting procedure is employed to derive the amplitudes of the measurement perturbations caused by embedded absorbers and scatterers located along a line of sight between the source and detector. Images are presented of a highly scattering, solid plastic phantom with optical properties closely matched to those of human breast tissue at near-infrared wavelengths. The phantom is a 54-mm-thick slab, containing four small cylinders of contrasting scatter and absorption. Results show that embedded absorbers can be distinguished from embedded scatterers, and that the diffusion perturbation amplitude provides inherently greater spatial resolution than the absorption perturbation amplitude.

Heckmeier, M.; S. E. Skipetrov; G. Maret; R. Maynard; "Imaging of dynamic heterogeneities in multiple-scattering media," *J. Opt. Soc. Am. A*, Vol. 14, pg. 185 (1997).

A new method of visualizing objects with distinct internal dynamics of the constituent scattering particles embedded in a liquid multiple-scattering medium is presented. We report dynamic multiple-light-scattering experiments and a theoretical model, based on diffusing photon-density waves for concentrated colloidal suspensions in Brownian motion, as a background medium into which is inserted a capillary containing (i) the same suspension under flow, or (ii) suspensions of different particle sizes in Brownian motion. These model objects, with purely dynamic but no static scattering contrast, can be visualized by space-resolved measurements of the time autocorrelation function g_2(tau) of the scattered light intensity at the sample surface. Maximum contrast occurs at a parameter-dependent finite correlation time tau. The physical origin of this effect is outlined. Our data are in excellent quantitative agreement with the model, with no adjustable parameter.

Hyde, S. C. W.; N. P. Barry; R. Jones; J. C. Dainty; P. M. W. French; "Sub-100- μm depth-resolved holographic imaging through scattering media in the near infrared," *Opt. Lett.*, Vol. 20, pg. 2330 (1995).

We discuss the compromise between depth and transverse spatial resolution for photorefractive holographic imaging through turbid media. Results from an optimized geometry for a 45ø-cut rhodium-doped barium titanate photorefractive crystal are presented, demonstrating two-dimensional imaging through turbid media with both sub-100- μm depth and transverse spatial resolution.

Hyde, S. C. W.; N. P. Barry; R. Jones; J. C. Dainty; P. M. W. French; M. B. Klein; B. A. Wechsler; "Depth-resolved holographic imaging through scattering media by photorefraction," *Opt. Lett.*, Vol. 20, pg. 1331 (1995).

A depth-resolved near-infrared imaging system has been demonstrated for recording three-dimensional images of objects embedded in diffuse media. Time-gated holographic imaging employing rhodium-doped barium titanate as the recording medium is used to acquire whole depth-resolved two-dimensional images in 1 s. Millimeter depth resolution has been achieved with a transverse resolution of \similar 30 μm .

Kalpaxis, L. L.; Wang, L. M.; Galland, P.; Liang, X.; Ho, Ping Pei; Alfano, R. R.; "Three-dimensional temporal image reconstruction of an object hidden in highly scattering media by time-gated optical tomography," *Opt. Lett.*, Vol. 18, pg. 1691 (1993).

Katsev, Iosif L.; Eleonora P. Zege; Alexander S. Prikhach; Igor N. Polonsky; "Efficient technique to determine backscattered light power for various atmospheric and oceanic sounding and imaging systems," *J. Opt. Soc. Am. A*, Vol. 14, pg. 1338 (1997).

A new technique that efficiently simplifies the estimations of backscattered light power is presented. This technique allows one to consider real inhomogeneous scattering media with an underlying surface and to account for all the features both of media phase functions and of the reflection of underlying surfaces. It is assumed that the phase function has a sharp forward peak. Different sounding, vision, and location systems can be considered. The image transfer equation is derived for any observed object, including three-dimensional and non-Lambertian objects. A few examples of the successful use of the developed approach to solve various problems are given.

Kempe, M.; W. Rudolph; E. Welsch; "Comparative study of confocal and heterodyne microscopy for imaging through scattering media," *J. Opt. Soc. Am. A*, Vol. 13, pg. 46 (1996).

We investigate limits of the confocal microscope when applied to imaging through scattering layers and compare its performance with that of a correlation (heterodyne) microscope. Confocal laser scanning microscopy is shown to make possible imaging through scattering media owing to the spatial filtering of the signal backreflected from the sample. Its performance is limited by the noise of the detection system and/or insufficient rejection of scattered light, depending on the sample under investigation. Correlation microscopy with narrow or broad bandwidth light can extend these limits by selective, optical amplification of the image information.

Le Tolguenec, Gaelle; Eric Lantz; Fabrice Devaux; "Imaging through scattering media by parametric amplification of images: study of the resolution and the signal-to-noise ratio," *Appl. Opt.*, Vol. 36, pg. 8292 (1997).

An imaging scheme through scattering media in which parametric image amplification is used is presented. An image of a resolution chart through a solution of latex microspheres with an attenuation of 22 mean free paths is obtained with a resolution of 20 mum. The evolution of the signal-to-noise ratio with respect to the medium attenuation is studied and compared with a rough modeling of the imaging process.

Leith, Emmett N.; Chen, C.; Chen, Y.; Lopez, Joaquin L.; Sun, Pang-Chen; Dilworth, David S.; "Imaging through scattering media using spatial incoherence techniques," *Opt. Lett.*, Vol. 16, pg. 1820 (1991).

Leith, E.; P. Naulleau; D. Dilworth; "Ensemble-averaged imaging through highly scattering media," *Opt. Lett.*, Vol. 21, pg. 1691 (1996).

A method for imaging through highly scattering media is described that consists of forming a multiplicity of holograms and performing an extensive averaging process. This process produces an estimate of the phase distribution across the exiting surface of the medium. This information is combined with the available magnitude data to form an ensemble-averaged wave front that can be backprojected to form an image of absorbers within or behind the scattering medium.

Liang, X.; L. Wang; P. P. Ho; R. R. Alfano; "Time-resolved polarization shadowgrams in turbid media," *Appl. Opt.*, Vol. 36, pg. 2984 (1997).

The use of the polarization of light as a parameter to discriminate against multiple-scattered light for transillumination imaging through random scattering media is examined. Time-resolved two-dimensional images of submillimeter test bars immersed in 5-cm-thick Intralipid solutions with different micelle dilutions were measured for two orthogonal polarizations (parallel and perpendicular) of light emerging from the turbid medium by using a picosecond Kerr-Fourier (KF) imaging system. The measured contrast and intensity of parallel-polarized KF shadowgrams decreased as the concentration of the scattering medium was increased, whereas the behavior of the perpendicular-polarized KF shadowgrams varied in an opposite matter to the micelle concentration.

Model, Regine; Matthias Orlt; Monika Walzel; Rolf Hunlich; "Reconstruction algorithm for near-infrared imaging in turbid media by means of time-domain data," *J. Opt. Soc. Am. A*, Vol. 14, pg. 313 (1997).

Light propagation in highly scattering media as a result of the injection of short laser pulses is described by the diffusion model, with absorption and scattering being considered as parameter functions. Different boundary conditions are discussed. The resulting parabolic differential equation with boundary conditions of the third kind has been integrated by the finite-element method, which allows different geometries and various embedded objects to be taken into account. For image reconstruction we introduce an iterative method based on the finite-element method forward model and on an optimization strategy that uses the full information contained in the time-resolved measurements. The algorithm includes a regularization strategy so that it is specifically fit for solving the ill-posed problem. Furthermore, it is shown that considerable improvements in reconstruction results can be achieved by adaptation of the detector-source arrangement.

Morgan, Stephen P.; Man P. Khong; Michael G. Somekh; "Effects of polarization state and scatterer concentration on optical imaging through scattering media," *Appl. Opt.*, Vol. 36, pg. 1560 (1997).

The imaging resolution in turbid media is severely degraded by light scattering. Resolution can be improved if the unscattered or weakly scattered light is extracted. Here the state of polarization of the emerging light is used to discriminate photon path length, with the more weakly scattered photons maintaining their original polarization state. It is experimentally demonstrated that over a wide range of scatterer concentrations there exist three distinct imaging regimes. It is also shown that within the intermediate regime one of two distinct imaging techniques is appropriate, depending on the particle size.

Naulleau, P. ; D. Dilworth, E. Leith, and J. Lopez; "Detection of moving objects embedded within scattering media by use of speckle methods," *Opt. Lett.*, Vol. 20, pg. 498 (1995).

Speckle-pattern subtraction methods are used for the detection of moving objects embedded in scattering media. Results show that the ability to detect small nonstationary objects is greatly enhanced.

Naulleau, P.; D. Dilworth; "Motion-resolved imaging of moving objects embedded within scattering media by the use of time-gated speckle analysis," *Appl. Opt.*, Vol. 35, pg. 5251 (1996).

Spectral analysis of a time-evolving speckle pattern is used to provide motion-resolved detection of moving objects embedded within scattering media. Results show that the ability to detect small nonstationary scattering objects and to discriminate between objects moving at different rates is greatly enhanced.

Naulleau, P.; D. Dilworth; E. Leith; J. Lopez; "Resolution-enhanced detection of moving objects embedded within scattering media using time-gated speckle methods," *Appl. Opt.*, Vol. 35, pg. 3065 (1996).

The holographic first-arriving-light method in combination with the speckle differencing method is used to provide resolution-enhanced detection of moving objects embedded in scattering media. Results show that the first-arriving-light technique provides significant resolution improvements over standard speckle differencing.

Naulleau, P.; E. Leith; H. Chen; B. Hoover; J. Lopez; "Time-gated ensemble-averaged imaging through highly scattering media," *Appl. Opt.*, Vol. 36, pg. 3889 (1997).

A previously described ensemble-averaged imaging method [Opt. Lett. 21, 1691 (1996)] is extended by its combination with holographically implemented time-gated imaging. This combined method is shown to extend the effectiveness of the ensemble-averaged method by permitting imaging through thicker diffusers. Experimental results are presented.

Paithankar, D. Y.; A. U. Chen; B. W. Pogue; M. S. Patterson; E. M. Sevick-Muraca; "Imaging of fluorescent yield and lifetime from multiply scattered light reemitted from random media," *Appl. Opt.*, Vol. 36, pg. 2260 (1997).

The feasibility of employing fluorescent contrast agents to perform optical imaging in tissues and other scattering media has been examined through computational studies. Fluorescence lifetime and yield can give crucial information about local metabolite concentrations or environmental conditions within tissues. This information can be employed toward disease detection, diagnosis, and treatment if noninvasively quantitated from reemitted optical signals. However, the problem of inverse image reconstruction of fluorescence yield and lifetime is complicated because of the highly scattering nature of the tissue. Here a light propagation model employing the diffusion equation is used to account for the scattering of both the excitation and fluorescent light. Simulated measurements of frequency-domain parameters of fluorescent modulated ac amplitude and phase lag are used as inputs to an inverse image-reconstruction algorithm, which employs the diffusion model to predict frequency-domain measurements resulting from a modulated input at the phantom periphery. In the inverse image-reconstruction algorithm, a Newton-Raphson technique combined with a Marquardt algorithm is employed to converge on the fluorescent properties within the medium. The successful reconstruction of both the fluorescence yield and lifetime in the case of a heterogeneous fluorophore distribution within a scattering medium has been demonstrated without a priori information or without the necessity of obtaining absence images.

Pan, Yingtian; Reginald Birngruber; Ralf Engelhardt; "Contrast limits of coherence-gated imaging in scattering media," *Appl. Opt.*, Vol. 36, pg. 2979 (1997).

The fundamental difference between time-resolved and coherence-gated imaging modalities in scattering media is analyzed in terms of their optical transfer functions. The effectiveness of coherence gating for multiple-scattering rejection is shown by imaging a 100-mum-thick razor blade hidden in the scattering phantoms (i.e., Intralipid suspensions) with different scattering coefficients. We found that the imaging contrast is limited by multiple scattering and speckle effects in high-scattering media, and the measured effective penetration depth of optical coherence tomography is approximately equal to six mean free paths under the experimental conditions of a numerical aperture of less than 0.1 and a scattering anisotropy of approximately 0.8.

Papazoglou, Theodore G.; W. Q. Liu; Alex Vasiliou; Ralf Grassmel; Costas Kalpouzos; Costas Fotakis; "Effect of diffraction on early-arriving photons during femtosecond laser transillumination of highly scattering media of biological significance," *Appl. Opt.*, Vol. 35, pg. 3759 (1996).

Early-arriving photons of 100-fs laser pulses transmitted through highly scattering media have been detected by a streak camera. Because of their partial spatial coherence, they are affected by diffraction from small hidden discontinuities. The experimental data of the patterns are analyzed with Fresnel diffraction theory and then corrected accordingly. Submillimeter hidden objects were scanned and imaged. Diffraction correction resulted in a significantly improved contrast in the hidden object's image. Key words: Ultrafast transillumination, diffraction correction, image-contrast improvement.

Reynolds, Jeffery S.; Charles A. Thompson; Kevin J. Webb; Fred P. LaPlant; Dor Ben-Amotz; "Frequency domain modeling of reradiation in highly scattering media," *Appl. Opt.*, Vol. 36, pg. 2252 (1997).

We present a straightforward procedure for frequency domain modeling of reradiation in a highly scattering medium with an arbitrary, finite three-dimensional geometry. We use a finite difference numerical solver to determine the fluence distribution at the excitation wavelength, which is then coupled to the emission wavelength with an array of equivalent reradiating sources. We then calculate the fluence distribution at the emission wavelength with a second, independent numerical simulation with new optical parameters appropriate to the emission wavelength, using the distributed reradiating sources as the excitation. We compare three-dimensional simulations of a fluorophore distributed in a scattering medium with experimental data. We also compare simulations of the Raman reradiation of small diamonds in a scattering medium with experiment.

Rowe, M. P. and E. N. Pugh, Jr.; J. S. Tyo and N. Engheta; "Polarization-difference imaging: a biologically inspired technique for observation through scattering media," *Opt. Lett.*, Vol. 20, pg. 608 (1995).

Many animals have visual systems that exploit the polarization of light, and some of these systems are thought to compute difference signals in parallel from arrays of photoreceptors optimally tuned to orthogonal polarizations. We hypothesize that such polarization-difference systems can improve the visibility of objects in scattering media by serving as common-mode rejection amplifiers that reduce the effects of background scattering and amplify the signal from targets whose polarization-difference magnitude is distinct from the background. We present experimental results obtained with a target in a highly scattering medium, demonstrating that a manmade polarization-difference system can render readily visible surface features invisible to conventional imaging.

Sevick, E. M.; Frisoli, J. K.; Burch, C. L.; Lakowicz, Joseph R.; "Localization of absorbers in scattering media by use of frequency-domain measurements of time-dependent photon migration," *Appl. Opt.*, Vol. 33, pg. 3562 (1994).

Thompson, C. A.; K. J. Webb; A. M. Weiner; "Imaging in scattering media by use of laser speckle," *J. Opt. Soc. Am. A*, Vol. 14, pg. 2269 (1997).

We present a method for the detection and localization of inhomogeneities embedded within highly scattering media that employs speckle statistics with a partially coherent light source. Variations in speckle contrast as a function of position are used to interrogate inhomogeneities deeply embedded within scattering media. A numerical model based on photon diffusion theory is introduced to predict speckle contrast as a function of scan position. This model uses measured speckle data to determine scattering and absorption parameters.

Thompson, Charles A.; Jeffery S. Reynolds, and Kevin J. Webb; Fred P. LaPlant and Dor Ben-Amotz; "Raman spectroscopic studies of diamond in Intralipid," *Opt. Lett.*, Vol. 20, pg. 1195 (1995).

Raman spectroscopic measurements were performed to determine the presence of a diamond located within an Intralipid scattering medium. The Raman signature of the diamond was recorded for varying thicknesses and concentrations of the Intralipid by use of a 514-nm argon-ion laser and a cw 785-nm Ti:sapphire laser. Results show how Raman spectroscopy can be used as a diagnostic tool for identifying objects embedded within scattering media. In particular, future applications could include tissue diagnostics performed *in vivo*.

Tyo, J. S.; M. P. Rowe; E. N. Pugh; N. Engheta; "Target detection in optically scattering media by polarization-difference imaging," *Appl. Opt.*, Vol. 35, pg. 1855 (1996).

Polarization-difference imaging (PDI) was recently presented by us as a method of imaging through scattering media [Opt. Lett. 20, 608 (1995)]. Here, PDI is compared with conventional, polarization-blind imaging systems under a variety of conditions not previously studied. Through visual and numerical comparison of polarization-difference and polarization-sum images of metallic targets suspended in scattering media, target features initially visible in both types of images are shown to disappear in polarization-sum images as the scatterer concentration is increased, whereas these features remain visible in polarization-difference images. Target features producing an observed degree of linear polarization of less than 1% are visible in polarization-difference images. The ability of PDI to suppress partially polarized background variations selectively is demonstrated, and discrimination of target features on the basis of polarization information is discussed. Our results show that, when compared with conventional imaging, PDI yields a factor of 2-3 increase in the distance at which certain target features can be detected.

Wabnitz, H.; H. Rinneberg; "Imaging in turbid media by photon density waves: spatial resolution and scaling relations," *Appl. Opt.*, Vol. 36, pg. 0064 (1997).

Spatial resolution of transillumination imaging through highly scattering media normalized to sample thickness d depends on only the normalized wavelength Lambda/d of photon density waves and normalized penetration depth delta/d. This was concluded theoretically and verified experimentally by the derivation of edge-spread functions from measured time-resolved transmittance and its Fourier transform by the use of dilute milk at various concentrations as scatterer in cuvettes of d = 2 cm and d = 4 cm. In the frequency domain and the time domain, spatial resolution was found experimentally to be given by approximately 0.3d obtained at Lambda \approximately d or when only the first arriving 1% of all photons detected were taken into account.

Walker, Scott A.; Sergio Fantini; Enrico Gratton; "Image reconstruction by backprojection from frequency-domain optical measurements in highly scattering media," *Appl. Opt.*, Vol. 36, pg. 170 (1997).

The reconstruction of the location and optical properties of objects in turbid media requires the solution of the inverse problem. Iterative solutions to this problem can require large amounts of computing time and may not converge to a unique solution. Instead, we propose a fast, simple method for approximately solving this problem in which calculated effective absorption and reduced scattering coefficients are backprojected to create an image of the objects. We reconstructed images of objects with centimeter dimensions embedded in a diffusive medium with optical characteristics similar to those of human tissue. Data were collected by a frequency-domain spectrometer operating at 120 MHz with a laser diode light source emitting at 793 nm. Intensity and phase of the incident photon density wave were collected from linear scans at different projection angles. Although the positions of the objects are correctly identified by the reconstructed images, the optical parameters of the objects are recovered only qualitatively.

Wang, L.; Ho, Ping Pei; Alfano, R. R.; "Time-resolved Fourier spectrum and imaging in highly scattering media," *Appl. Opt.*, Vol. 32, pg. 5043 (1993).

Wang, Lihong and Xuemei Zhao, "Ultrasound-modulated optical tomography of absorbing objects buried in dense tissue-simulating turbid media," Appl. Opt. 36, No. 28, pp. 7277-7282 (1997).

Continuous-wave ultrasonic modulation of scattered laser light was used to image objects buried in tissue-simulating turbid media. The buried object had an absorption coefficient greater than the background turbid medium. The ultrasonic wave that was focused into the turbid media modulated the laser light that passed through the ultrasonic field. The modulated laser light that was collected by a photomultiplier tube reflected the local mechanical and optical properties in the zone of ultrasonic modulation. Objects buried in the middle plane of 5-cm-thick dense turbid media were imaged with millimeter resolution through the scanning and detecting alterations of the ultrasound-modulated optical signal. The optical properties of the dense turbid media included an absorption coefficient of 0.1 cm-1 and a reduced scattering coefficient of 10 cm-1 and were comparable with those of biological tissues in the visible and near-IR ranges. The dependence of the ultrasound-modulated optical signal on the off-axis distance of the detector from the optic axis and the area of the detector was studied as well.

Yoo, K. M.; Xing, Qirong; Alfano, R. R.; "Imaging objects hidden in highly scattering media using femtosecond second-harmonic-generation cross-correlation time gating," *Opt. Lett.*, Vol. 16, pg. 1019 (1991).

Yoo, K. M.; Zang, Zhi-Wei; Ahmed, Samir A.; Alfano, R. R.; "Imaging objects hidden in scattering media using a fluorescence-absorption technique," *Opt. Lett.*, Vol. 16, pg. 1252 (1991).

Zhu, Wenwu; Yao Wang; Yuqi Yao; Jenghwa Chang; Harry L. Graber; Randall L. Barbour; "Iterative total least-squares image reconstruction algorithm for optical tomography by the conjugate gradient method," *J. Opt. Soc. Am. A*, Vol. 14, pg. 799 (1997).

We present an iterative total least-squares algorithm for computing images of the interior structure of highly scattering media by using the conjugate gradient method. For imaging the dense scattering media in optical tomography, a perturbation approach has been described previously [Y. Wang et al., Proc. SPIE 1641, 58 (1992); R. L. Barbour et al., in Medical Optical Tomography: Functional Imaging and Monitoring (Society of Photo-Optical Instrumentation Engineers, Bellingham, Wash., 1993), pp. 87-120], which solves a perturbation equation of the form $W\delta x=\delta I$. In order to solve this equation, least-squares or regularized least-squares solvers have been used in the past to determine best fits to the measurement data δI while assuming that the operator matrix W is accurate. In practice, errors also occur in the operator matrix. Here we propose an iterative total least-squares (ITLS) method that minimizes the errors in both weights and detector readings. Theoretically, the total least-squares (TLS) solution is given by the singular vector of the matrix $[W|\delta I]$ associated with the smallest singular value. The proposed ITLS method obtains this solution by using a conjugate gradient method that is particularly suitable for very large matrices. Simulation results have shown that the TLS method can yield a significantly more accurate result than the least-squares method.

Medical imaging

"Endoscope with traced raster and elemental photodetectors," *Appl. Opt.*, Vol. 30, pg. 361 (1991).

Andersson-Engels, S.; Berg, Roger; Persson, A.; Svanberg, Sune; "Multispectral tissue characterization with time-resolved detection of diffusely scattered white light," *Opt. Lett.*, Vol. 18, pg. 1697 (1993).

Andersson-Engels, S.; Johansson, J.; Svanberg, S.; "Medical diagnostic system based on simultaneous multi-spectral fluorescence imaging," *Appl. Opt.*, Vol. 33, pg. 8022 (1994).

Bashkansky, M.; Reintjes, John F.; "Nonlinear-optical field cross-correlation techniques for medical imaging with lasers," *Appl. Opt.*, Vol. 32, pg. 3842 (1993).

Berg, Roger; Jarlman, Olof; Svanberg, Sune; "Medical transillumination imaging using short-pulse diode lasers," *Appl. Opt.*, Vol. 32, pg. 574 (1993).

Beuthan, J.; Minet, O.; Muller, G.; "Observations of the fluorescence response of the coenzyme NADH in biological samples," *Opt. Lett.*, Vol. 18, pg. 1098 (1993).

Chen, Hsuan; Chen, Y.; Dilworth, David S.; Leith, Emmett N.; Lopez, Joaquin L.; Valdmanis, J. A.; "Two-dimensional imaging through diffusing media using 150-fs gated electronic holography techniques," *Opt. Lett.*, Vol. 16, pg. 487 (1991).

Chen, Hsuan; Shih, M.; Arons, E.; Leith, Emmett N.; Lopez, Juaquin; Dilworth, David S.; Sun, Pang-Chen; "Electronic holographic imaging through living human tissue," *Appl. Opt.*, Vol. 33, pg. 3630 (1994).

Chen, Y.; Chen, H.; Dilworth, David; Leith, Emmett N.; Lopez, Joaquin L.; Shih, M.; Sun, Pang-Chen; Vossler, G.; "Evaluation of holographic methods for imaging through biological tissue," *Appl. Opt.*, Vol. 32, pg. 4330 (1993).

Chen, Ye; "Characterization of the image resolution for the first-arriving-light method," *Appl. Opt.*, Vol. 33, pg. 2544 (1994).

Clymer, Bradley D.; DeVore, Thomas B.; Jagadeesh, Jogikal; Tomei, L. David; "Spatial resolution of images reconstructed from a bulk-detection scanning-laser microscope," *Appl. Opt.*, Vol. 30, pg. 5056 (1991).

Colak, S. B.; D. G. Papaioannou; G. W. 't Hooft; M. B. van der Mark; H. Schomberg; J. C. J. Paasschens; J. B. M. Melissen; N. A. A. J. van Asten; "Tomographic image reconstruction from optical projections in light-diffusing media," *Appl. Opt.*, Vol. 36, pg. 0180 (1997).

The recent developments in light generation and detection techniques have opened new possibilities for optical medical imaging, tomography, and diagnosis at tissue penetration depths of \similar10 cm. However, because light scattering and diffusion in biological tissue are rather strong, the reconstruction of object images from optical projections needs special attention. We describe a simple reconstruction method for diffuse optical imaging, based on a modified backprojection approach for medical tomography. Specifically, we have modified the standard backprojection method commonly used in x-ray tomographic imaging to include the effects of both the diffusion and the scattering of light and the associated nonlinearities in projection image formation. These modifications are based primarily on the deconvolution of the broadened image by a spatially variant point-spread function that is dependent on the scattering of light in tissue. The spatial dependence of the deconvolution and nonlinearity corrections for the curved propagating ray paths in heterogeneous tissue are handled semiempirically by coordinate transformations. We have applied this method to both theoretical and experimental projections taken by parallel- and fan-beam tomography geometries. The experimental objects were biomedical phantoms with multiple objects, including *in vitro* animal tissue. The overall results presented demonstrate that image-resolution improvements by nearly an order of magnitude can be obtained. We believe that the tomographic method presented here can provide a basis for rapid, real-time medical monitoring by the use of optical projections. It is expected that such optical tomography techniques can be combined with the optical tissue diagnosis methods based on spectroscopic molecular signatures to result in a versatile optical diagnosis and imaging technology.

Das, B. B.; Yoo, K. M.; Alfano, R. R.; "Ultrafast time-gated imaging in thick tissues: a step toward optical mammography," *Opt. Lett.*, Vol. 18, pg. 1092 (1993).

den Outer, P. N.; Nieuwenhuizen, Th. M.; Lagendijk, Ad; "Location of objects in multiple-scattering media," *J. Opt. Soc. Am. A*, Vol. 10, pg. 1209 (1993).

Hebden, Jeremy C.; "Time-resolved attenuation of transmitted laser pulses by a homogenous scattering medium," *Opt. Lett.*, Vol. 17, pg. 444 (1992).

Hee, Michael R.; Izatt, Joseph A.; Swanson, Eric A.; Fujimoto, James G.; "Femtosecond transillumination tomography in thick tissues," *Opt. Lett.*, Vol. 18, pg. 1107 (1993).

Knuttel, A.; Schmitt, J. M.; Knutson, J. R.; "Spatial localization of absorbing bodies by interfering diffusive photon-density waves," *Appl. Opt.*, Vol. 32, pg. 381 (1993).

Leith, Emmett N.; Chen, Hsuan; Chen, Y.; Dilworth, David S.; Lopez, Joaquin L.; Masri, R.; Rudd, J. V.; Valdmanis, J. A.; "Electronic holography and speckle methods for imaging through tissue using femtosecond gated pulses," *Appl. Opt.*, Vol. 30, pg. 4204 (1991).

Milbocker, Michael T.; Feke, Gilbert T.; "Intensified charge-coupled-device-based eyetracker image stabilizer," *Appl. Opt.*, Vol. 31, pg. 3719 (1992).

Milner, Thomas E.; Goodman, Dennis M.; Tanenbaum, B. Samuel; Nelson, J. Stuart; "Depth profiling of laser-heated chromophores in biological tissues by pulsed photothermal radiometry," *J. Opt. Soc. Am. A*, Vol. 12, pg. 1479 (1995).
A solution method is proposed to the inverse problem of determining the unknown initial temperature distribution in a laser-exposed test material from measurements provided by infrared radiometry. A Fredholm integral equation of the first kind is derived that relates the temporal evolution of the infrared signal amplitude to the unknown initial temperature distribution in the exposed test material. The singular-value decomposition is used to demonstrate the severely ill-posed nature of the derived inverse problem. Three inversion methods are used to estimate solutions for the initial temperature distribution. A nonnegatively constrained conjugate-gradient algorithm using early termination is found superior to unconstrained inversion methods and is applied to image the depth of laser-heated chromophores in human skin. Key words: constrained conjugate gradients, ill-posed problem, infrared radiometry, laser surgery, nonnegative, singular-value decomposition.

Mitic, Gerhard; Kolzer, Jochen; Otto, Johann; Plies, Erich; Solkner, Gerald; Zinth, Wolfgang; "Time-gated transillumination of biological tissues and tissuelike phantoms," *Appl. Opt.*, Vol. 33, pg. 6699 (1994).

Ono, Kimizo; Kanda, Masahiko; Hiramoto, J.; Yotsuya, Kohro; Sata, Nobuhiro; "Fiber optic reflectance spectrophotometry system for *in vivo* tissue diagnosis," *Appl. Opt.*, Vol. 30, pg. 98 (1991).

Schiller, Stephan; Byer, Robert L.; "Subwavelength optical magnetic-resonance imaging," *J. Opt. Soc. Am. A*, Vol. 9, pg. 683 (1992).

Wang, L.; Ho, Ping Pei; Liang, X.; Dai, H.; Alfano, R. R.; "Kerr-Fourier imaging of hidden objects in thick turbid media," *Opt. Lett.*, Vol. 18, pg. 241 (1993).

Optical coherence tomography

Bashkansky, M.; M. D. Duncan; M. Kahn; D. Lewis III; J. Reintjes; "Subsurface defect detection in ceramics by high-speed high-resolution optical coherent tomography," *Opt. Lett.*, Vol. 22, pg. 61 (1997).
We use optical coherence tomography with a new configuration to determine the size and location of subsurface defects in solid ceramic and composite ceramic materials. Cross-sectional subsurface regions either parallel or perpendicular to the surface were examined. We present experimental results showing that the size and distribution of small subsurface defects can be determined with depth and lateral resolutions of 10 and 4 , respectively.

Bhmer, Michael ; Heiko Hinrichs; Klaus D. Hinsch; Jan Kickstein; Ren Netter; Christof Surmann; "Optical Coherence Tomography for the Analysis of Three-dimensional Flow Fields," *Opt. Phot. News*, Vol. 8, pg. 19 (1997).

Boppart, S. A.; B. E. Bouma; C. Pitris; G. J. Tearney; J. G. Fujimoto; M. E. Brezinski; "Forward-imaging instruments for optical coherence tomography," *Opt. Lett.*, Vol. 22, pg. 1618 (1997).
We discuss the design and implementation of forward-imaging instruments for optical coherence tomography (OCT), which require the delivery, scanning, and collection of single-spatial-mode optical radiation. A hand-held surgical probe for use in open surgery can provide cross-sectional images of subsurface tissue

before surgical incisions are made. A rigid laparoscope for minimally invasive surgical OCT imaging provides a simultaneous en face view of the area being imaged. OCT imaging is demonstrated on *in vitro* human specimens.

Bouma, B. E.; G. J. Tearney; I. P. Bilinsky; B. Golubovic; J. G. Fujimoto; "Self-phase-modulated Kerr-lens mode-locked Cr:forsterite laser source for optical coherence tomography," *Opt. Lett.*, Vol. 21, pg. 1839 (1996).

An all-solid-state Kerr-lens mode-locked Cr:forsterite laser operating at 1.28 um is demonstrated as a short-coherence-length, high-average-power source for optical coherence tomographic (OCT) imaging. We achieve ultrahigh resolution by spectrally broadening the laser pulses, using self-phase modulation in a dispersion-shifted single-mode fiber. OCT imaging with a resolution of 6 μm and a dynamic range of 115 dB is achieved.

Bouma, B.; G. J. Tearney; S. A. Boppart; M. R. Hee, M. E. Brezinski; J. G. Fujimoto; "High resolution optical coherence tomographic imaging using a modelocked Ti:Al2O3 laser source," *Opt. Lett.,* Vol. 20, pg. 1 (1995).

Chen, Zhongping; Thomas E. Milner; Shyam Srinivas; Xiaojun Wang; Arash Malekafzali; Martin J. C. van Gemert; J. Stuart Nelson; "Noninvasive imaging of *in vivo* blood flow velocity using optical Doppler tomography," *Opt. Lett.*, Vol. 22, pg. 1119 (1997).

We report the development of an optical technique for noninvasive imaging of *in vivo* blood flow dynamics and tissue structures with high spatial resolution (2-10) in biological systems. The technique is based on optical Doppler tomography (ODT), which combines Doppler velocimetry with optical coherence tomography to measure blood flow velocity at discrete spatial locations. The exceptionally high resolution of ODT permits noninvasive *in vivo* imaging of both blood microcirculation and tissue structures surrounding the vessel, which has significance for biomedical research and clinical applications. Tomographic imaging of *in vivo* blood flow velocity in the chick chorioallantoic membrane and in rodent skin is demonstrated.

Chinn, S. R.; E. A. Swanson; J. G. Fujimoto; "Optical coherence tomography using a frequency-tunable optical source," *Opt. Lett.*, Vol. 22, pg. 340 (1997).

We have developed a simple, wide-optical-bandwidth, high-resolution system for performing rapid optical frequency domain reflectometry measurements and applied it to multidimensional tomographic imaging. The source is a grating-tuned external cavity semiconductor laser with a tuning capability of 25 nm in 100 ms. We discuss system performance and show a two-dimensional optical coherence tomography image of a thin glass sandwich structure as a preliminary demonstration of the systems depth and resolution capabilities.

de Boer, Johannes F.; Thomas E. Milner; Martin J. C. van Gemert; J. Stuart Nelson; "Two-dimensional birefringence imaging in biological tissue by polarization-sensitive optical coherence tomography," *Opt. Lett.*, Vol. 22, pg. 934 (1997).

Using a low-coherence Michelson interferometer, we measure two-dimensional images of optical birefringence in bovine tendon as a function of depth. Polarization-sensitive detection of the signal formed by interference of backscattered light from the sample and a mirror in the reference arm give the optical phase delay between light that is propagating along the fast and slow axes of the birefringent tendon. Images showing the change in birefringence in response to laser irradiation are presented. The technique permits rapid noncontact investigation of tissue structural properties through two-dimensional imaging of birefringence.

Fujimoto, J. G.; S. DeSilvestri; E. P. Ippen; C. A. Puliafito; R. Margolis; A. Oseroff; "Femtosecond optical ranging in biological systems," *Opt. Lett.*, Vol. 11, pg. 150 (1986).

Golubovic, B.; B. E. Bouma; G. J. Tearney; J. G. Fujimoto; "Optical frequency domain reflectometry using rapid wavelength tuning of a Cr^{4+}:forsterite laser," *Opt. Lett.,* Vol. 22, pg. 1704 (1997).

Hee, M. R.; D. Huang; E. A. Swanson; J. G. Fujimoto; "Polarization sensitive low coherence reflectometer for birefringence characterization and ranging," *J. Opt. Soc. Am. B,* Vol. 9, pg. 903 (1992).

Hee, Michael R.; Joseph A. Izatt; Joseph M. Jacobson; James G. Fujimoto; Eric A. Swanson; "Femtosecond transilluminaton optical coherence tomography," *Opt. Lett.*, Vol. 18, pg. 950 (1993).

Hee, Michael R.; Izatt, Joseph A.; Swanson, Eric A.; Fujimoto, James G.; "Femtosecond transillumination tomography in thick tissues," *Opt. Lett.*, Vol. 18, pg. 1107 (1993).

Izatt, Joseph A.; Manish D. Kulkarni; Siavash Yazdanfar; Jennifer K. Barton; Ashley J. Welch; "*In vivo* bidirectional color Doppler flow imaging of picoliter blood volumes using optical coherence tomography," *Opt. Lett.*, Vol. 22, pg. 1439 (1997).

We describe a novel optical system for bidirectional color Doppler imaging of flow in biological tissues with micrometer-scale resolution and demonstrate its use for *in vivo* imaging of blood flow in an animal model. Our technique, color Doppler optical coherence tomography (CDOCT), performs spatially localized optical Doppler velocimetry by use of scanning low-coherence interferometry. CDOCT is an extension of optical coherence tomography (OCT), employing coherent signal-acquisition electronics and joint time-frequency analysis algorithms to perform flow imaging simultaneous with conventional OCT imaging. Cross-sectional maps of blood flow velocity with spatial resolution and velocity precision were obtained through intact skin in living hamster subdermal tissue. This technology has several potential medical applications.

Izatt, Joseph; Manish D. Kulkarni; Kenji Kobayashi; Michael V. Sivak; Jennifer K. Barton; Ashley J. Welch; "Optical Coherence Tomography for Bio-diagnostics," *Opt. Phot. News*, Vol. 8, pg. 41 (1997).

Pan, Yingtian; Reginald Birngruber; Ralf Engelhardt; "Contrast limits of coherence-gated imaging in scattering media," *Appl. Opt.*, Vol. 36, pg. 2979 (1997).
The fundamental difference between time-resolved and coherence-gated imaging modalities in scattering media is analyzed in terms of their optical transfer functions. The effectiveness of coherence gating for multiple-scattering rejection is shown by imaging a 100-μm-thick razor blade hidden in the scattering phantoms (i.e., Intralipid suspensions) with different scattering coefficients. We found that the imaging contrast is limited by multiple scattering and speckle effects in high-scattering media, and the measured effective penetration depth of optical coherence tomography is approximately equal to six mean free paths under the experimental conditions of a numerical aperture of less than 0.1 and a scattering anisotropy of approximately 0.8.

Podoleanu, Adrian Gh.; George M. Dobre; David J. Webb; David A. Jackson; "Simultaneous en-face imaging of two layers in the human retina by low-coherence reflectometry," *Opt. Lett.*, Vol. 22, pg. 1039 (1997).
We show that with a fiberized multiple Michelson-interferometer-type configuration, transverse images from several layers in the human eye can be simultaneously obtained. We demonstrate the principle by producing simultaneous pixel en-face images of a region on a postmortem retina for two depth positions 250 apart.

Schmitt, J. M.; A. Knuttel; "Model of optical coherence tomography of heterogeneous tissue," *J. Opt. Soc. Am. A*, Vol. 14, pg. 1231 (1997).
A comprehensive model of optical coherence tomography (OCT) is described that includes the interference effects that produce speckle in images of dense heterogeneous tissue. It is based on the extended Huygens-Fresnel formulation of beam propagation in a turbulent atmosphere, adapted to the analysis of OCT. Incorporated in the model is a fractal description of the size distribution of scatterers in tissue. We demonstrate its application in the simulation of images of tissue volumes containing high-contrast targets embedded in a mixture of two sizes of particles. The simulated images show the degradation of image quality caused by speckle noise, along with the benefits of employing a light source with a short coherence time and an objective lens with a high numerical aperture. Based on model results, an estimate of the maximum probing depth is given in terms of the design variables of an OCT scanner and the optical properties of the tissue.

Sergeev, A. M.; V. M. Gelikonov; G. V. Gelikonov; F. I. Feldchtein; R. V. Kuranov; N. D. Gladkova; N. M. Shakhova; L. B. Snopova; A. V. Shakov; I. A. Kuzn; "*In vivo* endoscopic OCT imaging of precancer and cancer states of human mucosa," *Opt. Express*, Vol. 1, pg. 432 (1997).
First results of endoscopic applications of optical coherence tomography for *in vivo* studies of human mucosa in respiratory, gastrointestinal, urinary and genital tracts are presented. A novel endoscopic OCT (EOCT) system has been created that is based on the integration of a sampling arm of an all-optical-fiber interferometer into standard endoscopic devices using their biopsy channel to transmit low-coherence radiation to investigated tissue. We have studied mucous membranes of esophagus, larynx, stomach, urinary bladder, uterine cervix and body as typical localization for carcinomatous processes. Images of tumor tissues versus healthy tissues have been recorded and analyzed. Violations of well-defined stratified healthy mucosa structure in cancered tissue are distinctly seen by EOCT, thus making this technique promising for early diagnosis of tumors and precise guiding of excisional biopsy.

Swanson, E. A.; D. Huang; M. R. Hee; J. G. Fujimoto; C. P. Lin; C. A. Puliafito, "High speed optical coherence domain reflectometry," *Opt.* Lett., Vol. 17, pg. 151 (1992).

Swanson, Eric A.; J. A. Izatt; Michael R. Hee; David Huang; C. P. Lin; J. S. Shuman; C. A. Puliafito; James G. Fujimoto; "In-vivo retinal imaging by optical coherence tomography," *Opt. Lett.*, Vol. 18, pg. 1864 (1993).

Tearney, G. J.; B. E. Bouma; J. G. Fujimoto; "High-speed phase- and group-delay scanning with a grating-based phase control delay line," *Opt. Lett.*, Vol. 22, pg. 1811 (1997).
A rapid-scanning optical delay line that employs phase control has several advantages, including high speed, high duty cycle, phase- and group-delay independence, and group-velocity dispersion compensation, over existing optical delay methods for interferometric optical ranging applications. We discuss the grating-based phase-control delay line and its applications to interferometric optical ranging and measurement techniques such as optical coherence domain reflectometry and optical coherence tomography. The system performs optical ranging over an axial range of 3 mm with a scanning rate

of and a repetition rate of 2 kHz. The device is especially well suited for applications such as optical coherence tomography that require high-speed, repetitive, linear delay line scanning with a high duty cycle.

Tearney, G. J.; B. E. Bouma; S. A. Boppart; B. Golubovic; E. A. Swanson; J. G. Fujimoto; "Rapid acquisition of *in vivo* biological images by use of optical coherence tomography," *Opt. Lett.*, Vol. 21, pg. 1408 (1996).
The development of techniques for high-speed image acquisition in optical coherence tomography (OCT) systems is essential for suppressing motion artifacts when one is imaging living systems. We describe a new OCT system for performing micrometer-scale, cross-sectional optical imaging at four images / s. To achieve OCT image-acquisition times of less than 1 s, we use a piezoelectric fiber stretcher to vary the reference arm delay. A Kerr-lens mode-locked chromium-doped forsterite laser is employed as the low-coherence source for the high-speed OCT system. Dynamic, motion-artifact-free *in vivo* imaging of a beating Xenopus laevis (African frog) heart is demonstrated.

Tearney, G. J.; M. E. Brezinski; J. F. Southern; B. E. Bouma; M. R. Hee; J. G. Fujimoto; "Determination of the refractive index of highly scattering human tissue by optical coherence tomography," *Opt. Lett.*, Vol. 20, pg. 2258 (1995).
We describe two new techniques, based on optical coherence tomography (OCT), for determining the refractive index of highly scattering human tissue. We obtained refractive indices of *in vitro* human tissue, using OCT to measure the physical and optical path lengths of the sample. We obtained measurements of the refractive index of *in vitro* and *in vivo* human tissue, using OCT to track the focal length shift that results from translating the focus along the optic axis within the tissue. The refractive indices of human skin, adipose, and muscle were measured and compared with previously published estimates.

Tearney, G. J.; S. A. Boppart; B. E. Bouma; M. E. Brezinski; N. J. Weissman; J. F. Southern; J. G. Fujimoto; "Scanning single-mode fiber optic catheter-endoscope for optical coherence tomography," *Opt. Lett.*, Vol. 21, pg. 543 (1996).
We describe a new optical coherence tomography catheter-endoscope for micrometer-scale, cross-sectional imaging in internal organ systems. The catheter-endoscope uses single-mode fiber optics with a novel transverse scanning design. The distal end of the catheter-endoscope uses a gradient-index lens with a microprism to emit and collect a single spatial-mode optical beam with specific focusing characteristics. The beam is scanned in a circumferential pattern and can image transverse cross sections through the structure into which it is inserted. A device with a diameter as small as 1.1 mm has been achieved, and imaging of *in vitro* human venous morphology is demonstrated.

Tearney, G. J.; S. A. Boppart; B. E. Bouma; M. E. Brezinski; N. J. Weissman; J. F. Southern; J. G. Fujimoto; "Scanning single-mode fiber optic catheter-endoscope for optical coherence tomography: erratum," *Opt. Lett.*, Vol. 21, pg. 912 (1996).

Thompson, C. A. ; K. J. Webb; A. M. Weiner; "Imaging in scattering media by use of laser speckle," *J. Opt. Soc. Am. A*, Vol. 14, pg. 2269 (1997).
We present a method for the detection and localization of inhomogeneities embedded within highly scattering media that employs speckle statistics with a partially coherent light source. Variations in speckle contrast as a function of position are used to interrogate inhomogeneities deeply embedded within scattering media. A numerical model based on photon diffusion theory is introduced to predict speckle contrast as a function of scan position. This model uses measured speckle data to determine scattering and absorption parameters.

Yazdanfar, Siavash; Manish D. Kulkarni; Joseph A. Izatt; "High resolution imaging of *in vivo* cardiac dynamics using color Doppler optical coherence tomography," *Opt. Express*, Vol. 1, pg. 424 (1997).
Color Doppler optical coherence tomography (CDOCT) is a noninvasive technique for simultaneous high spatial resolution (~20 \microm) imaging and high velocity resolution (~500 \microm/s) imaging flowmetry in living tissues. In this paper, we demonstrate a reconstruction method which overcomes fundamental limitations on Doppler flow mapping associated with both high- and low-speed imaging. This algorithm is successful in retaining the high velocity resolution of CDOCT while eliminating motion artifact caused by slow image acquisition in samples which exhibit repetitive motion. We demonstrate reconstruction of blood flow throughout a beating Xenopus laevis heart and surrounding vasculature using gated CDOCT reconstruction.

Optical imaging

Alfano, R. R.; James Fujimoto; "Advances in Optical Imaging and Photon Migration," *Opt. Phot. News*, Vol. 7, pg. 37 (1996).

Barter, J. D.; P. H. Y. Lee; "Polarimetric optical imaging of scattering surfaces," *Appl. Opt.*, Vol. 35, pg. 6015 (1996).
A polarimetric optical specular event detector (OSED) has been developed to provide spatially and temporally resolved polarimetric data of backscattering in the visible from water wave surfaces. The OSED acquires simultaneous, two-dimensionally resolved images of the remote target in two orthogonal planes of polarization. With the use of plane-polarized illumination the OSED

presently can measure, in an ensemble of breaking waves, the equivalent four-element polarization matrix common to polarimetric radars. Upgrade to full Stokes parameter state of polarization measurements is straightforward with the use of present single-aperture, multi-imager CCD camera technology. The OSED is used in conjunction with a coherent pulse-chirped radar (PCR), which also measures the four-element polarization matrix, to provide direct time-correlated identification of backscattering mechanisms operative during wave-breaking events which heretofore have not been described theoretically. We describe the instrument and its implementation, and examples of spatially resolved polarimetric data are displayed as correlated with the PCR backscatter cross section and polarization ratio records.

Betzig, E.; "Proposed method for molecular optical imaging," *Opt. Lett.*, Vol. 20, pg. 237 (1995).
We can resolve multiple discrete features within a focal region of m spatial dimensions by first isolating each on the basis of n>=1 unique optical characteristics and then measuring their relative spatial coordinates. The minimum acceptable separation between features depends on the point-spread function in the (m+n) - dimensional space formed by the spatial coordinates and the optical parameters, whereas the absolute spatial resolution is determined by the accuracy to which the coordinates can be measured. Estimates of each suggest that near-field fluorescence excitation microscopy / spectroscopy with molecular sensitivity and spatial resolution is possible.

Chan, Kin Pui; Makoto Yamada, and B. Devaraj; Humio Inaba; "Optical imaging through highly scattering media by use of heterodyne detection in the 1.3-mu m wavelength region," *Opt. Lett.*, Vol. 20, pg. 492 (1995).
through highly scattering media was studied in the 1.3-μm wavelength region by use of a continuous-wave Nd:YAG laser and an optical heterodyne detection technique. We measured and compared the extinction coefficients of the fat emulsion Intralipid-10% at 0.80, 1.064, and 1.319 μm and demonstrated that the low scattering at 1.319 μm will permit optical imaging through highly scattering media, which otherwise may not be achieved. A possible use of water absorption at 1.319 μm to image the interior structure of biological tissues is also presented and discussed.

Clark, S. E.; Jones, Linda Ramball; DeSandre, Lewis F.; "Coherent array optical imaging," *Appl. Opt.*, Vol. 30, pg. 1804 (1991).

Cubeddu, R.; A. Pifferi; P. Taroni; A. Torricelli; G. Valentini; "Time-resolved imaging on a realistic tissue phantom: μs\prime and mua images versus time-integrated images," *Appl. Opt.*, Vol. 35, pg. 4533 (1996).
A method is proposed by which we construct images through turbid media, plotting directly either the transport-scattering coefficient μs\prime or the absorption coefficient mua. These optical parameters are obtained from the best fit of the time-resolved transmittance curves with a diffusion model. Measurements were performed with a time-correlated single-photon counting system on realistic tissue phantoms simulating a tumor mass within a breast. Images were obtained with an incident power of <1 mW and an acquisition time of 1 s/point. Comparison of mus\prime and mua images with time-integrated images constructed from the same experimental data shows that the fitting method discriminates between scattering and absorption inhomogeneities and improves image quality for scattering but not for absorption inhomogeneities.

Demos, S. G. ; R. R. Alfano; "Optical polarization imaging," *Appl. Opt.*, Vol. 36, pg. 0150 (1997).
The temporal profiles of the parallel and perpendicular polarization components of a light pulse backscattered from a scattering medium are different. The depth of penetration into the tissue and depolarization of the backscattered light depend on the scattering and absorption characteristics of the tissue. Based on these facts, a novel technique is demonstrated for noninvasive surface and beneath-the-surface imaging of biological systems.

Erickson, Michael G.; Jeffery S. Reynolds; Kevin J. Webb; "Comparison of sensitivity for single-source and dual-interfering-source configurations in optical diffusion imaging," *J. Opt. Soc. Am. A*, Vol. 14, pg. 3083 (1997).
A numerical study of single-source and dual-interfering-source modalities for optical imaging through biological media with a view to practical implications is given. Our comparisons are based on multigrid finite-difference frequency-domain solutions of the diffusion equation, for example, inhomogeneous imaging problems. Detector noise levels, the influence of boundaries, and ease of measurement are considered. This investigation indicates that a dual source, or dual-source data synthesized from single-source measurements, offers significant imaging opportunities when compared with single-source data.

Gandjbakhche, A. H.; R. F. Bonner; R. Nossal; G. H. Weiss; "Absorptivity contrast in transillumination imaging of tissue abnormalities," *Appl. Opt.*, Vol. 35, pg. 1767 (1996).
We calculate the time-resolved flux of photons transmitted across an optically turbid slab containing a partially absorbing inclusion. An analytical expression is obtained for the flux at a detector positioned opposite a point source (at a distance equal to the thickness of the slab) when the center of the inclusion lies on the line connecting those points. The calculation employs a discrete-time lattice random-walk model of photon

transport. The resulting expression is used to assess the affects of time resolution on the detectability of the inclusion. Key words: Random walks, optical imaging, detectability, optical contrast.

Hatano, Hiroshi; Yasushi Inouye; Satoshi Kawata; "Near-field optical microscope with a multiheight scanning-imaging mode," *Opt. Lett.*, Vol. 22, pg. 1532 (1997).

A near-field scanning optical microscope has been developed to yield optical images with various gap distances between the probe and the sample surface. The microscope uses an apertureless metallic probe, the position of which is controlled by regulation of the tunneling-electron current from the probe to the sample and by computer-generated bias voltage. Experimental results of near-field optical imaging with the developed microscope at different gap distances are shown. Thirteen images at gap distances of 0 to 500 nm demonstrate that the near-field image depends strongly on the gap distance. The imaging characteristics of a near-field imaging system are shown with the spatial-frequency spectra of images. Future investigation of the developed microscope is also discussed.

Hebden, Jeremy C.; David J. Hall; Michael Firbank; David T. Delpy; "Time-resolved optical imaging of a solid tissue-equivalent phantom," *Appl. Opt.*, Vol. 34, pg. 8038 (1995).

A solid plastic phantom has been developed with optical properties that closely match those of human breast tissue at near-IR wavelengths. The phantom is a 54-mm-thick slab containing four small cylinders of contrasting scatter and absorption. A detailed description of the phantom is followed by an account of an attempt to image the phantom by a time-resolved imaging technique. Images generated with transmitted light with the shortest flight times revealed the embedded cylinders with greater visibility than images obtained with continuous light transillumination. However, images corresponding to flight times of less than approximately 700 ps were severely degraded from a lack of detected photons. An attempt was made to overcome this degradation by extrapolating the measured temporal distributions with an analytic model of photon transport. Results suggest that subcentimeter resolution imaging of low-contrast tumors in the breast is scientifically possible. Our phantom is available to any other research groups wishing to evaluate their systems.

Holmes, Timothy J.; "Maximum-likelihood image restoration adapted for noncoherent optical imaging," *J. Opt. Soc. Am. A*, Vol. 5, pg. 666 (1988).

Hoshimiya, Tsutomu; "Simulation of photothermally diffused images and a concept of holoscan in photoacoustics and optical-probe imaging," *Appl. Opt.*, Vol. 33, pg. 7251 (1994).

Idell, Paul S.; Webster, Arthur; "Resolution limits for coherent optical imaging: signal-to-noise analysis in the spatial-frequency domain," *J. Opt. Soc. Am. A*, Vol. 9, pg. 43 (1992).

Matson, Charles L.; Natalie Clark; Lenore McMackin; Janet S. Fender; "Three-dimensional tumor localization in thick tissue with the use of diffuse photon-density waves," *Appl. Opt.*, Vol. 36, pg. 0214 (1997).

A new approach to three-dimensional tumor localization in turbid media with the use of measurements in a single plane is presented. Optical diffuse photon-density waves are used to probe the turbid medium. Relative amplitudes and phases are measured in the detection plane. Lateral localization is accomplished in the detection plane. With a Fourier optics approach, the scattered wave is reconstructed throughout the volume to provide depth localization. Computer-simulation results that validate this technique are presented. Applications of this technique to multiple tumors and to optical mammography are discussed.

McLean, E. A.; H. R. Burris, Jr.; M. P. Strand; "Short-pulse range-gated optical imaging in turbid water," *Appl. Opt.*, Vol. 34, pg. 4343 (1995).

Using a very short laser pulse and camera gate in a range-gated optical imaging system, we have demonstrated an improvement in imaging in turbid water. A 0.5-ns FWHM laser pulse at 532 nm and a camera gate time ranging down to 120 ps were used to record images of 0.635-cm-bar-resolution targets at 6.5 attenuation lengths through 4.572 m of turbid water. With a four-channel version of this camera, coupled to a CCD camera, we can observe a quasi-three-dimensional image in real time.

Morgan, Stephen P.; Man P. Khong; Michael G. Somekh; "Effects of polarization state and scatterer concentration on optical imaging through scattering media," *Appl. Opt.*, Vol. 36, pg. 1560 (1997).

The imaging resolution in turbid media is severely degraded by light scattering. Resolution can be improved if the unscattered or weakly scattered light is extracted. Here the state of polarization of the emerging light is used to discriminate photon path length, with the more weakly scattered photons maintaining their original polarization state. It is experimentally demonstrated that over a wide range of scatterer concentrations there exist three distinct imaging regimes. It is also shown that within the intermediate regime one of two distinct imaging techniques is appropriate, depending on the particle size.

Paithankar, D. Y.; A. U. Chen; B. W. Pogue; M. S. Patterson; E. M. Sevick-Muraca; "Imaging of fluorescent yield and lifetime from multiply scattered light reemitted from random media," *Appl. Opt.*, Vol. 36, pg. 2260 (1997).

The feasibility of employing fluorescent contrast agents to perform optical imaging in tissues and other scattering media has been examined through computational studies. Fluorescence lifetime and yield can give crucial information about local metabolite concentrations or environmental conditions within tissues. This information can bc employed toward disease detection, diagnosis, and treatment if noninvasively quantitated from reemitted optical signals. However, the problem of inverse image reconstruction of fluorescence yield and lifetime is complicated because of the highly scattering nature of the tissue. Here a light propagation model employing the diffusion equation is used to account for the scattering of both the excitation and fluorescent light. Simulated measurements of frequency-domain parameters of fluorescent modulated ac amplitude and phase lag are used as inputs to an inverse image-reconstruction algorithm, which employs the diffusion model to predict frequency-domain measurements resulting from a modulated input at the phantom periphery. In the inverse image-reconstruction algorithm, a Newton-Raphson technique combined with a Marquardt algorithm is employed to converge on the fluorescent properties within the medium. The successful reconstruction of both the fluorescence yield and lifetime in the case of a heterogeneous fluorophore distribution within a scattering medium has been demonstrated without a priori informa-tion or without the necessity of obtaining absence images.

Roggemann, Michael C.; Byron M. Welsh; Dennis Montera; Troy A. Rhoadarmer; "Method for simulating atmospheric turbulence phase effects for multiple time slices and anisoplanatic conditions," *Appl. Opt.*, Vol. 34, pg. 4037 (1995).

Simulating the effects of atmospheric turbulence on optical imaging systems is an important aspect of understanding the performance of these systems. Simulations are particularly important for understanding the statistics of some adaptive-optics system performance measures, such as the mean and variance of the compensated optical transfer function, and for understanding the statistics of estimators used to reconstruct intensity distributions from turbulence-corrupted image measurements. Current methods of simulating the performance of these systems typically make use of random phase screens placed in the system pupil. Methods exist for making random draws of phase screens that have the correct spatial statistics. However, simulating temporal effects and anisoplanatism requires one or more phase screens at different distances from the aperture, possibly moving with different velocities. We describe and demonstrate a method for creating random draws of phase screens with the correct space-time statistics for arbitrary turbulence and wind-velocity profiles, which can be placed in the telescope pupil in simulations. Results are provided for both the von Karman and the Kolmogorov turbulence spectra. We also show how to simulate anisoplanatic effects with this technique.

Sevick, E. M.; Frisoli, J. K.; Burch, C. L.; Lakowicz, Joseph R.; "Localization of absorbers in scattering media by use of frequency-domain measurements of time-dependent photon migration," *Appl. Opt.*, Vol. 33, pg. 3562 (1994).

Sevick-Muraca, E. M.; D. L. Heintzelman; J. Lee; T. L. Troy; D. Y. Paithankar; "Role of higher-order scattering in solutions to the forward and inverse optical-imaging problems in random media," *Appl. Opt.*, Vol. 36, pg. 9058 (1997).

From analytical and numerical solutions that predict the scattering of diffuse photon density waves and from experimental measurements of changes in phase shift theta and ac amplitude demodulation M caused by the presence of single and double cylindrical heterogeneities, we show that second- and higher-order perturbations can affect the prediction of the propagation characteristics of diffuse photon density waves. Our experimental results for perfect absorbers in a lossless medium suggest that the performance of fast inverse-imaging algorithms that use first-order Born or Rytov approximations might have inherent limitations compared with inverse solutions that use iterative solutions of a linear perturbation equation or numerical solutions of the diffusion equation.

Simarmata, Theodorus T. H.; Timothy J. Schulz; "Statistical analysis for coherent optical-imaging systems with registration errors," *J. Opt. Soc. Am. A*, Vol. 14, pg. 1439 (1997).

Coherent images of diffuse objects exhibit two fundamental noise artifacts: speckle noise that is due to the use of coherent illumination and photon noise that is due to fundamental limitations in the detection of electromagnetic radiation. Because of the nature of speckle noise, slight changes in the source-scene-sensor geometry can give rise to independent speckle-noise realizations, which can, in turn, be used to form ensemble averages and reduce the speckle-induced degradations. The very nature of this procedure, however, can result in a collection of images that are not registered relative to one another, and new distortions, unrelated to the speckle effect, will be seen in the resulting images. Our goal, therefore, is to present a statistical analysis of coherent optical-imaging systems that collect and average independent, photon-limited speckle images with registration errors and to use this analysis in the derivation of mean-square-error and signal-to-noise ratio expressions that are applicable for both image-domain and spatial-frequency-domain analyses.

Tearney, G. J.; B. E. Bouma; S. A. Boppart; B. Golubovic; E. A. Swanson; J. G. Fujimoto; "Rapid acquisition of *in vivo* biological images by use of optical coherence tomography," *Opt. Lett.*, Vol. 21, pg. 1408 (1996).

The development of techniques for high-speed image acquisition in optical coherence tomography (OCT) systems is essential for suppressing motion artifacts when one is imaging living systems. We describe a new OCT system for performing micrometer-scale, cross-sectional optical imaging at four images / s. To achieve OCT image-acquisition times of less than 1 s, we use a piezoelectric fiber stretcher to vary the reference arm delay. A Kerr-lens mode-locked chromium-doped forsterite laser is employed as the low-coherence source for the high-speed OCT system. Dynamic, motion-artifact-free *in vivo* imaging of a beating Xenopus laevis (African frog) heart is demonstrated.

von Bally, Gert; Brune, E.; Mette, W.; "Holographic endoscopy with gradient-index optical imaging systems and optical fibers," *Appl. Opt.*, Vol. 25, pg. 3425 (1986).

Wang, L. M.; Ho, Ping Pei; Alfano, R. R.; "Double-stage picosecond Kerr gate for ballistic time-gated optical imaging in turbid media," *Appl. Opt.*, Vol. 32, pg. 535 (1993).

Wang, L.; Ho, Ping Pei; Liang, X.; Dai, H.; Alfano, R. R.; "Kerr-Fourier imaging of hidden objects in thick turbid media," *Opt. Lett.*, Vol. 18, pg. 241 (1993).

Wang, Lihong; Xuemei Zhao; "Ultrasound-modulated optical tomography of absorbing objects buried in dense tissue-simulating turbid media," *Appl. Opt.*, Vol. 36, pg. 7277 (1997).
Continuous-wave ultrasonic modulation of scattered laser light was used to image objects buried in tissue-simulating turbid media. The buried object had an absorption coefficient greater than the background turbid medium. The ultrasonic wave that was focused into the turbid media modulated the laser light that passed through the ultrasonic field. The modulated laser light that was collected by a photomultiplier tube reflected the local mechanical and optical properties in the zone of ultrasonic modulation. Objects buried in the middle plane of 5-cm-thick dense turbid media were imaged with millimeter resolution through the scanning and detecting alterations of the ultrasound-modulated optical signal. The optical properties of the dense turbid media included an absorption coefficient of 0.1 cm-1 and a reduced scattering coefficient of 10 cm-1 and were comparable with those of biological tissues in the visible and near-IR ranges. The dependence of the ultrasound-modulated optical signal on the off-axis distance of the detector from the optic axis and the area of the detector was studied as well.

Wu, Jun; "Convolution picture of the boundary conditions in photon migration and its implications in time-resolved optical imaging of biological tissues," *J. Opt. Soc. Am. A*, Vol. 14, pg. 280 (1997).
To model the effect of a tissue-air boundary on time-resolved optical measurements, a convolution picture is presented based on the photon migration picture. It is demonstrated that the boundary conditions (either index matched or index mismatched) can be formulated as a spatiotemporal convolution of two terms, with the first being identical to the solution in the infinite medium and the second being independent of the original light source. The conditions under which the spatial convolution part in the second term becomes negligible are also determined, thus permitting the complete separation of the two terms by use of a temporal Laplace or Fourier transformation. This result is promising, since it suggests a method of removing the effect of the boundary conditions in the applications of time-resolved optical imaging (in both the time domain and the frequency domain) in biological tissues.

Xiang, Yongqing; Theodore Raphan; X. Liang; L. Wang; P. P. Ho; R. R. Alfano; "Image-quality enhancement of objects in turbid media by use of a combined computational-photonics approach," *Appl. Opt.*, Vol. 36, pg. 1045 (1997).
Ultrafast time-gated optical imaging and computational image-enhancement techniques were combined to produce a robust system for viewing objects in turbid media. Image enhancement was implemented by use of images from the early light with a histogram contrast-enhancement algorithm. Image quality was assessed by use of the contrast radius of gyration and the contrast-to-noise ratio. The technique was applied to viewing the dispersion of water droplets emanating from a jet spray and to pictures of an object embedded in turbid media. In all instances there were substantial improvements in image quality at a given time delay.

Yao, Yuqi; Yao Wang; Randall L. Barbour; Harry L. Graber; Jenghwa Chang; "Sensitivity studies for imaging a spherical object embedded in a spherically symmetric, two-layer turbid medium with photon-density waves," *Appl. Opt.*, Vol. 35, pg. 0735 (1996).
We present analytic expressions for the amplitude and phase of photon-density waves in strongly scattering, spherically symmetric, two-layer media containing a spherical object. This layered structure is a crude model of multilayered tissues whose absorption and scattering coefficients lie within a range reported in the literature for most tissue types. The embedded object simulates a pathology, such as a tumor. The normal-mode-series method is employed to solve the inhomogeneous Helmholtz equation in spherical coordinates, with suitable boundary conditions. By comparing the total field at points in the outer layer at a fixed distance from the origin when the object is present and when it is absent, we evaluate the potential sensitivity of an optical imaging system to inhomogeneities in absorption and scattering. For four types of background media with different absorption and scattering properties, we determine the modulation frequency that achieves an optimal compromise between signal-detection reliability and sensitivity to the presence of an object, the minimum

detectable object radius, and the smallest detectable change in the absorption and scattering coefficients for a fixed object size. Our results indicate that (1) enhanced sensitivity to the object is achieved when the outer layer is more absorbing or scattering than the inner layer; (2) sensitivity to the object increases with the modulation frequency, except when the outer layer is the more absorbing; (3) amplitude measurements are proportionally more sensitive to a change in absorption, phase measurements are proportionally more sensitive to a change in scattering, and phase measurements exhibit a much greater capacity for distinguishing an absorption perturbation from a scattering perturbation.

Yao, Yuqi; Yao Wang; Yaling Pei; Wenwu Zhu; Randall L. Barbour; "Frequency-domain optical imaging of absorption and scattering distributions by a Born iterative method," *J. Opt. Soc. Am. A*, Vol. 14, pg. 325 (1997).

We present a Born iterative method for reconstructing optical properties of turbid media by means of frequency-domain data. The approach is based on iterative solutions of a linear perturbation equation, which is derived from the integral form of the Helmholtz wave equation for photon-density waves. In each iteration the total field and the associated weight matrix are recalculated based on the previous reconstructed image. We then obtain a new estimate by solving the updated perturbation equation. The forward solution, also based on a Helmholtz equation, is obtained by a multigrid finite difference method. The inversion is carried out through a Tikhonov regularized optimization process by the conjugate gradient descent method. Using this method, we first reconstruct the distribution of the complex wave numbers in a test medium, from which the absorption and the scattering distributions are then derived. Simulation results with two-dimensional test media have shown that this method can yield quantitatively (in terms of coefficient values) as well as qualitatively (in terms of object location and shape) accurate reconstructions of absorption and scattering distributions in cases in which the first-order Born approximation cannot work well. Both full-angle and limited-angle measurement schemes have been simulated to examine the effect of the location of detectors and sources. The robustness of the algorithm to noise has also been evaluated.

Zaccanti, Giovanni; Donelli, Paolo; "Attenuation of energy in time-grated transillumination imaging: numerical results," *Appl. Opt.*, Vol. 33, pg. 7023 (1994).

Photon migration

Alfano, R. R.; James Fujimoto; "Advances in Optical Imaging and Photon Migration," *Opt. Phot. News*, Vol. 7, pg. 37 (1996).

Bassani, Michele; Fabrizio Martelli; Giovanni Zaccanti; Daniele Contini; "Independence of the diffusion coefficient from absorption: experimental and numerical evidence," *Opt. Lett.*, Vol. 22, pg. 853 (1997).

We investigated the dependence of the diffusion coefficient on the absorption coefficient by studying the propagation of light emitted by an isotropic source in an infinitely extended medium. Comparisons with both experimental and numerical results showed that the diffusion equation gives a better description of photon migration when the diffusion coefficient is assumed to be independent of absorption.

Bevilacqua, Frederic; Pierre Marquet; Olivier Coquoz; Christian Depeursinge; "Role of tissue structure in photon migration through breast tissues," *Appl. Opt.*, Vol. 36, pg. 0044 (1997).

has been investigated experimentally *in vitro* on human breast tissues, bovine liver, tissue phantoms, and theoretically by Monte Carlo simulations and diffusion theory. The spatial intensity profiles have been measured at the output surface of a sample illuminated by a collimated beam. Experimental results have then been compared with simulations that assume the sample to be homogeneous. Measurements on phantoms, i.e., fat emulsion and microspheres suspension, and on liver are in good agreement with theory. On the other hand, the width of the intensity profiles measured on breast tissues (adipose and fibrous) are systematically larger than those measured on phantoms or calculated by simulations. The structure of these samples, not considered in simulations and not present in phantoms, explains these differences.

Boas, D. A.; "A fundamental limitation of linearized algorithms for diffuse optical tomography," *Opt. Express*, Vol. 1, pg. 404 (1997).

Diffuse Optical Tomography is rapidly developing as a new imaging modality for characterizing the spatially varying optical properties of media which strongly scatter light (e.g. tissue). Numerous imaging algorithms exist, and more are being developed. Many of these algorithms rely on assumptions which linearize the relationship between the optical contrast and the perturbed signal. We show that this linear approximation makes quantitative imaging of spatially varying optical properties impossible. The explanation for this result is presented and the implication for Diffuse Optical Tomography is discussed.

Boas, D. A.; M. A. O'Leary; B. Chance; A. G. Yodh; "Detection and characterization of optical inhomogeneities with diffuse photon density waves: a signal-to-noise analysis," *Appl. Opt.*, Vol. 36, pg. 0075 (1997).

Diffusing photons provide information about the optical properties of turbid media. In biological tissues these optical properties may be correlated to physiological parameters, enabling one to probe effectively the physiological states of tissue for abnormalities such as tumors and hemorrhages. We show that positional uncertainty in the source and detector lead to significant random errors that degrade the optical information available from diffusing photons. We investigate the limits for the detection, localization, and characterization of optical inhomogeneities by using diffusing photons as a probe. Although detection is sufficient for tumor screening, full characterization of the optical properties is desirable for specification of the tumor. Our findings in model breast systems with realistic signal-to-noise ratios indicate that tumors as small as 0.3 cm in diameter can be unambiguously detected; however, simultaneous determination of tumor size and optical properties is possible only if its diameter is of the order of 1.0 cm or larger. On the other hand, if a priori information about the size (optical properties) is available, then the optical properties (size) of tumors as small as 0.3 cm in diameter can be determined.

Bonner, R. F.; Nossal, Ralph; Havlin, Shlomo; Weiss, George H.; "Model for photon migration in turbid biological media," *J. Opt. Soc. Am. A*, Vol. 4, pg. 423 (1987).

Cerussi, Albert E.; John S. Maier; Sergio Fantini; Maria Angela Franceschini; William W. Mantulin; Enrico Gratton; "Experimental verification of a theory for the time-resolved fluorescence spectroscopy of thick tissues," *Appl. Opt.*, Vol. 36, pg. 0116 (1997).

Fluorescence spectroscopy provides potential contrast enhancement for near-infrared tissue imaging and physiologically correlated spectroscopy. We present a fluorescence photon migration model and test its quantitative predictive capabilities with a frequency-domain measurement that involves a homogeneous multiple-scattering tissue phantom (with optical properties similar to those of tissue in the near infrared) that contains a fluorophore (rhodamine B). After demonstrating the validity of the model, we explore its ability to recover the fluorophore's spectral properties from within the multiple-scattering medium. The absolute quantum yield and the lifetime of the fluorophore are measured to within a few percent of the values measured independently in the absence of scattering. Both measurements are accomplished without the use of reference fluorophores. In addition, the model accurately predicts the fluorescence emission spectrum in the scattering medium. Implications of these absolute measurements of lifetime, quantum yield, concentration, and emission spectrum from within multiple-scattering media are discussed.

Contini, Daniele; Fabrizio Martelli; Giovanni Zaccanti; "Photon migration through a turbid slab described by a model based on diffusion approximation. I. Theory," *Appl. Opt.*, Vol. 36, pg. 4587 (1997).

The diffusion approximation of the radiative transfer equation is a model used widely to describe photon migration in highly diffusing media and is an important matter in biological tissue optics. An analysis of the time-dependent diffusion equation together with its solutions for the slab geometry and for a semi-infinite diffusing medium are reported. These solutions, presented for both the time-dependent and the continuous wave source, account for the refractive index mismatch between the turbid medium and the surrounding medium. The results have been compared with those obtained when different boundary conditions were assumed. The comparison has shown that the effect of the refractive index mismatch cannot be disregarded. This effect is particularly important for the transmittance. The discussion of results also provides an analysis of the role of the absorption coefficient in the expression of the diffusion coefficient.

Cui, Weijia; Wang, Naiguang; Chance, Britton; "Study of photon migration depths with time-resolved spectroscopy," *Opt. Lett.*, Vol. 16, pg. 1632 (1991).

Dogariu, A.; M. Dogariu; K. Richardson; S. D. Jacobs; G. D. Boreman; "Polarization asymmetry in waves backscattering from highly absorbant random media," *Appl. Opt.*, Vol. 36, pg. 8159 (1997).

Within the range in which light penetration depth is approximately the same as or less than the diameter of the particles in the medium, particulate media with considerable absorption behave as two-dimensional, rough-surface structures. As penetration depth increases, a complicated transition between volume and surface effects is seen. For these media, low-order scattering sequences have small spatial extent, making observation of polarization characteristics difficult. We present an experimental technique to access the low-order scattered photons by artificially reinjecting them through total internal reflections. Using a dielectric layer in contact with the high-absorption medium, we are able to observe fourfold polarization asymmetry in backscattering from highly absorbant media. We discuss the origin of the polarization patterns in a ray-optics approximation and suggest possibilities for solving practical problems encountered in characterizing composites with appreciable absorption.

Durduran, T.; A. G. Yodh; B. Chance; D. A. Boas; "Does the photon-diffusion coefficient depend on absorption?," *J. Opt. Soc. Am. A*, Vol. 14, pg. 3358 (1997).

We investigate the controversy over the precise form of the photon diffusion coefficient and suggest that it is largely independent of absorption, i.e., D_0=v/3mu$_s$^{\prime}. After presentation of the general theoretical arguments underlying this assertion, Monte Carlo simulations are performed and explicitly reveal that the absorption-independent diffusion coefficient gives better agreement with theory than the traditionally accepted photon diffusion coefficient, D_{mua}=v/3(mu$_s$^{\prime}+mu $_a$). The importance of resolving this controversy for the proper characterization of the material optical properties is discussed.

Durian, D. J.; J. Rudnick; "Photon migration at short times and distances and in cases of strong absorption," *J. Opt. Soc. Am. A*, Vol. 14, pg. 235 (1997).

Diffusion theory is widely used to describe photon migration in turbid media because it is simple and, in certain cases, can be accurate to within a few percent. However, it neglects the ballistic nature of photon propagation between successive scattering events and hence entirely breaks down for short times and distances, as well as for strong absorption. Here we generalize on the exact two-stream theory of one-dimensional photon migration to obtain a telegrapher equation that accounts for both diffusive and ballistic aspects of propagation in three-dimensional media. At long times and distances the standard diffusion theory is recovered, whereas at short times and distances we find improved predictions for such phenomena as pulse spreading, diffuse photon-density wave dispersion, transmission through a slab, and pulse reflection from a semi-infinite medium. Our theory should be useful for accurately characterizing turbid media, such as biological tissues, and may also aid in improving the spatial resolution of images made with diffuse light.

Durian, D. J.; J. Rudnick; "Photon migration at short times and distances and in cases of strong absorption: erratum," *J. Opt. Soc. Am. A*, Vol. 14, pg. 940 (1997).

Fantini, Sergio; Maria Angela Franceschini; Enrico Gratton; "Effective source term in the diffusion equation for photon transport in turbid media," *Appl. Opt.*, Vol. 36, pg. 0156 (1997).

The Green's function for the diffusion equation is widely used to describe photon transport in turbid media. We have performed a series of spectroscopy experiments on a number of uniform turbid media with different optical properties (absorption coefficient in the range 0.03-0.14 cm-1, reduced scattering coefficient in the range 5-22 cm-1). Our experiments have been conducted in the frequency domain, where the measured parameters are the dc intensity (Idc), ac amplitude (Iac), and phase (Phi) of the light intensity wave. In an infinite medium, the Green's function predicts a linear dependence of ln(rIdc) and Phi on the source-detector separation r. Our measurements show that the intercepts of these straight lines predicted by the Green's function do not agree with the experimental results. To reproduce the experimental results, we have introduced an effective photon source whose spatial extent and source strength depend on the optical properties of the medium. This effective source term has no effect on the slopes of the straight lines predicted by the Green's function at large values of r.

Feng, Shechao; Fan-An Zeng; Britton Chance; "Photon migration in the presence of a single defect: a perturbation analysis," *Appl. Opt.*, Vol. 34, pg. 3826 (1995).

We present an analytical perturbation analysis for studying the sensitivity of diffusive photon flux to the addition of a small spherical defect object in multiple-scattering media such as human tissues. As a first simple application of our perturbation method, we derive analytically the photon migration path distributions and the shapes of the so-called banana regions in which the photon migration paths are concentrated. We then derive analytically the sensitivity of detected photon flux densities to the inclusion of small spherical defects in the multiple-scattering medium for both single-source and two-source configurations, at both steady-state (dc) and frequency-modulation conditions, and compare the results with Monte Carlo simulations.

Fishkin, Joshua B.; Olivier Coquoz; Eric R. Anderson; Matthew Brenner; Bruce J. Tromberg; "Frequency-domain photon migration measurements of normal and malignant tissue optical properties in a human subject," *Appl. Opt.*, Vol. 36, pg. 0010 (1997).

A 1-GHz multifrequency, multiwavelength frequency-domain photon migration instrument is used to measure quantitatively the optical absorption (μa) and effective optical scattering (μs\prime) of normal and malignant tissues in a human subject. Large ellipsoidal (\similar10-cm major axis, \similar6-cm minor axes) subcutaneous malignant lesions were compared with adjacent normal sites in the abdomen and back. Absorption coefficients recorded at 674, 811, 849, and 956 nm were used to calculate tissue hemoglobin concentration (oxyhemoglobin, deoxyhemoglobin, and total), water concentration, hemoglobin oxygen saturation, and blood volume fraction *in vivo*. Our results show that the normal and the malignant tissues measured in the patient have clearly resolvable optical and physiological property differences that may be broadly useful in identifying and characterizing tumors.

Gandjbakhche, Amir H.; Robert F. Bonner; Ralph Nossal; George H. Weiss; "Effects of multiple-passage probabilities on fluorescent signals from biological media," *Appl. Opt.*, Vol. 36, pg. 4613 (1997).

By applying a random-walk model for photon migration, we find an exact expression for fluorescent signals emitted from a homogeneous, optically turbid medium containing a single fluorescent mass. In contrast to diffusion-theory-based models, our analysis accounts for multiple photon passages through the fluorophore sites and allows for a variable degree of fluorescent absorptivity. We particularly discuss effects on the amplitudes and phase shifts of frequency-domain fluorescent signals.

Hielscher, Andreas H.; Angela A. Eick; Judith R. Mourant; Dan Shen; James P. Freyer; Irving J. Bigio; "Diffuse backscattering Mueller matrices of highly scattering media," *Opt. Express*, Vol. 1, pg. 441 (1997).

We report on the development of a method that records spatially dependent intensity patterns of polarized light that is diffusely backscattered from highly scattering media. It is demonstrated that these intensity patterns can be used to differentiate turbid media, such as polystyrene-sphere and biological-cell suspensions. Our technique employs polarized light from a He-Ne laser ($\lambda = 543$ nm), which is focused onto the surface of the scattering medium. A surface area of approximately 4x4 cm centered on the light input point is imaged through polarization-analysis optics onto a CCD camera. One can observe a large variety of intensity patterns by varying the polarization state of the incident laser light and changing the analyzer configuration to detect different polarization components of the backscattered light. Introducing the Mueller-matrix concept for diffusely backscattered light, a framework is provided to select a subset of measurements that comprehensively describe the optical properties of backscattering media.

Ho, Ping Pei; Baldeck, Patrice L.; Wong, K. S.; Yoo, K. M.; Lee, Don; Alfano, R. R.; "Time dynamics of photon migration in semiopaque random media," *Appl. Opt.*, Vol. 28, pg. 2304 (1989).

Howard, Patrick M.; Robert V. Edwards; "Implementation of a likelihood ratio test for laser Doppler velocimeter burst detection," *Appl. Opt.*, Vol. 36, pg. 7629 (1997).

A likelihood ratio test for laser Doppler velocimeter burst detection is derived and implemented with the appropriate photon detection statistics. This detector utilizes the first point of the autocorrelation function of the laser Doppler velocimeter photodetector signal and relies on a simple identity for this point. The detector looks at the ratio of two probability functions of the data (burst and no burst) to make its decision. Because the detector does not depend on the absolute strength of the signal, detection down to very low (-10-dB) signal-to-noise levels can be achieved. Because the autocorrelation function is insensitive to the phase of the signal, the detector will operate reliably with signals containing multiple overlapping bursts.

Jiang, Huabei; Joseph Pierce; Jeffrey Kao; Eva Sevick-Muraca; "Measurement of particle-size distribution and volume fraction in concentrated suspensions with photon migration techniques," *Appl. Opt.*, Vol. 36, pg. 3310 (1997).

The determination of the particle size distribution and the volume fraction in concentrated suspensions from the multiwavelength measurement of isotropic-scattering coefficients by use of frequency-domain photon migration techniques is demonstrated for three different polydisperse polystyrene suspensions. When a Newton-type inverse algorithm is used, the successful recovery of the particle size distribution, in the form of a Weibull function, and the volume fraction of polystyrene suspensions is achieved. Our results are in excellent agreement with dynamic light-scattering size distribution measurements. On consideration of the particle mass conservation as an additional constraint penalty term in the inverse algorithm, it is shown that the quality of the particle size distribution reconstruction can be improved. Because no calibration is needed, photon migration techniques are especially suited for on-line measurement of the particle size distribution and the volume fraction in the chemical- and the pharmaceutical-based industries.

Madsen, Steen J.; erson, Eric R.; Haskell, Richard C.; Tromberg, Bruce J.; "Portable, high-bandwidth frequency-domain photon migration instrument for tissue spectroscopy," *Opt. Lett.*, Vol. 19, pg. (1994).

Martelli, Fabrizio; Daniele Contini; Adriana Taddeucci; Giovanni Zaccanti; "Photon migration through a turbid slab described by a model based on diffusion approximation. II. Comparison with Monte Carlo results," *Appl. Opt.*, Vol. 36, pg. 4600 (1997).

In our companion paper we presented a model to describe photon migration through a diffusing slab. The model, developed for a homogeneous slab, is based on the diffusion approximation and is able to take into account reflection at the boundaries resulting from the refractive index mismatch. In this paper the predictions of the model are compared with solutions of the radiative transfer equation obtained by Monte Carlo simulations in order to determine the applicability limits of the approximated theory in different physical conditions. A fitting procedure, carried out with the optical properties as fitting parameters, is used to check the application of the model to the inverse problem. The results show that significant errors can be made if the effect of the refractive index mismatch is not properly taken into account. Errors are more important when measurements of transmittance are used. The effects of using a receiver with a limited angular field of view and the angular distribution of the radiation that emerges from the slab have also been investigated.

Nossal, Ralph; Kiefer, J.; Weiss, George H.; Bonner, R. F.; Taitelbaum, Haim; Havlin, Shlomo; "Photon migration in layered media," *Appl. Opt.*, Vol. 27, pg. 3382 (1988).

Perelman, Lev T.; Joshua Winn; Jun Wu; Ramachandra R. Dasari; Michael S. Feld; "Photon migration of near-diffusive photons in turbid media: a Lagrangian-based approach," *J. Opt. Soc. Am. A*, Vol. 14, pg. 224 (1997).

We show that light transport in a turbid medium can be described by a path integral with a quasi-particle Lagrangian. The most probable trajectory over which photons can be found can be obtained from this Lagrangian. This approach extends the diffusion approximation to the near-diffusive regime, in which photons travel only a few transport mean free paths.

Pogue, Brian W.; Markus Testorf; Troy McBride; Ulf Osterberg; Keith Paulsen; "Instrumentation and design of a frequency-domain diffuse optical tomography imager for breast cancer detection," *Opt. Express*, Vol. 1, pg. 391 (1997).

The instrument development and design of a prototype frequency-domain optical imaging device for breast cancer detection is described in detail. This device employs radio-frequency intensity modulated near-infrared light to image quantitatively both the scattering and absorption coefficients of tissue. The functioning components of the system include a laser diode and a photomultiplier tube, which are multiplexed automatically through 32 large core fiber optic bundles using high precision linear translation stages. Image reconstruction is based on a finite element solution of the diffusion equation. This tool for solving the forward problem of photon migration is coupled to an iterative optical property estimation algorithm, which uses a Levenberg-Marquardt routine with total variation minimization. The result of this development is an automated frequency-domain optical imager for computed tomography which produces quantitatively accurate images of the test phantoms used to date. This paper is a description and characterization of an automated frequency-domain computed tomography scanner, which is more quantitative than earlier systems used in diaphanography because of the combination of intensity modulated signal detection and iterative image reconstruction.

Polishchuk, A. Ya. ; S. Gutman; M. Lax; R. R. Alfano; "Photon-density modes beyond the diffusion approximation: scalar wave-diffusion equation," *J. Opt. Soc. Am. A*, Vol. 14, pg. 230 (1997).

Various modifications of the diffusion model designed to describe photon migration in scattering and absorbing media are analyzed and compared in the nondiffusion region. A more accurate modified diffusion equation that has a broader region of applicability is introduced. The radiative-transfer equation is solved numerically to assess the validity range of different models tested and to select the best one.

Polishchuk, A. Ya.; J. Dolne; F. Liu; R. R. Alfano; "Average and most-probable photon paths in random media," *Opt. Lett.*, Vol. 22, pg. 430 (1997).

Time-resolved experiments have revealed that, in contrast to the predictions of conventional diffusion theory, photons select certain curvilinear paths to travel between a source and a detector. Concepts of the average photon paths and Fermat paths are introduced on the basis of the non-Euclidean diffusion equation (NED) to explain experimental results. Comparison of the theory and the experiment demonstrates the potential of the NED to describe nondiffusive features of photon migration in the multiple-scattering regime.

Polishchuk, A. Ya; R. R. Alfano; "Photon diffusion on the velocity sphere," *Opt. Lett.*, Vol. 21, pg. 916 (1996).

The concept of photon diffusion on the velocity sphere and the non-Euclidean diffusion equation (NED) are introduced to describe photon migration in highly forward-scattering random media. The NED covers the ballistic, transient, and developed diffusion modes of photon migration in random media. An approximate analytic solution to the NED is presented.

Reynolds, Jeffery S.; Andreas Przadka; Simon P. Yeung; Kevin J. Webb; "Optical diffusion imaging: a comparative numerical and experimental study," *Appl. Opt.*, Vol. 35, pg. 3671 (1996).

We present a low-cost photodiode-LED apparatus for making broadband frequency domain photon migration measurements. We compare measured data to finite-difference frequency domain solutions of the diffusion approximation of the Boltzmann transport equation. Specific comparisons include the influence of boundary conditions on simulations and the effect of finite source size on resolution.

Sevick, E. M.; Frisoli, J. K.; Burch, C. L.; Lakowicz, Joseph R.; "Localization of absorbers in scattering media by use of frequency-domain measurements of time-dependent photon migration," *Appl. Opt.*, Vol. 33, pg. 3562 (1994).

Sevick-Muraca, E. M.; D. L. Heintzelman; J. Lee; T. L. Troy; D. Y. Paithankar; "Role of higher-order scattering in solutions to the forward and inverse optical-imaging problems in random media," *Appl. Opt.*, Vol. 36, pg. 9058 (1997).

From analytical and numerical solutions that predict the scattering of diffuse photon density waves and from experimental measurements of changes in phase shift theta and ac amplitude demodulation M caused by the presence of single and double cylindrical heterogeneities, we show that second- and higher-order perturbations can affect the prediction of the propagation characteristics of diffuse photon density waves. Our experimental results for perfect absorbers in a lossless medium suggest that the performance of fast inverse-imaging algorithms that use first-order Born or Rytov approximations might have inherent limitations compared with inverse solutions that use iterative solutions of a linear perturbation equation or numerical solutions of the diffusion equation.

Soelkner, Gerald; Gerhard Mitic; Robert Lohwasser; "Monte Carlo simulations and laser Doppler flow measurements with high penetration depth in biological tissuelike head phantoms," *Appl. Opt.*, Vol. 36, pg. 5647 (1997).

Laser Doppler flow measurements on biological tissuelike phantoms have shown that penetration depths of 30 mm could be obtained, thus exceeding the penetration depth of commercial instruments for the measurement of skin perfusion by more than an order of magnitude. Monte Carlo simulations were performed and compared with measurement results obtained on a headlike tissue model to quantify the influence of perfusion of the scalp on the cortex perfusion results. We found Doppler frequency spectra to be independent of the mean scattering angle and could be fitted with a sum of Gaussian functions, using a simple analytical model.

Taitelbaum, Haim; Havlin, Shlomo; Weiss, George H.; "Approximate theory of photon migration in a two-layer medium," *Appl. Opt.*, Vol. 28, pg. 2245 (1989).

Walker, Scott A.; Sergio Fantini; Enrico Gratton; "Image reconstruction by backprojection from frequency-domain optical measurements in highly scattering media," *Appl. Opt.*, Vol. 36, pg. 0170 (1997).

The reconstruction of the location and optical properties of objects in turbid media requires the solution of the inverse problem. Iterative solutions to this problem can require large amounts of computing time and may not converge to a unique solution. Instead, we propose a fast, simple method for approximately solving this problem in which calculated effective absorption and reduced scattering coefficients are backprojected to create an image of the objects. We reconstructed images of objects with centimeter dimensions embedded in a diffusive medium with optical characteristics similar to those of human tissue. Data were collected by a frequency-domain spectrometer operating at 120 MHz with a laser diode light source emitting at 793 nm. Intensity and phase of the incident photon density wave were collected from linear scans at different projection angles. Although the positions of the objects are correctly identified by the reconstructed images, the optical parameters of the objects are recovered only qualitatively.

Wu, Jun; Partovi, Firooz; Feld, Michael S.; Rava, Richard P.; "Diffuse reflectance from turbid media: an analytical model of photon migration," *Appl. Opt.*, Vol. 32, pg. 1115 (1993).

Wu, Jun; "Convolution picture of the boundary conditions in photon migration and its implications in time-resolved optical imaging of biological tissues," *J. Opt. Soc. Am. A*, Vol. 14, pg. 280 (1997).

To model the effect of a tissue-air boundary on time-resolved optical measurements, a convolution picture is presented based on the photon migration picture. It is demonstrated that the boundary conditions (either index matched or index mismatched) can be formulated as a spatiotemporal convolution of two terms, with the first being identical to the solution in the infinite medium and the second being independent of the original light source. The conditions under which the spatial convolution part in the second term becomes negligible are also determined, thus permitting the complete separation of the two terms by use of a temporal Laplace or Fourier transformation. This result is promising, since it suggests a method of removing the effect of the boundary conditions in the applications of time-resolved optical imaging (in both the time domain and the frequency domain) in biological tissues.

Yaroslavsky, I. V.; A. N. Yaroslavsky; V. V. Tuchin; H.-J. Schwarzmaier; "Effect of the scattering delay on time-dependent photon migration in turbid media," *Appl. Opt.*, Vol. 36, pg. 6529 (1997).

We modified the diffusion approximation of the time-dependent radiative transfer equation to account for a finite scattering delay time. Under the usual assumptions of the diffusion approximation, the effect of the scattering delay leads to a simple renormalization of the light velocity that appears in the diffusion equation. Accuracy of the model was evaluated by comparison with Monte Carlo simulations in the frequency domain for a semi-infinite geometry. A good agreement is demonstrated for both matched and mismatched boundary conditions when the distance from the source is sufficiently large. The modified diffusion model predicts that the neglect of the scattering delay when the optical properties of the turbid material are derived from normalized frequency- or time-domain measurements should result in an underestimation of the absorption coefficient and an overestimation of the transport coefficient. These observations are consistent with the published experimental data.

Time-resolved imaging

Andersson-Engels, S.; Berg, Roger; Persson, A.; Svanberg, Sune; "Multispectral tissue characterization with time-resolved detection of diffusely scattered white light," *Opt. Lett.*, Vol. 18, pg. 1697 (1993).

Arons, Eric; Emmett N. Leith; An-Chen Tien; Robert Wagner; "High-resolution optical chirped pulse gating," *Appl. Opt.*, Vol. 36, pg. 2603 (1997).

We describe a system for achieving high-resolution range gating using optically chirped pulses. The technique converts signals from the time domain into signals in the frequency domain through a nonlinear, sum-frequency generation process. The technique is based on similar methods used in microwave radar. We draw analogies between our method and conventional and time-lens imaging processes, and present experimental results demonstrating the method.

Arridge, Simon R.; "Photon-measurement density functions. Part I: Analytical forms," *Appl. Opt.*, Vol. 34, pg. 7395 (1995).

This paper addresses the problem of tomographic reconstruction of absorption and scattering parameters in the optical region from measurements of transilluminated light. Specifically, the question of the sensitivity of different measurement schemes on the boundary of an object to perturbations of the optical parameters within the object are addressed. The concept of a photon-sampling volume [Appl. Opt. 33, 448 (1994)] and a photon-hitting density [Appl. Opt. 32, 448 (1993)] is extended to a photon-measurement density function (PMDF). The PMDF is derived from the Green's function of the diffusion equation and can be expressed for measurements such as the time-varying intensity, integrated intensity, temporal moments, and phase shift, as well as for both absorption and diffusion perturbations. Closed-form solutions are given for a number of these functions in infinite space, half-space, and slab geometries. Example results are given in terms of three-dimensional images.

Chang, Jenghwa; Wenwu Zhu; Yao Wang; Harry L. Graber; Randall L. Barbour; "Regularized progressive expansion algorithm for recovery of scattering media from time-resolved data," *J. Opt. Soc. Am. A*, Vol. 14, pg. 306 (1997).

Reconstructions of the absorption cross sections of dense scattering media from time-resolved data are presented. A progressive expansion (PE) algorithm, similar to a layer-stripping approach, is developed to circumvent the underdeterminedness of the inverse problem. An overlapping scheme, which uses detector readings from several consecutive time intervals, is introduced to reduce the propagation of reconstruction errors that occur at shallower depths. To reduce the sensitivity of the PE algorithm to noise, a regularized progressive expansion (RPE) algorithm is proposed, which incorporates regularization techniques into the PE algorithm. The PE and the RPE algorithms are applied to the problem of image reconstruction from time-resolved data. The test media were isotropically scattering slabs containing one or two compact absorbers at different depths below the surface. The data were corrupted by additive white Gaussian noise with various strengths. The reconstruction results show that the PE and the RPE algorithms, when they are combined by proper overlapping, can effectively overcome the underdeterminedness of the inverse problem. The RPE algorithm yields reconstructions that are more accurate and more stable under the same noise level.

Cubeddu, R.; A. Pifferi; P. Taroni; A. Torricelli; G. Valentini; "Time-resolved imaging on a realistic tissue phantom: mu $_{s}$ \prime and mu $_{a}$ images versus time-integrated images," *Appl. Opt.*, Vol. 35, pg. 4533 (1996).

A method is proposed by which we construct images through turbid media, plotting directly either the transport-scattering coefficient mu $_{s}$` or the absorption coefficient mu $_{a}$. These optical parameters are obtained from the best fit of the time-resolved transmittance curves with a diffusion model. Measurements were performed with a time-correlated single-photon counting system on realistic tissue phantoms simulating a tumor mass within a breast. Images were obtained with an incident power of less than 1 mW and an acquisition time of 1 s/point. Comparison of mu $_{s}$` and mu $_{a}$ images with time-integrated images constructed from the same experimental data shows that the fitting method discriminates between scattering and absorption inhomogeneities and improves image quality for scattering but not for absorption inhomogeneities.

Grosenick, D.; H. Wabnitz; H. Rinneberg; "Time-resolved imaging of solid phantoms for optical mammography," *Appl. Opt.*, Vol. 36, pg. 0221 (1997).

We have recorded time-resolved transillumination images of solid phantoms with objects embedded that differ in their scattering and absorption coefficients from those of the bulk material, simulating a compressed human breast with a tumor inside. Employing time-correlated single photon counting at rates of up to 1 MHz, we recorded distributions of times of flight of photons at 1369 scan positions within 2.5 min. Several quantities, such as fractional transmittance, first moments, Fourier amplitudes, phase shifts, and frequency-dependent effective transport scattering and absorption coefficients, have been derived from experimental data to form two-dimensional images. By recording such images at a selected total number of photons detected, we have determined the contrast and effective signal-to-noise ratio in each case.

Hall, David J.; Jeremy C. Hebden; David T. Delpy; "Imaging very-low-contrast objects in breastlike scattering media with a time-resolved method," *Appl. Opt.*, Vol. 36, pg. 7270 (1997).

An investigation was performed of the effectiveness of a time-resolved method for imaging very-low-contrast features embedded in highly scattering media. Experiments employed slabs of breastlike material into which were inserted small cylindrical objects having either a scattering or an absorption coefficient of 4, 2, 1.5, and 1.1 times greater than the surrounding medium. An attempt was made to quantify the degree of contrast produced by each object. The results indicate that time-gating is far more effective at enhancing the contrast of the scattering inhomogeneities than of the absorbing inhomogeneities. This observation is shown to agree with a diffusion-based model, which also predicts that time-gating can decrease the contrast of absorbing inhomogeneities unless very short time-gates can be employed.

Hebden, Jeremy C.; "Time-resolved imaging of opaque and transparent spheres embedded in a highly scattering medium," *Appl. Opt.*, Vol. 32, pg. 3837 (1993).

Hebden, Jeremy C.; Delpy, David T.; "Enhanced time-resolved imaging with a diffusion model of photon transport," *Opt. Lett.*, Vol. 19, pg. (1994).

Hebden, Jeremy C.; Kruger, Robert A.; Wong, K. S.; "Time resolved imaging through a highly scattering medium," *Appl. Opt.*, Vol. 30, pg. 788 (1991).

Hebden, Jeremy C.; Wong, K. S.; "Time resolved optical tomography," *Appl. Opt.*, Vol. 32, pg. 372 (1993).

Hebden, Jeremy C.; David J. Hall; Michael Firbank; David T. Delpy; "Time-resolved optical imaging of a solid tissue-equivalent phantom," *Appl. Opt.*, Vol. 34, pg. 8038 (1995).

A solid plastic phantom has been developed with optical properties that closely match those of human breast tissue at near-IR wavelengths. The phantom is a 54-mm-thick slab containing four small cylinders of contrasting scatter and absorption. A detailed description of the phantom is followed by an account of an attempt to image the phantom by a time-resolved imaging technique. Images generated with transmitted light with the shortest flight times revealed the embedded cylinders with greater visibility than images obtained with continuous light transillumination. However, images corresponding to flight times of less than approximately 700 ps were severely degraded from a lack of detected photons. An attempt was made to overcome this degradation by extrapolating the measured temporal distributions with an analytic model of photon transport. Results suggest that subcentimeter resolution imaging of low-contrast tumors in the breast is scientifically possible. Our phantom is available to any other research groups wishing to evaluate their systems.

Hebden, Jeremy C.; Maria Tziraki; David T. Delpy; "Evaluation of the temporally extrapolated absorbance method for dual-wavelength imaging through tissuelike scattering media," *Appl. Opt.*, Vol. 36, pg. 3802 (1997).

An independent assessment is described of a dual-wavelength imaging technique, known as the temporally extrapolated absorbance method (TEAM), proposed by Yamada et al. [Opt. Eng. 32, 634-641 (1993)]. The technique involves recording the temporal distribution of light transmitted across a scattering medium at two carefully chosen wavelengths at which the scattering properties of the medium are assumed to be identical. The objective is to image internal structure that absorbs more strongly at one wavelength than it does at the other. A simple theoretical treatment of TEAM is presented that employs a perturbation model of photon transport. This indicates that despite the lack of a secure theoretical basis, the technique may provide a potentially effective ad hoc method of generating images of highly scattering media. The method was also evaluated experimentally by using, for the first time to our knowledge, a single object and two wavelengths. A single-projection, two-dimensional image was obtained of a solid phantom with optical properties representative of breast tissue. The results exhibited good agreement with the theoretical model, and a small embedded feature that absorbs 3.5 times as strongly as the surrounding medium at one wavelength was revealed successfully.

Hebden, Jeremy C.; Simon R. Arridge; "Imaging through scattering media by the use of an analytical model of perturbation amplitudes in the time domain," *Appl. Opt.*, Vol. 35, pg. 6788 (1996).

A method of generating images through highly scattering media is presented that involves comparing measurements of the time-dependent intensity of transmitted light with an analytical model describing the sensitivity of that intensity on localized changes in optical properties. A least-squares fitting procedure is employed to derive the amplitudes of the measurement perturbations caused by embedded absorbers and scatterers located along a line of sight between the source and detector. Images are presented of a highly scattering, solid plastic phantom with optical properties closely matched to those of human breast tissue at near-infrared wavelengths. The phantom is a 54-mm-thick slab, containing four small cylinders of contrasting scatter and absorption. Results show that embedded absorbers can be distinguished from embedded scatterers, and that the diffusion perturbation amplitude provides inherently greater spatial resolution than the absorption perturbation amplitude.

Joblin, Anthony; "Tumor contrast in time-domain, near-infrared laser breast imaging," *Appl. Opt.*, Vol. 36, pg. 9050 (1997).

The contrast of breast disease in an image produced by a time-domain, near-infrared laser, imaging system was calculated by use of a finite-difference method. The contrast was investigated for tumors of different sizes, in a range of positions within the breast, and in a range of breast thicknesses. Contrast is greatest for large tumors that are near the breast surfaces and that are measured at short times of flight. The contrast was shown to increase as the scattering and the absorption properties of the tumor increased. The scattering properties of the tumor were shown to make the biggest contribution to the contrast at short times of flight. Both pointlike and large extended source and detector configurations were investigated. The effect of the system size on the signal-to-noise ratio was investigated.

Kalpaxis, L. L.; Wang, L. M.; Galland, P.; Liang, X.; Ho, Ping Pei; Alfano, R. R.; "Three-dimensional temporal image reconstruction of an object hidden in highly scattering media by time-gated optical tomography," *Opt. Lett.*, Vol. 18, pg. 1691 (1993).

Liang, X.; L. Wang; P. P. Ho; R. R. Alfano; "Time-resolved polarization shadowgrams in turbid media," *Appl. Opt.*, Vol. 36, pg. 2984 (1997).

The use of the polarization of light as a parameter to discriminate against multiple-scattered light for trans-illumination imaging through random scattering media is examined. Time-resolved two-dimensional images of submillimeter test bars immersed in 5-cm-thick Intralipid solutions with different micelle dilutions were measured for two orthogonal polarizations (parallel and perpendicular) of light emerging from the turbid medium by using a picosecond Kerr-Fourier (KF) imaging system. The measured contrast and intensity of parallel-polarized KF shadowgrams decreased as the concentration of the scattering medium was increased, whereas the behavior of the perpendicular-polarized KF shadowgrams varied in an opposite matter to the micelle concentration.

Mahon, Rita; Duncan, Michael D.; Tankersley, Lawrence L.; Reintjes, John F.; "Time-gated imaging through dense scatterers with a Raman amplifier," *Appl. Opt.*, Vol. 32, pg. 7425 (1993).

Mitic, Gerhard; Kolzer, Jochen; Otto, Johann; Plies, Erich; Solkner, Gerald; Zinth, Wolfgang; "Time-gated transillumination of biological tissues and tissuelike phantoms," *Appl. Opt.*, Vol. 33, pg. 6699 (1994).

Pan, Yingtian; Reginald Birngruber; Ralf Engelhardt; "Contrast limits of coherence-gated imaging in scattering media," *Appl. Opt.*, Vol. 36, pg. 2979 (1997).

The fundamental difference between time-resolved and coherence-gated imaging modalities in scattering media is analyzed in terms of their optical transfer functions. The effectiveness of coherence gating for multiple-scattering rejection is shown by imaging a 100-μm-thick razor blade hidden in the scattering phantoms (i.e., Intralipid suspensions) with different scattering coefficients. We found that the imaging contrast is limited by multiple scattering and speckle effects in high-scattering media, and the measured effective penetration depth of optical coherence tomography is approximately equal to six mean free paths under the experimental conditions of a numerical aperture of less than 0.1 and a scattering anisotropy of approximately 0.8.

Schweiger, Martin; Simon R. Arridge; "Direct calculation with a finite-element method of the Laplace transform of the distribution of photon time of flight in tissue," *Appl. Opt.*, Vol. 36, pg. 9042 (1997).

Reconstruction methods for optical tomographic imaging require the development of models of light transport in highly scattering materials. While the simulation of the full temporal response function arising from a short source light pulse is computationally expensive, there are methods to evaluate efficiently certain transforms of the temporal profile. We previously presented methods to obtain directly the Mellin transform, which is related to the moments of the temporal intensity distribution; We introduce a similar method to calculate directly the Laplace transform. This method provides an additional, largely independent measurement type that can be combined with the moments to improve image quality in optical tomography, in particular with respect to the simultaneous reconstruction of absorption and scattering distributions.

Wabnitz, H.; H. Rinneberg; "Imaging in turbid media by photon density waves: spatial resolution and scaling relations," *Appl. Opt.*, Vol. 36, pg. 0064 (1997).

Spatial resolution of transillumination imaging through highly scattering media normalized to sample thickness d depends on only the normalized wavelength λ/d of photon density waves and normalized penetration depth δ/d. This was concluded theoretically and verified experimentally by the derivation of edge-spread functions from measured time-resolved transmittance and its Fourier transform by the use of dilute milk at various concentrations as scatterer in cuvettes of d = 2 cm and d = 4 cm. In the frequency domain and the time domain, spatial resolution was found experimentally to be given by approximately 0.3d obtained at λ \approximately d or when only the first arriving 1% of all photons detected were taken into account.

Wang, L.; X. Liang, P. Galland, P. P. Ho, and R. R. Alfano; "True scattering coefficients of turbid matter measured by early-time gating," *Opt. Lett.*, Vol. 20, pg. 913 (1995).

The true scattering coefficients of turbid matter have been determined by use of picosecond time-resolved imaging. The scattering coefficients measured by the conventional cw collimation method were found to be smaller than those obtained from the early-time-sliced ballistic photons.

Wu, Jun; "Convolution picture of the boundary conditions in photon migration and its implications in time-resolved optical imaging of biological tissues," *J. Opt. Soc. Am. A*, Vol. 14, pg. 280 (1997).

To model the effect of a tissue-air boundary on time-resolved optical measurements, a convolution picture is presented based on the photon migration picture. It is demonstrated that the boundary conditions (either index matched or index mismatched) can be formulated as a spatiotemporal convolution of two terms, with the first being identical to the solution in the infinite medium and the second being independent of the original light source. The conditions under which the spatial convolution part in the second term becomes negligible are also determined, thus permitting the complete separation of the two terms by use of a temporal Laplace or Fourier transformation. This result is promising, since it suggests a method of removing the effect of the boundary conditions in the applications of time-resolved optical imaging (in both the time domain and the frequency domain) in biological tissues.

Xiang, Yongqing; Theodore Raphan; X. Liang; L. Wang; P. P. Ho; R. R. Alfano; "Image-quality enhancement of objects in turbid media by use of a combined computational-photonics approach," *Appl. Opt.*, Vol. 36, pg. 1045 (1997).

Ultrafast time-gated optical imaging and computational image-enhancement techniques were combined to produce a robust system for viewing objects in turbid media. Image enhancement was implemented by use of images from the early light with a histogram contrast-enhancement algorithm. Image quality was assessed by use of the contrast radius of gyration and the contrast-to-noise ratio. The technique was applied to viewing the dispersion of water droplets emanating from a jet spray and to pictures of an object embedded in turbid media. In all instances there were substantial improvements in image quality at a given time delay.

Tissue Optics

Andersson-Engels, S.; Berg, Roger; Persson, A.; Svanberg, Sune; "Multispectral tissue characterization with time-resolved detection of diffusely scattered white light," *Opt. Lett.*, Vol. 18, pg. 1697, (1993).

Bahman Anvari; B. Samuel Tanenbaum; Thomas E. Milner; Kimberly Tang; Lih-Huei Liaw; Ken Kalafus; Sol Kimel; J. Stuart Nelson; "Spatially selective photocoagulation of biological tissues: feasibility study utilizing cryogen spray cooling," *Appl. Opt.*, Vol. 35, pg. 3314, (1996).

Successful laser treatment of selected dermatoses such as hemangiomas requires thermally induced damage to blood vessels while protecting the epidermis. We present and test a procedure in a rabbit liver tissue model that utilizes cryogen spray cooling during continuous Nd:YAG laser irradiation to induce deep photocoagulation necrosis while protecting superficial tissues from thermal injury. Gross and histologic observations are consistent with calculated thicknesses of protected and photocoagulated tissues and demonstrate the feasibility of inducing spatially selective photocoagulation when cryogen spray cooling is used in conjunction with laser irradiation. This procedure may be useful in the thermal treatment of some pathological conditions for which it is desired that deep photocoagulation be induced while protecting superficial tissues. Key words: Chlorodifluoromethane, hemangioma, infrared radiometry, laser, microwave, thermal-damage confinement.

Roland Bays; Georges Wagnieres; Dimitri Robert; Daniel Braichotte; Jean-Francois Savary; Philippe Monnier; Hubert van den Bergh; "Clinical determination of tissue optical properties by endoscopic spatially resolved reflectometry," *Appl. Opt.*, Vol. 35, pg. 1756, (1996).

A noninvasive method to measure the optical properties of a diffusing and absorbing medium is described. Based on the spatially resolved measurement of diffuse reflectance at the sample surface, this method is particularly suitable for investigating the *in vivo* optical properties of biological tissues endoscopically in a clinical context. The sensitivity of the measurement is discussed, and two optical probes for two different clinical applications are presented. Preliminary measurements are performed on a nonbiological medium, which illustrate the possibilities of the proposed method. Finally, we report on *in vivo* measurements of the optical properties of the human esophageal wall at 630 nm. Key words: Diffuse reflectance, tissue optics, optical probe, optical properties, scattering.

Beek, Johan F.; Sterenborg, Henricus J. C. M.; van Gemert, Martin J. C.; "Continuous pneumothorax monitoring by remittance measurement," *Appl. Opt.*, Vol. 32, pg. 454, (1993).

Sabrina Beretta; Lorenzo Lunelli; Giuseppe Chirico; Giancarlo Baldini; "Dynamic light scattering from small particles: expected accuracy in hemoglobin data reduction," *Appl. Opt.*, Vol. 35, pg. 3763, (1996).

Dynamic light scattering from protein solutions can be applied to the detection of conformational changes and to the measurement of particle size. Because accurate results can be adversely affected by experimental perturbations, careful procedures are necessary both in data acquisition and in data analysis. Autocorrelation functions (ACF's) of scattered light are simulated here to evaluate the role of the perturbations affecting measured signals from dilute protein solutions. The analysis of measured and simulated ACF's has been performed both by cumulant expansion and by a nonlinear least-squares fit, thereby allowing the definition of criteria for the optimization of the fitting parameters and of the measuring conditions. Moreover, by comparing experimental data from hemoglobin solutions and computer simulations, we show how to evaluate the contributions of polydispersity and statistical noise affecting the measurements. Key words: Dynamic light scattering, polydispersity, small particle sizing, autocorrelation function.

R. Berg; S. Andersson-Engels; O. Jarlman; S. Svanberg; "Time-gated viewing studies on tissuelike phantoms," *Appl. Opt.*, Vol. 35, pg. 3432, (1996).

A time-gated technique to enhance viewing through highly scattering media such as tissue is discussed. Experiments have been performed on tissuelike plastic phantoms to determine the possibilities and limitations of the technique. The effects of the time-gate width and the localization, size, and optical properties of hidden objects have been studied. A computer model to simulate light propagation in tissue is also presented. The predictions of the model are compared with experimental results. Key words: Medical diagnostics, time-gated viewing, tissue optics, transillumination, scattering media, diffusion equation, numerical modeling.

Frederic Bevilacqua; Pierre Marquet; Olivier Coquoz; Christian Depeursinge; "Role of tissue structure in photon migration through breast tissues," *Appl. Opt.*, Vol. 36, pg. 0044, (1997).

Photon migration has been investigated experimentally *in vitro* on human breast tissues, bovine liver, tissue phantoms, and theoretically by Monte Carlo simulations and diffusion theory. The spatial intensity profiles have been measured at the output surface of a sample illuminated by a collimated beam. Experimental results have then been compared with simulations that assume the sample to be homogeneous. Measurements on phantoms, i.e., fat emulsion and microspheres suspension, and on liver are in good agreement with theory. On the other hand, the width of the intensity profiles measured on breast tissues (adipose and fibrous) are systematically larger than those measured on phantoms or calculated by simulations. The structure of these samples, not considered in simulations and not present in phantoms, explains these differences.

D. A. Boas; "A fundamental limitation of linearized algorithms for diffuse optical tomography," *Opt. Express*, Vol. 1, pg. 404, (1997).

Diffuse Optical Tomography is rapidly developing as a new imaging modality for characterizing the spatially varying optical properties of media which strongly scatter light (e.g. tissue). Numerous imaging algorithms exist, and more are being developed. Many of these algorithms rely on assumptions which linearize the relationship between the optical contrast and the perturbed signal. We show that this linear approximation makes quantitative imaging of spatially varying optical properties impossible. The explanation for this result is presented and the implication for Diffuse Optical Tomography is discussed.

Bosman, Saskia; "Heat-induced structural alterations in myocardium in relation to changing optical properties," *Appl. Opt.*, Vol. 32, pg. 461, (1993).

Bouma, B.; G. J. Tearney; S. A. Boppart; M. R. Hee; M. E. Brezinski; J. G. Fujimoto; "High-resolution optical coherence tomographic imaging using a mode-locked Ti:Al2O3 laser source," *Opt. Lett.*, Vol. 20, pg. 1486, (1995).

A Kerr-lens mode-locked Ti: Al_2O_3 oscillator, optimized for minimal coherence length, is demonstrated as a high-power source for high-resolution optical coherence tomographic imaging. Dispersion compensation and heterodyne noise rejection are demonstrated to yield in situ images of biological tissues with 3.7- μm resolution and 93-dB dynamic range.

James F. Brennan III; Mark E. Beattie; Yang Wang; Michael J. Cantella; Bor-Yeu Tsaur; Ramachandra R. Dasari; Michael S. Feld; "PdSi focal-plane array detectors for short-wave infrared Raman spectroscopy of biological tissue: a feasibility study," *Appl. Opt.*, Vol. 35, pg. 5736, (1996).

We have used a PdSi focal-plane array detector to measure short-wave infrared Raman spectra of pure compounds and human tissue. Raman bands of the pure compounds are clearly visible in the spectra, and a calcification feature at 960 cm^{-1} is readily identifiable in the spectra of diseased human aorta. The performance characteristics of our detection device were good; dark noise contributed approximately 60 (electrons/s)/pixel, and the read noise was approximately 50 rms electrons/pixel. The primary noise in the spectra was due to fixed-pattern noise, which is the variation in measured signal across a detector when it is uniformly illuminated.

J. T. Bruulsema; J. E. Hayward; T. J. Farrell; M. S. Patterson; L. Heinemann; M. Berger; T. Koschinsky; J. Sandahl-Christiansen; H. Orskov; M. Essenpreis; G. Schmelzeisen-Redeker; D. Bocker; "Correlation between blood glucose concentration in diabetics and noninvasively measured tissue optical scattering coefficient," *Opt. Lett.*, Vol. 22, pg. 190, (1997).

Diabetics would benefit greatly from a device capable of providing continuous noninvasive monitoring of their blood glucose levels. The optical scattering coefficient of tissue depends on the concentration of glucose in the extracellular fluid. A feasibility study was performed to evaluate the sensitivity of the tissue reduced scattering coefficient in response to step changes in the blood glucose levels of diabetic volunteers. Estimates of the scattering coefficient were based on measurements of the diffuse reflectance on the skin at distances of 1-10 mm from a point source. A correlation was observed between step changes in blood glucose concentration and tissue reduced scattering coefficient in 30 out of 41 subjects measured.

Stephen A. Burns; Marwood N. Ediger; Rebecca R. Richards-Kortum; "Optical technology in medicine and biology: Introduction," *Appl. Opt.*, Vol. 35, pg. 3313, (1996).

Cerussi, Albert E.; John S. Maier; Sergio Fantini; Maria Angela Franceschini; William W. Mantulin; Enrico Gratton; "Experimental verification of a theory for the time-resolved fluorescence spectroscopy of thick tissues," *Appl. Opt.*, Vol. 36, pg. 0116, (1997).

Fluorescence spectroscopy provides potential contrast enhancement for near-infrared tissue imaging and physiologically correlated spectroscopy. We present a fluorescence photon migration model and test its quantitative predictive capabilities with a frequency-domain measurement that involves a homogeneous multiple-scattering tissue phantom (with optical properties similar to those of tissue in the near infrared) that contains a fluorophore (rhodamine B). After demonstrating the validity of the model, we explore its ability to recover the fluorophore's spectral properties from within the multiple-scattering medium. The absolute quantum yield and the lifetime of the fluorophore are measured to within a few percent of the values measured independently in the absence of scattering. Both measurements are accomplished without the use of reference fluorophores. In addition, the model accurately predicts the fluorescence emission spectrum in the scattering medium. Implications of these absolute measurements of lifetime, quantum yield, concentration, and emission spectrum from within multiple-scattering media are discussed.

Eric Chan; Thomas Menovsky; Ashley J. Welch; "Effects of cryogenic grinding on soft-tissue optical properties," *Appl. Opt.*, Vol. 35, pg. 4526, (1996).

Optical properties obtained from spectrophotometer measurements of reflectance and transmittance were determined for both frozen-ground and intact soft tissues. The tissues used in these experiments were calf aorta, rat jejunum, and rabbit sciatic nerve. Tissue specimens from each tissue type were frozen in liquid nitrogen and then ground with a pestle and mortar into a fine powder. A tissue paste formed once the powder returned to room temperature. The tissue paste was then sandwiched between glass slides for spectrophotometer measurements. For comparison, the optical properties of the intact specimens were also measured. Total transmission and diffuse reflection were obtained on a Varian Cary 5E spectrophotometer (400-850 nm). Absorption and reduced scattering coefficients of the tissues were determined with the Inverse Adding Doubling method. Our results suggested that within the 400-nm to 850-nm spectrum, optical properties of the ground tissue approximated intact tissue within limits of experimental error.

Chen, Hsuan; Shih, M.; Arons, E.; Leith, Emmett N.; Lopez, Juaquin; Dilworth, David S.; Sun, Pang-Chen; "Electronic holographic imaging through living human tissue," *Appl. Opt.*, Vol. 33, pg. 3630, (1994).

Chen, Y.; Chen, H.; Dilworth, David; Leith, Emmett N.; Lopez, Joaquin L.; Shih, M.; Sun, Pang-Chen; Vossler, G.; "Evaluation of holographic methods for imaging through biological tissue," *Appl. Opt.*, Vol. 32, pg. 4330, (1993).

Chen, Ye; "Characterization of the image resolution for the first-arriving-light method," *Appl. Opt.*, Vol. 33, pg. 2544, (1994).

Cilesiz, I.; Welch, A. J.; "Light dosimetry: effects of dehydration and thermal damage on the optical properties of human aorta," *Appl. Opt.*, Vol. 33, pg. 3571, (1994).

Cilesiz, Inci F.; Welch, Ashley J.; "Light dosimetry: effects of dehydration and thermal damage on optical properties of human aorta," *Appl. Opt.*, Vol. 32, pg. 477, (1993).

S. B. Colak; D. G. Papaioannou; G. W. 't Hooft; M. B. van der Mark; H. Schomberg; J. C. J. Paasschens; J. B. M. Melissen; N. A. A. J. van Asten; "Tomographic image reconstruction from optical projections in light-diffusing media," *Appl. Opt.*, Vol. 36, pg. 0180, (1997).

The recent developments in light generation and detection techniques have opened new possibilities for optical medical imaging, tomography, and diagnosis at tissue penetration depths of \similar10 cm. However, because light scattering and diffusion in biological tissue are rather strong, the reconstruction of object images from optical projections needs special attention. We describe a simple reconstruction method for diffuse optical imaging, based on a modified backprojection approach for medical tomography. Specifically, we have modified the standard backprojection method commonly used in x-ray tomographic imaging to include the effects of both the diffusion and the scattering of light and the associated nonlinearities in projection image formation. These modifications are based primarily on the deconvolution of the broadened image by a spatially variant point-spread function that is dependent on the scattering of light in tissue. The spatial dependence of the deconvolution and nonlinearity corrections for the curved propagating ray paths in heterogeneous tissue are handled semiempirically by coordinate transformations. We have applied this method to both theoretical and experimental projections taken by parallel- and fan-beam tomography geometries. The experimental objects were biomedical phantoms with multiple objects, including *in vitro* animal tissue. The overall results presented demonstrate that image-resolution improvements by nearly an order of magnitude can be obtained. We believe that the tomographic method presented here can provide a basis for rapid, real-time medical monitoring by the use of optical projections. It is expected that such optical tomography techniques can be combined with the optical tissue diagnosis methods based on spectroscopic molecular signatures to result in a versatile optical diagnosis and imaging technology.

R. Cubeddu; A. Pifferi; P. Taroni; A. Torricelli; G. Valentini; "Time-resolved imaging on a realistic tissue phantom: μ_s\prime and μ_a images versus time-integrated images," *Appl. Opt.*, Vol. 35, pg. 4533, (1996).

A method is proposed by which we construct images through turbid media, plotting directly either the transport-scattering coefficient μ_s` or the absorption coefficient μ_a. These optical parameters are obtained from the best fit of the time-resolved transmittance curves with a diffusion model. Measurements were performed with a time-correlated single-photon counting system on realistic tissue phantoms simulating a tumor mass within a breast. Images were obtained with an incident power of less than 1 mW and an acquisition time of 1 s/point. Comparison of μ_s` and μ_a images with time-integrated images constructed from the same experimental data shows that the fitting method discriminates between scattering and absorption inhomogeneities and improves image quality for scattering but not for absorption inhomogeneities.

Cummings, Joseph P.; Walsh, Joseph T.; "Tissue tearing caused by pulsed laser-induced ablation pressure," *Appl. Opt.*, Vol. 32, pg. 494, (1993).

Das, B. B.; Yoo, K. M.; Liu, Feng; Cleary, J.; Prudente, R.; Celmer, E.; Alfano, R. R.; "Spectral optical-density measurements of small particles and breast tissues," *Appl. Opt.*, Vol. 32, pg. 549, (1993).

de Mul, F. F. M.; Koelink, M. H.; Weijers, A. L.; Greve, Jan; Aarnoudse, J. G.; Graaff, R.; Dassel, A. C. M.; "Self-mixing laser-Doppler velocimetry of liquid flow and blood perfusion of tissue," *Appl. Opt.*, Vol. 31, pg. 5844, (1992).

Domankevitz, Yacov; Lee, Margaret S.; Nishioka, Norman S.; "Effects of irradiance and spot size on pulsed holmium laser ablation of tissue," *Appl. Opt.*, Vol. 32, pg. 569, (1993).

Dunn, Andrew; Charles DiMarzio; "Efficient computation of time-resolved transfer functions for imaging in turbid media," *J. Opt. Soc. Am. A*, Vol. 13, pg. 65, (1996).

The transfer function of a turbid medium such as biological tissue provides a method of analyzing the spatial resolution of a time-resolved tissue imaging system. A method is presented of calculating the transfer function with the use of a Monte Carlo simulation. The model allows the computation of the time-resolved line-spread function of a sample of thickness d from a simulation of thickness d/2 by use of reciprocity under certain conditions, and the transfer function can then be computed from the line-spread function. Results with this method agree with previously published theoretical and experimental results.

Dunn, Andrew K.; Colin Smithpeter; Ashley J. Welch; Rebecca Richards-Kortum; "Sources of contrast in confocal reflectance imaging," *Appl. Opt.*, Vol. 35, pg. 3441, (1996).

The relationship between optical properties and image contrast in confocal imaging is investigated. A Monte Carlo simulation has been developed to analyze the effects of changes in scattering, index of refraction, and absorption in a three-layer medium. Contrast was calculated from the computed signal-to-background ratios for changes in tissue optical properties. Results show that the largest source of contrast is changes in refractive index.

T. Durduran; A. G. Yodh; B. Chance; D. A. Boas; "Does the photon-diffusion coefficient depend on absorption?," *J. Opt. Soc. Am. A*, Vol. 14, pg. 3358, (1997).

We investigate the controversy over the precise form of the photon diffusion coefficient and suggest that it is largely independent of absorption, i.e., $D_0=v/3\mu_s'$. After presentation of the general theoretical arguments underlying this assertion, Monte Carlo simulations are performed and explicitly reveal that the absorption-independent diffusion coefficient gives better agreement with theory than the traditionally accepted photon diffusion coefficient, $D_{\mu a}=v/3(\mu_s'+\mu_a)$. The importance of resolving this controversy for the proper characterization of the material optical properties is discussed.

Essenpreis, Matthias; Elwell, C. E.; Cope, M.; van der Zee, P.; Arridge, S. R.; Delpy, D. T.; "Spectral dependence of temporal point spread functions in human tissues," *Appl. Opt.*, Vol. 32, pg. 418, (1993).

Joshua B. Fishkin; Olivier Coquoz; Eric R. Anderson; Matthew Brenner; Bruce J. Tromberg; "Frequency-domain photon migration measurements of normal and malignant tissue optical properties in a human subject," *Appl. Opt.*, Vol. 36, pg. 10, (1997).

A 1-GHz multifrequency, multiwavelength frequency-domain photon migration instrument is used to measure quantitatively the optical absorption (μa) and effective optical scattering (μs\prime) of normal and malignant tissues in a human subject. Large ellipsoidal (\similar10-cm major axis, \similar6-cm minor axes) subcutaneous malignant lesions were compared with adjacent normal sites in the abdomen and back. Absorption coefficients recorded at 674, 811, 849, and 956 nm were used to calculate tissue hemoglobin concentration (oxyhemoglobin, deoxyhemoglobin, and total), water concentration, hemoglobin oxygen saturation, and blood volume fraction *in vivo*. Our results show that the normal and the malignant tissues measured in the patient have clearly resolvable optical and physiological property differences that may be broadly useful in identifying and characterizing tumors.

Joshua B. Fishkin; Peter T. C. So; Albert E. Cerussi; Enrico Gratton; Sergio Fantini; Maria Angela Franceschini; "Frequency-domain method for measuring spectral properties in multiple-scattering media: methemoglobin absorption spectrum in a tissuelike phantom," *Appl. Opt.*, Vol. 34, pg. 1143, (1995).

We have measured the optical absorption and scattering coefficient spectra of a multiple-scattering medium (i.e., a biological tissue-simulating phantom comprising a lipid colloid) containing methemoglobin by using frequency-domain techniques. The methemoglobin absorption spectrum determined in the multiple-scattering medium is in excellent agreement with a corrected methemoglobin absorption spectrum obtained from a steady-state spectrophotometer measurement of the optical density of a minimally scattering medium. The determination of the corrected methemoglobin absorption spectrum takes into account the scattering from impurities in the methemoglobin solution containing no lipid colloid. Frequency-domain techniques allow for the separation of the absorbing from the scattering properties of multiple-scattering media, and these techniques thus provide an absolute measurement of the optical absorption spectra of the methemoglobin/lipid colloid suspension. One accurately determines the absolute methemoglobin absorption spectrum in the frequency domain by extracting the scattering and absorption coefficients from the phase shift Phi and average light intensity DC (or Phi and the amplitude of the light-intensity oscillations AC) data with relationships provided by diffusion theory, but one determines it less accurately by using the Phi and modulation M (M identically equals AC/DC) data and the diffusion theory relationships. In addition to the greater uncertainty in the absorption and scattering coefficients extracted from the Phi and M data, the optical parameters extracted from the Phi and M data exhibit systematically inaccurate behavior that cannot be explained by random noise in the system. Possible reasons for the systematically lower accuracy of the methemoglobin absorption spectrum obtained from Phi and M data are discussed.

Fukshansky, Leonid; Fukshansky-Kazarinova, Nina; von Remisowsky, Alexander Martinez; "Estimation of optical parameters in a living tissue by solving the inverse problem of the multiflux radiative transfer," *Appl. Opt.*, Vol. 30, pg. 3145, (1991).

Gandjbakhche, A. H.; R. F. Bonner; R. Nossal; G. H. Weiss; "Absorptivity contrast in transillumination imaging of tissue abnormalities," *Appl. Opt.*, Vol. 35, pg. 1767, (1996).

We calculate the time-resolved flux of photons transmitted across an optically turbid slab containing a partially absorbing inclusion. An analytical expression is obtained for the flux at a detector positioned opposite a point source (at a distance equal to the thickness of the slab) when the center of the inclusion lies on the line connecting those points. The calculation employs a discrete-time lattice random-walk model of photon transport. The resulting expression is used to assess the affects of time resolution on the detectability of the inclusion.

Gandjbakhche, A. H.; Nossal, R.; Bonner, R. F.; "Scaling relationships for theories of anisotropic random walks applied to tissue optics," *Appl. Opt.*, Vol. 32, pg. 504, (1993).

Craig M. Gardner; Steven L. Jacques; Ashley J. Welch; "Fluorescence spectroscopy of tissue: recovery of intrinsic fluorescence from measured fluorescence," *Appl. Opt.*, Vol. 35, pg. 1780, (1996).

We present a method for recovering the intrinsic fluorescence coefficient, defined as the product of the fluorophore absorption coefficient and the fluorescence energy yield, of an optically thick, homogeneous, turbid medium from a surface measurement of fluorescence and from knowledge of medium optical properties. The measured fluorescence signal is related to the intrinsic fluorescence coefficient by an optical property dependent path-length factor. A simple expression was developed for the path-length factor, which characterizes the penetration of excitation light and the escape of fluorescence from the medium. Experiments with fluorescent tissue phantoms demonstrated that intrinsic fluorescence line shape could be recovered and that fluorophore concentration could be estimated within ñ15%, over a wide range of optical properties. Key words: Fluorescence, diagnostics, light transport, photodynamic therapy, photosensitizer.

R. Graaff; J. G. Aarnoudse; F. F. M. de Mul; M. H. Koelink; M. L. Kok; P. J. Harmsma; J. Greve; "Laser Doppler velocimetry and Monte Carlo simulations on models for blood perfusion in tissue," *Appl. Opt.*, Vol. 34, pg. 6595, (1995).

Laser Doppler flow measurements and Monte Carlo simulations on small blood perfusion flow models at 780 nm are presented and compared. The dimensions of the optical sample volume are investigated as functions of the distance of the laser to the detector and as functions of the angle of penetration of the laser into the sample. The effects of homodyne and heterodyne scattering are investigated.

D. Grosenick; H. Wabnitz; H. Rinneberg; "Time-resolved imaging of solid phantoms for optical mammography," *Appl. Opt.*, Vol. 36, pg. 0221, (1997).

We have recorded time-resolved transillumination images of solid phantoms with objects embedded that differ in their scattering and absorption coefficients from those of the bulk material, simulating a compressed human breast with a tumor inside. Employing time-correlated single photon counting at rates of up to 1 MHz, we recorded distributions of times of flight of photons at 1369 scan positions within 2.5 min. Several quantities, such as fractional transmittance, first moments, Fourier amplitudes, phase shifts, and frequency-dependent effective transport scattering and absorption coefficients, have been derived from experimental data to form two-dimensional images. By recording such images at a selected total number of photons detected, we have determined the contrast and effective signal-to-noise ratio in each case.

Yici Guo; P. P. Ho; A. Tirksliunas; Feng Liu; R. R. Alfano; "Optical harmonic generation from animal tissues by the use of picosecond and femtosecond laser pulses," *Appl. Opt.*, Vol. 35, pg. 6810, (1996).

Second- and third-harmonic generations of femtosecond and picosecond laser pulses have been measured from chicken skin, muscle, and fat tissues. The magnitude of the harmonic signals showed a strong structural dependence with the signal from skin interface being the strongest. The polarization dependence of the signal was also measured and found to be consistent with the fact that the tissue samples were highly scattering random media. The second-harmonic- and third-harmonic-generation conversion efficiencies were found to be in the range of approximately 10^{-7} to approximately 10^{-10}.

Havlin, Shlomo; Kiefer, J. E.; Trus, B.; Weiss, George H.; Nossal, Ralph; "Numerical method for studying the detectability of inclusions hidden in optically turbid tissue," *Appl. Opt.*, Vol. 32, pg. 617, (1993).

Jeremy C. Hebden; David J. Hall; Michael Firbank; David T. Delpy; "Time-resolved optical imaging of a solid tissue-equivalent phantom," *Appl. Opt.*, Vol. 34, pg. 8038, (1995).

A solid plastic phantom has been developed with optical properties that closely match those of human breast tissue at near-IR wavelengths. The phantom is a 54-mm-thick slab containing four small cylinders of contrasting scatter and absorption. A detailed description of the phantom is followed by an account of an attempt to image the phantom by a time-resolved imaging technique. Images generated with transmitted light with the shortest flight times revealed the embedded cylinders with greater visibility than images obtained with continuous light transillumination. However, images corresponding to flight times of less than approximately 700 ps were severely degraded from a lack of detected photons. An attempt was made to overcome this degradation by extrapolating the measured temporal distributions with an analytic model of photon transport. Results suggest that subcentimeter resolution imaging of low-contrast tumors in the breast is scientifically possible. Our phantom is available to any other research groups wishing to evaluate their systems.

Jeremy C. Hebden; Simon R. Arridge; "Imaging through scattering media by the use of an analytical model of perturbation amplitudes in the time domain," *Appl. Opt.*, Vol. 35, pg. 6788, (1996).

A method of generating images through highly scattering media is presented that involves comparing measurements of the time-dependent intensity of transmitted light with an analytical model describing the sensitivity of that intensity on localized changes in optical properties. A least-squares fitting procedure is employed to derive the amplitudes of the measurement perturbations caused by embedded absorbers and scatterers located along a line of sight between the source and detector. Images are presented of a highly scattering, solid plastic phantom with optical properties closely matched to those of human breast tissue at near-infrared wavelengths. The phantom is a 54-mm-thick slab, containing four small cylinders of contrasting scatter and absorption. Results show that embedded absorbers can be distinguished from embedded scatterers, and that the diffusion perturbation amplitude provides inherently greater spatial resolution than the absorption perturbation amplitude.

Hielscher, Andreas H.; Hanli Liu; Britton Chance; Frank K. Tittel; Steven L. Jacques; "Time-resolved photon emission from layered turbid media," *Appl. Opt.*, Vol. 35, pg. 0719, (1996).

We present numerical and experimental results of time-resolved emission profiles from various layered turbid media. Numerical solutions determined by time-resolved Monte Carlo simulations are compared with measurements on layered-tissue phantoms made from gelatin. In particular, we show that in certain cases the effects of the upper layers can be eliminated. As a practical example, these results are used to analyze *in vivo* measurements on the human head. This demonstrates the influence of skin, skull, and meninges on the determination of the blood oxygenation in the brain.

Hielscher, Andreas H.; Judith R. Mourant; Irving J. Bigio; "Influence of particle size and concentration on the diffuse backscattering of polarized light from tissue phantoms and biological cell suspensions," *Appl. Opt.*, Vol. 36, pg. 0125, (1997).

We present experimental results that show the spatial variations of the diffuse-backscattered intensity when linearly polarized light is incident upon highly scattering media. Experiments on polystyrene-sphere and Intralipid suspensions demonstrate that the radial and azimuthal variations of the observed pattern depend on the concentration, size, and anisotropy factor g of the particles that constitute the scattering medium. Measurements performed on biological-cell suspensions show the potential of this method for cell characterization.

Christina L. Hutchinson; Tamara L. Troy; Eva M. Sevick-Muraca; "Fluorescence-lifetime determination in tissues or other scattering media from measurement of excitation and emission kinetics," *Appl. Opt.*, Vol. 35, pg. 2325, (1996).

Measurements of nanosecond and subnanosecond fluorescence lifetimes are restricted to dilute, nonscattering systems since excitation and emission photon times of flight significantly affect measured fluorescent decay kinetics. We provide the theoretical rationale for frequency-domain measurements of phase-shift and amplitude demodulation made at excitation and emission wavelengths for direct determination of lifetimes in tissues and other scattering media. We confirm our analytical expressions using standard laser dyes such as 3,3`-diethylthiatricarbocyanine iodide, IR-125, and IR-140 in polystyrene suspensions with similar scattering properties as tissues. Our results have significant implication for lifetime-based spectroscopy in tissues and other scattering media.

Joseph A. Izatt; Manish D. Kulkarni; Siavash Yazdanfar; Jennifer K. Barton; Ashley J. Welch; "*In vivo* bidirectional color Doppler flow imaging of picoliter blood volumes using optical coherence tomography," *Opt. Lett.*, Vol. 22, pg. 1439, (1997).

We describe a novel optical system for bidirectional color Doppler imaging of flow in biological tissues with micrometer-scale resolution and demonstrate its use for *in vivo* imaging of blood flow in an animal model. Our technique, color Doppler optical coherence tomography (CDOCT), performs spatially localized optical Doppler velocimetry by use of scanning low-coherence interferometry. CDOCT is an extension of optical coherence tomography (OCT), employing coherent signal-acquisition electronics and joint time-frequency analysis algorithms to perform flow imaging simultaneous with conventional OCT imaging. Cross-sectional maps of blood flow velocity with spatial resolution and velocity precision were obtained through intact skin in living hamster subdermal tissue. This technology has several potential medical applications.

Jansen, E. Duco; Le, Tuong H.; Welch, Ashley J.; "Excimer, Ho:YAG and Q-switched Ho:YAG ablation of aorta: a comparison of temperatures and tissue damage *in vitro*," *Appl. Opt.*, Vol. 32, pg. 526, (1993).

Jarry, Gilbert; Ludovic Poupinet; Jennifer Watson; Thierry Lepine; "Extinction measurements in diffusing mammalian tissue with heterodyne detection and a titanium:sapphire laser," *Appl. Opt.*, Vol. 34, pg. 2045, (1995).
Total optical absorption in mammalian tissues is measured in the near infrared by the use of heterodyne detection and a Ti:sapphire laser. Because of the high sensitivity, directivity, and signal-to-noise ratio of the setup, we were able to detect coherent photons after attenuation by more than 9 optical densities. This method allows us to detect unscattered photons that are passing through more than 7 mm of various tissues such as brain, muscle, liver, skin, and fat selectively.

Huabei Jiang; Keith D. Paulsen; Ulf L. Osterberg; Brian W. Pogue; Michael S. Patterson; "Simultaneous reconstruction of optical absorption and scattering maps in turbid media from near-infrared frequency-domain data," *Opt. Lett.*, Vol. 20, pg. 2128, (1995).
Simultaneously reconstructed images of both absorption and scattering coefficients in optically heterogeneous media based on frequency-domain measurements are presented. The images are obtained from absolute measured data by use of a diffusion approximation to light propagation in tissue that is modeled with a finite-element method. Images that have different contrast levels are demonstrated.

Joblin, Anthony J.; "Method of calculating the image resolution of a near-infrared time-of-flight tissue-imaging system," *Appl. Opt.*, Vol. 35, pg. 0752, (1996).
A new method for calculating the image resolution for a near-infrared time-of-flight tissue-imaging system is presented. The image resolution is calculated from the full width at half-maximum of the photon-path function along the midplane of the medium, integrated over all times of flight, and weighted by the time-resolved detector response. Detailed treatment of the optical gating mechanism shows that for some types of gating mechanism, there exists an optimal gating time beyond which the image resolution is not improved by arbitrarily decreasing the gating time. This theory predicts a limiting image resolution of approximately 20% of the medium thickness, which is consistent with the research of others.

Kienle, Alwin; Lothar Lilge; I. Alex Vitkin; Michael S. Patterson; Brian C. Wilson; Raimund Hibst; Rudolf Steiner; "Why do veins appear blue? A new look at an old question," *Appl. Opt.*, Vol. 35, pg. 1151, (1996).
We investigate why vessels that contain blood, which has a red or a dark red color, may look bluish in human tissue. A CCD camera was used to make images of diffusely reflected light at different wavelengths. Measurements of reflectance that are due to model blood vessels in scattering media and of human skin containing a prominent vein are presented. Monte Carlo simulations were used to calculate the spatially resolved diffuse reflectance for both situations. We show that the color of blood vessels is determined by the following factors: (i) the scattering and absorption characteristics of skin at different wavelengths, (ii) the oxygenation state of blood, which affects its absorption properties, (iii) the diameter and the depth of the vessels, and (iv) the visual perception process.

Kienle, Alwin; Lothar Lilge; Michael S. Patterson; Raimund Hibst; Rudolf Steiner; Brian C. Wilson; "Spatially resolved absolute diffuse reflectance measurements for noninvasive determination of the optical scattering and absorption coefficients of biological tissue," *Appl. Opt.*, Vol. 35, pg. 2304, (1996).
The absorption and transport scattering coefficients of biological tissues determine the radial dependence of the diffuse reflectance that is due to a point source. A system is described for making remote measurements of spatially resolved absolute diffuse reflectance and hence noninvasive, noncontact estimates of the tissue optical properties. The system incorporated a laser source and a CCD camera. Deflection of the incident beam into the camera allowed characterization of the source for absolute reflectance measurements. It is shown that an often used solution of the diffusion equation cannot be applied for these measurements. Instead, a neural network, trained on the results of Monte Carlo simulations, was used to estimate the absorption and scattering coefficients from the reflectance data. Tests on tissue-simulating phantoms with transport scattering coefficients between 0.5 and 2.0 mm^{-1} and absorption coefficients between 0.002 and 0.1 mm^{-1} showed the rms errors of this technique to be 2.6% for the transport scattering coefficient and 14% for the absorption coefficients. The optical properties of bovine muscle, adipose, and liver tissue, as well as chicken muscle (breast), were also measured ex vivo at 633 and 751 nm. For muscle tissue it was found that the Monte Carlo simulation did not agree with experimental measurements of reflectance at distances less than 2 mm from the incident beam. Key words: Tissue optics, reflectance, Monte Carlo, neural network.

Kienle, Alwin; Michael S. Patterson; "Improved solutions of the steady-state and the time-resolved diffusion equations for reflectance from a semi-infinite turbid medium," *J. Opt. Soc. Am. A*, Vol. 14, pg. 246, (1997).

Improved solutions of the diffusion equation for time-resolved and steady-state spatially resolved reflectance are investigated for the determination of the optical coefficients of semi-infinite turbid media such as tissue. These solutions are derived for different boundary conditions at the turbid-medium-air interface and are compared with Monte Carlo simulations. Relative reflectance data are fitted in the time domain, whereas relative and absolute reflectance are investigated in the steady-state domain. It is shown that the error in deriving the optical coefficients is, especially for steady-state spatially resolved reflectance, considerably smaller for the solutions under study than for the commonly used solutions. Analysis of experimental measurements of absolute steady-state spatially resolved reflectance confirms these results.

Koelink, M. H.; de Mul, F. F. M.; Greve, J.; Graaff, R.; Dassel, A. C. M.; Aarnoudse, J. G.; "Laser doppler blood flowmetry using two waveguides: Monte Carlo simulations and measurements," *Appl. Opt.*, Vol. 33, pg. 3549, (1994).

Matthias Kohl; Russell Watson; Mark Cope; "Optical properties of highly scattering media determined from changes in attenuation, phase, and modulation depth," *Appl. Opt.*, Vol. 36, pg. 0105, (1997).

The optical properties of scattering media determine the attenuation (A) and the transit time (\leftanglet\rangle) of light reflected from the medium as well as the phase (Phi) and modulation depth (M) of an intensity-modulated lightwave. Our primary finding is that the ratio of changes in A, Phi, and M is approximately independent of the scattering properties and gives a good estimate of the absorption coefficient. These changes can be induced either by small changes in the absorption coefficient of the medium, by the tuning of the wavelength, or by changes in the light source-detector distance. The application for the *in vivo* monitoring of hemoglobin and oxyhemoglobin concentrations in human tissue is discussed.

Matthias Kohl; Russell Watson; Mark Cope; "Determination of absorption coefficients in highly scattering media from changes in attenuation and phase," *Opt. Lett.*, Vol. 21, pg. 1519, (1996).

The accurate, quantitative analysis of absorption and scattering properties in tissue is a central problem in biochemical optics, in particular for the determination of hemoglobin and oxyhemoglobin concentrations. Because of light scattering, the absolute concentrations of these chromophores (i.e., the absorption coefficient) cannot easily be inferred. A new method for the estimation of the absorption coefficients in scattering media, based on measurements obtained from an intensity-modulated optical spectrometer, is proposed. The ratios of the changes in attenuation and phase that are induced by changes in the absorption coefficient give a good approximation of the absorption coefficient.

Vladimir G. Kolinko; Frits F. M. de Mul; Jan Greve; Alexander V. Priezzhev; "Probabilistic model of multiple light scattering based on rigorous computation of the first and the second moments of photon coordinates," *Appl. Opt.*, Vol. 35, pg. 4541, (1996).

We consider a concise method based on recurrent relations that permit rigorous computing of the first and the second moments of the components of the vector locating a randomly walking photon in an infinite homogeneous light-scattering medium. On assumption that the components obey a three-dimensional Gaussian distribution a probability density for the photon locations at the Nth scattering event can readily be written down and the light-intensity distribution in the medium may be calculated. The results from theoretical analyses are compared with the solution of a light-diffusion equation and with results of Monte Carlo simulations and show a better fit with simulated data than the diffusion approximation.

G. Kumar; J. M. Schmitt; "Optimal probe geometry for near-infrared spectroscopy of biological tissue," *Appl. Opt.*, Vol. 36, pg. 2286, (1997).

The results of this study clarify the influence of probe geometry on spectroscopic measurements obtained from the surface of a turbid biological tissue. We show that the transition between the measurement of the predominantly backward-propagating and the predominantly forward-propagating photon fluxes is marked by the separation between the source probe and the detector probes at which the dependence of the fluence on small changes in scattering coefficient vanishes. This is the probe separation at which a variable scattering background has the least influence on the measurement of optical absorption in turbid materials. Estimates of the optimum probe spacing for typical values of absorption and scattering coefficients of soft tissue in the near-infrared spectral region (800-2500 nm) are derived from an analytical solution of the diffusion equation. The estimates were verified by Monte Carlo simulations and experiments on particle suspensions with optical properties similar to those of skin tissue.

Leith, Emmett N.; Chen, Hsuan; Chen, Y.; Dilworth, David S.; Lopez, Joaquin L.; Masri, R.; Rudd, J. V.; Valdmanis, J. A.; "Electronic holography and speckle methods for imaging through tissue using femtosecond gated pulses," *Appl. Opt.*, Vol. 30, pg. 4204, (1991).

Hui Li; Shusen Xie; "Measurement method of the refractive index of biotissue by total internal reflection," *Appl. Opt.*, Vol. 35, pg. 1793, (1996).

A simple method based on total internal reflection is presented for measuring the refractive index of biological tissue. A narrow laser beam and a semicylindrical lens in contact with a tissue specimen are used in the experimental apparatus. The dependence of the internal reflectivity on the angle of incidence is measured to determine the critical angle and therefore the refractive index of tissue. The experimental results for several tissue samples show that the method is reliable and useful for studying tissue optics. The principal advantages of the method are its elimination of scattering effects, suitability to a small *in vitro* sample, and excellent accuracy.

X. D. Li; M. A. O'Leary; D. A. Boas; Britton Chance; A. G. Yodh; "Fluorescent diffuse photon density waves in homogeneous and heterogeneous turbid media: analytic solutions and applications," *Appl. Opt.*, Vol. 35, pg. 3746, (1996).

We present analytic solutions for fluorescent diffuse photon density waves originating from fluorophores distributed in thick turbid media. Solutions are derived for a homogeneous turbid medium containing a uniform distribution of fluorophores and for a system that is homogeneous except for the presence of a single spherical inhomogeneity. Generally the inhomogeneity has fluorophore concentration, and lifetime and optical properties that differ from those of the background. The analytic solutions are verified by numerical calculations and are used to determine the fluorophore lifetime and concentration changes required for the accurate detection of inhomogeneities in biologically relevant systems. The relative sensitivities of absorption and fluorescence methods are compared.

Shao-Pow Lin; Lihong Wang; Steven L. Jacques; Frank K. Tittel; "Measurement of tissue optical properties by the use of oblique-incidence optical fiber reflectometry," *Appl. Opt.*, Vol. 36, pg. 0136, (1997).

Fiber-optic-based oblique-incidence reflectometry is a simple and accurate method for measuring the absorption and reduced scattering coefficients mua and mu\primes of semi-infinite turbid media. Obliquely incident light produces a spatial distribution of diffuse reflectance that is not centered about the point of light entry. The amount of shift in the center of diffuse reflectance is directly related to the medium's diffusion length D. We developed a fiber-optic probe to deliver light obliquely and sample the relative profile of diffuse reflectance. Measurement in absolute units is not necessary. From the profile, it was possible to measure D, perform a curve fit for the effective attenuation coefficient mueff, and then calculate mua and mu\primes. This method was verified with Monte Carlo simulations and tested on tissue phantoms. Our measurements of D and mueff had an accuracy of approximately 5%, thus giving us 10% and 5% accuracy for mua and mu\primes, respectively.

Wei-Chiang Lin; Massoud Motamedi; Ashley J. Welch; "Nonlinear optical behavior of ocular tissue during laser irradiation," *Appl. Opt.*, Vol. 34, pg. 7979, (1995).

A pump (cw Ho-YAG laser) and probe (He-Ne laser) system was used to study the dynamics of the optical behavior of ocular tissue during laser heating. The nonlinear optical behavior of porcine corneal and vitreous-humor tissue was characterized *in vitro* by means of measurements of the radial profile of a He-Ne laser beam transmitted through the tissue. Temperature gradients in the tissue created by the absorption of pump radiation caused the probe beam to diverge. For constant laser power, the rate of divergence was made dependent on the spot size of the pump beam. The profile of the transmitted probe beam returned to its original magnitude and shape after the tissue was permitted to cool. This reversible change in optical behavior was attributed to the formation of a negative lens owing to thermally induced local gradients in the refractive index of the tissue.

Wei-Chiang Lin; Massoud Motamedi; Ashley J. Welch; "Dynamics of tissue optics during laser heating of turbid media," *Appl. Opt.*, Vol. 35, pg. 3413, (1996).

The dynamics of the optical behavior of tissue during the photothermal interaction of laser radiation with tissue could significantly affect the optimization of light doses for effective and safe applications of lasers in medicine. Characterization of the dynamics of tissue optics during laser heating was performed by means of simultaneous measurements of the total transmittance, diffuse reflectance, and surface temperature of fresh and thermally coagulated human skin and canine aorta during long-pulsed Nd:YAG laser heating with a double integrating-sphere system and an infrared camera. Thermally induced changes in the optical properties of tissue caused a decrease in the total transmittance and an increase in the diffuse reflectance of both fresh and precoagulated skin and aorta samples. For fresh tissue, these changes were primarily reversible until photocoagulation occurred, then both the reversible, as well as the irreversible, changes were observed. However, for precoagulated tissue the reversible changes in the optical properties were dominant, whereas the irreversible changes were insignificant. Results from this study indicate the existence of the nonlinear behavior in the optics of turbid biological media during pulsed laser heating. Possible mechanisms responsible for this nonlinear optical behavior are discussed. Key words: Tissue optics, reversible optical behavior, tissue coagulation, pulsed laser, laser heating, total transmittance, diffuse reflectance.

Maitland, Duncan J.; Walsh, Joseph T.; Prystowsky, Jay B.; "Optical properties of human gallbladder tissue and bile," *Appl. Opt.*, Vol. 32, pg. 586, (1993).

Manni, Jeffrey G.; "High Powered Delivery: Fiberoptic Laser Surgery," *Opt. Phot. News*, Vol. 7, pg. 23, (1996).

Marble, David R.; Burns, David H.; Cheung, Peter W.; "Diffusion-based model of pulse oximetry: *in vitro* and *in vivo* comparisons," *Appl. Opt.*, Vol. 33, pg. 1279, (1994).

Guillermo Marquez; Lihong V. Wang; "White light oblique incidence reflectometer for measuring absorption and reduced scattering spectra of tissue-like turbid media," *Opt. Express*, Vol. 1, pg. 454, (1997).

We designed a quick and inexpensive system for spectral measurements of optical properties including absorption coefficient and reduced scattering coefficient. The system was based on oblique incidence reflectometry.[1] A broad band light source was coupled into an optic fiber to deliver light obliquely to turbid media. Nine detection optic fibers were used to collect the diffuse reflectance as a function of source-detector distance. The relative diffuse reflectance profile was used to deduce the absorption and reduced scattering spectra. The system was able to acquire data in a wavelength range of 256 nm within a fraction of a second.

S. J. Matcher; "Closed-form expressions for obtaining the absorption and scattering coefficients of a turbid medium with time-resolved spectroscopy," *Appl. Opt.*, Vol. 36, pg. 8298, (1997).

A simple closed-form relation for extracting the absorption coefficient of a multiply scattering medium from a measured temporal impulse response is pointed out. Such a relation eliminates the need for time-consuming iterative curve-fitting techniques and may also be of use for calculating the regional sensitivity of the time-resolved spectroscopy technique. Equations for three types of measurement geometry (reflection from an infinite half-space, transmission through an infinite slab, and reflection from an infinite slab) are described. The effects of detector temporal resolution and detector intensity linearity are also discussed.

S. J. Matcher; M. Cope; D. T. Delpy; "*In vivo* measurements of the wavelength dependence of tissue-scattering coefficients between 760 and 900 nm measured with time-resolved spectroscopy," *Appl. Opt.*, Vol. 36, pg. 0386, (1997).

We present *in vivo* values for the optical transport coefficients (μa, μs\prime) of the adult human forearm, calf, and head from 760 to 900 nm measured with time-resolved spectroscopy. The accuracy of the method is tested with tissue-simulating phantoms. We obtain μs\prime(λ) \approximately 1.1 - (5.1 \times 10-4 λ) mm-1 (forearm), 1.6 - (8.9 \times 10-4 λ) mm-1 (calf), and 1.45 - (6.5 \times 10-4 λ) mm-1 (head), where λ is measured in nanometers. At 800 nm we obtain μa = 0.023 -+ 0.004 mm-1 (forearm), 0.017 -+ 0.005 mm-1 (calf), and 0.016 -+ 0.001 mm-1 (head). Our values differ substantially from published *in vitro* data. In particular, our transport coefficients for the adult head are substantially lower than previously reported values for adult human cerebral matter and pig skull cortical bone measured *in vitro*.

Charles L. Matson; Natalie Clark; Lenore McMackin; Janet S. Fender; "Three-dimensional tumor localization in thick tissue with the use of diffuse photon-density waves," *Appl. Opt.*, Vol. 36, pg. 0214, (1997).

A new approach to three-dimensional tumor localization in turbid media with the use of measurements in a single plane is presented. Optical diffuse photon-density waves are used to probe the turbid medium. Relative amplitudes and phases are measured in the detection plane. Lateral localization is accomplished in the detection plane. With a Fourier optics approach, the scattered wave is reconstructed throughout the volume to provide depth localization. Computer-simulation results that validate this technique are presented. Applications of this technique to multiple tumors and to optical mammography are discussed.

Milner, Thomas E.; Goodman, Dennis M.; Tanenbaum, B. Samuel; Nelson, J. Stuart; "Depth profiling of laser-heated chromophores in biological tissues by pulsed photothermal radiometry," *J. Opt. Soc. Am. A*, Vol. 12, pg. 1479, (1995).

A solution method is proposed to the inverse problem of determining the unknown initial temperature distribution in a laser-exposed test material from measurements provided by infrared radiometry. A Fredholm integral equation of the first kind is derived that relates the temporal evolution of the infrared signal amplitude to the unknown initial temperature distribution in the exposed test material. The singular-value decomposition is used to demonstrate the severely ill-posed nature of the derived inverse problem. Three inversion methods are used to estimate solutions for the initial temperature distribution. A nonnegatively constrained conjugate-gradient algorithm using early termination is found superior to unconstrained inversion methods and is applied to image the depth of laser-heated chromophores in human skin. Key words: constrained conjugate gradients, ill-posed problem, infrared radiometry, laser surgery, nonnegative, singular-value decomposition.

Mitic, Gerhard; Kolzer, Jochen; Otto, Johann; Plies, Erich; Solkner, Gerald; Zinth, Wolfgang; "Time-gated transillumination of biological tissues and tissuelike phantoms," *Appl. Opt.*, Vol. 33, pg. 6699, (1994).

Judith R. Mourant; Irving J. Bigio; Darren A. Jack; Tamara M. Johnson; Heather D. Miller; "Measuring absorption coefficients in small volumes of highly scattering media: source-detector separations for which path lengths do not depend on scattering properties," *Appl. Opt.*, Vol. 36, pg. 5655, (1997).

The noninvasive measurement of variations in absorption that are due to changes in concentrations of biochemically relevant compounds in tissue is important in many clinical settings. One problem with such measurements is that the path length traveled by the collected light through the tissue depends on the scattering properties of the tissue. We demonstrate, using both Monte Carlo simulations and experimental measurements, that for an appropriate separation between light-delivery and light-collection fibers the path length of the collected photons does not depend on scattering parameters for the range of parameters typically found in tissue. This is important for developing rapid, noninvasive, and inexpensive methods for measuring absorption changes in tissue.

Judith R. Mourant; James Boyer; Andreas H. Hielscher; Irving J. Bigio; "Influence of the scattering phase function on light transport measurements in turbid media performed with small source-detector separations," *Opt. Lett.*, Vol. 21, pg. 546, (1996).

Many methods of optical tissue diagnosis require that measurements be performed with small source-detector separations in a backscatter geometry. Monte Carlo simulations are used to demonstrate that for these situations light transport depends on the exact form of the angular scattering probability distribution, $P(\theta)$. Simulations performed with different forms of $P(\theta)$ with the same value of $<\cos\theta>$ result in the collection of significantly different fractions of the incident photons, particularly when small-numerical-aperture delivery and collection fibers are employed. More photons are collected for the distribution that has a higher probability of scattering events with theta >125 degrees. For the clinically relevant optical parameters employed here, the differences in light collection are >60%.

Judith R. Mourant; Tamika Fuselier; James Boyer; Tamara M. Johnson; Irving J. Bigio; "Predictions and measurements of scattering and absorption over broad wavelength ranges in tissue phantoms," *Appl. Opt.*, Vol. 36, pg. 0949, (1997).

Predictions from Mie theory regarding the wavelength dependence of scattering in tissue from the near UV to the near IR are discussed and compared with experiments on tissue phantoms. For large fiber separations it is shown that rapid, simultaneous measurements of the elastic scatter signal for several fiber separations can yield the absorption coefficient and reduced scattering coefficient. With this information, the size of the scattering particles can be estimated, and this is done for Intralipid. Measurements made at smaller source detector separations support Mie theory calculations, demonstrating that the sensitivity of elastic scatter measurements to morphological features, such as scatterer size, is enhanced when the distance between the source and detector fibers is small.

Egberto Munin; "Picosecond magneto-optical phenomena in turbid media: toward magneto-optical characterization of highly scattering biological samples," *Appl. Opt.*, Vol. 36, pg. 2990, (1997).

The behavior of a Faraday-active turbid medium under ultrafast optical excitation is investigated. As the degree of polarization of early arriving photons is mostly preserved during the first 100 ps after the arrival of the ballistic component, the possibility of using the magneto-optical rotation of the light polarization as a new tool for tissue characterization is addressed. A technique is proposed for determining photon-scattering statistics in turbid biological media. The analysis is performed on the basis that photons trapped in multiple-scattering events leave the medium with larger induced rotation angles. A measurement of the magneto-optical rotatory power of turbid biological samples is possible if only the early arriving unscattered photons are probed.

Babak Nemati; H. Grady Rylander III; Ashley J. Welch; "Optical properties of conjunctiva, sclera, and the ciliary body and their consequences for transscleral cyclophotocoagulation," *Appl. Opt.*, Vol. 35, pg. 3321, (1996).

A number of recent studies have demonstrated the success of Nd:YAG and diode laser transscleral cyclophotocoagulation in the treatment of advanced glaucoma. Wavelength selection, however, has seldom been based on a clear understanding of the optical properties of tissues involved. The optical properties of conjunctiva, sclera, and the ciliary body adjacent to the limbus were investigated to find an optimal wavelength range for transscleral cyclophotocoagulation. The absorption and scattering coefficients of these layers were determined in the 300-1200-nm wavelength range by the use of a one-dimensional inverse adding-doubling method. The measured optical properties of conjunctiva, sclera, and the ciliary body provide a basis for a comparative analysis of the laser wavelengths used clinically for transscleral cyclophotocoagulation.

Babak Nemati; H. Grady Rylander III; Ashley Welch; "Optical properties of conjunctiva, sclera, and the ciliary body and their consequences for transscleral cyclophotocoagulation: erratum," *Appl. Opt.*, Vol. 36, pg. 0416, (1997).

Corrections are given to the reported [Appl. Opt. 35, 3321 (1996)] units of the absorption and scattering coefficients mua and mus, respectively. The corrections pertain to two figures, several tables, and the text describing results.

Michael G. Nichols; Edward L. Hull; Thomas H. Foster; "Design and testing of a white-light, steady-state diffuse reflectance spectrometer for determination of optical properties of highly scattering systems," *Appl. Opt.*, Vol. 36, pg. 0093, (1997).

We present a steady-state radially resolved diffuse reflectance spectrometer capable of measuring the absorption and transport scattering spectra of tissue-simulating phantoms over an adjustable 170-nm wavelength interval in the visible and near infrared. Measurements in a variety of phantoms are demonstrated over the relevant range of tissue optical properties, and the accuracy of the instrument is found to be approximately 10% in both scattering and absorption. Monte Carlo simulations designed to test the accuracy of the instrument are presented that support the experimental findings.

Nilsson, Annika M. K.; Roger Berg; Stefan Andersson-Engels; "Measurements of the optical properties of tissue in conjunction with photodynamic therapy," *Appl. Opt.*, Vol. 34, pg. 4609, (1995).

A simple optical dosimeter was used to measure the light intensity in rat liver and muscle *in vivo* with fibers positioned at different depths to investigate whether the light penetration changed during photodynamic therapy (PDT). The results were then correlated with measurements of the three optical-interaction coefficients μ_s, μ_a, and g for wavelengths in the range 500-800 nm for PDT-treated and nontreated rat liver and muscle tissue *in vitro*. A distinct increase in the absorption coefficient was seen immediately after treatment, in agreement with the decreasing light intensity observed during the treatment, as measured with the optical dosimeter. The collimated transmittance was measured with a narrow-beam setup, and an optical integrating sphere was used to measure the diffuse reflectance and total transmittance of the samples. The corresponding optical properties were obtained by spline interpolation of Monte Carlo-simulated data. To ensure that the measured values were correct, we performed calibration measurements with suspensions of polystyrene microspheres and ink.

M. A. O'Leary, D. A. Boas, B. Chance, and A. G. Yodh; "Experimental images of heterogeneous turbid media by frequency-domain diffusing-photon tomography," *Opt. Lett.*, Vol. 20, pg. 426, (1995).

We present images of heterogeneous turbid media derived from measurements of diffuse photon-density waves traveling through highly scattering tissue phantoms. To our knowledge, the images are the first experimental reconstruction based on data collected in the frequency domain. We demonstrate images of both absorbing and scattering heterogeneities and show that this method is sensitive to the optical properties of the heterogeneity. The algorithm employs a differential measurement scheme that reduces the effect of errors resulting from incorrect estimation of the background optical properties. The relative advantages of sources with low and high modulation frequency are discussed within this context.

M. A. O'Leary; D. A. Boas; X. D. Li; B. Chance; A. G. Yodh; "Fluorescence lifetime imaging in turbid media," *Opt. Lett.*, Vol. 21, pg. 158, (1996).

The lifetime of a fluorophore generally varies in different environments, making the molecule a sensitive indicator of tissue oxygenation, pH, and glucose. However, lifetime measurements are complicated when the fluorophore is embedded in an optically thick, highly scattering medium such as human tissue. We formulate the inverse problem for fluorescence lifetime tomography using diffuse photon density waves, and we demonstrate the technique by deriving spatial images of heterogeneous fluorophore distribution and lifetime, using simulated measurements in heterogeneous turbid media.

Eiji Okada; Martin Schweiger; Simon R. Arridge; Michael Firbank; David T. Delpy; "Experimental validation of Monte Carlo and finite-element methods for the estimation of the optical path length in inhomogeneous tissue," *Appl. Opt.*, Vol. 35, pg. 3362, (1996).

To validate models of light propagation in biological tissue, experiments to measure the mean time of flight have been carried out on several solid cylindrical layered phantoms. The optical properties of the inner cylinders of the phantoms were close to those of adult brain white matter, whereas a range of scattering or absorption coefficients was chosen for the outer layer. Experimental results for the mean optical path length have been compared with the predictions of both an exact Monte Carlo (MC) model and a diffusion equation, with two differing boundary conditions implemented in a finite-element method (FEM). The MC and experimental results are in good agreement despite poor statistics for large fiber spacings, whereas good agreement with the FEM prediction requires a careful choice of proper boundary conditions. Key words: Light propagation, time-of-flight measurement, Monte Carlo method, finite-element method.

Eiji Okada; Michael Firbank; Martin Schweiger; Simon R. Arridge; Mark Cope; David T. Delpy; "Theoretical and experimental investigation of near-infrared light propagation in a model of the adult head," *Appl. Opt.*, Vol. 36, pg. 0021, (1997).

Near-infrared light propagation in various models of the adult head is analyzed by both time-of-flight measurements and mathematical prediction. The models consist of three- or four-layered slabs, the latter incorporating a clear cerebrospinal fluid (CSF) layer. The most sophisticated model also incorporates slots that imitate sulci on the brain surface. For each model, the experimentally measured mean optical path length as a function of source-detector spacing agrees well with predictions from either a Monte Carlo model or a finite-element method based on diffusion theory or a hybrid radiosity-diffusion theory. Light propagation in the adult head is shown to be highly affected by the presence of the clear CSF layer, and both the optical path length and the spatial sensitivity profile of the models with a CSF layer are quite different from those without the CSF layer. However, the geometry of the sulci and the boundary between the gray and the white matter have little effect on the detected light distribution.

Ono, Kimizo; Kanda, Masahiko; Hiramoto, J.; Yotsuya, Kohro; Sata, Nobuhiro; "Fiber optic reflectance spectrophotometry system for *in vivo* tissue diagnosis," *Appl. Opt.*, Vol. 30, pg. 98, (1991).

Oraevsky, Alexander A.; Steven L. Jacques; Frank K. Tittel; "Measurement of tissue optical properties by time-resolved detection of laser-induced transient stress," *Appl. Opt.*, Vol. 36, pg. 0402, (1997).

We report on a technique utilizing time-resolved detection of laser-induced stress transients for the measurement of optical properties in turbid media specifically suitable for biological tissues. The method was tested initially in nonscattering absorbing media so that it could be compared with spectrophotometry. The basis of this method is provided by the conditions of temporal stress confinement in the irradiated volume where the pressure generated in tissues heated instantly by laser pulses is proportional to the absorbed laser energy density, and the exponential profile of the initial stress distribution in the irradiated volume corresponds to the z-axial distribution of the absorbed laser fluence. Planar thermoelastic waves can propagate in water-containing media with minimal distortion, and their axial profiles can be detected by an acoustic transducer with sufficient temporal resolution. The acoustic waves induced by 14-ns laser pulses in nonscattering media, turbid gels, and tissues were measured by a piezoelectric transducer with a 3-ns response time. Temporal profiles of stress transients yielded z-axial distributions of the absorbed laser energy in turbid and opaque media, provided that the speed of sound in these media was known. The absorption and effective scattering coefficients of beef liver, dog prostate, and human aortic atheroma at three wavelengths, 1064 nm (in near infrared), 532 nm (visible), and 355 nm (near UV), were deduced from laser-induced stress profiles with additional measurements of total diffuse reflectance.

Martin R. Ostermeyer; Steven L. Jacques; "Perturbation theory for diffuse light transport in complex biological tissues," *J. Opt. Soc. Am. A*, Vol. 14, pg. 255, (1997).

A perturbation theory for the forward problem of optical transport in turbid media is developed. It is applicable to media with scattering and absorption inhomogeneities and steady-state and modulated light. Absorbing perturbations can be described by a volume distribution of virtual sources that primarily causes a monopole perturbation light field. Scattering objects have an additional contribution that, in the limiting case of sharply bounded objects, is represented by a surface distribution of virtual sources and causes a dipolelike perturbation pattern. Using the concept of virtual sources, we discuss a possible ambiguity between the perturbations from scattering and absorbing inhomogeneities and the implications for the source-detector placement in inverse problems. We show that the surface effects due to sharp boundaries of scattering objects pose both a numerical problem and a chance to improve the resolution of inverse algorithms.

D. Y. Paithankar; A. U. Chen; B. W. Pogue; M. S. Patterson; E. M. Sevick-Muraca; "Imaging of fluorescent yield and lifetime from multiply scattered light reemitted from random media," *Appl. Opt.*, Vol. 36, pg. 2260, (1997).

The feasibility of employing fluorescent contrast agents to perform optical imaging in tissues and other scattering media has been examined through computational studies. Fluorescence lifetime and yield can give crucial information about local metabolite concentrations or environmental conditions within tissues. This information can be employed toward disease detection, diagnosis, and treatment if noninvasively quantitated from reemitted optical signals. However, the problem of inverse image reconstruction of fluorescence yield and lifetime is complicated because of the highly scattering nature of the tissue. Here a light propagation model employing the diffusion equation is used to account for the scattering of both the excitation and fluorescent light. Simulated measurements of frequency-domain parameters of fluorescent modulated ac amplitude and phase lag are used as inputs to an inverse image-reconstruction algorithm, which employs the diffusion model to predict frequency-domain measurements resulting from a modulated input at the phantom periphery. In the inverse image-reconstruction algorithm, a Newton-Raphson technique combined with a Marquardt algorithm is employed to converge on the fluorescent properties within the medium. The successful reconstruction of both the fluorescence yield and lifetime in the case of a heterogeneous fluorophore distribution within a scattering medium has been demonstrated without a priori information or without the necessity of obtaining absence images.

Yingtian Pan; Reginald Birngruber; Jurgen Rosperich; Ralf Engelhardt; "Low-coherence optical tomography in turbid tissue: theoretical analysis," *Appl. Opt.*, Vol. 34, pg. 6564, (1995).

On the basis of white-light interferometry and statistical optics, a theoretical model for low-coherence optical tomography is presented that establishes the relation of interference modulation with path-length-resolved reflectance and that can provide analytical expressions and numerical solutions by means of a Fourier transform. The Monte Carlo technique is used to simulate the path-length-resolved reflectance from different multilayer tissue phantoms. Theoretical analyses and preliminary experimental results suggest that, unlike time-resolved spectroscopy, low-coherence optical tomography detects the local relative variations of path-length-resolved reflectance from the turbid tissues.

Yingtian Pan; Reginald Birngruber; Ralf Engelhardt; "Contrast limits of coherence-gated imaging in scattering media," *Appl. Opt.*, Vol. 36, pg. 2979, (1997).

The fundamental difference between time-resolved and coherence-gated imaging modalities in scattering media is analyzed in terms of their optical transfer functions. The effectiveness of coherence gating for multiple-scattering rejection is shown by imaging a 100-mum-thick razor blade hidden in the scattering phantoms (i.e., Intralipid suspensions) with different scattering coefficients. We found that the imaging contrast is limited by multiple scattering and speckle effects in high-scattering media, and the measured effective penetration depth of optical coherence tomography is approximately equal to six mean free paths under the experimental conditions of a numerical aperture of less than 0.1 and a scattering anisotropy of approximately 0.8.

Patterson, Michael S.; Moulton, J. David; Wilson, Brian C.; Berndt, Klaus W.; Lakowicz, Joseph R.; "Frequency-domain reflectance for the determination of the scattering and absorption properties of tissue," *Appl. Opt.*, Vol. 30, pg. 4474, (1991).

Patterson, Michael S.; Pogue, Brian W.; "Mathematical model for time-resolved and frequency-domain fluorescence spectroscopy in biological tissues," *Appl. Opt.*, Vol. 33, pg. 1963, (1994).

Keith D. Paulsen; Huabei Jiang; "Enhanced frequency-domain optical image reconstruction in tissues through total-variation minimization," *Appl. Opt.*, Vol. 35, pg. 3447, (1996).

Optical image reconstruction in heterogeneous turbid media is sensitive to noise, especially when the signal-to-noise ratio of a measurement system is low. A total-variation-minimization-based iterative algorithm is described in this paper that enhances the quality of reconstructed images with frequency-domain data over that obtained previously with a regularized least-squares approach. Simulation experiments in an 8.6-cm-diameter circular heterogeneous region with low- and high-contrast levels between the target and the background show that the quality of the reconstructed images can be improved considerably when total-variation minimization is included. These simulated results are further verified and confirmed by images reconstructed from experimental data by the use of the same geometry and optically tissue-equivalent phantoms. Measures of imaging performance, including the location, size, and shape of the reconstructed heterogeneity, along with absolute errors in the predicted optical-property values are used to quantify the enhancements afforded by this new approach to optical image reconstruction with diffuse light. The results show improvements of up to 5 mm in terms of geometric information and an order of magnitude or more decrease in the absolute errors in the reconstructed optical-property values for the test cases examined.

G. H. Pettit; M. N. Ediger; "Corneal-tissue absorption coefficients for 193- and 213-nm ultraviolet radiation," *Appl. Opt.*, Vol. 35, pg. 3386, (1996).

The small-signal absorption coefficients of 193- and 213-nm nanosecond laser pulses in bovine corneal tissue have been studied. The absolute reflectance of a planar quartz-cornea interface was measured at various angles of incidence for low-intensity laser irradiation (i.e., pulse fluences 3 orders of magnitude below the ablation threshold). The reflectance-versus-angle data were analyzed by the use of Fresnel theory to estimate the effective complex index of refraction of the tissue. This analysis indicated corneal absorption coefficients of 39,900 ñ 9800 cm^{-1} at 193 nm and 21,400 ñ 4900 cm^{-1} at 213 nm. Key words: Cornea, tissue optics, photorefractive keratectomy.

Pettit, G. H.; Ediger, M. N.; Weiblinger, R. P.; "Dynamic optical properties of collagen-based tissue during ArF excimer laser ablation," *Appl. Opt.*, Vol. 32, pg. 488, (1993).

Pickering, John W.; Bosman, Saskia; Posthumus, Paul; Blokland, Peter; Beek, Johan F.; van Gemert, Martin J. C.; "Changes in the optical properties (at 632.8 nm) of slowly heated myocardium," *Appl. Opt.*, Vol. 32, pg. 367, (1993).

Pickering, John W.; Prahl, Scott A.; van Wieringen, Niek; Beek, Johan F.; Sterenborg, Henricus J. C. M.; van Gemert, Martin J. C.; "Double-integrating-sphere system for measuring the optical properties of tissue," *Appl. Opt.*, Vol. 32, pg. 399, (1993).

Brian W. Pogue; Markus Testorf; Troy McBride; Ulf Osterberg; Keith Paulsen; "Instrumentation and design of a frequency-domain diffuse optical tomography imager for breast cancer detection," *Opt. Express*, Vol. 1, pg. 391, (1997).

The instrument development and design of a prototype frequency-domain optical imaging device for breast cancer detection is described in detail. This device employs radio-frequency intensity modulated near-infrared light to image quantitatively both the scattering and absorption coefficients of tissue. The functioning components of the system include a laser diode and a photomultiplier tube, which are multiplexed automatically through 32 large core fiber optic bundles using high precision linear translation stages. Image reconstruction is based on a finite element solution of the diffusion equation. This tool for solving the forward problem of photon migration is coupled to an iterative optical property estimation algorithm, which uses a Levenberg-Marquardt routine with total variation minimization. The result of this development is an automated frequency-domain optical imager for computed tomography which produces quantitatively accurate images of the test phantoms used to date. This paper is a description and characterization of an automated frequency-domain computed tomography scanner, which is more quantitative than earlier systems used in diaphanography because of the combination of intensity modulated signal detection and iterative image reconstruction.

Hans Pratisto; Martin Frenz; Michael Ith; Hans J. Altermatt; E. Duco Jansen; Heinz P. Weber; "Combination of fiber-guided pulsed erbium and holmium laser radiation for tissue ablation under water," *Appl. Opt.*, Vol. 35, pg. 3328, (1996).

Because of the high absorption of near-infrared laser radiation in biological tissue, erbium lasers and holmium lasers emitting at 3 and 2 μm, respectively, have been proven to have optimal qualities for cutting or welding and coagulating tissue. To combine the advantages of both wavelengths, we realized a multiwavelength laser system by simultaneously guiding erbium and holmium laser radiation by means of a single zirconium fluoride (ZrF_4) fiber. Laser-induced channel formation in water and poly(acrylamide) gel was investigated by the use of a time-resolved flash-photography setup, while pressure transients were recorded simultaneously with a needle hydrophone. The shapes and depths of vapor channels produced in water and in a submerged gel after single erbium and after combination erbium-holmium radiation delivered by means of a 400- μm ZrF_4 fiber were measured. Transmission measurements were performed to determine the amount of pulse energy available for tissue ablation. The effects of laser wavelength and the delay time between pulses of different wavelengths on the photomechanical and photothermal responses of meniscal tissue were evaluated *in vitro* by the use of histology. It was observed that the use of a short (200-μs, 100-mJ) holmium laser pulse as a prepulse to generate a vapor bubble through which the ablating erbium laser pulse can be transmitted (delay time, 100 μs) increases the cutting depth in meniscus from 450 to 1120 μm as compared with the depth following a single erbium pulse. The results indicate that a combination of erbium and holmium laser radiation precisely and efficiently cuts tissue under water with 20-50- μm collateral tissue damage. Key words: Multiwavelength laser system, pressure wave, cavitation, channel formation, infrared-fiber-delivery system, tissue damage, cartilage.

Jannick P. Rolland; Robin N. Strickland; "An approach to the synthesis of biological tissue," *Opt. Express*, Vol. 1, pg. 414, (1997).

Mathematical phantoms developed to synthesize realistic complex backgrounds such as those obtained when imaging biological tissue, play a key role in the quantitative assessment of image quality for medical and biomedical imaging. We present a modeling framework for the synthesis of realistic tissue samples. The technique is demonstrated using radiological breast tissue. The model employs a two-component image decomposition consisting of a slowly, spatially varying mean-background and a residual texture image. Each component is synthesized independently. The approach and results presented here constitute an important step towards developing methods for the quantitative assessment of image quality in medical and biomedical imaging, and more generally image science.

Sadhwani, Ajay; Kevin T. Schomacker; Guillermo J. Tearney; Norman S. Nishioka; "Determination of Teflon thickness with laser speckle. I. Potential for burn depth diagnosis," *Appl. Opt.*, Vol. 35, pg. 5727, (1996).

A quantitative method for determining the depth of burn eschar would aid surgeons in determining whether to excise and subsequently graft a burn wound. We hypothesize that tissue viability could be assessed by an analysis of the spatial modulation of near-field laser speckle by flowing blood. A feasibility study of the technique was performed with two-layer tissue phantoms used to simulate a burn wound. A sheet of polytetrafluoroethylene (PTFE) was used to simulate nonperfused burn eschar, and tissue perfusion within deeper layers was represented by Brownian motion from a scattering solution. A low-power He-Ne laser was focused onto the target, and the resulting speckle image was captured with a CCD camera and stored on a computer for further processing. The diameter of the speckle pattern was found to be directly proportional to the thickness of the overlying layer. These data suggest that the thickness of PTFE can be determined to ñ100- μm accuracy with 95% confidence and may be suitable for burn depth detection *in vivo*.

J. M. Schmitt; A. Knuttel; "Model of optical coherence tomography of heterogeneous tissue," *J. Opt. Soc. Am. A*, Vol. 14, pg. 1231, (1997).

A comprehensive model of optical coherence tomography (OCT) is described that includes the interference effects that produce speckle in images of dense heterogeneous tissue. It is based on the extended Huygens-Fresnel formulation of beam propagation in a turbulent atmosphere, adapted to the analysis of OCT. Incorporated in the model is a fractal description of the size distribution of scatterers in tissue. We demonstrate its application in the simulation of images of tissue volumes containing high-contrast targets embedded in a mixture of two sizes of particles. The simulated images show the degradation of image quality caused by speckle noise, along with the benefits of employing a light source with a short coherence time and an objective lens with a high numerical aperture. Based on model results, an estimate of the maximum probing depth is given in terms of the design variables of an OCT scanner and the optical properties of the tissue.

J. M. Schmitt; G. Kumar; "Turbulent nature of refractive-index variations in biological tissue," *Opt. Lett.*, Vol. 21, pg. 1310, (1996).

Phase-contrast microscopy shows that the structure of the refractive-index inhomogeneities in a variety of mammalian tissues resembles that of frozen turbulence. Viewed over a range of scales, the spectrum of index variations exhibits a power-law behavior for spatial frequencies spanning at least a decade (5-50 μm^{-1}) and has an outer scale in the range of 4-10 μm , above which correlations are no longer seen. The observed structure function fits the classical Kolmogorov model of turbulence. These observations are fundamental to understanding light propagation in tissue and may provide clues about how tissues develop and organize.

J. M. Schmitt; K. Ben-Letaief; "Efficient Monte Carlo simulation of confocal microscopy in biological tissue," *J. Opt. Soc. Am. A*, Vol. 13, pg. 952, (1996).

A variance-reduction technique is described that greatly improves the efficiency of Monte Carlo simulations of reflection-mode confocal microscopy in anisotropically scattering media. The efficiency gain is large enough that the performance of confocal microscopes probing as deep as 5 scattering lengths can be simulated with a desktop computer. We use the technique to simulate the response of a true confocal microscope probing biological tissue, a problem that has been impractical to undertake by using conventional Monte Carlo methods. Our most important finding is that operation of a confocal microscope in the true confocal mode enables much more effective rejection of undesired scattered light than operation in the partially coherent mode, but the maximum probing depths of microscopes operated in either mode are similar (2-3 scattering lengths) in practice because of sensitivity limitations.

Schmitt, J. M.; Knuttel, A.; Bonner, R. F.; "Measurement of optical properties of biological tissues by low-coherence reflectometry," *Appl. Opt.*, Vol. 32, pg. 6032, (1993).

Sergeev, A. M.; V. M. Gelikonov; G. V. Gelikonov; F. I. Feldchtein; R. V. Kuranov; N. D. Gladkova; N. M. Shakhova; L. B. Snopova; A. V. Shakov; I. A. Kuzn; "*In vivo* endoscopic OCT imaging of precancer and cancer states of human mucosa," *Opt. Express*, Vol. 1, pg. 432, (1997).

First results of endoscopic applications of optical coherence tomography for *in vivo* studies of human mucosa in respiratory, gastrointestinal, urinary and genital tracts are presented. A novel endoscopic OCT (EOCT) system has been created that is based on the integration of a sampling arm of an all-optical-fiber interferometer into standard endoscopic devices using their biopsy channel to transmit low-coherence radiation to investigated tissue. We have studied mucous membranes of esophagus, larynx, stomach, urinary bladder, uterine cervix and body as typical localization for carcinomatous processes. Images of tumor tissues versus healthy tissues have been recorded and analyzed. Violations of well-defined stratified healthy mucosa structure in cancered tissue are distinctly seen by EOCT, thus making this technique promising for early diagnosis of tumors and precise guiding of excisional biopsy.

Sevick-Muraca, Eva M.; Christina L. Hutchinson; Dilip Y. Paithankar; "Optical Tissue Biodiagnostics Using Fluorescence Lifetime," *Opt. Phot. News*, Vol. 7, pg. 24, (1996).

Sevick, E. M.; Frisoli, J. K.; Burch, C. L.; Lakowicz, Joseph R.; "Localization of absorbers in scattering media by use of frequency-domain measurements of time-dependent photon migration," *Appl. Opt.*, Vol. 33, pg. 3562, (1994).

HanQun Shangguan; Lee W. Casperson; Alan Shearin; Kenton W. Gregory; Scott A. Prahl; "Drug delivery with microsecond laser pulses into gelatin," *Appl. Opt.*, Vol. 35, pg. 3347, (1996).

Photoacoustic drug delivery is a technique for localized drug delivery by laser-induced hydrodynamic pressure following cavitation bubble expansion and collapse. Photoacoustic drug delivery was investigated on gelatin-based thrombus models with planar and cylindrical geometries by use of one microsecond laser pulses. Solutions of a hydrophobic dye in mineral oil permitted monitoring of delivered colored oil into clear gelatin-based thrombus models. Cavitation bubble development and photoacoustic drug delivery were visualized with flash photography. This study demonstrated that cavitation is the governing mechanism for photoacoustic drug delivery, and the deepest penetration of colored oil in gels followed the bubble collapse. Spatial distribution measurements revealed that colored oil could be driven a few millimeters into the gels in both axial and radial directions, and the penetration was less than 500 μm when the gelatin structure was not fractured. Key words: Ablation, photoacoustic transients, localized drug delivery, cavitation bubble, laser thrombolysis.

HanQun Shangguan; Lee W. Casperson; Scott A. Prahl; "Pressure impulses during microsecond laser ablation," *Appl. Opt.*, Vol. 36, pg. 9034, (1997).

The collapse of laser-induced cavitation bubbles creates acoustic transients within the surrounding medium and also pressure impulses to the ablation target and light-delivery fiber during microsecond laser ablation. The impulses are investigated here with time-resolved flash photography, and they are found to occur whether or not the light-delivery fiber is in contact with the target. We demonstrate that the impulses depend primarily on the energy stored in the cavitation bubble. They are not directly dependent on the mode of light delivery (contact versus noncontact), and they are also not directly correlated to the other acoustic transients. The pressure impulses do seem to be associated with the bubble-driven jet formation caused by the bubble collapse.

Gerald Soelkner; Gerhard Mitic; Robert Lohwasser; "Monte Carlo simulations and laser Doppler flow measurements with high penetration depth in biological tissuelike head phantoms," *Appl. Opt.*, Vol. 36, pg. 5647, (1997).

Laser Doppler flow measurements on biological tissuelike phantoms have shown that penetration depths of 30 mm could be obtained, thus exceeding the penetration depth of commercial instruments for the measurement of skin perfusion by more than an order of magnitude. Monte Carlo simulations were performed and compared with measurement results obtained on a headlike tissue model to quantify the influence of perfusion of the scalp on the cortex perfusion results. We found Doppler frequency spectra to be independent of the mean scattering angle and could be fitted with a sum of Gaussian functions, using a simple analytical model.

Paul T. Staveteig; Joseph T. Walsh, Jr.; "Dynamic 193-nm optical properties of water," *Appl. Opt.*, Vol. 35, pg. 3392, (1996).

Previous assumptions that water is not a 193-nm chromophore during ArF excimer laser tissue ablation are based on room-temperature data and ignore spectroscopic literature that suggests a strong temperature dependence of far-ultraviolet water absorption. By the use of a Q-switched Er:YAG laser as a pump source and an ArF excimer laser as a probe source, thermal generation and relaxation of 193-nm water absorption were characterized under nonequilibrium high-temperature and high-pressure conditions. At volumetric energy densities as small as 2 kJ/cm^3 relative to room temperature, the 193-nm absorption coefficient of water was measured to increase by more than 5 orders of magnitude. These results are consistent with the hypothesis that the absorption of 193-nm radiation by water may play a role in ArF excimer laser ablation of tissue. Key words: Water, hydrogen bonding, laser-tissue interaction, dynamic optical properties, nonlinear absorption, excimer laser ablation.

Kazunori Tanaka; Marcos T. T. Pacheco; James F. Brennan III; Irving Itzkan; Andrew J. Berger; Ramachandra R. Dasari; Michael S. Feld; "Compound parabolic concentrator probe for efficient light collection in spectroscopy of biological tissue," *Appl. Opt.*, Vol. 35, pg. 0758, (1996).

We describe a compound parabolic concentrator (CPC)-based probe for enhanced signal collection in the spectroscopy of biological tissues. Theoretical considerations governing signal enhancement compared with conventional collection methods are given. A ray-tracing program was used to analyze the throughput of CPC's with shape deviations and surface imperfections. A modified CPC shape with 99% throughput was discovered. A 4.4-mm-long CPC was manufactured and incorporated into an optical fiber-based near-infrared Raman spectrometer system. For human tissue samples, light collection was enhanced by a factor of 7 compared with collection with 0.29-NA optical fibers.

Tromberg, Bruce J.; Svaasand, Lars O.; Tsay, Tsong-Tseh; Haskell, Richard C.; "Properties of photon density waves in multiple-scattering media," *Appl. Opt.*, Vol. 32, pg. 607, (1993).

Verkruysse, Wim; Pickering, John W.; Beek, Johan F.; Keijzer, Marleen; van Gemert, Martin J. C.; "Modeling the effect of wavelength on the pulsed dye laser treatment of port wine stains," *Appl. Opt.*, Vol. 32, pg. 393, (1993).

I. Alex Vitkin; Brian C. Wilson; R. Rox Anderson; "Analysis of layered scattering materials by pulsed photothermal radiometry: application to photon propagation in tissue," *Appl. Opt.*, Vol. 34, pg. 2973, (1995).

A model of pulsed photothermal radiometry (PPTR) based on optical diffusion theory is presented for a turbid, two-layer, semi-infinite medium containing a surface layer whose optical absorption and scattering properties differ from that of the underlying layer. Assuming one-dimensional geometry, we develop expressions for the depth-dependent fluence distributions and radiant-energy-density profiles and for the time dependence of the PPTR signal. Experimental tests of the PPTR model in a series of layered phantoms of varying optical properties are described. The results of these tests are consistent with the model predictions.

H. Wabnitz; H. Rinneberg; "Imaging in turbid media by photon density waves: spatial resolution and scaling relations," *Appl. Opt.*, Vol. 36, pg. 0064, (1997).

Spatial resolution of transillumination imaging through highly scattering media normalized to sample thickness d depends on only the normalized wavelength Λ/d of photon density waves and normalized penetration depth δ/d. This was concluded theoretically and verified experimentally by the derivation of edge-spread functions from measured time-resolved transmittance and its Fourier transform by the use of dilute milk at various concentrations as scatterer in cuvettes of d = 2 cm and d = 4 cm. In the frequency domain and the time domain, spatial resolution was found experimentally to be given by approximately 0.3d obtained at Λ\approximately d or when only the first arriving 1% of all photons detected were taken into account.

Georges A. Wagnieres; Andre P. Studzinski; Daniel R. Braichotte; Philippe Monnier; Christian Depeursinge; Andre Chatelain; Hubert E. van den Bergh; "Clinical imaging fluorescence apparatus for the endoscopic photodetection of early cancers by use of Photofrin II," *Appl. Opt.*, Vol. 36, pg. 5608, (1997).

A fluorescence imaging device applied to the detection of early cancer is described. The apparatus is based on the imaging of laser-induced fluorescence of a dye that localizes in a tumor with a higher concentration than in the surrounding normal tissue after iv injection. Tests carried out in the upper aerodigestive tract, the tracheobronchial tree, and the esophagus with Photofrin II (1 mg/kg of body weight) as the fluorescent agent are reported as examples. The fluorescence is induced by violet (410-nm) light from a continuous-wave (cw) krypton-ion laser. The fluorescence contrast between tumor and surrounding tissue is enhanced by real-time image processing. This is done by the simultaneous recording of the fluorescence image in two spectral domains (470-600 and 600-720 nm), after which these two images are digitized and manipulated with a mathematical operator (look-up table) at video frequency. Among the 7 photodetections performed in the tracheobronchial tree, 6 were successful, whereas it was the case for only 5 of the 15 lesions investigated in squamous mucosa (upper aerodigestive tract and esophagus). The sources of false positives and false negatives are evaluated in terms of the fluorescent dye, tissue optical properties, and illumination optics.

Scott A. Walker; Sergio Fantini; Enrico Gratton; "Image reconstruction by backprojection from frequency-domain optical measurements in highly scattering media," *Appl. Opt.*, Vol. 36, pg. 0170, (1997).

The reconstruction of the location and optical properties of objects in turbid media requires the solution of the inverse problem. Iterative solutions to this problem can require large amounts of computing time and may not converge to a unique solution. Instead, we propose a fast, simple method for approximately solving this problem in which calculated effective absorption and reduced scattering coefficients are backprojected to create an image of the objects. We reconstructed images of objects with centimeter dimensions embedded in a diffusive medium with optical characteristics similar to those of human tissue. Data were collected by a frequency-domain spectrometer operating at 120 MHz with a laser diode light source emitting at 793 nm. Intensity and phase of the incident photon density wave were collected from linear scans at different projection angles. Although the positions of the objects are correctly identified by the reconstructed images, the optical parameters of the objects are recovered only qualitatively.

Lihong Wang, Steven L. Jacques, and Xuemei Zhao; "Continuous-wave ultrasonic modulation of scattered laser light to image objects in turbid media," *Opt. Lett.*, Vol. 20, pg. 629, (1995).

Continuous-wave ultrasonic modulation of scattered laser light has been used to image objects in tissue-simulating turbid media for what is to our knowledge the first time. The ultrasound wave focused into the turbid media modulates the laser light passing through the ultrasonic focal zone. The modulated laser light collected by a photomultiplier tube reflects the local mechanical and optical properties in the focal zone. Buried objects are located with millimeter resolution by scanning and detecting alterations of the modulated optical signal. This technique has the potential to provide a noninvasive, nonionizing, inexpensive diagnostic tool for diseases such as breast cancer.

Lihong Wang; Da Liu; Nancy He; Steven L. Jacques; Sharon L. Thomsen; "Biological laser action," *Appl. Opt.*, Vol. 35, pg. 1775, (1996).

The narrowing of the spectral linewidth and the increasing of the peak intensity characteristic of laser action were observed in emission spectra of dye-infused biological tissues. The tissue was infused with a solution of Rhodamine 640 perchlorate in ethanol and then excited with frequency-doubled Q-switched Nd:YAG laser pulses. The dependence of emission linewidth on the excitation radiant exposure and dye concentration was investigated. Laser action was also observed in biologically compatible fluorescein sodium dye dissolved in phosphate-buffered saline mixed with scattering polystyrene spheres. The sharp spectral peaks of laser action in tissues may find applications in the detection of superficial disease. Key words: Tissue optics, laser action, fluorescence, amplification of spontaneous emission, turbid media.

Yamada, Yukio; Hasegawa, Yasuo; Yamashita, Yutaka; "Simulation of fan-beam-type optical computed-tomography imaging of strongly scattering and weakly absorbing media," *Appl. Opt.*, Vol. 32, pg. 4808, (1993).

Yuqi Yao; Yao Wang; Randall L. Barbour; Harry L. Graber; Jenghwa Chang; "Sensitivity studies for imaging a spherical object embedded in a spherically symmetric, two-layer turbid medium with photon-density waves," *Appl. Opt.*, Vol. 35, pg. 0735, (1996).

We present analytic expressions for the amplitude and phase of photon-density waves in strongly scattering, spherically symmetric, two-layer media containing a spherical object. This layered structure is a crude model of multilayered tissues whose absorption and scattering coefficients lie within a range reported in the literature for most tissue types. The embedded object simulates a pathology, such as a tumor. The normal-mode-series method is employed to solve the inhomogeneous Helmholtz equation in spherical coordinates, with suitable boundary conditions. By comparing the total field at points in the outer layer at a fixed distance from the origin when the object is present and when it is absent, we evaluate the potential sensitivity of an optical imaging system to inhomogeneities in absorption and scattering. For four types of background media with different absorption and scattering properties, we determine the modulation frequency that achieves an optimal compromise between signal-detection reliability and sensitivity to the presence of an object, the minimum detectable object radius, and the smallest detectable change in the absorption and scattering coefficients for a fixed object size. Our results indicate that (1) enhanced sensitivity to the object is achieved when the outer layer is more absorbing or scattering than the inner layer; (2) sensitivity to the object increases with the modulation frequency, except when the outer layer is the more absorbing; (3) amplitude measurements are proportionally more sensitive to a change in absorption, phase measurements are proportionally more sensitive to a change in scattering, and phase measurements exhibit a much greater capacity for distinguishing an absorption perturbation from a scattering perturbation.

I. V. Yaroslavsky; A. N. Yaroslavsky; T. Goldbach; H.-J. Schwarzmaier; "Inverse hybrid technique for determining the optical properties of turbid media from integrating-sphere measurements," *Appl. Opt.*, Vol. 35, pg. 6797, (1996).

We have combined the Monte Carlo method with the small-angle approximation of the radiative transfer theory to derive the optical properties (the absorption coefficient, the scattering coefficient, and the anisotropy factor) of turbid materials from integrating-sphere measurements (the total transmittance and the diffuse reflectance) and the collimated transmittance. Unlike one-dimensional models, the technique accounts for the side losses of light at the edges of the sample. In addition, it enables the correction of the measured collimated signal for the contribution of multiply scattered light. On the other hand, the hybrid technique allows a significant reduction in calculation time compared with inverse methods based on a pure Monte Carlo technique. Numerical tests and experimental results from a phantom material (milk) as well as samples of biological tissue (porcine myocardium) confirmed the feasibility of applying this technique to the determination of the optical properties of turbid media.

I. V. Yaroslavsky; A. N. Yaroslavsky; T. Goldbach; H.-J. Schwarzmaier; "Inverse hybrid technique for determining the optical properties of turbid media from integrating-sphere measurements," *Appl. Opt.*, Vol. 35, pg. 6797, (1996).

We have combined the Monte Carlo method with the small-angle approximation of the radiative transfer theory to derive the optical properties (the absorption coefficient, the scattering coefficient, and the anisotropy factor) of turbid materials from integrating-sphere measurements (the total transmittance and the diffuse reflectance) and the collimated transmittance. Unlike one-dimensional models, the technique accounts for the side losses of light at the edges of the sample. In addition, it enables the correction of the measured collimated signal for the contribution of multiply scattered light. On the other hand, the hybrid technique allows a significant reduction in calculation time compared with inverse methods based on a pure Monte Carlo technique. Numerical tests and experimental results from a phantom material (milk) as well as samples of biological tissue (porcine myocardium) confirmed the feasibility of applying this technique to the determination of the optical properties of turbid media.

I. V. Yaroslavsky; A. N. Yaroslavsky; V. V. Tuchin; H.-J. Schwarzmaier; "Effect of the scattering delay on time-dependent photon migration in turbid media," *Appl. Opt.*, Vol. 36, pg. 6529, (1997).

We modified the diffusion approximation of the time-dependent radiative transfer equation to account for a finite scattering delay time. Under the usual assumptions of the diffusion approximation, the effect of the scattering delay leads to a simple renormalization of the light velocity that appears in the diffusion equation. Accuracy of the model was evaluated by comparison with Monte Carlo simulations in the frequency domain for a semi-infinite geometry. A good agreement is demonstrated for both matched and mismatched boundary conditions when the distance from the source is sufficiently large. The modified diffusion model predicts that the neglect of the scattering delay when the optical properties of the turbid material are derived from normalized frequency- or time-domain measurements should result in an underestimation of the absorption coefficient and an overestimation of the transport coefficient. These observations are consistent with the published experimental data.

Yoon, Gilwon; Roy, Dilip N. Ghosh; Straight, Richard C.; "Coherent backscattering in biological media: measurement and estimation of optical properties," *Appl. Opt.*, Vol. 32, pg. 580, (1993).

Renato Amaro Zangaro; Landulfo Silveira, Jr.; Ramasamy Manoharan; George Zonios; Irving Itzkan; Ramachandra R. Dasari; Jacques Van Dam; Michael S. Feld; "Rapid multiexcitation fluorescence spectroscopy system for *in vivo* tissue diagnosis," *Appl. Opt.*, Vol. 35, pg. 5211, (1996).

We have designed, fabricated, and tested a compact, transportable, excitation-emission spectrofluorimeter with optical-fiber light delivery and collection for use in rapid analysis of tissues in a clinical setting. This system provides up to eleven different excitation wavelengths, permitting collection of all the corresponding emission spectra in approximately 600 ms. It uses a N_2 laser that pumps a sequence of dyes placed in cuvettes on a rotating wheel. A white-light excitation source permits acquisition of the tissue's diffuse reflectance spectrum on each cycle. Return fluorescence and reflected light are dispersed by a small spectrograph and detected by a photodiode-array detector. The system can collect a single-shot spectrum from biological tissue with a signal-to-noise ratio in excess of 50:1.

Qienyuan Zhou; Robert W. Knighton; "Light scattering and form birefringence of parallel cylindrical arrays that represent cellular organelles of the retinal nerve fiber layer," *Appl. Opt.*, Vol. 36, pg. 2273, (1997).

The retinal nerve fiber layer (RNFL) comprises bundles of unmyelinated axons that run across the surface of the retina. The cylindrical organelles of the RNFL (axonal membranes, microtubules, neurofilaments, and mitochondria) as seen by electron microscopy were modeled as parallel cylindrical arrays in order to gain insight into their optical properties. Arrays of thin fibrils were used to represent organelles that are thin relative to wavelength, and the model took into account interference effects that may arise from spatial order. Angular and spectral light-scattering functions were calculated for the backscattering hemisphere. Scattering was much larger from axonal membranes than from microtubules or neurofilaments. Spectra from 400 to 700 nm show that scattering increases at shorter wavelengths for both axonal membranes and microtubules. At 560 nm, scattering from mitochondria modeled as thick cylinders was approximately the same as that from microtubules but showed little wavelength dependence. The model reveals differences in backscattered polarization ratios that may permit experimental discrimination between microtubule and membrane mechanisms for the RNFL reflectance. Calculated backscattering exceeds measured values by at least 1 order of magnitude, but calculated form birefringence for microtubule arrays is approximately the same as measured birefringence.

Jaap R. Zijp; Jaap J. ten Bosch; "Anisotropy of volume-backscattered light," *Appl. Opt.*, Vol. 36, pg. 1671, (1997).

In many measurements used for the determination of optical properties of biological tissues, a detector with a small cone of acceptance is used. Therefore a small part of the anisotropically distributed backscattered light is detected. We present the results of Monte Carlo calculations and study the anisotropy of the direction distribution of volume-backscattered photons. Close to the source it is impossible to give a general correction factor to relate the measurements to theoretical results. Far from the source, the measured radiation divided by the square of the sine of the accepting angle of the detector is a relatively good approximation of the radiant exitance at the surface of the sample under investigation. We recommend the use of a detector with a large angle of acceptance for these kinds of measurements.

Zijp, Jaap R.; ten Bosch, Jaap J.; "Theoretical model for the scattering of light by dentin and comparison with measurements," *Appl. Opt.*, Vol. 32, pg. 411, (1993).

D. A. Zimnyakov; V. V. Tuchin; A. A. Mishin; "Spatial speckle correlometry in applications to tissue structure monitoring," *Appl. Opt.*, Vol. 36, pg. 5594, (1997).

Asymptotic behavior of temporal autocorrelation functions of speckle intensity fluctuations induced by tissue scanning with a focused probe beam is experimentally studied for the transition from a single-scattering to a multiple-scattering mode. Such parameters as the exponential factor (or the Hurst coefficient) and the Hausdorff dimension are proposed for the characterization and the visualization of the variations of the studied tissues' optical properties in generalized form. We studied reversible transition between various scattering modes stimulated by the application of certain chemical agents to the human sclera samples using speckle intensity correlation analysis; corresponding results are presented. The possibilities of the scattering structures imaging with local estimations of the exponential factor of speckle intensity fluctuations are shown in *in vitro* experiments with samples of human skin epidermis.

Author Index

Subject Index

Optics Classification and Indexing Scheme (OCIS) codes supplied in parenthesis, where applicable.